MW00846252

Physical Chemistry of Biological Interfaces

Physical Chemistry of Biological Interfaces

edited by

Adam Baszkin

Laboratoire de Physico-Chimie des Surfaces,
UMR 8612 CNRS
Université Paris-Sud
Châtenay-Malabry, France

Willem Norde

Laboratory of Physical Chemistry and Colloid Science
Wageningen University
Wageningen, The Netherlands

MARCEL DEKKER, INC.   NEW YORK · BASEL

ISBN: 0-8247-7581-3

This book is printed on acid-free paper.

Headquarters
Marcel Dekker, Inc.
270 Madison Avenue, New York, NY 10016
tel: 212-696-9000; fax: 212-685-4540

Eastern Hemisphere Distribution
Marcel Dekker AG
Hutgasse 4, Postfach 812, CH-4001 Basel, Switzerland
tel: 41-61-261-8482; fax: 41-61-261-8896

World Wide Web
http://www.dekker.com

The publisher offers discounts on this book when ordered in bulk quantities. For more information, write to Special Sales/Professional Marketing at the headquarters address above.

Copyright © 2000 by Marcel Dekker, Inc. All Rights Reserved.

Neither this book nor any part may be reproduced or transmitted in any form or by any means, electronic or mechanical, including photocopying, microfilming, and recording, or by any information storage and retrieval system, without permission in writing from the publisher.

Current printing (last digit):
10 9 8 7 6 5 4 3 2

PRINTED IN THE UNITED STATES OF AMERICA

Preface

Biological systems are complex supramolecular assemblies and constitute reaction sites with highly accumulated functions. The great advances made during the last several decades in our understanding of biological cell and biological membrane functions have been due mainly to the interest in this area of a number of scientific disciplines ranging from chemistry and physics to molecular biology, stimulated by the need for a multidisciplinary approach to investigating these supramolecular functional architectures. The unifying power of physical surface chemistry in the development of this research area, as witnessed by the great number of publications, reviews, congresses, and schools, has enabled us to approach complex biological problems in a logical stepwise fashion. The characterization of biological interfaces often requires modeling of interfacial phenomena encountered in these systems by the concepts of surface science in order to analyze them in terms of basic molecular features. Major research interests center on extensive studies of adsorption phenomena that are analogous to biological situations. They are focused on the concepts of forces operating across interfaces, of electrical double layers at charged interfaces, and of the structure of water adjacent to (bio)interfaces.

The adsorption or adhesion to the surface of a biocomponent more often than not triggers a change in its physicochemical properties which, in turn, affects its biological functioning. Hence, the presence of interfaces plays a crucial role in biomedicine, food processing, environmental sciences, and various other biotechnological applications.

In the past decades, it has been fully recognized that, in addition to specific interactions, generic physical–chemical interactions are omnipresent in various biological systems. This especially applies to mechanisms that control bio-interfacial phenomena such as immobilization of enzymes in biocatalysis, biofouling, biofilm formation, phagocytosis, membrane fusion, diffusion, and transport in biomembranes. Application of interfacial physicochemistry principles—initially developed for studies of relatively simple, well-defined inorganic or synthetic molecules—to biological systems has revolutionized an understanding of molecular recognition interactions.

The purpose of *Physical Chemistry of Biological Interfaces* is twofold: 1) to introduce the most important fundamental concepts of physical-chemical interface science applied to systems of biological origin and 2) to serve as a comprehensive state-of-the-art reference that addresses experimental techniques and applications of surface science to biological systems. The book contains contributions from outstanding specialists in physical chemistry who have an open eye for biological applications, as well as contributions from more biologically oriented scientists who employ physicochemical concepts.

The first three chapters refer to the theories and principles of thermodynamics, interfacial interactions, and electrical properties of charged interfaces and interphases described at the fundamental level. They introduce the most important concepts of interface science and provide essential background for approaching biological systems with a straightforward logic.

Chapters 4—6 develop the topics of protein and polysaccharide adsorption with a heavy emphasis on the physical aspects of the subject. Chapters 7—10 present an incisive review of biochemical and biophysical properties of biomembranes followed by a survey of important recent experimental works designed to mimic biomembrane processes. Such artificial assemblies of lipids, proteins, and polysaccharides are expected to perform novel functions that are not efficiently exercised by biological systems and enable the development of receptor-ligand interaction studies at a molecular level.

In Chapter 11, the text moves to enzymatic reactions at interfaces involving thorough studies of their mechanisms. The material presented in Chapters 12—14 is dedicated to comprehension of cell adhesion with solid surfaces and cell-cell interactions. Special emphasis is given to the role of the bending elasticity concept of shape transitions of free and adhering cells and to short- and long-range forces that control cell adhesion to solid surfaces. An overview of the main classes of cell adhesion molecules and of the role of cell glycocalyx in establishing the first bond and extension of the contact area between adhering cells is given in Chapter 14. Thermodynamic and kinetic models for the extension of contact area during cell adhesion, as well as the different methods of measuring adhesive strength in cell detachment experiments, are largely developed.

In Chapters 15—24 the reader will find a detailed description of recently developed methods and techniques akin to accurate measurements of biointerfacial phenomena, particularly those applied to quantitative understanding of structure-property relationships and molecular recognition interactions. These chapters cover axisymmetric drop shape analysis, Brewster angle microscopy, ellipsometry, neutron and x-ray reflectivity, time-resolved fluorescence techniques, circular dichroism and infrared spectroscopy, and scanning force and cryotransmission microscopies.

We expect that *Physical Chemistry of Biological Interfaces* will be valuable to anyone who wants to appreciate the wide range of biointerfaces from fundamentals to applications. Owing to its interdisciplinary character, the book will be appreciated by physical chemists and biologists from academia and industrial communities and also by graduate students who need a conceptual framework showing how studies in these diverse areas are related to one another.

Finally, our deep gratitude goes to the excellent secretarial services provided by Théresia Fraboulet at Laboratoire de Physico-Chimie des Surfaces, UMR 8612 CNRS, Université Paris-Sud, Châtenay-Malabry and by Yvonne, Wil, and Josie at the Laboratory of Physical Chemistry and Colloid Science of Wageningen University. The book could not have been edited without their efficient typing, checking of references and illustrations, and maintaining our correspondence with the authors and the publisher.

Adam Baszkin
Willem Norde

Contents

Contributors

H. Arwin Laboratory of Applied Optics, Department of Physics and Measurement Technology, Linköping University, Linköping, Sweden

Y. Barenholz Laboratory of Membrane and Liposome Research, Department of Biochemistry, Hebrew University, Hadassah Medical School, Jerusalem, Israel

A. Baszkin Laboratoire de Physico-Chimie des Surfaces UMR 8612 CNRS, Université Paris-Sud, Châtenay-Malabry, France

A. Benoliel Laboratoire d'Immunologie, INSERM U387, Hôpital de Sainte-Marguerite, Marseille, France

P. Bongrand Laboratoire d'Immunologie, INSERM U387, Hôpital de Sainte-Marguerite, Marseille, France

R. Bos Department of Biomedical Engineering, University of Groningen, Groningen, The Netherlands

H. J. Busscher Department of Biomedical Engineering, University of Groningen, Groningen, The Netherlands

G. Cevc Department of Biophysics, Klinikum r.d.I., The Technical University of Munich, Munich, Germany

P. Chen Department of Mechanical and Industrial Engineering, University of Toronto, Toronto, Ontario, Canada

D. Danino Department of Chemical Engineering, Technion-Israel Institute of Technology, Haifa, Israel

H. H. J. de Jongh Centre for Protein Technology, Wageningen University and Research Centre, Wageningen, The Netherlands

O. I. del Río Department of Mechanical and Industrial Engineering, University of Toronto, Toronto, Ontario, Canada

R. A. Dluhy Department of Chemistry, University of Georgia, Athens, Georgia

P. S. Handley   Department of Biological Sciences, University of Manchester, Manchester, United Kingdom

S. Hénon   Laboratoire de Biorhéologie et d'Hydrodynamique Physicochimique, Universités Paris VI et Paris VII, Paris, France

D. J. Keller   Department of Chemistry, University of New Mexico, Albuquerque, New Mexico

J. M. Kleijn   Laboratory of Physical Chemistry and Colloid Science, Wageningen University, Wageningen, The Netherlands

T. Kunitake   Department of Chemistry and Biochemistry, Kyushu University, Fukuoka, Japan

C. Lheveder   Laboratoire de Physique Statistique de l'ENS, UMR 8550 du CNRS, associé aux Universités Paris VI et VII, Paris, France

J. R. Lu   Department of Chemistry, University of Surrey, Guildford, United Kingdom

J. Lyklema   Laboratory of Physical Chemistry and Colloid Science, Wageningen University, Wageningen, The Netherlands

F. MacRitchie   Department of Grain Science and Industry, Kansas State University, Manhattan, Kansas

J. Meunier   Laboratoire de Physique Statistique de l'ENS, UMNR 8550 du CNRS, associé aux Universités Paris VI et VII, Paris, France

A. W. Neumann   Department of Mechanical and Industrial Engineering, University of Toronto, Toronto, Ontario, Canada

W. Norde   Laboratory of Physical Chemistry and Colloid Science, Wageningen University, Wageningen, The Netherlands

I. Panaiotov   Biophysical Chemistry Laboratory, University of Sofia, Sofia, Bulgaria

A. Pierres   Laboratoire d'Immunologie, INSERM U387, Hôpital de Sainte-Marguerite, Marseille, France

H. Ringsdorf   Institute of Organic Chemistry, University of Mainz, Mainz, Germany

V. Rosilio   Laboratoire de Physico-Chimie des Surfaces UMR 8612 CNRS, Université Paris-Sud, Châtenay-Malabry, France

E. Sackmann   Department of Physics, The Technical University of Munich, Munich, Germany

R. Simson   Software Design and Management AG, Munich, Germany

K. J. Stine   Department of Chemistry and Center for Molecular Electronics, University of Missouri—St. Louis, St. Louis, Missouri

J. Sunamoto   Niihama National College of Technology, Niihama, Japan

Y. Talmon   Department of Chemical Engineering, Technion-Israel Institute of Technology, Haifa, Israel

R. K. Thomas   Physical and Theoretical Chemistry Laboratory, University of Oxford, Oxford, United Kingdom

H. C. van der Mei   Department of Biomedical Engineering, University of Groningen, Groningen, The Netherlands

A. van Hoek   Micro-Spectroscopy Centre, Wageningen University and Research Centre, Wageningen, The Netherlands

H. P. van Leeuwen   Laboratory of Physical Chemistry and Colloid Science, Wageningen University, Wageningen, The Netherlands

C. P. M. van Mierlo   Centre for Protein Technology and Department of Biomolecular Sciences, Laboratory of Biochemistry, Wageningen University and Research Centre, Wageningen, The Netherlands

R. Verger   Laboratoire de Lipolyse Enzymatique, CNRS, Marseilles, France

A. J. W. G. Visser   Micro-Spectroscopy Centre, Wageningen University and Research Centre, Wageningen, The Netherlands

H. Wennerström   Center for Chemistry and Chemical Engineering, Lund University, Lund, Sweden

1

Interfacial Thermodynamics with Special Reference to Biological Systems

J. Lyklema

Wageningen University, Wageningen, The Netherlands

Thermodynamics is powerful but sometimes seems discouraging. It is powerful because relations of general validity can be derived or, for that matter, relations for which the validity is specified. At the same time it may appear discouraging because the formalism seems abstract and beyond everyday reality. However, as biological interfaces, whatever their complexity, are also subject to the laws of thermodynamics, it is appropriate to start the present book with a review of interfacial thermodynamics. We shall have to discuss what thermodynamics has to say about such features as interfacial energies, interfacial tension, adhesion, adsorption, and particle-particle interaction and about relations between them.

For this first chapter it is assumed that the reader is familiar with the basics of bulk thermodynamics but not with the idiosyncrasies of the thermodynamics of heterogeneous systems containing phase boundaries. We shall also introduce some aspects of interfacial statistical thermodynamics that may be useful background reading for some later chapters. Sometimes digressions beyond the domain of strict thermodynamics will be made, doing our best to delineate the borderline (and the trespassing!).

In order to remain concrete, we consider only smooth, homogeneous, not strongly curved interfaces; i.e., we shall not include porous surfaces or microemulsions. Although most of the material to be presented will deal with solid-liquid interfaces, much of it also applies to liquid-liquid interfaces. In view of the biological interest, we shall emphasize aqueous solutions. The term "interface" refers to the boundary between two condensed phases, whereas "surface" applies to the border between a condensed phase and vapor. When specifically referring to the outer side of a system, the term "surface" is also used (e.g., "the particle surface").

I. ON THERMODYNAMIC PRINCIPLES

A. Functions of State

Basically, classical thermodynamics deals with measurable macroscopic quantities. By "measurable" it is understood that an unambiguous experiment can be carried out to obtain a value for the required quantity at given ambient conditions, say p and T. When the outcome of such measurements is independent of the path taken, data for such quantities can be tabulated, either as absolute values or with respect to an agreed reference state. Examples of quantities that can thus be tabulated are the energy U, the entropy S, the enthalpy H, the Helmholtz energy F (formerly called free energy), the Gibbs energy G (formerly called free enthalpy), and the chemical potential μ_i of component i. These so-called characteristic functions are *functions of state*: when the macroscopic state of the system is fully specified (nature of the system, amount, composition, p and T, etc.), the functions of state are also unambiguously fixed.

For fluid-fluid interfaces interfacial and surface tensions γ can be measured by a host of methods. The measurability stems from the possibility of extending those interfaces reversibly and isothermally. Therefore, for such interfaces interfacial tensions are also functions of state. We shall use the superscripts S, L, and G to indicate a solid, liquid, or gaseous phase, respectively, and combinations to denote the nature of the interface. So, the preceding can be condensed to stating that γ^{LL} and γ^{LG} are functions of state.

A problem is posed for γ^{SG} and γ^{SL} because it is next to impossible to extend a solid-gas or solid-liquid interface reversibly. The work to be done involves not only the creation of a new surface but also breaking or cleaving of the solid. Apart from the difficulties incurred in doing this in a reversible way, it is also cumbersome to account quantitatively for its contribution to the work. Only for ductile solids can reliable tensions be obtained. In practice, γ^{SG} and γ^{SL} values reported in the literature have been obtained by various indirect means. Between different authors and different methods, the outcome may vary substantially.

Notwithstanding the experimental inaccessibility of γ^{SG} and γ^{SL}, the statement that these quantities are functions of state can be defended on the grounds that at a given fully relaxed state of the system γ^{SG} and γ^{SL} are fully determined. The consequence of this is that $d\gamma^{SG}$ and $d\gamma^{SL}$ are complete differentials, just like dU, dH, dS, etc., so that cross-differentiation is allowed to obtain a variety of useful relationships. We shall present an additional argument in Sec. I.B. For LL and LG interfaces this problem does not arise.

As is known, work and heat are not functions of state because these quantities depend on the path along which they are obtained. One can therefore not speak of "the work of a given system" and the differential is not complete. This also applies to surface work and surface heat.

B. Reversibility, Hysteresis, and the Deborah Number

We have anticipated another important issue by stating that processes carried out to obtain values for functions of state should be reversible. A process is called reversible if at any stage the system is at equilibrium, so that at that stage the process can be reversed by an infinitesimal change in the state variables. Strictly speaking, this implies that a reversible process is infinitely slow, because the driving force for it must be infinitesimally small.

However, "infinitesimally slow" is a relative notion. What counts is not the absolute rate of the process but its rate *relative* to the rate at which the system can adjust itself to the new situation (i.e., *relax* to the new equilibrium situation). It is enlightening for this to introduce the *Deborah number De* as the ratio

$$De = \frac{\tau_{process}}{\tau_{meas}} \tag{I.1}$$

where $\tau_{process}$ is the time scale on which the system relaxes and τ_{meas} the time scale of the measurement or observation. Processes for which $De \ll 1$ are reversible.

Illustration: LL interfaces, in the absence of adsorbed substances, relax very rapidly. "Process" now stands for the adjustment of the boundary layer structure after bringing about an extension. Typically, $\tau_{process}$ is of the order 10^{-4} s, so that reversibility is ensured if the time scale of the extension process is longer than, say, 10^{-2} s. Actually, most interfacial tension measurement techniques are much slower than that. However, solid surfaces relax much more slowly. In fact, in many situations they do not relax at all during a measurement. This observation gives rise to consideration of the notion of *frozen equilibrium*.

Frozen equilibrium implies a situation in which at least one of the processes does not proceed at all. Strictly speaking, we are dealing with frozen nonequilibrium. The typical illustration is that of detonation gas: two volumes of hydrogen plus one of oxygen, mixed at room temperature. This mixture is not at thermodynamic equilibrium: the Gibbs energy of the water formed after the reaction is lower than that of the gas mixture. One little spark suffices to complete the reaction explosively. However, kinetically it is at equilibrium, and one can for instance use this mixture to find out whether the ideal gas law $pV = nRT$ applies. Speaking in time scales, adjustment of p after changing V is for all practical purposes instantaneous; for this process $De \ll 1$, it is reversible. However, for the reaction $De \gg 1$ and hence this reaction can, for the present purpose, be ignored.

The issue is directly relevant to processes involving solid surfaces, in particular adsorption phenomena. Most solid surfaces are not flat and homogeneous but rather contain asperities, dislocations, crevices, and other irregularities. These leftovers of the formation process are not yet relaxed to the equilibrium state, which would call for a surface with minimum Gibbs energy, that is, having minimum area and the crystal plane with the lower indices exposed. Similar things can be stated about the surfaces of latex particles: these may also be nonrelaxed with respect to the distribution of polymer chains because after solidification, diffusion of chains or chain elements is very slow. The analogy with the detonation gas case enters when we are studying adsorption-desorption cycles. For the limiting, but not uncommon, case in which the solid does not relax during these cycles and the adsorption-desorption processes are themselves reversible, the analogy with the detonation gas case is complete. In terms of Deborah numbers, we have to distinguish two of them, De(surf) for rearrangements in the solid and De(ads) for adsorption. For the usual adsorption time scales De(surf) $\gg$ 1 whereas De(ads) $\ll$ 1.

With this in mind, we can return to the statement made in Sec. I.A, that for SG and SL interfaces $d\gamma$ is also a complete differential. Even when the solid is not at equilibrium, $d\gamma$ is complete with respect to changes in p, T, and μ_i's, and hence cross-differentiations are allowed. This condition underlies the validity and application of the Gibbs adsorption equation.

Especially for adsorption phenomena involving biological systems it is appropriate to discuss the validity of what has been said a little further. Applicability of the pair of conditions De(surf)$\gg$1 and De(ads)$\ll$1 can be inferred when adsorption-desorption, temperature, and pressure cycles are entirely reversible. (Incidentally, complete reversibility is also expected with both Deborah numbers $\ll$1.) Absence of full reversibility implies *hysteresis*.

Sometimes hysteresis is a virtue (because certain fine touches of a process can be studied, as in capillary condensation) but mostly it is a nuisance, because its origin is hard to identify, let alone quantify, and hence the applicability of equilibrium thermodynamics becomes debatable.

Experience has shown that the adsorption of low-molecular-weight species is often hysteresis free over large ranges of the solute concentration. The most general reason is that De(surf)$\gg$1, but in a few special cases the structure of the solid surface may vary in phase with the surface concentration (reversible *reconstruction*). Because of its phenomenological nature, to be addressed in Sec. II, thermodynamics does not have the means of distinguishing between the solid-side and the solution-side contributions to $d\gamma$ and to other complete interfacial differentials.

Many biological and other surfaces in an aqueous medium carry a charge. The question must be asked whether the formation of such charges is reversible. For systems of constant charge (sulfonated polystyrene latices, the plate charges on clay particles) this poses no problem. For oxides, proteins, biomembranes, etc., the *surface charge density* σ^o, usually (but not thermodynamically, Sec. IIB) defined as

$$\sigma^o \equiv F(\Gamma_{H^+} - \Gamma_{OH^-}) \tag{I.2}$$

where F is the Faraday, is a function of pH. For sufficiently low pH $\sigma^o > 0$, for sufficiently high pH $\sigma^o < 0$, and at a given pH^o, the *point of zero charge (p.z.c.)*, characteristic of the nature of the oxide, $\sigma^o = 0$. So σ^o depends on pH and the issue is therefore how reversible σ^o(pH) cycles are. Collected evidence from potentiometric acid-base titrations indicates that reversibility is usually ensured provided a variety of other sources of hysteresis are suppressed (i.e., the titrations are carried out sufficiently slowly and the systems are sufficiently aged). Sometimes it is found that the first back-and-forth pH cycle contains some hysteresis, after which the subsequent ones are free of it; the probable explanation is that the first cycle gives rise to some reconstruction of the surface after which De(surf)$\gg$1.

At given pH, σ^o also depends on the concentration of electrolytes and on the presence of adsorbates. In many cases changes of σ^o then remain reversible.

In conclusion, for most cases of adsorption, including those leading to electric double layers, the application of Gibbs' law is validated.

When hysteresis does occur, making general statements becomes difficult and impossible on thermodynamic grounds. Hysteresis may have a variety of origins, and no progress can be made unless these are identified. Strictly speaking, distinction has to be made between real hysteresis and lack of patience. The former implies that

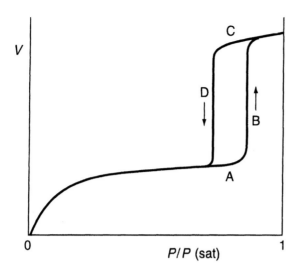

Fig. 1 Example of a hysteresis loop for gas adsorption to a porous surface (capillary condensation).

the state of the system is physically well defined but different between the way "up" and the way "down." Thermodynamics may then be applied for not too large changes in the state of either system. "Lack of patience" is the case we have discussed so far. No hysteresis is found when one of the processes does not relax during our measurement because we prefer not to wait so long. (De≫1 for the process under consideration.) Thermodynamics may be applied, but with the caveat that the results obtained refer to a system that is not fully relaxed. Because of that, different authors may obtain different results: it is unlikely that they are working with exactly identically frozen nonequilibria. Two examples from interface science may serve to illustrate this.

This first concerns capillary condensation of gases in pores. See Fig. 1. Hysteresis loops like this one are known to occur on adsorbates consisting of consolidated aggregates. On the adsorption branch, after a certain adsorbed volume V has been attained, the narrow slits or pores between the solids become filled with gas condensing to liquid, giving rise to a steep ascent in the isotherm. The sharpness of the rise is related to the size distribution of the pores: the narrower this distribution, the steeper the slope. On the desorption branch, liquid is retained longer in the capillaries than on the way up because of the stabilizing effect of the Laplace pressure across the (condensed) fluid-vapor interface. So points A and C refer to completely different states of the system and thermodynamics can be applied to either of them if short cycles are reversible.

Our second example, adsorption-desorption hysteresis like that in Fig. 2 is very common in the adsorption of macromolecules in general and of proteins in particular. Upon adsorption, proteins tend to undergo some irreversible unfolding, to an extent that is determined by the nature of the molecule and with a time scale that is commensurate with that of the adsorption process. Hence De is neither ≪1 nor ≫1. The process on the ascending branch is therefore very difficult to control: the amount adsorbed will depend on the number of molecules arriving by diffusion and convection, on the interaction of molecules with the surface and with each other, and,

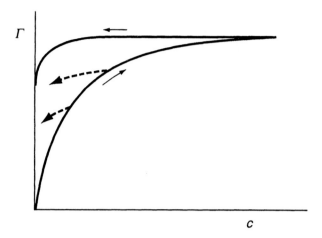

Fig. 2 Example of a hysteresis loop in protein adsorption. See discussion in the text.

last but not least, on the rate of unfolding in the adsorbed state as compared with the rate of desorption of neighboring molecules. However, once molecules have adsorbed and unfolded they are progressively difficult to remove by dilution, because the Gibbs energy of adsorption increases very strongly with the number of protein-surface contacts. Hence the tendency of virtually no desorption from the plateau and, for that matter, not much either from other points in the acclivity of the isotherm (dashed in Fig. 2). It is beyond this chapter to analyze further the many interesting features involved. Let it suffice to state that the ascending branch is probably defined more poorly than the descending one (because of the improbability of controlling the formation of nonrelaxed states) and that *scanning* (short excursions up and down) may reveal interesting features.

In conclusion of this section, the paradox may be noted that, although the discussion concerned rates and time scales, the purpose was to establish the range of applicability of classical thermodynamics, in which the time does not occur. In fact, the term "thermodynamics" ought to be reserved for the thermodynamics of irreversible processes, calling classical thermodynamics "thermostatics."

II. IMPLICATIONS OF THE MACROSCOPIC-PHENOMENOLOGICAL NATURE OF THERMODYNAMICS

A. When Is a System "Macroscopic"?

Thermodynamics deals with *macroscopic* amounts of matter, be it moles or tiny fractions of it, but still containing gigantic numbers of molecules, say of order 10^{16}–10^{23}. This is in contrast to molecular theories, such as molecular (or statistical) thermodynamics. The macroscopic phases are so large that they seem to be continuous. In fact, no molecular picture is needed: one can derived the entire formula set without believing that matter is composed of molecules. As a result, by purely thermodynamic arguments, one can never derive molecular properties.

The obvious next question is how many molecules a system should minimally contain to still be called "macroscopic." No sharp boundary can be given, but a useful distinction can be made on the basis of the relative size of *fluctuations* in quantities that are not fixed by the choice of the state of the system. When fluctuations are relatively insignificant, the system may be considered macroscopic. Let us make this concrete.

Consider a macroscopic system, containing a fixed number of molecules N in a volume V, placed in a thermostat of temperature T. The energy U is, macroscopically speaking, fixed and measurable (it is a function of state; see Sec. I.A). Nevertheless, this energy is subject to small random thermal fluctuations that one could measure if one had an extremely sensitive energy meter, responding at time scales of fractions of fractions of seconds. Taking longer measuring times than that—the usual procedure, that is—an *average* energy $\langle U \rangle$ is obtained, which is well defined and identified as *the* energy U. By the same token, in an open system at equilibrium with its surroundings, N can fluctuate and *the* number N is virtually the average $\langle N \rangle$.

Averaged values of U, N, etc. start to lose their physical meaning when the fluctuations are no longer small relative to the absolute value, that is, when the system becomes small. Then the system should no longer be called "macroscopic" and classical thermodynamics starts to lose its applicability.

How small? The answer to that follows from the theory of *random fluctuations*. According to this theory, the spread in a quantity is quantified through its *standard deviation* σ, defined as*

$$\sigma_U^2 = \langle U^2 \rangle - \langle U \rangle^2 \qquad \sigma_N^2 = \langle N^2 \rangle - \langle N \rangle^2 \tag{II.1a, b}$$

Elaborating the right-hand sides, it follows for the relative size of the fluctuations that

$$\frac{\sigma_U}{\langle U \rangle} = O(\langle U \rangle)^{-1/2} \qquad \frac{\sigma_N}{\langle N \rangle} = O(\langle N \rangle)^{-1/2}, \text{ etc.} \tag{II.2a, b}$$

where $O(x)$ stands for "of order." So, even if the absolute value of σ increases with the size of the system, its relative value decreases and it is the relative value that counts. Quantitatively, for $\langle N \rangle = 10^6$, 10^4, and 10^2 molecules $\sigma_N / \langle N \rangle$ is 1‰, 1%, and 10%, respectively. Hence we may say that systems containing at least $\sim 10^3$ molecules are "macroscopic." In fact, classical thermodynamics is always applied to, say, $O(\gtrsim 10^{16})$ molecules; then nobody worries about fluctuations and tacitly sets $\langle U \rangle = U$, $\langle N \rangle = N$, etc.

For interface science the consequence is that thermodynamically only macroscopic interfaces will be considered. The surface of a molecule cannot be thermodynamically identified, let alone its surface tension measured (how would one measure the force to expand such a surface reversibly?). Amounts of molecules in an adsorbate may be so low that classical thermodynamics becomes less appropriate; at any rate, statistical thermodynamics is the technique required for deriving adsorption isotherm equations. Likewise, polymers in solution and in the adsorbed state (relevant for steric stabilization) are preferably statistically treated. Isolated protein molecules in solution may have enough molecules (or rather atoms) to warrant being considered macroscopic, but they are not homogeneous phases and the fluctuations are not random.

---

* Standard deviations and surface charge densities have the same symbol, but confusion is unlikely because of the subscript in the former.

B. Consequences of the Phenomenological Nature

By "phenomenological" is meant "based on an actual measurement". Phenomenological relations relate experimentally measurable variables. We shall use the terms "operational" and "inoperational" for quantities for which a process can or cannot be designed to measure them, respectively.

The advantages of keeping these features in mind are threefold.

1. As long as one remains within the domain of phenomenological relations between operational variables, these are universally valid, because they are based on the first and second law of thermodynamics. Whatever nonthermodynamic relation is derived, it should never violate these phenomenological laws.
2. By considering the operationality of (steps in) the processes, one becomes conscious of having made nonthermodynamic steps in the reasoning.
3. The notion of driving force can be generalized.

Let us elaborate these points by considering some aspects of (biological) interfaces.

One illustration of (1) regards the question of whether or not an osmotic pressure across a membrane depends on the nature of that membrane. The answer is "no," the argument being that if it did one could arrange two different membranes in series, separating the same solution from pure water, and in this way let water flow without an external source. Putting a membrane in a system can never lead to flow of fluid.

Applications of point (2) can, for instance, be encountered in the domain of electric double layers, with obvious interest for biological interfaces.

One example is that it is impossible to establish the absolute potential difference between two adjacent phases of different compositions. Otherwise stated, such a potential difference is thermodynamically inoperational. To realize that, one has to go back to the definition. The potential $\psi(r)$ at a certain position r is the electric work needed to transfer a unit charge from an agreed reference position to r. In actual systems ions are the charge carriers. If an ion has to be transferred across a phase boundary, not only electric work but also nonelectric, or chemical, work has to be performed. There is no operational way to separate the total isothermic reversible work (i.e., the change in Helmholtz or Gibbs energy, depending on whether V or p is kept constant) into its two parts.

The other involves the charge formation in double layers. Let us consider an insoluble oxide, for which we have defined the surface charge σ^0 through (I.2). There, we stated that this definition is not thermodynamic. The reason is that one cannot let single ionic species adsorb as the sole outcome of a process. Because of the phenomenological nature of thermodynamics, one cannot thermodynamically distinguish between ions adsorbed on the solid and those adsorbed on the solution side of the double layer. Thermodynamically, double layers are *electroneutral*. Hence, only electroneutral combinations of ions can be adsorbed or desorbed, and that we have the habit of calling some ions surface ions and others counterions is just our choice. One could equally well have identified the surface with the slip plane so that the charge on the solid plus the charge in the stagnant layer would become the "surface" charge. This alternative definition would not be at variance with thermodynamics.

Let us elaborate this for the preceding oxide in an aqueous electrolyte containing HNO_3, KOH, and KNO_3. For these three components the surface concentrations, Γ_{HNO_3}, Γ_{KOH}, and Γ_{KNO_3} are operable. Each proton that adsorbs must be accompanied by an NO_3^- ion, etc. Γ_{KNO_3} is always negative (except at the p.z.c., where it is zero): double layers expel (electroneutral!) electrolyte (*negative adsorption*, or the *Donnan effect*). In operational language we can now define the following three charges:

$$\sigma^o \equiv F(\Gamma_{HNO_3} - \Gamma_{KOH}) \tag{II.3}$$

$$\sigma_{K^+} \equiv F(\Gamma_{KOH} + \Gamma_{KNO_3}) \tag{II.4}$$

$$\sigma_{NO_3^-} \equiv F(\Gamma_{HNO_3} + \Gamma_{KNO_3}) \tag{II.5}$$

of which the sum is zero. σ_{K^+} and $\sigma_{NO_3^-}$ are the ionic components of charge: these charges together constitute the countercharge, but thermodynamics does not tell us so much.

In conclusion, operationally it must always be possible to describe charge formation and adsorption in electric double layers in terms of electroneutral components.

Now consider the notion of *driving force*. Especially for interactions involving biological surfaces, it is necessary to extend the restrictive notion of *mechanical force* to a *thermodynamic force*. Mechanical forces are purely energetic and conservative, whereas thermodynamic forces have an entropic component and are not necessarily conservative. The term "conservative" means that if work is done by that force, the energy resulting from the process remains stored in the system; it can be completely released if the process is reversed. Examples: the electric force of moving an ion toward a surface carrying fixed charges and the force needed to move an object up against gravity. In both cases the energy is released if the process is reversed. These two examples are illustrations of mechanical forces, characterized by the fact that they can be described without introducing the notion of temperature. Another aspect of conservative forces is that energies can be related to *potentials*. For instance, in electric fields the potential, $\psi(r)$ at a position r equals the electric energy $u_{el}(r)$ minus the electric energy in the reference state.

However, for our systems entropic contributions can rarely be ignored. Consequently, it is not enough to consider only the energy of a process; rather the Helmholtz or Gibbs energy is needed and the forces are temperature dependent. Typical examples include double-layer overlap and steric interaction upon adhesion of, say, bacterial cells. In the same vein, interfacial tensions are Helmholtz energies, so they have an energetic and an entropic contribution, and when processes take place under the influence of a gradient in the interfacial tension (Marangoni phenomena), the driving force has a thermodynamic nature.

Whether or not thermodynamic forces are conservative depends on the extent to which energy is dissipated. This, in turn, depends on the rate of the process or, more precisely, on De.

The upshot for the present chapter is that quantities and interactions must generally be treated thermodynamically rather than mechanically.

III. THE EXCESS NATURE OF INTERFACIAL QUANTITIES. THE GIBBS DIVIDING PLANE

Let us now consider interfacial energies, enthalpies, entropies, Helmholtz and Gibbs energies, and interfacial tensions and their relationships. We shall call them

U^σ, H^σ, S^σ, F^σ, G^σ, and γ, in this order. Thermodynamically, one cannot specify what the origin of, say, U^σ is or where this energy is exactly situated; neither is it possible to make any statements about the thickness of the interfacial layer. [We know from experience that GS interfaces are relatively thin and that charged interfaces may have thicknesses of $O(\kappa^{-1})$, but those are nonthermodynamic facts.] Nevertheless it is possible to derive the required relationships.

Consider first a two-phase system, containing the phases α and β in contact, separated by a flat interface of which the energy is U^σ. See Fig. 3. The total energy of the system U is measurable. We can write this as

$$U = U^\alpha + U^\beta + U^\sigma \tag{III.1}$$

Similar equations can be written for the other characteristic functions. Basically, U^σ is obtainable if U^α and U^β can be subtracted. For the isolated homogeneous phases α and β the energies can also be measured. These are extensive quantities, i.e., $U = VU_V$ if U_V is the energy per unit volume (J m^{-3} in SI units). U_V is also measurable and the subtraction becomes

$$U^\sigma = U - V^\alpha U_V^\alpha - V^\beta U_V^\beta \tag{III.2}$$

with $V^\alpha + V^\beta = V$ (known). Now it is appreciated where the problem lies: there is no way of telling where the one phase ends and the other starts because there is a certain transition range between them, however narrow it might be. Some convention is required to locate *the* interface. Hence, the value obtained for U^σ is subject to this choice, but as long as we remain in the domain of interfacial thermodynamics this does not matter, provided we adhere consistently to the same convention. Relations between measurable quantities should of course be unaffected by this choice.

We shall adhere to the *Gibbs convention*. Gibbs reasoned that, if we cannot thermodynamically state where the various interfacial excesses of molecules (n_i^σ) are,

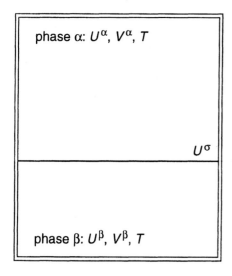

Fig. 3   Two phases in contact.

we can simply assign all these excesses to one mathematically thin plane whose position is determined by letting the interfacial excess of a major component (water in our case) be zero by definition:

$$n_w^\sigma \equiv 0 \qquad\qquad\qquad\text{(III.3)}$$

The interface defined in this way is called the *Gibbs dividing plane* and Fig. 4 illustrates what it means. The symbols $\rho_{N,s}$ and $\rho_{N,w}$ stand for the number densities (m^{-3}) of the solid and water, respectively. Over a (very) narrow range they change from their homogeneous bulk values to zero. The Gibbs dividing plane is thus situated so that the two hatched areas are identical. If we introduce the surface excess concentration of water (i.e., number of moles in the interface with respect to bulk water) this is also zero:

$$\Gamma_w = \frac{n_w^\sigma}{A} = 0 \qquad\qquad\qquad\text{(III.4)}$$

Here, A is the interfacial area. All other surface excesses, covering those of all dissolved components and the solid, are assumed to be located at the mathematically thin Gibbs dividing plane.

For our case at hand this convention is realistic. It means that all adsorptions, i.e., all surface excesses, are counted with respect to water, the density of which is assumed to stay constant up to the dividing plane. It can be proved that slight displacements of this plane have a negligible effect on the other Γ_i's. At the same time the Γ's in the thermodynamic equations may be identified with the analytically determined surface concentrations. Adsorption becomes more cumbersome to handle when the solutions are not dilute (i.e., when the mole fractions x_i of other components besides water are not $\ll x_w$), or when the surfaces are strongly curved. However, we have decided not to treat such systems.

In conclusion, interfacial excesses can be defined in terms of the Gibbs dividing plane, that is, with respect to adsorption of the solvent water. For all surface excesses the same convention must be maintained to ensure consistency. If that is done, there is thermodynamically no need to specify the structure of the interface further.

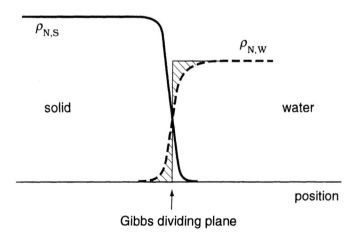

Fig. 4 Positioning of the Gibbs dividing plane between a solid and an aqueous solution.

IV. RELATIONS BETWEEN INTERFACIAL CHARACTERISTIC FUNCTIONS

A. Integral Relations

In this and the following section we shall derive integral and differential relationships between the interfacial excess characteristic functions. By way of a starting point, let us first recapitulate those for two homogeneous phases separated by an interface. In these equations U, H, S, F, G, V, and n_i apply to the entire system. Equilibrium is supposed, meaning that p, T, and all μ_i's are the same throughout the system.

The integral equations are

$$U = TS - pV + \gamma A + \sum_i \mu_i n_i \tag{IV.1}$$

$$H \equiv U + pV = TS + \gamma A + \sum_i \mu_i n_i \tag{IV.2}$$

$$F \equiv U - TS = -pV + \gamma A + \sum_i \mu_i n_i \tag{IV.3}$$

$$G \equiv U + pV - TS = \gamma A + \sum_i \mu_i n_i \tag{IV.4}$$

These equations, but without the γA term, can be found in all thermodynamics textbooks.

The corresponding differentials are

$$dU = T\,dS - p\,dV + \gamma\,dA + \sum_i \mu_i dn_i \tag{IV.5}$$

$$dH = T\,dS + V\,dp + \gamma\,dA + \sum_i \mu_i dn_i \tag{IV.6}$$

$$dF = -S\,dT - p\,dV + \gamma\,dA + \sum_i \mu_i dn_i \tag{IV.7}$$

$$dG = -S\,dT + V\,dp + \gamma\,dA + \sum_i \mu_i dn_i \tag{IV.8}$$

Let us assume that U^σ is known from the determination outlined in Sec. III. The next step to be taken is to define H^σ, F^σ, and G^σ. For bulk phases the energy is converted into an enthalpy via $H = U + pV$. In the Gibbs convention $V^\sigma = 0$, so the definition $H^\sigma \equiv U^\sigma + pV^\sigma$ makes no sense, although one could of course always decide that by definition H^σ and U^σ are identical. However, as we are dealing with interfacial work, the most logical interfacial equivalent is

$$H^\sigma \equiv U^\sigma - \gamma A \tag{IV.9}$$

In fact, this definition is recommended by the International Union of Pure and Applied Chemistry (IUPAC). A consequence of this way of defining H^σ is that this quantity is given not by $H^\sigma = H - H^\alpha - H^\beta$ (which would render H^σ and U^σ identical) but by

$$H^\sigma = H - H^\alpha - H^\beta - \gamma A \qquad\qquad (IV.10)$$

Physically, (IV.10) means that H^σ stands for the enthalpy that is left after subtracting the two bulk parts and the surface work contribution.

In line with this, the definitions of F^σ and G^σ follow as

$$F^\sigma \equiv U^\sigma - TS^\sigma \qquad\qquad (IV.11)$$

$$G^\sigma \equiv U^\sigma - TS^\sigma - \gamma A \qquad\qquad (IV.12)$$

where G^σ does not follow from $G^\sigma = G - G^\alpha - G^\beta$ but from

$$G^\sigma = G - G^\alpha - G^\beta - \gamma A \qquad\qquad (IV.13)$$

The second step is writing U^σ explicitly. To that end, starting from (III.1), substituting (IV.1) for U, and using the corresponding bulk expressions for U^α and U^β, we find

$$U^\sigma = TS - pV + \gamma A + \sum_i \mu_i n_i - TS^\alpha + pV^\alpha - \sum_i \mu_i n_i^\alpha - TS^\beta + pV^\beta$$
$$- \sum_i \mu_i n_i^\beta$$
$$= T[S - S^\alpha - S^\beta] - p[V - V^\alpha - V^\beta] + \gamma A - \sum_i \mu_i\left[n_i - n_i^\alpha - n_i^\beta\right]$$

or

$$U^\sigma = TS^\sigma + \gamma A + \sum_i \mu_i n_i^\sigma \qquad\qquad (IV.14)$$

Hence,

$$H^\sigma = TS^\sigma + \sum_i \mu_i n_i^\sigma \qquad\qquad (IV.15)$$

$$F^\sigma = \gamma A + \sum_i \mu_i n_i^\sigma \qquad\qquad (IV.16)$$

$$G^\sigma = \sum_i \mu_i n_i^\sigma \qquad\qquad (IV.17)$$

Equations (IV.14–17) complete the required set of integral expressions for surface excess characteristic functions. An alternative way to write these is by dividing by

A, writing U_a^σ for the interfacial excess energy per unit area (J m^{-2}), etc.

$$U_a^\sigma = TS_a^\sigma + \gamma + \sum_i \mu_i \Gamma_i \tag{IV.18}$$

$$H_a^\sigma = TS_a^\sigma + \sum_i \mu_i \Gamma_i \tag{IV.19}$$

$$F_a^\sigma = \gamma + \sum_i \mu_i \Gamma_i \tag{IV.20}$$

$$G_a^\sigma = \sum_i \mu_i \Gamma_i \tag{IV.21}$$

So, in this convention γ equals the excess Helmholtz energy per unit area for a pure liquid ($\Gamma_i = 0$). Equation (IV.21) is the interface equivalent of the three-dimensional $G = \Sigma_i \mu_i n_i$ for a homogeneous bulk phase [(IV.4) without the γA term]. Note that in these equations Γ_i and the excesses depend on the choice of the dividing plane, whereas γ is independent of it.

B. Differential Relations

For the differentials of the characteristic functions of the total energy and Helmholtz and Gibbs energy of a two-phase system with an interface the expressions have already been given; see (IV.5–8). These are the "differential counterparts" of the four integral expressions (IV.1–4). Now the corresponding differential equations between the surface excess functions are required. To that end, we proceed in the same way as in Sec. IV.A; i.e., dU^σ is obtained as $dU - dU^\alpha - dU^\beta$, with dU given by (IV.5), $dU^\alpha = TdS^\alpha - pdV^\alpha + \Sigma_i\mu_i dn_i^\alpha$, and similarly for dU^β. The following set results

$$dU^\sigma = T\,dS^\sigma + \gamma\,dA + \sum_i \mu_i dn_i^\sigma \tag{IV.22}$$

$$dH^\sigma = T\,dS^\sigma - A\,d\gamma + \sum_i \mu_i dn_i^\sigma \tag{IV.23}$$

$$dF^\sigma = -S^\sigma\,dT + \gamma\,dA + \sum_i \mu_i dn_i^\sigma \tag{IV.24}$$

$$dG^\sigma = -S^\sigma\,dT - A\,d\gamma + \sum_i \mu_i dn_i^\sigma \tag{IV.25}$$

This completes our sets of relations between interfacial excess characteristic functions. Together, these equations contain very much information and can be used for a variety of purposes. A few of these will now be mentioned.

1. There are two thermodynamic interpretations of the interfacial tension in terms of the interfacial Helmholtz energy, the integral one, derived from (IV.20),

$$\gamma = F_a^\sigma - \sum_i \mu_i \Gamma_i \qquad \text{(for solutions)} \qquad \text{(IV.26)}$$

$$\gamma = F_a^\sigma \qquad \text{(for a pure liquid)} \qquad \text{(IV.27)}$$

or the differential definition, from (IV.24),

$$\gamma = \left(\frac{\partial F^\sigma}{\partial A}\right)_{T,n_i^{\sigma'}\text{s}} \qquad \text{(for solutions)} \qquad \text{(IV.28)}$$

$$\gamma = \left(\frac{\partial F^\sigma}{\partial A}\right)_T \qquad \text{(for a pure liquid)} \qquad \text{(IV.29)}$$

Equations (IV.27) and (IV.29) are identical because $F_a^\alpha = F^\sigma/A$. Interpreting γ according to (IV.28) requires extension of the interface, keeping its composition constant, which is not an easy constraint. If the surface tension is expressed in terms of the excess energy, using (IV.22)

$$\gamma = \left(\frac{\partial U^\sigma}{\partial A}\right)_{S^\sigma,n_i^\sigma} \qquad \text{(IV.30)}$$

it is realized that area extension should take place at constant excess entropy, which is a very awkward procedure. Of course, γ can also be defined in terms of differentials of the entire system. For instance, from (IV.7) or (IV.8),

$$\gamma = \left(\frac{\partial F}{\partial A}\right)_{T,V,n_i'\text{s}} \qquad \text{(IV.31)}$$

$$\gamma = \left(\frac{\partial G}{\partial A}\right)_{p,T,n_i'\text{s}} \qquad \text{(IV.32)}$$

Operationally speaking, (IV.31) means that the area should be isothermally increased by a unit amount and the corresponding rise in the Helmholtz energy recorded, but the composition of the system has to remain unaltered and no volume work on the system is done; hence all the increase of F is associated with the interface.

2. Useful cross-differentiations can be carried out. For instance, from (IV.24)

$$\left(\frac{\partial \gamma}{\partial T}\right)_{A,n_i^\sigma} \left[= \left(\frac{\partial \gamma}{\partial T}\right)_{A,\Gamma_i} \right] = -\left(\frac{\partial S^\sigma}{\partial A}\right)_{T,n_i^\sigma} \qquad \text{(IV.33)}$$

This equation demonstrates what has stated in Sec. II.B regarding the fact that γ also had an entropic part: had the interfacial tension been purely energetic (i.e., purely mechanical), the right-hand side of (IV.33) would have been zero and γ would not vary with the temperature.

3. By combining integral and differential expressions interesting new information is obtainable. For instance, from (IV.17) it follows that $dG^\sigma = \Sigma_i \mu_i dn_i^\sigma + \Sigma_i n_i^\sigma d\mu_i$, which may be equated to (IV.25) to yield $A\,d\gamma + S^\sigma dT + \Sigma_i n_i^\sigma d\mu_i = 0$, or

$$d\gamma = -S_a^\sigma\,dT - \sum_i \Gamma_i d\mu_i \qquad\qquad (IV.34)$$

which is the famous *Gibbs adsorption equation*, to which we shall return in Sec. V because there is more to be said about it.

4. Equations (IV.22–25) can be used to introduce new characteristic functions, just as is routinely done for their three-dimensional counterparts.

V. THE GIBBS-DUHEM AND GIBBS (ADSORPTION) EQUATIONS

The *Gibbs-Duhem* equation is a familiar expression relating to the chemical potentials of the components in a homogeneous bulk mixture. We shall now derive it and then demonstrate that the Gibbs (adsorption) equation is virtually its interfacial equivalent.

A. The Gibbs-Duhem Equation

Only a limited number of state variables (p, T, μ_i's, ...) suffice to define completely the thermodynamic state of a system. Thermodynamics sets rules for this number. This rule is the sought Gibbs-Duhem equation and it is embodied in the integral and differential expressions we have for the characteristic functions, i.e., (IV.1–8) without the γA and γdA terms.

It is typical of functions of state that their integral equations completely specify them. Hence, if the differential of these is taken it is complete and identical to the differential as it occurs in the differential expression. The two may therefore be equated, and in doing so the sought additional relation is obtained. It does not matter for which characteristic function the exercise is done; for each the result is the same. We already used this strategy to obtain (IV.34). Let us take the energy as the example. For a bulk phase $U = TS - pV + \Sigma_i \mu_i n_i$ [(IV.1) without the interface term]. So $dU = T\,dS + S\,dT - p\,dV - Vdp + \Sigma_i \mu_i dn_i + \Sigma_i n_i d\mu_i$, which may be equated to dU in (IV.5), again without the $\gamma\,dA$ term, resulting in

$$SdT - V\,dp + \sum_i n_i d\mu_i = 0 \qquad\qquad (V.1)$$

or, on a molar basis,

$$S_m\,dT - V_m\,dp + \sum_i x_i d\mu_i = 0 \qquad\qquad (V.2)$$

where the mole fraction of i is $x_i = n_i/\Sigma_i n_i$. Expression (V.1) or (V.2) is the required *Gibbs-Duhem relation*.

Equation (V.2) is the basis of the phase rule, in that it tells us how many variables can be independently varied. For a homogeneous bulk phase at fixed p and T, containing j components, only the chemical potential of $j-1$ can be freely chosen.

For interfacial thermodynamics the relevance is that interfaces are not autonomous, they can exist only by virtue of the two adjoining phases, and the two Gibbs-Duhem relations in these phases automatically fix the number of variables that can be independently varied.

B. On the Derivation of the Gibbs Adsorption Equation

We have had this equation before [see (IV.34)] and derived it by equating dG^σ from (IV.17) to dG^σ according to (IV.25). The derivation is the two-dimensional analogue of that leading to (V.2) so the Gibbs adsorption equation can be viewed as the two-dimensional Gibbs-Duhem relation. Instead of the volume work term, $-V_m\,dp$, we now have the surface work term, $\gamma\,dA$. One difference is that in (IV.34) the sum over i extends over all components except water, because we have used $\Gamma_w = 0$ 0 to locate the dividing plane, subsequently referring all other surface concentrations to that of water. In fact, when working isothermally (usually the case in many biological experiments) μ^S is constant, so that the contribution of the solid also drops out. Then, more simply,

$$dy = -\sum_i \Gamma_i d\mu_i \qquad (T \text{ constant}) \tag{V.3}$$

where the sum extends over all dissolved components.

The fact that the Gibbs equation was derived from dG^σ, that is, the differential of a characteristic function requiring some convention, did not matter. Only the property of being a complete differential was needed. It is also possible to derive it without the introduction of G^σ, starting from any of the bulk characteristic functions, but now with the interfacial work term included. For example, choosing the energy function to elaborate this, we take the differential from (IV.1) and equate it to (IV.5). Gibbs' adsorption equation follows immediately.

C. Application of the Gibbs Equation

For fluid interfaces, for which γ is measurable, the Gibbs equation is often used to establish adsorbed amounts. The advantage is that this procedure may be applied for very small interfacial areas, where not enough solute adsorbs to estimate Γ analytically. For various kinds of surfactants this has been carried out to find Γ_i as a function of μ_i. Otherwise stated, in this way adsorption isotherms can be obtained.

Considering biological surfaces implies that γ is often not measurable. However, the property that dy is a complete differential enables one to relate different surface concentrations to each other. The great advantage of this is that coadsorption processes can be identified and quantified and/or that the adsorption of secondary components can be computed from that of the primary one, even if (again) the former is analytically not measurable.

A few cases are discussed in the following.

1. Electric Double Layers

Consider an insoluble oxide in an aqueous solution of HNO_3, KOH, and KNO_3, already discussed in Sec. IIB. Let the temperature be constant and the oxide inert

$(d\mu_s = 0)$. Then four components are left in the Gibbs equation, which we write phenomenologically in terms of electroneutral components

$$d\gamma = -\Gamma_{HNO_3}d\mu_{HNO_3} - \Gamma_{KOH}d\mu_{KOH} - \Gamma_{KNO_3}d\mu_{KNO_3} - \Gamma_w d\mu_w \qquad (V.4)$$

where the subscript w stands for water. In practice, usually the difference $(\Gamma_{HNO_3} - \Gamma_{KOH})$ is measured by potentiometric acid-base titration and related to the surface charge, according to (II.3). More rarely, the negative adsorption Γ_{KNO_3} is also determined, but because of the coupling the latter can also be obtained from the former.

Equation (V.4) is still redundant in the number of variables, because of the acid-base reaction $HNO_3 + KOH \rightleftharpoons KNO_3 + H_2O$, leading to

$$d\mu_{KOH} + d\mu_{HNO_3} = d\mu_{KNO_3} + d\mu_w \qquad (V.5)$$

The terms containing $d\mu_w$ in (V.4) and (V.5) can be eliminated using the Gibbs-Duhem relation (V.2): at given p and T, $d\mu_w = -\Sigma_{j \neq i}(x_i/x_w)d\mu_i$ which for the dilute solutions we are considering is negligible. Hence, all $d\mu_w$ terms ≈ 0. With this equation we can eliminate either $d\mu_{KOH}$ or $d\mu_{HNO_3}$. If we get rid of $d\mu_{KOH}$ the result is

$$d\gamma = -(\Gamma_{HNO_3} - \Gamma_{KOH})d\mu_{HNO_3} - (\Gamma_{KOH} + \Gamma_{KNO_3})d\mu_{KNO_3} \qquad (V.6a)$$

Had we eliminated $d\mu_{HNO_3}$ we would have obtained

$$d\gamma = (\Gamma_{HNO_3} - \Gamma_{KOH})d\mu_{KOH} - (\Gamma_{HNO_3} + \Gamma_{KNO_3})d\mu_{KNO_3} \qquad (V.6b)$$

These two expressions are equivalent. Using (II.3–5) they can also be written as

$$Fd\gamma = \pm\sigma^\circ d\mu_{HNO_3} - \sigma_{K^+}d\mu_{KNO_3} \qquad (V.7a)$$

$$Fd\gamma = \sigma^\circ d\mu_{KOH} + \sigma_{NO_3^-}d\mu_{KNO_3} \qquad (V.7b)$$

Experimentally σ° is measured as a function of pH at various electrolyte concentrations. The conversions are easy and thermodynamically correct provided $a_{KNO_3} \gg a_{HNO_3}, a_{KOH}$, a condition that is always met in the practice of titrations. Then, to a very good approximation $d\mu_{HNO_3} = d\mu_{H^+} + d\mu_{NO_3^-} = d\mu_{H^+}$ because $\mu_{NO_3^-}$ is dominated by the nitrate originating from the KNO_3, which is kept constant when HNO_3 is added. Further, $d\mu_{H^+} = RTd \ln a_{H^+} = -2.303RTd$pH and $d\mu_{KNO_3} = d\mu_{K^+} + d\mu_{NO_3^-} = RTd \ln a_{K^+} + RTd \ln a_{NO_3^-} \approx 2RTd \ln a_{KNO_3}$.

The last transition involves a nonthermodynamic step of which we must be aware. Single ionic activities (a_+, a_-) are not operational, nor are single ion chemical potentials (μ_+, μ_-). It is impossible to add single ions to an aqueous phase as the sole outcome of an unambiguous process. However, it is always possible to *define* single ionic activities and chemical potentials, provided these nonthermodynamic definitions are consistent with phenomenological behavior. The customary way of treating this, also by us, is to assume that the relations $\mu_+(a_+)$ and $\mu_-(a_-)$ are exactly identical to those of uncharged compounds, i.e., $\mu_+ = \mu_+^\circ + RT \ln a_+$, etc. This relation is at least correct in the limit of very low concentrations, where the ions behave as if they are on their own in the system. In the limit of extreme dilution $d\mu_+$, $d\mu_-$, and $d\mu$

are related through $d\mu = d\mu_+ + d\mu_- = RTd \ln c_+ + RTd \ln c_- = 2RTd \ln c$ and the assumption is made that this remains valid when activity coefficients cannot be ignored, i.e., one writes

$$\mu_{KNO_3} = \mu^o_{KNO_3} + 2RT \ln a_{\pm} = \mu^o_{KNO_3} + RT \ln a_{K^+}a_{NO_3^-} \tag{V.8}$$

with $\mu^o_{KNO_3} = \mu^o_{K^+} + \mu^o_{NO_3^-}$ and the *mean activity* $a_{\pm}$ is defined via

$$a^2_{\pm} \equiv a_{K^+}a_{NO_3^-} \tag{V.9}$$

Similarly, introducing individual ionic activity coefficients y via $a_{K^+} = y_{K^+}c_{K^+}$, etc., we may also write

$$\mu_{KNO_3} = \mu^o_{KNO_3} + 2RT \ln y_{\pm}c_{KNO_3} \tag{V.10}$$

with

$$y^2_{\pm} \equiv y_{K^+}y_{NO_3^-} \tag{V.11}$$

Mean ionic activities and mean ionic activity coefficients are measurable.

Returning after this digression to the elaboration of (V.7a,7b), we eventually obtain

$$\begin{cases} Fd\gamma = 2.303RT\sigma^o d\text{pH} - 2\sigma_{K^+}RTd \ln y_{\pm}c_{KNO_3} \\ Fd\gamma = -2.303RT\sigma^o d\text{pH} + 2\sigma_{NO_3^-}RTd \ln y_{\pm}c_{KNO_3} \end{cases} \tag{V.12a, b}$$

This pair contains the variables that are experimentally controlled.

2. Ionic Components of Charge

One of the interesting features of (V.7a,b) is that they show that the *ionic components of charge*, σ_{K^+} and $\sigma_{NO_3^-}$, are thermodynamically accessible, except for a constant. This must mean that the contributions of positive adsorption (ions accompanying adsorbing H^+ and OH^- ions) and negative adsorption (Donnan expulsion) are fully defined by the titration curves. Figure 5 gives a sketch of such a curve for an oxidic surface in an electrolyte solution. For other systems the thermodynamics remain the same, mutatis mutandis. The information that we need is the slope of the various curves at each pH and the distance between the curves at constant pH or at constant σ^o. So, for good accuracy a fine-grained set of curves is required.

Assuming such a set to be available, the thermodynamic analysis starts again from the Gibbs equation. For instance, starting from (V.7a) and cross-differentiating

$$\left(\frac{\partial\sigma^o}{\partial\mu_{KNO_3}}\right)_{\mu_{HNO_3}} = \left(\frac{\partial\sigma_{K^+}}{\partial\mu_{HNO_3}}\right)_{\mu_{KNO_3}} \tag{V.13}$$

Following the same reasoning as in the previous section, for $c_{KNO_3} \gg c_{HNO_3}$ the right-hand side (r.h.s.) may be written as $-(\partial\sigma_{K^+}/2.303RTd\text{pH})_{a_{\pm}}$ considering that at constant T the only way to keep μ_{KNO_3} fixed is to keep $a_{\pm}$ constant. In the left-hand side (l.h.s.) we want pH constant rather than μ_{HNO_3}; this conversion can be

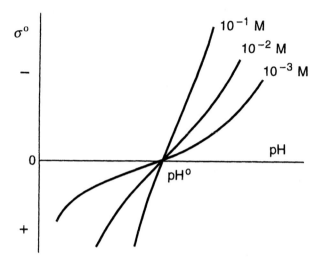

Fig. 5   Sketch of a typical experimental set of σ° (pH, c_{salt}) curves for an insoluble oxide in a solution of an indifferent electrolyte, of which the concentration (mol dm^{-3}) is given. The point of zero charge coincides with the common intersection point (for electrolytes containing specifically adsorbing ions this is no longer the case, although the thermodynamics of Sec. V. C1 and 2 remains valid). The first and second terms on the r.h.s. of (V.16) are identified as the vertical distance between and the slope of the curves, respectively.

accomplished using the so-called extended chain rule, which for the case at hand reads

$$\left(\frac{\partial \sigma^\circ}{\partial \mu_{KNO_3}}\right)_{pH} = \left(\frac{\partial \sigma^\circ}{\partial \mu_{KNO_3}}\right)_{\mu_{HNO_3}} + \left(\frac{\partial \sigma^\circ}{\partial \mu_{HNO_3}}\right)_{\mu_{KNO_3}} \left(\frac{\partial \mu_{HNO_3}}{\partial \mu_{KNO_3}}\right)_{pH} \qquad (V.14)$$

$$\frac{1}{2RT}\left(\frac{\partial \sigma^\circ}{\partial \ln a_\pm}\right)_{pH} = \left(\frac{\partial \sigma^\circ}{\partial \mu_{KNO_3}}\right)_{\mu_{HNO_3}} - \frac{1}{2.303RT}\left(\frac{\partial \sigma^\circ}{\partial pH}\right)_{a_\pm} \cdot \frac{1}{2} \qquad (V.15)$$

Hence,

$$\left(\frac{\partial \sigma_{K^+}}{\partial pH}\right)_{a_\pm} = -\frac{1}{2}\left(\frac{\partial \sigma^\circ}{\partial \log a_\pm}\right)_{pH} - \frac{1}{2}\left(\frac{\partial \sigma^\circ}{\partial pH}\right)_{a_\pm} \qquad (V.16)$$

The r.h.s. contains measurable quantities only. Titrations produce σ° (pH) curves at various $a_\pm$. The first term on the r.h.s. gives the cross section at given pH, the second the slope at given electrolyte activity, that is, at given electrolyte concentration, because $c_{KNO_3} \gg c_{HNO_3}, c_{KOH}$. The slope is proportional to the double layer capacitance; this capacitance increases with c_{KNO_3} because screening allows more charges to be adsorbed for a given pH increment.

   Absolute values of σ_{K^+} cannot be obtained in this way. A reference value is needed, say σ_{K^+} at the point of zero charge. If from other experience it is known that K$^+$ does not adsorb specifically, σ_{K^+} (pH) = 0. Generally σ_{K^+}(pH) is obtained with respect to that reference by integration of (IV.16)

$$\sigma_{K^+}(pH) = \sigma_{K^+}(pH^\circ) - \frac{1}{2}\int_{pH^\circ}^{pH}\left(\frac{\partial\sigma^\circ}{\partial\ln a_\pm}\right)_{pH} dpH - \frac{1}{2}\sigma^\circ(pH) \qquad (V.17a)$$

Similarly,

$$\sigma_{NO_3^-}(pH) = \sigma_{NO_3^-}(pH^\circ) + \frac{1}{2}\int_{pH^\circ}^{pH}\left(\frac{\partial\sigma^\circ}{\partial\ln a_\pm}\right)_{pH} dpH - \frac{1}{2}\sigma^\circ(pH) \qquad (V.17b)$$

It is readily verified that (V.17a and b) are equivalent, using the electroneutrality condition $\sigma^\circ = -(\sigma_{K^+} + \sigma_{NO_3^-})$ at any pH.

3. Double Layer in the Presence of an Uncharged Organic Adsorptive

Let us call the organic molecule A. It may be an alcohol, urea, or any low-molecular-weight compound. To the Gibbs equation a term $-\Gamma_A d\mu_A$ now has to be added. Extending (V.7a,b),

$$Fd\gamma = -\sigma^\circ d\mu_{HNO_3} - \sigma_{K^+}d\mu_{KNO_3} - F\Gamma_A d\mu_A \qquad (V.18)$$

Cross-differentiation between the first and third terms on the r.h.s. at constant μ_{KNO_3} yields

$$\frac{1}{F}\left(\frac{\partial\sigma^\circ}{\partial\mu_A}\right)_{\mu_{HNO_3},\mu_{KNO_3}} = \left(\frac{\partial\Gamma_A}{\partial\mu_{HNO_3}}\right)_{\mu_{KNO_3},\mu_A} \qquad (V.19)$$

Transformation is simple; in line with Sec. V.C.1

$$\frac{1}{F}\left(\frac{\partial\sigma^\circ}{\partial\ln c_A}\right)_{pHc_{KNO_3}} = -0.434\left(\frac{\partial\Gamma_A}{\partial pH}\right)_{c_{KNO_3},c_A} \qquad (V.20)$$

Nothing can be inferred from the titration curves if the organic does not adsorb at all or if its adsorption does not change with pH. However, if Γ_A does depend on pH, this leads to a coupled c_A dependence of σ° at given pH, from which $\Gamma_A(pH)$ can be computed except for a constant. As the relation between σ° and pH is also known, the dependence of Γ_A on σ° can be established, even if this could not be directly measured, say, for analytical reasons.

For a variety of organic adsorptives (mostly on AgI as the adsorbate rather than oxides) it was thus found that $\Gamma_A(\sigma^\circ)$ curves passed through a maximum. This maximum is a thermodynamic fact, but interpretation requires a molecular model. Most probably the maximum finds its origin in the competition between water dipoles and A molecules (polar or not) for the surface. At very high positive or very high negative surface charge the water molecules may be preferred because of their higher dipole moment, so that somewhere between these extremes the organic molecules can compete most successfully. Usually the resulting maximum does not coincide with the point of zero charge.

Phenomena such as this one, in which adsorption and double-layer phenomena influence each other, are collectively known as *electrosorption*. Interfacial thermodynamics is very helpful in analyzing the corresponding titration data.

4. Adsorption of Ionic Surfactants

As another illustration of the application of Gibbs' adsorption law consider the surface tension lowering caused by ionic surfactants. This is of course a very important topic for wetting, emulsification, foaming, etc. In the (older) literature there has been a discussion about the need to have a factor of 2 in the simplified Gibbs equation $d\gamma = -2RT\Gamma_{surf}d \ln c_{surf}$, where the subscript surf refers to the surface and bulk concentrations of the surfactant. The absence or presence of the 2 leads to an uncertainty of the same amount in Γ if determined from the slopes of $\gamma(\ln c)$ curves. Using Gibbs' law it is immediately appreciated that the factor is related to the absence or presence of electrolyte.

On any surface in an aqueous solution containing, say, NaDS (sodium dodecyl sulfate) and NaCl at fixed temperature

$$d\gamma = -\Gamma_{NaDS}d\mu_{NaDS} - \Gamma_{NaCl}d\mu_{NaCl} \tag{V.21}$$

The phenomenology of this equation implies that at this level we cannot worry about problems such as "does Γ_{NaDS} include only the surfactant or the surfactant with its counterion?" Splitting the chemical potentials into their ionic parts,

$$d\gamma = -\Gamma_{NaDS}\left(d\mu_{Na^+} + d\mu_{DS^-}\right) - \Gamma_{NaCl}\left(d\mu_{Na^+} + d\mu_{Cl^-}\right) \tag{V.22}$$

$$= -RT\left[\Gamma_{NaDS}\left(\frac{da_{Na^+}}{a_{Na^+}} + \frac{da_{DS^-}}{a_{DS^-}}\right) + \Gamma_{NaCl}\left(\frac{da_{Na^+}}{a_{Na^+}} + \frac{da_{Cl^-}}{a_{Cl^-}}\right)\right] \tag{V.23}$$

Now it is seen that the factor 2 depends on the salt concentration relative to the surfactant concentration. If $c_{NaCl} \gg c_{NaDS}$, da_{Na^+}/a_{Na^+} is negligible upon increase of c_{NaDS}, so the first term becomes $RT\Gamma_{NaDS}da_{DS^-}/aDS^- \approx RT\Gamma_{NaDS}d \ln c_{NaDS}$. However, in the opposite case the first term RT $\Gamma_{NaDS}(da_{Na^+}/a_{Na^+} + da_{DS^-}/a_{DS^-}) \approx 2RT\Gamma_{NaDS}d \ln c$ NaDS. For the negative adsorption of NaCl the factors are just the other way around. For intermediate cases (V.23) requires further elaboration and nonthermodynamic assumptions have to be made. Obviously, there is a gradual transition between 1 and 2 for the surfactant term with increasing c_{NaDS}/c_{NaCl} ratio and the reverse for the negative adsorption contribution. We shall not elaborate this.

5. Adsorption of Polymers and Proteins

Unfortunately, the Gibbs equation cannot be applied to these systems because of the irreversibility of the adsorption of these components. As discussed in connection with Fig. 2, there is no unique way of linking $d\gamma$ to Γ and $d \ln c$. If we applied this law to the range where the hysteresis loop is found, we would conclude that γ decreases more on the way up than it increases on the way down. In fact, this is sometimes observed experimentally, but then the conclusion is drawn that for the dashed parts in the figure, for which $d\gamma/d \ln c$ is found to be low, the surface concentration is also low. And for the drawn "return loop" Γ is even found to be zero. All of this is clearly in conflict with reality. Computations of Γ from $d\gamma/d \ln c$ for polymers also lead to erroneous results. Apart from this, there is the practical problem that polymers are usually heterodisperse. The longer molecules adsorb more strongly, but the smaller ones adsorb faster. Hence, with time there is a gradual increase of Γ, which is experimentally not reflected in a decrease of γ. In fact, from statistical thermodynamics it can be concluded that γ is related to the density profile of adsorbed macromolecules in such a way that the segments that are in the interface proper (the so-called train segments) more strongly contribute to the reduction of γ than those farther out. Recall the adsorption of surfactants: long-chain and short-chain molecules having the same surface concentration in moles m^{-2} lead to the same reduction in γ, although for the former the adsorbed amount in mg m^{-2} is higher. So, even if there were equilibrium, the conversion of moles in weight, or in segments, has its subtleties and so has the issue of relating $d\mu$ to $d \ln c$. Mutatis mutandis, the same can be stated for proteins.

What *is* possible, however, is to study the adsorption of a minor component on top of that of an adsorbed macromolecule, provided the equilibrium of the latter is frozen. This is an illustration of a situation where $De_1 \ll 1$ and $De_2 \gg 1$, as discussed in Sec. I.B. An example is the Donnan exclusion of electrolyte from a polyelectrolyte or protein adsorbate as a function of c_{salt}.

VI. INTERFACIAL STATISTICAL THERMODYNAMICS

In Sec. II we discussed the macroscopic-phenomenological nature of classical thermodynamics. The lower size borderline of its applicability was related to the relative magnitude of the fluctuations in extensive variables. Underneath that border one can apply *statistical thermodynamics*, also called *molecular thermodynamics* or *statistical mechanics*. The two names do not reflect entirely equivalent approaches but rather differences in the way the statistics are handled. Common to all of these disciplines is that they start from a specific model of interactions and phenomena at the molecular level, thereafter invoking statistical arguments to predict macroscopic properties. One of these arguments is that probabilities of finding systems in a given molecular state become progressively more accurate when the numbers of molecules in the system grow. So, by statistical thermodynamics it is possible to give molecular interpretations of macroscopic quantities (the entropy and interfacial tension, for example) and to describe the properties of systems such as adsorbates and polymer coils.

Regarding adsorption isotherms, classical thermodynamics *can* be involved to describe them, but in this approach molecular properties cannot be made explicit. For example, in deriving the *Langmuir adsorption isotherm* (equation) for (sub-)monolayer adsorption of small uncharged molecules

$$\frac{\theta}{1-\theta} = k_L c \tag{VI.1}$$

where θ is the fraction of the monolayer covered $[= \Gamma / \Gamma(max)]$, one could start by assuming that for the adsorbing component i the chemical potential in the solution is

$$\mu_i = \mu_i^o + RT \ln c_i \tag{VI.2}$$

and that the corresponding expression for the adsorbate is

$$\mu_i^\sigma = \mu_i^{\sigma o} + RT \ln\left[\frac{\theta}{(1-\theta)}\right] \tag{VI.3}$$

Equating μ_i and μ_i^σ gives (VI.1) and relates k_L to $\mu_i^{\sigma o} - \mu_i^\sigma$. The advantage of such a derivation is that it shows that (VI.1) must have general validity if the solution is ideal and the adsorbate has an ideal entropy of mixing between covered ($\sim \theta$) and empty $[\sim(1-\theta)]$ parts, because under these premises (VI.2 and 3) do apply. The drawback is that little can be stated about k_L. First, there is a standard state problem: in (VI.2) the logarithm is taken of a dimension-having quantity, implying that μ_i^0 contains a $-RT \log c_i^*$ contribution, where c_i^* is a standard concentration. Otherwise, k_L contains a standard Gibbs energy of transfer, but one cannot say whether this term contains entropic contributions (besides the configurational parts, which are already accounted for in the $RT \log$ terms). By statistical means all of this can be made explicit, but then it should be kept in mind that the outcome is as good as the premises are.

This introductory example illustrates that classical thermodynamics and statistical thermodynamics are each other's counterparts. In the following we shall briefly review some basic aspects of statistical thermodynamics, avoiding most of the mathematics, and thereafter concentrate on interfaces, adsorption, and two-dimensional equations of state in particular. The reader is not assumed to be (very) familiar with this.

A. Some General Principles

In statistical thermodynamics and statistical mechanics the two laws of thermodynamics are replaced by two basic postulates of general validity. These postulates are different between statistical mechanics and thermodynamics and so are the elaborations, but for both approaches they refer to counting probabilities in multimolecular systems. We shall essentially follow the thermodynamic approach, largely developed by Gibbs. In passing, it is mentioned that Monte Carlo and Molecular Dynamics simulations also invoke some of these principles.

Classically, the thermodynamic state of a system is described by a limited number of macroscopic variables. For instance, for a closed isothermal system these are the n_i's, V, and T; for an open system the μ_i's, V, and T. For interfacial systems (adsorbates) we choose the n_i^σ's, A, T or the μ_i^σ's, A, and T, respectively. An interfacial system is closed/open when it cannot/can exchange matter with the surroundings, i.e., with the solution. Examples: spread and adsorbed monolayers, respectively. Molecularly, the system is interpreted as consisting of molecules. Their numbers being N_i, the macroscopic state is given by N_i, V, T; N_i^σ, A, T; μ_i^σ, A, T; etc. To each thermodynamic state belong a vast number Ω of *molecular states*, all of them differing with respect to positions and interactions of molecules. Ω is called the *degeneracy* of a system; its order of magnitude is $O(10^N)$, i.e., $O(10^{10^{23}})$ for molar amounts of matter. To each molecular state j a *probability* P_j can be assigned. These P_j's are very small numbers but the sum over all states is not small and by definition *normalized* to unity:

$$\sum_j P_j = 1 \tag{VI.4}$$

Not every molecular state is equally probable, and the basic issue is to count and weigh these probabilities using the postulates. In a closed adsorbate (N_i^σ, A, and T fixed), the energy can fluctuate. For each state j, U_j^σ has to be evaluated and averaged to give the average energy $\langle U^\sigma \rangle$, which may be identified with the thermodynamic U^σ. In the

Gibbs approach the problem of following the fluctuations in U, U^σ, N, etc. with time is avoided by defining *ensembles*, imaginary collections of a very large number L of systems, that are all in the same thermodynamic state, but in which the molecular state may vary from system to system, subject to the macroscopic constraints (say, V, T constant). As a whole, the ensemble is isolated from the environment, so for each system in the ensemble all the remaining $L-1$ systems together constitute *the* surroundings. Depending on the thermodynamic state of the systems in it, different ensembles can be constructed. Some of the familiar ones have special names. Of these, we shall consider only the ensemble consisting of closed isothermal systems, called *canonical*, and that composed of open isothermal systems, called *grand canonical*, or just *grand*. These names have been coined by Gibbs. For systems in a canonical ensemble the energy can fluctuate; for those in a grand canonical ensemble the energy and the numbers of molecules can both fluctuate.

Using the ensemble method it is possible to derive *partition functions*. They play a central role in statistical theories because they allow the determination of thermodynamic and mechanical quantities of the entire system by weighting and summing the contributions of the various molecular states: partition functions are *weighted sums*. In German and Dutch they are called "state sums" and in Russian "statistical sums." The next step is to make this quantitative for the canonical and grand canonical ensembles.

B. Partition Functions

The *canonical partition functions* $Q(N_i$'s, V, $T)$ and $Q(N_i^\sigma$'s, A, $T)$ are defined as

$$Q(N, V, T) = \sum_j \Omega_j e^{-U_j(N,V)/kT} \tag{VI.5a}$$

and

$$Q(N^\sigma, A, T) = \sum_j \Omega_j e^{-U_j^\sigma(N^\sigma,A)/kT} \tag{VI.5b}$$

respectively. Equation (VI.5b) applies to a system containing N^σ molecules on an area A, i.e., having a surface concentration of N^σ/A per area or $\Gamma = N^\sigma/AN_{Av}$, where N_{Av} is Avogadro's number. For more types of molecules one writes N_1^σ, N_2^σ, ... and for bulk phases A is replaced by V. The degeneracy Ω_j counts how often the energy U_j^σ occurs in the probability distribution. When the various energies U_j^σ are very close to each other, the sum can be replaced by an integral.

A basic feature of (VI.5) is that $U_j^\sigma(N, A)$ is a *purely mechanical quantity*; i.e., it can be calculated without invoking the notion of temperature. For instance, one could compute it on the basis of Schrödinger's equation. It is because of the kT in the denominator of the exponent and the way of summing (Boltzmann factors) that Q becomes a *thermodynamic quantity*. For example, from Q the entropy or the Helmholtz energy can be obtained. One can also find the thermodynamic energy U of the system.

For a surface phase the grand *(canonical)* partition function is

$$\Xi(\mu, A, T) = \sum_{N^\sigma} \sum_j \Omega_j(N^\sigma) e^{-U_j^\sigma(N^\sigma,A)/kT} e^{N^\sigma \mu/kT} \tag{VI.6}$$

or

$$\Xi(\mu, A, T) = \sum_{N^\sigma} Q(N^\sigma, A, T) e^{N^\sigma \mu / kT} \tag{VI.7}$$

and similarly for a bulk phase.

The transition from $Q(N^\sigma, A, T)$ to $\Xi(\mu, A, T)$ in (VI.7) is the statistical way of changing variables; it may be compared with the way in which this is done in classical thermodynamics.

Partition functions contain all the molecular information of the system and that is why they are so important. From them it is possible to find the *standard deviations* in U and N. In fact, quantitative expressions for σ_U and σ_N in (II.1) can be obtained from the canonical and grand canonical partition functions, respectively. It is also possible to quantify the probabilities P_j. For instance, canonically

$$P_j = P(U_i) = \frac{\Omega_j e^{-U_j/kT}}{\sum_j \Omega_j e^{-U_j/kT}} = \frac{\Omega_j e^{-U_j/kT}}{Q(N, V, T)} \tag{VI.8}$$

which satisfies the normalization requirement (VI.4).

Most important for our purposes is how mechanical and thermodynamic quantities are obtained. There are fixed rules for that; they read as follows.

Canonically, for three-dimensional systems

$$F = -kT \ln Q \tag{VI.9}$$

$$S = -k \sum_j P_j \ln P_j \tag{VI.10}$$

$$S = k \ln Q + kT \left(\frac{\partial \ln Q}{\partial T} \right)_{N,V} \tag{VI.11}$$

$$U = kT^2 \left(\frac{d \ln Q}{dT} \right)_{N,V} \tag{VI.12}$$

$$H = kT^2 \left(\frac{\partial \ln Q}{\partial T} \right)_{N,V} + kTV \left(\frac{\partial \ln Q}{\partial V} \right)_{T,N} \tag{VI.13}$$

$$G = -kT \ln Q + kTV \left(\frac{\partial \ln Q}{\partial V} \right)_{T,N} \tag{VI.14}$$

$$p = kT \left(\frac{\partial \ln Q}{\partial V} \right)_{T,N} \tag{VI.15}$$

$$\mu = -kT \left(\frac{\partial \ln Q}{\partial N} \right)_{T,V} \tag{VI.16}$$

This chemical potential is a factor of N_{Av} lower than that in classical thermodynamics.

For two-dimensional systems the required alterations are that F in (VI.9) becomes F^σ, etc., and (VI.15) becomes

$$\pi = kT\left(\frac{\partial \ln Q}{\partial A}\right)_{T,N^\sigma} \tag{VI.17}$$

where π is the *two-dimensional* or *interfacial pressure*. The quantity $-kT \ln(\text{partition function})$ is called the *characteristic function* of the ensemble. For the canonical ensemble this is the Helmholtz energy; see (VI.9). Differentiations like the one in (VI.15) are therefore equivalent to the thermodynamic $p = (\partial F/\partial V)_{T,N}$.

One of the postulates states that when the energy of the system is fixed all molecular states are equally probable: $P_1 = P_2 = \cdots P_j \cdots = \Omega^{-1}$, so that in that case (VI.10) reduces to

$$S = -k\Omega\left(\frac{1}{\Omega}\ln\frac{1}{\Omega}\right) = k \ln \Omega \tag{VI.18}$$

which is the famous Boltzmann equation. It is a particular case of (VI.10).

Grand canonically the corresponding expressions become

$$F = -kT \ln \Xi + kT\mu\left(\frac{\partial \ln \Xi}{\partial \mu}\right)_{V,T} \tag{VI.19}$$

$$S = k \ln \Xi + kT\left(\frac{\partial \ln \Xi}{\partial T}\right)_{V,\mu} \tag{VI.20}$$

$$U = kT^2\left(\frac{\partial \ln \Xi}{\partial T}\right)_{V,\mu} + kT\mu\left(\frac{\partial \ln \Xi}{\partial \mu}\right)_{V,T} \tag{VI.21}$$

$$H = kT^2\left(\frac{d \ln \Xi}{dT}\right)_{V,\mu} + kT\mu\left(\frac{\partial \ln \Xi}{\partial \mu}\right)_{V,T} + kT \ln \Xi \tag{VI.22}$$

$$G = N\mu \tag{VI.23}$$

$$pV = kT \ln \Xi \tag{VI.24}$$

$$p = kT\left(\frac{\partial \ln}{\partial V}\right)_{T,\mu} \tag{VI.25}$$

$$N = \langle N \rangle = kT\left(\frac{\partial \ln \Xi}{\partial \mu}\right)_{T,V} \tag{VI.26}$$

In this case $-pV$ is the characteristic function, i.e., the (volume) work term. For a two-dimensional system this becomes

$$\pi A = kT \ln \Xi \tag{VI.27}$$

and for a system containing a bulk phase and an interface

$$\pi A + pV = kT \ln \Xi \tag{VI.28}$$

The two-dimensional pressure can also be obtained from (VI.25), after replacing V by A, whatever is more convenient. The two-dimensional equivalent of (VI.26) is

$$N^\sigma = \langle N^\sigma \rangle = kT \left(\frac{\partial \ln \Xi}{\partial \mu} \right)_{A,T} \tag{VI.29}$$

With this, the strategy for obtaining mechanical and thermodynamic quantities can be completed. It consists of three steps. First, a *subsystem* is selected, which is such a small building brick of the system that it can be completely treated mechanically. For interfacial science it could be an empty or filled adsorption site (in the case of localized adsorption) or the two-dimensional equivalent of a "particle in a box." Sometimes it is expedient to evaluate first the partition function of the subsystem only. We shall use small letters to indicate subsystem properties. Canonically,

$$q(T) \equiv \sum_j \omega_j e^{-u_j/kT} \tag{VI.30}$$

where u_j is the energy of the subsystem in state j. When the subsystems are dependent, q becomes $q(N, T)$. Second, the complete system is built from these building bricks, considering their interactions; this must lead to a partition function. Finally, the required mechanical and/or thermodynamic characteristics are derived.

C. Derivation of the Langmuir Adsorption Isotherm Equation

It is beyond the scope of this chapter to give a full account of all available isotherm equations. Instead, we shall derive one of the most important ones, the *Langmuir equation*, in some detail to show how the preceding statistical principles can be made to work and what assumptions have to be made. In Sec. D these assumptions and alternative equations will be discussed.

Isotherms can be derived canonically or grand canonically. Canonically, the adsorbate is considered to be a two-dimensional phase having a certain N^σ, A, and T. Then (N^σ, A, T) is formulated, and μ^σ found from (VI.16). Obviously, μ^σ also is a function of N^σ, A, and T. Next, μ^σ is equated to the bulk chemical potential, which is a function of c. In this way the required relation between N^σ and c is found, that is, the isotherm equation.

Alternatively, the derivation is carried out grand canonically. Now the adsorbate is considered open, the amount adsorbed being determined by the chemical potential in bulk, using (VI.26). For a given system and given model the two methods must give the same outcome, because in both equilibrium is assumed to be established. Choosing to solve a problem canonically or grand canonically is therefore a matter not of principle but of convenience. Usually, grand canonical partition functions are more complex than canonical ones, see (VI.7), but elaboration of the sums in them is sometimes easier in the latter case and, of course, $\langle N^\sigma \rangle$ is directly found.

For a Langmuir adsorbate the following assumptions are made:

1. The adsorption is *localized*, meaning that the surface contains a number of adsorption sites onto which a molecule can adsorb.
2. Lateral interaction between adsorbed molecules is absent; in this respect the adsorption is ideal. Another consequence is that the total energy of the adsorbate is the sum of the energies of the molecules.
3. The surface is homogeneous; all sites are identical.
4. Adsorption is limited to a monolayer.
5. Adsorption is reversible.
6. The adsorption sites contain either an adsorbed molecule or a molecule of the solvent (water): the two need not necessarily have the same size.

As we are interested in the properties of the adsorbate and not those of the solvent, we consider water as a continuum. (Strictly speaking, consistency of the analysis would have called for a lattice-type treatment). In this way we can distinguish two types of sites, "filled" (with adsorbate) and "empty" (without adsorbate). We count the adsorption energy, $\Delta_{ads}u$ per adsorbed molecule *with respect to the energy in the bulk of the water*; i.e., it is the energy of exchanging a water molecule at the surface against an adsorptive molecule. This is the energy ($\approx$ enthalpy) of adsorption as it is experimentally measured. It may consist of several contributions and even contain some entropic contribution, e.g., from the vibrational degrees of freedom, which may differ between the adsorbed molecule and the water. In this case the quantity is no longer purely energetic, but for the moment we ignore these. As a result, q has only one term, namely

$$q_{loc}(T) = e^{-\Delta_{ads}u/kT} \tag{VI.31}$$

For two filled sites the energy is $2\Delta_{ads}u$, etc., and for N^σ adsorbed molecules we obtain a factor of $q_{loc}^{N^\sigma}$ in the partition function, because U_j^σ becomes $N^\sigma \Delta_{ads}u$ because all filled sites are equal. To make (VI.5b) complete, we must substitute for Ω_j the degeneracy of the state having N^σ molecules on N_s^σ sites. This number is known from permutation science. There are

$$\Omega(N^\sigma, N_s^\sigma) = \frac{N_s^\sigma!}{N^\sigma!(N_s^\sigma - N^\sigma)!} \tag{VI.32}$$

possibilities of putting N^σ molecules on N_s^σ sites.

With all of this, we find for the canonical partition function (VI.5) of a Langmuir adsorbate

$$Q(N^\sigma, N_s^\sigma, T) = \frac{N_s^\sigma!}{N^\sigma!(N_s^\sigma - N^\sigma)!} q_{loc}^{N^\sigma} \tag{VI.33}$$

Here $Q(N^\sigma, A, T)$ has been replaced by $(N^\sigma, N_s^\sigma, T)$, because for a surface, characterized only by the number of sites, A has no physical meaning. The factorials $N^\sigma!$ and $(N_s^\sigma - N^\sigma)!$ in the denominator account for the indistinguishability of the molecules on the filled and open sites. The permutation factor accounts for the entropic part; in this case it is typical for ideal systems that (and other partition functions) can be written as the *product* of an entropic and an energetic term. For nonideal systems this is no longer the case; then the entropy depends on the (interaction) energy. As we always need the logarithm of, we obtain from (VI.33) using Stirling's approximation

$$\ln N! = N \ln N - N \tag{VI.34}$$

valid for large N,

$$\ln Q\left(N^\sigma, N_s^\sigma, T\right) = N_s^\sigma \ln N_s^\sigma - N^\sigma \ln N^\sigma - \left(N_s^\sigma - N^\sigma\right) \ln\left(N_s^\sigma - N^\sigma\right) + N \ln q_{\text{loc}} \tag{VI.35}$$

which now consists of the *sum* of an entropic and an energetic part (a so-called separable partition function). Using (VI.16),

$$\mu^\sigma = \pm kT \ln q_{\text{loc}} + kT \ln\left(\frac{N^\sigma}{N_s^\sigma - N^\sigma}\right) \tag{VI.36}$$

$$\mu^\sigma = \mu^{\sigma o} + kT \ln\left(\frac{\theta}{1 - \theta}\right) \tag{VI.37}$$

where the occupied fraction $\theta = N^\sigma / N_s^\theta$. What has been accomplished in this way is that (IV.3) has been derived and that $\mu^{\sigma o}$ has received a very clear physical meaning: in this approximation it is just the adsorption energy with respect to the solution:

$$\mu^{\sigma o} = \Delta_{\text{ads}} u \tag{VI.38}$$

To obtain the isotherm equation we can write for the present approximation

$$\mu = kT \ln x \tag{VI.39}$$

for the chemical potential of the solution. The μ^o is lacking because we want to equate μ^o to μ and $\mu^{\sigma o}$ has been referred to the bulk as the reference point. Instead of the concentration c the mole fraction x is chosen, which is dimensionless, so that this part of the standard state problem is avoided. We take the solution to be dilute ($x \ll 1$), otherwise we would need an $\ln[x(1-x)]$ factor. The adsorbate is, of course, not dilute. The resulting Langmuir equation is

$$\frac{\theta}{1 - \theta} = k_L x \tag{VI.40}$$

with

$$k_L = \exp\left(-\frac{\Delta_{\text{ads}} u}{kT}\right) \tag{VI.41}$$

    Let us postpone the discussion of the quality of this derivation to Sec. VI.D and now finish the methodical part by giving the grand canonical derivation. The partition function for the adsorbate follows from (VI.7) and (VI.33):

$$\Xi(\mu, N_s^\sigma, T) = \sum_{N^\sigma} \frac{N_s^\sigma}{N^\sigma!(N_s^\sigma - N^\sigma)!} q_{loc}^{N^\sigma} e^{N^\sigma \mu/kT} \tag{VI.42}$$

$$= \sum_{N^\sigma} \frac{N_s^\sigma}{N^\sigma!(N_s^\sigma - N^\sigma)!} (q_{loc}x)^{N^\sigma} \tag{VI.42A}$$

where we used (VI.39). The r.h.s. is a well-known binominal expansion:

$$\Xi(\mu, N_s^\sigma, T) = (1 + q_{loc}x)^{N_s^\sigma} \tag{VI.43}$$

$$\ln \Xi(\mu, N_s^\sigma, T) = N_s^\sigma \ln(1 + q_{loc}x) \tag{VI.44}$$

Because of (VI.31) and (VI.41) this may also be written as

$$\ln \Xi(\mu, N_s^\sigma, T) = N_s^\sigma \ln(1 + k_L x) \tag{VI.45}$$

Using (VI.26) in the form

$$N^\sigma = \langle N^\sigma \rangle = kT(\partial \ln \Xi / \partial \mu)_{N_s^\sigma, T} = kT(\partial \ln \Xi / \partial \ln x)_{N_s^\sigma, T}$$
$$= kTx(\partial \ln \Xi / \partial x)_{N_s^\sigma, T}$$

(VI.40) is immediately obtained.

D. Model Assumptions and Alternative Isotherm Equations

The derivations of the previous section served the dual purpose of (a) indicating how isotherm equations can be derived by statistical means and (b) showing which physical assumptions and approximations had to be made to obtain the required result. Let us consider item (a) sufficiently discussed and now turn to (b).

The Langmuir equation is perhaps the simplest and most widely used equation with a sound physical background. The only obviously more simple example is the *Henry isotherm equation.*

$$q = k_L x \tag{VI.46}$$

to which the Langmuir equation reduces for $\theta \ll 1$, i.e., for the initial part of the isotherm. This limit is not discriminative because most theories predict linear initial isotherm parts, sometimes with the same constant, sometimes with another.

Given the relatively large number of restrictive assumptions that had to be made to arrive at the Langmuir equation, it is surprising how often this equation appears to be obeyed by experiment, even for cases where the premises do not seem to apply.

The reason is that relaxing some of the constraints still keeps the *shape* of the isotherm intact. If the water is not treated as a continuum, but a similar kind of lattice statistics is applied as for the adsorbate, the isotherm shape remains unaltered. Adsorption enthalpies can be handled in a more proficient way, say by accounting for the solute-solvent, solute-surface, and solvent-surface interactions in more molecular detail. Of particular interest is the adsorption of hydrophobic molecules (or hydrophobic parts of more complex molecules), for which the increase in the configurational entropy of water molecules released upon the binding is the driving force (*hydrophobic bonding*). In that case the adsorption energy $\Delta_{ads}u$ has to be replaced by a Helmholtz energy, $\Delta_{ads}f$. In fact, experiments have shown that at low temperatures the adsorption is endothermic; i.e., had there been no entropy rise, no adsorption would take place. The upshot is that as long as ideality is maintained μ_i and μ_i^σ retain their composition dependence so that Langmuir behavior follows, albeit with different values for the constant.

Langmuir behavior breaks down when the adsorbate is not localized but mobile and when it is not ideal. Obviously, it also fails if the solution is nonideal, but as the intermolecular interactions in an adsorbate are so much stronger (because of the shorter average distances) we shall not consider this situation. Surface heterogeneity also alters the shape of the isotherm.

Mobile adsorbates require a very different kind of statistics. The three-dimensional analogue is that of a "particle in a box," which for an ideal gas can be elaborated rigorously. Not surprisingly, for an ideal adsorbate (point molecules without interaction) the model leads to a Henry-type isotherm

$$q = k_{mob}x \tag{VI.47}$$

However, to obtain an equation that is on the same level of nonideality as the Langmuir isotherm, the cross section of the adsorbed molecule (or rather the excluded area, a_m) cannot be neglected. The isotherm equation for this case is

$$\frac{\theta}{1-\theta}e^{\theta/(1-\theta)} = k_V x \tag{VI.48}$$

with

$$k_V = \left(\frac{h^2}{2\pi mkT}\right)^{1/2} a_m e^{-\Delta_{ads}u/kT} \cdot V_m^{-1} \tag{VI.49}$$

where h is Planck's constant, m the mass of the molecule (these parameters stem from kinetic gas theory), and V_m the molar volume of the solution (required to obtain the dimensionless mole fraction x on the r.h.s). Equation (VI.48) is the *Volmer isotherm equation*. Its shape is different from that of the Langmuir isotherm, and to show this difference the two isotherms are compared in Fig. 6. According to Langmuir, half-coverage is attained for $k_L x = 1$, but for mobile adsorbates $\theta = 0.5$ corresponds to $k_V x = e$; therefore we have scaled the abscissa axis for the localized adsorption by e to let the two isotherms cross at $\theta = 0.5$. Volmer isotherms start more rapidly [their Henry coefficient is larger: k_{mob} in (VI.47) $< k_L$ in (VI.46)] but they approach saturation more slowly. The reason for this difference is the translational entropy that molecules have in a mobile adsorbate. At low θ this favors the mobile adsorbate, but at high θ mobility is much more restricted because of the competition for space. Localization, with its inherent configurational entropy, may then become more favorable.

When in an experiment a number of data points are available and one wants to find out whether the data obey Langmuir or Volmer, it is expedient to make a discriminative plot that magnifies the difference. In particular, data for high θ should be involved. One possibility is plotting θ^{-1} as a function of x^{-1}. For localized

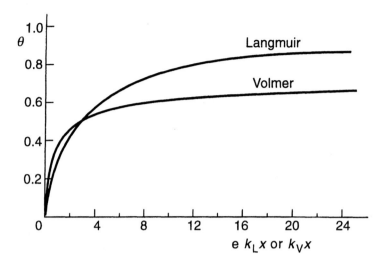

Fig. 6 Comparison of isotherms for localized (Langmuir) and mobile adsorption (Volmer). The abscissa axes are scaled to let the isotherms cross at $\theta = 0.5$.

adsorbates this results in a linear graph over the entire range, the slope of which gives k_L. However, for a mobile adsorbate such a plot is not linear, particularly not in the range of low x^{-1} (= high adsorption). Of course, this method of discrimination remains viable when no data are available for the plateau; then one simply plots Γ^{-1} versus x^{-1} or c^{-1}. In practice, such linearizations are often not made over a sufficiently long range of data points (if they are made at all). It may well be that reportedly localized adsorbates are in reality mobile, at least over part of the range.

Nonideality has two causes: finite ion size and intermolecular interaction. Recall the van der Waals equation of state for a gas,

$$\left(p + \frac{an^2}{V^2}\right)(V - nb) = nRT \tag{VI.50}$$

where the constants b and a account for these two nonidealities, respectively. Of these, the finite size is already accounted for in the Langmuir and Volmer isotherms, in the former because of the limit put on the number of sites, N_s^σ, in the latter explicitly through the excluded area a_m. Let us now relax the other nonideality feature and consider lateral interactions in the adsorbate. They can be repulsive or attractive. Repulsion may, for instance, occur between parallel oriented dipolar adsorbate molecules or between ions. Its consequence is most noticeable in the upper parts of the isotherms because of the increased resistance against filling up. In fact, because of that it is sometimes not so easy to distinguish between localized adsorption with lateral repulsion and mobile adsorption without. Additional information may be required (say, electrolytes would screen lateral ionic repulsion but would have no effect on uncharged species). Lateral attraction may lead to two-dimensional phase separation, quite analogous to the three-dimensional equivalent of the condensation of vapors. For bulk phases this phenomenon can be quantitatively accounted for with (VI.50).

Statistically speaking, this nonideality poses the problem that the logarithm of the partition function is no longer the sum of an entropic and an energetic contribution, as was the case in (VI.36). "Mixed" terms appear. Mathematical problems involved in rigorous treatments have led to a number of fairly generally accepted approximations, of which two will be mentioned.

The first or zeroth-order interaction correction is the *Bragg-Williams approximation*, in which next neighbor pair interactions are taken into account but it is assumed that this interaction has no consequences for the configurational entropy. Otherwise stated, the adsorbate is forced to remain ideal, so that the partition function is still separable. The resulting localized isotherm is the *Frumkin-Fowler-Guggenheim (FFG) isotherm equation*.

$$\frac{\theta}{1-\theta} e^{zw\theta/kT} = k_{\text{L}} x \tag{VI.51}$$

where z is the coordination number of the lattice and w the pair energy. For lateral repulsion ($w > 0$) the isotherm is flattened as compared with Langmuir, $w = 0$ results in (VI.1), and $w < 0$ (lateral attraction) leads to steeper isotherms and even, if w/kT is high enough, to phase separation or *two-dimensional condensation*. Qualitatively this is a new feature. The critical point can be obtained from (VI.51) as the condition for which $d\theta/dx \to \infty$. Quantitatively, the critical point is poorly predicted, which is not unexpected for such a simple model.

The next better (first-order) interaction correction is the *quasi-chemical* or *Guggenheim-Bethe* approximation, in which pairs of interactions are considered independent ("Bragg-Williams of pairs"). Now the influence of interaction on the entropy is also accounted for. The isotherm equation is

$$\frac{\theta}{1-\theta} \left[\frac{(\beta - 1 + 2\theta(1 - \theta))}{(\beta + 1 - 2\theta)\theta} \right]^{z/2} e^{w/kT} = k_{\text{L}} x \tag{VI.52}$$

where

$$\beta^2 = 1 - 4\theta(1 - \theta)\left(1 - e^{w/kT}\right) \tag{VI.53}$$

This equation predicts the same qualitative features as (VI.51) but is quantitatively better.

For mobile adsorbates lateral interaction cannot be well described with a lattice approach. Rather a two-dimensional equivalent of the van der Waals theory is used, which leads to the introduction of a^σ, the surface van der Waals constant [the analogue of a in (VI.50)]. The isotherm equation is

$$\frac{\theta}{1-\theta} e^{\theta/(1-\theta)} e^{-2a^\sigma \theta/a_{\text{m}} kT} = k_{\text{V}} x \tag{VI.54}$$

that is, the *Hill–de Boer* or *two-dimensional van der Waals equation*. The reason for the second name will become clear after (VI.62). Obviously, this equation also predicts two-dimensional condensation. Comparison of experimental data for the two-dimensional critical point and those predicted by (VI.52) and (VI.54) is another, more sophisticated way to discriminate between localized and mobile adsorbates.

Surface heterogeneity is a very real but compounding difficulty. In practice, most surfaces have a certain distribution of $\Delta_{ads}u$ over the sites or over the surface. The trend is that the more "high energy" sites are occupied first, but this is not strictly sequential for entropic reasons. The impact that heterogeneity has on shapes of isotherms depends on such issues as the mode of the heterogeneity (patchwise or random), the width of the distribution, and the mobility of the adsorbate (mobile adsorbed molecules can become localized on high-energy patches). Another feature is that the extent of heterogeneity "seen" by a certain molecular species is not necessarily identical to that "seen" by another. A systematic and comprehensive description would require a book on its own and is beyond our present scope.

Still another type of nonideality is encountered in adsorbates of molecules having different sizes and/or shapes. In lattice models this feature can be accounted for by allowing certain species to occupy more sites than others. Usually it gives rise to modifications of the Langmuir bahavior. For instance, the $(1-\theta)$ in the denominator tends to change into $(1-\theta)^r$ for an r-mer. In connection with molecules of biological origin, the situation for very large r is relevant. This will be considered in the next section.

E. On Adsorption of Polymers, Polyelectrolytes, and Proteins

This is a vast scientific territory with obvious biophysical implications, such as to bacterial adhesion, immunochemistry, biofouling, and enzyme immobilization, of which several aspects will recur in this book. The present subsection is intended to review some basic principles. Let us first consider uncharged, linear homopolymers and summarize what has been achieved by statistical thermodynamics in combination with experimental observations.

Polymers in solution have a high conformational entropy, determined by the many molecular states the chain elements can assume in a polymer coil. The expansion of such a coil is determined by the molecular mass M and the quality of the solvent, quantified by the Flory-Huggins χ-parameter. The lower χ, the better the solvent and the more expanded the coil is. Upon adsorption, a reduction of this conformational entropy takes place. Hence, adsorption will not take place unless there is compensation in the attraction between segments and the surface. This attraction is described by Silverberg's χ^s-parameter, which is a measure of the excess binding (Gibbs or Helmholtz) energy of a segment over that of a solvent molecule. To compensate for the entropy loss, χ^s should be above a certain critical value, $\chi^s(crit)$, typically 0.2–$0.5kT$. When χ^s is well above $\chi^s(crit)$, the adsorption is very strong (because there are so many segments involved) and seemingly irreversible. The isotherms obtain a high-affinity character. Figures 7 and 8 illustrate some basic features of polymer adsorption.

Figure 7 shows that the segments in an adsorbed polymer may be distributed over *trains, loops,* and *tails.* The length distributions and numbers of each of these groups of segments depend on M, χ^s, and χ, but their roles in polymer adsorption are generic. Trains account for the attached segments. For entropic reasons they are rarely very long and do not cover the entire surface, leaving about 5–15% space free. Loops account for most of the adsorbed weight. They are present for entropic reasons, but their extension is determined by χ: a high loop density is tolerated close to the surface only if the solvent is relatively poor. This accounts for the trend of Fig. 8. Once χ^s is well above $\chi^s(crit)$, further increase of χ^s no longer has consequences for the

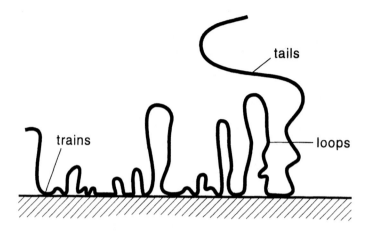

Fig. 7 Distribution of segments over trains, loops, and tails in a polymeric adsorbate.

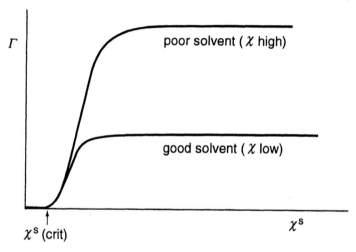

Fig. 8 Basic trend of the effects of χ^s and χ on polymer adsorption.

adsorption, but the solvent quality has, because loops are entirely surrounded by solvent. For a poor solvent Γ is typically around 2 mg m^{-2}, for a good solvent Γ is around 0.5 mg m^{-2}. So the loops typically account for about 75% of the adsorbed weight. Tails are also there for entropic reasons. They contribute negligibly to Γ but play a dominant role in steric interaction because of their extension. The experimental thickness of a polymer layer depends on the method. If it is obtained from ellipsometry ($\delta_{ell.}$) essentially the loop layer is counted, but if deduced from dynamic light scattering or steric interaction the tail extension (δ_h) is counted. Invariably, $\delta_h > \delta_{ell.}$.

With this in mind it is relatively straightforward to predict *polyelectrolyte* behavior. Because of the charge they carry, polyelectrolytes form very expanded coils in solution. Phenomenologically speaking, water is an excellent solvent for polyelectrolytes. As a consequence, the formation of loops is suppressed. Typically, polyelectrolytic adsorbates are thin and Γ is a few tenths of an mg m^{-2}. There may be an occasional tail dangling out far into the solution. Such tails will dominate steric repulsion, but in their absence particle-particle interaction between surfaces with a polyelectrolyte adsorbate is essentially electrostatic.

As for uncharged polymers, polyelectrolytes also require χ^s above χ^s(crit) to become adsorbed. This adsorption (Gibbs) energy may have an electrostatic and/or a nonelectrostatic contribution and, depending on the charge signs of polyelectrolyte and surface, the electrostatic contribution may be attractive or repulsive; it may or may not outweigh the nonelectrostatic component. If too strongly repulsive, it may prevent the polyelectrolyte from adsorbing. To obtain some feeling for the relative strengths of these forces, the electric energy of a monovalent ion or ionic group in an electric field is about $1kT$ for every 25 mV, and the Gibbs energy per CH$_2$ segment adsorbed on a hydrophobic surface is about $1.2kT$. Hence, molecules with some hydrophobic groups in their chain may readily adsorb on a hydrophobic surface against an adversary potential.

It is typical for polyelectrolytes that their adsorption is very sensitive to indifferent low-molecular-weight *electrolytes*. In this respect they can be distinguished from uncharged polymers. Electrolytes exert a dual effect:

1. They screen the segment-segment repulsion, which phenomenologically means that water becomes a poorer solvent. Addition of salt therefore promotes loop formation and hence leads to an increase of Γ.
2. They also screen segment-adsorbent interactions, as far as these are of an electrostatic origin. Attachment is promoted/inhibited by salts if this interaction is electrically repulsive/attractive.

So, trend (1) is a χ effect, trend (2) a χ^s effect.

Along similar lines, the pH influence on the adsorption of weak polyelectrolytes may be interpreted. Consider for instance polyacrylic acid (PAA). At very low pH it is uncharged, and it adsorbs with thick layers; Γ is high and does not depend on ionic strength. However, at pH ≥ 7 it is fully dissociated, Γ is low but increases with salt concentration.

For similar reasons, *amphoteric polyelectrolytes* tend to have a plateau adsorption with a maximum around their isoelectric point. With increasing electrolyte content, this maximum becomes less pronounced, mainly because the levels on both sides rise.

At this point it is appropriate to consider also the influence of the *molecular weight*. Theory predicts that in a poor solvent Γ increases with M and that in (very) good solvents there is little effect of M. The former trend is essentially of entropic origin. Upon adsorption, larger coils would lose more conformational entropy than smaller ones, so they would not adsorb unless there is entropic compensation by making longer loops. In a good solvent the adsorbate layer is thin; in the limiting case it is entirely flat and there is no influence of M because 10 molecules of $M = 1000$ lead to the same Γ as 1 molecule of $M = 10,000$.

The combined electrolyte and M effect is again characteristic of polyelectrolytes. In the absence of salt, Γ is low and there is no or little M dependence. However, upon addition of electrolyte Γ not only rises but simultaneously becomes dependent on M.

All the foregoing features have been theoretically predicted and experimentally confirmed for model systems.

In Sec. I.B the slowness of the *rate* of adsorption was discussed. With the present information one of the more familiar reasons can be given. Most polymers are heterodisperse. The larger molecules adsorb more strongly (more train segments per molecule), but the smaller ones diffuse more rapidly toward the surface, so initially they enrich the adsorbate. Replacement of small molecules by large ones will take place, but this is a slow process. At the same time it is realized that the hysteresis increases with the age of the adsorbate.

Regarding protein adsorption, the application of statistical thermodynamics is just embryonic because it is next to impossible to account for all the fine structure in the molecule.

F. Two-Dimensional Equations of State

In three-dimensional systems an equation of state is a relation between p, V, n, and T. Equation (VI.50) is an example, and so is the ideal gas law $pV = nRT$. The two-dimensional equivalent relates πA to $N^\sigma kT$. Equations of state contain properties of the adsorbate only, so they really describe the state of the system. This is in contradistinction to adsorption isotherms, which relate N^σ/A to bulk properties (such as c or x) and properties related to transport from the bulk solution to the surface (like $\Delta_{ads}u$). Given N^σ, π, A, and T, the state of the system should be fully specified, irrespective of whether the adsorbate is isolated from or in equilibrium with the substrate.

Partition functions contain all the information to describe isotherms, fluctuations, and probabilities and to obtain the equation of state. For a Langmuir adsorbate, if the derivation is carried out canonically (i.e., considering the adsorbate closed) we replace A in (VI.17) by $a_s N_s^\sigma$, where a_s is the area per site:

$$\pi a_s = kT \left(\frac{\partial \ln Q}{\partial N_s^\sigma} \right)_{T, N^\sigma} \tag{VI.55}$$

which must be combined with (VI.35). The result is

$$\pi a_s = -kT \ln(1 - \theta) \tag{VI.56}$$

This equation of state can also be obtained immediately from the grand canonical partition function, using (VI.27) and (VI.45), the work term being the characteristic function in this ensemble. $\pi(\theta)$ is an ascending function. Starting from zero, the first part is linear (Henry)

$$\pi a_s = kT\theta \qquad \text{or} \qquad \pi A = N^\sigma kT \tag{VI.57A, B}$$

but for $\theta \to 1$, $\pi \to \infty$. It is a measure of the work to be done to compress the area or, for that matter, to reduce the number of sites at fixed number of molecules. The Henry equation of state is entirely congruent with its three-dimensional analogue for ideal gases.

The equations of state in the Bragg-Williams and quasi-chemical approximations read

$$\pi a_s = -kT \ln(1 - \theta) + \frac{1}{2} zw\theta^2 \tag{VI.58}$$

and

$$\pi a_s = -kT \ln(1 - \theta) - \frac{zkT}{2} \ln\left[\frac{\beta + 1 - 2\theta}{(\beta + 1)(1 - \theta)}\right] \tag{VI.59}$$

respectively. These pressures also run from 0 to ∞, but below the critical point they may contain loops, indicating two-dimensional condensation, phenomenologically equivalent to the "van der Waals loops" in vapor-liquid coexistence. Once again it is noted that none of these equations contain k, $\Delta_{ads}u$, or x, but lateral, i.e., internal interaction is accounted for, in terms of either w or β.

For mobile adsorbates without lateral interaction (Volmer) the two-dimensional equation of state reads

$$\pi a_{mob} = kT\left(\frac{\theta}{1 - \theta}\right) \tag{VI.60}$$

The trend $\pi(\theta)$ is qualitatively the same as for its localized equivalent, (VI.56), but at higher θ the rise is steeper. Colloquially stated, at high θ it is more difficult to compress a mobile than a localized adsorbate. The reasons for this have already been discussed in connection with Fig. 6.

For mobile adsorbates with lateral interaction

$$\pi a_{mob} = kT\left(\frac{\theta}{1 - \theta}\right) - \frac{a^\sigma}{a_{mob}}\theta^2 \tag{VI.61}$$

where the parameter a^σ has already been met in the corresponding isotherm (VI.54). Equation (VI.61) also predicts condensation loops below the critical point, and because of the mobility the similarity to the three-dimensional van der Waals analogue is even greater than it is for localized adsorbates. In fact, (VI.61) can be converted into

$$\left(\pi + \frac{a^\sigma(N^\sigma)^2}{A^2}\right)(A - N^\sigma a_{mob}) = N^\sigma kT \tag{VI.62}$$

completing the analogy with the three-dimensional van der Waals equation of state, (VI.50).

VII. APPLICATION TO SOME CURRENT ISSUES IN BIOSURFACE SCIENCE

There are a number of ongoing disputes regarding important notions, interpretations, and approximations related to biophysical interfaces. They involve such important issues as the interpretation of interfacial tensions and adhesion, the distinction between high- and low-energy surfaces or between hydrophobic and hydrophilic surfaces, and the notion of surface equation of state. On the basis of the presented thermodynamics and statistical thermodynamics we shall now try to contribute to defining and, where possible, solving these issues, hoping to help distinguish opinions from facts.

A. Surface Energy. High and Low Energy Surfaces

The distinction between "high-energy" and "low-energy" surfaces can be found in several places in the literature. However, this distinction is sometimes made loosely, that is, without having a clear picture in mind about its background. Do we really mean the (excess) energy of a surface or are we referring to the energy of binding of certain molecules to surfaces? The former is an intrinsic property of each surface and interface, but the latter resembles the adsorption energy and is a property of the interaction of that surface with other molecules and therefore dependent on the nature of the probe.

At any rate, it is a thermodynamic tenet that phenomenological physical properties should be defined on the basis of an unambiguous measurement. Only when such a process can be designed and carried out is the property operational.

Let us consider this principle for the surface excess energy U^σ. The defining equation is (III.2), but a process to determine the energy of two phases in contact is not so simple to envisage for solid phases [it is of course possible to measure the energy of adhesion of phases α and β, but that gives us the energy difference $U^{\sigma(\alpha\beta)} - U^{\sigma(\alpha)} - U^{\sigma(\beta)}$, which is something else (see Sec. VII.C)]. However, for fluids we are better off because interfacial tensions can be unambiguously measured. The question is then whether it is possible to split $\gamma^{\alpha\beta}$ into its entropic and energetic parts. It appears that for water this can be done in a satisfactory (although not rigorous) way for the temperature range 0–40°C over which biophysical studies are usually carried out. Over this range surface tension decreases linearly with the temperature. The rule also applies to other fluids just above their melting points. As $\gamma = F_a^\sigma$ [(IV.27)] and $F_a^\sigma = U_a^\sigma - TS_a^\sigma$ [(IV.11)],

$$\gamma = U_a^\sigma - TS_a^\sigma \tag{VII.1}$$

and linearity suggests that U_a^σ and S_a^σ may be obtained from intercept and slope. This is operational and no assumptions about the dividing plane have to be made. The surface excess energy is *larger than* the surface tension. For example, for water $U_a^\sigma \approx 118$ mJ m^{-2} and γ decreases from 72 mN m^{-1} at 25°C to 60 mN m^{-1} at 100°C, so the entropic contribution TS_a^σ to γ increases from about 46 mJ m^{-2} at 25°C to 58 mJ m^{-2} at 100°C. Let us conclude that for pure fluids U_a^σ is accessible but that this is not the case for pure solids. Obviously, this U_a^σ may not be identified with its counterpart for solids, because crystallization gives rise to a drastic rearrangement in the interface.

When the interface carries an adsorbate, from (IV.26) and (IV.11)

$$\gamma = U_a^\sigma - TS_a^\sigma - \sum_i \mu_i \Gamma_i \tag{VII.2}$$

Hence, no general statements can be made, because the adsorption term is also temperature dependent. The chemical potentials are Gibbs energies, so they contain an entropic contribution, mainly in μ. It is feasible that, after appropriate $\Gamma(T)$ measurements, the contribution of the adsorbate to the surface energy can be sequestered and added to U_a^σ for the pure surface.

However, for most purposes the energy obtained in this way is not the most interesting. One is rather interested in its *change* when one phase is replaced by another (say, air by water as in *adsorption* or in *wetting*) or when the water is replaced by other molecules, as in *adsorption* from aqueous solution. The energies of these processes, which for practical purposes are identical to the heats or enthalpies, are measurable. For biological purposes the drawback of probe dependence does not count because one is exclusively dealing with aqueous systems. The energies of these three processes are different. Adsorption energies refer to the interaction between individual water molecules and a surface, whereas energies of wetting also involve water-water interactions. So the former gives a more absolute measure of the affinity of the surface for water, whereas the latter rather informs us about the affinity for (bulk) water over the affinity of bulk water over itself. It depends on the problem at hand which one of the two is the more relevant. Adsorption from solution does not give this primary information because a third component is involved.

A property that can be reasonably well measured is the *enthalpy of immersion*, $\Delta_{imm}H_a$, which is equivalent to the *enthalpy of wetting*, $\Delta_w H_a$. The experiment is carried out in a calorimeter, but it is not easy to account for numerous experimental problems. Accepting proper data to be available, they can be carried out for a variety of solids or, for one solid, for pure water and water-containing adsorptives. The absolute values obtained with pure water vary between several hundreds of mJ m^{-2} (equivalent to mN m^{-1}) for various inorganic oxides or clay minerals, through about 30 for graphon (a hydrophobic carbon) to as low as 6 mJ m^{-2} for Teflon. All these wetting enthalpies are exothermal; i.e., they have a minus sign. Such measurements offer a number of possibilities:

1. They offer a way to quantify the notions "high" versus "low" energy surfaces on the basis of $\Delta_{imm}H_a$. As water is used as the probe, these notions are equivalent to quantifying the notions *hydrophobic* versus *hydrophilic*. To that end the data are preferentially written as molar quantities ($\Delta_w H_m$ or $\Delta_{imm}H_m$), allowing comparison with other types of measurement. Values of $\Delta_w H_m$ below about 40 kJ mol^{-1} indicate physical binding of water to the surface, from about 50 to 70 kJ mol^{-1} to hydrogen bond formation (for instance, with surface hydroxyls), and above that to stronger adsorption (say on Lewis acid sites) or even chemisorption. The heat (enthalpy) of condensation for water is 44 kJ mol^{-1}, so one could put the borderline there: if $|\Delta_w H_m| \gtrsim |44$ kJ mol$^{-1}|$ the surface is called *hydrophilic/hydrophobic*. This quantification makes sense; it measures by how much, energetically speaking, water prefers the surface over itself. The caveat should be made that the discussion refers only to the energy and not to the entropy. Typically, for hydrophobic interactions the process is entropically driven, so that under the appropriate conditions even the sign of ΔH does not determine the affinity.

2. The difference between the heat of attachment to a solid and the heat of condensation also occurs in the Brunauer-Emmet-Teller (BET) theory of multilayer gas adsorption. In this case energies are counted but energies, enthalpies, and heats may also be taken as identical. The difference $\Delta_{ads}U_m - \Delta_{cond}U_m$ is obtainable from the isotherms. For "low-energy" surfaces the isotherms are convex with respect to the pressure axis (molecules are reluctant to adsorb, but once they are adsorbed there is a trend of other molecules to condense on those present). See Fig. 9. For "high-energy" surfaces it is the other way around: the initial part is concave, and then there is a knee bend after which the adsorption rises again.

3. Heat of immersion measurements may be carried out for aqueous solutions of different compositions to obtain information on heats of adsorption. For instance,

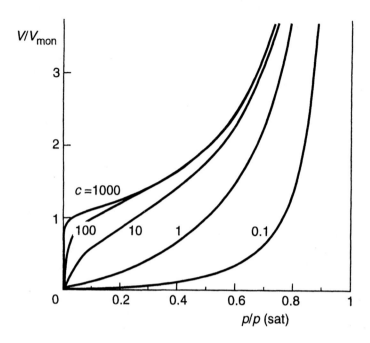

Fig. 9 Example of a set of BET isotherms. V is the volume of gas adsorbed, V_{mon} that in a monolayer. The constant C equals $(\Delta_{ads}U_m - \Delta_{cond}U_m)/RT$. For low C, $|\Delta_{ads}U_m| \ll |\Delta_{cond}U_m|$, the isotherm is convex; for large C, $|\Delta_{ads}U_m| \gg |\Delta_{cond}U_m|$, the isotherm is concave until multilayer adsorption sets in.

by comparing the immersional enthalpy of oxides in solutions of different pH one can obtain the enthalpy of double-layer formation.

So, it may be concluded that useful operational thermodynamic information is or can be made available to estimate the "energetics" of surfaces.

It is repeated that these measurements provide information only about *energies*, not about *Helmholtz energies*. It is one of the basic tenets of thermodynamics that one can obtain energies and entropies from Helmholtz energies (as a function of temperature) but not the other way around. The interaction of water with many surfaces, particularly low-energy ones, is to a large extent entropically determined and the entropic part of the Helmholtz energy is not available calorimetrically.

Entropic and energetic contributions are included in interfacial tensions but, as explained before, these are inoperable for solid-liquid interfaces. The closest one can approach the affinity of a surface for water is to measure the *contact angle* of a water droplet on that surface. Using the Young equation, the difference $\gamma^{SL} - \gamma^{SG}$ is obtainable, but not the absolute values, unless a nonthermodynamic argument is invoked to establish an additional relationship between these two tensions. The enthalpic part of $\gamma^{SL} - \gamma^{SG}$ is identical to $\Delta_w H_a$, as discussed before:

$$\Delta_w H_a = \left(\gamma^{SL} - T\frac{\partial \gamma^{SL}}{\partial T} \right) - \left(\gamma^{SG} - T\frac{\partial \gamma^{SG}}{\partial T} \right) \qquad \text{(VII.3)}$$

or, using the Young equation,

$$\Delta_{\mathrm{w}} H_{\mathrm{a}} = -\left(\gamma^{\mathrm{LG}} \cos\alpha - T\frac{\partial(\gamma^{\mathrm{LG}} \cos\alpha)}{\partial T}\right) \tag{VII.4}$$

where α is the contact angle. For three-liquid systems such as oil-water-mercury, the agreement between the enthalpy of wetting obtained in this way and that obtained microcalorimetrically has been verified.

B. Molecular Interpretation of Interfacial Tensions

Surface and interfacial tensions are determined by the (excess) molecular interactions in the interface and by the distribution of molecules in that interface. The former gives the energetic, the latter the entropic contribution. A number of theories have been proposed to account quantitatively for these two terms and from there interpret interfacial tensions. Obviously, such theories rest on statistical mechanics or statistical thermodynamics. They are intrinsically complicated because partition functions cannot be separated into purely energetic and purely entropic contributions and the distributions are asymmetrical. Alternatively, various simplifying models have been proposed in which approximations are made. Some of these are rather drastic (say, the entropy is ignored) but they can be readily used with experimental data. There seems to be not much available on the "middle" level.

Rigorous statistical mechanical interpretation would require the introduction of pressure tensors, of which the horizontal ($\parallel$ interface) component determines the tension, and of radial distribution functions and elaboration of them for heterogeneous systems. Such approaches are beyond the scope of the present chapter. Other approaches involve a mean field approximation. These theories do not allow direct application to predict interfacial tensions, let alone those in the presence of adsorbates, because they contain parameters that are hard to obtain experimentally. However, to illustrate the lines along which a statistical mean field theory might be developed,

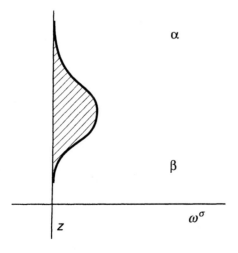

Fig. 10   Excess profile of the function ω^{σ} across an interface separating phases α and β.

consider the pictorial representation of Fig. 10, which is in line with the statistics and gives at least the essentials.

In this figure the quantity ω^σ stands for the *excess of the grand canonical characteristic function density* [see (VI.27)] in an interface. It is plotted as a function of position z across that interface. To the lower side there is phase β, above it phase α. From β to α a certain transition takes place in composition and local entropy, which gives rise to a local excess in $\omega^\sigma(z)$. For high z we are in the bulk of α and $\omega^\sigma = 0$; similarly for phase β at low z. The width of the interface is given by the range over which $\omega^\sigma \neq 0$. Usually it amounts to a few molecular layers; in situations of surfactant or polymer adsorption it may be more than that. The shape depends on the orientation and nature of molecules in the interfacial region. The sketch in Fig. 10 is a very simple one and would apply to the interface between two simple pure liquids. When, say, surfactants are adsorbed the profile may have positive and negative regions, reflecting attractive and repulsive parts. There is no need to define the position of the interface. It is enough to know $\omega^\sigma(z)$ from a molecular model.

Figures like this one are characteristic; they are found from both statistical mechanics (where rather the tangential pressure profile is plotted) and lattice theories (extensions of the Langmuir approach in which the excesses are counted in lattice layers parallel to the surface).

Once such a profile has been established, the next step is simple:

$$\gamma^{\alpha\beta} = \int_{z=-\infty}^{+\infty} \omega^\sigma(z)\,dz \tag{VII.5}$$

i.e., the interfacial tension corresponds to the hatched area in Fig. 10. In this manner the relation between interfacial structure and tension is established in a general way.

From this molecular interpretation it follows that γ is a *macroscopic quantity*: it makes sense only at a scale much greater than the thickness of the ω^σ-excess layer. When the surface is curved on a scale comparable to this thickness, drastic modifications have to be brought about, and for stronger curvature all physical meaning is lost. Models in which an interfacial tension is assigned to the "interface" between a hydrocarbon tail of a surfactant and the water around it have no physical basis.

Let us now briefly discuss a model often used for the interpretation of interfacial tensions. It was proposed by Fowkes and is perhaps the best available. According to this approximation, the interfacial tension $\gamma^{\alpha\beta}$ is related to the two surface tensions γ^α and γ^β via

$$\gamma^{\alpha\beta} = \gamma^\alpha + \gamma^\beta - 2(\gamma_d^\alpha \gamma_d^\beta)^{1/2} \tag{VII.6}$$

Here γ_d^α and γ_d^β are the dispersion contributions to γ^α and γ^β, respectively.

The idea behind this equation is the following. When the surfaces of phases α and β are brought from infinity to contact, each brings its own surface tension along, but $\gamma^{\alpha\beta}$ is less than this sum because of the gain in attraction between the two phases. This attraction is attributed to dispersion (London–van der Waals) forces, the only force having a long enough range to act far across the interface. For paraffin γ_d is a large fraction of γ, but for water γ_d is a relatively minor part ($\gamma_d^w \sim 22$ mN m^{-1}, i.e., 30% of γ^w) because the main role is played by hydrogen bonding. The fact that the geometric mean is subtracted is based on the general experience that Hamaker constants for heteroattraction are also to a good approximation geometric averages of those for homointeraction:

$$A^{\alpha\beta} \approx \left(A^{\alpha\alpha} A^{\beta\beta}\right)^{1/2} \tag{VII.7}$$

How good are (VII.6) and expressions derived from it? The main problem is that an energetic argument is applied to a Helmholtz energy; entropic contributions to γ are ignored. Otherwise stated, any structural changes occurring in surfaces if brought in contact with a condensed phase are neglected. Had (VII.6) been applied to U_a^σ, the approximation would hold within about 10–15%, that is, the uncertainty incurred in (VII.7). (This equation stems from the molecular or microscopic theory of van der Waals forces: it involves the Berthelot principle for interactions between two different molecules across vacuum and assumes additivity of London forces. The results of this theory agree within 10–15% with those of the macroscopic theory.) As discussed in Sec. VII.A, the absolute error made in γ by ignoring the TS term is about 20–30% for water. However, the problem is that the *shape* of the combination rule (VII.6) also changes. Entropies of mixing are not geometric means but rather go with $x^{\sigma\alpha} \ln x^{\sigma\alpha} + x^{\sigma\beta} \ln x^{\sigma\beta}$ for ideal mixing or in a more complicated way if the mixing is nonideal. The overall conclusion is that (VII.6) is a reasonable approximation if only the energetics of interfaces are involved and no adsorption takes place. Variants of it have been used to obtain a quick estimate of Hamaker constants.

Equation (VII.6) cannot deal well with adsorption. When phase α (water) contains a low concentration of an adsorptive that does not adsorb on its surface (say sodium biphosphate from a buffer) but adsorbs very strongly at the $\alpha\beta$ interface (e.g., if β is an iron oxide) $\gamma^{\alpha\beta}$ should be lower than in the absence of the adsorptive but the equation cannot account for that. When adsorption by hydrophobic bonding is involved, i.e., entropically driven processes, it fails completely. Procedures to "determine" interfacial tensions for solid-liquid interfaces from contact angles and combination rule (VII.6) are therefore suspect.

C. Adhesion

In the domain of particle and cell adhesion, a controversy is sometimes noted between Deryagin-Landau-Verwey-Overbeek (DLVO)–type interpretations and the so-called surface free energy method. The distinction is ill posed because DLVO theory also gives a free energy of interaction—more specifically, a Helmholtz energy, because the isothermal reversible work is computed to bring particles to the surface. Nevertheless, it is worthwhile to describe the principles of the two approaches and compare their applications.

To that end, consider Fig. 11, showing three stages in the adhesion of a particle p from water to a solid surface s. To keep the considerations simple, the particle is assumed to have a cubic shape with each face having unit area. In (a) the particles are so far from the surface that they do not interact (the initial stage), in (b) the particle has come very close and experiences the influence of the solid, and in (the final) state (c) intimate contact has been established. It is characteristic in situation (c) that a unit area of pw interface and a unit area of sw interface have been annihilated and a unit area of sp interface has been formed. In adhesion science the *work of*

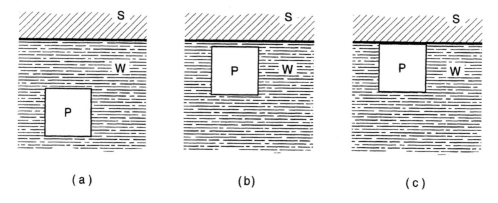

Fig. 11   Three stages in the adhesion of a particle p from water on a solid s.

adhesion

$$W_{\text{a}}^{\text{pws}} \equiv \gamma^{\text{sw}} + \gamma^{\text{pw}} - \gamma^{\text{ps}} \qquad\qquad\qquad (\text{VII.8})$$

is defined. It is minus the Helmholtz energy of the process envisaged in Fig. 11. Otherwise stated, it is the work to be performed to loosen the particle from the substrate.

Equation (VII.8) is not at all useful. The r.h.s. contains three nonmeasurable quantities, and $W_{\text{a}}^{\text{pws}}$ is not so easy to obtain either. Calorimetry gives the enthalpy of adhesion rather than the Helmholtz energy. Standard Helmholtz energies can be obtained from an adhesion isotherm (the number of particles adhered as a function of those in solution) but that requires a model.

The situation becomes more amenable when p or s is a fluid. Then the Young equation may be used, say if s is an oil phase $\gamma^{\text{ps}} = \gamma^{\text{pw}} + \gamma^{\text{sw}} \cos \alpha$, where α is measurable. The result is

$$W_{\text{a}}^{\text{pws}} = \gamma^{\text{sw}}(1 - \cos \alpha) \qquad\qquad\qquad (\text{VII.9})$$

so the number of unmeasurable quantities on the r.h.s. is reduced to one. It is again concluded that γ^{sw} cannot be obtained unless a nonthermodynamic assumption is made (to obtain $W_{\text{a}}^{\text{pws}}$).

Another limitation of definition (VII.8) is that the r.h.s. can be replaced by a Helmholtz energy balance only for pure solids and liquids. Then, form (IV.20)

$$W_{\text{a}}^{\text{pws}} = F_{\text{a}}^{\sigma(\text{sw})} + F_{\text{a}}^{\sigma(\text{pw})} - F_{\text{a}}^{\sigma(\text{ps})} \qquad\qquad\qquad (\text{VII.10})$$

but when adsorptives are present in one or more of the three phases the three $\Sigma_i \mu_i \Gamma_i$ terms also have to be considered.

In conclusion, an analysis along the lines of (VII.8) or (VII.10) or variants of them appears cumbersome. In addition, the question has to be asked whether complete annihilation of interfaces occurs in practice. For biological surfaces this hardly seems the case.

These problems are taken care of in colloid stability theories in which the Helmholtz energy of interaction is computed as a function of distance (DLVO for only electrical repulsion and van der Waals attraction; for biological systems usually steric and solvent structure–induced forces also have to be considered). So these theories give $F_a(h)$, the Helmholtz energy of bringing the surfaces from $h = \infty$ to h. All three situations of Fig. 11 are considered, with $F_a(\infty) \equiv 0$.

Colloid stability theories are more versatile and general than the preceding adhesion method, but at a price. This price is that all contributions to $F_a(h)$ have to be evaluated on the basis of some model assumptions. A concomitant issue is the extent to which the overlapping electric and steric layers do come to equilibrium during the adhesion event. This problem takes us back to the Deborah number issue (Sec. I.B) with questions such as "do adsorbed large biological molecules have time enough to escape from the narrowing gap between p and s during the encounter?"

Much interesting work remains to be done.

VIII. LITERATURE

Most of the material presented in this review can be found in *Fundamentals of Interface and Colloid Science* (FICS) by the same author (although with less emphasis on biological systems). Specifically, for interfacial thermodynamics see FICS I, Chap. 2, FICS II, Chap. 3, statistical thermodynamics, FICS I, Chap. 3, adsorption FICS II, Chap. 1 and 2, electric double layers, FICS II, Chap. 3, polymer adsorption FICS II, Chap. 5.

FICS is published by Academic Press. Volume I (Fundamentals) appeared in 1993 (second printing 1995), Vol. II (Solid-Liquid Interfaces) in 1995, and Vol. III (Fluid-Liquid Interfaces) is in preparation and will contain chapters on interfacial tensions, monolayers, and wetting.

ACKNOWLEDGMENT

The author is indebted to the University of Florida, Gainesville, for a Visiting Eminent Scholarship, during which most of this chapter was written.

2

Electrostatic and Electrodynamic Properties of Biological Interphases

J. M. Kleijn and H. P. van Leeuwen
Wageningen University, Wageningen, The Netherlands

I. INTRODUCTION

In biological systems electric fields are invariably associated with interfaces—or better, interphases*—and they play an important role in many biological processes. To illustrate this we have only to pick at random a few examples from the wealth of bio(electro)chemical information available. A range of important regulatory events in cells, such as the activation of particular enzymes, may be mediated by electrostatic interactions with membrane surfaces (1,2). Electrostatic interactions affect the binding, insertion, and folding of proteins into membranes (3,4) and the fusion of phospholipid bilayers (5,6). The generation and rapid transmission of nerve impulses are based on changes in ionic permeability of membranes resulting from changes in electric fields (7–9). Both the photosynthetic apparatus and the respiratory chain involve extensive membrane-bound enzyme systems in which charge separation across the membrane is coupled to the production of ATP (9–11).

   This chapter is concerned with fundamental physicochemical aspects of electric properties of biological interphases. It is not our objective to dig into the huge amount

---

* In the context of this chapter we prefer the term "interphase" over "interface," because the word interface implies a dividing plane (intrinsically two-dimensional) between two (bulk) phases. However, biological structures that form boundaries between two compartments are in themselves distinct phases and are often inhomogeneous and multilayered. Their three-dimensional structure and charge distribution have important consequences for their electric properties.

of biological and biochemical data that may be of relevance; we shall confine ourselves to discussing general principles and—whenever possible—assessing their implications. In particular, we shall focus on the electric properties of (bacterial) cell walls and biological membranes. In a number of cases we will consider bilayer lipid membranes as a model for biological membranes, of which they constitute the basic element. Surface charges, dipolar effects, and ionic distributions of different types are taken as the primary building blocks for the operational electric potential profile across the interphase. Not only the equilibrium electrochemistry of the biological interphase but also some fundamental aspects of its electrodynamics will be treated. Thus we shall discuss the elementary time characteristics of the formation and relaxation of biological electric double layers. Subsequently, we concentrate on electrokinetics and dielectric relaxation techniques as applied to biological interphases. In this, there is special attention to surface conduction, i.e., lateral ionic migration. The impact of interphasial electrodynamics on practical properties of biophysical systems is briefly outlined. Finally, we will pay attention to the much more classical subject of permeation of ions through biological membranes and lipid bilayers. In connection with this, the generation of the transmembrane potential is briefly recalled.

II. ELECTROSTATICS OF BIOLOGICAL INTERPHASES

A. Potential-Generating Phenomena

Biological interphases such as membranes and cell walls usually contain numerous charges, which, as far as these are not internally compensated, give rise to the formation of a diffuse electric double layer in the surrounding solution. The extension of the double layer into the surrounding solution depends on the electrolyte concentration. In aqueous systems of low salt concentration, for example, streams and freshwater lakes, the diffuse countercharge around the cell wall or cell membrane of a microorganism may extend over distances of the order of 10 nm, whereas in more saline systems, such as seawater and most body fluids, the compensating charge will be at distances as close as less than a nanometer from the membrane or cell.

The charge on bacterial cell walls mostly originates from carboxyl, phosphate, and amino groups (12,13). The degree of (de)protonation of these anionic and cationic groups is determined by the pH and the ionic composition of the surrounding electrolyte solution. At neutral pH almost all bacterial cells are negatively charged because the number of carboxyl and phosphate groups is generally higher than that of amino groups. The compensating charge consists mainly of counterions that can penetrate into the porous cell wall and to a minor extent of coions that are expelled from it. The thickness of bacterial cell walls is in the range of several tens of nanometers (13).

Biological membranes consist of lipoic materials (mostly phospholipids and, except for plant cells, cholesterol), proteins, and also saccharides, sometimes mutually bound in the form of lipoproteins, glycolipids, and glycoproteins. The specific composition of a membrane varies depending on its function. Protein-to-lipid ratios are thus not fixed, but they are usually higher than 1 : 1 by weight (14). In a biological membrane the molecules are organized in such a way that the apolar (hydrophobic) parts escape the water phase, clustering together, while the polar parts reside at or near the boundaries with the surrounding aqueous solutions. Singer and Nicolson (15) were the first to propose that a bilayer structure of lipids forms the membrane

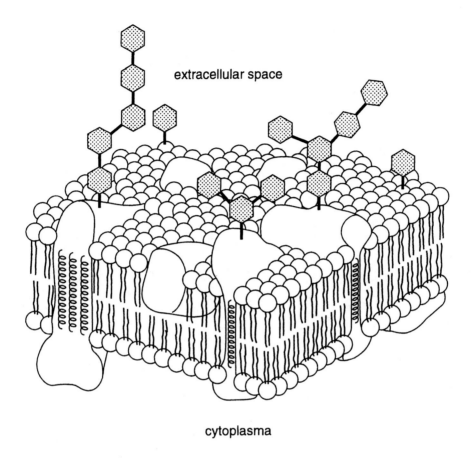

extracellular space

cytoplasma

Fig. 1 Schematic representation of a typical eukaryotic plasma membrane, based on the Singer and Nicolson fluid mosaic model for biological membranes. Carbohydrate moieties on lipids and proteins face the extracellular space (the "carbohydrate glycocalyx").

matrix in which proteins are embedded to varying degrees of penetration (Fig. 1). The thickness of a bilayer lipid membrane is between 6 and 9 nm. However, proteins and other moieties may stick out at both sides of the lipid matrix as far as 20 nm. An extensive treatment of the structure of and structure-function relations in biological membranes is given in Ref. 16.

In understanding the properties of biomembranes it is essential to take into account that the polar headgroups of the lipids are not located as neatly in a plane as suggested by Fig. 1. From molecular dynamics simulations (17–20) and self-consistent field modeling (21–24), it has been found that the boundary between lipid and polar components is rather diffuse. Results obtained with these computational techniques also indicate a relatively large degree of disorder in the lipid tail region (see Fig. 2). Neutron diffraction measurements have revealed that water molecules penetrate into the bilayer up to the level of the glycerol backbone of the lipids (25). Another important aspect is that, at physiological temperatures, most biomembranes are in a liquid crystalline state. This implies that the lateral mobility of the membrane components is high, whereas their tangential movements are highly

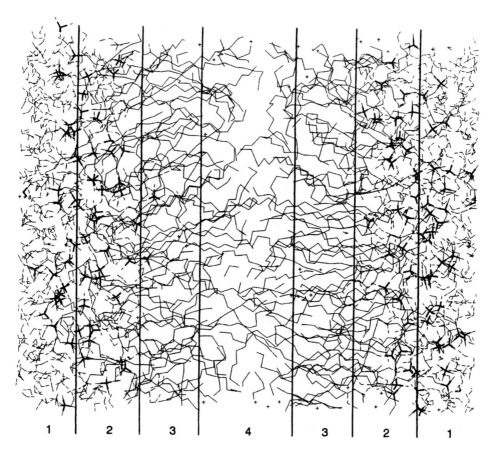

Fig. 2 Snapshot of a dipalmitoylphosphatidylcholine (DPPC) bilayer obtained by MD simulation. Dashed lines are used for water molecules, bold lines for choline and phosphate groups. Vertical lines indicate different regions. Crosses are artifacts resulting from bonds cut by the periodic boundaries. (From Ref. 18.)

restricted. As a result of this restriction in tangential movements, membrane asymmetry is maintained. This is a basic characteristic of biological membranes, reflected not only in variously arranged proteins but also in different lipid compositions at the two sides of the membrane. The lipid chains are highly flexible, assuming many different conformations within a very short period of time (25). Like the membrane lipids, membrane proteins can show high degrees of lateral mobility. Sometimes they cluster together or are immobilized in lateral direction by adsorption on structures outside the membrane, e.g., the cytoskeleton, a network of fibers in the cell anchored to the membrane at specific sites.

Electrostatic fields that are associated with biomembranes arise from several sources. The net charge on the membrane generates a potential at the surface relative to the bulk aqueous phase (the *surface* or *double-layer potential*). This charge resides in the outer parts of the membrane and arises from charged headgroups of phospholipids, adsorbed and penetrated ions, and proteins. It is dependent on the pH and ionic composition of the adjacent solution phase. In the hydrophobic core of the membrane the net charge density is essentially zero. In virtually all biological systems the double-layer potential of membranes is negative owing to the predominance of negatively charged lipids (for mammalian cells mainly phosphatidylserines). These typically constitute 10–20% of the effective lipid area of the membrane (about one elementary charge per 10 nm^2, corresponding to about -0.02 C/m^2) (2,9,26). Furthermore, most of the membrane proteins have isoelectric points below neutral pH. Lipid, protein, and ion contributions together result in potentials of -8 to -30 mV as found from electrophoretic mobility measurements (2,26).

Internal membrane potentials are due to the inhomogeneous distribution of charges within the membrane. Lipid bilayers possess a substantial *dipole potential* arising from the structural organization of dipolar groups and molecules, primarily the ester linkages of the phospholipids and water. These groups are oriented in such a way that the hydrocarbon phase is positive with respect to the outer membrane regions. The magnitude of the dipole potential is usually large, typically several hundreds of millivolts (2,26,27). Other internal potentials arise from membrane asymmetry, including differences in adsorption at the two sides of the membrane, and from differences in penetration of ions.

Finally, separation of charge across the membrane gives rise to a *transmembrane potential*. The transmembrane potential is defined as the electric potential difference between the bulk aqueous phases at the two sides of the membrane and results from selective transport of charged molecules across the membrane. In biological systems it is typically of the order of 10–100 mV (9,26). As a rule, the potential at the cytoplasm side of cell membranes is negative relative to the extracellular physiological solution. We will return to the issues of ion permeation and transmembrane potential in Sec. III.E.

In the following sections we will first concentrate on the electrochemical double-layer properties of biological interphases and then consider the full potential profile of cell walls and lipid bilayer membranes. We do not pursue the full coverage of all modern theoretical approaches to electric double layers but rather try to extend first principles to the complex situation met in biological systems.

B. Basic Electrostatics of Interphasial Electric Double Layers

The mathematical description of an interphasial double layer starts from the basic laws of electrostatics. The Poisson equation gives the relation between the divergence of the electric field gradient (grad ψ) and the space charge density ρ:

$$\text{div} \cdot \text{grad}\, \psi = -\frac{\rho}{\varepsilon\varepsilon_0} \tag{1}$$

where $\varepsilon\varepsilon_0$ denotes the (electric) permittivity. Diffuse parts of double layers are conveniently treated by the Gouy-Chapman (GC) approach (28,29). This is based on the primary premises that ions are point charges, the solvent is a structureless continuum, and interactions are purely electrostatic. The GC model was originally applied to flat geometry, which in the case of bilayer membranes and cell walls is a good approach because the extension of the double layer is usually small compared with the radii of curvature of the membrane or cell wall. The one-dimensional Poisson equation is a simplification of Eq. (1) to

$$\frac{d^2\psi(x)}{dx^2} = -\frac{\rho(x)}{\varepsilon\varepsilon_0} \tag{2}$$

The GC model further assumes that the surface charges are not discrete but uniformly smeared out over a plane, adjacent to the diffuse part of the double layer. This is a rather crude approach for biological interphases, and we will come back to the issue in the next section. The space charge density ρ counts all ions:

$$\rho(x) = F \sum_i z_i c_i(x) \tag{3}$$

where z_i and c_i represent the valence and concentration of ions of type i, respectively. In the case of thermodynamic equilibrium, local concentrations are related to the electric potential field through Boltzmann's law

$$c_i(x) = c_{i,\infty} \exp[-z_i F\psi(x)/RT] \tag{4}$$

in which $c_{i,\infty}$ represents the concentration of ion i in the bulk solution. F, R, and T have their usual meanings as the Faraday constant, the gas constant, and the absolute temperature, respectively. Combination of Eqs (2)–(4) leads to the well-known Poisson-Boltzmann equation, which we write here for a symmetrical electrolyte $(z_+ = z_- = z)$ as

$$\frac{d^2 y(x)}{dx^2} = \frac{\kappa^2}{z} \sinh[zy(x)] \tag{5}$$

where y is the normalized potential $F\psi/RT$ and κ is the reciprocal Debye length, defined by

$$\kappa^2 = \frac{2F^2 c z^2}{\varepsilon\varepsilon_0 RT} \tag{6}$$

With the usual boundary condition that $\psi = 0$ in the bulk of the solution phase $(x = \infty)$, the relation between the surface charge density σ_0 and the potential at the surface ψ_0 becomes

$$\sigma_0 = (8\varepsilon\varepsilon_0 RTc)^{1/2} \sinh(zy_0/2) \tag{7}$$

which for potentials less than 25 mV (corresponding to $y_0 < 1$) approaches

$$\sigma_0 = \frac{2z^2 Fc}{\kappa} y_0 = \varepsilon\varepsilon_0 \kappa\psi_0 \tag{8}$$

The quantity $\varepsilon\varepsilon_0\kappa$ is the diffuse double-layer capacitance, C_d. It is a useful term in dealing with dynamic electric properties of the interphase (see Sec. III.A). The potential profile $y(x)$ follows from

$$\tanh[zy(x)/4] = \tanh(zy_0/4)\exp(-\kappa x) \tag{9}$$

which comes to

$$y(x) = y_0 \exp(-\kappa x) \tag{10}$$

for sufficiently low potentials ($y_0 < 1$).

Biological solutions, of course, normally contain both monovalent and divalent cations as counterions of the negatively charged membranes and cell walls. For aqueous solutions containing a mixture of electrolytes, Grahame (30) derived a general equation that describes the surface potential analogous to Eq. (5) and which is given in, e.g., Ref. 31.

In many systems (including biological ones) the electrostatic potential near the surface is less than predicted by the GC theory. A simple and frequently used way to deal with this problem is to introduce a *Stern layer* in the double-layer model (32). This extended model, the Gouy-Chapman-Stern (GCS) model, accounts for the finite size of the ions (only near the surface) and for specific ion adsorption. The Stern model defines a region adjacent to the surface, which has a permittivity ε_S different from that of the bulk solution (Fig. 3). Its thickness δ is taken as the radius of the hydrated counterion. At a distance δ from the surface, the diffuse double layer can be thought to begin with a potential ψ_d. In the absence of specific adsorption the Stern layer is taken to be free of charges and thus, according to the Poisson equation [Eq. (2)], the potential in this layer drops off linearly:

$$\psi_0 - \psi_d = \sigma_0 \frac{\delta}{\varepsilon_S} = \sigma_0 C_S \tag{11}$$

with C_S the capacitance of the Stern layer. Frequently, ψ_d is low enough to allow

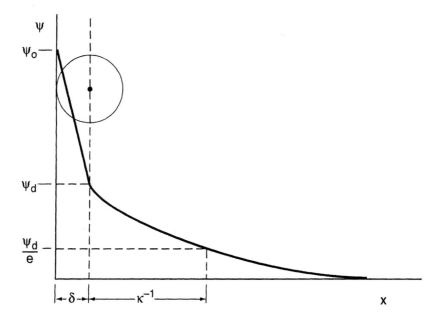

Fig. 3 Potential in the electric double layer according to the Gouy-Chapman-Stern model in the absence of specific adsorption. The size of the hydrated counterion is indicated.

approximation of the potential in Eq. (10), which now becomes

$$y(x) = y_d \exp[-\kappa(x - \delta)] \tag{12}$$

with $y_d = F\psi_d/RT$. Specifically adsorbing ions are thought to penetrate the Stern layer; i.e., they lose part of their hydration shell. In their presence the Stern layer is no longer charge free and generally the potential drops off more strongly than indicated in Fig. 3. For strong specific adsorption it is even possible for the sign of ψ_d to become opposite to that of ψ_0 ('superequivalent adsorption'). The specific adsorption in the Stern layer can be described in several ways. Stern applied the Langmuir adsorption isotherm, whereas later many authors [e.g., Healy and White (33)] used so-called site-binding models in which surface ions are thought to be bound to surface groups with an intrinsic chemical association constant K_a.

For detailed treatments of the GC and GCS models and for more refined double-layer models the reader is referred to the extensive literature, of which overviews can be found in Refs. 2, 26, and 31.

C. Electric Double Layers in Biological Systems

Clearly, the primary shortcoming of GC and GCS in their application to biological interphases is that these models completely ignore the interphasial structure with its spatial distribution of charged sites, counterions, and coions. Nevertheless, in a number of cases GCS adequately describes the dependence of potential on distance in close proximity to a lipid membrane surface (1,34,35). Langner et al. (35) have shown, on the basis of experimental results as well as statistical mechanical calculations, that deviations from GCS are negligible for a bilayer membrane consisting of monovalent lipids but are significant for trivalent lipids and larger for coions than for counterions. Of course, extension of the GC model with an additional layer with a permittivity different from that of the bulk aqueous phase will better fit experimental results than the pure GC model. It should be stressed, however, that the molecular picture of a Stern layer as described in the previous section is far from the situation encountered at the biological membrane–solution interface.

An obvious first attempt to take into account the three-dimensional distribution of charges in the interphase is to extend the double-layer model with a Donnan layer. A Donnan layer contains a number of fixed charges and is accessible for water and dissolved ions. For a bilayer lipid membrane, for example, the Donnan layer would encompass the headgroup charges of the membrane lipids and part of the counter-charge. We emphasize that, generally, the countercharge is distributed over the Donnan layer and the diffuse layer, the distribution ratio being primarily dependent on the electrolyte concentration in the medium.

A theoretical derivation of the potential profile in a system with a combination of a Donnan layer and a diffuse layer has been given by Ohshima and Kondo (36). The basic step is that Poisson's equation is modified to [compare Eq. (2)]

$$\frac{d^2\psi}{dx^2} = -\frac{(\rho_h + \rho_e)}{\varepsilon_D\varepsilon_0} \tag{13}$$

where ρ_h and ρ_e represent the space charge densities due to the lipid headgroups (or, alternatively, the charged groups of the cell wall) and the electrolyte ions, respectively, and ε_D is the relative permittivity of the Donnan layer phase. For the adjacent diffuse layer in the aqueous phase the GC approach is maintained. The potential profile can then be derived. Under conditions of continuity of potential and potential gradient across the interphase (Gauss), and with ε_D identical to the relative permittivity ε in solution, i.e.,

$$\psi_{+Ds} = \psi_{-Ds} \tag{14a}$$

$$(d\psi/dx)_{x=+Ds} = (d\psi/dx)_{x=-Ds} \tag{14b}$$

Ohshima and Kondo (36) finally obtained

$$\psi_{Ds} = \psi_D - \frac{RT}{zF}\tanh(zF\psi_D/2RT) \tag{15}$$

where ψ_{Ds} is the potential at the boundary between Donnan layer and solution and $+Ds$ and $-Ds$ denote the solution side and the Donnan layer side, respectively. The potential ψ_D is the Donnan potential, corresponding to the situation of a pure Donnan profile (the limit of no diffuse layer at all):

$$\psi_D = \frac{RT}{zF}\sinh(\rho_h/2zFc) \tag{16}$$

Figure 4 shows a set of calculated potential profiles for different electrolyte concentrations. For large c the profile approaches the step functionality of the pure Donnan model. Note that for low potentials ψ_{Ds}/ψ_D does not vary with c, in accordance with Eq. (15). As in the GCS model, the diffuse double-layer potential is lower than in the pure GC model, because part of the "fixed" charge is already compensated within the Donnan layer phase. In the Ohshima-modified Donnan approach it is assumed that the distribution of fixed charges in the Donnan layer, ρ_h, is homogeneous and that the thickness of this layer is larger than κ^{-1}, resulting in an exponential variation of the potential in the interphase. This feature makes the Ohshima-modified Donnan approach suitable primarily for describing the double layer at the cell wall–solution interface and less appropriate for the lipid bilayer membrane. Reviews concerning the electric double layer in biological systems are given in Refs. 1, 2, and 37.

D. The Total Potential Profile of Biological Interphases

For the construction of the full electric potential profile of a biological interphase, the preceding approach is a good starting point, i.e., dividing the interphase into different regions, each with its own thickness, permittivity, and fixed charge density, and using a Poisson-Boltzmann type of approach. At all boundaries between distinguishable layers we should have the Gauss conditions of continuity taking into account the differences in permittivity of the various layers $[\psi_+ = \psi_-;$ $\varepsilon_+(d\psi/dx)_+ = \varepsilon_-(d\psi/dx)_-]$. For the diffuse part of the double layer in the solution the GC model may be used again.

Total potential profiles for bacterial cell walls at various electrolyte concentrations calculated by van der Wal (38) are shown in Fig. 5. Here, the complete cell wall has been modeled as one layer; the "fixed" charge density is dependent on

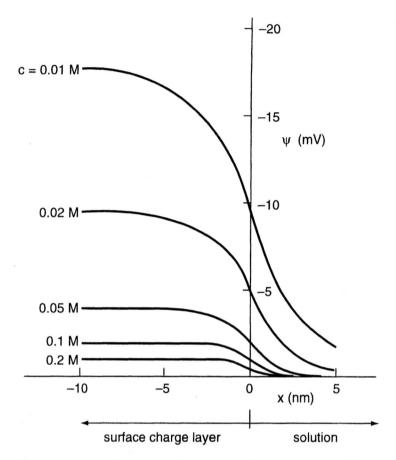

Fig. 4 Potential distribution $\psi(x)$ across the Donnan layer and diffuse double layer for different electrolyte concentrations c in solution, calculated with $z = -1$, $\rho_h = 0.015$ M (for comparison given in the same unit as c), thickness of the Donnan layer $= 10$ nm, $\varepsilon_D = \varepsilon = 78.5$, and $T = 298$ K. (Adapted from Ref. 36.)

the pH and the electrolyte concentration. The Poisson equation [Eq. (13)] has been solved numerically with the boundary condition $d\psi/dx = 0$ for the inner core and Gauss conditions on the solution–cell wall boundary. The relative permittivity of the cell wall material was chosen equal to 60.

The total electric potential profile of a biological membrane is the result of the various potential-generating phenomena as discussed in Sec. II.A. In Fig. 6 the potential profile for a lipid bilayer is presented schematically. In the inner core the charge density is zero and, consequently, the potential varies linearly in this region. The dipole potential profoundly affects the total potential profile. Because the transmembrane potential is the result of selective ion transport across the membrane, it is not possible to calculate this contribution to the potential profile from purely electrostatic principles as described earlier. For this we need a model that describes the ion permeation and the mobilities of the various ions in the membrane, as will be shown in Sec. III.E.

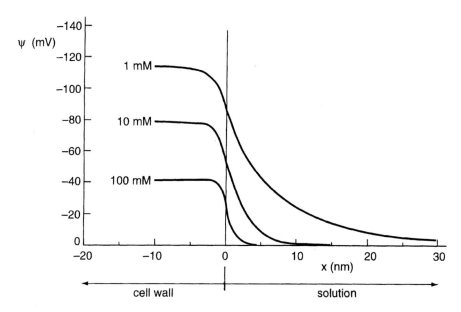

Fig. 5 Potential profiles across the cell wall of *Corynebacterium* and the surrounding solution at neutral pH and various electrolyte concentrations. Results of calculations by van der Wal et al. (38).

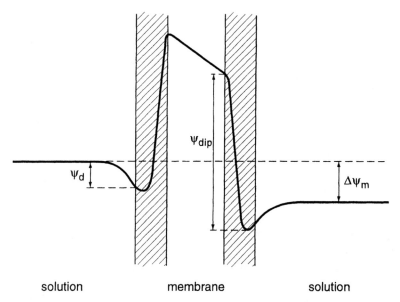

Fig. 6 Schematic representation of the potential profile across a bilayer lipid membrane. The charge density in the hydrophobic inner part of the membrane is zero. The shaded (outer) regions of the membrane contain membrane charges and dipoles, as well as adsorbed and penetrated counterions. The net charge in the membrane is compensated by diffuse countercharges in the solution phases; ψ_d is the potential of the diffuse double layer. The structural organization of the dipoles produces a large internal dipole potential, ψ_{dip}, which renders the hydrophobic interior positive with respect to each of the bulk solutions. Separation of charge across the membrane gives rise to the transmembrane potential $\Delta\psi_m$.

It is important to realize that for the electrostatic properties of molecules in and near membranes the actual local electric field rather than the transmembrane potential is important. Surface charge asymmetry, proteins, and adsorbed and penetrating (hydrophobic) ions may substantially change the internal potential profile of the membrane without affecting $\Delta\psi_m$ but with important consequences for membrane electric phenomena.

As already mentioned in Sec. II.A, the structure of bilayer lipid membranes has been investigated using computational techniques, e.g., molecular dynamics (MD) simulations (17–20) and self-consistent field (SCF) calculations (21–24). These methods also allow calculation of the electric fields across the membrane. Figure 7 shows the segment density, total charge, and potential profiles for a dimyristoylphosphatidylcholine (DMPC) bilayer surrounded by a 50 mM NaCl aqueous solution as calculated using the SCF method (24). DMPC is a zwitterionic phospholipid and its chemical structure is given in Fig. 7. From Fig. 7b it can be seen that the nitrogen resides in the same layers as the phosphate group, indicating that on average the phosphocholine headgroup is almost parallel to the interface, in accordance with experimental data (25). However, the profiles of N and P do not match perfectly and the net charge separation leads to a potential profile across the membrane. In Fig. 7c the volume fraction profiles of the Na^+ and Cl^- ions are given. The ions prefer the water phase over the hydrocarbon core of the bilayer; in the headgroup region they locally follow the electrostatic field of the P-N dipoles and contribute to the total potential profile.

For comparison, in Fig. 8 the interphasial atom distribution and potential profile for a dipalmitoylphosphatidylcholine (DPPC) bilayer in pure water as obtained by MD simulation is presented (18). Apart from the total potential, the contributions to the potential from the lipid and the water molecules are also given (Fig. 8b). The potential arising from the lipid partial charges has a positive sign at the outside of the membrane, in line with the distributions of the P and N atoms depicted in Fig. 8a. On average, the N atom is located closer to the water phase than the P atom, just the other way around as given by the SCF results described above. However, the water dipoles, for which the potential profile is almost a mirror image of the lipid potential, overcompensate for the lipid dipoles. As a result, it is found that the total potential drop across the interphase between the membrane core and the bulk electrolyte is negative (about -200 mV).

Figures 7 and 8 are just given as examples and it is too early to discuss in detail the differences between the results of the two methods. Of course, in both approaches the results depend heavily on the parameters chosen. Apart from this, an obvious cause of differences is that in the SCF method the water molecules have been modeled as isotropic monomers with only nearest neighbour (contact) interactions, whereas in the MD simulations their dipole character is maintained. In virtually

Fig. 7 Segment density, electrostatic potential, and charge profiles through a DMPC bilayer membrane according to SCF calculations. The chemical structure of DMPC is indicated. The bulk volume fraction of NaCl, φ^b_{NaCl}, is 0.002 (corresponding to about 50 mM NaCl). The layer thickness of the lattice used was set to 0.3 nm; layer $z = 0$ corresponds to the center of the bilayer. (a) Overall segment distribution of the main components: the water, hydrocarbon, and headgroup segments, as well as the oxygen from the myristoyl ester. (b) Detailed segment distributions for the phosphocholine headgroup. (c) Distributions of the electrolyte ions. (d) Electrostatic potential and total charge profiles. (From Ref. 24.)

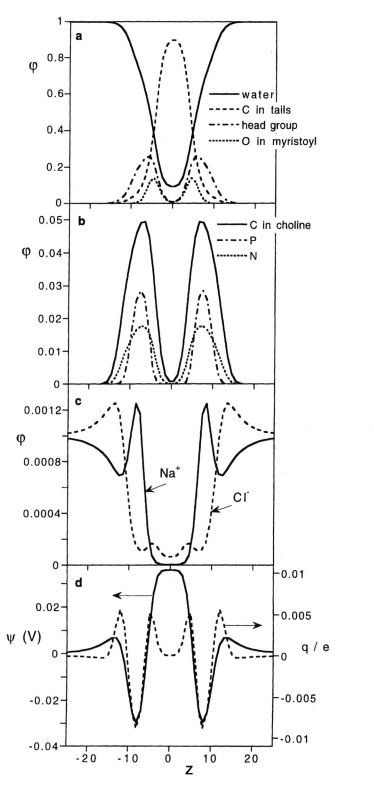

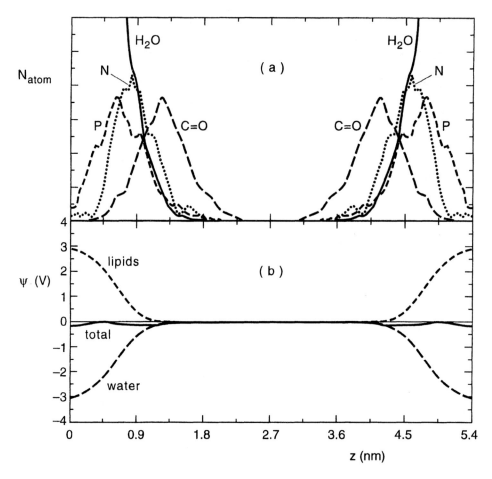

Fig. 8 Structure and electrostatic potential profiles of a DPPC bilayer membrane from MD simulations. (a) Interfacial atom distributions. (b) Electrostatic potential profile across the bilayer, split up in a water and a phospholipid contribution. Potentials are given relative to the potential in the center of the bilayer. (Redrawn from Ref. 18.)

all recent MD studies it has been found that the polarization due to orientation of water molecules contributes to a net negative potential in the water relative to the bilayer interior (20). The different simulations vary, however, in their predictions of the range of the orientational polarization and the magnitude of the resulting dipole potential.

III. ELECTRODYNAMICS OF BIOLOGICAL INTERPHASES

A. Elementary Dynamics of Electric Double Layers

How fast does an electric double layer in a biological interphase and the adjacent solution build itself up or, equivalently, adjust itself to changing conditions? In other words, what are the characteristic time constants for formation of elements of the interphasial double layer?

The time constant τ_d for relaxation of the diffuse part of the double layer is solely determined by *bulk* properties of the electrolyte solution:

$$\tau_d = \frac{\varepsilon\varepsilon_0}{K} \tag{17}$$

with K the specific conductivity of the solution. It is useful to realize that in terms of electric equivalent circuitry the relaxation process corresponds to the discharge of the geometric capacitance $\varepsilon\varepsilon_0$ (i.e., the dielectric formed by the bulk electrolyte solution) across the solution resistivity K^{-1}. In electric jargon, τ_d represents the *RC* time constant of the solution, and this indeed governs interfacial double-layer formation. Equation (17) can be rewritten in terms of the ionic diffusion coefficients D_i because

$$K = F\sum_i |z_i|u_ic_i = \frac{F^2}{RT}\sum_i z_i^2 D_ic_i \tag{18}$$

in which u_i represents the ionic mobilities. For a symmetrical electrolyte solution with $D_{\text{cation}} = D_{\text{anion}} = D$, and using Eq. (6), it follows that

$$\tau_d = \frac{1}{D\kappa^2} \tag{19}$$

Thus, the time constant τ_d is directly related to the time necessary for ions to migrate over a distance equal to the Debye length κ^{-1}. For example, for a 10^{-3} M aqueous electrolyte solution with $\varepsilon = 80$, D of the order of 10^{-9} m^2 s^{-1}, and κ of the order of 10^8 m^{-1}, Eq. (19) yields a τ_d value around 10^{-7} s. Note that κ is proportional to the square root of the electrolyte concentration [Eq. (6)], so that τ_d is inversely proportional to c. Thus, for electrolyte concentrations of the order of 0.1 M, τ_d reaches values as low as 10^{-9} s.

Lipid bilayer membranes, with their apolar cores, generally have extremely poor conductivities. K can be as low as 10^{-6} Ω^{-1} m^{-2} (39), also depending on the type of ions transferring the charge across the membrane (see Sec. III.E). The time constant τ_m to relax from electric polarization by conduction through the bilayer is correspondingly large. For a membrane with a relative permittivity of the order of order 10 and a thickness of about 10^{-8} m the geometric capacitance is of order 10^{-2} F m^{-2}. Thus, the corresponding time constant τ_m is of order 10^4 s. This means that on the relevant time scales of biological processes, the double layer at the solution side adjusts instantaneously (τ_d typically 10^{-9}–10^{-7} s), whereas the apolar core of the lipid bilayer behaves as a dielectric that does not allow appreciable passage of charge on time scales below 10^4 s.

Obviously, the picture changes when the bilayer contains ion transport channels or other transport mediators. Measurements on individual ion channels have shown that their conductivities are typically of order 10^{-11} Ω^{-1} (40). Realizing that such a channel takes a part of the membrane surface area of the order of 10 nm^2, this corresponds to a local K of about 10^6 Ω^{-1} m^{-2}. Thus, locally, the relaxation time constant comes down to around 10^{-8} s, which logically is again in the range of that for electrolyte solutions.

Lateral transfer of ionic species through the biointerphasial double layer has only recently received attention. Yet it is a subject of significant relevance because it may play a crucial role in the interactions of organisms with their surroundings. We shall consider the subject in some detail in the following. In connection with this, we shall go through the principles of electrokinetics, which are so helpful in measuring static as well as dynamic double-layer properties of biointerphases.

B. Principles of Electrokinetics and Dielectric Relaxation

Charged colloidal particles migrate under the influence of an applied electric field. The phenomenon has been called *electrophoresis*. The nature of the electrophoretic behavior of suspended particles is much more complicated than that of individual ions. This is due to the more involved double-layer situation: the particle has a certain surface charge and some countercharges (ions within some surface layer or Donnan layer), which generally have a certain mobility. On the solution side of the electric double layer there are only the individual ions with mobilities equal to their bulk values. This distribution of charges and mobilities has consequences for the overall response to an externally applied electric field, as we shall treat in some detail in the following.

Generally speaking, electrokinetics covers phenomena due to the movement of charged particles (or a charged surface) and surrounding liquid with respect to each other. For a quantitative interpretation of electrokinetic data it is essential to know the position of the so-called *slip plane* (or the hydrodynamic plane of shear). The slip plane is the hypothetical boundary plane that separates a stagnant liquid layer at the interface from the bulk liquid. Although neither a discrete slip plane nor a discrete interfacial plane exists, theory is usually based on infinitely sharp transitions. Only modern molecular dynamics simulations relax from this. A vast amount of experimental data indicates that the slip plane does not coincide with the particle-solution interface but that it is situated at some finite distance in the solution. Apparently, there is a more or less stagnant layer of liquid immediately adjacent to the surface. The electric potential at the slip plane, the so-called ζ-*potential*, is an important experimentally accessible parameter. On the basis of double-layer data, colloidal stabilities, and electrokinetic information, the ζ-potential has often been identified with ψ_d, the potential at the Stern layer–solution boundary (see Fig. 3). This expresses current belief that the stagnant layer between the slip plane and the surface has a thickness of only a few molecular layers of water.

1. Electrophoresis of Spherical Particles

In a very first approximation one could express the electrophoretic velocity v in terms of the electric force on the particle (equal to the particle charge Q times the field strength E) and the friction coefficient f (for spherical particles f equals $6\pi\eta a$, where η is the viscosity of the medium and a the particle radius). Thus we introduce the particle mobility as

$$u = \frac{v}{E} = \frac{Q}{6\pi\eta a} \tag{20}$$

The oversimplification in this approach is that it ignores the presence of the countercharge $-Q$ at the solution side of the double layer. The coun025charge is subjected to an opposite electric force $-QE$, and this results in substantial hydrodynamic retardation of the particle movement. In addition, there is an electric retardation effect that is related to the asymmetric nature of the double layer around the migrating particle. We shall return to this later in this section.

Hückel (41) and Onsager (42) derived the classical theory for stationary migration of a spherical particle surrounded by a diffuse countercharge layer. This countercharge atmosphere is divided into infinitesimally thin concentric shells at distance r from the center of the particle and with thickness dr. The external field E exerts a force $4\pi r^2 \rho E dr$ (ρ being the local volume charge density) on the shell. Using Eq. (20), this results in a migration velocity of the shell

$$v(r, dr) = \frac{4\pi r^2 dr \rho E}{6\pi \eta r} \tag{21}$$

Integration over the total countercharge atmosphere yields the velocity of the diffuse double layer with respect to the innermost layer and the particle

$$v_{dl} = \int_a^\infty v(r, dr) = \int_a^\infty \frac{2r\rho E}{3\eta} dr \tag{22}$$

The velocity of the particle with respect to its immediate surroundings is given by Eq. (20), so the *net* velocity of the particle with respect to the bulk solution is the algebraic sum of the velocities given by Eqs. (20) and (22)

$$v = \left[\frac{Q}{6\pi \eta a} + \int_a^\infty \frac{2r\rho}{3\eta} dr \right] E \tag{23}$$

Overall electroneutrality implies the identity of Q and the integral of ρ:

$$Q = -\int_a^\infty 4\pi r^2 \rho \, dr \tag{24}$$

Using Poisson's laws for spherical symmetry,

$$\rho = -\frac{\varepsilon}{r} \frac{\partial^2 r\psi}{\partial r^2} \tag{25}$$

Equation (23) is transformed into

$$\frac{v}{E} = \frac{2\varepsilon}{3\eta} \left\{ -\frac{1}{a} \int_a^\infty (r - a) \frac{\partial^2 r\psi}{\partial r^2} dr \right\} \tag{26}$$

For $\kappa a \ll 1$ and $\psi_{r=a} = \zeta$, this becomes

$$u = \frac{2\varepsilon}{3\eta} \zeta \qquad (\kappa a \ll 1) \tag{27}$$

The treatment is justified for such low electrolyte concentrations and/or such small particles that κa is indeed well below unity. According to Smoluchowski's theory (43) for double layers that are relatively thin compared with the particle radius ($\kappa a \gg 1$) the result just differs by a numerical factor:

$$u = \frac{\varepsilon}{\eta} \zeta \qquad (\kappa a \gg 1) \tag{28}$$

In generalizing the theory, Henry (44) demonstrated that Eqs. (27) and (28) give the limiting values for u for the extremes of the double-layer characteristic κa. Henry computed a factor $f(\kappa a)$ such that

$$u = f(\kappa a)\frac{\varepsilon}{\eta}\zeta \tag{29}$$

and showed that it gradually changes from 1 to 2/3 on going from high to low κa values.

2. Double Layer Dynamics in an Alternating-Current Field

The migration of a charged particle in an applied electric field is generally accompanied by a distortion of the counterionic atmosphere. The originally symmetric charge distribution is modified as a result of the opposite movements of particle and counterions. Figure 9 illustrates that the ionic atmosphere is dispersed at one side of the migrating particle and rebuilt at the other side. The resulting effect can be described in terms of an induced dipole moment (formed by the displacement between the centers of positive charge and negative charge) or an electric polarization field opposing the applied field. The resulting retardation is often denoted as a "relaxation effect" because a certain time is necessary to generate an asymmetric ion distribution around a moving particle. In an a.c. electric field it will depend on the frequency, usually expressed by the angular frequency ω, whether polarization effects are important or not. The elementary time constant τ for building an asymmetric double layer around a particle with radius a is [and compare Eq. (19)]

$$\tau = \frac{a^2}{2D} \tag{30}$$

which means that for frequencies around τ^{-1} the polarization field is just able to partly follow the externally applied field.

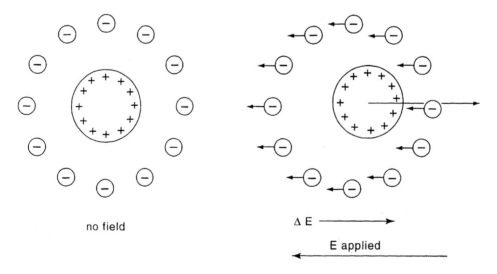

no field $\Delta E \longrightarrow$

E applied

Fig. 9   Development of the polarization field ΔE in electrophoresis.

The technique dielectric relaxation spectroscopy measures the complex conductivity $\hat{K}$ of the particle dispersion as a function of frequency and detects the double-layer polarization as a relaxation process at its characteristic frequency τ^{-1}. The complex conductivity $\hat{K}$ is composed of a real component K (the purely conductive element) and an imaginary component ε (the purely dielectric or capacitive component):

$$\hat{K}(\omega) = K(\omega) - i\omega\varepsilon(\omega) \tag{31}$$

where i is the imaginary unit. The conductivity $K(\omega)$ is an ascending function of ω: at low ω double-layer relaxation hinders ion movement around the particle, whereas at high ω polarization is not generated because the field changes sign many times within a period τ. The dielectric permittivity $\varepsilon(\omega)$ is a descending function of ω because at sufficiently low ω double-layer polarization represents strong dipoles that enhance the ε of the dispersion. As an example, Fig. 10 shows $\varepsilon(\omega)$ and $K(\omega)$ of a hematite sol in an aqueous KCl solution.

C. The Role of Surface Conduction

Anticipating the experimental evidence in Sec. III.D, electrokinetic and electrodynamic techniques have already proved to be powerful tools in the electrochemical characterization of bacterial cell surfaces. An essential feature appeared to be a high degree of surface conductance, i.e., conductance due to mobile counterions within the stagnant cell wall layer (45,46). The phenomenon of high lateral mobility of surface ions *within* the stagnant layer has been observed for several systems, including well-defined model colloids (47–50). It therefore seems desirable to outline the extensions of electrodynamic theory to explicitly include the effects of surface conduction.

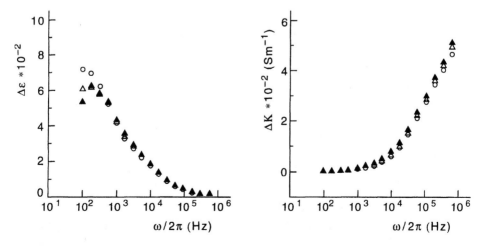

Fig. 10 Dielectric (left) and conductivity (right) dispersion of a hematite sol in 1.3×10^{-3} M KCl. Volume fractions: ($\bigcirc$) $\varphi = 0.04$, ($\triangle$) $\varphi = 0.03$, and ($\blacktriangle$) $\varphi = 0.02$. (Redrawn from Ref. 48.)

1. Particle Mobility

Under conditions where tangential ionic transfer within the slip plane is significant, the electrophoretic mobility takes the form (51)

$$u = \frac{\varepsilon}{\eta}\zeta + \frac{\varepsilon kT}{\eta e}\frac{Du}{1 + Du(1 - \frac{1}{2}\gamma)}\left\{\frac{2\ln 2}{z} - \frac{e\zeta}{kT} - \gamma\left(\frac{2\ln 2}{z} - \frac{e\zeta}{2kT}\right)\right\} \tag{32}$$

where γ is a complex functions of the nondimensional angular frequency $\omega\tau$ [τ as defined by Eq. (30)]:

$$\gamma = \frac{(\omega\tau)^{2/3} + i\omega\tau}{(1 + (\omega\tau)^{1/2})(1 + \omega\tau)} \tag{33}$$

and Du is the relative surface conductivity (31)

$$Du = \frac{2}{\kappa a}\left(1 + \frac{3m}{z^2} + \theta\right)\left[\exp\left(\frac{ze\zeta}{2kT}\right) - 1\right] \tag{34}$$

The drag coefficient m in the electro-osmotic term $3m/z^2$ is defined by

$$m = \frac{2\varepsilon k^2 T^2}{3\eta e^2 D} \tag{35}$$

and the parameter θ reflects the ratio between conduction behind the slip plane and conduction beyond it (see Refs. 52 and 53 for details).

The first term on the right-hand side of Eq. (32) is identical to Smoluchowski's result, Eq. (28). The second term expresses the retardation due to polarization of the double layer and becomes unimportant for $Du \ll 1$, i.e., for systems without significant surface conduction. It is useful to mention two other limiting cases that immediately follow from Eq. (32). The low-frequency limit is

$$u = \frac{\varepsilon\zeta}{\eta} + \frac{\varepsilon kT}{\eta e}\frac{Du}{1 + Du}\left(\frac{2\ln 2}{z} - \frac{\varepsilon\zeta}{kT}\right) \qquad (\omega\tau \ll 1) \tag{36}$$

which is equivalent to the Dukhin-Semenikhin expression (54) for the steady-state mobility. On the high-frequency side we have

$$u = \frac{\varepsilon\zeta}{\eta}\left(1 - \frac{Du}{2 + Du}\right) \qquad (\omega\tau \gg 1) \tag{37}$$

which reduces to what is known as the Henry-Booth equation (55,56) for $\theta = 0$. Around $\omega = \tau^{-1}$ the mobility shows dispersion and the motion is out of phase compared with the applied field.

2. Relaxation Spectrum

The full dynamic spectrum of a particle dispersion is given by $\hat{K}(\omega)$. For a sufficiently dilute dispersion of spherical particles with relatively thin double layers ($\kappa a \gg 1$) fairly rigorous theory has been developed. The results are conveniently expressed in terms of conductivity and permittivity increments relative to some reference value (see Fig. 10). The conductivity increment $\Delta\hat{K}(\omega)$ is the actual conductivity $K(\omega)$ minus K at frequencies much *smaller* than τ^{-1}:

$$
\begin{aligned}
\Delta K(\omega) &= K(\omega) - K(\omega \ll \tau^{-1}) \\
&= K^\infty \frac{9}{8} \varphi (1 - h^2) \left(\frac{Du}{1 + Du} \right)^2 \\
&\quad \times \frac{(\omega\tau)^{3/2}}{(1 + \omega\tau)(1 + (\omega\tau)^{1/2}) - (1-h)Du(2+2Du)^{-1}(\omega\tau)^{3/2}}
\end{aligned}
\tag{38}
$$

and the permittivity increment $\Delta\varepsilon(\omega)$ is the actual $\varepsilon(\omega)$ minus ε at frequencies much *higher* than τ^{-1}:

$$
\begin{aligned}
\Delta\varepsilon(\omega) &= \varepsilon(\omega) - \varepsilon(\omega \gg \tau^{-1}) \\
&= \varepsilon^\infty \frac{9}{16} \varphi (\kappa a)^2 (1 - h^2) \left(\frac{Du}{1 + Du} \right)^2 \frac{1}{(1 + \omega\tau)(1 + (\omega\tau)^{1/2})}
\end{aligned}
\tag{39}
$$

where K^∞ and ε^∞ are the bulk conductivity and permittivity, respectively, φ is the particle volume fraction of the dispersion, and h is a parameter defined by

$$
h = \frac{z_2 D_2 + z_1 D_1}{z_2 D_2 - z_1 D_1}
\tag{40}
$$

where the index 2 refers to the counterion and the index 1 to the coion. Note that h may vary between -1 and $+1$ and that it is close to zero for electrolytes with $D_{cation} \approx D_{anion}$. Equations (38) and (39) have been applied to the ε, K spectra of well-defined model dispersions and work reasonably well for a variety of electrolytes (52).

D. Lateral Electrodynamics of Biological Interphases

The quantitative understanding of physicochemical properties of (micro)biological particles and surfaces requires thorough knowledge of the electric double-layer properties, static as well as dynamic ones. In this section we concentrate on lateral migration effects of surface ions. The measurement and interpretation of lateral ion movements are a new and rewarding development, which represents a forefront topic in both (micro)biology and colloid chemistry.

As an example, we shall discuss a study of bacterial cells of the genus *Corynebacterium*. The cells are more or less spherical, with diameters of 1.1 and 0.8 μm for the longer and shorter axes, and can be prepared as fairly homodisperse suspensions. Measured electrophoretic mobilities of *Corynebacterium* cells are displayed as a function of pH in Fig. 11. The mobility in some way represents the particle charge via the ζ-potential and other double-layer features [see Eqs (27)–(29), (32), (36), and (37)]. Thus, the mobility increases with increasing charge in the cell wall and with decreasing electrolyte concentration. For the *Corynebacterium* cells the mobility versus charge function levels off and becomes quite insensitive to the electrolyte concentrations at higher pH, i.e., at increasing magnitude of the (negative) charge. This behavior could be indicative of surface conduction effects because Du increases with decreasing electrolyte concentration. [Note in Eq. (34) the occurrence of κ, which is proportional to $c^{1/2}$.]

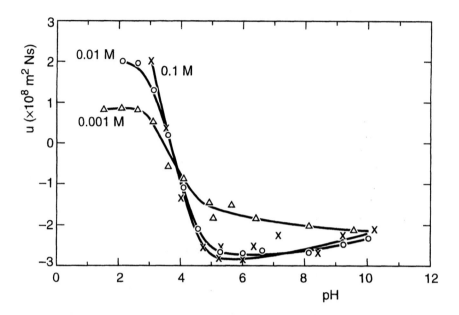

Fig. 11 Electrophoretic mobility of *Corynebacterium* as a function of pH in KNO_3 solutions of various concentrations. (Redrawn from Ref. 38.)

Consistent and unambiguous interpretation of the mobility data has been achieved by combining them with the full dielectric relaxation spectra, $\hat{K}(\omega)$, of the bacteria suspension. The experimental data, in the form of ΔK and $\Delta \varepsilon_r (= \Delta \varepsilon / \varepsilon_0)$, are collected in Fig. 12. The spectra show the expected behavior with ΔK increasing and $\Delta \varepsilon$ decreasing with increasing frequency. The figure also shows the fits of the spectra to the theoretical expressions for ΔK and $\Delta \varepsilon$ as given by Eqs (38) and (39). The quality of the coincidence between experiment and theory is remarkably satisfactory. For the particular case presented in Fig. 12, the best fits are obtained (and note the consistency between K and &varepsilon) for $\zeta = -70$ mV and $\theta = 16$. For such a set of data, the relative surface conductivity Du takes values of order unity [see Eq. (34)] and, consequently, the second term in the rigorous mobility expression [Eq. (32)] is most significant at these low concentration levels. Reconstruction of the dependence of the mobility on the ζ-potential for various electrolyte concentrations (Fig. 13) shows regular behavior, in contrast to the results obtained on the basis of the classical Smoluchowski equation [Eq. (28)], which ignores lateral mobility of double-layer ions. We emphasize that the findings on the parameter θ confirm the mobility of counterions *within the cell wall*. It turns out that these ions have lateral mobilities that are only about a factor 2 lower than the mobilities in aqueous bulk solution. This is a remarkable and important finding and has been confirmed by results on several model colloids (47–50).

The electrodynamic results obtained for microorganisms and their similarity to findings in colloid chemical studies call for some preliminary extrapolation to bacterial aggregation and adhesion processes. We have seen that lateral charge redistribution processes are based on surface ion mobilities almost equal to bulk mobilities. Thus for bacterium-bacterium interaction or bacterium–macroscopic surface interaction, estimation of lateral charge redistribution times can be based

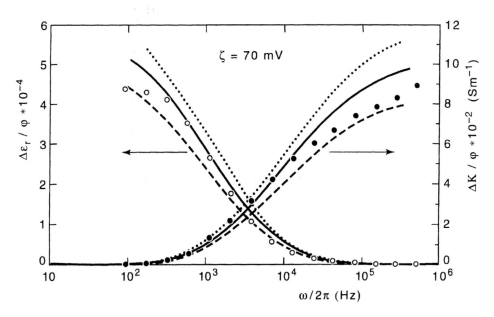

Fig. 12 Dielectric dispersion of *Corynebacterium* in 10^{-3} M KNO_3. Measured values are indicated by points: ($\square$) $\Delta\varepsilon_r/\varphi$ and ($\bullet$) $\Delta K/\varphi$, with φ the volume fraction of bacteria. Curves have been calculated with $a = 570$ nm, $\zeta = -70$ mV, and $\theta = 13$ (– – –), 16 (————), and 19 (······). (Adapted from Ref. 46.)

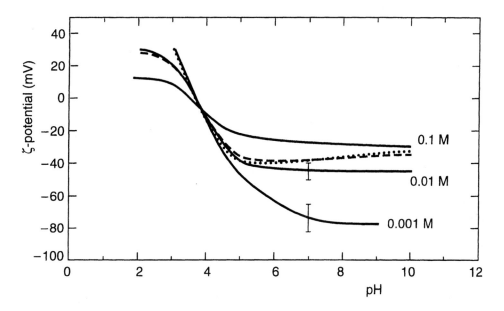

Fig. 13 The ζ-potentials of *Corynebacterium* calculated from the electrophoretic mobility as given in Fig. 11. The continuous curves are obtained using Eq. (32), i.e., taking surface conduction into account. The dashed curves are computed using the Smoluchowski equation (······, 10^{-3} M and – – –, 10^{-2} M). (Redrawn from Ref. 38.)

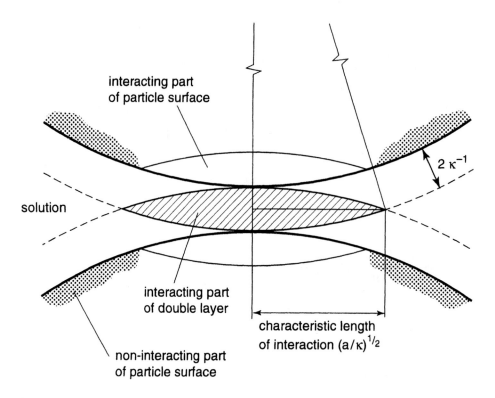

Fig. 14   Double-layer interaction between two spherical colloidal particles of radius a in dispersion. The lens-shaped interaction region has a radius of $(a/\kappa)^{1/2}$. (From Ref. 57.)

on characteristic interaction distances and bulk mobilities. For spherical particles with radius a the interaction distance is typically of order $(a/\kappa)^{1/2}$ (57). Figure 14 illustrates this. The time constant for lateral surface charge relaxation, τ_l, would therefore be

$$\tau_l = \frac{a}{2D_l \kappa} \tag{41}$$

where D_l is the lateral diffusion coefficient of the surface ions. For $a = 500$ nm, $\kappa^{-1} = 10$ nm, and $D_l = 5 \times 10^{-10}$ m^2 s^{-1} we get τ_l of order 10^{-5} s. This should be compared with the characteristic time of Brownian encounter between two bacterial particles or the time of approach of one particle encountering a macroscopic surface, τ_{Br}

$$\tau_{Br} = \frac{1}{2D_p \kappa^2} \tag{42}$$

in which D_p is the diffusion coefficient of the particle. For the mentioned particle of 500 nm, D_p would be of order 10^{-12} m^2 s^{-1}, so that with $\kappa^{-1} = 10$ nm we get a τ_{Br} of order 10^{-4} s. This means that lateral ion dynamics can effectively minimize the electrostatic repulsion in interaction processes. The resulting practical aggregation or adhesion kinetics will derive heavily from this.

E. Ionic Membrane Permeation

Because of the nonpolar nature of its inner core, the lipid membrane is intrinsically impermeable to polar and charged molecules. Only small hydrophobic molecules and ions (e.g., tetraphenylborate, picrate, and tetraalkylammonium ions) are able to pass the membrane by themselves. For other molecules and ions, the lipid membrane presents too much resistance and special regulatory mechanisms are required to allow their permeation. Such regulatory mechanisms involve more or less specific *transport mediators* (see Fig. 15) such as transport proteins, which generally have high specificity for the transported species. Biological membranes contain membrane-spanning protein channels for ion transport. These channels are dynamic structures that change their properties in response to outer influences, such as radiation, changes in electric potential, and bulk ion concentrations. Rearrangement of the channel structure caused by such perturbations is called "channel gating" if the different conformational states are characterized by different electric conductivities "open" and "closed" states). Generally, channel gating is a quite complex process involving various intermediate states (26,58). A well-known example of channel gating is the voltage-dependent opening and closing of sodium nerve membrane channels. Studies concerning voltage-dependent channel gating can be found, for example, in Refs. 59 and 60.

Ion-selective transmembrane channels may also be formed by molecules other than proteins, having a very specific structure. Characteristics are a lyophilic exterior, a flexible conformation (so that conformational changes connected with the presence of an ion readily occur), sufficient length (so that the channel connects both sides of the membrane that are accessible for ions), and a homogeneous (not too strong) binding of ions throughout the channel. An example is the pentadecapeptide gramicidin A, of which two molecules in the α-helix conformation associate to form an ion-selective channel. Several approaches to preparing artificial ion channels in lipid membranes have been reported in the literature (see, e.g., Refs. 61–63).

Other mediators include relatively small carrier molecules (ionophores) that characteristically facilitate the transport of ions across biological membranes. The ways in which these mediators bring about transport of ions across biological membranes are schematically given in Fig. 15. In cases where the bond between the transported ion and the carrier is weak, the ion may jump from one ionophore molecule to another

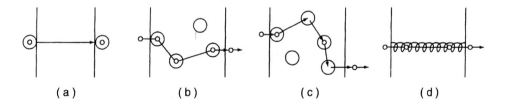

(a)　　　　　(b)　　　　　(c)　　　　　(d)

Fig. 15 Mechanisms for ion transport across biological membranes. (a) Transport of lyophilic ions. (b) Carrier-mediated transport; the ion-carrier complex travels from one side of the membrane to the other. (c) Carrier-mediated transport, in which the ion jumps over from one carrier to another ("carrier relay"). (d) Transport through an ion-specific transmembrane channel.

during transport across the membrane. The binding between ion and carrier can be rather selective and forms the basis of modern ion-selective electrode potentiometry. Characteristic carriers are macrocyclic molecules with a hydrophobic exterior and a hydrophilic interior in which an ion fits. These carriers specifically uncouple oxidative phosphorylation in mitochondria, which led to their discovery in the 1950s. This property is also connected with their antibiotic action. A well-known example is the antibiotic valinomycin, a cyclic oligopeptide. Also, hydrophobic ions, such as tetraphenylborate and picrate-containing compounds, may act as ion carriers.

Before we discuss the electrodynamic aspects of ion permeation across biological membranes in more detail, we will briefly treat the driving force for this process.

1. Passive and Active Transport

The chemical potential of a species i, μ_i, thermodynamically indicates the free energy associated with it. In case of a charged molecule the term *electrochemical potential*, denoted as $\tilde{\mu}_i$, is used. Under conditions of constant temperature and pressure and in the absence of force fields other than an electric field, the electrochemical potential is expressed as

$$\tilde{\mu}_i = \mu_i^0 + RT \ln a_i + z_i F \psi \tag{43}$$

where μ_i^0 is the standard chemical potential, a_i the activity of the ion, z_i its valence (including sign), and ψ the electric potential.

If the electrochemical potential of a type of ion depends on its position, a net movement or flux of those ions will tend to occur toward the region where its electrochemical potential is lower (so that its free energy is minimized). The negative gradient of its electrochemical potential $[-\mathrm{grad}(\tilde{\mu}_i)$, or $-\partial\tilde{\mu}_i/\partial x$ for the one-dimensional case] acts as the driving force for this flux. This process is called *passive transport*.

Active transport is a membrane transport process with a source of energy other than the (electro)chemical potential gradient of the transported substance. This source of energy can be either a metabolic reaction or an (electro)chemical potential gradient of another substance (10). Active transport is of basic importance for life processes.

2. A Simple Kinetic Model for Ion Permeation

A relatively simple way to investigate ion permeation through a membrane is to measure the electrical resistance across the membrane between two electrolyte solutions. Generally, analysis of the membrane conductivity K_m is based on a kinetic model for ion permeation discriminating between an adsorption, a permeation, and a desorption step. The permeation step corresponds to crossing an energy barrier that is highest in the apolar heart of the bilayer lipid membrane. From the laws of conservation for the permeant ion it follows that

$$\frac{d\Gamma_1}{dt} = k_{\mathrm{ads},1}c_1 - k_{\mathrm{des},1}\Gamma_1 - \vec{k}\Gamma_1 + \overleftarrow{k}\Gamma_2 \tag{44a}$$

$$\frac{d\Gamma_2}{dt} = k_{\mathrm{ads},2}c_2 - k_{\mathrm{des},2}\Gamma_2 + \vec{k}\Gamma_1 - \overleftarrow{k}\Gamma_2 \tag{44b}$$

with $\Gamma_{1(2)}$ the adsorbed amount at side 1 (resp. 2) of the membrane, $k_{\mathrm{ads},1(2)}$ the rate constant for adsorption from the solution, and $k_{\mathrm{des},1(2)}$ the desorption rate constant;

$\vec{k}$ and $\overleftarrow{k}$ are the rate constants for transfer through the membrane, from side 1 to side 2 and vice versa, respectively. The charge separation between 1 and 2 is given by the net ion flux $J_{1\rightarrow 2}$:

$$J_{1\rightarrow 2} = \vec{k}\Gamma_1 - \overleftarrow{k}\Gamma_2 \quad (= -J_{2\rightarrow 1}) \tag{45}$$

The rate of charge transfer depends on the electric potential profile across the membrane: internal potential differences affect the rate constants for permeation $\vec{k}$ and $\overleftarrow{k}$, whereas the adsorption and desorption rate constants are functions of the potential difference between the membrane surface and the corresponding bulk solutions (the "adsorption potential"). Let us consider the simple case of a symmetrical membrane with identical electrolyte solutions at both sides, so that $\Gamma_1 = \Gamma_2 = \Gamma$. If under these conditions K_m is measured using an a.c. field ($\Delta\psi_{a.c.}$), $\vec{k}$ and $\overleftarrow{k}$ can be expressed as

$$\vec{k} = k^{\theta} \exp\left(\frac{-\alpha zF\Delta\Psi_{a.c.}}{RT}\right) \tag{46a}$$

$$\overleftarrow{k} = k^{\theta} \exp\left(\frac{(1-\alpha)zF\Delta\psi_{a.c.}}{RT}\right) \tag{46b}$$

where k^{θ} is the exchange rate constant for permeation in the given system (with defined Γ and c). These equations are based on the concept that a "cooperative" potential difference lowers the height of the energy barrier with an amount of $\alpha zF\Delta\psi_{a.c.}$, which

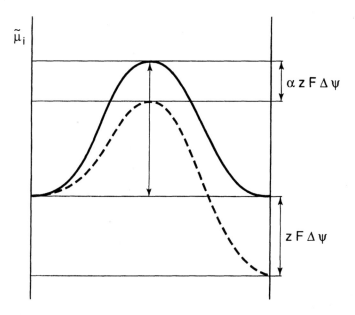

Fig. 16   Electrochemical potential profiles for an ion (valence z) in a lipid membrane. To pass the membrane the ion has to cross an energy barrier, which has its top in the center of the membrane; ———— $\tilde{\mu}_i$ for $\Delta\psi = 0$ and – – – $\tilde{\mu}_i$ for some finite $\Delta\psi$. In this picture the transfer coefficient α has a value of 0.5.

is illustrated in Fig. 16. Therefore, permeation of ions is accelerated in one direction by a factor of $\exp(-\alpha z F \Delta \psi_{a.c.})$ and slowed down in the other direction by a complementary factor of $\exp((1-\alpha)z F \Delta \psi_{a.c.})$. The factor α is referred to as the transfer coefficient; for a symmetrical situation it has a value of 0.5 (64).

From Eqs. (45) and (46) we get for small $\Delta \psi_{a.c.}$

$$J_{1 \to 2} = -\frac{zF}{RT} \Delta \psi_{a.c.} k^\theta \Gamma \tag{47}$$

For low concentrations of the permeant ion the adsorbed amount is linear with its concentration (the "Henry region" of the adsorption isotherm): $\Gamma = (k_{ads}/k_{des})c$. Then it follows from Eq. (47) that

$$K_m = \left| \frac{j}{\Delta \psi_{a.c.}} \right| = \frac{z^2 F^2}{RT} Pc \tag{48}$$

in which j is the current density $(j = zFJ)$ and $P = k^\theta k_{ads}/k_{des}$. Note that P has the dimension of m s^{-1} because k^θ is in s^{-1} and k_{ads}/k_{des}, the so-called Henry coefficient, is in m. P can be interpreted as an effective permeability (or exchange rate constant) of the membrane for the ion under consideration. As expected, it is generally found that Eq. (48) is applicable for relatively low concentrations of the permeant ion (39).

A number of ionic permeability values for phosphatidylcholine (lecithin) bilayers are listed by de Gier (65). For example, for Cl$^-$ ions the permeability is 7.6×10^{-13} m s^{-1}. For a concentration of 0.01 M Cl$^-$ this corresponds to a K_m value of 2.8×10^{-8} Ω^{-1} m^{-2}. For Na$^+$ ions P is as low as 9.5×10^{-15} m s^{-1}, in line with the extremely high energy barrier for Na$^+$ in the inner region of a lipid bilayer ($\sim 100kT$) as calculated by Pethig (9). For protons P is about 10^{-5} m s^{-1}; mechanisms that may explain the relatively rapid transport of H$^+$ in lipid bilayers are discussed in Ref. 9.

In the case of ion transport through transmembrane channels, discrete conductivity phenomena may be observed. If the channel-forming compound is present in only a small amount and the conductivity of the membrane is measured by a sensitive, fast instrument, one is able to see the effect of the separate channels. This is illustrated in Fig. 17. K_m takes values of $n \times K_{m,min}$ (with $n = 0, 1, 2, \ldots$). $K_{m,min}$ is in the order of 10^{-11} Ω^{-1} and the lifetime of a conductive channel is in the order of 1 s. Of course, for mediated ion transfer across membranes, the concept of a high energy barrier in the center of the membrane is not applicable.

3. The Transmembrane Potential

Selective ion permeation across membranes results in the generation of an electric potential difference between the bulk solutions at the two sides of the membrane, the transmembrane potential $\Delta \psi_m$. In his treatment of membrane potentials Nernst (66) assumed that the membrane is permeable for only one type of ion. In equilibrium, the membrane potential follows from the condition for equality of the electrochemical potential of the permeant ion i at both sides of the membrane [see Eq. (43)], yielding

$$\Delta \psi_m \equiv \psi_2 - \psi_1 = \frac{RT}{z_i F} \ln \frac{a_{i,1}}{a_{i,2}} \tag{49}$$

in which z_i is the valence (including sign) of the permeant ion and $a_{i,1}$ and $a_{i,2}$ are its

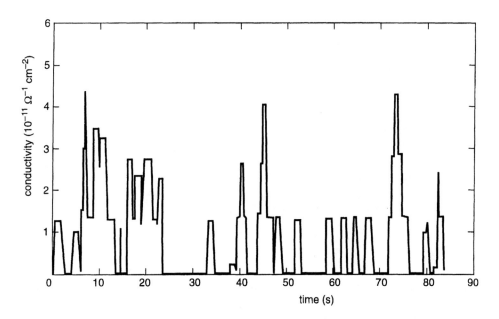

Fig. 17 Fluctuation of the conductivity of a bilayer lipid membrane in the presence of gramicidin A. The discrete steps correspond to the opening and closing of individual transmembrane channels. (From Ref. 40.)

activities in the respective bulk solutions. The membrane potential expressed by this equation is called the *Nernst potential*.

For the more general case that there are various permeant ions with different mobilities in the membrane, the starting point for the description of membrane potentials is usually the one-dimensional *Nernst-Planck equation* for the flux J_i of each permeant ion i as a result of both a concentration gradient and an electric potential gradient*:

$$J_i = -D_i \frac{dc_i}{dx} - \frac{|z_i|}{z_i} u_i c_i \frac{d\psi}{dx} \tag{50}$$

D_i is the diffusion coefficient in the membrane, z_i the valence, and u_i the (electric) mobility. D_i and u_i are related according $u_i = |z_i|FD_i/RT$. Note that the first term on the right-hand side corresponds to Fick's first law of diffusion. In the steady state of constant membrane potential no further charge separation takes place, i.e.,

$$\sum_i z_i J_i = 0 \tag{51}$$

---

* Equation (50) can be obtained, e.g., from the general expression for the contribution of the passive transport of ion i to the current density: $j_i = -u_i c_i (\partial \tilde{\mu}_i / \partial x)$.

Combination of Eqs. (50) and (51) gives

$$\frac{d\psi}{dx} = -\frac{RT}{F}\frac{\sum_i \frac{|z_i|}{z_i} u_i \frac{dc_i}{dx}}{\sum_i |z_i| u_i c_i} \tag{52}$$

To integrate Eq. (52), expressions for either the electric potential ψ or the concentrations c_i as a function of position x in the membrane are needed or, alternatively, equations that determine the relation between $c_i(x)$ and $\psi(x)$. In early studies of ion permeation through transmembrane channels Goldman (67) and later Hodgkin and Katz (68) assumed a linear potential drop along the channel, which results in the following expression for the transmembrane potential:

$$\Delta\psi_m = \frac{RT}{F}\ln\frac{\sum_i^+ u_i c_{i,1} + \sum_i^- u_i c_{i,2}}{\sum_i^+ u_i c_{i,2} + \sum_i^- u_i c_{i,1}} \tag{53}$$

in which Σ^+ and Σ^- are summations over all abundant cations and anions, respectively. The potential $\Delta\psi_m$ thus arises from differences in ion mobilities in the membrane (channel) and concentrations in the bulk solutions. Equation (53) is known as the *Goldman-Hodgkin-Katz equation.*

Generally, when ions pass through a complex inhomogeneous medium such as a biological membrane, diffusion coefficients and electric mobilities are spatial variables. Therefore, it is better to replace the mobility u_i in Eq. (53) by an effective permeability of the system for each ionic species, P_i. Earlier, we presented an expression for the permeability P based on a simple kinetic model [Eq. (48)]. In Ref. 69 a more general integral expression for P_i is given and introduced into the Nernst-Planck equation. For the specific case for which Eq. (53) was originally derived, i.e., transport of K^+, Na^+, and Cl^- ions across nerve cell membranes, it is found that

$$\Delta\psi_m \equiv \psi_{in} - \psi_{out} = \frac{RT}{F}\ln\frac{P_K[K^+]_{out} + P_{Na}[Na^+]_{out} + P_{Cl}[Cl^-]_{in}}{P_K[K^+]_{in} + P_{Na}[Na^+]_{in} + P_{Cl}[Cl^-]_{out}} \tag{54}$$

where the indexes "out" and "in" refer to the extracellular and intracellular liquid, respectively.

Based on their experimental results and using Eq. (54), Hodgkin and Huxley (7) proposed a mechanism of nerve excitation and conduction. The permeability of an axon for K^+ and Na^+ varies with the membrane potential $\Delta\psi_m$. At "rest" (i.e., when the axon is not excited) the permeability for K^+ is much higher than that for Na^+. Typically, in animal cells, the concentration of K^+ in the extracellular liquid is about 0.01 M, while in the axon its concentration is about 0.4 M. Therefore, the rest potential amounts to about -80 mV (approximately equal to the Nernst potential for the K^+ ion); the potential in the axon is negative relative to that of the extracellular liquid. Positive excitation pulses locally lead to depolarization of the membrane. If the depolarization pulse exceeds a certain threshold value, the Na^+ channels open and transport of Na^+ starts to contribute to the membrane potential. Because the concentration of Na^+ in the extracellular liquid (~ 0.46 M) is higher than that in the axon (~ 0.05 M), $\Delta\psi_m$ suddenly increases (the *action potential*), resulting in opening of nearby Na^+ channels and propagation of the pulse along the axon. The concentration gradients of K^+ and Na^+ over the axon membrane are maintained because of the action of Na^+,K^+-ATPase, transferring K^+ ions from the extracellular liquid into the axon and Na^+ ions in the opposite direction, through hydrolysis of ATP. For more details see, e.g., Refs. 8, 10, and 70.

To solve the Nernst-Planck equation under steady-state conditions for mediated as well as nonmediated ion permeation through a lipid bilayer membrane, a number of specific models have been developed. For example, Syganow and Von Kitzing (71) have presented a theory to calculate integral properties of biological ion channels. The approach is directly related to physical parameters such as the channel length and the minimal ionic number density inside the channel. It explains current-voltage and conductance-concentration relations found in experiments. Aguilella et al. (69) have modeled nonmediated transport of ions across lipid bilayers. In their treatment the membrane has been divided into three layers: an inner hydrophobic region and two hydrophilic outer regions with dielectric properties different from those of the bulk solutions. All fixed charges and dipoles are assumed to be on the boundaries between the polar zones and the surrounding electrolyte solutions. For the case in which the inner hydrophobic layer is the rate-controlling step for ion transfer, it was found that the relative selectivity of the membrane for cation or anion transfer depends not only on membrane characteristics such as the thickness of the polar zones and the surface dipole density but also on the electrolyte concentration in the bulk solutions.

IV. SUMMARY AND CONCLUDING REMARKS

In this chapter we have explored some of the physical concepts of relevance to the electrostatic and electrodynamic properties of biological interphases. The discussion has been particularly focused on cell walls and biological membranes. We have considered electric double layers and effects of electric fields, as well as some of the physical factors involved in ion transport in membrane structures, a process that is intimately associated with the transmembrane potential. We trust that this chapter presents a functional mix of "old" and "new" elements and that it will underscore the value of electrochemistry in the study of biological systems.

Biological interphases generally contain numerous charges and dipoles that give rise to the formation of electric double layers and internal electric fields. Their description usually starts from the classical Poisson equation. For a number of properties the Gouy-Chapman (or Gouy-Chapman-Stern) model provides a reasonable first approximation to the potential profile near a biological surface. However, this model is too simple to account for the three-dimensional distribution of charges in the interphase. A logical further step is to extend the double-layer model with a Donnan layer. Generally, for the construction of the total electric potential profile one may divide the interphase into different regions, each with its own thickness, permittivity, and charge density, and calculate the ion distribution by a Poisson-Boltzmann type of approach. Alternatively, computational techniques such as molecular dynamics simulations or self-consistent field lattice modeling may be used to obtain the potential profile and the ionic distributions. There are still, however, important differences between the results of these techniques, and further development and verification are needed. Theoretical analysis of the interphasial electric potential profile is valuable for analysis and determination of the electrokinetic potential derived from cell electrophoresis, for the interpretation of studies of ion adsorption and ion transport across membranes, and for many other electrochemical phenomena related to biological interphases.

In this chapter we have not considered lateral inhomogeneity of biological interphases. Still, the presence of proteins and other specialized molecules with charged groups may have a large effect on the local potential. Consequently, such local electric fields may be expected to have physiological significance, for example, in recognition, binding, and channel-gating processes.

Interphasial double layers adjust themselves to changing conditions. On the relevant time scale of biological processes the part of the double layer in solution adjusts instantaneously, whereas the apolar core of lipid membranes generally does not allow passage of charge unless special transport mediators are present. Ions inside bacterial cell walls appear to have high lateral mobilities, almost equal to their bulk solution mobilities. This is important because lateral transfer of ions may play a crucial role in the interactions of organisms with their surroundings: it effectively minimizes the electrostatic repulsion in interaction processes, with drastic consequences for aggregation and adhesion. Lateral electrodynamics of interphasial double layers thus promises to develop to a new basic factor in the modeling of cell surface interactions.

For ion permeation through biological membranes various models have been developed. The usual starting point is the one-dimensional Nernst-Planck equation, which describes the permeation as a result of diffusion and conduction. Selective ion transport results in an electric potential difference between the bulk phases at the two sides of the membrane, the so-called transmembrane potential. However, it is the total potential profile across the interphase, rather than the transmembrane potential, that is a determining factor for ion permeation processes.

The field of electrochemistry of biointerphases has grown rapidly over the last decades. Apparently, electrochemists have seized the opportunity to extend their area of interest to the challenging realms of biological systems; in turn, biologists, biochemists, and physiologists have integrated electrochemical principles into their fields of research. It may be expected that this mutual interest will continue to provide better understanding of the physiological features of biological and biotechnological systems.

REFERENCES

1. S McLaughlin. The electrostatic properties of membranes. Annu Rev Biophys Biophys Chem 18:113–136, 1989.
2. G Cevc. Membrane electrostatics. Biochim Biophys Acta 1031:311–382, 1990.
3. DS Cafiso. Lipid bilayers: Membrane-protein electrostatic interactions. Curr Opin Struct Biol 1:185–190, 1991.
4. S High. Membrane protein insertion into the endoplasmatic reticulum—Another channel tunnel? Bioessays 14:535–540, 1992.
5. A Raudino. Lateral inhomogeneous lipid membranes: Theoretical aspects. Adv Colloid Interface Sci 57:229–285, 1995.
6. MN Jones. Surface properties and interactions of vesicles. Curr Opin Colloid Interface Sci 1:91–100, 1996.
7. AL Hodgkin, AF Huxley. A quantitative description of membrane current and its application to conduction and excitation in nerve. J Physiol (Lond) 177:500–544, 1952.

8. PKJ Kinnunen, JA Virtanen. A qualitative, molecular model of the nerve impulse. Conductive properties of unsaturated lyotropic liquid crystals. In: F Gutmann, H Keyzer, eds. Modern Bioelectrochemistry. New York: Plenum, 1985, pp 457–479.

9. R Pethig. Ion, electron, and proton transport in membranes: A review of the physical processes involved, In: F Gutmann, H Keyzer, eds. Modern Bioelectrochemistry. New York: Plenum, 1985, pp 199–239.

10. C Tanford. Mechanism of free energy coupling in active transport. Annu Rev Biochem 52:379–409, 1983.

11. R van Grondelle. Excitation energy transfer, trapping and annihilation in photosynthetic systems. Biochim Biophys Acta 811:147–195, 1985.

12. VP Harden, JO Harris. The isoelectric point of bacterial cells. J Bacteriol 65:198–202, 1952.

13. A van der Wal, W Norde, B Bendinger, AJB Zehnder, J Lyklema. Chemical analysis of isolated cell walls of gram-positive bacteria and the determination of the cell wall to cell mass ratio. J Microbiol Methods 28:147–157, 1997.

14. PF Luckham, PG Hartley. Interactions between biosurfaces. Adv Colloid Interface Sci 49:341–386, 1994.

15. SJ Singer, GL Nicolson. The fluid mosaic model of the structure of cell membranes. Science 175:720–731, 1972.

16. P Yeagle, ed. The Structure of Biological Membranes. Boca Raton, FL: CRC Press, 1991.

17. E Egberts, HJC Berendsen. Molecular dynamics simulation of a smectic liquid crystal with atomic detail. J Chem Phys 89:3718–3732, 1988.

18. E Egberts, SJ Marrink, HJC Berendsen. Molecular dynamics simulation of a phospholipid membrane. Eur Biophys J 22:423–436, 1994.

19. RW Pastor. Molecular dynamics and Monte Carlo simulations of lipid bilayers. Curr Opin Struct Biol 4:486–492, 1994.

20. DJ Tobias, K Tu, ML Klein. Atomic-scale molecular dynamics simulations of lipid membranes. Curr Opin Colloid Interface Sci 2:15–26, 1997.

21. FAM Leermakers, JMHM Scheutjens. Statistical thermodynamics of associated colloids. I. Lipid bilayer membranes. J Chem Phys 89:3264–3274, 1988.

22. LA Meijer, FAM Leermakers, J Lyklema. Head-group conformations in lipid bilayer membranes. Recl Trav Chim Pays-Bas 113:167–175, 1994.

23. LA Meijer, FAM Leermakers, A Nelson. Modelling of the electrolyte ion-phospholipid layer interaction. Langmuir 10:1199–1206, 1994.

24. LA Meijer, FAM Leermakers, J Lyklema. Modelling the interactions between phospholipid bilayer membranes with and without additives. J Phys Chem 99:17282–17293, 1995.

25. J Seelig, A Seelig. Lipid conformation in model membranes and biological membranes. Q Rev Biophys 13:19–61, 1980.

26. BH Honig, WL Hubbell, RS Flewelling. Electrostatic interactions in membranes and proteins. Annu Rev Biophys Biophys Chem 15:163–193, 1986.

27. K Gawrisch, D Ruston, J Zimmerburg, VA Parsegian, RP Rand, N Fuller. Membrane dipole potentials, hydration forces, and the ordering of water at membrane surfaces. Biophys J 61:1213–1223, 1992.

28. M Gouy. Sur la constitution de la charge électrique à la surface d'un électrolyte. J Phys Paris 9:457–468, 1910.

29. DL Chapman. A contribution to the theory of electrocapillarity. Philos Mag 25:475–481, 1913.

30. DC Grahame. The electrical double layer and the theory of electrocapillarity. Chem Rev 41:441–501, 1947.

31. J Lyklema. Fundamentals of Interface and Colloid Science. Vol II. Solid-Liquid Interfaces. London: Academic Press, 1995, Chap. 3 and Chap. 4.

32. O Stern. Zur Theorie der Elektrolytischen Doppelschicht. Z Elektrochem 30:508–516, 1924.

33. TW Healy, LR White. Ionizable surface group models of aqueous interfaces. Adv Colloid Interface Sci 9:303–345, 1978.
34. G Cevc, D Marsh. Phospholipid Bilayers. New York: Wiley-Interscience, 1987.
35. M Langner, D Cafiso, S Marcelja, S McLaughlin. Electrostatics of phosphoinositide bilayer membranes—Theoretical and experimental results. Biophys J 57:335–349, 1990.
36. HO Ohshima, T Kondo. On the electrophoretic mobility of biological cells. Biophys Chem 39:191–198, 1991.
37. M Blank, ed. Electrical Double Layers in Biology. New York: Plenum, 1985.
38. A van der Wal, M Minor, W Norde, AJB Zehnder, J Lyklema. Electrokinetic potential of bacterial cells. Langmuir 13:165–171, 1997.
39. N Lakshminarayanaiah. Transport Phenomena in Membranes. New York: Academic Press, 1969.
40. DA Haydon, SB Hladky. Ion transport across thin lipid membranes: A critical discussion of mechanisms in selected systems. Q Rev Biophys 5:187–282, 1972.
41. E Hückel. Die Kataphorese der Kugel. Phys Z 25:204–210, 1924.
42. L Onsager. The theory of electrolytes I. Phys Z 27:388–392, 1926.
43. M von Smoluchowski. Contribution à la théorie de l'endosmose électrique et de quelques phénomès corrélatifs. Bull Intern Acad Sci Cracovie 182–199, 1903.
44. DC Henry. The cataphoresis of suspended particles. I. The equation of cataphoresis. Proc R Soc Lond A133:106–129, 1931.
45. J Kijlstra, A van der Wal. Electrokinetic behaviour of bacterial suspensions. Bioelectrochem Bioenerg 37:149–151, 1995.
46. A van der Wal, M Minor, W Norde, AJB Zehnder, J Lyklema. Conductivity and dielectric dispersion of gram-positive bacterial cells. J Colloid Interface Sci 186:71–79, 1997.
47. J Kijlstra, HP van Leeuwen, J Lyklema. Low-frequency dielectric relaxation of hematite and silica sols. Langmuir 9:1625–1633, 1993.
48. J Kijlstra, RAJ Wegh, HP van Leeuwen. Impedance spectroscopy of colloids. J Electroanal Chem 366:37–42, 1994.
49. M Minor, A van der Wal, J Lyklema. Dielectric spectroscopy of model colloids, and the role of conduction behind the plane of shear. In: E Pelizzetti, ed. Fine Particles Science and Technology. Boston: Kluwer Academic, 1996, pp 225–238.
50. J Lyklema, M Minor. On surface conduction and its role in electrokinetics. Colloids Surf A Physicochem Eng Aspects 140:33–41, 1998.
51. EJ Hinch, JD Sherwood, WC Chew, PN Sen. Dielectric response of a dilute suspension of spheres with thin double layers in an asymmetric electrolyte. J Chem Soc Faraday Trans 2 80:535–551, 1984.
52. M Minor, AJ van der Linde, HP van Leeuwen, J Lyklema. Dynamic aspects of electrophoresis and electroosmosis: A new fast method for measuring particle mobilities. J Colloid Interface Sci 189:370–375, 1997.
53. M Minor, HP van Leeuwen, J Lyklema. Low-frequency dielectric response of polystyrene latex dispersions. J Colloid Interface Sci 206:397–406, 1998.
54. SS Dukhin, NM Semenikhin. Theory of double layer polarization and its influence on the electrokinetic and electrooptical phenomena and the dielectric permeability of disperse systems. Calculation of the electrophoretic and difusiophoretic mobility of solid spheres. Russ Colloid J 32:298–305, 1970.
55. DC Henry. The electrophoresis of suspended particles. IV. The surface conductivity effect. Trans Faraday Soc 44:1021–1026, 1948.
56. F Booth. Surface conductance and cataphoresis. Trans Faraday Soc 44:955–959, 1948.
57. HP van Leeuwen, J Lyklema. Interfacial electrodynamics of interacting colloidal particles. Geometrical aspects. Ber Bunsenges Phys Chem 91:288–291, 1987.
58. KD Schulze. Investigations of the channel gating influence on the dynamics of biomembranes. Z Phys Chem 186:47–63, 1994.

59. JI Kourie. Vagaries of artificial bilayers and gating modes of the SCI channel from the sarcoplasmic reticulum of skeletal muscle. J Membr Sci 116:221–227, 1996.
60. CM Jones, DM Taylor. Voltage gating of porin channels in lipid bilayers. Thin Solid Films 285:748–751, 1996.
61. KS Åkerfeldt, JD Lear, ZR Wasserman, LA Chung, WF DeGrado. Synthetic peptides as models for ion channel proteins. Acc Chem Res 26:191–197, 1993.
62. T Kinoshita. Biomembrane mimetic systems. Prog Polym Sci 20:527–583, 1995.
63. N Voyer, L Potvin, E Rousseau. Electrical activity of artificial ion channels incorporated into planar lipid bilayers. J Chem Soc Perkin Trans 2:1469–1471, 1997.
64. AJ Bard, LR Faulkner. Electrochemical Methods—Fundamentals and Applications. New York: Wiley, 1980.
65. J de Gier. Permeability barriers formed by membrane lipids. Bioelectrochem Bioenerg 27:1–10, 1992.
66. W Nernst. Über die Löslichkeit von Mischkrystallen. Z Phys Chem 9:137–142, 1892.
67. DE Goldman. Potential, impedance and rectification in membranes. J Gen Physiol 27:37–60, 1944.
68. AL Hodgkin, B Katz. The effect of sodium ions on the electrical activity of the giant axon of the squid. J Physiol (Lond) 108:37–77, 1949.
69. V Aguilella, M Belaya, V Levadny. Ion permeability of a membrane with soft polar interfaces. 1. The hydrophobic layer as the rate-determining step. Langmuir 12:4817–4827, 1996.
70. M Blank. Membrane transport: Insights from surface science. In: M Bender, ed. Interfacial Phenomena in Biological Systems. Surfactant Science Series 39. New York: Marcel Dekker, 1991, pp 337–365.
71. A Syganow, E von Kitzing. Integral weak diffusion and diffusion approximations applied to ion transport through biological ion channels. J Phys Chem 99:12030–12040, 1995.

3

Interfacial Interactions

H. Wennerström
Lund University, Lund, Sweden

I. IMPORTANCE OF INTERFACIAL INTERACTIONS IN LIVING SYSTEMS

From the point of view of a chemist, the living organism displays an incredibly complex maze of parallel and sequential chemical transformations that in the common language we refer to as birth, life, and death. Every day biochemical and molecular biology research reveals more about the crucial molecular events in these series of chemical transformations. There is another aspect of the living system that has so far received less attention. How is the multitude of individual enzymatic transformations organized into a coherent functional system? This is a major challenge that is solved by a combination of compartmentalization and signaling.

The chemistry of the living cell occurs in a number of organelles and (colloidal) "particles" that have their specific functional roles: mitochondria, ribosomes, nuclei, lyzosomes, chloroplasts, etc. The more developed the organism, the more organelles there tend to be in a given cell or the more specialized its function. Furthermore, cells exist in an environment of other cells, and these can either be part of the same organism or represent other uni- or multicellular individuals.

Communication between cells or between organelles in a cell can occur either through individual signal molecules or through the surface interaction between entities of colloidal size. This chapter deals primarily with the latter route of communication of interaction. The approach we adopt is to describe general mechanisms for the interactions distinguished by their molecular origin. In a real system there are always several of these mechanisms operating at the same time, but the particular combination varies from application to application and a detailed discussion of that aspect easily leads from the general to the very specific. The present chapter is aimed at serving as a guide for doing this transition to the specific case, but it deals mainly with the general mechanisms of interactions.

In this chapter we concentrate on the conceptual issues of surface interactions. A number of reviews and textbooks discuss the quantitative aspects of surface forces in greater detail. For a more thorough discussion of these issues we refer primarily to the textbooks by Israelachvili (1) and by Evans and Wennerström (2,3).

II. BASIC INTERMOLECULAR INTERACTIONS

The interaction between two colloidal entities in a solvent is determined by the combined action of a large number of molecular contributions. Today we know a great deal about how molecules interact pairwise, and it is essential for the conceptual understanding of colloidal systems to utilize the accumulated knowledge about interactions on the more fundamental molecular level. The intermolecular, noncovalent, interaction between a pair of isolated (no solvent present) molecules has a conceptually relatively simple description. Let us denote the interaction energy between two molecules A and B at a separation R_{AB} and with the orientations Ω_A and Ω_B, respectively, as $V(R_{AB}, \Omega_A, \Omega_B)$. For reasonably small molecules (currently less than 20 atoms) quantum chemical methods are good enough to allow a reliable calculation of the potential V. However, in addition to this quantitative information it is essential to have a more conceptual description of the interaction. It is possible to separate the potential V into five (2) different contributions when considering the interaction between closed-shell molecules.

1.   There is a short-range repulsion due to the energy increase when the electron clouds of two molecules are forced to overlap. This contribution can be traced back to the action of the Pauli exclusion principle.

2.   There is an attraction between electron clouds that are not overlapping but in close proximity due to correlations between electrons located at the different molecules. This attraction, called the dispersion or London interaction, varies as the inverse sixth power of the separation R_{AB} and normally represents the major contribution to the van der Waals interaction. The magnitude of the dispersion force increases with the number of electrons in the molecules and is only weakly orientation dependent.

3.   The nuclei and the electrons of the molecules together give rise to a charge distribution, and this generates an electric field outside the molecule. We can see the charge distribution as a single entity and analyze how it interacts with the charge distribution of another molecule. This strategy leads, for example, to the hydrogen bond. Another strategy is to divide the charge distribution into multipole moments and describe the electrostatic interaction between molecules in terms of their multipole moments. The simplest illustration of this method is the ion-ion interaction

$$V_{\text{ion-ion}} q_A q_B / (4\pi\varepsilon_0 R_{AB}) \tag{2.1}$$

where we say that the interaction is dominated by the first moment of the charge distribution, which we normally call the (net) charge. Equation (2.1) illustrates a virtue of the multipole expansion: that we can write the interaction potential in an analytical form where the only system-specific parameter is the magnitude of the multipole moment, which is a property of the isolated molecule.

The next two terms in the multipole expansion after the ion-ion interaction are the ion-dipole and dipole-dipole interactions. Explicit expressions for these interactions can be found in textbooks (1–4); here we simply note that for each step one goes up in the multipole hierarchy, the range of the interaction is shortened by one inverse power of R. In addition, the interaction becomes increasingly orientation dependent. A further general property is that these orientation-dependent interactions average to zero when one of the molecules is rotated isotropically.

For cases in which the multipole expansion converges slowly, it is not very useful, and one adopts a conceptually different description of the electrostatic intermolecular interaction. The best known representative of this approach is the hydrogen bond. In this case it is simpler to analyze the interaction in terms of the full charge distributions. For the hydrogen bond it is not sufficient to include only the dipole-dipole interaction because higher order terms also contribute substantially. Thus there is no general functional form for either the distance or the orientation dependence. However, because of the substantial contributions from higher order multipole moments, it follows that the hydrogen bond potential is of relatively short range and in the region of optimal separations it is highly orientation dependent.

4. There is an additional interaction term of electrostatic origin that can sometimes be significant. Even a nonpolar molecule (or atom) can respond to the electric field from a polar molecule or ion. The electric field induces a dipole moment because all molecules are polarizable, and this results in an attractive interaction of relatively short range.

5. For a pair of molecules of which one has a high electron affinity and the other a low ionization potential, one can have a charge transfer interaction. This occurs under rather specific circumstances but is important when it occurs. The interaction is attractive, strong, and of short range.

As an illustration of the subdivision of the pair intermolecular interaction into the different contributions, Fig. 1 shows the calculated water-water interaction for a fixed orientation but with a varying oxygen-oxygen separation. At long range an attractive dipole-dipole term dominates, and in the region of optimal separation the electrostatic interaction still dominates but the dipole-dipole character is lost. At the minimum there is also a smaller attractive contribution from the dispersion interaction and there is a overlap repulsion of the same magnitude as the dispersion term. At closer separations the overlap repulsion starts to dominate.

From a molecular perspective, function in a biological system is coupled to molecular recognition and organization through intermolecular interactions. How is this recognition accomplished? The short-range repulsion determines size and shape, and these properties are always one element in recognition events of the lock-and-key type. The dispersion interaction is common to all molecules, and although the magnitudes vary it cannot contribute in any essential degree to recognition and can contribute only moderately to organization. The intermolecular interaction of absolutely dominant importance in biological systems is the electrostatic one. It has molecular specificity and there is also a strong orientational variation. Usually these specific interactions are discussed in terms of electrical double-layer force, hydrophobic intractions, or hydrogen bonds, but one should be aware of the fact that electrostatics determine the underlying intermolecular interaction in all these cases.

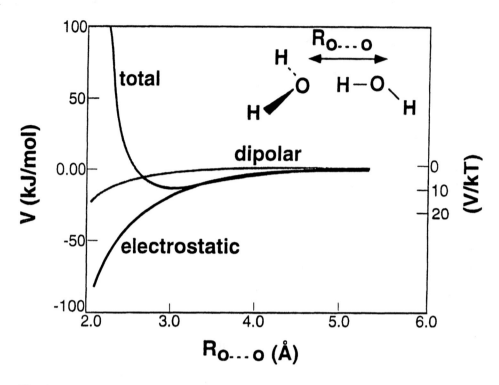

Fig. 1 Calculated interaction energy between two water molecules in a hydrogen bond configuration. The three lines show the total interaction, the total electrostatic contribution, and the dipolar part of the electrostatic contribution as a function of the oxygen-oxygen separation. The equilibrium separation is at 2.8 Å; for separations exceeding 3.2 Å the interaction is purely electrostatic, and at shorter separations there is a contribution from both the dispersion interaction and overlap repulsion. (Courtesy of G. Karlström.)

III. THE SINGLE SURFACE. WATER AS A SOLVENT

Water is the natural solvent and most interactions between aggregates or macromolecules in a biological system occur in an aqueous environment. To set the scene for a detailed discussion of interaction forces between these entities, we first discuss the properties of the single aggregate or macromolecule. This generally consists of a more or less apolar interior and a polar surface. The surface shows a large variation in chemical composition, and this will to a large extent determine the nature of the interactions with other aggregates or macromolecules.

To exist as an identifiable object, the aggregate or macromolecule has to show stability relative to association with the majority of other aggregates or macromolecules in the system, and this can be achieved by a few different strategies. The most direct is charge stabilization. Nucleic acids and most globular proteins carry charges localized to groups fixed at the surface. Nature uses negative charge to provide stability relative to association. Positively charged species, at neutral pH, typically carry this charge for specific association purposes related to the biological function.

For lipid membranes colloidal stability is partly achieved by the charge mechanism, but uncharged lipid bilayers also show limited swelling in water and then stability also emerges from local flexibility of the surface groups at either side of the lipid membrane. For systems in which being inert relative to association over a wide range of conditions is important, the polyelectrolyte brush is the preferred arrangement. In this case the surface contains flexible and partly charged polymeric groups. These are usually polysaccharides but they can also be proteins. This arrangement is found in the glycocalyx of cell surfaces and in mucous tissue as well as in the casein micelles of milk. Having a charged but flexible surface combines the first two mechanisms of stabilization.

The polar surfaces exist in an aqueous medium and their properties also depend on those of the medium. The medium effects can be discussed on two different levels: thermodynamic and mocular. Let us first discuss the former. A most important characteristic of the medium is the osmotic pressure or, more appropriately, the chemical potential of the water. The osmotic pressure is affected by all types of solutes. It will turn out to be of some importance to have a proper understanding of the different molecular causes of osmotic pressure. We can distinguish between four major contributions. First, there are dissolved electrolytes in the medium, with Na^+, K^+, Cl^-, and HCO_3^- as typically the main ionic species. Second, there are low-molecular-weight neutral solutes such as sugars, urea, and amino acids. The third contribution comes from the (positive) counterions of polyelectrolytes and other charged aggregates or macromolecules. Finally, there can also be a contribution from flexible polymers. The interior of a cell is a crowded system and there are no large aqueous domains, so even surface polymeric groups can contribute to the osmotic pressure.

Of all the molecular species in a living cell, we expect that water is the one that equilibrates fastest, and one should expect a constant water chemical potential (or constant osmotic pressure) throughout an organism, except for regions with fast turnover of water molecules. There is an intimate coupling between osmotic pressure and interaggregate forces. The functioning of a living system is strongly influenced by the osmotic pressure through its effect on association processes. For example, simple preservatives such as salt and sugar owe their physiological effect to the high osmotic pressure they induce.

To obtain a reference frame for the discussion of the molecular mechanisms behind forces between colloidal particles, it is very useful to review some basic thermodynamic relations for (aqueous) solutions, focusing on the solvent. The basic quantity is the chemical potential of the water, $\mu(H_2O)$. By an old convention this is expressed in terms of the osmotic pressure Π_{osm}, and for an incompressible liquid there is a simple relation between the chemical potential and the osmotic pressure:

$$\mu°(H_2O) - \mu(H_2O) = V(H_2O)\Pi_{osm} \tag{3.1}$$

where $\mu°(H_2O)$ is the chemical potential of pure water and $V(H_2O)$ is its molar volume. Because the chemical potential of water in a solution is smaller than that of the pure solvent, the osmotic pressure is a positive quantity. For the remaining discussion in this chapter we will stick to the assumption of incompressibility, so Eq. (3.1) is considered valid.

One virtue of using the osmotic pressure concept is that there is a close analogy between the behavior of a one-component real gas and that of a two-component incompressible solution. In the latter case the osmotic pressure plays the same role as the pressure of the gas in a statistical mechanical description. Here we cite one relation obtained through this analogy, namely the virial expansion of the osmotic pressure:

$$\Pi_{osm} = RTc(1 + B_2 c + \cdots) \tag{3.2}$$

where c is the solute concentration and B_2 is the so-called second virial coefficient, which, for example, can be measured in a light scattering experiment. This coefficient is determined by the pair solute-solute interaction, when this is attractive the osmotic pressure is smaller, meaning B_2 negative, whereas for a repulsive interaction B_2 is positive.

There exists a range of solution models that provide explicit relations for the chemical potential and thus the osmotic pressure in terms of the solute concentration as in Eq. (3.2). These are typically not quantitatively accurate when applied to a real system, but they are most important for the development of a conceptual understanding. For an ideal solution

$$\mu - \mu° = RT \ln(1 - X) \approx -RTX(1 - X/2 + \cdots) \tag{3.3}$$

$$\Pi_{osm} = RTc(1 - 1 V_s c/2 + \cdots) \tag{3.4}$$

where X is the mole fraction of the solute and V_s the molar volume of the solvent.

For an electrolyte there are two species present, so the expression for the water chemical potential is

$$\mu(H_2O) - \mu°(H_2O) = -RT(z_+ + z_-)/\min\{z_+, z_-\} V(H_2O)c + \mu(\text{elec}) \tag{3.5}$$

$$\Pi_{osm} = RT(z_+ + z_-)/\min\{z_+, z_-\}c + \mu(\text{elec})/V(H_2O) \tag{3.6}$$

where $(z_+ + z_-)/\min \{z_+, z_-\}$ is two for a symmetrical electrolyte and three for a 2:1 or 1:2 electrolyte. Here $\mu(\text{elec})$ denotes the correction to the chemical potential from the ion-ion interactions, which are usually described in terms of Debye-Hückel correction factors. We will typically operate at concentrations and at a level of approximation where these can be neglected.

The simplest description of a nonideal solution of a noncharged species is through the regular solution model (5). Here the effective solute-solute interaction w is written as a combination of the bare solute-solvent interaction W_{AB}, the solute-solute interaction W_{BB}, and the solvent-solvent interaction W_{AA} so that

$$w = N_{Av} z_b (W_{AB} - W_{AA}/2 - W_{BB}/2) \tag{3.7}$$

where N_{Av} is Avogadro's number and z_b is the number of near neighbors in the liquid. The total free energy of mixing of a regular solution is obtained by having an ideal entropy of mixing and an interaction with randomly distribution near neighbors. This yields the solvent (A) chemical potential

$$\mu_A - \mu_A° = RT \ln(1 - X_B + w X_B^2) \approx -RTX_B + RTX_B^2/2 + w X_B^2 \tag{3.8}$$

$$\Pi_{osm} = RTc_B - RTc_B^2 V_A(1/2 - w/RT) + \cdots \tag{3.9}$$

and by comparison with Eq. (3.2) we see that the second virial coefficient is

$$B_2 = V_A(1/2 - w/RT) \tag{3.10}$$

One application of the regular solution model is to understanding the properties of aqueous sugar solutions. In this case the interaction parameter is unusually small and can even take negative values.

Another merit of the regular solution model is that it forms the basis for the Flory-Huggins model of polymer solutions. Having the solute units on large strings as in a homopolymer clearly affects the solution thermodynamics but, as we will see, the difference relative to the regular solution is basically quantitative rather than qualitative. For a polymer system

$$\mu_A - \mu_A^\circ = -RT\Phi_p/N_p + (w - RT/2)\Phi_p^2 \tag{3.11}$$

$$\Pi_{osm} = (RT/V_A)\{\Phi_p/N_p + (1/2 - w/RT)\Phi_p^2\} \tag{3.12}$$

where the polymer concentration enters through its volume fraction Φ_p rather than the molar concentration. Thus, counted on a per weight basis, a flexible polymer is only moderately less effective in increasing the osmotic pressure than the monomer molecules. The main difference is that because of the large degree of polymerization, N_p, the term linear in Φ_p is small in Eqs. (3.11) and (3.12). These equations are valid in the semi-dilute regime, where the polymer coils overlap forming a dynamic network. Although it is far from obvious, this effect on the osmotic pressure in Eq. (3.12) comes from the entropy associated with conformational degrees of freedom of the polymer molecules. For a stiff polymer as in an ordered helix, the contribution to the osmotic pressure is virtually negligible. Similarly, a denatured protein contributes much more to the osmotic pressure than the native globular form.

The preceding expressions for the chemical potentials and osmotic pressures are obtained for general solution models that make no specific reference to water as the solvent. However, there is a consensus that water has some rather unique features as a solvent. Unfortunately, the consensus ceases when one goes into details about these unique properties. The following is a short version of the author's view on this subject, but the reader should be warned that conflicting views exist.

From a molecular perspective, water is an unusually polar molecule considering its size. There are few liquids (HCl, HF, etc.) that can produce a higher dipolar polarization, i.e., a higher dipole moment per unit volume. There are other solvents that have higher molecular dipole moments, such as dimethyl sulfoxide (DMSO), and even those that have a higher dielectric permittivity, such as methyl formamide. Short-range electrostatic interactions dominate in liquid water and these are usually called hydrogen bonds, so the liquid has strong cohesion as manifested in the high melting and boiling points and also in a high surface tension (≈ 72 mJ/m^2).

The strong cohesion due to electrostatic forces results in the ability to solvate other species with strong electrostatic cohesion, most notably electrolyte ions. Water is furthermore miscible with many less polar solvents such as acetone and alcohols (methanol, ethanol, propanol, and *t*-butanol). Sugars can be used to illustrate the versatility of water as a solvent. Individual water molecules can interact with the OH groups on the sugar ring, and due to the structural flexibility coming from having small separate entities this interaction is favorable relative to the direct sugar-sugar interaction, which is hampered by the stiffness of the ring system. Thus, optimal sugar-sugar contact can be realized only over small parts of the molecules.

For apolar solutes such as hydrocarbons or simple gases that have not much to offer in terms of intermolecular electrostatic interactions, the cohesion in water makes the solubility low. We can illustrate the mechanisms with reference to the regular solution model, where in Eq. (3.7) we wrote the effective interaction w in terms of three contributions. For a weakly interacting species both W_{AB} and W_{BB} are small, but the strong cohesion makes W_{AA} large for water. This is the basic mechanism behind the important hydrophobic interaction in water. This matter has been the subject of scientific discussion over half a century. Why all this fuss if the explanation is so simple? The answer is that this basic mechanism is obscured by some complicating but highly significant features. These complications concern both the thermodynamics and the molecular structure in the liquid.

The thermodynamic observation is that for the transfer of a hydrophobic molecule from an apolar solvent to water the enthalpy change is close to zero at room temperature while the *free energy* change is strongly positive as illustrated in Table 1. The regular solution model would have equal changes in standard free energy and enthalpy. Furthermore, there is an anomalously large difference in the heat capacity so that at 100°C the free energy and enthalpy changes match. In fact, at higher temperatures the thermodynamic data for alkane solubility approach those expected from the regular solution model (6). The anomalous behavior at room and lower temperatures shows that it is too simplistic to model the aqueous system in terms of the regular solution. The most reasonable interpretation of the unusual properties is that they are caused by some structural reorganization among the water molecules in the liquid when the apolar solutes are introduced. At higher temperatures, where liquid water is not significantly more ordered than other polar liquids, we observe normal, i.e., regular solution, behavior.

Unfortunately, it is a widespread view in the biochemical literature that the hydrophobic interaction arises because a hydrophobic molecule structures water, making the transfer from an apolar environment unfavorable. This argument, originally due to Frank and Evans (7), confuses cause and effect and is fundamentally wrong. Yes, there is a structural response in the liquid water when an apolar molecule is introduced, but this is an effect induced by introducing the apolar solute and by the principle of Le Chatelier this effect causes a decrease in the free energy, not an

Table 1   Thermodynamic Data for the Transfer of Some Alkanes from Water to the Neat Liquid at 25°C

Compound	$\mu^0(l)-\mu^0(aq)$ (kJ/mol)	$H^0(l) - H^0(aq)$ (kJ/mol)	$S^0(l) - S^0(aq)$ (kJ/mol)	$C_p^0(aq)$ (J/mol K)	$C_p^0(l)$ (J/mol K)
C_4H_{10}	−25.0	3.3	96	414	142
C_6H_{14}	−32.4	0	108	635	196
C_6H_6	−19.2	−2.1	59	359	134
$C_6H_5CH_3$	−22.6	−1.7	71	418	155
C_4H_9OH	−10.0	9.4	65	478	178

increase. Consequently, water is in fact a much better solvent for apolar solutes than a similarly cohesive but less flexible solvent. Experimental studies using hydrazine as a reference solvent have demonstrated this point (8). Thus, using Eqs. (3.7) and (3.8) predicts too low solubility at room temperature but they work reasonably well for the solubility at high temperatures. In fact, the solubility of alkanes actually increases as the temperature is decreased below $\approx +20°C$ and this can in some cases lead to cold denaturation of proteins.

These thermodynamic properties of the hydrophobic interaction highlight another unique feature of liquid water. For most liquids the structure is determined by the short-range repulsive interactions present for all molecules. For water there is a more delicate balance between the short-range repulsion, the dispersion attraction, and the electrostatic interactions. The last is optimal at rather specific orientations of the molecules relative to one another, but the former two are more isotropic. The structure of ice at normal pressures shows that the minimum in energy occurs for a structure of lower density where the electrostatic interactions are optimized. On melting, local structures with shorter molecular separations but higher energies are also populated. The increase in electrostatic energy is compensated by an increase in entropy due to the disorder in both the orientational and translational degrees of freedom. There is also a decrease in the dispersion energy in the liquid state due the shorter intermolecular separations.

This balance between the two types of local structures, the disordered denser and the ordered less dense one, is easily affected by an external perturbation such as the introduction of an apolar molecule. Thus, in this case the density of water decreases locally and this leads to a decrease in enthalpy. At room temperature this decrease of the enthalpy from the water-water interactions fully compensates for the enthalpy increase due to the weak water-solute interaction. At still lower temperatures the enthalpy decrease due to the restructuring in the water is so large that the solubility of liquid alkanes actually increases with decreasing temperature. For apolar gases the solubility increases strongly with decreasing temperature because there is an additional effect due to the general feature that the stability of the gas decreases relative to a liquid when the temperature is lowered.

It is a common suggestion that because water is a structured liquid it has difficulties in responding to the perturbation caused by a solute. The argument presented earlier indicates that in fact the opposite is the case. Water is an unusually versatile solvent, a property that is utilized in biological systems to solvate polar surface groups, and this provides a first basis for obtaining colloidal stability for a wide range of systems (9).

IV. SURFACE INTERACTIONS. FREE ENERGY, ENERGY AND ENTROPY

In our theoretical analysis of forces between colloidal particles we will mainly use the force between two planar parallel surfaces as an illustration. Mechanistically there is no fundamental difference between forces in the planar case and forces between spherical particles as long as the surfaces are the same and homogeneous. In biological systems surface inhomogeneities are commonplace, but we will postpone a discussion of this complication. The restriction to the planer case is less severe than one might think due to the existence of the so-called Derjaguin approximation, which will be discussed at the end of this section and which relates the behavior of the planar system to that for curved surfaces.

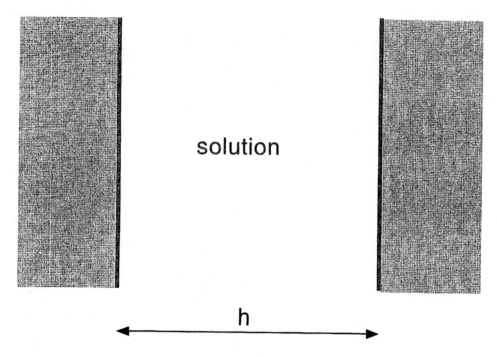

Fig. 2 Illustration of the planar geometry used for the discussion of surface forces.

Consider a system of two planar surfaces with an intervening solution in the gap as shown in Fig. 2. This solution is in equilibrium with a bulk system and we will assume, as long as the opposite is not explicitly stated, that the solution in the gap can maintain equilibrium with all the components in the solution. Furthermore, we will maintain the approximation of an incompressible system. In this case the Helmholtz, A and Gibbs, G, free energies differ by only a trivial constant and we can write the force F as the change in free energy A or G with changes in the gap size h so that

$$F = -\partial G/\partial h = -\partial A/\partial h = -\partial U/\partial h + T\partial S/\partial h \qquad (4.1)$$

where $G, A, U,$ and S refer to the total system gap plus bulk. Here the sign convention is such that a repulsive force is positive and an attractive force is negative.

A change in h implies a change in the volume of the gap and, for the case in which the solvent (water) is in excess of other components, the volume change implies a transport of solvent between bulk and gap. The free energy change in adding or removing solvent from the bulk solution is by definition related to the chemical potential of the solvent or, as discussed in Sec. III, to the osmotic pressure in the bulk. For the gap a change dh in separation implies a volume changes dV for a given area, and this in turn implies a change in the number of solvent molecules $dn_s = dV/V_s = \text{area } dh/V_s$, where V_s is the molar volume of the solvent. This leads to the force per unit area of

$$
\begin{aligned}
F/\text{area} - \Pi_{\text{osm}}(\text{bulk}) &= -(\partial A/\partial n_s)_{\text{gap}}/V_s \\
&= -[(\partial U/\partial n_s)_{\text{gap}} - T(\partial S/\partial n_s)_{\text{gap}}]/V_s
\end{aligned}
\qquad (4.2)
$$

Table 2 The Geometry-Dependent Function $f(R_1, R_2)$

Sphere-plane	Sphere-sphere	Crossed cylinders
$2\pi R$	$2\pi\left(\dfrac{R_1 R_2}{R_1 + R_2}\right)$	$2\pi\dfrac{\sqrt{R_1 R_2}}{\sin\theta}$

One way to interpret this equation is to say that the resulting force is the difference between the osmotic pressure in the gap and in the bulk. In Eq. (4.1) we have explicitly divided the force into an energy and an entropy contribution. It will turn out that this way of looking at the force provides a useful basis for the conceptual understanding of different mechanisms.

We have formulated the force in a thermodynamic framework based on free energy changes. A force also has a mechanical interpretation and it is sometimes tempting to refer to this mechanical picture when discussing forces on a conceptual level. However, in most cases this mechanical picture leads to difficulties that far exceed the intended simplification. Making the assumption of incompressibility for the liquid between the surfaces, we have in fact eliminated the possibility of describing the force in a mechanical way. For hard objects in contact the actual force can be anything between zero and infinity. Thus there is no way to make the connection between structure and the magnitude of the force.

Equation (4.2) gives a formal expression for the force per unit area between two planar surfaces. In applications to biological systems this can be relevant as it stands when considering the interaction between planar areas of membranes, but in most cases we are interested in the interaction between particles of finite size and with curved surfaces. This situation can often be simply related to the planar case by the so-called Derjaguin approximation (10), which states that the force between curved surfaces is related to the interaction free energy $A(h)$ for equivalent planar surfaces

$$F(h) \approx f(R_1, R_2)(A(h))_{\text{plane}} \tag{4.3}$$

where the function f depends only on the geometry. Table 2 shows the value of $f(R_1, R_2)$ for a number of cases.

The range of validity of the Derjaguin approximation has been a matter of some debate, but it reflects a fundamental relation. The first condition for its validity is that the surfaces are laterally homogeneous, but for the intervening solution it is required only that it adopts its equilibrium properties. There is also a limit at small radii where the approximation loses its validity. There is no generally valid rule for the breakdown of the Derjaguin approximation. A minimum condition is that the range of the force has to be smaller than the radii of curvature of the surfaces.

V. FORCE MEASUREMENTS

Forces in colloidal systems can be measured in several ways. One can identify three different principles of force measurements. Most direct is mechanical measurement. In this case the magnitude of the force is obtained by bringing two colloidal entities to a controlled separation and simultaneously measuring the force mechanically, for example, by the deflection of a spring. Using this principle, the most precise data are obtained using the surface force apparatus (SFA) (1) or one of its variants, for example, MASSIF (11). The principle of the SFA is simple. The mechanical force is determined between two crossed cylinders of macroscopic size through the deflection of a spring. The related distance is measured by an interferometric technique with a precision that can be as high as 0.1 nm. There are two main experimental limitations. The method requires macroscopically smooth surfaces, and these can easily be obtained only with a few materials, mainly mica. However, one can make surface modifications to allow for broader chemical applications. The second limitation is that the observed force is caused by a contact over a large area, so large quantities of material are needed and there is no lateral resolution.

During the past few years the atomic force microscopic (AFM) has been developed into a versatile instrument that allows explicit force measurements in addition to a determination of surface morphology (12). In the standard setup the force is measured between a surface and a sharp tip of an inorganic material such as silicone nitride. However, with current techniques the tip can be modified chemically in such a variety of ways that the method also appears very promising for studies of specific interactions between biological molecules (13). The AFM force measurement is less accurate than the SFA version but the method has several advantages. It is more versatile when it comes to the nature of the sample and there is very good lateral resolution. One can also foresee further development of the method.

A number of other methods can be utilized for mechanical force measurements. One that has received attention recently is the use of optical tweezers (14) to manipulate particles. Here a particle of high refractive index is captured at the cross point of two focused laser beams. For example in trying to remove a particle, such as a bacterial cell, from a surface it is possible to measure an adhesion force provided one knows the intensity of the laser beams and the contrast in refractive index.

The second principle for measuring forces is through the combination of thermodynamic control of osmotic pressure, or equivalently solvent chemical potential, and simultaneous measurement of distances in a macroscopic sample. Currently, the version of this strategy most commonly used is the so-called osmotic stress technique (15), where one has an ordered sample, preferably liquid crystalline, and measures the characteristic distances by X-ray diffraction for varying osmotic pressures. For well-ordered samples this yields very accurate force-distance relations. The limitation is that one can measure only under circumstances dictated by bulk equilibrium conditions. For example, attractive forces are not easily measured by this method. A historically older method based on the same idea is the determination of equilibrium thicknesses of soap films (16). In this case the distance measurement is performed optically using interference, but the variation of the osmotic pressure in the bulk phase is analogous (17).

The third class of methods we denote as indirect ones. Here one measures some physical property that is affected by interparticle forces and then deduces the force from a more or less established model. The weakness of these methods is often that the analysis of the experimental data requires some deconvolution to give an explicit force. The most important representative of this class of methods is scattering of light, X-rays, or neutrons. In this case the measured structure factor contains information on particle interactions, but it is a somewhat complex process to extract explicit force data. This involves the solution of the statistical mechanical problem of deducing a pair potential from a known pair correlation function or its Fourier transform (18). This problem has been solved for some simple forms of pair potentials, but the general case provides a rather formidable challenge.

VI. DISPERSION AND VAN DER WAALS FORCES

The dispersion interaction provides an attraction between all molecules as discussed in Sec. III. This interaction thus also operates between all particles of colloidal dimensions. There are two qualitative features to discuss that are different from the interactions between a pair of small molecules although the basic mechanism is the same.

The first new aspect is the range of the interaction. One can see the colloidal particle as built from a large number of small molecular units, amino acids, nucleotides, or lipids in the biological case. In a first, rather accurate description, we can see the total dispersion force between two particles as the sum of the pair interactions between molecular units on each particle. The simplest case is to consider the interaction between two half-planes separated by a gap of width h. The addition of all the pair interactions can be accomplished through an integration over volume elements dV and dV' in the two half-spaces as shown in Fig. 3. Integrating over two of the variables, $dx'dy'$, parallel to the gap yields an area, while the integration over the remaining four, $dx\,dy\,dz\,dz'$, changes the original R^{-6} distance dependence

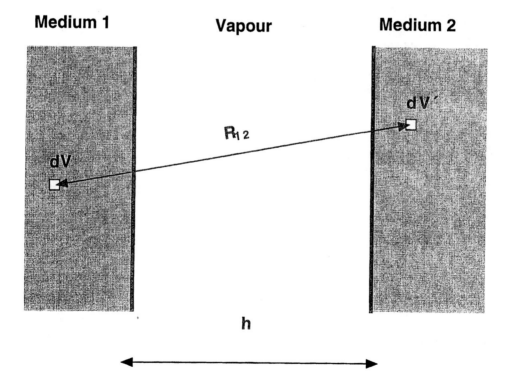

Fig. 3 Illustration of the Hamaker approach to calculate the dispersion interaction between two bodies by intergrating the intermolecular interaction between volume elements dV and dV'.

by one power unit for each integration so that the interaction energy is

$$V(h)/\text{area} = -H/(6\pi h^2) \tag{6.1}$$

where H is called the Hamaker constant and is a material constant that depends on the chemical nature of the particles. The corresponding force is obtained from the derivative of Eq. (6.1),

$$F(h)/\text{area} = -H/(12\pi h^3) \tag{6.2}$$

Thus, by summing over many molecular contributions, the force has become substantially more long range.

The second complication arises when we want to determine the effective dispersion interaction across a medium. Equations (6.1) and (6.2) refer to the direct interaction between the two particles. However, in practice we have an (aqueous) medium between the particles and there is also a particle-medium and medium-medium dispersion interaction that should enter the energy balance. The effect of the medium is typically to reduce the dispersion interaction substantially, and in extreme cases it can even change the sign when the two particles are very different. The presence of the medium does not change the functional forms of Eqs. (6.1) and (6.2), so it is only the magnitude of the Hamaker constant that is affected. In a biological system we typically have an apolar bulk of a mainly hydrocarbon character interacting across water. Because water and hydrocarbon have relatively similar polarizabilities, there is a substantial cancellation and the effective Hamaker constant is reduced by nearly an order of magnitude. Some representative values of the Hamaker constant are shown in Table 3.

The dispersion interaction arises from correlations between electrons in molecules in the two media, and since electrons move even at zero temperature this correlation effect is largely temperature independent except for an indirect effect due

Table 3 Values of Hamaker Constants ($\times 10^{20}$/J) for Some Common Combinations of Materials M

Material, M	M|air|M	M|water|M	M|water|air	M|air|water
Water	3.7	0	0	3.7
Alkanes				
$n = 5$	3.8	0.3	0.15	3.6
$n = 6$	4.1	0.4	0	3.8
$n = 10$	4.8	0.5	−0.3	4.1
$n = 14$	5.1	0.5	−0.5	4.2
$n = 16$	5.2	0.5	−0.5	4.3
Fused quartz	6.5	0.8	−1.0	4.8
Fused silica	6.6	0.8	−1.0	4.8
Sapphire	16	5	−3.8	7.4
Polymethyl methacrylate	7.1	1.1	−1.3	5.0
Polystyrene	6.6	1.0	−1.1	4.8
Polyisoprene	6.0	0.7	−0.8	4.6
Polytetrafluoroethylene	3.8	0.3	0.1	3.7
Mica (green)	10	2.1		

to changes in the density. There is another molecular contribution to the interaction between colloidal particles that gives rise to the same distance dependence as the dispersion interaction. At finite temperatures polar molecules rotate and dipole-dipole correlations can develop between molecular dipoles of two particles. These correlations will be temperature dependent and their effect can be included in the general formalism by writing

$$F(h)/\text{area} = -H_{\text{eff}}(T)/(12\pi h^3) \tag{6.3}$$

The contribution from the dipolar correlations is usually relatively small, but for the special case of hydrocarbon-hydrocarbon interaction across water it contributes approximately half of the value of the Hamaker constant. When one includes this dipolar correlation contribution it is customary to refer to the interaction as a van der Waals force, although this nomenclature is not firmly established.

Even though the Hamaker constant is relatively small for typical aggregates in a biological system, the van der Waals force is still sufficiently strong to cause aggregation of virtually all particles larger than a few nanometers in radius. The larger the particle, the larger is the interaction, and this general sticking mechanism provides one limitation for operating with large aggregates in the living system. To prevent the van der Waals attraction leading to aggregation it is necessary for some of the repulsive forces to be operating, which we discuss later. It is common in technological applications of colloids to balance attractive and repulsive forces, but this balance operates in a much more sophisticated way in the living cell, although this aspect has received less attention.

VII. THE ELECTRIC DOUBLE-LAYER FORCE

Practically all entities of colloidal size in a biological system carry charge, and this has a large influence on the interactions and the function of the system. The standard method for describing the interaction between two charged surfaces is through the Poisson-Boltzmann equation. Let us consider a model system of two homogeneously charged walls separated by a distance h and with counterions and other small electrolyte ions in the gap as shown in Fig. 4. The solution in the gap is typically in contact with a bulk solution of given osmotic pressure and electrolyte concentration. The solvent is seen as a dielectric continuum with a dielectric permittivity $\varepsilon_r \varepsilon_0$. The Poisson-Boltzmann equation describes how the ions are distributed in the gap as a response to the variation of the electrical potential $\phi(z)$.

The discussion of the interaction between two charged surfaces is greatly facilitated by first specifying the properties of a single charged surface in an electrolyte medium. (This case is discussed in more detail in Chapter 2.) The characterizing parameters are the surface charge or alternatively the surface potential and the bulk concentration c_{salt} of the electrolyte, which for simplicity we will assume to be of the 1:1 type as in NaCl. For this case, which is usually referred to as the Gouy-Chapman theory, the Poisson-Boltzmann equation has a transparent analytical solution even in its nonlinearized form. This gives a relation between surface charge density σ and electric surface potential ϕ_0, which for the linearized care is

$$\sigma = \varepsilon_r \varepsilon_0 \kappa \phi_0 \tag{7.1}$$

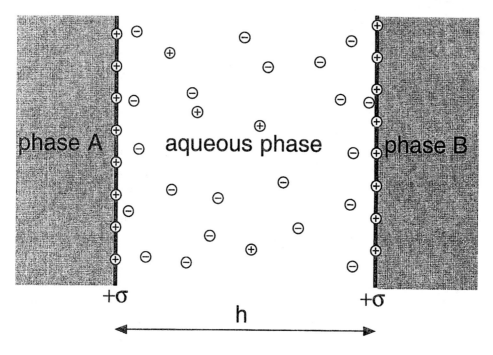

Fig. 4 Two charged surfaces with neutralizing counterions and salt in the intervening solution.

where κ^{-1} is the Debye screening length and is determined by the electrolyte concentration c_i and

$$\kappa = \left[\varepsilon_r \varepsilon_0 RT / \sum (z_i e)^2 c_i \right]^{1/2} \tag{7.2}$$

In addition, the Gouy-Chapman solution provides a prediction of the ion concentration profiles, the Fig. 5 shows some examples. Away from the surface these concentration profiles decay exponentially with a decay length κ^{-1} toward the bulk electrolyte concentration.

For biological systems, a significant consequence of the existence of a charged surface is that surface groups such as COO^- ($COOH$) and NH_2 (NH_3^+) titrate at a different pH than in bulk solution. Figure 6 illustrates this effect, showing the degree of ionization of a surface consisting of $COOH$ groups with an intrinsic pK_a of 4.8. In elementary texts this phenomenon is sometimes explained in terms of a surface pH different from that of bulk. However, it is obviously the carboxylic groups that experience a different environment from the bulk due to the charge-charge interactions, and this causes the shift in the pK_a value. At equilibrium the pH is homogeneous throughout the solution.

Let us now turn to the interaction between two identical charged surfaces. At separations much larger than the Debye length the two surfaces experience bulk conditions and there is no interaction. However, as the surfaces approach, the inhomogeneous ion distributions start to overlap, and at this point one surface begins

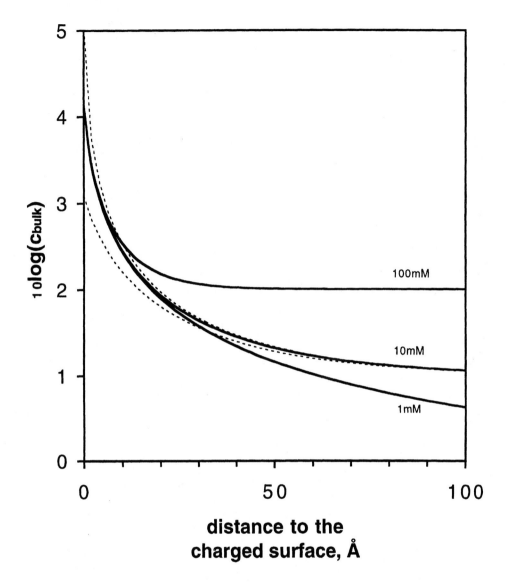

Fig. 5 Counterion concentration (in mM) outside a single charged wall. Solid lines: surface charge density 0.2 C/m^2. Dashed lines: bulk concentration 10 mM and surface charge 0.07 C/m^2 and 0.6 C/m^2. Note that for a given (high) charge density the counterion concentration close to the wall is independent of the bulk concentration.

to sense the presence of the other. It is possible to evaluate the derivative in Eq. (5.2) for a free energy expression consistent with the Poisson-Boltzmann equation. This yields a conceptually simple form and the force is determined solely by the ion concentrations in the midplane ($z = 0$) where the electric field is zero so that

$$F(h)/\text{area} = RT \sum c_i(0) - \Pi_{\text{osm}}(\text{bulk}) \tag{7.3}$$

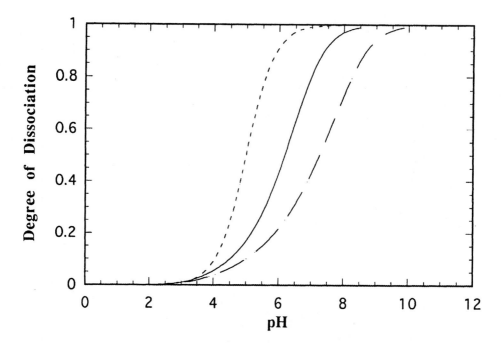

Fig. 6 Calculated degree of dissociation of an acid group of intrinsic $pK_a = 5$ located at a surface with a density of one site per 1.0 nm². Dotted line: electrostatic interactions ignored. Dashed line: electrostatic interactions calculated using the Gouy-Chapman theory with a smeared-out surface charge. Solid line: calculation using the Poisson-Boltzmann equation with discrete surface charges. (From Wägnerud, thesis, Lund University, 1995.)

where we can see using Eq. (3.6) that the first term represents the osmotic pressure at the midplane. Equation (7.3) is exact within the Poisson-Boltzmann approximation and for large $h \geq \kappa^{-1}$ the concentration profiles decay exponentially and are furthermore not greatly affected by the interaction between the surfaces. Then one can derive the approximate formula

$$F(h)/\text{area} = 64RTc_i\Gamma_0^2\exp(-\kappa h) \qquad (7.4)$$

where Γ_0 is related to the surface potential ϕ_0 through

$$\Gamma_0 = [\exp(ze\phi_0/2kT) - 1]/[\exp(-ze\phi_0/2kT) + 1] \qquad (7.5)$$

The force between similar surfaces is always repulsive, and one can extend this to show that it is in fact always repulsive between surfaces carrying the same sign of the charge even if the charge densities are very different. The mechanism behind the repulsion is that when the two surfaces approach the counterions have less available space and their entropy is decreased, which leads to a repulsion. It is thus the entropy term in the general expression for the force that dominates, and it is important to realize that the ion-ion interaction actually tends to a lower internal energy as the two surfaces approach. In the presence of divalent ions, because of correlation effects, this energy decrease can dominate the entropic repulsion with the result that similarly charged surfaces experience an attractive interaction (19,20).

For an asymmetric situation in which the two surfaces have charges of the same sign but different magnitude we can still use the exact expression of Eq. (7.3), but in this case it is more problematic to locate the position where the electric field is zero. In the weak overlap limit one can derive an expression similar to Eq. (7.4),

$$F(h)/\text{area} = 64RTc_i\Gamma_{0,1}\Gamma_{0,2}\exp(-\kappa h) \tag{7.6}$$

where Γ_0^2 is simply replaced by the product $\Gamma_{0,1}\Gamma_{0,2}$ for the two different surfaces according to Eq. (7.5).

In classical colloidal systems the basic description of stability is built from a competition between the attractive van der Waals force promoting association and the repulsive double-layer force providing stability. The combination of these two forces is considered in the DLVO theory, named after the originators Derjaguin, Landau, Verwey, and Overbeek. This theory adequately describes the force between two similar surfaces at long range, $h \geq \kappa^{-1}$. It forms a most useful basis for the description of forces between colloidal entities, but one should always keep in mind that there are many additional force contributions. In the following we discuss a few of these that we judge particularly relevant to biological systems.

An aspect of double-layer interaction that has only recently received attention concerns the interaction between oppositely charged surfaces. Because these are obviously different, the magnitudes of the surface charge densities are usually different. At long range there is always an attraction between a positively and a negatively charged surface. One way to see the mechanism behind the attraction is to consider the surface with the smaller magnitude of the charge density to act as a giant counterion of the one with the higher charge. As the surfaces begin to approach, small counterions are released from both surfaces and join the bulk solution. This amounts to a decrease in the ion entropy and thus an attractive interaction. However, it also follows that when the gap has become sufficiently narrow there are no more superfluous counterions to be released. The remaining ones are necessary to maintain neutrality, and they will experience a confinement with a concomitant decrease in counterion entropy. Thus, at some separation the attraction turns into a repulsion by this mechanism. One can obtain an approximate expression for the force by solving the linearized Poisson-Boltzmann equation,

$$A(h)/\text{area} = \{[(\sigma_1^2+\sigma_2^2)\exp(-\kappa h)+2\sigma_1\sigma_2]/[\exp(\kappa h)-\exp(-\kappa h)]\}/(\kappa\varepsilon_1\varepsilon_0) \tag{7.7}$$

where σ_i is the surface charge density of surface i. Figure 7 shows the calculated interaction potential for some combinations of surface charge densities. The more similar the magnitudes of the two charge densities, the lower is the free energy and the shorter is the separation at the minimum. The existence of a minimum in the interaction at some short separation $\leq$the Debye length is very significant. This can, for example, be used to understand ion exchange chromatography of proteins (21, 22).

The argument giving a minimum away from contact between the surfaces is based on a situation in which the surface charge densities remain unchanged. A minimum of the double-layer interaction at contact can be achieved by one of the surfaces adjusting its charge density so that a perfect match is obtained. However, there is a free energy cost associated with such an adjustment of the charge, and it is a delicate

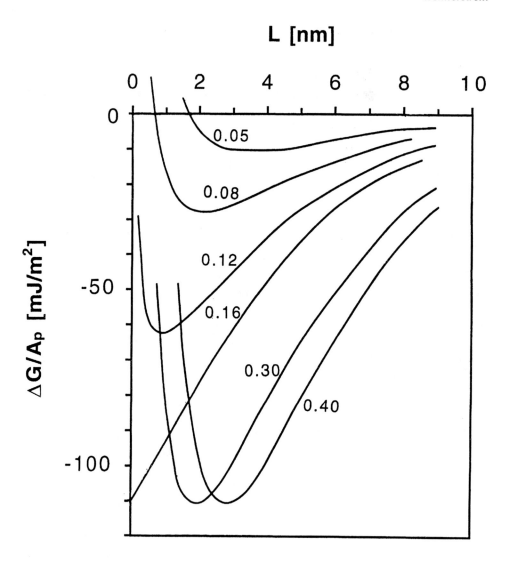

Fig. 7 Calculated interaction between two oppositely charged surfaces with one surface at a constant charge density of -0.16 C/m^2 while the surface charge density of the other is varied from 0.05 to 0.40 C/m^2. It is only when the charge densities match that the curve is monotonically attractive. (From Ref. 21.)

question to identify the global free energy minimum. At short range it is necessary to introduce another complication in the description. We have so far seen the surfaces as homogeneously charged. For a metal electrode this is in fact the case, but in biological systems the charges are carried by ionized groups and this becomes an important issue at separations smaller than the average lateral distance between the surface charges. Thus, to have the minimum at contact it is necessary either to generate surfaces with no charge or alternatively to match them in so-called salt bridges. If the minimum is at some separation between the particles there is room for dynamic processes. For example, spherical particles can slide and roll on an oppositely charged surface, which could clearly have some interesting functional consequences. For example, repressor proteins are thought to find their target sequences on a DNA strand by using this sliding-rolling motion in an electrostatically bound state (23).

This electrical double-layer attraction is widely used in the cell to promote and enhance specific association processes. Most proteins that function by interaction with nucleic acids are positively charged at neutral pH. This applies, for example, to histones, repressors, and nucleases. Presumably the basic myelin protein has a function of associating with the negatively charged membrane to keep the myelin structure intact. It is more of a curiosity that snake venoms are rich in positively charged polypeptides that promote lysis by binding to cell membranes.

VIII. THE HYDROPHOBIC INTERACTION

The hydrophobic interaction plays a most fundamental role in the molecular organization in the cell. It provides the stabilizing interaction for the membrane lipid bilayer and the associated membrane proteins. Similarly, globular proteins and nucleic acid helices owe their conformational stability to the hydrophobic interaction. It also plays a significant role in a large number of functions triggered by temporary association processes.

In Sec. III we pointed out that the molecular basis for the hydrophobic interaction is found in the strong cohesion between water molecules in the bulk (24). We thus repeat that the hydrophobic interaction is the effective interaction formed by a combination of different molecular contributions, where the contribution from the electrostatic water-water interaction dominates. That hydrophobic molecules tend to associate in water in thus due not to a strong intrinsic attraction between the molecules but rather to the fact that they are expelled from the aqueous environment because they are unable to form strong interactions in that environment.

It can sometimes be revealing to see the hydrophobic interaction as belonging to the general class of solvophobic interactions. When the solvent is not compatible with the solute, there is always a tendency for association. The reverse of the hydrophobic interaction is found inside the lipid bilayers. There polar groups are out of place and tend to associate, which provides one mechanism for the interaction between α-helices of membrane proteins. Also, exposed polar amino acid side chains prefer to associate with one another.

The hydrophobic interaction can be seen to act on two different levels. On the one hand, single molecules or patches of molecules come together from an aqueous environment. There exist a number of semiempirical or empirical estimates of the free energy changes for these processes. These can be based on the number of CH_2 groups or on the molecular area withdrawn from the aqueous environment or based simply on solubility data for suitable reference compounds. It would lead outside the scope of this chapter to analyze critically the merits of these different approaches to the description of the hydrophobic interaction between single molecules. Here we focus primarily on the interactions on a colloidal rather than a molecular level and consider the other extreme of the hydrophobic interaction, namely that between two planar hydrophobic surfaces separated by an aqueous liquid film.

The alkane-water interfacial free energy is ≈ 50 mJ/m^2, and bringing two surfaces of a hydrocarbon character from a large separation in water into contact involves a decrease in energy of $2 \times 50 = 100$ mJ/m^2. This is in fact a large number; as a comparison, the corresponding process in a medium of a low-density gas instead of water is only half as large although the contribution from the van der Waals force is close to 10 times larger in this case. The difference is due to the contribution from the cohesion in the water, as we have already discussed. From the knowledge of the surface free energies we can thus obtain a reliable estimate of the total interaction in bringing two completely hydrophobic surfaces from a large separation into contact. However, the argument does not provide any direct information on the distance dependence.

During the past 10 years many reports of observations of attractive forces between hydrophobic surfaces have appeared (25–28). This has led to a widespread belief that the hydrophobic interaction has a long-range character. The experimental observations are genuine, but it is far from obvious that the actual mechanism behind the observed forces is properly identified as a hydrophobic force (29–31). It is one of the major current challenges in fundamental colloid science to resolve this issue, and a detailed discussion in this context would lead too far. The present author's opinion is that the "true" hydrophobic force between surfaces is of a short-range character (32) and that the experimental observations of a longer range interaction are manifestations of other mechanisms. It might be that the true hydrophobic interaction can be measured only between small surface patches because otherwise the system is so far from equilibrium that some instability occurs. It is in general more difficult to measure attractive than repulsive forces experimentally, and strong attractions easily generate mechanical or structural instabilities.

The conclusion from this section is that there exist reasonable estimates of the magnitude of the hydrophobic interaction between both molecules and surfaces. On the other hand, there is little reliable information on the distance dependence of the force and further work is needed to clarify this aspect of the problem.

IX. REPULSIVE ENTROPIC FORCES

So far we have discussed the van der Waals, the electrical double-layer, and the hydrophobic interactions, which all play a role in virtually all biologically relevant systems. In addition, there are a number of effects of a more specific nature that are absolutely essential in some functions but are absent in many other cases. Organization within a cell and also between cells is based on a controlled aggregation of macromolecules, organelles, and cells. To be able to control association, a first prerequisite is that the basic rule is nonassociation. It is thus essential to have strategies for forming nonassociating colloidal entities. In this section we discuss the basic strategy with three examples.

Although difficult to prove in a fundamental way, it seems that a repulsive interaction between two similar surfaces is always caused by a decrease in entropy as the two surfaces approach (33). One basis for this statement is that the energy minimum is always at contact between the surfaces. As we discussed in Sec. VII, the electrical double-layer force is of such a character, but there are several other examples.

In technical applications surface-adsorbed polymers are often used to generate stability in suspensions of colloidal particles. An adsorbed polymer typically has high flexibility on the surface, and there is thus a substantial entropic contribution to the free energy from the large number of allowed configurations. When two similar surfaces approach and the polymers on opposite surfaces start to overlap, less configurations are possible, resulting in a repulsive interaction of entropic origin (34). There is typically a compensating decrease in energy due to the increased monomer-monomer contacts, but in a good solvent the entropic term wins out. Figure 8 shows a typical interaction between two surfaces with an adsorbed noncharged polymer in a good solvent. The molecular mechanism behind the repulsion is thus quite

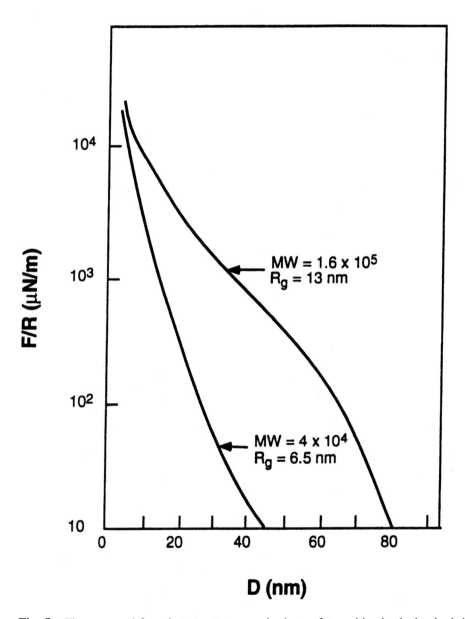

Fig. 8 The measured force between two curved mica surfaces with adsorbed polyethylene oxide in water. Water is a good solvent for this polymer and the force is monotonically repulsive with a range given by a few multiples of the radius of gyration of the polymer. (From Ref. 50.)

analogous to the mechanism leading to the relatively high osmotic pressure of polymer solutions in a good solvent as in Eq. (3.11).

In the literature the repulsive interaction between surfaces covered with polymers is usually called the "steric interaction," but we (3) propose the nomenclature "steric entropic interaction" for two reasons. First, the interaction is of an entropic origin and the simple name "steric interaction" invites conceptual confusion between the thermodynamic and mechanical interpretations of surface interactions. From a mechanical point of view all interactions are steric in the approximation of incompressibility. A second reason for the name "steric entropic force" is to distinguish it from the "steric energetic interaction," which we will discuss in Sec. XII.

In biological systems the steric entropic interaction usually occurs for extracellular surfaces that are covered with charged polysaccharide chains. Such a layer of flexible partly charged polymer segments acts as very effective protection for adsorption of other cells, bacteria, or viruses. Infectious bacteria can, of course, find counterstrategies that allow them to penetrate this protective shield, but the general steric entropic repulsion sets the scene for this host-guest battle.

Chemically, the flexible polymer character is realized in the living system by either having a glycosylated (sphingo)lipid or by glycosylating a membrane protein. Glycoproteins of another type are found in the so-called mucins of mucous tissue (35). These are giant proteins apparently built as a linear polymer of polypeptide units with polysaccharide side chains. The stability of casein micelles in milk is caused by the polyelectrolyte character of κ-casein covering the micelles. The rennet used in cheese making acts by hydrolyzing the κ-casein and thus inducing colloidal aggregation and the formation of cheese curd (36).

Within a cell the molecular conditions are more constant, but a wide spectrum of colloidal entities is present and these should be protected from nonspecific association. The lipid membrane carries a negative charge because approximately 15 to 20% of the membrane lipids are ionic. However, even the pure lipid systems of noncharged lipids show an unusually weak van der Waals attraction that is not compatible with the DLVO theory. In these systems a strong repulsive force appears at separations shorter than 2 nm (37) as shown in Fig. 9. The molecular source of this repulsion has been a matter of some debate. According to one widespread view, it is caused by some structural effect in the aqueous solvent (38). This author finds such a proposition untenable and feels that the source of the repulsion should instead be found in entropic confinement of the lipid configurations, similarly to the case of the adsorbed polymers (39). However, because of the much smaller spatial extent of the lipid headgroups, the force is of shorter range. Relevant degrees of freedom are both headgroup conformations and out-of-plane protrusions of lipid molecules. Also, in this case the result of large conformational freedom is a repulsion, or reduced attraction, of an unspecified macromolecule or aggregate.

A third repulsive entropic mechanism is the undulation force (40). This operates only for bilayers and the relevant degrees of freedom are the collective bending modes of the bilayer. Two bilayers in proximity restrict one another and allow fewer bilayer conformations. This is similar to the protrusion case just discussed, but here we are considering the limit where the bilayer can be considered as a flexible sheet and the molecular detail is incorporated in a bending stiffness only. Consequently, the force is operating on a larger scale and the force law was derived from a continuum

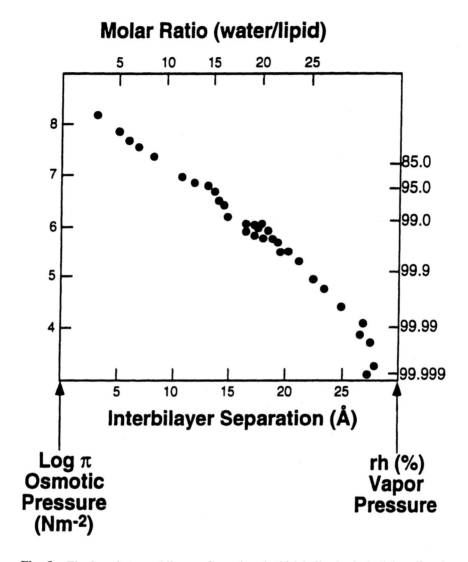

Fig. 9 The force between bilayers of egg phosphatidylcholine in the bulk lamellar phase as a function of interbilayer separation of alternatively water content. (From Ref. 38.)

description by Helfrich,

$$F(h)/\text{area} = \text{constant}(kT)^2/\kappa h^{-3} \tag{9.1}$$

where the value of the constant is somewhat debated but is of order 0.03. This undulation force has the same distance dependence as the van der Waals attraction between half-spaces and is thus dominant for the interaction between bilayers at separations larger than their thickness. It is an important force in model bilayer systems (41) but it remains to be seen how significant the effect is in a living system.

X. PACKING AND DEPLETION FORCES

The presence of a surface will affect the density in the liquid immediately outside this surface. We have briefly discussed the case of adsorption from solution onto the surface. Here we will consider two different situations. In a solution one can also have the opposite of adsorption, with a solute avoiding the surface and being depleted from it. Such a depletion or excluded volume provides a positive contribution to the surface free energy, and by this thermodynamic argument we see that the free energy change in bringing two similar surfaces into contact is larger in the presence of the solute than for the pure liquid. Lowering of the free energy implies the presence of an additional attractive component of the force (42,43). The range of the force is determined by the thickness of the depletion layer, h_{dep}. When there is no room for the solute in the gap, a further decrease of the separation implies moving solvent from a pure state to a solution with a finite osmotic pressure. Comparison with Eq. (4.2) shows that the force in this case is

$$F(h)/\text{area} = -\Pi_{osm} \tag{10.1}$$

Depending on the nature of the concentration profile outside a single surface, this depletion force either decays monotonically to zero or oscillates as illustrated in Fig. 10. To obtain the force for curved surfaces we can use the Derjaguin approximation, and in the simplest case the force varies linearly with separation with a bend around $h \approx 2h_{dep}$. One should note that the same type of force curve is also observed for the more general case of capillary phase separation (44–46).

The preceding argument has for simplicity presupposed one solute only, but particularly in biological systems there is a mix of solutes. However, as long as these do not interfere with one another the effect is additive in the different solute concentrations; for example, with only one solute being surface depleted the force is the same as for the pure binary system.

Depletion interactions can in principle be induced by any solute, but from a practical point of view nonadsorbing polymers provide the most important applications. For example, high concentrations of polyethylene oxide are used in biological systems to promote cell-cell aggregation and DNA transfection. One could be tempted to think that the role of the polymer is to take some active part in the critical process, but the most likely explanation is that it acts by its absence from the scene inducing an depletion attraction by affecting the osmotic pressure of the solvent. A more esoteric application of the depletion attraction that has had practical importance is that in infected patients erythrocytes aggregate and have a high sedimentation rate (SR) in a test tube. This effect provides a simple assay for the occurrence of an infection. It is due to enhanced concentration of fibrinogen and γ-globulins, which presumably cause aggregation of the erythrocytes by a depletion attraction (47).

XI. THE STERIC ENERGETIC FORCE

In Sec. II we pointed out that the size and shape of a molecule can be related to the overlap repulsion force between molecules. We then took the existence of this repulsive force to account for the practical incompressibility of liquids that has been used extensively up to this point. When one considers solids, the overlap repulsion has somewhat different significance. The solid body remains intact in shape on reasonable time

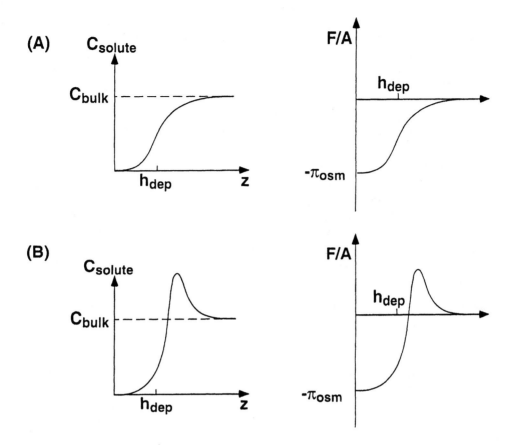

Fig. 10 Illustration of the relation between the solute concentration profile outside a surface repelling the solute and the interaction force. (A) A monotonically decaying concentration profile yields a monotonically attractive force. (B) An accumulation outside the depletion zone gives rise to a force barrier before the depletion attraction sets in.

scales. So even if the equilibrium shape is different, this is of no practical consequence. Thus, as two solid bodies approach there is a very strong repulsion when they start to overlap. We call this the steric energetic interaction and it occurs only in rigid systems. For example, planar solid bodies should adhere strongly through the van der Waals force, yet we know that the book normally does not stick to the table. The reason is that unless extreme care is taken, solid surfaces are rough and two such surfaces can establish close molecular contact only in a few positions, leading to weak adhesion. A material, such as a glue, that could float would fill the initial voids and establish strong adhesive contact.

Even though solid structures are rare in biological systems, proteins and nucleic acids can be highly structured with definite shapes. Thus the steric energetic interaction is a major player in recognition events. The lock-and-key analogy is based on this steric picture, but as we will discuss in the following section the steric energetic interaction is only one element in a highly specific association equilibrium.

When dealing with interactions between bodies that are rigid on the molecular level, it is also necessary to consider the molecularity of the solvent. Let us for simplicity use a perfectly planar hard surfaces as an example. The solvent density outside such a wall will vary in a periodic way with distance because of the molecularity of the solvent. On a purely packing basis, one can identify a first, a second, a third, and so on solvation layer. When two such surfaces approach, the oscillations in density will generate an oscillation in the force as solvation layers are successively removed (48). Particularly significant is the removal of the last solvation layer, because the barrier can be rather high. Conversely, separating two such surfaces that have established contact can be difficult because of the barrier involved in bringing in a first solvation layer. In practice, this packing effect is just one of several contributions to the total free energy and it can happen that the global minimum is with the first solvation layer intact. This is, for example, seen in clathrates, which are crystals formed by hydrophobic solutes with a retained first solvation layer of water (49).

XII. MOLECULAR RECOGNITION

The detailed function of a biological system is fundamentally based on molecular recognition processes. This holds for a range of phenomena such as replication, transformations, assembly, and regulation. In these cases the appropriate molecular pair is formed out of a large selection of other possibilities by intermolecular interactions. The present chapter is not aimed at analyzing this central problem, but the knowledge presented up to now could serve as a very useful background in the analysis of the recognition problem. There are two aspects that we will address here.

The first aspect is that the specific interactions occur in an environment determined by the general colloidal interactions. Consider as an illustration of this concept the action of a repressor protein regulating a specific gene by selective binding to the DNA double helix. Such proteins tend to be positively charged, and this cases general electrostatic binding to the negatively charged DNA helix. The next stage in the binding then involves specific matching between groups on the protein and groups at the specific recognition site on the helix. Repressor proteins have very high binding constants, and in this picture we can see the total binding constant K_{ass} as a product

$$K_{ass} = K_{unsp}K_{spec} \tag{12.1}$$

of one term for the unspecific binding to the DNA helix and one specific term representing the free energy gain in going from the generally bound state to the specific one. Often the free energy changes are of the same magnitude in the two steps. It is thus essential for a quantitative understanding of the specific association to have a proper description of the more unspecific effects that are generally present.

The separation of the interaction in terms of an energetic and an entropic term can also be used to understand the basis of a specific recognition. For a system having large configurational freedom a specific interaction implies a selection of one or a few of these configurations that can interact specifically with the binding molecule. Thus the energy gain in binding is partly compensated by a corresponding decrease in entropy. The energy term could still be the largest, but it is difficult to achieve a very high binding constant and the system could bind a range of similar molecules with moderate strength. The high specificity is obtained preferentially with stiff conformationally locked units where a combination of the steric energetic effect and an attractive force an result in high binding constants.

REFERENCES

1. J Israelachvili. Intermolecular & Surface Forces. 2nd ed. New York: Academic Press, 1992.
2. DF Evans, H Wennerström. The Colloidal Domain. Where Physics, Chemistry, Biology and Technology Meet. New York: VCH, 1994.
3. DF Evans, H Wennerström. The Colloidal Domain. Where Physics, Chemistry, Biology and Technology Meet. 2nd ed. New York: Wiley-VCH, 1999.
4. H Margenau, NR Kestner. Theory of Intermolecular Forces. 2nd ed. Oxford: Pergamon, 1969.
5. See Ref. 3, Secs. 1.4 and 1.5.
6. K Shinoda. "Iceberg" formation and solubility. J Phys Chem 81:1300, 1977.
7. HS Frank, MW Evans. Free volume and entropy in condensed systems. III. Entropy in binary liquid mixtures; partial molal entropy in dilute solutions; structure and thermodynamics in aqueous electrolytes. J Chem Phys 13:507, 1945.
8. DF Evans. Self-organization of amphiphiles. Langmuir 4:6, 1988.
9. J Israelachvili, H Wennerström. Role of hydration and water structure in biological and colloidal interactions. Nature 379:219, 1996.
10. B Derjaguin. Untersuchungen über die Reibung and Adhäsion. IV. Theorie des Anhaften kleiner Teilchen. Kolloid Z 69:155, 1934.
11. JL Parker. Surface force measurements in surfactant systems. Prog Surf Sci 47:205, 1994.
12. H-J Butt, M Jaschke, W Ducker. Measuring surface forces in aqueous solution with the atomic force microscope. Bioelectrochem Bioenergetics 38:191, 1995.
13. A Noy, CD Frisbie, LF Rozsnyai, MS Wrighton, CM Lieber. Chemical force microscopy: Exploiting chemically-modified tips to quantify adhesion, friction and functional group distribution in molecular assemblies. J Am Chem Soc 117:7943, 1995.
14. A Ashkin, JM Dziedzic. Optical trapping of viruses and bacteria. Science 235:1517, 1987.
15. VA Parsegian, RP Rand, NL Fuller, DC Rau. Osmotic stress for direct measurement of intermolecular forces. Methods Enzymol 127:400, 1986.
16. J Lyklema, KJ Mysels. A study of double layer repulsion and van der Waals attraction in soap films. J Am Chem Soc 87:2539, 1965.
17. V Bergeron, Å Waltermo, PM Claesson. Disjoining pressure measurements for foam films stabilized by a nonionic sugar-based surfactant. Langmuir 12;1336, 1996.
18. RH Ottewill. Colloidal stability and instability: "Order disorder." Langmuir 5:4, 1989.
19. L Guldbrand, B. Jönsson, H Wennerström, P Linse. Electrical double layer forces. A Monte Carlo study. J Chem Phys 88:2221, 1984.
20. A Khan, B Jönsson, H Wennerström. Phase equilibria in the mixed sodium and calcium di-2-ethylsulphosuccinate aqueous system. An illustration of repulsive and attractive double layer forces, J Phys Chem 89:5180, 1985.
21. J Ståhlberg, B Jönsson, C Horváth. Theory for electrostatic interaction chromatography of proteins. Anal Chem 63:1867, 1991.
22. J Ståhlberg, B Jönsson, C Horváth. Combined effect of Coulombic and van der Waals interactions in the chromatography of proteins. Anal Chem 64:3118, 1992.
23. OG Berg, C Blomberg. Association kinetics with coupled diffusion. An extension to coiled-chain macromolecules applied to the LAC repressor-operator system. Biophys Chem 7:33, 1977.
24. C Tanford. The Hydrophobic Effect: Formation of Micelles and Biological Membranes. New York: Wiley, 1980.
25. J Israelachvili, RM Pashley. The hydrophobic interaction is long range, decaying exponentially with distance. Nature 300:341, 1982.

26. RM Pashley, PM McGuiggan, BW Ninham, DF Evans. Attractive forces between uncharged hydrophobic surfaces: Direct measurements in aqueous solution. Science 229:1088, 1985.

27. PM Claesson, CE Blom, PC Herder, BW Ninham. Interactions between water-stable hydrophobic Langmuir-Blodgett monolayers on mica. J Colloid Interface Sci 114:234, 1986.

28. JL Parker, PM Claesson. Forces between hydrophobic silanated glass surfaces. Langmuir 10:635, 1994.

29. Y-H Tsao, SX Yang, DF Evans, H Wennerström. Interactions between hydrophobic surfaces. Dependence on temperature and alkyl chain length. Langmuir 7:3154, 1991.

30. Y-Y Tsao, DF Evans, H Wennerström. Long range attraction between hydrophobic surfaces observed by atomic force microscopy. Science 262:547, 1993.

31. JL Parker, PM Claesson, P Attard. Bubbles, cavities and the long-ranged attraction between hydrophobic surfaces. J Phys Chem 98:8468, 1994.

32. J Forsman, B Jönsson, CE Woodward, H Wennerström. Attractive surface forces due to liquid density depression. J Phys Chem B 101:4253, 1997.

33. J Israelachvili, H Wennerström. Entropic forces between amphiphilic surfaces in liquids. J Phys Chem 96:520, 1992.

34. DH Napper. Polymeric Stabilization of Colloidal Dispersions. New York: Academic Press, 1983.

35. J Hilkens, MJ Ligtenberg, HL Vos, SV Litviniov. Cell membrane–associated micins and their adhesion-modulating property. Trends Biochem Sci 17:359, 1992.

36. P Walstra, R Jenness. Dairy Chemistry and Physics. New York: Wiley, 1984, Chap. 13.

37. DM LeNeveu, RP Rand, VA Parsegian, D Gingell. Measurement and modification of forces between lecithin bilayers. Biophys J 18:209, 1977.

38. RP Rand, VA Parsegian. Hydration forces between phospholipid bilayers. Biochim Biophys Acta 988:351, 1989.

39. J Israelachvili, H Wennerström. Hydration or steric forces between amphiphilic surfaces? Langmuir 6:873, 1990.

40. W Helfrich. Steric interaction of fluid membranes in multilayer systems. Z Naturforsch 33:305, 1978.

41. H Bagger-Jörgensen, U Olsson. Experimental study of undulation forces in a nonionic lamellar phase. Langmuir 12:4057, 1996.

42. S Asakura, F Oosawa. On interaction between two bodies immersed in a solution of macromolecules. J Chem Phys 22:1255, 1954.

43. A Vrij. Polymers at interfaces and the interactions in colloidal dispersions. Pure Appl Chem 48:471, 1976.

44. See Ref. 3, Sec. 5.6.

45. P Petrov, U Olsson, H Wennerström. Surface forces in bicontinuous microemulsions: Water capillary condensation and lamellae formation. Langmuir 13:3331, 1997.

46. H Wennerström, K Thuresson, P Linse, E Freyssingeas. Long range attractive surface forces due to capillary induced polymer-polymer phase separation. Langmuir, 14:5664, 1998.

47. R Fåhreus. The suspension stability of the blood. Physiol Rev 9:241, 1929.

48. See Ref. 1, Chap. 13.

49. DW Davidson. Clathrate hydrates. In: F Franks, ed. Water: A Comprehensive Treatise. Vol 2. New York: Plenum, 1973, Chap. 3.

50. J Klein, PF Luckham. Forces between two adsorbed polyethylene oxide layers immersed in a good aqueous solvent. Nature 300:429, 1982.

4

Proteins at Solid Surfaces

W. Norde
Wageningen University, Wageningen, The Netherlands

I. INTRODUCTION

Protein adsorption is a widespread event occurring at biological interfaces. Wherever and whenever a protein-containing (aqueous) solution is exposed to a (solid) surface, it results in the spontaneous accumulation of protein molecules at the solid-water interface, thereby altering the characteristics of the sorbent surface and, in most cases, also of the protein molecules. It is, therefore, not surprising that the interaction between proteins and surfaces has attracted attention from various disciplines, ranging from soil and food science to biotechnology and (biomedical) materials science. Apart from its impact in these practical domains, the interaction between proteins and interfaces is of academic interest. When a protein molecule adsorbs from solution at an interface it changes its environment, which, more often than not, is accompanied by structural rearrangements. Hence, studying the interaction between proteins and interfaces may contribute to the understanding of the mechanism that determines the three-dimensional (3-D) structure of protein molecules.

Knowledge of the adsorption behavior of proteins has largely progressed over the past few decades but a unified predictive theory is still lacking; at best a number of principles that are common in most protein-interface systems can be indicated.

Proteins are biopolymers, and it may therefore be useful to discuss their adsorption behavior starting from the general trends observed for the adsorption of more simple polymers, in particular water-soluble flexible polyelectrolytes.

Theoretical aspects of polymer adsorption (that have been confirmed experimentally) are briefly considered in Chap. 1 of this book. The main features may be summarized as follows:

1. Adsorption of a flexible polymer molecule results in reduction of its con-
    formational entropy. Therefore, if adsorption occurs, the loss in con-
    formational entropy is compensated by sufficient attractive interaction
    between segments of the polymer and the surface. Even if this attractive
    interaction is only slightly higher than the energy of thermal motion (one
    kT unit for a segment moving in two directions at a sorbent surface), the
    whole polymer molecule adsorbs with a high affinity. This is because the
    contribution from each adsorbing segment adds to the Gibbs energy of
    adsorption of the whole polymer molecule.
2. A flexible polymer molecule typically adsorbs in a train-loop-tail–like con-
    formation (see Fig. 7 of Chap. 1). The extension of the loops and, to a lesser
    extent, the tails is determined by the solubility of the polymer in the solvent.
    A high loop density in the adsorbed layer is tolerated only in the case of
    a poor solvent.
3. Generally, because of their ionic groups, polyelectrolytes are quite soluble in
    water and therefore the formation of loops in the adsorbed layer is strongly
    suppressed; they adsorb in a relatively flat conformation.
4. Depending on the charge contrast between the polyelectrolyte and the
    sorbent surface, their electrostatic interaction is attractive or repulsive. If
    repulsive, the polyelectrolyte still adsorbs if the nonelectrostatic contribution
    to the Gibbs energy of adsorption (over)compensates the electrostatic one.
    Consequently, polyelectrolytes containing apolar groups may readily adsorb
    under electrostatically adverse conditions.
5. Unlike the case of uncharged polymers, the adsorption of polyelectrolytes is
    strongly influenced by the presence of low-molecular-weight ions. Because
    they screen electrostatic interactions, ions reduce the quality of water as
    a solvent for polyelectrolytes so that the formation of loops in adsorbed
    layers of polyelectrolytes is promoted. Furthermore, the ions weaken the
    electrostatic interaction between a polyelectrolyte and a charged sorbent
    surface, leading to a stronger or weaker adsorptive bond, depending on
    the charge constrast between the interacting components. The pH,
    determining the charge of the polyelectrolyte and, often, the sorbent surface,
    influences the adsorption process in a similar way. Thus, at a pH at which the
    polyelectrolyte is nearly uncharged, it adsorbs in a relatively thick loopy layer
    that is not or only slightly dependent on ionic strength. At a pH at which the
    polyelectrolyte is (fully) charged, it adsorbs in a flat conformation that
    becomes more loopy with increasing ionic strength.
6. According to these principles, polyampholytes, i.e., polyelectrolytes that
    contain both anionic and cationic groups, show maximum adsorption at
    the isoelectric point, and the reduced adsorption at either side of the
    isoelectric point becomes less pronounced with increasing ionic strength.

Proteins are copolymers of some 22 different amino acids linked together in a linear
polypeptide chain. A number of amino acid residues along the polypeptide contain an
anionic or cationic group. This makes the protein a polyampholyte. The various amino
acid side groups differ greatly in polarity, so the protein is amphiphilic. The sequence
of the amino acids in the polypeptide chain ultimately determines the folded, spatial
architecture, i.e., the 3-D structure of the protein molecule. The 3-D structure is the

net result of interactions between segments within the protein molecule but also between segments of the protein and the environment, which is usually an aqueous medium. Based on the 3-D structure, the following division may be made:

1.   Protein molecules that are highly solvated and flexible, resulting in a disordered coil-like structure. This group comprises some proteins whose natural function is nutritional, such as glutens in wheat grains and caseins in milk. The adsorption characteristics of such proteins are more or less similar to those of the flexible polyampholytes mentioned earlier.
2.   Protein molecules that have a very regular structure (e.g. helices, pleated sheets), the so-called fibrous proteins. These proteins are usually insoluble in water; they are mainly found in muscle and connective tissues.
3.   By far the greatest proportion of the protein species (but only a small fraction of the protein mass on earth) contains different structural elements, i.e., α-helices, β-pleated sheets, and parts that are unordered, which are folded together into a compact dense globule: the globular proteins. In an aqueous environment the apolar amino acid residues are mainly located in the interior of the molecule and the polar residues are found primarily at the periphery of the molecule. The globular proteins have evolved to fulfill specific functions, such as biocatalysis (enzymes) and immunologic reactions (antibodies). An almost countless number of different kinds of globular proteins exist, each kind having its own specific biological function related to its own characteristic 3-D structure.

With respect to practical applications, the adsorption of globular proteins is most relevant. Examples can be found in biomedical engineering, biosensors, immobilized enzymes in bioreactors, immunological diagnostic tests, drug targeting and drug delivery systems, stabilizing agents of dispersions in foodstuffs, pharmaceutics and cosmetics, wastewater and soil treatments, etc. Therefore, this chapter focuses on the adsorption of globular proteins. It may be understood that, because of the complicated 3-D structures that globular proteins molecules adopt, their interaction with interfaces is complicated as well. The occurrence of structural rearrangements and the resulting effects on biological activity are among the most intriguing and challenging aspects of protein adsorption. Therefore, before treating protein adsorption in more detail, the major factors determining the 3-D structure of a globular protein molecule will be briefly discussed.

II.   THE THREE-DIMENSIONAL STRUCTURE OF GLOBULAR PROTEINS

Folding of a polypeptide into a compact globular structure occurs at the expense of the conformational entropy of the polypeptide chain. In particular, hydrogen bonds formed between the peptide units in α-helices and β-sheets largely reduce the rotational freedom of the bonds in the polypeptide chain. For instance, the folding of a polypeptide of molar mass 10,000 Da into a compact structure with 50% ordered (α-helix and/or β-sheet) structure involves a loss of conformational entropy of several hundreds of $J\ K^{-1}$, which, at room temperature, corresponds to a Gibbs energy increase of a few hundreds of kJ per mol (1). Hence, a stable globular structure exists only if this entropy loss is compensated by favorable interactions within the protein

molecule and/or between the protein molecule and its environment. The major interactions for a protein molecule in an aqueous medium are indicated next.

Hydrophobic interaction. The favorable dehydration of the apolar amino acid
   residues promotes association of these residues in the interior of the globular
   protein molecule, thereby contributing to the stability of a compact structure.
   This contribution is estimated to amount to 9.2 kJ mol^{-1} per nm^2 reduction
   of water-accessible surface area (2). Hence, folding of a completely hydrated
   polypeptide of molar mass 10,000 Da into a compact globular structure
   of which the interior is 60% apolar residues lowers the Gibbs energy at room
   temperature by 500 kJ per mol.

Coulomb interaction. The majority of the ionic groups of a protein reside at the
   aqueous exterior of the molecule. When these charged groups are more or
   less homogeneously distributed their interactions may or may not stabilize
   a compact globular structure, depending on the pH relative to the isoelectric
   point of the protein. Under isoelectric conditions, when the protein has
   as many positive as negative charges, the overall intramolecular Coulombic
   interaction is attractive and therefore favors a compact conformation. Away
   from the isoelectric point the excess of either positive or negative charge
   results in intramolecular repulsion, which promotes a more expanded
   structure. Moreover, ionization of residues that are in the nonionized form
   buried in the low-dielectric interior of the compact protein (e.g., histidine
   and tyrosine) stimulates the protein to unfold (3). When ionic groups are present in the interior of a compact protein molecule they usually occur as
   ion pairs. Disruption of such electrostatically favorable bonds, as would occur
   when the protein unfolds, is more or less compensated by hydration of the two
   ionic groups in the unfolded state (4).

Lifshitz – van der Waals interaction. These are interactions between fixed and/or
   induced dipoles. They are highly sensitive to the separation distance between
   the dipoles. When a highly hydrated polypeptide chain folds into a compact
   globular structure, dipolar interactions between the polypeptide and water
   are broken but new dipolar interactions inside the protein molecule and
   between water molecules are formed. The net effect of dipolar interactions
   on the stability of the protein structure is difficult to estimate (it would require
   accurate knowledge of the exact positions of the dipolar groups in the protein
   molecule), but it is generally assumed that because of the relatively high volume density in globular proteins dipolar interactions slightly favor a compact
   structure (5).

Hydrogen bonding. As with dipolar interactions, folding of the polypeptide chain
   implies a loss of hydrogen bonds between peptide units and water molecules
   on the one hand and the formation of intramolecular hydrogen bonds between
   peptide units (and possibly other groups) and among water molecules on the
   other hand. In the nonaqueous interior of a compact protein molecule,
   hydrogen bonds between peptide units stabilize secondary structures as
   α-helices and β-sheets. However, because of the compensating effects due
   to loss and creation of hydrogen bonds, it is not clear whether hydrogen
   bonding promotes a compact or an unfolded structure (6, 7).

Under not too extreme conditions of pH, temperature, etc., hydrophobic inter-action is the major factor counteracting the loss of conformational entropy upon folding the polypeptide in an aqueous environment. Thus, a stable, low-entropy 3-D structure of the protein is achieved by virtue of entropy gain of water molecules that are released from contact with apolar amino acid residues. Because of these opposite effects that are of comparable magnitude, the folded compact structure is only marginally supported with a Gibbs energy of stabilization of, typically, a few tens of kJ per mole of protein. Hence, it should be realized that the other factors (i.e., Coulomb interaction, Lifshitz–van der Waals interaction, hydrogen bonding) may be decisive as to the structure the polypeptide chain adopts. As a consequence, even mild changes in the environment, such as changes in the temperature, pH, ionic strength, addition of other solutes, and, exposure to an interface, may induce structural rearrangements in the protein molecule.

III. THE ADSORPTION PROCESS

Protein adsorption comprises various aspects: kinetics, type of the binding, adsorbed amount, and structure of the adsorbed layer and of the individual molecules therein. Figure 1 schematically depicts the various steps through which an adsorbing and desorbing protein molecule passes: transport toward the sorbent surface (1), deposition at the surface (2), relaxation of the adsorbed molecule (3), detachment from the surface (4), transport away from the surface (5), and possible restructuring of the desorbed protein molecule (6). The asterisks indicate the degree of relaxation of the adsorbed molecule. Each of these steps will be considered in the following discussion.

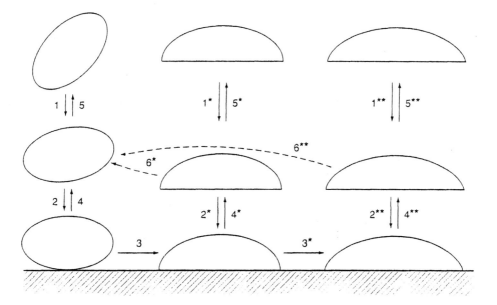

Fig. 1   Schematic presentation of the protein adsorption process. For an explanation of the numbers see the text.

A. Transport Toward the Surface

The basic mechanisms of protein transport toward the sorbent surface are diffusion and convection by laminar or turbulent flow. In the absence of convection, the transfer of the protein molecules is a stochastic process. As adsorption proceeds the solution near the sorbent surface becomes progressively depleted, so that for the flux J toward a smooth sorbent surface

$$J = (c_b - c_s)\left(\frac{D}{\pi t}\right)^{1/2} \tag{1}$$

where c_b and c_s are the protein concentrations in the bulk solution and at the surface, respectively, D is the diffusion coefficient, and t is the incubation time.

However, most transport processes take place under steady-state convective diffusion driven by a fixed concentration gradient. This results in

$$J = k_1(c_b - c_s) \tag{2}$$

where k_1 is a transport rate constant that depends on the hydrodynamic conditions and the diffusion coefficient. Expressions for k_1 in different geometries have been given by Adamczyk et al. (8).

B. Deposition at the Sorbent Surface

Deposition of the protein at the sorbent surface may be considered as a first-order process

$$\frac{d\Gamma}{dt} = k_2 c_s \tag{3}$$

where Γ is the adsorbed mass per unit sorbent surface area and k_2 is the deposition rate constant. The value of k_2 decreases with increasing coverage of the surface by the protein (i.e., with decreasing space available for the arriving protein to adsorb) and k_2 is also lowered by any repulsive barrier for attachment. The origin of such a barrier might be electrostatic repulsion (9, 10), a hydration effect (11), or the fact that a fraction of molecular collisions with the surface do not lead to attachment (9).

As the surface coverage increases, the rate of detachment increases. Eventually, equilibrium is reached, which implies that

$$\frac{d\Gamma}{dt} = k_2(c_s - c_{eq}) \tag{4}$$

where c_{eq} is the concentration in bulk solution corresponding to the equilibrium value for Γ, as given by the adsorption isotherm (see Sec. III.D). When the sorbent surface acts as a perfect sink, i.e., when $d\Gamma/dt = J$ (that is, at low surface coverage and in the absence of a deposition barrier), combining Eqs. (2), (3), and (4) gives

$$\frac{d\Gamma}{dt} = \frac{c_b - c_{eq}(\Gamma)}{(1/k_1) + (1/k_2)} \tag{5}$$

To obtain an explicit expression for $d\Gamma/dt$ the adsorption isotherm $c_{eq}(\Gamma)$ must be known (see Secs. III.D and E).

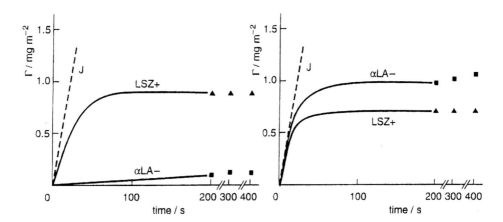

Fig. 2 Adsorption of positively charged lysozyme, LSZ+, and negatively charged α-lactalbumin, αLA−, on surfaces of negatively charged polar silica (left) and negatively charged apolar polystyrene (right). The dashed line indicates the flux of the proteins toward the sorbent surface. (Data taken from Ref. 9.)

Figure 2 shows experimental results for the rate of adsorption of lysozyme and α-lactalbumin. For the polar surface $d\Gamma/dt$ reflects electrostatic interaction between the protein and the sorbent: $d\Gamma/dt$ largely decreases under electrostatically repulsive conditions. Even when the interaction is attractive the rate of adsorption is significantly lower than the flux toward the surface, indicating some (nonelectrostatic) adsorption barrier. With the apolar surface the rates of adsorption are higher and, remarkably, the influence of electrostatic interaction is masked. Apparently, a factor or factors other than electrostatic ones, probably dehydration of the apolar surface and/or structural rearrangements in the adsorbing protein molecules, dominate the adsorption process.

C. Relaxation of the Adsorbed Layer

Once the protein molecule has attached, it relaxes toward its equilibrium structure, which, because of the altered environment is as a rule different from the (native) structure in solution. Structural relaxation is represented by the steps 3 in Fig. 1. Relaxation becomes more difficult as the protein-sorbent interaction is stronger because nonequilibrium states tend to become quenched. Furthermore, relaxation is retarded as the protein molecule has a strong internal coherence. Structural relaxation implies optimization of protein-surface interaction and it normally involves a certain degree of "spreading" of the protein molecule over the sorbent surface, developing a larger number of protein-surface contacts. As a consequence, after relaxation it becomes more difficult for the protein to detach from the surface. This may lead to a structural heterogeneity in the adsorbed layer (12, 13): molecules arriving at an early stage of the adsorption process find sufficient area available for spreading, whereas this is not the case for the molecules that arrive when the surface is already (partially) covered with protein. Another consequence is that the outcome of the adsorption process depends on the rate of attachment and the rate of spreading relative to each other. When spreading occurs relatively quickly, the adsorbed molecules are more flattened.

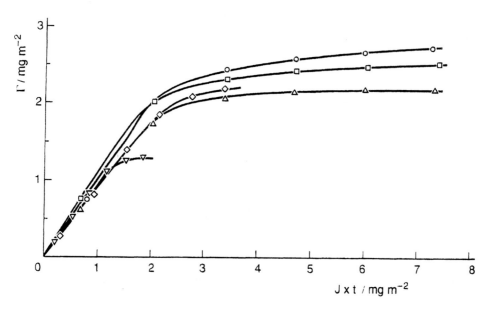

Fig. 3 The effect of the protein concentration in solution, c_b, on the adsorption of IgG. c_b (mg dm^{-3}): ($\bigcirc$) 40; ($\square$) 25; ($\diamondsuit$) 7.5; ($\triangle$) 5; ($\triangledown$) 1.

If, however, the protein flux to the surface increases, the adsorbed molecules retain a more globular conformation and, therefore, the adsorbed mass per unit surface area is higher. Hence, when the rates of supply and spreading are of the same order, saturation adsorption increases with increasing flux. Figure 3, showing the adsorption kinetics for immunoglobulin G (IgG) on a silica surface, illustrates this phenomenon. In this figure Γ is plotted against $J \times t$. As long as the adsorption is controlled by the transport toward the surface, i.e., at low values for $J \times t$, the slopes of the curves coincide. Adsorption saturation increases with increasing flux and by considering the conditions in which the curves deviate from unity, a relaxation time in the range of about 1000 s is estimated for the spreading of IgG (14).

D. Adsorption Isotherm

From a compilation of $\Gamma(t \rightarrow \infty)$ obtained at various supply rates the adsorption isotherm, $\Gamma(c_{eq})$, can be derived. See Fig. 4. Adsorption isotherms for globular proteins usually display well-defined plateau values that are reached at c_{eq} typically less than a few tenths of a g dm^{-3}. Usually, the plateau adsorption is lower than or, maximally, comparable to the adsorbed amount in a closely packed monolayer of more or less native-like molecules. In view of the discussion in Sec. III.C, it will be understood that $\Gamma(c_{eq})$ depends on the mode of supply, more specifically, on the relaxation rate relative to the supply rate (the latter being higher at higher c_{eq}). Indeed, using various experimental techniques it has been concluded that the degree of structural rearrangements in adsorbed proteins decreases with increasing c_{eq} (13, 15–17). Hence, as at a given Γ value c_{eq} may depend on the history (i.e., the supply mode), $\Gamma(c_{eq})$ may not reflect true thermodynamic equilibrium.

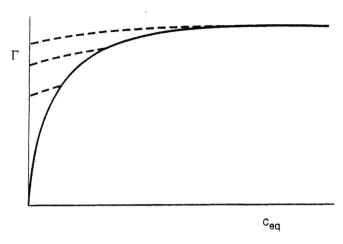

Fig. 4 Typical shapes of protein adsorption (————) and desorption (– – –) isotherms.

Just like other polymers, protein molecules attach via multiple contacts to a sorbent surface and, as explained in Chapter 1, one would therefore expect high-affinity adsorption, i.e., merging of the initial part of the isotherm with the Γ axis. Such isotherms are indeed often observed, but isotherms showing a more gentle slope are not exceptional. It is remarkable that even in the case of non–high-affinity adsorption the desorption isotherm (dashed curve in Fig. 4) usually has a high-affinity character. Such a hysteresis between adsorption and desorption is a manifestation of (apparent) irreversibility of the protein adsorption process. This feature will be dealt with further in Sec. III.F.

E. Adsorption and Desorption Rates

When the adsorption isotherm, $\Gamma(c_{eq})$, is available, it can be inverted in $c_{eq}(\Gamma)$ and this may be inserted in Eq. (5) to give an explicit expression for $d\Gamma/dt$. For the rising part of the isotherm c_b is considerably larger than c_{eq}, resulting in a relatively high rate of adsorption, but as soon as the plateau value is reached c_{eq} rapidly approaches c_b and $d\Gamma/dt$ becomes very small. In the case of a high-affinity isotherm c_{eq} is almost zero up to reaching the $\Gamma(c_{eq})$ plateau value. Hence, desorption into the pure buffer ($c_b = 0$) proceeds too slowly to be detected. This is consistent with the experimental observation that adsorbed proteins (and other polymers) cannot be removed from surfaces by simply rinsing with the solvent. In the case of hysteresis both the adsorption and the desorption isotherms reflect (meta-)stable states. Estimation of adsorption and desorption rates should then be based on the corresponding isotherms.

F. Reversibility of the Adsorption Process

Phenomenologically, a system is in equilibrium if no changes take place at constant surroundings. At constant pressure p and temperature T the equilibrium state of a system is characterized by a minimum value of the total Gibbs energy G. Any other state, away from this minimum, is in nonequilibrium and there will be a spontaneous transition toward the equilibrium state provided that the Gibbs energy barriers along this transition are not prohibitively large. By definition, a process is reversible if during the whole trajectory of the process the departure from equilibrium is infinitesimally small, so that in the reverse process the variables characterizing the state of the system return through the same values but in the reverse order. Because a finite amount of time is required for the system to relax to its equilibrium state, investigating the reversibility of a process requires that the time of observation exceeds the relaxation time.

If both the adsorption and desorption isotherms are of the high-affinity type it is difficult, if not impossible, to verify reversibility, because, apart from the slowness of the desorption process (as discussed in Sec. III.E), it would require a method for determining the protein concentration in an almost infinitely diluted solution. However, as mentioned before, diluting a non–high-affinity isotherm usually does not lead to detectable desorption; in other words, $\Gamma(c)$ rarely, if ever, follows the same path backward. The deviation between adsorption and desorption remains even when the observation time is extended to several days and is therefore much longer than the relaxation time of the protein at the surface (14, 18–20). The occurrence of such hysteresis indicates that at a given protein concentration in solution two metastable states exist, one on the adsorption isotherm and one on the desorption isotherm, each being characterized by local minima of G, which are separated by a Gibbs energy barrier that prevents the transition from the one state into the other. The fact that the adsorption and desorption isotherms represent different metastable states implies that between adsorption and desorption a physical change has occurred in the system that may not be restored after desorption. Because of the poor desorbability of proteins upon dilution, it is virtually impossible to investigate desorbed molecules. However, adsorbed proteins may readily be exchanged against other surface-active components of the same or of another kind (16, 21–23). Permanent structural changes are indeed established in such exchanged molecules (10, 15, 22).

IV. DRIVING FORCES FOR PROTEIN ADSORPTION

The tendency of proteins to accumulate at interfaces is determined not only by properties of the protein molecules and the sorbent surface but also by the nature of the solvent, the presence of other solutes, pH, ionic strength, and temperature. Whatever the mechanism, at constant temperature T and pressure p adsorption proceeds spontaneously if the Gibbs energy G of the system decreases.

$$\Delta_{ads}G = \Delta_{ads}H - T\Delta_{ads}S < 0 \tag{6}$$

where H and S are the enthalpy and the entropy of the system and where Δ_{ads} indicates the change invoked by the adsorption process. The more negative the value of $\Delta_{ads}G$, the higher the adsorption affinity is.

One way to analyze the overall adsorption process is to establish how various interactions contribute to $\Delta_{ads}G$. Of course, such a thermodynamic approach can be successful only if the system is well characterized. Therefore, the forthcoming discussion will be restricted to systems composed of one type of protein and one type of sorbent in a well-defined environment.

A. Interaction Between Electrical Double Layers

Generally, both the protein molecule and the sorbent surface are electrically charged. In an aqueous medium charged sorbents and proteins are surrounded by counterions and coions, together accounting for the so-called countercharge that neutralizes the surface charge. A fraction of the countercharge may be bound to the sorbent surface and/or the protein molecule and the other part is diffusely distributed in the solution. The surface charge and the countercharge together form the electrical double layer.

The Gibbs energy G_{el} to invoke a charge distribution can be calculated as the isothermal, isobaric reversible work (24)

$$G_{el} = \int_0^{\sigma_0} \phi_0' d\sigma_0' \qquad (7)$$

where ϕ_0' and σ_0' are the variable surface potential and surface charge density, respectively, during the charging process. Solving this equation requires knowledge of $\phi_0'(\sigma_0')$ and this functionality can be derived from models for the electrical double layer. For the bare sorbent surface the Gouy-Stern model may be adopted (25). For the dissolved protein molecule a discrete-charge model, as developed by Kirkwood (26), may be more appropriate and for the protein-covered sorbent surface different models have been proposed (27–31). Charge distributions for the system before and after adsorption are schematically depicted in Fig. 5.

Under most (practical) conditions the Debye length, i.e., the separation distance over which charges interact (25), is considerably smaller than the thickness of the adsorbed protein layer. For instance, in a solution of 0.1 M ionic strength the Debye length is about 1 nm, whereas the thickness of the adsorbed protein layer in which the molecules retain a compact conformation (see Sec. IV.D) is at least a few nanometers. Hence, such a compact protein layer shields the protein-sorbent contact region from electrostatic interaction with the solution so that the net charge density in that contact region is essentially zero. Using phenomenological (thermodynamic) linkage relations, this charge regulation can be derived from the electrolyte depen-

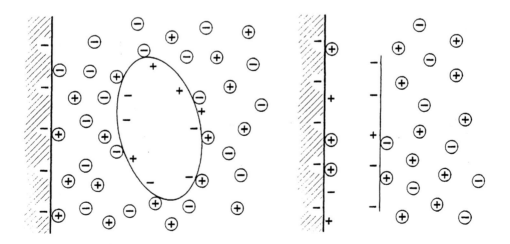

Fig. 5 Schematic representation of charge distributions before (left) and after (right) protein adsorption. The charge on the sorbent surface and the protein molecule are indicated by $+/-$. The low-molecular-weight electrolyte ions are indicated by $\oplus/\ominus$.

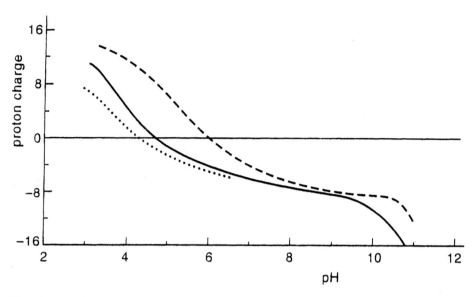

Fig. 6 Protein charge (net number of charged groups per protein molecule) as a function of pH for α-lactalbumin in solution (————) and adsorbed on a negatively (– – –) and a positively (·····) charged surface. (Data taken from Ref. 35.)

dence of the protein adsorption isotherm (32). Charge neutralization has been confirmed experimentally by the shift in proton charge of the protein upon adsorption at a charged surface (33–35). Examples are shown in Fig. 6.

Adjustment of the sorbent surface charge may occur as well, as has been demonstrated for, e.g., silver halide and clay particles (36, 37). Apart from adjustments in the protein and the sorbent, charge neutralization may be further regulated by the incorporation of indifferent ions from solution into the protein-sorbent contact region. This has been demonstrated by electrokinetic data (38, 39) and, more directly, by tracing radiolabeled ions (40). Trends, derived from electrokinetics, are shown in Fig. 7; they clearly follow the charge antagonism between the protein molecules in solution and the bare sorbent surface.

As a consequence of the charge regulation, $\Delta_{ads} G_{el}$ is not very sensitive to the charge densities of the protein and the sorbent and it usually does not exceed a few tens of RT per mole of adsorbing protein. Its sign and value depend on the charge distributions and the dielectric constants of the electrical double layers before and after adsorption, respectively (31, 39).

It should be realized that the transfer of ions from the aqueous solution into the nonaqueous protein-sorbent environment is chemically unfavorable. In other words, the chemical effect of ion incorporation opposes the overall adsorption process. This explains why maximum protein adsorption affinity is reached when the charge density on the protein just matches that on the sorbent surface so that no additional ions have to be incorporated (36).

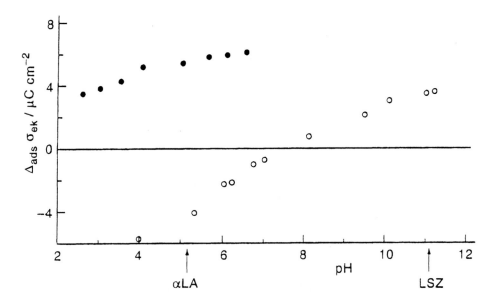

Fig. 7 Variation in the electrokinetic charge density resulting from plateau adsorption of lysozyme (○) and α-lactalbumin (●) on negatively charged polystyrene particles. The arrows indicate the isoelectric points of the two proteins. (Data taken from Ref. 35.)

B. Changes in the State of Hydration

In liquid water of not too high temperature, say <80°C, the water molecules are strongly hydrogen bonded, causing strong internal coherence.

Polar groups interact favorably with water molecules, mainly through electrostatic interaction, including hydrogen bonding. These interactions overcompensate the strong cohesion between the water molecules, rendering the polar components readily soluble in water. Apolar groups, which do not offer the possibility for such favorable interactions with water, are expelled from an aqueous environment. This mechanism is at the basis of hydrophobic interaction. A detailed discussion of water as a solvent and the hydrophobic interaction is given by Wennerström in Chapter 3 of this book.

If the surfaces of the protein molecule and the sorbent are polar, their hydration is favorable. In that case it is probable that some hydration water is retained between the sorbent surface and the adsorbed protein molecule. However, if (one of) the surfaces are (is) apolar, dehydration would be a driving force for adsorption.

Although the apolar residues of a globular protein in water tend to be buried in the interior of the molecule, the water-accessible surface of the protein may still constitute a significant apolar fraction, even up to 40–50% (41). For water-soluble, nonaggregating proteins the apolar residues are more or less evenly distributed over the exterior of the molecules. The presence of pronounced apolar patches may lead to protein aggregation.

In several studies (42–44) the influence of protein polarity on adsorption has been experimentally confirmed. In this context it should be realized that apart from the polarity of the outer shell the overall polarity of the protein could be relevant for its adsorption behavior. The overall polarity influences the protein structural stability (see section II) and, hence, the extent of structural perturbation upon adsorption. This, in turn, affects the adsorption affinity, as will be discussed in Sec. IV.D.

The effect of sorbent surface polarity is also difficult to assess, because modifying the polarity usually involves a change in the surface electrostatic potential as well. A fair estimate of the contribution from hydration changes to the Gibbs energy of adsorption, $\Delta_{ads}G_{hydr}$, may be inferred from partitioning (model) components in water-nonaqueous two-phase systems (45, 46). It has thus been estimated that, at room temperature, dehydration of apolar surfaces involves a lowering of the Gibbs energy of about 10–20 mJ m^{-2}. For a protein of molar mass 15,000 Da that adsorbs to about 1 mg m^{-2} it results in $\Delta_{ads}G_{hydr}$ ranging between -60 and $-120RT$ per mole of adsorbed protein. It is obvious that apolar hydration dominates over the effects from overlapping electrical double layers and dispersion interaction (see Secs. IV.A and C).

C. Dispersion Interaction

Dispersion (or London–van der Waals) interaction results from attraction between electron "clouds" that are in close proximity but do not overlap. This interaction is attractive. For a sphere interacting with a planar surface the contribution from dispersion interaction to the Gibbs energy of adsorption, $\Delta_{ads}G_{disp}$, can be approximated by (47)

$$\Delta_{ads}G_{disp} = -\frac{A_{132}}{6}\left(\frac{a}{h} + \frac{a}{h+2a} + \ln\frac{h}{h+2a}\right)$$ (8)

where A_{132} is the Hamaker constant for the interaction between the flat sorbent (1) and the spherical protein molecule (2) across the (aqueous) medium (3), a the radius of the sphere, and h the distance of closest approach between the sphere and the surface. Under most conditions $h \ll a$ so that Eq. (8) simplifies to

$$\Delta_{ads}G_{disp} = -\frac{A_{132}a}{6h}$$ (9)

Values for Hamaker constants (of the individual components) are given in several references (e.g., 5, 25, 48, 49). The Hamaker constant for the system can be derived from the individual ones according to the following rules.

$$A_{132} = \left(A_1^{1/2} - A_3^{1/2}\right)\left(A_2^{1/2} - A_3^{1/2}\right)$$ (10)

and

$$A_{132} = (A_{131}A_{232})^{1/2}$$ (11)

In aqueous media usually $A_1 > A_3$ and $A_2 > A_3$, so that, according to Eq. (10), $A_{132} > 0$ and, hence, $\Delta_{ads}G_{disp} < 0$, which implies attraction. The Hamaker constant for interaction across water is about 6.6×10^{-21} J for globular proteins (5), $(1–3) \times 10^{-19}$ J for metals (25), and $(4–12) \times 10^{-21}$ J for synthetic polymers such as polystyrene or Teflon (25). According to Eq. (9), $\Delta_{ads}G_{disp}$ varies proportionally with the dimensions of the protein molecule and it drops off hyperbolically with increasing distance between the protein and the sorbent surface. Thus, for a globular protein molecule of 3 nm radius at 0.15 nm distance from the surface $\Delta_{ads}G_{disp}$ amounts to $-(1-3)RT$ per mole at a synthetic polymer surface and to $-(4-7)RT$ at a metal surface. Because of the various approximations involved, these values are only indicative. Deriving more accurate values is practically impossible because of the irregular shape protein molecules and, possibly, sorbent surfaces adopt in real systems. Moreover, rearrangements in the protein structure induced by adsorption may affect the Hamaker constant in an unknown way.

D. Rearrangements in the Protein Structure

As discussed in Sec. II, the three-dimensional structure of a globular protein molecule is only marginally stable, so interaction with a sorbent surface may induce changes in that structure. However, as compared with flexible polymers the conformational changes in adsorbing protein molecules are usually small. It is generally observed that the thickness of a monolayer of adsorbed protein, determined, for example, by ellipsometry, light scattering, viscometry, scanning probe microscopy, or the surface force technique, is comparable to the dimensions of the native protein molecule (50–52). Structural rearrangements do not lead to unfolding into a loose, highly hydrated "loop-and-tail" structure. After adsorption, at one side of the protein molecule the aqueous environment is replaced by the sorbent material. As a consequence, *intra*molecular hydrophobic interaction becomes less important as a structure-stabilizing factor; i.e., apolar parts of the protein that are buried in the interior of the dissolved molecule may become exposed to the sorbent surface without making contact with water. Hydrophobic interactions between amino acid residues support the formation of α-helices and β-sheets; hence, a reduction of these interactions tends to destabilize such secondary structures. A reduction of the α-helix and/or β-sheet content is indeed expected to occur if the peptide units released from the helices and sheets can form hydrogen bonds with the sorbent surface, as is the case at polar surfaces. Then a decrease in ordered (secondary) structure would result in an increased conformational entropy of the protein (see Sec. II) and, hence, an increased adsorption affinity. The contribution from increased conformational entropy to a negative value for $\Delta_{ads}G$ may amount to some tens of RT per mole of protein (13, 15, 17, 53). However, adsorption at an apolar, non–hydrogen-bonding surface may stimulate intramolecular peptide-peptide hydrogen bonding, resulting in increased order in the protein's structure. Whether or not extra hydrogen bonding within the protein molecule occurs depends on the outcome of the opposing effects of the energetically favorable hydrogen bonds and the unfavorable change in the conformation entropy (13, 54).

E. Sorbent Surface Morphology

Many surfaces are not completely smooth and rigid. For instance, at surfaces of polymeric materials polymer chains may protrude into the solution to some extent. At natural surfaces such as those of, e.g., biological membranes and bacterial cell walls, natural polymers such as proteins and/or polysaccharides are often present. When these surface polymers extend into the surrounding medium with some flexibility, the surface will respond dynamically to protein adsorption. On the one hand, this would offer the possibility of optimizing contact by confirming to the shape of the adsorbing protein molecule. On the other hand, squeezing the surface polymers

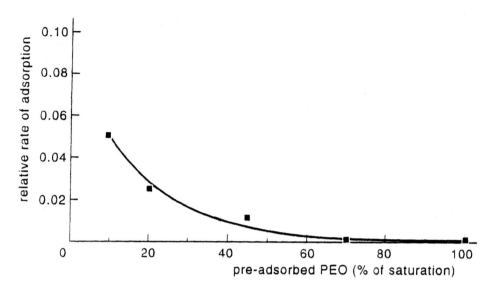

Fig. 8 Relative rate of adsorption of lipase on a surface as a function of the preadsorbed amount of polyethylene oxide, PEO. (From Ref. 60.)

between the sorbent and the adsorbed protein layer would cause steric repulsion because of increased osmotic pressure and decreased conformational entropy.

Grafting or preadsorbing water-soluble oligomers or polymers has been used to tune protein adsorption. In particular, polyethylene oxide (PEO) has been proved to be successful in producing protein-repelling surfaces (55). The protein resistance is determined by the length of the PEO chains and their density at the sorbent surface (56, 57). Thus, relatively short chains of PEO, consisting of, say, less than 10 monomer units, do not severely hamper protein adsorption but they do prevent intimate contact between the protein and the underlying sorbent surface. As a consequence, less structural perturbation occurs so that the adsorbed molecules retain more biological activity (58, 59). Longer PEO chains more effectively repel proteins. As an example, Fig. 8 gives the relative rate of adsorption of a lipase as a function of the coverage of the sorbent by PEO chains that contain an average number of 127 monomers (60). The figure shows a strong retardation of the adsorption process and almost complete suppression at surface coverages beyond 50%.

In view of the discussion given in Secs. IV.A–E, it is to be expected that proteins do adsorb from an aqueous solution onto apolar surfaces, even under conditions of electrostatic repulsion. With polar surfaces a distinction must be made between structurally stable ("hard") and structurally labile ("soft") proteins. The hard proteins adsorb at polar surfaces only if they are electrostatically attracted. The soft proteins undergo more severe structural rearrangements (i.e., a decrease in ordered secondary structure) resulting in an increase in conformational entropy large enough to make them adsorb on a polar, electrostatically repelling surface.

The mutual influences of these effects on protein adsorption are clearly demonstrated in a few systematic studies using well-defined proteins and sorbent materials (9, 13, 61, 62).

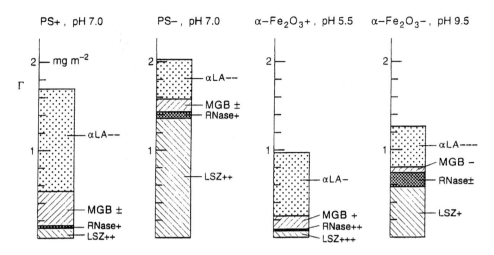

Fig. 9 Competitive adsorption between lysozyme (LSZ), ribonuclease (RNase), myoglobin (MGB), and α-lactalbumin (αLA) on surfaces of apolar polystyrene (PS) and polar iron oxide (α-Fe₂O₃). The "+" and "−" signs indicate the electrical charge of the components. The solutions supplied to the sorbent surfaces contain equal concentrations (0.3 g dm⁻³) of each of the proteins. (Data taken from Ref. 75.)

V. COMPETITIVE PROTEIN ADSORPTION

Most practical systems, e.g., essentially all biological fluids, are multiprotein systems. The various proteins compete with each other (and with other surface-active components) for adsorption at any interface present. As a rule, the sorbent surface will at first become covered by the molecules that have the highest rate of arrival (i.e., the smaller ones that have the highest diffusion coefficient and the ones that occur most abundantly in the solution). At later stages the initially adsorbed molecules may be displaced in favor of other molecules that have a higher affinity for the surface.

Competitive adsorption between monomers, dimers, or higher aggregates of the same type of protein is easily understood in terms of more anchoring segments per adsorbing entity as the number of segments in that entity increases. Thus, preferential adsorption of aggregates, relative to monomers, has been reported for, e.g., serum albumin (63, 64), fibrinogen (65), and insulin (66).

Competition between proteins of different types is more complicated. The molecular size may still play an important role but other variables such as polarity, electrical charge density, and structural stability should be taken into account as well.

Most of the literature on competitive protein adsorption refers to blood proteins (67, 68). In mixtures containing albumin, γ-globulins, and fibrinogen it has been observed that albumin adsorbs first, followed by γ-globulin, and that fibrinogen finally takes over (12, 69–72). This sequence corresponds to that of the molar masses (67,000, 150,000, and 340,000 Da, respectively) and, hence, to that of the diffusion coefficients. The same sequence has been found with blood plasma and with whole blood, where high-molecular-weight kininogen, which occurs in relatively low concentration in blood, covers the sorbent surface in the end (72–74). It is to be expected that the rate (and the extent) of these transitions depends on the affinity between the adsorbed proteins and the surface. Such a dependence is not always unambiguously found experimentally (73). More systematic research is needed to solve that matter. To that end, adsorption from solutions containing equal concentrations of lysozyme, ribonuclease, myoglobin, and α-lactalbumin on sorbent surfaces of different charges and polarities has been investigated (75). These proteins have very similar molar masses and dimensions and therefore a difference in transport rate is practically eliminated, so adsorption preference is determined by the affinity for the sorbent surface. At both the polar and the apolar sorbent surface the adsorption preference tends to reflect electrostatic interaction, except for the relatively structurally labile α-lactalbumin, which adsorbs in disproportionately high amounts at surfaces of the same charge sign. See Fig. 9. This result is in line with the result for competitive adsorption between β-lactoglobulin and κ-casein (76). Thus, it is concluded that protein structure stability is an important and often even a dominating factor in competitive protein adsorption, resulting in preferential adsorption of the soft proteins over the hard ones.

REFERENCES

1. TE Creighton. Proteins: Structures and Molecular Properties. 2nd ed. New York: WH Freeman, 1993, Chap. 5.
2. FM Richards. Areas, volumes, packing and protein structure. Annu Rev Biophys Bioeng. 6:151–176, 1977.
3. JA Schelman. Solvent denaturation. Biopolymers 17:1305–1322, 1978.
4. DJ Barlow, JM Thornton. Ion pairs in proteins. J Mol Biol 168:867–885, 1983.
5. S Nir. van der Waals interactions between surfaces of biological interest. Prog Surf Sci 8:1–58, 1977.
6. GC Kresheck, IM Klotz. The thermodynamics of transfer of amides from an apolar to an aqueous solution. Biochemistry 8:8–11, 1969.
7. C Tanford. Protein denaturation. Part C. Theoretical models for the mechanism of denaturation. Adv Protein Chem 24:1–95, 1970.
8. Z Adamczyk, B Siwek, M Zembala, P Belouschek. Kinetics of localized adsorption of colloidal particles. Adv Colloid Interface Sci 48:151–280, 1994.
9. W Norde, T Arai, H Shirahama. Protein adsorption in model systems. Biofouling 4:37–51, 1991.
10. W Norde, ACI Anusiem. Adsorption, desorption and re-adsorption of proteins on solid surfaces. Colloids Surf 66:299–306, 1992.
11. M Hasegawa, H Kitano. Adsorption kinetics of proteins onto polymer surfaces as studied by the multiple internal reflection fluorescence method. Langmuir 8:1582–1586, 1992.
12. WG Pitt, K Park, SL Cooper. Sequential protein adsorption and thrombus deposition on polymeric biomaterials. J Colloid Interface Sci 111;343–362, 1986.
13. T Zoungrana, GH Findenegg, W Norde. Structure, stability, and activity of adsorbed enzymes. J Colloid Interface Sci 190:437–448, 1997.
14. MCP van Eijk. Semi-flexible polymers near interfaces. Equilibrium aspects and adsorption kinetics. PhD. thesis Wageningen Agricultural University, Wageningen, The Netherlands, 1998.
15. W Norde, JP Favier. Structure of adsorbed and desorbed proteins. Colloids Surf 64:87–93, 1992.
16. V Ball, A Bentaleb, J Hemmerle, JC Voegel, P Schaaf. Dynamic aspects of protein adsorption onto titanium surfaces: Mechanism of desorption into buffer and release in the presence of proteins in the bulk. Langmuir 12:1614–1621, 1996.

17. A Kondo, F Murakami, K Higashitani. Circular dichroism studies on conformational changes in protein molecules upon adsorption on ultrafine polystyrene particles. Biotechnol Bioeng 40:889–894, 1992.

18. HP Jennissen. Protein adsorption hysteresis. In: JD Andrade, ed. Surface and International Aspects of Biomedical Polymers. Vol 2. New York: Plenum, 1985, pp 295–320.

19. R Barbucci, A Casolaro, A Magnani. Characterization of biomaterial surfaces: ATR-FTIR, potentiometric and calorimetric analysis. Clin Mater 11:37–51, 1992.

20. MCP van Eijk, MA Cohen Stuart. Polymer adsorption kinetics: Effects of supply rate. Langmuir 13:5447–5450, 1997.

21. JL Brash, QM Samak. Dynamics of interactions between human albumin and polyethylene surface. J Colloid Interface Sci 65:495–504, 1978.

22. BMC Chan, JL Brash. Conformational change in fibrinogen desorbed from glass surface. J Colloid Interface Sci 84:263–265, 1981.

23. PR van Tassel, P Viot, G Tarjus. A kinetic model of partially reversible protein adsorption. J Chem Phys 106:761–770, 1997.

24. AJ Babchin, Y Gur, IJ Lin. Repulsive interface forces in overlapping electric double layers in electrolyte solutions. Adv Colloid Interface Sci 9:105–141, 1978.

25. J Lyklema. Fundamentals of Interface and Colloid Science. Vol 1. Fundamentals. London: Academic Press, 1991.

26. JG Kirkwood. Theory of solutions of molecules containing widely separated charges with special application to zwitterions. J Chem Phys 2:351–361, 1934.

27. J Ståhlberg, U Appelgren B Jönsson. Electrostatic interactions between a charged sphere and a charged planar surface in an electrolyte solution. J Colloid Interface Sci 176:397–407, 1995.

28. J Ståhlberg, B Jönsson. Influence of charge regulation in electrostatic interaction chromatography of proteins. Anal Chem 68:1536–1544, 1996.

29. BJ Yoon, AM Lenhoff. Computation of electrostatic interaction energy between a protein and a charged surface. J Chem Phys 96:3130–3134, 1992.

30. W Norde, J Lyklema. The adsorption of human plasma albumin and bovine pancreas ribonuclease at negatively charged polystyrene surfaces. IV. The charge distribution in the adsorbed state. J Colloid Interface Sci 66:285–294, 1978.

31. W Norde, J Lyklema. Thermodynamics of protein adsorption. Theory with special reference to the adsorption of human plasma albumin and bovine pancreas ribonuclease at polystyrene surfaces. J Colloid Interface Sci 71:350–366, 1979.

32. JGEM Fraaije, RM Murris, W Norde, J Lyklema. Interfacial thermodynamics of protein adsorption, ion co-adsorption and ion binding in solution. I. Phenomenological linkage relations for ion exchange in lysozyme chromatography and titration in solution. Biophys Chem 40:303–315, 1991.

33. CA Haynes, E Sliwinski, W Norde. Structural and electrostatic properties of globular proteins at a polystyrene-water interface. J Colloid Interface Sci 164:394–409, 1994.

34. F Galisteo, W Norde. Adsorption of lysozyme and α-lactalbumin on poly(styrene sulphonate) latices. II. Proton titrations. Colloids Surf B Biointerfaces 4:389–400, 1995.

35. W Norde, F Galistéo Gonzalez, CA Haynes. Protein adsorption on polystyrene latex particles. Polym Adv Techn 6:518–524, 1995.

36. JGEM Fraaije. Interfacial thermodynamics and electrochemistry of protein partitioning in two-phase systems. PhD thesis Wageningen Agricultural University, Wageningen, The Netherlands, 1987.

37. H Quicampoix, RG Ratcliffe. A ^{31}P NMR study of the adsorption of bovine serum albumin on montmorillonite using phosphate and the paramagnetic cation Mn^{2+}: Modification of conformation with pH. J Colloid Interface Sci 148:343–352, 1992.

38. W Norde, J Lyklema. The adsorption of human plasma albumin and bovine pancreas ribonuclease at negatively charged polystyrene surfaces. III Electrophoresis. J Colloid Interface Sci 66:277–284, 1978.

39. W Norde, J Lyklema. Why proteins prefer interfaces. J Biomater Sci Polym Ed 2:183–202, 1991.

40. P van Dulm, W Norde, J Lyklema. Ion participation in protein adsorption at solid surfaces. J Colloid Interface Sci 82:77–82, 1981.

41. B Lee, FM Richards. The interpretation of protein structures: Estimation of static accessibility. J Mol Biol 55:379–400, 1971.

42. CA Haynes, W Norde. Globular proteins at solid-liquid interfaces. Colloids Surf B Biointerfaces 2:517–566, 1994.

43. FE Regnier. The role of protein structure in chromatographic behavior. Science 238:319–323, 1987.

44. AA Gorbunov, AY Lukyanov, VA Pasechnik, AV Vakrushev. Computer simulation of protein adsorption and chromatography. J Chromatogr 365:205–212, 1986.

45. G Némethy, HA Scheraga. Structure of water and hydrophobic bonding in proteins. J Chem Phys 36:3401–3417, 1962.

46. FM Richards. Areas, volumes, packing and protein structure. Annu Rev Biophys Bioeng 6:151–176, 1977.

47. J Visser. Adhesion of colloidal particles. In: E Matijeuic, ed. Surface and Colloid Science. Vol 89. New York: Academic Press, 1976, Chap. 1.

48. J Visser. Hamaker constants. Comparison between Hamaker constants and Lifshitz–van der Waals constants. Adv Colloid Interface Sci 3:331–363, 1972.

49. JN Israelachvili. Intermolecular and Surface Forces, 2nd ed., Academic Press, 1992.

50. JA Reynaud, I Tavernier, LT Yu, JM Cochet. The adsorption of RNase, BSA and cytochrome c at the graphite powder/liquid interface using in parallel the adsorption isotherm plot and linear sweep voltammetry on graphite paste electrode. Bioelectrochem Bioenerg 15:103–112, 1986.

51. P Schaaf, P Dejardin. Structural changes within an adsorbed fibrinogen layer during the adsorption process: A study by scanning angle reflectometry. Colloids Surf 19:89–103, 1988.

52. E Blomberg, PM Claesson, JC Fröberg, RD Tilton. Interaction between adsorbed layers of lysozyme studied with the surface force technique. Langmuir 10:2325–2334, 1994.

53. A Kondo, S Oku, K Higashitani. Structural changes in protein molecules adsorbed on ultrafine silica particles. J Colloid Interface Sci 143:214–221, 1991.

54. MCL Maste, W Norde, AJWG Visser. Adsorption induced conformational changes in the serine proteinase savinase: A tryptophan fluorescence and circular dichroism study. J Colloid Interface Sci 196:224–230, 1997.

55. JD Andrade, V Hlady, SI Jeon. Polyethylene oxide and protein resistance. In: JE Glass, ed. Hydrogels, Biocompatible and Biodegradable Polymers and Associating Polymers. Advances in Chemistry Series 248:51–59. Washington, DC; American Chemical Society, 1996.

56. PG de Gennes. Flexible polymers at S/L interfaces. Ann Chim 77:389–410, 1987.

57. SI Jeon, HJ Lee, JD Andrade, PG de Gennes. Protein surface interactions in the presence of polyethylene oxide. J Colloid Interface Sci 142:149–166, 1991.

58. MCL Maste, HA Rinia, CMJ Brands, MR Egmond, W. Norde. Inactivation of a subtilisin in colloidal systems. Biochim Biophys Acta 1252:261–268, 1995.

59. W Norde, T Zoungrana. Surface-induced changes in the structure and activity of enzymes physically immobilized at solid-liquid interfaces. Biotech Appl Biochem, 28:133–143, 1998.

60. CGPH Schroën, K van der Voort Maarschalk, MA Cohen Stuart, A van der Padt, K van 't Riet. Influence of pre-adsorbed block co-polymers on protein adsorption: Surface properties, layer thickness and surface coverage. Langmuir 11:3068–3074, 1995.

61. T Arai, W Norde. The behavior of some model proteins at solid-liquid interfaces. 1. Adsorption from single protein solutions. Colloids Surf 51:1–16, 1990.
62. CA Haynes, W Norde. Structures and stabilities of adsorbed proteins. J Colloid Interface Sci 169:313–328, 1995.
63. RLJ Zsom. Dependence of preferential serum albumin oligomer adsorption on the surface properties of monodisperse polystyrene latices. J Colloid Interface Sci. 111:434–445, 1986.
64. H Shirahama T Susawa. Adsorption of heat-denatured albumin onto polymer latices. J Colloid Interface Sci 126:269–277, 1988.
65. E Brynda, M Houska, F Lednicky. Adsorption of human fibrinogen onto hydrophobic surfaces: The effect of concentration in solution. J Colloid Interface Sci 113:164–171, 1986.
66. T Nylander, P Kékicheff, BW Ninham. The effect of solution behavior of insulin on interactions with adsorbed layers of insulin. J Colloid Interface Sci 164:136–150, 1994.
67. TA Horbett, JL Brash. Proteins at interfaces: Current issues and future prospects. ACS Symp Ser 343:1–33, 1987.
68. L Vroman. Proteins from blood plasma at interfaces. In: M Bender, ed. Interfacial Phenomena in Biological Systems. New York: Marcel Dekker, 1992, pp 137–150.
69. BK Lok, YL Chang, CR Robertson. Protein adsorption on cross-linked polydimethylsiloxane using total internal reflection fluorescence. J Colloid Interface Sci 91:104–116, 1983.
70. A Baszkin, MM Boissonnade. Competitive adsorption of albumin and fibrinogen at solution-air and solution-polyethylene interfaces. In situ measurements. ACS Symp Ser 602:209–227, 1995.
71. HGW Lensen, D Bargeman, P Bergveld, CA Smolders, J Feijen. High-performance liquid chromatography as a technique to measure the competitive adsorption of plasma proteins onto latices. J Colloid Interface Sci 99:1–8, 1984.
72. L Vroman, AL Adams. Adsorption of proteins out of plasma and solutions in narrow spaces. J Colloid Interface Sci 111:391–402, 1986.
73. JL Brash, P ten Hove. Transient adsorption of fibrinogen on foreign surfaces: Similar behavior in plasma and whole blood. J Biomed Mater Res 23:157–169, 1989.
74. P Wojciechowski, P ten Hove, JL Brash. Phenomenology and mechanism of the transient adsorption of fibrinogen from plasma (Vroman effect). J Colloid Interface Sci 111:455–465, 1986.
75. T Arai, W Norde. The behavior of some model proteins at solid-liquid interfaces. 2. Sequential and competitive adsorption. Colloids Surf 51:17–28, 1990.
76. T Arnebrant, T Nylander. Sequential and competitive adsorption of β-lactoglobulin and κ-casein on metal surfaces. J Colloid Interface Sci 111:529–533, 1986.

5

Proteins at Liquid Interfaces

F. MacRitchie
Kansas State University, Manhattan, Kansas

I. INTRODUCTION

The adsorption of proteins at liquid interfaces has relevance to biological problems such as the role of the pulmonary surfactant as well as industrial systems such as food foams and emulsions. The dual hydrophobic-hydrophilic nature of proteins causes them to concentrate at interfaces such as those between air and water and between oil and water. Protein molecules tend to adopt globular conformations in aqueous solution in which nonpolar groups are congregated in the centers and polar groups concentrate at the periphery. In this way, the free energy of the system is minimized by reducing interactions between nonpolar groups and water molecules. In practice, this is a somewhat idealized structure and there is evidence that, because of steric constraints, the periphery of the molecule contains an appreciable proportion of nonpolar groups, which therefore interact with the aqueous phase. When this type of molecule reaches a polar-nonpolar interface, it can adopt a conformation in which polar groups predominantly interact with the polar phase (usually water) and the nonpolar groups can escape. This, together with the elimination of an area of interface of high free energy (e.g., approximately 72 mN m^{-1} for air with pure water at 25°C), means a considerable lowering of the free energy of the system. This, in simple terms, accounts for the surface activity of proteins at liquid interfaces.

Some of the questions that have been controversial in the area concern the molecular conformational changes that occur at the interface and whether the adsorption process is reversible or irreversible. Such questions have not been completely resolved and will be among those to be considered in the following discussion.

II. REVERSIBILITY

In the earliest studies of protein adsorption at liquid interfaces, many results were interpreted to mean that the process is irreversible. These results will be summarized and analyzed in the light of subsequent studies.

A. Gibbs Adsorption Equation Anomaly

Once adsorbed or spread at an interface, proteins are usually very difficult to desorb, even though many are highly soluble in water. This apparent anomaly led Langmuir and Schaeffer (1) to apply the Gibbs adsorption equation in its simple form to calculate the increase in solubility that should accompany compression of a surface film of a protein, i.e.,

$$d\Pi/d\ln c_{\text{b}} = c_{\text{s}}kT \tag{1}$$

where Π is the surface pressure, c_{b} the concentration in solution, and c_{s} the number of molecules per unit area of surface. Application of this equation to the protein ovalbumin (molecular weight 35,000) showed that an increase in surface pressure of 15 mN m^{-1} should increase the solubility of ovalbumin in the solution by a factor of 10^{95}. Because no tendency for the ovalbumin film to desorb was observed for compressions of this magnitude, it was interpreted to mean that the adsorption process must be an irreversible one. Although in principle this analysis is valid, there are some other considerations to take into account as a result of experimental work since that time. First, it has been established that the behavior of long-chain linear polymers reflects the properties of segments of the molecules. For example, the viscosity of long-chain hydrocarbons becomes independent of molecular size above a certain chain length (2). This is because these long-chain flexible molecules can be considered to behave as a series of segments that move, at least to some extent, independently of each other. The behavior of proteins at liquid interfaces has been interpreted in terms of this effect. When a protein monolayer is compressed, a decrease of area (at constant surface pressure) or decrease of surface pressure (at constant area) is observed. This is consistent with displacement of portions of molecules (segments) from the interface into the adjacent aqueous subphase (3), a process that is reversible on expanding the surface. At a given surface pressure, the distribution of segments between interface (N_{i}) and bulk solution (N_{b}) has been calculated from the equation

$$N_{\text{i}}/N_{\text{b}} = \exp[(\Delta G - \Pi\Delta A)/kT] \tag{2}$$

where ΔG is the free energy difference between adsorbed and displaced segments at $\Pi = 0$, ΔA is the average area per unit segment of the adsorption-displacement process, k is Boltzmann's constant, and T is the absolute temperature. Some results for several proteins under different conditions are summarized in Table 1. Values of ΔA calculated from Eq. (2) range from 0.9 to 1.6 nm^2, consistent with segment sizes corresponding to 6–10 amino acid residues. The general conclusion is that the pressure increment of solubility is considerably less than predicted from Eq. (1) because the displacement involves segments rather than whole molecules.

The other relevant point discussed in the next section is that since the analysis by Langmuir and Schaeffer (1), proteins have been found to desorb under certain conditions.

Table 1 Values of ΔG and ΔA for Segments of Proteins Calculated from the Pressure Displacement Equilibria

Protein	Subphase	Temperature (°C)	ΔG ($\Pi = 0$)	ΔA (nm^2)
γ-Globulin	Buffer, pH 7.3	25	$7.3kT$	1.3
	Buffer, pH 4.4	25	$6.4kT$	1.3
	Buffer, pH 3.5	25	$5.5kT$	1.4
γ-Globulin	Pure water, pH 5.4	5	$4.6kT$	0.9
		25	$5.1kT$	1.0
		45	$5.8kT$	1.1
Catalase	Pure water, pH 5.4	25	$7.7kT$	1.5
	Buffer, pH 7.3	25	$8.8kT$	1.5
	Buffer, pH 4.4	25	$8.2kT$	1.6
	Buffer, pH 3.5	25	$7.5kT$	1.5
Ferritin	Pure water, pH 5.4	25	$7.8kT$	1.2

From Ref. 3.

B. Stability of Monolayers

The early studies of protein monolayers highlighted their stability. Many proteins could be compressed to moderate surface pressures (20 mN m^{-1}) without observable losses of area that would result from desorption. However, Langmuir and Waugh (4), in a detailed study of the effect of compression, distinguished between pressure displacement and pressure solubility. Pressure displacement corresponded to the removal of molecular segments as already discussed, whereas pressure solubility was equivalent to the desorption of whole molecules that was observed at high pressures. Their work also illustrated that fragments of molecules produced by proteolysis desorbed at greater rates than the whole protein molecules. Later work has confirmed that proteins can be desorbed from their monolayers under certain conditions (5–7). Kinetics of desorption have been measured for different proteins (5,8) at constant surface pressures by rates of decrease of surface area.

C. Molecular Conformation

The first measurements of spread protein monolayers at air-water interfaces showed that their Π-A curves were characterized by similarity. Sigmoidal curves were obtained in which the approximately linear region extrapolated to areas close to 1 m^2 mg^{-1} at zero surface pressure. This indicated that the molecules were unfolded and that the structure was similar for different proteins, independent of their size and shape. The conclusion reached was that the globular conformation taken up by protein molecules in solution was lost on adsorption at the interface. Such a large change in molecular conformation has been interpreted as a denaturation and therefore an irreversible effect. On the other hand, these conformational changes can be rationalized on the basis of free energy considerations as discussed in Sec. I and have been well established for polymers when transferred from one phase to another. For example, polymers tend to fold up in a poor solvent and unfold in a good solvent. A liquid interface may be considered as a good solvent for proteins. There is thus no justification for interpreting molecular unfolding in terms of irreversibility. Some qualification should be mentioned, however. The unfolding and high concentration of molecules at the interface may lead to chemical reactions such as disulfide-sulfhydryl interchange, which could be superimposed and transform an otherwise reversible process into an irreversible one.

D. Interfacial Coagulation

The observation that proteins precipitate when their solutions are shaken has been considered to be further evidence that interfaces produce irreversible changes in proteins. The surface-coagulated material has properties different from those of the original crystalline protein, being characterized by an apparent loss of solubility. The nature of the process has been investigated by a number of workers, by following the formation of coagulated protein as a result of shaking solutions and as a result of compression of surface films (9). The effects of different variables on protein precipitation by shaking have been reported by Reese (10) and Reese and Robbins (11). Of particular interest are their studies on solubility properties of surface-coagulated protein. For β-lactoglobulin, partial dissolution (up to 40%) was found on placing the suspended material in a bath at 50°C for several hours. The rate of dissolution was higher at 50°C than at 30°C. It was found that surface-coagulated protein redissolved more easily than heat-coagulated protein (11).

These results may be rationalized by taking into account the change in molecular conformation of proteins on adsorption. Surface coagulation is a two--dimensional–three-dimensional phase change and the protein precipitate is formed from molecules in extended conformations and in the same plane as each other. In order to redissolve, the extended protein molecule would need to pass through a transition state in which it leaves the precipitate and passes through a high free energy state involving contact of many nonpolar groups with water molecules before passing to a folded conformation in which the nonpolar groups can interact with each other in the interior of the molecule. For a heat-coagulated protein, molecules are highly entangled, making it even more difficult for them to attain the transition state. Reese and Robbins (11) also found that practically all the surface-coagulated β-lactoglobulin could be redissolved if suspended at pH 2.0 or pH 7.0 and the protein remained in solution on readjusting the pH to 5.0. The properties of the redissolved protein were closely similar to those of the original protein. For example, it could be coagulated by shaking at the same rate and was hydrolyzed by pepsin at similar rates, whereas the surface-coagulated (unfolded) protein was hydrolyzed at a much higher rate. The evidence suggests that surface coagulation may not be irreversible in principle and that the changes in solubility properties could be interpreted on the basis of a high activation energy for return to the solution state, although the topic remains to be firmly resolved. This question apart, surface coagulation is a secondary process and not directly related to the question of whether or not protein adsorption is reversible.

III. KINETICS OF ADSORPTION

As a result of the uncertainty associated with the reversibility of adsorption and the measurement of adsorption isotherms, there has been a preference to study kinetics.

Study of adsorption kinetics at liquid interfaces has several advantages over that at solid interfaces. Most important, it is possible to ensure the purity of the system by measuring interfacial tension and checking for buildup of surface-active contaminants with time. In this way, measurements can be made with the assurance that no artifacts are introduced through the presence of extraneous surface-active impurities. Furthermore, films can be manipulated at liquid interfaces by means of barriers, allowing accurate determinations of area while surface tension is monitored. This facility has been used to obtain accurate measurements of the kinetics of adsorption, preferably by the rate of change in area with the interfacial pressure held constant. Using this approach, MacRitchie and Alexander (12–14) found three main energy barriers to protein adsorption, one associated with diffusion from the bulk solution, another related to the interfacial pressure, and a third to the interfacial electrical potential. The first two barriers are always present and the third acts only when the adsorbing protein molecule carries an effective electrical charge.

A. Effect of Bulk Diffusion on Adsorption

When a clean interface is formed on a protein solution (or any surfactant solution), all molecules close to the interface adsorb, leaving a sublayer of several molecular diameters thickness depleted of protein. Provided that there is no back-diffusion from the interface, the rate of adsorption is equal to the rate of diffusion from the bulk solution (concentration c_b) to the sublayer at zero concentration, c_0. The number of molecules, N, adsorbing in time t is then given by diffusion theory as

$$N = 2c_b(Dt/\Pi)^{1/2} \tag{3}$$

where D is the diffusion coefficient of the protein and $\Pi = 3.14$.

Application of this equation to the adsorption of bovine serum albumin (BSA) from solutions of different concentrations gave results that are summarized in Table 2. Here, experimentally measured times for a BSA film to reach a concentration of 0.7 mg m^{-2} (corresponding to a surface pressure of 0.1 mN m^{-1}) are compared with those calculated from Eq. (3). At short times (higher BSA concentrations), agreement between experiment and theory is remarkably good. At lower BSA concen-

Table 2 Measured and Calculated Times for BSA Films to Reach a Pressure of 0.1 mN m^{-1} from Solutions of Different Concentrations After Creation of a Fresh Surface

Concentration (g l^{-1})	t measured (s)	t calculated (s)	Width of stationary layer at t calculated (cm)
0.03	6±1	7	0.004
0.02	15±2	16	0.005
0.01	55±5	64	0.011
0.005	210±12	252	0.022
0.003	540±12	714	0.037
0.002	1140±60	1602	0.055
0.001	2700±300	6420	0.11

From Ref. 12.

trations and therefore longer times, the experimentally measured times deviate, becoming shorter than predicted. Evidently, the effects of convection cause molecules to reach the interface from outside the stationary or diffusion layer at those longer times.

B. Interfacial Pressure Barrier

If the experimental times are compared with those predicted from Eq. (3) as surface pressures increase above 0.1 mN m^{-1}, the two values deviate, showing that the adsorption can no longer be explained by diffusion alone (13). This is illustrated in Table 3, where the experimentally determined times for adsorbed films of BSA to reach surface pressures from 0.1 to 2.0 mN m^{-1} are compared with the times predicted from Eq. (3). The results indicate that not all molecules reaching the surface are adsorbed and that there must be an energy barrier present. It was also found in this study that, provided electrical charge effects were minimized, the rate of adsorption of protein into monolayers of different compounds spread at the surface was independent of the nature of the compound and depended only on the surface pressure. These results, taken together, suggest that the energy barrier is associated with the surface pressure. A theory of adsorption developed by Ward and Tordai (15) postulates that, for a molecule to adsorb at an interface from solution, it must do work against the surface pressure (Π) in order to create a hole of area A for it to move into. This amount of work (ΔG) is equal to

$$\Delta G = \int \Pi dA \tag{4}$$

If the adsorption step (for a molecule) is so rapid that the surface pressure does not change significantly, then $\int \Pi \, dA$ is simply equal to ΠA. The rate of adsorption (dn/dt) is given by

$$dn/dt = kc_0 \exp\text{-}(\Pi A/kT) \tag{5}$$

In many studies, Eq. (5) has been applied to the rate of adsorption of proteins as a

Table 3 Measured and Calculated Times for BSA Adsorbed Films to Reach Various Pressures from Solutions of Two Concentrations

Pressure (mN m^{-1})	Concentration = 0.02 g l^{-1}		Concentration = 0.03 g l^{-1}	
	t measured (s)	t calculated (s)	t measured (s)	t calculated (s)
0.1	15	16	6	7
0.2	22	17.5	10	7.5
0.3	32	18.5	15	8
0.5	40	19.5	22	8.5
1.0	65	23	50	10
2.0	135	26.5	120	11.5

From Ref. 12.

Table 4 Values of A for Different Proteins Calculated from Rates of Adsorption

Protein	Concentration (g l^{-1})	A (nm^2)	Molecular weight$\times 10^{-3}$
Myosin	0.03	1.45	600
Human γ-globulin	0.01	1.30	160
Human albumin	0.02	1.00	67
Ovalbumin	0.03	1.75	40
Lysozyme	0.01	1.00	15

From Ref. 13.

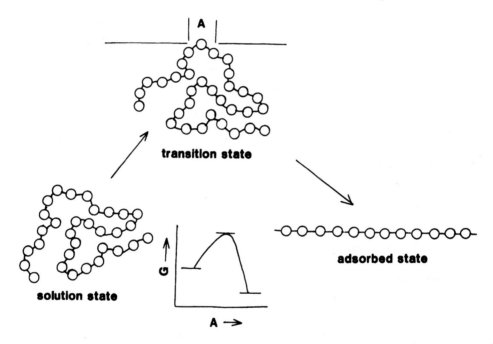

Fig. 1 Schematic illustration of model for transition state in adsorption of a protein molecule at a liquid interface. The inserted graph shows how the free energy G of the system changes as the area of penetration A of the molecule in the surface increases. When the maximum value of G is reached, the molecule then adsorbs spontaneously. (From Ref. 16.)

function of pressure. From plots of the logarithm of the rate against Π, values of A calculated from the slope have invariably been found to be similar for different proteins and to be much smaller than those of whole molecules. Some values are summarized in Table 4. As seen from the table, the areas correspond to segments of molecules of roughly 6–10 amino acid residues. This general result indicates that adsorption is an activated process in which the transition state is one in which only a small portion of the protein molecule needs to penetrate the surface for adsorption to proceed spontaneously. The process is illustrated schematically in Fig. 1.

C. Electrical Energy Barrier

When a protein carries a net electrical charge q, its adsorption produces an electrical potential ψ at the interface, which exerts a repulsive force on approaching molecules. If the molecule carries a charge q and the electrical potential at the interface is ψ, the rate of adsorption is given by the equation

$$dn/dt = Kc_0 \exp\text{-}\left(\int qd\psi/kT \right) \tag{6}$$

Using the same assumption as in Eq. (5) for the rapidity of the adsorption step, the energy term is simply $q\psi$. Some results for the initial rates of adsorption of negatively charged lysozyme into differently charged monolayers are summarized in Table 5. The variable of surface pressure was eliminated by comparing results at the same value, 6 mN m^{-1}. Because the adsorption of lysozyme into the monolayers changes the electrical potential at the surface and therefore the rate, rates of adsorption were measured from the initial slopes of the log A–time curves. Equation (6) was tested by plotting the logarithm of the rate against the surface electrical potential, which had been calculated independently from the electrophoretic mobility of oil droplets in water on which the monolayer compounds had been adsorbed. The value of q for lysozyme calculated from the slope was much less than the net charge and was thought to correspond to the electrophoretic charge under the conditions of the experiment (pH 6.5, μ 0.01).

IV. DESORPTION

Some of the important questions about proteins at liquid interfaces pertain to their desorption behavior. For example, the question of reversibility revolves around whether proteins can really be desorbed from the interface and whether, on desorption, they can recover their original solution conformation.

Table 5   Relative Initial Rates of Adsorption at a Constant Surface Pressure of 6 mN m^{-1} of Lysozyme into Monolayers Having Different Electrical Potentials from a Solution at pH 6.5

Monolayer	Initial rate[a]	ζ-Potential (mV)
Cephalin	34.5	−68.1
Polyglutamic acid	13.2	−51.5
Pepsin	5.4	−16.7
BSA	4.5	−14.1
Octadecanol	5.0	Not measured
Trypsin	3.0	+3.2
Lysozyme	(1.0)	+12.2
Polylysine	0.3	+38.5

[a]   Rates are relative to a rate of 1.0 for adsorption into a lysozyme monolayer.
From Ref. 14.

A. Confirmation of Desorption

If desorption of protein occurs from a monolayer, it should manifest itself by a loss of area with time if the monolayer is held at constant pressure. The problem with proteins and many polymers is that losses of area can occur because of other mechanisms. One of these is the pressure displacement of segments already recognized by Langmuir and Waugh (4). As this is a reversible process, it is necessary, in order to measure desorption, to measure only the permanent loss of area. This can be achieved by measuring the area at a low pressure at which no permanent area loss is observed (e.g., 5 mN m^{-1}) and using this value as a benchmark. The monolayer can then be compressed at a higher pressure for a given time and then expanded and the permanent loss of area measured with reference to the benchmark pressure. In this way, Gonzalez and MacRitchie (5) measured the permanent losses of monolayer for BSA under different conditions to ascertain whether the losses were consistent with desorption. Before discussing these results, let us first consider the mechanism of the desorption process.

If a partially soluble monolayer is placed on the surface of a solution containing none of the monolayer substance, an equilibrium is rapidly reached between the surface and a thin adjacent region of the bulk solution. The rate of desorption is then equal to the rate of diffusion from this equilibrium sublayer into the bulk solution initially at zero concentration, provided the transport of molecules occurs only through the stationary or diffusion layer. This is the layer that is adjacent to all interfaces and its thickness depends on the type of interface and the degree of stirring. At an air-water interface, under conditions of no stirring, this can be of the order of 1 mm, much greater than the thickness of the equilibrium sublayer, which is assumed to be only a few molecular diameters. Under these conditions, the number of molecules desorbing in a time t is given by an expression similar to Eq. (3). The rate of desorption is then given by differentiation of this equation as

$$dn/dt = c_0(D/\Pi)^{1/2}t^{-1/2} \tag{7}$$

Gonzalez and MacRitchie (5) used two variables to test whether their effects would alter the rate of area loss of BSA in the way expected for desorption. The effects are shown in Fig. 2. First, they varied the concentration of BSA in the solution. Increase of solution concentration should reduce the rate of diffusion from the sublayer from that predicted by Eq. (7). This was what was observed as seen in Fig. 2. The second variable was the stirring conditions. When the solution below the monolayer (of zero BSA concentration) was stirred, the rate of area loss first followed the same kinetics as those without stirring but, after a given time, deviated and became constant. This is consistent with a diffusion layer of lesser thickness so that when the diffusion front reached the limit of this layer, molecules were carried away by convection.

B. Kinetics of Desorption

Kinetics of desorption of BSA have been measured at different surface pressures based on the procedure described earlier (5). Plots of rate as a function of $t^{-1/2}$ are shown in Fig. 3. They are linear as predicted by Eq. (7), although there

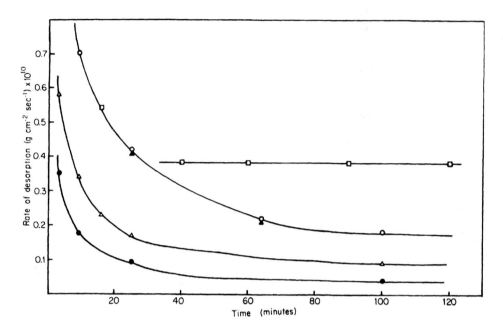

Fig. 2 Rate of desorption–time curves for a BSA monolayer at 25.6 mN m^{-1}. (○) No protein in subphase, no stirring; (□), stirring; (▲) 0.008% BSA in subphase; (△) 0.05% BSA in subphase; (●) 0.10% BSA in subphase. (From Ref. 5.)

is an indication that the lines of best fit do not extrapolate to zero, indicative of a surface barrier to desorption. Molecular weight is an important parameter in determining adsorption rates. It is generally found that, as the molecular weight increases, the surface pressure at which desorption becomes measurable correspondingly increases. Some examples in which this effect is demonstrated are summarized in Table 6. Another approach to measuring protein desorption was used by MacRitchie and Ter-Minassian-Saraga (6) using radiolabeled protein. Monolayers of ^{125}I-labeled BSA were held at several surface pressures for 20-min periods. The permanent loss of monolayer was calculated from the area loss at 10 mN m^{-1}. As a further check, samples of the monolayer were removed on glass slides after expansion of the monolayer to 10 mN m^{-1} and the radioactivity was measured in a scintillation counter. Examples of the compression cycles are shown in Fig. 4. The specific radioactivity (in counts min^{-1} cm^{-2}) is noted at each expansion stage. Radioactivity and film compressibility remained constant within experimental error with loss of film showing that there was no change in the monolayer, consistent with a desorption process. Measurement of the radioactivity in the solution at the end of the experiment was also of the same order as expected from the estimated loss of monolayer by desorption into the subphase.

C. Separation of Reversible and Irreversible Processes

This type of experiment also allows the kinetics of the two processes, displacement of segments and desorption, to be separated. Figure 5 illustrates how this

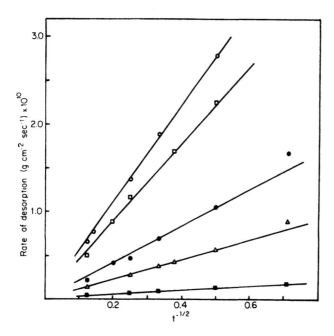

Fig. 3 Rate of desorption versus $t^{-1/2}$ for BSA monolayers at different surface pressures. ($\bigcirc$) 28.8 mN m^{-1}; ($\square$) 27.2 mN m^{-1}; ($\bullet$) 25.6 mN m^{-1}; ($\triangle$) 24.0 mN m^{-1}; ($\blacksquare$) 22.4 mN m^{-1}. (From Ref. 5.)

Table 6 Rates of Desorption of Proteins at Different Surface Pressures

Protein	Molecular weight ($\times 10^{-3}$)	Rates of desorption (min$^{-1} \times 10^4$)						
		15[a]	20[a]	25[a]	30[a]	35[a]	40[a]	45[a]
Insulin	6	56	530					
β-Lactoglobulin	17.5		20	50	90			
Myoglobin	17			34	67	144		
γ-Globulin	160				9	20	40	
Catalase	230					30	70	110

[a] Surface pressure (mN m^{-1}).
From Ref. 8.

is done for a monolayer of β-lactoglobulin that was held at 30 mN m^{-1} for different times (16). The initial area of the monolayer corresponds to point A in the left diagram. The film is compressed to C and the decrease of area monitored for a measured time. The change in area from C to D then represents the total loss of area. The loss of area from A to B at the benchmark pressure is the permanent area loss due to desorption. Subtraction of the fraction AB from the fraction DC then gives the reversible area loss due to pressure displacement. The kinetics of the reversible and irreversible processes are plotted in the right diagram.

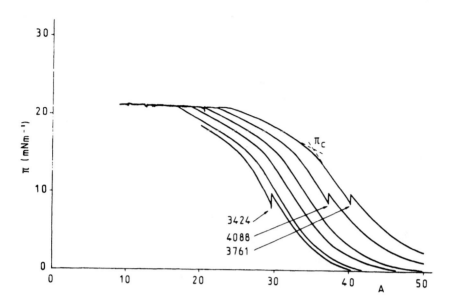

Fig. 4 Compression-expansion (not shown) cycles for ^{125}I-labeled BSA monolayer held for 20-min periods at 21 mN m^{-1}. Surface area may be calculated from the formula area $= (8.33A + 10.7)$ cm^2. The numbers correspond to measured counts (min^{-1} cm^{-2}) for films transferred onto glass slides at 10 mN m^{-1}. (From Ref. 6.)

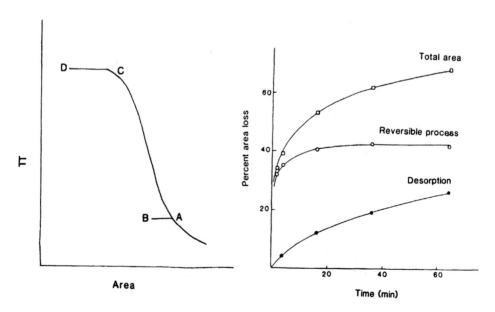

Fig. 5 Procedure for separating reversible (segment displacement) and irreversible (desorption) processes in a monolayer of β-lactoglobulin held at 30 mN m^{-1}. In the left diagram, CD is the total area loss at the high pressure and AB is the irrecoverable area loss measured at the benchmark pressure. (From Ref. 16.)

D. Model for Desorption

The apparent paradox noted by Langmuir and Schaeffer (1) that proteins are highly soluble in aqueous solution yet are extremely difficult to desorb may be rationalized by the relaxation process that occurs on compression of their monolayers. Thus, molecules of proteins and other polymers can reduce their free energy by displacing segments from the interface without the need for whole molecules to be desorbed in one process. As described earlier, the activated complex for adsorption of proteins at liquid interfaces is a segment of some 6–10 amino acid residues. This should also be the activated state for desorption because, once the area occupied by a molecule at the interface is reduced to this value, it will no longer be stable. A protein molecule can reach this state either by compression of the surface or by fluctuations in free energy about the equilibrium conformation, governed by the Boltzmann distribution. A model for the desorption process is depicted in Fig. 6. It can be appreciated that, at low surface pressures, the probability of a statistical fluctuation that enables the conformation to reach that of the transition state is infinitesimally low. As the surface pressure is increased and the equilibrium conformation is shifted toward displaced segments, the probability of a fluctuation attaining the transition state increases until at high pressures (above 20 mN m^{-1} for most proteins), desorption becomes measurable.

An estimate of the free energy barrier to desorption can be deduced from the Π-A curve of the protein monolayer. Figure 7 shows the Π-A curve for BSA (17). If the monolayer were to be compressed very rapidly (with no time for segments to be displaced), the isotherm would follow the dashed line. However, because of displacement of segments during compression, this relaxation process causes the curve to be shifted to lower areas. The full line in Fig. 7 is that for the equilibrium Π-A curve, i.e., that which is obtained when attached and displaced segments have reached equilibrium with each other at each surface pressure. Because the compression process along this curve can be carried out reversibly, we can use the equilibrium Π-A curve to calculate the free energy for desorption at any value of the surface pressure. If Π^* is the pressure required to bring

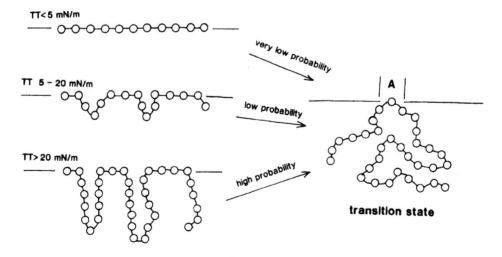

Fig. 6 Schematic illustration of how the probability of a fluctuation attaining the transition state for desorption increases with increasing surface pressure. (From Ref. 16.)

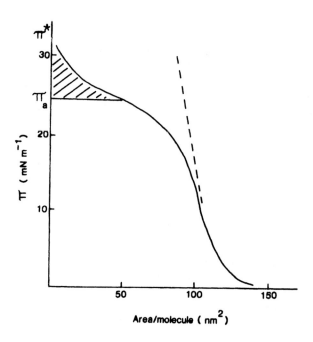

Fig. 7 Equilibrium (full line) and instantaneous (dashed line) Π-*A* curves for a BSA monolayer. Hatched area is a measure of the free energy of desorption estimated at a surface pressure of Π_a. Π^* is the transition state pressure for desorption. (From Ref. 17.)

the monolayer to its transition state and Π_a is the pressure being considered, then the free energy of activation for desorption (ΔG^*) is given by the simple expression

$$\Delta G^* = \int_{\Pi_a}^{\Pi^*} A \, d\Pi \tag{8}$$

This term may be evaluated from the area under the equilibrium Π-*A* curve between the limits Π_a and Π^*. Some values for ΔG^* for BSA evaluated from its equilibrium Π-*A* curve are summarized in Table 7 together with the corresponding desorption rates, which were

Table 7 Free Energies of Activation for Desorption (ΔG^*) Calculated from the Equilibrium Π-*A* Curve of BSA, Rate Constants for Desorption from BSA Monolayers and Calculated Subsurface Concentrations at Different Surface Pressures

Π (mN m^{-1})	$\Delta G^*/kT$	Rate constant for desorption (g cm^{-2} s$^{-1/2} \times 10^{12}$)	c_0 (from desorption rate) g cm^{-3}	c_0 (from adsorption isotherm) g cm^{-3}
0	650	—	—	—
20	106	—	—	—
22.4	64	0.42	5.7×10^{-8}	1.6×10^{-3}
24.0	42	1.83	2.62×10^{-7}	3.8×10^{-3}
25.6	24	3.50	4.86×10^{-7}	9.1×10^{-3}
27.2	14	7.50	1.04×10^{-6}	2.2×10^{-2}
28.8	9	9.17	1.28×10^{-6}	5.5×10^{-2}

evaluated from the experimental data taken from Fig. 3. It is seen that the values of ΔG^* relative to kT are prohibitively high for desorption to occur until the surface pressure rises above 20 mN m^{-1}. This is when the rates become detectable. Therefore, at least qualitatively, the experimental results are in accord with theory.

V. ADSORPTION ISOTHERMS

Despite the uncertainty associated with possible irreversibility of adsorption and the question of the applicability of the Gibbs adsorption equation, a number of workers have reported measurements of adsorption isotherms. In some cases, the Gibbs adsorption equation has been applied in its simple form:

$$c_s = (1/kT)(d\Pi/d \ln c_b) \tag{9}$$

where c_s is the interfacial concentration, c_b the bulk solution concentration, and Π the equilibrium interfacial pressure. By substituting $c_s = 1/A$, where A is the area per molecule of the adsorbed species, Eq. (9) can be written as

$$d\Pi/d \ln c_b = kT/A \tag{10}$$

This equation predicts a linear relationship between Π and $\ln c_b$ from which a value of A may be calculated. Table 8 summarizes data at fluid interfaces for adsorption of proteins where Π was found to be an approximately linear function of $\ln c_b$. Corresponding values of A calculated from the slopes are included. There is some variation between the calculated values of A. However, two conclusions may be drawn. First,

Table 8 Data from Adsorption Isotherms of Proteins at Liquid Interfaces

Protein	Conditions	Pressure range where Π–ln c_b was linear (mN m^{-1})	A (nm^2)	G_{ads}/kT per segment	Ref.
BSA	Air-water, pH 4.9, 0.1 N HCl	0–20	1.6	19	18
BSA	Air-water, pH 4.9, no salt	0–12	2.3	18	18
BSA	Air-water, pH 5.5, no salt	14–18	2.5	20	5
BSA	Heptane-water, pH 4.9, 0.1 N HCl	0–30	0.6	16	18
Pepsin	Air-water, pH 1.57	12–18	6.72	38	18
Trypsin	Air-water, pH 3.0	9–18	2.09	21	18
α-Chymotrypsin	Air-water, pH 3.0	5–12	1.0	14	18
Insulin	Air-water, 0.01 N HCl	12–18	2.9	21	19
β-Lactoglobulin	Air-water, pH 5.2, 0.01 N salt	12–16	3.3	23	19
β-Lactoglobulin	Air-water, no salt	15–25	2.1	18	21
Lysozyme	Air-water, pH 7	0–20	1.4	17	20

the values are very much less than would be expected for a whole molecule at the interface. For example, the molecular area for BSA would be expected to be in the order of 100 nm^2. Second, there is no apparent relation between the molecular size of the protein and the values of A. These results are consistent with the concept that interfacial behavior of proteins is a function of molecular segments rather than whole molecules.

It should be possible to use the equilibrium data from adsorption isotherms to predict the rates of desorption of proteins from interfaces. Where this has been done, it has been found that the desorption rates are less than those predicted because the sublayer concentrations are several orders of magnitude lower than the equilibrium concentrations calculated from the adsorption isotherm. This is consistent with a surface energy barrier to the desorption step. Values of the subsurface concentrations of BSA at different surface pressures were evaluated from the adsorption isotherm and from desorption rates [calculated from Eq. (7)] and the results are summarized in Table 7. It can be concluded from these data that the rate-determining step in desorption is not the diffusion from the sublayer but a surface barrier preventing the protein from attaining its equilibrium concentration in the sublayer.

An estimate of the magnitude of this surface barrier can be made by assuming that the rate of desorption is controlled by two barriers in parallel. These are the diffusional resistance R_1 (equal to δ/D, where δ is the thickness of the diffusion layer near the surface and D is the diffusion coefficient) and the surface resistance R_2. R_2 is associated with the desorption step and will thus be related to the magnitude of the activation energy barrier for a molecule to pass from the monolayer state to the solution. By substituting the equilibrium concentration obtained from the adsorption isotherm, it is possible to calculate the magnitude of R_2 from the equation

$$dn/dt = c_0/(R_1 + R_2) \tag{11}$$

From the data for BSA desorbing under steady-state conditions and using the appropriate value of c_0 from the adsorption isotherm, a value of 2.4×10^8 s cm^{-1} was calculated for R_2 (5). It may be concluded that, although the kinetics of desorption show the influence of the diffusional resistance (calculated to be 1.3×10^4 s cm^{-1}), the absolute rate is governed by the much larger surface resistance.

VI. CONCLUDING REMARKS

In spite of the many studies of proteins at liquid interfaces, many fundamental aspects are still controversial. In particular, there is disagreement about the reversibility of adsorption and the degree of molecular unfolding at the interface. To resolve these problems adequately will require well-designed and ingenious experimental work. It is fairly well established that proteins can desorb from liquid interfaces, but little work has been done to determine whether the molecules revert to their original solution conformation after desorption. There appears to be a need to carry out careful experiments in which the structure or conformation of protein that is known to have been desorbed is monitored and compared with the original protein in solution.

REFERENCES

1. I Langmuir, VJ Schaeffer. Chem Rev 24:181, 1939.
2. AL Van Geet, AW Adamson. J Phys Chem 68:238, 1964.
3. F MacRitchie. J Colloid Interface Sci 79:461, 1981.
4. I Langmuir, DF Waugh. J Am Chem Soc 62:2771, 1940.
5. G Gonzalez, F MacRitchie. J Colloid Interface Sci 32:55, 1970.
6. F MacRitchie, L Ter-Minassian-Saraga. Prog Colloid Polym Sci 68:14, 1983.
7. DE Graham, MC Phillips. J Colloid Interface Sci 70:415, 1979.
8. F MacRitchie. J Colloid Interface Sci 105:119, 1985.
9. F MacRitchie, NF Owens. J Colloid Interface Sci 29:66, 1969.
10. ET Reese. J Appl Biochem 2:36, 1980.
11. ET Reese, FM Robbins. J Colloid Interface Sci 83:393, 1981.
12. F MacRitchie, AE Alexander. J Colloid Sci 18:453, 1963.
13. F MacRitchie, AE Alexander. J Colloid Sci 18:458, 1963.
14. F MacRitchie, AE Alexander, J Colloid Sci 18:464, 1963.
15. AFH Ward, L Tordai. Rec Trav Chim Pays-Bas 71:572, 1952.
16. F MacRitchie. Colloids and Surfaces 76:159, 1993.
17. F MacRitchie. Anal Chim Acta 249:241, 1991.
18. P Joos. Proc Int Congr Surf Act 5th, 1968, Vol 2, p 513.
19. N Benhamou, J Guastalla. J Chim Phys 57:745, 1960.
20. MC Phillips, MTA Evans, DE Graham, D Oldani. Colloid Polym Sci 253:424, 1975.
21. F MacRitchie. Colloids Surf 41:25, 1989.

6

Polysaccharides at Interfaces

V. Rosilio and A. Baszkin
Université Paris-Sud, Châtenay-Malabry, France

A known feature of living systems is that they are constructed from self-assembled, hierarchically ordered microstructures of many different size scales. As self-organizing noncovalent aggregates, biological membranes have evolved to facilitate rapid and specific physiologic responses in multicellular systems. Connected into a complex meshwork, cell membrane proteins and a great number of membrane lipids are associated with polysaccharides, which ensure membrane structural organization and intercellular adhesion and play an essential role in modulating interfacial membrane junctions (1–4).

The various modes of self-organization in such systems are dictated by thermodynamics, and the alternation of system properties as well as a transition from one type of morphology to another may occur as a result of subtle changes in a balance of intermolecular forces. Control over the amphiphile self-organization is therefore crucial in maintaining the specified properties of the systems in which biological amphiphiles are present (5,6).

Synthetic oligo- and polysaccharide amphiphiles are of considerable interest primarily because they are a natural alternative to existing surfactants for many industrial applications but also because they offer a wide range of unexplored possibilities in the biomedical-pharmaceutical field. In this regard, expanding our knowledge of oligo- and polysaccharide amphiphile behavior at interfaces is important for the development of biosensors, drug delivery systems, surface-modified implants, and experimental models for understanding the biological function of cell surface receptors and receptor-ligand interactions. This may be accomplished by a strategy of chemical conjunction of selected carbohydrates to a lipid-modified building block component or by their incorporation in lipid membrane structures by passive adsorption (7–14).

Because two monolayers make up the membrane lipid bilayer, many properties of biological membranes can be derived from monolayer studies. Lipid monolayers spread at the air-water interface are, in fact, a good model system for studying interactions of many ordered molecules of biological significance at a molecular level. The technique of injecting soluble active species under preformed lipid monolayers, as pioneered by Schulman and coworkers for systems composed of proteins and lipids (15,16), has proved to be a convenient method for obtaining information that leads to a fundamental understanding of the relationship between the molecule structure and the macromolecular architecture at the different length scales, time scales, and levels of interaction that it entails.

We review here the use of two classical physicochemical techniques (surface pressure and surface potential) for determining the orientations, in their adsorbed state, of the two semisynthetic polysaccharides cholesteryl-pullulan (CHP) and cholesteryl-amylopectin (CHA). These polysaccharide derivatives have been shown to be particularly efficient in penetrating lipid bilayers and thus have potential in the preparation of coated liposomes (17). Because of the presence of sugar determinants, they are also capable of playing an important role in various biological recognition processes. The specific saccharide determinants conjugated to polysaccharide determinants may, therefore, act as sensory recognition devices for specific cells. This would explain why polysaccharide-coated liposomes have found successful applications in many specific cases of chemo- and immunotherapies (18–20).

I. CHOLESTERYL-AMYLOPECTIN AND CHOLESTERYL-PULLULANS. SYNTHESIS AND CHEMICAL STRUCTURES

Cholesterol derivatives of polysaccharides have been synthesized by two different routes. Cholesteryl-amylopectin was obtained by reacting amylopectin with sodium monochloracetate in an aqueous alkaline solution and then with ethylene diamine in the presence of 1-ethyl-3-(3-dimethylamino) propyl carbodiimide to give [(aminoethyl) carbamoyl] methyl-amylopectin (AECM-amylopectin). Finally, a cholesteryl derivative of amylopectin was obtained through the reaction between cholesteryl chloroformate and AECM-amylopectin (21). This synthesis procedure was also used to obtain cholesteryl-pullulan from pullulan with a molecular weight of 52,000.

Other cholesterol derivatives of pullulan (for pullulans with molecular weights 45,000 and 50,000) were prepared using a modified synthetic route. According to this procedure, cholesterol is reacted with 1,6-hexyl-diisocyanate in dry toluene containing pyridine. The cholesterol derivative obtained was reacted with pullulan in dry dimethyl sulfoxide (DMSO) containing pyridine and ethanol was added to the reaction mixture. The products were purified by dialysis and then lyophilized (22).

Chemical structures of cholesterol-substituted amylopectin and pullullans are shown in Fig. 1. The modified polysaccharides may be considered as random copolymers consisting of two comonomers: unmodified, native polysaccharide (A) and its hydrophobized, cholesterol derivative (B).

Table 1 summarizes the main chemical characteristics of these derivatives. The degree of their substitution by cholesterol residues was determined by ^{1}H nuclear magnetic resonance (NMR) and element analysis. $CHA_{113-1.0}$, for example, denotes

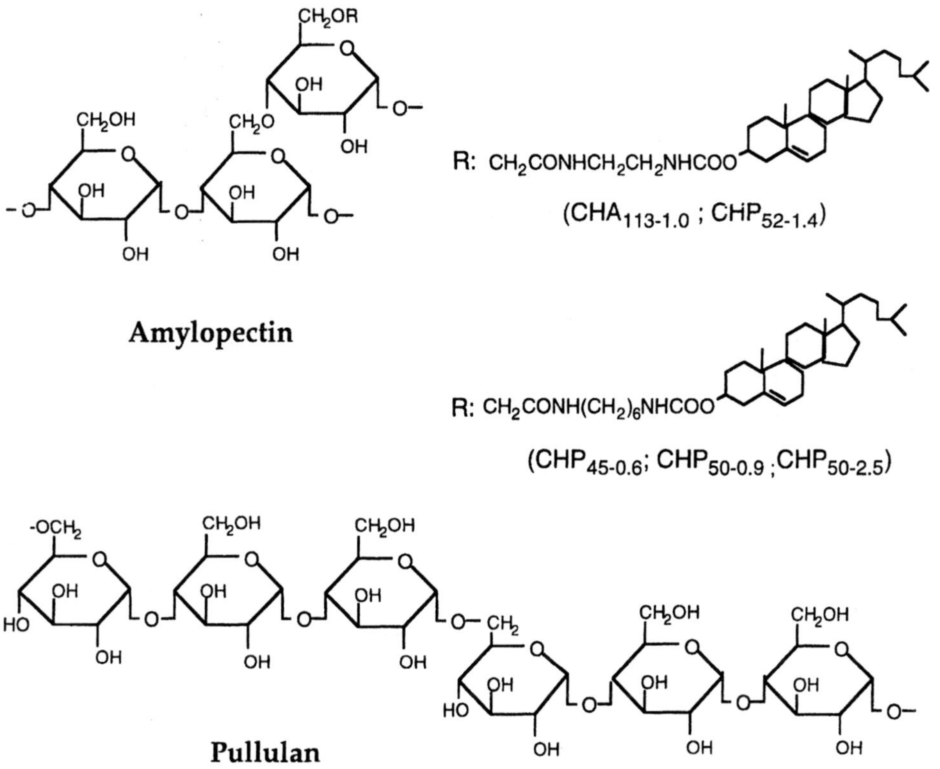

Fig. 1 Chemical structures of cholesteryl-amylopectin and cholesteryl-pullulans.

an amylopectin derivative with a molecular weight of 113,00 (polysaccharide part) and with a degree of substitution equal to 1.0 cholesteryl group per 100 glucose units.

Despite their low degree of substitution by hydrophobic cholesteryl groups, varying from 0.6 to 2.5 mol % for comonomer B, CHPs are barely soluble in water and their solubility decreases with an increase in the degree of cholesterol substitution. Whereas the unsubstituted pullulan is very soluble in water, CHPs

Table 1 Chemical Characteristics of Cholesteryl-Polysaccharide Derivatives

Derivative	M_w [a]	N_g	A/B	N_c	n
$CHP_{45-0.6}$	45,000	280	99.4/0.6	1.7	6
$CHP_{50-0.9}$	50,000	310	99.1/0.9	2.8	6
$CHP_{52-1.4}$	52,000	320	98.6/1.4	4.5	2
$CHP_{50-2.5}$	50,000	310	97.5/2.5	7.7	6
$CHA_{113-1.0}$	113,000	706	99.0/1.0	7.1	2

[a] M_w, molecular weight of the polysaccharide part; N_g, number of glucose units per molecule; A/B molar fraction of comonomers A and B; N_c, number of graft cholesterol units per molecule $= N_g B/100$; n, spacer length (see Fig. 1).

require simultaneous sonication and heating up to 50°C to form an optically clear solution.

II. ADSORPTION OF CHOLESTERYL-AMYLOPECTIN AND CHOLESTERYL-PULLULAN DERIVATIVES AT THE AIR-WATER INTERFACE

Adsorption of surfactants at the air-water interface is favored by a decrease of the free energy associated with the release of hydration water around their tail (hydrophobic moiety) when it moves from the bulk water out into the air. The free energy gain of removing surfactant hydrophobic tails from direct contact with water is also the driving force that leads surfactant molecules to aggregate and form micelles. Although large and negative, this hydrophobic dehydration contribution to the free energy of adsorption is not the only component; other components of the total free energy of adsorption and particularly the electrostatic contribution disfavor adsorption as it is essentially positive (repulsive). It results from repulsion between the charged polar groups exposed to water and is very large at low ionic strengths when the Debye length is large so that the screening of the electrostatic repulsion by counterions is weak.

For polymer surfactants such as polysaccharide derivatives, the free energy arising from steric repulsion due to polysaccharide chains immersed in the water phase and crowding at the interface would disfavor surfactant adsorption processes. From these considerations it follows that the larger the free energy of adsorption, the smaller the concentrations or the chemical potentials at which a surfactant monolayer will begin to adsorb.

At a fundamental level, but also in many practical applications associated with surfactants, accurate measurement of surface tension and surface potential provide extremely useful information regarding surfactant molecular configuration and orientation at interfaces. In the following we explain how for nonionic surfactants such as hydrophobically modified polysaccharides, these two experimental methods make it possible to quantify the number of hydrophobic moieties and their orientation at the air-solution interface.

A. Surface Potential

The surface potential at the air-water interface arises from polarized water molecules near the interface. The specific orientation of these molecules with their oxygen atoms preferentially oriented toward the air leads to potential jumps of the order of −100 to −200 mV (23). Because the randomly oriented water molecules in the bulk water do not contribute to this potential, the potential jump is essentially due to spontaneous orientation of these surface water molecules. When the surface potential is determined with the differential measuring device, using the ionizing electrode method (Fig. 2a), the electrodes are rotated and placed successively above either of the two cells. In the absence of a monolayer the difference in potential approaches the null value for two perfectly clean water interfaces.

The surface potential of the air-water interface is modified when surfactant molecules are spread or adsorbed from solution at this interface. The potential jump, ΔV, is then induced by formation of a monolayer and is proportional to the interfacial change of the normal component of the dipole density, $n\mu_{\perp}$, as defined by the Helmholtz

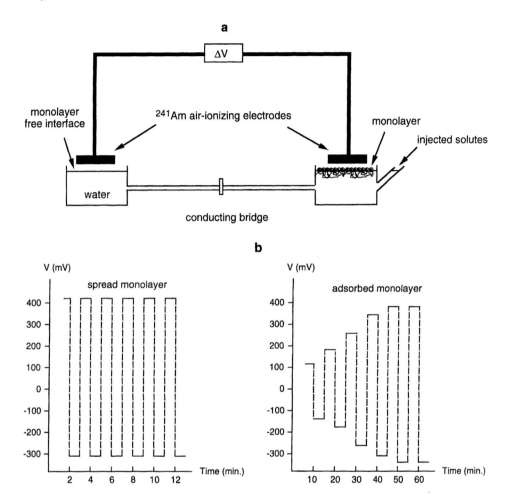

Fig. 2 (a) Schematic representation of the surface potential apparatus using two americium-241 electrodes. (b) A typical representation of the $\Delta V = f(t)$ relationship for adsorbed and spread monolayers.

equation.

$$\Delta V = \frac{1}{\varepsilon_0} n \mu_\perp$$

where n is the number of spread or adsorbed molecules per unit area ($n = 1/A$) and ε_0 the permittivity of vacuum. For charged monolayers ΔV also includes the contribution arising from the diffuse double layer, ψ_0. The factors contributing to $\mu_\perp$ are complex; they include the mean value of the normal components arising from (a) the change in effective dipole moments of monolayer polar headgroups, the change in dipole moment due to the reoriented water molecules near the monolayer-water interface, and changes in the ionic double layer, μ_ρ; and (b) the effective dipole moment of the hydrophobic part of surface-active molecules adsorbed at the interface, μ_λ. Because dipole fields superpose, the total effective dipole moment is $\mu_\perp = \mu_\rho + \mu_\lambda$.

For nonionic surfactants, the double-layer contribution is negligible and $\mu_\perp$, which is the vector sum of all the bond moments of the hydrocarbon chain, is oriented at an angle θ with respect to the vertical so that the effective dipole moment of the molecule is given by $\mu_\perp = \mu_\lambda \cos \theta$.

The method involves ionization of the air phase above the adsorbed film at constant area. The surface potential is measured by a 610C Keithley Instruments electrometer connected to two ^{241}Am air ionizing electrodes, supported by a rotating device suspended at about 2–3 mm above the reference (monolayer-free interface) and measurement (interface covered with an adsorbed or spread monolayer) cells. The two electrodes are made of ^{241}Am deposited on a silver disk between two thin layers of gold. The diameter of the disk is 15 mm. The two cells are connected by a glass tube with a Teflon stopcock. The stopcock is closed during the experiment to avoid any exchange of material, but the liquid bridge between the glass and the stopcock is ensured for the electrical conductivity (24,25) (Fig. 2a).

The potential difference measured when rotating the electrodes equals $2\Delta V$ (Fig. 2b). In the presence of a surfactant monolayer, the surface potential is

$$\Delta V = V_M - V_R$$

where V_M is the potential of the measuring cell and V_R that of the monolayer-free reference cell.

The sensitivities of the measurements are 5 mV in the range 0–100 mV and 10 mV in the range 100–400 mV.

Surface potentials of adsorbed polysaccharide derivatives at the monolayer-free interface were monitored at each given polysaccharide solution concentration. In such experiments a concentrated solution of a polysaccharide derivative was introduced in microliter quantities into the aqueous phase in the measurement cell through a side arm and the surface potential value was taken when its constant value was attained.

In the penetration experiments performed in the presence of a phospholipid monolayer spread at the air-water interface at a given surface coverage, the experimental procedure was identical to that described for the monolayer-free interface. The variations of the surface potential $\Delta(\Delta V)$ due to the interaction of the polysaccharides with spread phospholipid monolayers corresponded to the increments of surface potentials and were equal to

$$\Delta(\Delta V) = V_{PL-PS} - V_{PL}$$

where V_{PL-PS} denotes the equilibrium surface potential of a mixed phospholipid-polysaccharide monolayer and V_{PL} the surface potential of a phospholipid monolayer before the injection of a cholesterol derivative of pullulan or amylopectin to the subphase.

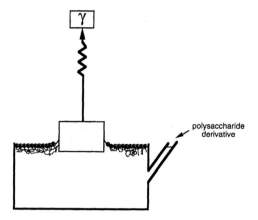

Fig. 3 Schematic representation of the Wilhelmy plate method used for surface tension measurements at constant area.

B. Surface Tension

The experiments are performed using the Wilhelmy plate method. A roughened platinum plate is attached by a thread to a force transducer (Krüss, digital tensiometer K10T). Measurements are taken without detaching the plate from the interface and the data are continuously monitored (Fig. 3).

To ascertain that the recorded surface tension data are reliable over long periods of time, a check is performed for each polysaccharide concentration studied. An experiment is run in which a polysaccharide is adsorbed at the bare air-water interface. After 20 h of adsorption, the platinum plate attached to the tensiometer is brought into contact with the aqueous phase and the surface tension is recorded. As the surface tension values recorded thereby did not differ from those obtained with the plate immersed in the aqueous phase for long periods of time, the data obtained were considered to correspond to the thermodynamic equilibrium values of the systems studied. The accuracy of the measurements was estimated to be 0.2 mN/m.

The surface pressure π of the polysaccharides adsorbed at the air-water interface are calculated for each polysaccharide solution concentration from the relationship $\pi = \gamma_0 - \gamma_{PS}$, where γ_0 corresponds to the surface tension of the pure water and γ_{PS} to the measured surface tension of polysaccharide solution at a given time.

In penetration experiments, the change in surface pressure, $\Delta\pi$, due to the interaction between adsorbing polysaccharide molecules and a spread phospholipid monolayer is calculated as the difference between the surface tension of the system in the presence of a phospholipid monolayer (γ_{PL}) and that in the presence of the mixed polysaccharide-phospholipid film after injection of a polysaccharide derivative into the water subphase (γ_{PL-PS})

$$\Delta\pi = \gamma_{PL} - \gamma_{PL-PS}$$

C. Surface Activity of Cholesteryl-Amylopectin (CHA) and Cholesteryl-Pullulan (CHP)

It may be noted from the data plotted in Fig. 4a that the increase in π is considerably higher for CHA than for CHP. The steady-state π values are achieved for CHA concentrations about three times lower than the CHP concentrations. Whereas for CHA this concentration is 9×10^{-7} M, for CHP the plateau value is achieved at concentrations higher than 3×10^{-6} M. Also, the corresponding plateau value of π is much higher for CHA (38 mN/m) than for CHP (24 mN/m). In other words, CHA is much more surface active than CHP.

By plotting the equilibrium surface tension of aqueous CHP and CHA solutions versus their logarithmic concentration and taking the $d\gamma/d \ln c$ slopes (for CHP within 2.7×10^{-7}–1.2×10^{-6} M and for CHA within 1.0×10^{-7}–3.4×10^{-7} M) prior to the appearance of the discontinuities in the slopes, the interfacial excess quantities, Γ, are obtained from the simplified Gibbs equation (Fig. 4b). In these calculations it is assumed that at low polysaccharide concentrations, their activities may be replaced by concentrations (26). They yield 19 and 40.5 Å^2 per adsorbing CHA and CHP, respectively. Evidently, these small areas do not correspond to the entire area of either a CHP or a CHA molecule at the interface. They clearly indicate that the hydrophilic sugar moieties of adsorbing cholesteryl derivatives are immersed in the solution. The surface pressure increase with concentration would thus result mainly from the increasing quantities of hydrophobic moieties arriving at the interface.

Careful examination of surface pressure data reveals that the mean molecular area found for CHP is almost the same as that reported for cholesterol molecules that orient vertically at the air-water interface giving a molecular area of 39 Å^2 (27). The smaller molecular area at saturation and the higher surface pressure observed with CHA relative to CHP suggest that cholesteryl moieties of CHA lie flat beneath the water-air interface. In such a horizontal orientation with respect to the interface, which may be favored by the branched structure of the CHA

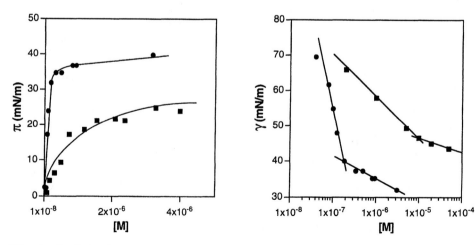

Fig. 4 (Left) Surface pressure dependence on solution concentration. (Right) Equilibrium surface tension (after 25 h) versus concentration; (■) CHP$_{52-1.4}$, (●) CHA$_{113-1.0}$.

molecule, only lateral CH_3 groups of cholesteryl moieties would appear at the interface. This would explain the good agreement between the value of the mean molecular area found for the CHA molecule and that of many fatty acid–type compounds, for which a typical cross-sectional area corresponding to a CH_3 group is close to 20 Å^2.

Support for the validity of these arguments is provided by the kinetics of adsorption of these two polysaccharide derivatives. At the same solution concentration, the time necessary for CHP to reach the steady-state value of the surface pressure is twice that for CHA. The adsorbing CHA reaches the saturation surface pressure (20 mN/m) 10 times faster than adsorbing CHP. Evidently, the time necessary to accommodate, at the air-solution interface, a much smaller CH_3 group of the CHA molecule is markedly shorter than that required for a whole cholesteryl moiety of CHP (26).

In contrast to the adsorption of proteins and flexible polymers, in which the penetration of the surface film is accomplished by spreading and unfolding of their adsorbed segments, CHP molecules penetrate the film only with their cholesteryl moieties. The rest of the polysaccharide molecule, which is connected to the hydrophobic anchors, more or less maintaining the original conformation in the aqueous medium, cannot be located in a two-dimensional surface lattice. Thus, if the segments of the adsorbed protein molecule can have many conformations within the lattice, the number of ways of arranging a molecule such as CHP is reduced. The interfacial flexibility of the CHP molecule becomes still smaller as a greater number of cholesteryl groups penetrate the air-water interface.

D. Quantitation of Surface Densities of Cholesteryl Groups of Adsorbed CHP Molecules

Surface tension and surface potential measurements performed with a series of cholesteryl-pullulan derivatives reveal that their surface properties are related to the degree of substitution of the polysaccharide by hydrophobic cholesteryl groups. Increasing the degree of cholesterol substitution of pullulan increases both the ability of CHP to decrease the solution surface tension and the value of the surface potential (28,29).

The surface potential data provide additional information that confirms the ordered structure of cholesteryl groups oriented toward the air phase. It is apparent from Fig. 5 that the maximum value of ΔV (305 mV) coincides with the surface potential value of spread cholesterol molecules in a tightly packed state.

From Fig. 5 it is also apparent that the maximum surface potential per cholesterol (330 mV) is reached at a surface density equal to 2.6×10^{14} molecules/cm^2, which corresponds to an area of 38.5 Å^2, in good agreement with the cross-sectional area of a cholesterol molecule obtained in surface pressure experiments.

The organization of cholesteryl groups extended toward the air phase makes it possible to assess their surface densities using a procedure in which the curve plotted for spread cholesterol molecules is used as a conversion curve for the maximum ΔV values for the CHP series.

Table 2 summarizes the areas per cholesteryl group for the series of CHP with varying cholesteryl content.

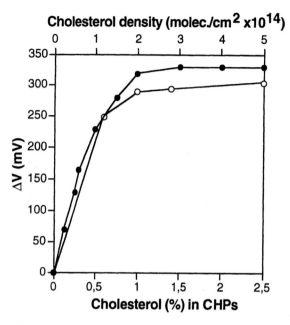

Fig. 5 Effect of cholesteryl content on the surface potential of hydrophobized pullulans (○) and surface potential versus surface density of spread cholesterol at constant surface area (●).

Table 2 Molecular Dimensions of Adsorbed CHP Molecules

CHP	ΔV_{max} (mV)	A_{chol} (Å^2)^a	n	A_{CHP} (Å^2)
CHP$_{45-0.6}$	249	86	1.7	146
CHP$_{50-0.9}$	290	59	2.8	162
CHP$_{52-1.4}$	295	57	4.5	256
CHP$_{50-2.5}$	305	51	7.7	393

^{a} A_{chol}, area per cholesteryl group; n, number of cholesteryl groups per pullulan chain; A_{CHP}, area per CHP chain.

E. Organization of Adsorbed CHP Molecules at the Air-Water Interface

Molecular dimensions of adsorbed pullulans with varying cholesteryl group contents make it possible to model their arrangements at the air-water interface (Fig. 6). It may be noted that the area per cholesteryl group decreases as the degree of grafting increases and that cholesteryl groups have enough room to occupy a tilted position with respect to the horizontal plane.

A marked difference in the orientation of the polysaccharide part between polymers is direct consequence of the varying degree of cholesterol grafting. An increase in cholesteryl content results in a decrease in the area occupied per cholesteryl group at the surface and in an increase in the overall area occupied per pullulan macromolecule. A higher number of cholesteryl anchoring points at the interface, which act as "polymer suspenders," reduce the thickness of the polysaccharide layer

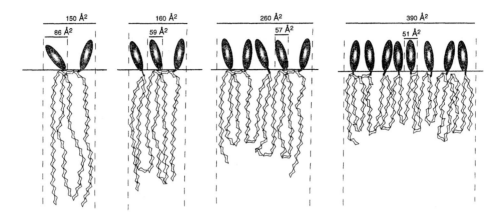

Fig. 6 Molecular arrangements of adsorbed pullulans with the varying grafted cholesteryl content.

attached to cholesteryl groups. It follows that polysaccharide chains have an increased degree of freedom and an increased density inward to the solution bulk when they are capped with only one or two cholesteryl groups. This configuration dependence of pullulan on the number of grafted chains has profound implications for its ability to interact sterically with proteins and cells.

III. PENETRATION OF EGG PHOSPHATIDYLCHOLINE MONOLAYERS BY CHOLESTERYL-AMYLOPECTIN AND CHOLESTERYL-PULLULAN DERIVATIVES

A classical technique is used to study the interaction of polysaccharide derivatives with lipids. A water-soluble surfactant is injected underneath a monolayer kept under a given surface pressure and the increments in surface pressure ($\Delta\pi$) and in surface potential $\Delta(\Delta V)$ are recorded. Their extent is directly related to the interaction of the reactant with a monolayer. The rate and the magnitude of penetration determine whether the reactant is entering the surface by solution in the monolayer (the case for many water-soluble proteins) or by molecular interaction resulting in association of the polar groups of the reactant and van der Waals attraction between aliphatic lipid chains and cholesterol moieties of polysaccharides (16).

As already pointed out for both CHA and CHP, the decrease in the surface tension at the monolayer-free interface takes place over a long time scale until the energetically most favorable orientation of cholesterol groups is attained. However, studies using the laser scanning confocal fluorescence microscopy (LSCM) technique clearly demonstrated that in the presence of a spread lipid monolayer, a CHP with a fluorescent label is very rapidly adsorbed at the interface (after 1 h of adsorption the fluorescence intensity no longer changes) (30). Because polysaccharide derivatives are so rapidly adsorbed at the interface with a spread lipid monolayer, the question arises of the origin of the marked dif-

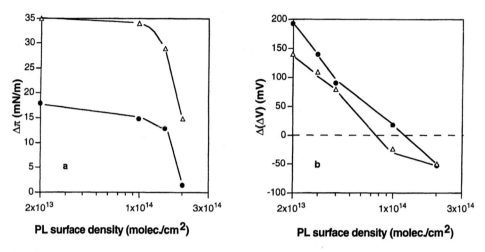

Fig. 7 Change in surface pressure (a) and in surface potential (b) as a function of phospholipid (PL) surface density after CHP or CHA injection into the aqueous subphase.

ferences in $\Delta\pi$ and $\Delta(\Delta V)$ values observed for CHP and CHA molecules (Fig. 7a and b).

From the data presented in Fig. 7a, it may be inferred that for both polysaccharide derivatives, the higher the monolayer surface density, the lower their penetration. This is quite obvious because an increase in surface density gives rise to reduced space between lipid molecules forming monolayers. It may also be noted that the $\Delta\pi$ values are always higher for CHA than for CHP. They approach zero at high surface coverages for CHP, but for CHA the $\Delta\pi$ values are always positive.

The smaller molecular area at saturation and higher surface pressure observed in the case of CHA, compared with that of CHP, suggest that cholesteryl moieties of CHA lie flat beneath the water-air interface. In such a horizontal orientation with respect to the interface, which may be favored by the branched structure of CHA molecules, only lateral groups of cholesteryl moieties would appear at the interface. This would explain good agreement between the mean molecular area found for CHA and that of many fatty acid–type compounds, for which a typical cross-sectional area corresponding to a closely packed linear aliphatic chain is about 20 Å². Because of their branched structure and horizontal orientation with respect to the monolayer plane, CHA molecules fully compensate lipid surface potential (Fig. 7b). Such an orientation of hydrophobized hydrophilic polymers with respect to phospholipid membranes has also been confirmed for the system of poly-(*N*-isopropylacrylamide) with grafted stearyl chains (31) and for dextrans bearing palmitoyl chains (32). In both cases, the hydrophobic parts were found to intercalate into the phospholipid bilayer while the hydrophilic polymer part covered the membrane.

Interfacial architectures of CHA and CHP molecules penetrating phospholipid monolayers are depicted in Fig. 8.

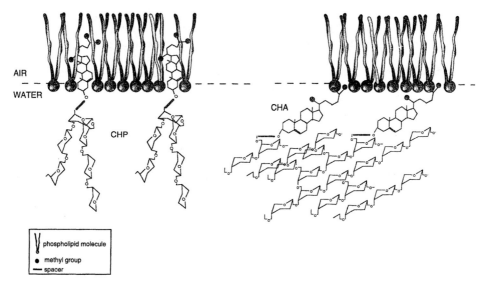

Fig. 8 Models of penetration of CHP and CHA molecules into phospholipid monolayers.

IV. CONCLUSIONS

It becomes apparent from the results discussed here that the specific mechanism behind the effectiveness of hydrophobized polysaccharides in stabilization of medical and pharmaceutical colloidal formulations is the high steric repulsion exerted by polysaccharide chains extended toward the aqueous environment as well as the uncompensated lipid potential remaining at the surface of phospholipid bilayers. Although the interrelations between molecular architecture and formulation properties need systematic work in all classes of amphiphilic copolymers, the experimental information and fundamental understanding emerging from this review can be used to differentiate the performance of naturally occurring polysaccharides in complex systems such as medical, pharmaceutical, cosmetic, and agricultural formulations.

REFERENCES

1. B Alberts, D Bray, J Lewis, M Raff, K Roberts, JD Watson. Molecular Biology of the Cell. 2nd ed. New York: Garland, 1989, p 298.
2. T Muramatsu. Les sucres de la membrane cellulaire. Recherche 20:624, 1989.
3. T Muramatsu. Developmentally regulated expression of cell surface carbohydrates during mouse embryogenesis. J Cell Biochem 36:1, 1988.
4. D Chapman. Biomembranes and new hemocompatible materials. Langmuir 9:39, 1993.
5. JN Israelachvili. Intermolecular and Surface Forces. 2nd ed. London: Academic Press, 1991, p 38.
6. H Ringsdorf, B Schlarb, J Venzmer. Molecular architecture and function of polymeric ordered systems: Model for the study of organization, surface recognition and dynamics of biomembranes. Angew Chem Int Ed Engl 27:113, 1988.
7. B Pfrannemüller, W Welte. Amphiphilic properties of synthetic glycolipids based on amide linkages. I. Electron microscopic studies on aqueous gels. Chem Phys Lipids 37:227, 1985.

8. M Hato, H Minamikawa. The effect of oligosaccharide sterochemistry on the physical properties of aqueous synthetic glycolipids. Langmuir 12:1658, 1996.

9. J Eastoe, P Rogueda, BJ Harrison, AM Howe, AR Pitt. Properties of a dichained sugar surfactant. Langmuir 10:4429, 1994.

10. P Boullanger, Y Chevalier. Surface active properties and micellar aggregation of alkyl 2-amino-2-deoxy-β-D-glucopyranosides. Langmuir 12:1771, 1996.

11. T Zhang, RE Marchant. Novel polysaccharide surfactants. The effect of hydrophobic and hydrophilic chain length on surface activity properties. J Colloid Interface Sci 177:419, 1996.

12. M Ruegsegger, T Zhang, RE Marchant. Surface activity of ABA-type nonionic oligosaccharide surfactants. J Colloid Interface Sci 190:152, 1997.

13. K Mishima, K Satoh, K Suzuki. Molecular order of lipid membranes modified with polysaccharide pullulan and a pullulan-derivative bearing a hydrophobic anchor. Colloids Surf B Biointerfaces 9:9, 1997.

14. K Akiyoshi, J Sunamoto. Physicochemical characterization of cholesterol-bearing polysaccharides in solution. In: SE Friberg, B Lindman, eds. Organized Solutions, Surfactants in Science and Technology. New York: Marcel Dekker, 1992, p 289.

15. P Doty, JH Schulman. Formation of lipo-protein monolayers. Part I. Premiminary investigation on the adsorption of proteins onto lipid monolayers. Discuss Faraday Soc. 6:21, 1949.

16. R Matalon, JH Schulman. Formation of lipo-protein monolayers. Part II. Mechanism of adsorption, solution and penetration. Discuss Faraday Soc 6:27, 1949.

17. T Sato, J Sunamoto. Site specific liposomes coated with polysaccharides. In: G Gregoriadis, ed. Liposome Technology, Interactions of Liposomes with the Biological Milieu. London: CRC Press, 1993, p 179.

18. J Sunamoto, M Goto, T Iida, K Hara, A Saito, A Tomonago. Unexpected tissue distribution of liposomes coated with amylopectin derivatives and successful use in the treatment of experimental Legionnaires' disease. In: G Gregoriadis, G Poste, J Senior, A Trouet, eds. Receptor-Mediated Targeting of Drugs New York: Plenum, 1984, p 359.

19. J Sunamoto, K Akiyoshi, T Sato. Cell Specific drug release from liposomes. In: KK Mitha, ed. Topics in Pharmaceutical Sciences. Noordwijk: Amsterdam Medical Press, 1989, p 25.

20. J Sunamoto. Cell specific transportation using functionalized artificial cell liposomes. In: S Yoshikawa, Y Murakami, eds. New Frontier in Supramolecular Chemistry. Tokyo: Mita Press, 1990, p 263.

21. T Sato, J Sunamoto. Recent aspects in the use of liposomes in biotechnology and medecine. Prog Lipid Res 31:345, 1992.

22. K Akiyoshi, S Deguchi, N Moriguchi, S Yamaguchi, J Sunamoto. Self aggregates of hydrophobized polysaccharides in water. Formation and characteristics of nanoparticles. Marcomolecules 26:3062, 1993.

23. V Vogel, D Mobius. Local surface potentials and electric dipole moments of lipid monolayers: Contributions of the water/lipid and lipid/air interfaces. J Colloid Interface Sci 126:408, 1988.

24. M Plaisance, L Ter-Minassian-Saraga. Ionized monolayers as models for polyelectrolytes. J Colloid Interface Sci 38:489, 1972.

25. A Baszkin, M Deyme, P Couvreur, G Albrecht. Surface pressure and surface potential studies of poly(isobutylcyanoacrylate)-ampicillin interactions at the water/air interface. J Bioactive Compat Polym 4:110, 1989.

26. A Baszkin, V Rosilio, G Albrecht, J Sunamoto. Cholesteryl-pullulan and cholesteryl-amylopectin interactions with egg-phosphatidyl-choline. J Colloid Interface Sci 145:502, 1991.

27. HE Ries, H Swift. Electron microscope and pressure-area studies on a granicidin and its binary mixtures with cerebronic acid, cholesterol and vilinomycin. Colloids Surf 40:145, 1989.

28. B Demé, V Rosilio, A Baszkin. Polysaccharides at interfaces. I. Adsorption of cholesteryl-pullulan derivatives at the solution/air interface. Kinetic study by surface tension measurements. Colloids Surf B Biointerfaces 4:357, 1995.
29. B Demé, V Rosilio, A Baszkin. Polysaccharides at interfaces. II. Surface potential of adsorbed cholesteryl-pullulan monolayers at the solution/air interface. Colloids Surf B Biointerfaces 4:367, 1995.
30. G Gluck, H Ringsdorf, Y Okumura, J Sunamoto. Vertical sectioning of molecular assemblies at air/water interface using laser scanning confocal fluorescence microscopy. Chem Lett 209, 1996.
31. H Ringsdorf, J Venzmer, FM Winnik. Interaction of hydrophobically modified poly-*N*-isopropylacrylamides with model membranes, or playing a molecular accordion. Angew Chem Int Ed Engl 30:315, 1991.
32. MGL Elferink, JG de Wit, G In't Weld, A Reichert, AJM Driessen, H Ringsdorf, WN Konings. The stability and functional properties of proteoliposomes mixed with dextran derivatives bearing hydrophobic anchor groups. Biochim Biophys Acta 1106:23, 1992.

7

Structure and Properties of Membranes

Y. Barenholz
Hebrew University, Hadassah Medical School, Jerusalem, Israel

G. Cevc
The Technical University of Munich, Munich, Germany

I. INTRODUCTION

Membranes abound in nature. Several billion years ago, primitive closed sacs first enshrined the droplets of dissolved primordial self-replicating molecules into an original "self." These membranes created a "milieu interne," which is a prerequisite for proper life (Fig. 1); this led to one of the theories of the origin of life, referred to as the "membrane came first" hypothesis. Thus, membranes have localized material into natural bioreactors by limiting and controlling diffusion into and from the surroundings. Equally important may have been the catalysis of reactions near the membrane surface, at the interface between the aqueous and lipidic compartment of each self-replicating unit (Deamer, 1997).

Membranes acting as selective barriers are obligatory for creating chemical-osmotic and electrical gradients that provide the basis for bioenergetics, neuronal action, and other kinds of signal transduction. Barrier function is crucial for many other processes in the cell as well, the most important of which are presented in Fig. 2.

All biological membranes on Earth are composed of two major types of molecules: proteins and lipids. The protein/lipid mass ratio in membranes varies greatly—from a value of ~ 0.23 for myelin sheets to a ratio of 3.2 in the mitochondrial inner membranes and many bacteria (Singer, 1975). In most biological membranes the protein quantity equals or exceeds that of lipids, but many membrane proteins have most of their mass outside the membrane bilayer. The membrane lipids and proteins normally act in cohorts and most often as functional and thermodynamic units, the large variability of individual molecules notwithstanding.

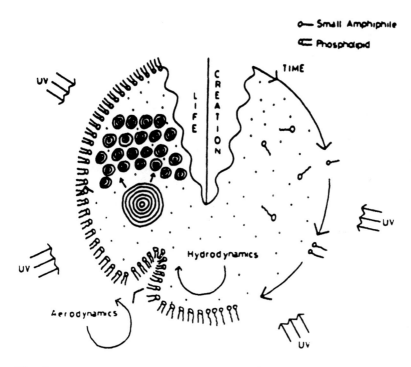

Fig. 1 Membranes came first! How hydrodynamic and aerodynamic forces coupled to solvation and entropic free energies generate clones of self-replicating, potassium-rich, closed membrane systems. Experiments with liposomes, reviewed by Bangham (1993), offer evidence to support the idea that lipid membranes preceded DNA or protein synthesis. Cloned liposomes would have provided catalytic surfaces within protected environments. (From Bangham, 1993.)

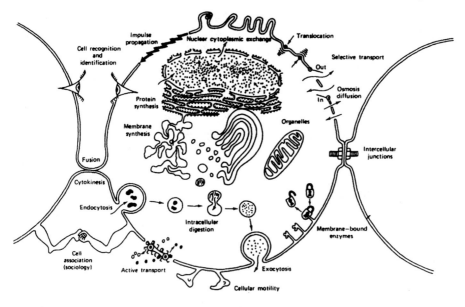

Fig. 2 Cartoon of a typical cell emphasizes the various processes modulated by the cell membrane. (From Jain and Wagner, 1980.)

The unique role that lipids play in membranes is due to the fact that no lipid is completely compatible with water. All lipids therefore spontaneously aggregate in water or related solvents: the hydrophilic ends (polar headgroups) are soluble in water to the limit of several moles per liter and thus typically seek contact with the solvent. The hydrophobic ends (fatty tails) prefer the relatively apolar media (methanol, chloroform, etc.) and typically have water solubility below 10^{-10} mol/L. Most biological lipids therefore (self-)aggregate in an aqueous environment. Their critical aggregation concentration (CAC) is in the range between 10^{-10} M (=moles per liter) and 10^{-5} M; only rarely are values above 10^{-4} M observed.

The current picture of membrane structure is a modification of the fluid-mosaic model proposed by Singer and Nicolson in 1972 (Fig. 3). In this model a lipid bilayer was seen as a continuum in which the protein molecules are either partially embedded or inserted or else associated with the lipid bilayer surface, for integral and peripheral proteins, respectively. More recently it became evident that an additional major group of membrane proteins is covalently attached to the lipids (Hooper and Turner, 1992); protein acylation, isoprenylation, and binding via glycerosyl phosphatidylinositol (GPI) are some of the possibilities for such "protein lipidation" (Hooper and Turner, 1992). The reason for having a very large repertoire of membrane proteins and a large heterogeneity of membrane protein composition is obviously the functional role that proteins play in membranes. The need for a large repertoire of membrane lipids [many hundred per eukaryotic cell (Marai and Kuksis, 1969; Rouser et al., 1968; White, 1973)] and for the great diversity in the lipid composition of various membranes (Barenholz, 1984; Shmeeda et al., 1994) is less clear, however.

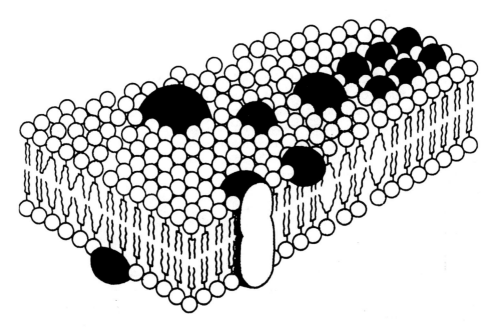

Fig. 3 A classical presentation of the fluid mosaic model of cells membranes, according to Singer and Nicolson (1972).

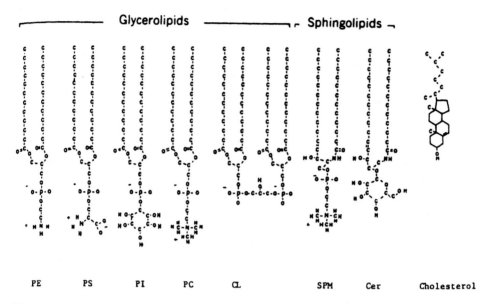

Fig. 4   Chemical structure of major membrane lipids. Acyl chains are schematically presented by C–C bonds with no reference to the double bonds (usually located on the acyl chain attached to the second carbon atom of the glycerophospholipid glycerol moiety). Abbreviated names of the various lipids are given in the text. (From Lichtenberg and Barenholz, 1988.)

In the membranes of most eukaryotic organisms four major lipid classes are found: glycerophospholipids, glyceroglycolipids, sphingolipids, and sterols (Fig. 4); most bacteria (prokaryotes) lack sterols and sphingolipids, however. Archaebacteria also contain unusual fatty ingredients (Gliozzi and Relini, 1996, and references listed therein) not found in other bacteria, with one or two polar heads, referred to as monopolar and bipolar (bolalipids), respectively (Fig. 5).

Mammalian membranes, which were the last to develop on Earth, have the most diverse lipid composition. The lipid composition of bacterial membranes is similar to that of mitochondrial and chloroplast membranes, which supports the theory of a close relationship between these organelles and bacteria.

Among the lipid classes encountered in eukaryotic and bacterial membranes, phospholipids form the largest group. They can make up 40–80% of mammalian and up to 40% of plant membrane mass but in the extreme case contribute less than 1% to the cell mass.

Phospholipids, glycolipids, and sterols build the matrix of each biological membrane, which is typically organized as a lipid bilayer (reviewed in Ansell et al., 1973; Weissmann and Clailborne, 1975; Yeagle, 1993; Mouritsen and Anderson, 1998). Locally, however, such molecules (sterols excepted) can also take the shape of inverse hexagonal (Seddon, 1990) or cubic (Landh, 1995) structures. The capability to undergo a transition between the different membrane forms is important for proper membrane functioning (Cevc, 1993e; Kinnunen, 1996). This review is focused on membrane (phospho)lipids, their interaction with the other membrane components (sterols, glycosphingolipids, proteins), and the contribution to physical and biological properties of cell membranes.

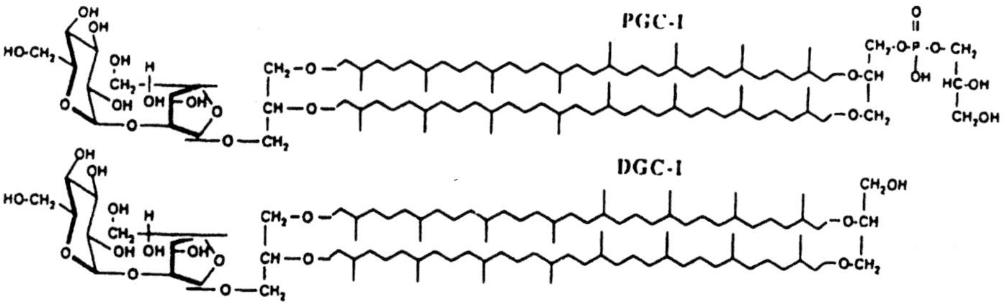

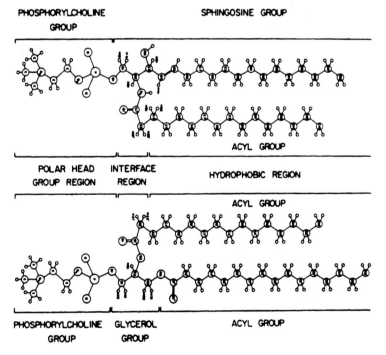

Fig. 5 Chemical structure of the membrane spanning (bolaform) lipids from *M. hungatei.* Notice that methanogens comprise monopolar as well as bolaform lipids. PGC-I, phosphoglycocaldarchaeol; DGC-I, diglycosylcaldarchaeol. (Based on Gliozzi and Relini, 1996.)

Phospholipids derived from eukaryotic cell membranes are divided into two main groups, which constitute 80–90% of the total membrane phospholipid pool. These groups are the choline phospholipids [as in phosphatidylcholine (PC) and sphingomyelin (SPM); see Fig. 6] and the amino phospholipids [such as phosphatidylethanolamine (PE) and phosphatidylserine (PS)].

The chemical classification of phospholipids is discussed in Sec. III. A more historical perspective is given in the review and reference books by Ansell et al. (1973), Danielli (1975), Bangham (1993), Cevc and Marsh (1987), Cevc (1993e), and Lasic and Barenholz (1996).

Fig. 6 Molecular structure of D-*erythro*-N-palmitoyl-sphingosylphosphorylcholine (top) and *sn*-L-dipalmitoyl-phosphatidylcholine (bottom) showing the amphipathic nature of the molecules. (From Barenholz and Gatt, 1982.)

II. LIPID DIVERSITY IN MEMBRANES

It is tempting to assume that lipid heterogeneity is a prerequisite for membrane function optimization. Indeed, this presumption is supported by numerous experimental findings (Roseman and Thompson, 1980; Wimley and Thompson, 1991; Pomorsky et al., 1996).

A.   Intermembrane Variability

1.   Lipid composition is often related to organ function (Rouser et al., 1968, 1972; White, 1973; Barenholz and Thompson, 1980; Barenholz and Gatt, 1982; Shmeeda et al., 1994).
2.   Membranes of various organelles have different lipid composition. An intracellular gradient exists, e.g., for sphingomyelin and cholesterol, with the highest content of these lipids found in the cell plasma membrane and the lowest amount in the nucleus and mitochondria (Fig. 7); (Rouser et al., 1968; White, 1973; Barenholz and Thompson, 1980; Barenholz and Gatt, 1982; Shmeeda et al., 1994).

Phospholipids spontaneously exchange very slowly between membranes, with $t_{\frac{1}{2}}$ of the order of many hours; cholesterol exchange is much faster ($t_{\frac{1}{2}}$ in the minutes to hours range). The bulk of transfer in either case depends on the activity of lipid transfer proteins, which have specific activities in the range 50–2500 nmol min^{-1} mg^{-1}; this should suffice for equalization of the composition of membrane lipids throughout the cell, but that is not observed. The mechanism counteracting expected equalization remains unknown but probably involves multifactorial, energy-dependent homeostatic processes involving both membrane lipid synthesis and vesicular transport (see Shmeeda et al., 1994).

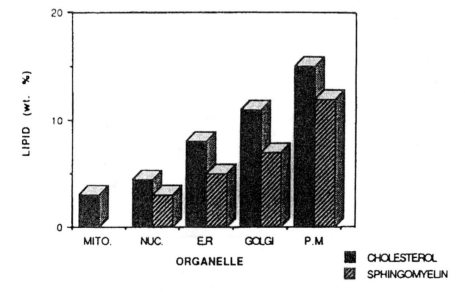

Fig. 7   Relative amount of sphingomyelin and cholesterol in membranes of rat hepatocyte organelles. (From Shmeeda et al., 1994.)

Intracellular lipid trafficking was partly elucidated by doing experiments with fluorescently labeled lipids (often using *N*-[5-(5,7-dimethyl BODIPY)-1-pentanoic acid as a probe). This revealed sphingolipid transport along the endocytic pathway in human skin fibroblasts. Qualitatively similar observations have also been made with labeled lactosylceramide and monosialoganglioside (GM_1; (Pagano et al., in press).

Membrane homeostasis deteriorates during cell aging. Increased cholesterol and sphingomyelin content and a higher degree of acyl chain saturation are the two most prominent results of this (Shmeeda et al., 1994).

B. Membrane Asymmetry

Membrane components are not randomly distributed. Proteins and lipids are found in different relative quantities in the outer and inner leaflets of most biological membranes (Rothman and Lenard, 1977; Etemadi, 1980a; 1980b). In the plasma membrane of eukaryotic cells, for example, phosphatidylserine, phosphatidylethanolamine, and phosphoinositides are mainly located in the inner monolayer, and phosphatidylcholine, sphingomyelin, and glycolipids reside chiefly in the outer bilayer half (Devaux, 1993; Devaux in Barenholz 1998a). Fatty acid composition, even for lipids with identical headgroups, is often different on either bilayer side as well (Patzer et al., 1978a; Crain and Zilversmit, 1980).

The ability of cells to sort apical and basolateral proteins and to target them to appropriate destinations helps to maintain membrane asymmetry. Such sorting involves two intracellular sites, by and large: the trans-Golgi network and the basolateral endosome. Constitutive protein traffic in the epithelial cells has been shown to be regulated via a classical signal transduction pathway involving heterotrimeric G proteins and protein kinases. The diversion of apical and basolateral proteins into specific pathways can be mediated by signals contained within those proteins (Le Gall et al., 1995). Apical sorting information is thought to be localized in the luminal domain of transmembrane proteins and, in the case of proteins anchored to the membrane, via a GPI anchor. Apical sorting information is provided by the lipid moiety. In contrast, basolateral signals have been identified in the cytoplasmic domain of transmembrane proteins. Similarities between basolateral signals and those required for endocytosis have suggested that the two sorting processes may be related mechanistically (Le Gall et al., 1995).

Maintenance of a stable membrane asymmetry is due in part to the activity of proteins responsible for the lipid synthesis and in part to the activity of specific proteins called "phospholipid flippases," which catalyze the exchange or "flip-flop" of the lipids between both membrane leaflets. Aminophospholipid translocases belong to the latter group (Devaux, 1993; Williamson and Shlegel, 1994). The best studied aminophospholipid translocases are those that selectively transport aminophospholipids from the outer to the inner monolayer of erythrocyte membranes (Devaux, 1993; Devaux in Barenholz 1998a). However, related activity has been found in the plasma membrane of many different cells, including yeast. Proteins of the multidrug resistance (MDR) family are also flippases, because they are able to translocate the amphiphilic drugs and the phospholipids from the inner to the outer monolayer of the plasma membrane. It has been shown that the expression of human *MDR3* and mouse *mdr2* genes promotes the translocation of long-chain phosphatidylcholines across the cell membrane. Conversely, expression of the *MDR1* gene stimulates the outward motion of phospholipids that possess at least one short chain (van Helvoort et al., 1996). Other proteins ("scramblases") may be involved in the rapid redistribution of phospholipids following calcium entry into the cell (Comfurius et al., 1996). Moreover, ATP-independent phospholipid transporters exist in the inner cell membranes. In particular, the diffusion of glycerophospholipids and of glycosphingolipids across the membrane of endoplasmic reticulum and of Golgi apparatus relies on the activity of such specific, ATP-independent proteins. The process is both saturable and competitive. Scramblases were invoked to explain the rapid redistribution of phospholipids after entry of calcium into the cell (Comfurius et al., 1996); Ca^{2+} therefore induces transbilayer redistribution of all major phospholipids in human erythrocytes (Williamson et al., 1992).

Although the reasons for lipid asymmetry are not fully understood, it is plausible that it provides an asymmetric environment for the membrane enzymes. Thus, PS and PE reorientation could be a way of controlling or triggering specific enzymes. Asymmetric distribution of phospholipids may also be important for the fusion of competent membranes and may govern the side on which the two membranes begin to fuse. Lipid pumps, as well as the enzymes responsible for the net transmembrane flux of phospholipids, may also provide the driving force for membrane bending. This arguably happens during the formation of endocytic vesicles (Devaux, 1991, 1992). Incorrect phospholipid asymmetry, conversely, may lead to pathologic situations (see further text and Zwaal and Schroit, 1997).

Few of the proteins mentioned so far have been identified unambiguously. Furthermore, all reconstitutions attempted so far with purified or partially purified proteins gave only low transport activity. In contrast, clear evidence for the importance of membrane asymmetry exists. For example, under normal conditions, phospholipid asymmetry in human red blood cells is well maintained during the cell circulation in the blood. In other situations, this asymmetry is impaired. Phosphatidylserine, which is normally confined to the inner membrane half, then becomes exposed on the outer leaflet of the erythrocyte membrane. This happens in sickle cell disease (Wood et al., 1996) and in β-thalassemia (Borenstain et al., 1993; Kuypers et al., 1998). The exposure of phosphatidylserine to the blood plays an important role in the pathology of these diseases, leads to anemia and shorter survival of adult red blood cells in the peripheral blood, but is also an early signal for cell apoptosis (Savill, 1997). Recognition of exposed phosphatidylserine by macrophages may also explain the shorter lifetime of adult red blood cells in these and other diseases (Kuypers et al., 1998) and the clearance of "old," senescent red blood cells (Schroit et al., 1985). Phosphatidylserine exposure, furthermore, may play a role in the hypercoaguable state associated with severe β-thalassemia intermedia (Borenstain et al., 1993). This illustrates the importance of membrane lipid asymmetry for our wellbeing.

C. Lipid Distribution in the Membrane Plane

Evidence suggests that membrane lipids are not randomly distributed but rather are organized laterally in distinct domains (Shmeeda et al., 1994; Barenholz, 1984; Lipka et al., 1991; Lin et al., 1991; Pike and Casey, 1996; Edidin, 1997). This is discussed in more detail later in the text.

One should distinguish between the long-range lateral organization, which can be observed in the membranes by methods such as fluorescence recovery after photobleaching (FRAP) (Peters, 1981), and the short-range order, which is normally explored by means of diffraction or spectroscopy. The latter order is related to the short-range interactions between membrane components. These properties determine microdomain formation and free volume in the membrane. The relation of long- and short-range organization to the membrane structure and function is another important topic of modern "membranology" (Shinitzky and Barenholz, 1978; Peters, 1981; Shinitzky and Yuli, 1982; Barenholz et al., 1996; Bar et al., 1997).

D. Relevance to Signal Transduction and to Gene Expression

Almost 20 years ago, phosphatidylinositol-4,5-biphosphate was found to be a key molecule in signal transduction (reviewed, e.g., by Nishizuka et al., 1990). Its hydrolysis by the specific phospholipase C releases two messengers: one of lipidic nature (diacylglycerol) and the other soluble in water (inositol triphosphate). Other membrane lipids, including sphingomyelin (Hannun, 1994; Kolesnick and Golde, 1994) and phosphatidic acid (Hannun, 1996; Spiegel and Milstien, 1995; Spiegel et al., 1996) and their metabolites (for a review see Gomez-Munoz, 1998), are important as well. It is noteworthy that the headgroup selectivity (for diacylglycerol and phosphatides) is complemented or supported by the selectivity for certain acyl chains (Hodgkin et al., 1998).

The connection between the membrane's physical state and gene expression is also interesting. Various studies of bacteria and eukaryotic organisms have shown, for example, that the fluidity and/or lateral phase separation in membranes (such as "raft" formation) can affect the expression of stress-related genes and of genes controlling membrane lipid composition (Vigh et al., 1998). This supports the notion that membrane lipid composition plays a role in the regulation of cell protein function.

E. Environmental Effects

The lipid composition of biological membranes is responsive to changes in environmental factors such as diet (Rintoul et al., 1979; Deuticke, 1977) and temperature (de Mendoza and Cronan, 1983; Reizer et al., 1985; Lindblom and Rilfors, 1989; Vigh et al., 1998). The response to temperature variation can be described in terms of homeoviscous adaptation (Cossins, 1977; Cossins and Prosser, 1978; Sinensky, 1974) and has also been studied in detail for bacteria (de Mendoza and Cronan, 1983; Reizer et al., 1985).

F. Effect of Development, Aging, and Disease

Membrane lipid composition also changes during the development and aging of the cell. This affects membrane and cell properties (Kutchai et al., 1976; Kapitulnick et al., 1979; Barenholz and Thompson, 1980; Barenholz and Gatt, 1982; Shmeeda et al., 1994; Yechiel et al., 1994), including cell elimination by the body or cell apoptosis. Various pathological conditions, such as malignant cell transformation or atherosclerosis, cause similar alterations as well (Rouser et al., 1968; 1972; Wood, 1973; Wallach, 1976; Barenholz and Gatt, 1982; Horrocks and Sharma, 1982; Shmeeda et al., 1994). Disturbed membrane homeostasis is one possible origin of this.

The relationship between (perturbed) membrane lipid composition and membrane function can be explained in different ways. It was first customary to consider membrane fluidity (Singer and Nicolson, 1972; Shinitzky and Barenholz, 1978; Sandermann, 1978) and viscotropic effect (Sandermann, 1978) for this purpose; more recently, the more detailed picture of a function-adapted, laterally and transversely organized membrane became fashionable. The latter approach is more reliable and is supported by the observation that lateral mobility of many membrane proteins is not diffusion limited (Peters, 1981); instead, the mobility of various membrane components depends on membrane "compartmentalization." (Zachowski, 1993; Simons and Ikonen, 1997).

III. CLASSIFICATION AND NOMENCLATURE OF PHOSPHOLIPIDS

There is more than one way to classify phospholipids. We will adopt here the classification of Strickland (1975). Accordingly (Table 1), we will first distinguish five phospholipid classes: (a) phosphoglycerides (all derivatives of *sn*-glycerol-3-phosphoric acid); (b) phosphoglycolipids; (c) phosphodiol lipids (monoacyl or monoalkyl derivatives of dihydroxy acetone phosphate); (d) phosphosphingolipids (all lipids containing a phosphorus attached to a sphingoacid base); (e) phosphonolipids, in which the phosphorus is attached to the nitrogenous base directly by a C–P bond (and not as in the phosphoglycerides, which are, with the exception of phosphatidic acids, chiefly phosphodiesters). The phosphonolipids include mainly glycero- and sphingolipids, which are present in some bacteria but mainly in protozoa (Rosenberg, 1975; Brown, 1998).

At the second level we will refer to the way or type of bond in or by which the aliphatic chain is bound to the phospholipid backbone. Most of the phosphoglycerides are mono-*O*-acyls or *O*-alkyls or *O*-alk-1-enyls, *O*-alk-1-enoyls, or the corresponding double-chain derivatives. Phosphosphingolipids contain a sphingoid base [(dihydrosphingosine (sphinganine), sphingosine, and phytosphingosine] instead (Barenholz, 1984).

The third level of lipid hierarchy pertains to the polar headgroups (as described in Table 1 and as illustrated in Fig. 4).

The fourth, and last, distinction can be made on the basis of acyl, alkyl, or sphingoid base chain variability in terms of the number of carbons, degree of unsaturation, type and location of double bond(s), degree and location of branching, presence of rings, etc. (Fig. 8). The sphingoid base may vary in type (sphingosine, dihydrosphingosine, phytosphingosine) and in chain length.

These four phospholipid levels of hierarchy or classification criteria, combined with the presence of sterols and glycero- and sphingoglycolipids, can be used to distinguish between numerous lipidic membrane ingredients, produced in a complex metabolic pathways (summarized in Fig. 9) (see Shmeeda et al., 1994, and recent biochemistry texts for more details). Lipid variability generates a broad spectrum of physicochemical membrane properties that are important for the proper cell function (Vigh et al., 1998).

A. Stereospecific Atom Numbering

All phosphoglycerides have at least one asymmetric carbon atom (carbon 2 of the glycerol), and sphingolipids have at least two [sphinganine (dihydrosphingosine)

Table 1 Phospholipid Classification[a]

Classes[b]	Classes continued[b]
A. Phosphoglycerides	(19) Monomannosyl-hexamannosyl inositol phosphoglycerides
(1) Phosphatidic acids[b]	(20) Glucose phosphoglyceride
(2) Cytidylic phosphoglycerides (CDP diglyceride)	(21) O-Diglucosylglycerol phosphoglyceride
(3) Choline phosphoglycerides[b]	
(4) Ethanolamine phosphoglycerides[b]	B. Phosphoglycolipids
(5) N-Methylethanolamine phosphoglycerides[b]	(1) Diacyl(glycerylphosphoryldiglucosyl) glycerol
(6) N,N-Dimethylethanolamine phosphoglycerides[b]	
(7) N-Acyl ethanolamine phosphoglyceride	C. Phosphodiol lipids
(8) Serine phosphoglycerides[b]	(1) Acyl dihydroxyacetone phosphate
(9) N-2-(Hydroxyethyl)alanine phosphoglyceride	(2) Alkyl dihydroxyacetone phosphate
(10) Glycerol phosphoglycerides	D. Phosphosphingolipids
(11) Glycerophosphate phosphoglycerides	(1) Sphingomyelin (ceramide phosphorylcholine)
(12) Phosphatidylglycerol phosphoglyceride (diphosphatidylglycerol)	(2) Ceramide phosphorylethanolamine
(13) Mono- and diacylglycerol phosphoglycerides (lyso bis phosphatidic acids)	(3) Ceramide phosphorylglycerol
(14) Glucosaminylglycerol phosphoglyceride	(4) Ceramide phosphorylglycerophosphate
(15) O-Amino acid esters of glycerol phosphoglycerides	(5) Ceramide phosphorylinositol–containing lipids
(16) Inositol phosphoglyceride	(6) Ceramide-1-phosphate
(17) Inositol monophosphate phosphoglyceride	(7) Sphingosine-1-phosphate
(18) Inositol diphosphate phosphoglyceride	E. Phosphonolipids
	Many types, including glycerolipids and phosphosphingolipids (Rosenberg, 1975)

[a] An attempt has been made to conform to the recommendations of the IUPAC-IUB Commission of Nomenclature.
[b] (a) exists in the form of l-acyl- or 2-acyl- (i.e., lyso); (b) 1,2-diacyl- (i.e., phosphatidyl); (c) 1-alk-1'-enyl-, 2-acyl- (i.e., plasmalogen); (d) 1-alkyl-, 2-acyl-; and (e) 2,3-dialkyl, also acyl and alkyl chains varied to a large extent.

and sphingosine (carbons 2 and 3 of this sphingoid base)] or three [phytosphingosines (carbons 2, 3, and 4)] such atoms. Biological phospholipids normally exist in only one enantiomeric form. In principle, however, several forms are possible, which can be described using the *R-S* or, D/L nomenclature, or stereospecific (*sn*) numbering.

Membrane phosphoglycerides are in 1-α phosphatidyl-X configuration, also referred to as 1-3-phosphatidyl-X or DL-phosphatidyl-X (which is confusing) or *R*-3-phosphatidyl- or 3-*sn*-phosphatidyl-. The nomenclature pertaining to membrane phosphosphingolipids, with two or three asymmetric carbon atoms, is even more confusing. The most common corresponding system refers to D-erythro (for dihydrosphingosine and sphingosine) and D-ribo- (for phytosphingosine) compounds; alternatively, the respective descriptors 2*S*, 3*R* and 2*S*, 3*R*, 4*R* can be used.

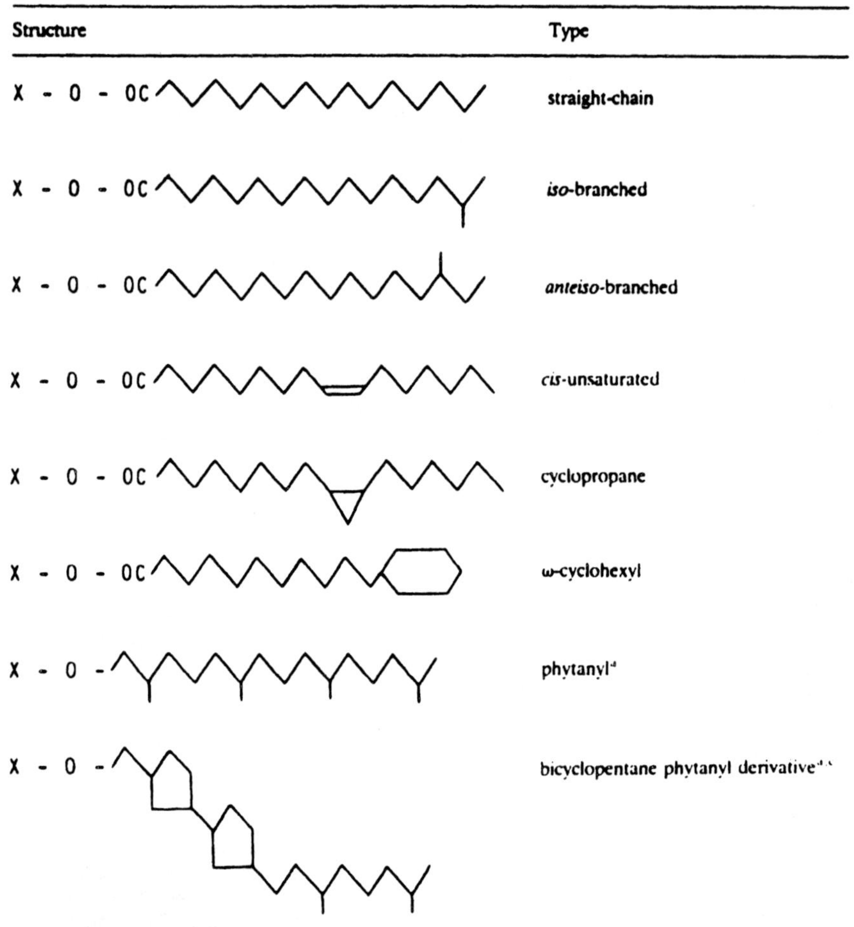

Structure	Type
X - 0 - OC	straight-chain
X - 0 - OC	*iso*-branched
X - 0 - OC	*anteiso*-branched
X - 0 - OC	*cis*-unsaturated
X - 0 - OC	cyclopropane
X - 0 - OC	ω-cyclohexyl
X - 0 -	phytanyl[a]
X - 0 -	bicyclopentane phytanyl derivative[a,c]

[a] In archaebacteria the phytanyl chains and their cyclopentane derivatives are in ether, not ester, linkage to the lipid head group.
[b] Only one acyl chain is shown, whereas two are usually present per lipid; the remainder of the molecule is represented by X.
[c] For ease of comparison, one half of the alkyl chain only is shown; *in vivo* they exist as 40C tetraethers that span the width of the membrane

Fig. 8 Diversity of membrane lipid chains. (From Russel, 1984.)

IV. (PHOSPHO)LIPID SELF-AGGREGATION AND ORGANIZATION

Membrane lipids are amphiphiles: they comprise at least one polar and one apolar region. Exposure of a hydrophobic moiety to water is thermodynamically unfavorable and enforces amphiphile self-aggregation in aqueous surroundings (reviewed, e.g., in Cevc and Marsh, 1987; Lichtenberg and Barenholz, 1988; Israelachvili, 1992). The resulting aqueous dispersion, or suspension, can comprise a variety of aggregate forms (Figs. 10 and 11). Their common feature is that the apolar portions of the amphiphiles form a hydrophobic core while the polar groups build an aggregate surface and interact with the aqueous medium.

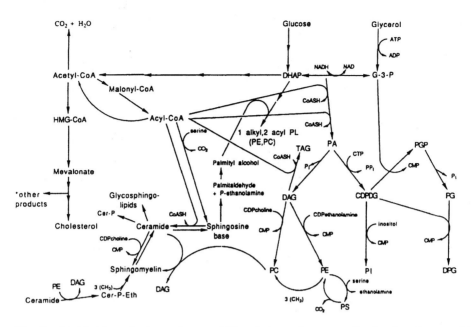

Fig. 9 Lipid biosynthetic pathways. (CE, cholesterol esters; Cer-P, phosphoceramide; Cer-P-Eth, ceramide phosporylethanolamine; Chol, cholesterol; CL, cardiolipin; DAG, diacylglycerol; DHAP, dihydroxyacetone phosphate; FA, fatty acid; G3P, glycerol-3-phosphate; HMG-CoA, 3-hydroxy-3-methylglutaryl-coenzyme A; LPC, lysophosphatidylcholine; LPE, lysophosphatidylethanolamine; PA; phosphatidic acid; PC, phosphatidylcholine; PE, phosphatidylethanolamine; P-ethanolamine, phosphorylethanolamine; PG, phosphatidyl-glycerol.)

Only exceptionally polar lipids, such as single-chain lysophospholipids and some charged lipids (such as phosphatidylglycerol), dissolve in water up to a limiting concentration of 10^{-5} to 10^{-2} mol/L. Above this limit (critical micelle concentration [CMC] or critical aggregation concentration [CAC]), such lipids form micellar (or other aggregate) suspensions. Typical small aggregates contain a few dozen up to several hundred molecules per unit (Fromherz et al., 1985) and have a shape that depends on the distribution of polar residues on the individual molecules (Helenius, 1979), as well as on the presence of other ingredients. Many lipid micelles, such as those consisting of sodium dodecyl sulfate (SDS) and other (longitudinal) amphipaths, are spherical. Disklike or cylindrical micelles are frequently observed with "facial" biosurfactants, such as bile salts; Fig. 11 illustrates this. Bigger aggregates often change their (rather irregular) shape rapidly. Such larger aggregates are at least locally of bilayer form.

Nonbilayer structures are highly curved (radius < 5 nm), at least locally. Structures in a normal, noninverted phase therefore resemble an open lipid bilayer, or an "edge". In the cells, inverse nonbilayer structures prevail in the mitochondria and thylakoids, where they form cubic phases (for further discussion see Landh, 1995).

Little hydrophilic, relatively "dry" but polar, lipids such as phosphatidyl-ethanolamines and many other phospholipids at low pH or in the presence of complexing ions form inverted nonbilayer structures, by and large. The reason for this is the relatively low polarity of protonated lipids or other lipid-ion complexes. Such

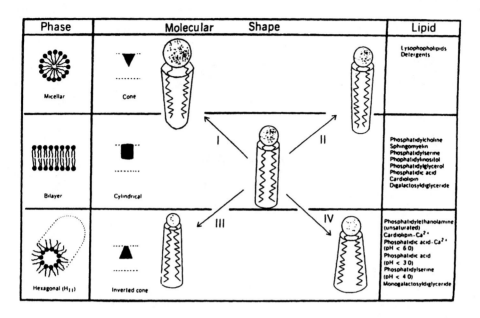

Fig. 10 Polymorphism and molecular shapes of major membrane lipids. Lipids with a balanced packing parameter ($PP = V/al_c \sim 1$) form bilayers. Increasing the headgroup repulsion path I, following charging or increasing molecular hydration and/or decreasing the total volume of lipid chains (path II, e.g., following hydrolysis of one of the chains, chain shortening, and/or lowering the temperature) will diminish the packing parameter; possibly resulting in transformation into a micellar phase. In contrast, headgroup area reduction (path III, e.g., after decreasing molecular charge and/or hydration, such as after ion binding), by increasing the volume of hydrophobic molecular part (path IV, e.g., after raising the temperature or increasing the lipid chain length and/or unsaturation), will enlarge the packing parameter value and favor transformation into a nonbilayer phase. (From Lichtenberg and Barenholz, 1988.)

aggregates resemble a "stack of spaghetti" (in an inverted hexagonal phase), "plumber's nightmare" (in one of the frequent inverted cubic phases), or other complex forms (Chapman, 1973; Seddon, 1990; Landh 1995; Delacroix et al., 1996). The lowest panel in Fig. 12 provides a nice example of this.

Can one predict, on the basis of thermodynamic considerations, the outcome of lipid aggregation? As a rule, amphiphiles form the smallest possible aggregate in which the hydrophobic portion of the molecule minimizes its contact with water; but this may yield different structures. In the absence of a universally accepted definition of the various structure types, the simplest categorization is into nonlamellar and lamellar aggregates. The latter are the only structural forms with an enclosed aqueous compartment and are found in plasma membranes, intracellular vesicles, or liposomes, for example. In contrast, nonlamellar phases do not often have a water core, are bicontinuous, or tend to have open ends.

Small (1970; 1986) suggested that only the nonsoluble, swelling amphiphiles that form stable monolayers at the water-air interface form lamellar structures. His empirical observation still holds, but over the years alternative, more quantitative approaches have been developed to predict the capability of different amphiphiles to form closed vesicles.

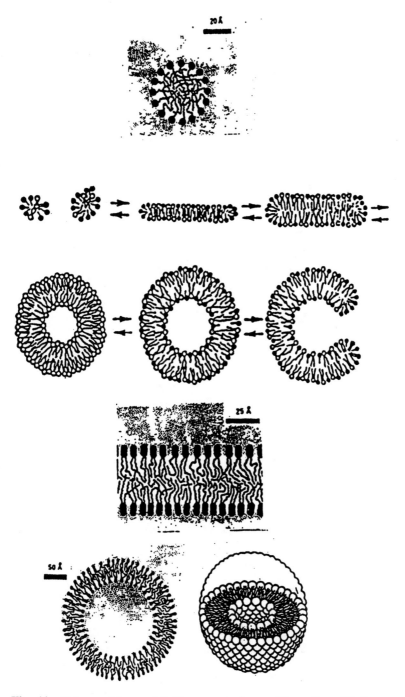

Fig. 11 Polymorphism and lipid aggregate forms. (From Cevc, 1996.)

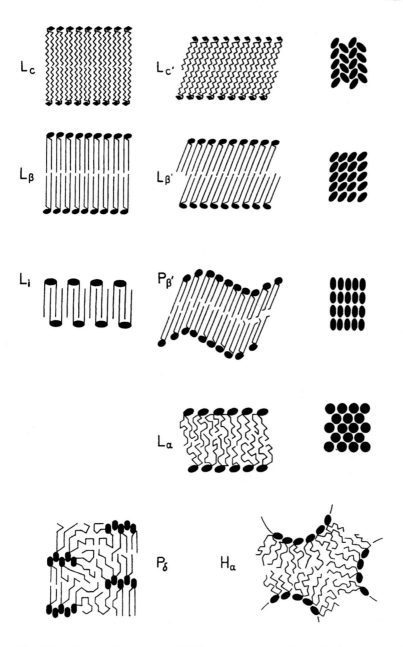

Fig. 12 Commonly observed lipid phases in suspension (side and top (right panel) views). Increasing temperature reduces the order in the hydrocarbon membrane core and sometimes induces nonbilayer (e.g., H_{II}) phases. The lipid chains expand laterally and the membrane thickness decreases in the less ordered phases, intermembrane separation simultaneously increases. All phases shown were observed with (partially) hydrated phosphatidylcholine. Biological membranes are typically in the L_α phase. (From Cevc, 1991a.)

The most popular approach has been proposed by Israelachvili and colleagues (Israelachvili et al., 1980; Israelachvili, 1992), who introduced a "packing parameter"

$$PP = \frac{V}{al_c} \tag{1}$$

based on simple geometric considerations and the structure of the amphiphiles considered. Geometric parameters are the hydrophobic volume (V), the average surface area occupied by the polar region of the amphiphile at the air-water interface (a), and the fully extended length of the hydrophobic region (l_c; in the case of phospholipids, glycolipids, and glycerides, the length of the hydrocarbon chain; for steroids, the steroid nucleus).

Amphiphiles with a value of the packing parameter close to 1, that is, cylinder-like molecules, tend to form bilayers; a majority of lipids under physiological conditions fulfill this criterion. Molecules with a much lower or a higher packing parameter value self-associate into micelles or form inverted hexagonal structures (H_{II} phase), respectively (Fig. 10). The smaller the packing parameter, the closer the middle shape is to a sphere.

However, changing the temperature, the medium pH, the ionic strength of the suspension medium, etc. may change the packing parameter and affect the aggregation state of an amphiphile. The concept of molecular shape, consequently, must be used with prudence, the idea of typical and constant PP for a given lipid being potentially misleading. The rich thermotropic (Cevc, 1991a) and lyotropic (Lichtenberg and Barenholz, 1988) polymorphism (Fig. 12) of biological amphiphiles proves this clearly.

It is noteworthy that gangliosides, despite having two hydrocarbon chains, have a rather low packing parameter value in a dilute aqueous suspension. This is due to the large, charged, polar headgroup. Such lipids therefore form large micelles rather than bilayer vesicles (Barenholz et al., 1980; Cantu et al., 1990; Felgner et al., 1981, 1983), but they may organize in a more complex and extended fashion as well. Membranes nevertheless can accommodate gangliosides at a level dependent on the headgroup length and charge, the bulkier heads being more destructive for the lipid bilayer. Saturation concentrations of 10 mol % and higher are possible (Barenholz et al., 1980) if the hydrocarbon chains are not too short.

Prediction of aggregate form for a mixture of amphiphiles requires detailed information on the molecular arrangement within the resulting assembly. For a rough estimate of the probability that a given mixture will form a bilayer, one can calculate the mean PP value from the weighted average of the PP values for the individual components.

Lipids or lipid mixtures that normally "dislike" lamellar phases and lipid vesicles, owing to too small or too large a packing parameter PP, can still be incorporated in a stable lipid bilayer. Good examples of this are the phosphatidylethanolamines with fluid chains combined with cholesterol (PP > 1.0) and lysophosphatidylcholines (lyso-PC) as well as gangliosides (PP < 1.0). The combination of individual amphiphile properties may affect the outcome of aggregation as well. Fluid-phase lyso-PC and fatty acids form micelles on their own, due to the small PP value, but yield lamellar phase vesicles when combined with each other in an equimolar mixture. On the other hand, diacyl-PC, which normally prefers lamellar phases, undergoes a transition in a fluid nonlamellar phase (most often of H_{II} type), when mixed with a twofold molar excess of homologous fatty acids or fatty alcohols (Seddon, 1990).

Amphiphile interaction with ions and polar molecules alters the PP value (Fig. 10). Such interaction, consequently, can enforce a transition from a lamellar phase into either a micellar or hexagonal structure. These possible effects of soluble compounds on the amphiphile packing must be kept in mind to understand the basic properties of biological membranes.

V. BILAYER MEMBRANE

The bilayer is the most common form of polar lipid aggregate with PP $\sim$1 in water. Its symmetry can be quasi-planar, cylindrical or periodic (e.g., of cubic type).

Each lipid bilayer consists of two opposing hydrocarbon monolayers. These can overlap but are always separated from the surrounding water by two layers of polar lipid headgroups (Fig. 13). Lipid headgroups can, but need not, attract each other

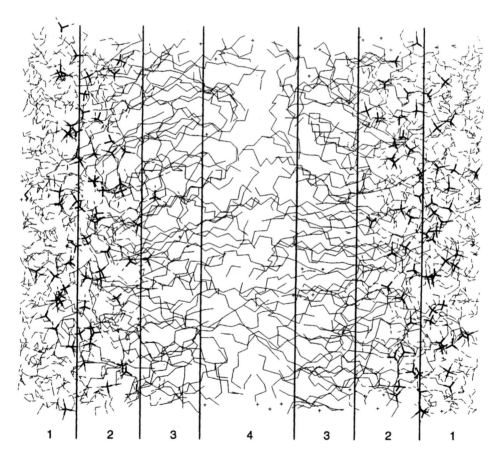

Fig. 13 Computer simulation of a hydrated phosphatidylcholine bilayer in the fluid phase. Chain disorder increases nonlinearly toward the membrane center (4) resulting in reduced density membrane core. Polar headgroups [2, thick: phosphate segments] are spread over substantial solute volume; however, they do not intercalate very deep into the membrane core (3) or reach too far into the solution (1). Solvent molecules (dashed), however, can penetrate the membrane in proportion to the number of defects or voids in the bilayer. (From Egberts et al., 1994.)

laterally via hydrogen bonds or ionic bridges. However, polar headgroups in most cases at least partially repel each other, e.g., via steric, electrostatic, or hydration forces.

When dispersed in water above the chain-melting phase transition temperature (T_m), most phospholipids are prone to form closed, spherical bilayer vesicles (liposomes) (Goll et al., 1982). This requires no external energy, if multilamellar liposomes are to be created (Bangham et al., 1974; Bangham, 1993; Cevc, 1992; Lasic, 1990). To make small lipid vesicles, however, energy must be put in the system to transform multilamellar into unilamellar and large into small liposomes. The radius of lipid vesicles always exceeds the corresponding lipid dimensions (< 3 nm), owing to the membrane packing constraints (Helfrich, 1986; Israelachvili, 1992). The minimum size for a reasonably stable, unstressed vesicle made of even the most polar lipids is greater than 40 nm; for the less polar species the size far exceeds 100 nm. Phosphatidylethanolamines in the gel phase, for example, do not form stable vesicles (Cevc, 1989).

Too small liposomes tend to fuse to relax packing stress. The equilibrium phospholipid vesicle radius is therefore greater than 100 nm and often exceeds the size of simple bacteria (> 0.25 μm).

Giant vesicles can be prepared, and maintained, with a radius greater than 50 μm. Such large liposomes normally do not arise spontaneously, but can be made by careful hydration of a thin lipid film or by means of an electrical field applied over a stack of partially oriented, surface-supported lipid membranes. Suspensions of unilamellar, relatively small lipid vesicles are most easily prepared by extrusion or dialysis. With the former method, care must be taken to eliminate oligolamellar liposomes, however.

A. Hydrocarbon Core

The core of a simple lipid membrane typically comprises aliphatic chains in different conformations: in the ordered or gel phase (L_β), fully saturated chains are completely extended in an all-*trans* configuration. Chains then occupy a layer of thickness $d_l = 1.27n_c$, perpendicular to the membrane surface, where n_c is the number of carbon atoms per chain. Tilting the chains relative to the membrane normal by an angle f_{tilt} reduces the hydrocarbon core thickness to $d_l = 1.27n_c \cos f_{tilt}$. Each double bond in the hydrocarbon chain shortens the effective chain length by approximately 0.15 nm in the gel phase. The chain unsaturation favors chain melting and transition into one of the (chain) disordered phases (see Sec. VII.D for a more detailed discussion).

In the lamellar fluid phase (L_α), which prevails in biological membranes, lipid chains are partly disordered, molten. The hydrophobic membrane part thickness in such a phase therefore depends less on the nominal chain length than in the gel phase, the incremental change for the former phase being only 60% of that for the latter. Orientational chain disorder increases nonlinearly toward the chains ends (Fig. 13). This gives rise to the so-called order- and polarity-parameter profile. The order-parameter profile exhibits a plateau if plotted along the chain axis (Fig. 14) but does not show this feature if plotted perpendicular to the membrane surface. The order parameter is sensitive not only to the degree of chain unsaturation but also to the total chain length (Fig. 13). Changes in the headgroup region are less important in this respect, if comparison is made at the same relative temperature ($T - T_c$) and corresponding phase. Headgroup changes strongly affect many other membrane parameters, however.

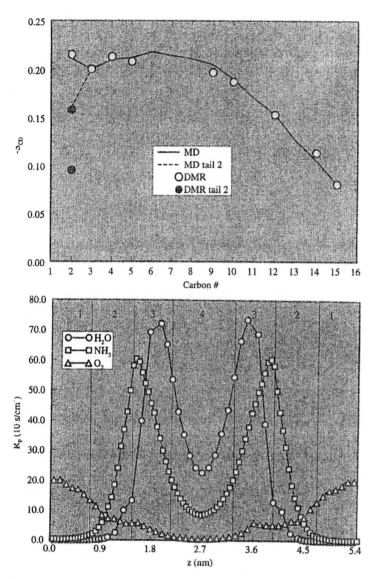

Fig. 14 Effect of lipid chain conformation on lipid bilayer properties. Order parameter profile in a fluid bilayer membrane perpendicular to the bilayer surface (upper panel) is indirectly reflected in the local resistance to various solute permeation (lower panel), which in turn is related to the dielectric constant profile across the membrane. (Modified from Egberts et al., 1994.)

Cholesterol, polypeptides, proteins, and other conformationally confined molecules affect the intramembrane order. Cholesterol typically stiffens chain ends but lowers the average chain order near the lipid headgroups (Straume and Litman, 1987; Ipsen et al., 1990). This happens preferentially near the saturated chains, which have a higher affinity for the cholesterol molecules than the more unsaturated lipids (Gawrisch and Huster, in Barenholz 1998a).

Proteins can increase or decrease the average chain order. The precise outcome of such "perturband" addition is sensitive to the protein shape and dimensions (relative to those of a typical lipid molecule); the result also depends on the extent and kind of the coupling with the lipid bilayer. In general, anything that will stretch or compress the lipid chains will increase the (local) degree of chain order in the membrane core. This includes modifications in or of lipid headgroups as well.

Completely hydrophobic molecules, such as alkanes with up to six carbon atoms per chain, partition into the core of lipid membranes. Quite frequently, even apolar molecules arrange themselves with the long axis perpendicular to the hydrocarbon chains.

Volume density of lipid bilayers decreases with increasing temperature and decreasing hydrocarbon chain length. So does the membrane thickness, whereas the area per lipid shows the opposite trend.

Some bilayer characteristics reflect the corresponding monolayer values directly. For example, the elastic curvature modulus of the bilayer is given by $B = K_A d_b^2/2$, where K_A is the membrane area compressibility modulus and $d_b/2$ is the lipid monolayer thickness. For many other membrane parameters no simple relations exist. The bulk membrane compressibility modulus is not given by the ratio $K_B = K_A/d_b = 3 \times 10^7$ N m^{-2}, as one might expect, but is much higher, $K_B = 1.3 \times 10^9$ N m^{-2} (Cevc and Marsh, 1987). This demonstrates that the surface of a fluid lipid bilayer is 100 times more compressible than that of a simple liquid. However, the temperature and the pressure responses to a periodic volume excitation are proportional to the equilibrium excess heat capacity and bulk modulus, respectively (van Osdol et al. 1991).

The main role of the hydrocarbon membrane core is to act as a barrier to the transport of polar solutes (Bangham et al., 1974; de Gier, 1993) and simultaneously to provide a good solvent for the less polar entities, such as lipids on the hydrophobic parts of proteins. The fluidity of membrane interior is essential in the latter respect.

Related lipids with very different hydrocarbon chain packing do not mix properly; typically, a difference of $n_c = 4$ causes lipid separation. The boundaries between the resulting domains are believed to be gradual and to involve several molecular layers in the transition region. The same is true for the transition region between the hydrophobic membrane core and the hydrophilic, or aqueous, surrounding of a membrane. Any such transition is gradual and takes place over a volume of finite thickness. It is therefore more appropriate to speak of a membrane-water interphase than of a membrane-water interface. Figure 13 illustrates this point.

B. Headgroup Region and Beyond

The membrane-solution "interface" (interphase) includes all molecular groups or molecules that do not fall clearly into the category of the membrane core and also do not belong directly to the membrane-bathing solution. Said interphase can therefore cover a substantial range if the polar parts of molecules in close proximity to nonpolar molecular parts extend over an appreciable distance. Even in the case of simple phospholipids, such as phosphatidylcholines, with a nominal headgroup length of 0.5 nm from the phosphate and 0.9 nm from the glycerol group, the interfacial region is at least 0.8 nm thick (Kirchner and Cevc, 1994); gangliosides span a distance of several nanometers; artificial lipopolymers, such as pegylated lipids used to suppress the adsorption of proteins to artificial membrane surfaces, have headgroup region 4–10 nm or 15 nm wide, in the case of poly(ethylene glycol) 2000 (PEG-2000) (Priev et al., 1998) and PEG-5000 (Vierl and Cevc, unpublished data), respectively. The carbohydrate-rich layer near the bacterial cell wall, called the glycocalyx, may be tens of nanometers wide in the case of gram-negative cells. The width of the glycocalyx of many eukaryotic cells also approaches 10 nm. Such large dimensions, together with the obvious variability in the interfacial properties of the different living systems, suggest that the membrane-solution interphase has an important functional role.

One feature of the protein- and lipid headgroup-rich region near the membrane surface is to ensure a smooth transition between the essentially charge-free membrane interior and the ion-rich bulk electrolyte with a high dielectric constant. The interphase takes the role of a localized buffer and can catalyze chemical reactions and lateral transport.

The continuous interfacial polarity profile of a phosphatidylcholine bilayer is shown in Fig. 14. Its thickness increases with progressive membrane hydration (Cevc et al., 1995) and with lowering of the bulk electrolyte concentration. Very polar lipids and substances with a long polar headgroup give raise to a wider interphase than substances with a less hydrophilic headgroup; the former consequently swell more strongly than the latter. Adsorption of water and other polar molecules at the membrane-solution interface generally widens the interphase; so does increased segmental mobility in the polar part of molecules forming a membrane. Increasing temperature enlarges the width of the interfacial region in the lamellar and undulated lamellar phases and in standard nonlamellar phases too. In the inverted nonlamellar phase the opposite trend is observed (see further discussion).

The membrane-solution interphase plays an important role in recognition as well. This can involve recognition between two membrane segments but also selective recognition of or adaptation to foreign molecules. Complex formation, molecular sorting, and immune responses exploit this principle (see Sec. VII for further discussion).

1. Membrane Charge Distribution

Structural charges are typically distributed nonuniformly throughout the interfacial region. Owing to the low polarity of the membrane core and the dielectric (polarity) profile across the interphase, the membrane-associated charges try to find their place as far away from the membrane core as possible. This (re)distribution is subject to mobility, entropy, and mutual electrostatic repulsion constraints, however. The actual charge distribution in the interphase therefore typically differs from the nominal location of charged groups on an isolated corresponding molecule: it is typically broader, especially toward the outer end of the interfacial region. A decaying exponential or a (skewed) Gaussian curve with corresponding width can be used to describe the resulting charge distribution on an artificial (phospholipid) membrane. To model the electrostatic properties of complex, biological, membranes with an extended interphase, several consecutive regions differing in charge density, width, polarity, and counterion concentration should be invoked. In the simpler approximation, a Donnan equilibrium approach can be taken (for more details see Cevc, 1990).

2. Electrostatic Membrane Properties

Most biological membranes are negatively charged. The properties of negatively charged lipid vesicles have therefore been widely studied. This has involved naturally occurring as well as synthetic phosphatidylserines (PSs), phosphatidylglycerols (PGs), phosphatidylinositols (PIs), phosphatidic acids (PAs), sulfatides, various gangliosides, and dialkyl phosphates. Until 10 years ago, however, few data were available on model membranes containing one of the few naturally occurring, but scarce, positively charged lipids (Bangham et al., 1974; Cullis et al., 1985; Lichtenberg and Barenholz, 1988). Since then, the growth of studies making use of lipids in cationic liposomes positively charged has been exponential, owing to the ability of cationic liposomes to complex with nucleic acids (DNA or oligonucleotides) and be used for gene or oligonucleotide delivery into cells (Lasic, 1997; Behr, 1994; Gershon et al. 1993).

One way to characterize the electrostatics of the membrane-solution interphase is to cite its so-called zeta potential (see Fig. 15a). This is defined as the potential in the zero-shear plane and is typically determined by the electrophoretic mobility measurements with vesicle suspensions. Simple microscopes or dedicated devices related to photon correlation spectrometers are used for the purpose (Bangham et al., 1974; for a review see Cevc, 1993b).

To study bilayer electrostatics in greater detail, fluorescent probes with a known location of the fluorophore relative to the membrane surface can be used. For example, 7-heptadecyl-7-hydroxy coumarin (HC) (Pal et al., 1985; Borenstain and Barenholz, 1993) has been used to monitor the electrostatic surface potential of viruses (Pal et al., 1983) and liposomes (Zuidam and Barenholz, 1997) (see Fig. 15b). Such a method was also used to estimate the pK_a or ΔpK_{el} of the label in charged membranes. For reference, the pK_a of HC in a neutral DOPC membrane was used, neglecting the effect of surface polarity (see Zuidam and Barenholz, 1997). Specifically, the pK_a values of HC in pure DOTAP or in DOTAP-DOPE (1:1) mixed membranes were found to be 4.2 and 3.8 pH units lower, respectively, than that of HC in a neutral DOPC membrane. The large negative ΔpK_{el} value implies that the electrostatic surface potential of such cationic membranes is highly positive (240 and 217 mV for DOTAP and DOTAP-DOPE large unilamellar vesicles, respectively). Distributing the surface charges over a wider region, for example, on the headgroups of ganglioside molecules, lowers the maximum ΔpK_{el} value.

Another way to study membrane electrostatics is to measure the charge- and salt-dependent phase transition shifts (Cevc and Marsh, 1987; Cevc, 1987; 1989; 1990; 1991a). Care must be taken, however, not to misinterpret any of the observed changes in terms of simple electrostatics rather than as a consequence of the latter and of membrane hydration, which often prevails (Cevc et al., 1980; Cevc, 1987, 1988).

3. Membrane Hydration

Membrane hydration follows essentially the distribution of hydrophilicity ("polarity") throughout the interfacial region. This is due to the relatively short range (~ 0.1–0.3 nm) of simple hydration phenomena but may be obscured by the interdependence of the surface hydration and electrostatics (Cevc et al., 1995). For a proper understanding of membrane hydration and its effects, it is essential to realize that the bilayer and its associated water are one thermodynamic unit that reacts consistently to external variations. This unit is narrowly defined, as the range of simple hydration

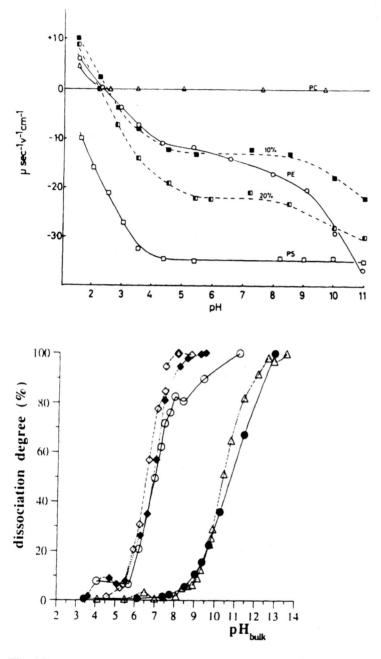

Fig. 15 Upper: Electrophoretic mobility of various phospholipids which is proportional to zeta-potential, as a function of pH. PC, phosphatidylcholine; PE, phosphatidylethanolamine; PS, phosphatidylserine (continuous curves). Dashed curves are for mixtures of 10% and 20% phosphatidylserine in phosphatidylcholine. (From Bangham et al., 1974.) Lower: Dissociation degree of HC in liposomes as translated into percentages of the maximum value against the pH_{bulk}. The liposomes were composed of DOTAP/DOPE (1:1) (○), DOTAP/DOPC (1:1) (◆), DOTAP (◇), DOPC (△), and DOPC/DOPE (1:1) (●). The curves shown are not fitted. (Adapted from Zuidam et al., 1997.)

phenomena does not exceed two to three water molecules (Vierl et al., 1994) or a few tenths of a nanometer (Cevc, 1993d).

Modifying membrane hydration is therefore prone to change the membrane structure and vice versa. Morphological adaptations of the whole membrane to osmotic stress were reported, including isothermal membrane fusion and phase transitions (Cevc, 1991a). At the level of lipid molecules, progressive hydration results in a concerted change in the lipid headgroup conformation, including an effective accessible headgroup volume expansion and turning of polar headgroups more toward the aqueous subphase. This increases the rate and the scope of lipid headgroup motion (Ulrich and Watts, 1994). The change is partially due to a shift in the lipid-phase transition temperature and is partly of entropic origin (Ulrich et al., 1994). Lipid and protein ionization also affects molecular conformation and interactions.

The importance of lipid bilayer hydration for physical and chemical membrane stabilization is well established. Solvent freezing, for example, minimizes lipid degradation induced by ionizing irradiation. This suggests that the aqueous subphase is the source of reactive oxygen species (ROSs) involved in the damaging process. A similar effect is achieved by introducing polyethylene glycol (PEG) covalently attached to lipids into the membrane. Even at low concentrations, this reduces radiation-induced oxidative damage (Tirosh et al., 1997; 1998).

Lipid headgroup dehydration increases acyl chain packing density and decreases the number of defects (Cevc, 1993d). Physical or chemical drying of the membrane, as caused by solute binding (Vierl et al., 1994) or by the presence of PEG-lipids in the membrane (Tirosh et al., 1998), has similar consequences; the dehydration of phospholipid headgroups by the latter in conjuction with increased hydration of the layer of the grafted PEG moiety (with molar mass > 750) provides an explanation for this. The lipid-bound water, moreover, contributes to the stability of membrane suspensions, and the (rare) water molecules inside a bilayer destabilize the membrane structure (Fig. 16) and (Tirosh et al., 1997; 1998; Torchillin, 1996).

Studies of pure phospholipid membranes or of various well-defined lipid mixtures have greatly contributed to our current knowledge of the relationships between membrane lipid composition and membrane properties. This includes membrane thermotropic polymorphism (summarized by Marsh, 1990; Cevc, 1989; 1991a; Caffrey et al., 1996), lipid bilayer permeability (Bangham 1974; de Gier, 1993), membrane electrostatics (Cevc, 1990), protein binding (Cevc et al., 1990), and other membrane properties (Lasic and Barenholz, 1996).

VI. MEMBRANE PACKING

A. Membrane Curvature

A mixture of different lipid molecules in a single membrane will form a uniform bilayer only if their characteristics are compatible. Otherwise, lipids will phase separate laterally, sometimes in a time- and location-dependent manner (Seddon et al., 1997). The reason for this is the coupling of local membrane composition and local bilayer curvature. Membrane lateral nonuniformity thus induces local transformations of membrane or vesicle shape (Lipowsky, 1992, 1993; Jülicher and Lipowsky, 1993). The reverse is also true: highly polar lipids with a relatively strong repulsion between

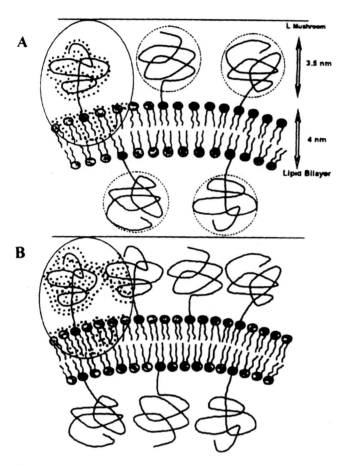

Fig. 16 Schematic diagrams of the changes in the outer layer and bilayer of PEG-grafted liposomes at different concentrations of PEG-lipid. (From Priev et al., 1998.)

the hydrophilic molecular parts will concentrate at sites with a high negative local curvature. Lipids with a predominantly hydrophobic character, on the contrary, accumulate in membrane regions with a positive surface curvature or at sites of quasi-planar geometry. Enforced vesicle shape transformations in external gradients therefore go hand in hand with molecular rearrangements in the lipid bilayer (Cevc, 1995); both may change the characteristics of an entire lipid vesicle. Modifying the membrane lipid composition may change the appearance of the organism as well. Incorporation of synthetic lipids with different chain compositions into *Acholeplasma laidlawii* was shown, for example, to affect de novo synthesis of endogenous lipids. The average cell diameter, consequently, increased with increasing acyl chain length and saturation, which correlated qualitatively with the packing properties of the resulting lipids (Wieslander et al., 1995).

B. Membrane Separation

Bilayer-layer, bilayer membrane-cell, or cell-cell separation is determined by the free energy minimum. The membranes thus take the position at which the attractive intermembrane forces are balanced by the intersurface repulsion. Consequently, increasing repulsion between the bilayers (by membrane charging, promoting membrane hydration, incorporation of long, sterically active molecules into the bilayer, making membranes more flexible, etc.) increases the separation between two membranes. Conversely, membrane rigidification, lowering the net charge density on or the hydrophilicty of a membrane, increasing the bulk salt concentration, partial membrane surface digestion, or other means for thinning the membrane-solution interface pushes the membranes closer together. An applied osmotic pressure has the same effect. The relative probability of such changes can be judged on the basis of Fig. 17, which illustrates the range of different repulsive forces between the surfaces.

It is noteworthy that increasing the membrane polarity increases the surface hydration and leads to greater intermembrane separation only initially; further hydrophilicity increase is nearly useless unless the limited accessibility, that is, the restricted interfacial volume constraint, is relieved by swelling of the interface in the third dimension. Such swelling for simple lipid bilayers is limited by the size of the polar headgroups: molecules with long and/or mobile polar headgroups typically enlarge the intermembrane repulsion, as does irreversible adsorption of polar molecules onto a membrane or an increase in the density or width of the charge distribution in the interfacial region. "Protrusion force" is a misnomer, however, as is the term "fluctuation force." (This notwithstanding, the former term is frequently

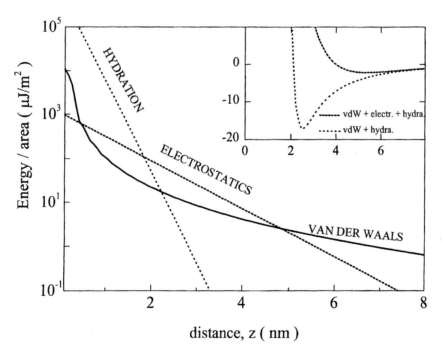

Fig. 17 Main forces that act between two polar (membrane) surfaces as a function of intermembrane separation. Fluctuation of the soft, fluid membranes extends the range of repulsion farther. Inset gives the result of attractive van der Waals and repulsive electrostatic and/or hydration force, showing that the minimum is flatter and farther from the membrane when membranes are charged.

used to describe the various interactions originating from steric interactions or else from the fluctuation-dependent forces.) In reality, the "fluctuation repulsion" is not a force in its own right but rather represents an entropic correction to the other (calculated) repulsive forces. Such correction is required when the amplitude and the wavelength of undulations in the membrane surface topography are comparable to the separation between the investigated membranes. Interfacial softness and membrane flexibility, consequently, tend to enlarge the repulsion between membranes by extending the range of primitive (hydration, electrostatic, steric) forces beyond the reach of interactions between the corresponding fixed, rigid surfaces (Helfrich, 1973; Cevc et al., 1995).

VII. LIPID POLYMORPHISM

A. Lamellar Phase

At relatively low temperatures, hydrated lipids typically adopt a densely packed, crystalline structure (L_c or $L_{c'}$ phase, prime denoting phases with tilted chains) (Mulukutla and Shipley, 1984; Wong and Mantsch, 1982); lipid chains then typically form an orthorhombic hybrid chain-subcell lattice (see Fig. 12, right); (Mulukutla and Shipley, 1984; Ruocco and Shipley, 1982; Harlos, 1978; Gudgin et al., 1981). Temperature increase first induces rotational chain excitations. (Two-dimensional) lipid crystalline structures hereby revert to a more expanded, gel or L_β or $L_{\beta'}$ phase at subtransition temperature, $T = T_s$. Aliphatic chains are now packed in a hexagonal lattice and are often tilted ($L_{\beta'}$ phase) (Hosemann et al., 1980) but may remain untilted as well (L_β) (Mulukutla and Shipley, 1984) (Fig. 12).

Heating speeds up the hydrocarbon chain oscillations. This is mirrored in an essentially unhindered, long-axis chain rotation (Füldner, 1981; Trahms et al;, 1983) at pretransition temperature, $T = T_p$. Heating also increases lipid headgroup mobility, most notably the rotation of lipid headgroups around the P–O bond to the glycerol backbone (Shepherd et al., 1978). The interfacial area per molecule increases during pretransition as well. It is very probable that during lipid pretransition individual chains shift along their long axes mutually to stay in close contact; this is believed to be the (a) reason for the bilayer surface breaking up into a series of periodic, asymmetric, quasi-lamellar bilayer segments, or ripples (Zasadzinski et al., 1988). Lipid chains in the resulting (P_β or $P_{\beta'}$ phase are at least partly molten, perhaps at the junctions between different quasi-lamellar segments of the ripples (Fig. 12). Owing to their all-*trans* configuration, in the lamellar gel phase, lipid chains have maximum extension (see, e.g., Ruocco and Shipley, 1982; Tardieu et al., 1973). Higher temperatures result in effective chain shortening and membrane thinning, owing to the orientational chain excitations and the concomitant loss of chain order. A cooperative chain melting (order-disorder, gel-to-fluid) phase transition at $T = T_m$ is therefore associated with a great change in system entropy and enthalpy.

B. Interdigitated Phase

When lipid chains are highly asymmetric (Hui et al., 1984; Li et al., 1990) or when the tendency of lipid headgroups, with their associated solvent and solutes, to expand laterally is greater than the original tendency of chains to stick together in two separate bilayer halves (Rowe, 1987; Vierl et al., 1994; Löbbecke and Cevc,

1995), interdigitated lamellar phases (L_i) form (Fig. 12). In such phases the chains of lipids from both membrane leaflets intercalate to increase the effective area per molecule in the membrane (Slater and Huang, 1988). High hydrostatic pressure also favors the formation of interdigitated lipid lamellae (Braganza and Worcester, 1986), probably because of release of the strain in the glycerol backbone. Furthermore, this strain is eliminated by replacement of the ester bond in the aliphatic chains with an ether bond. The carbonyl oxygen on the *sn*-1 chain plays a pivotal role in this.

Dialkyl- and alkyl-acyl chains of phosphatidylcholines, consequently, inter-digitate spontaneously; so does diacyl-phosphatidylcholine (such as dipalmitoyl-PC) at pressures higher than approximately 150 MPa. Moreover, bilayers of lipids physically similar to phosphatidylcholine interdigitate upon the addition of a sufficient, chain length-dependent amount of fatty alcohols (Rowe, 1987; Löbbecke and Cevc, 1995) or amines with two to eight carbon atoms per chain. The presence of certain antibiotics or of certain anions (Br^-, I^-, ClO_4^-, IO_4^-, KCN^-, etc.) has a similar effect, as do some buffers (such as Tris), peptides (e.g., mellitin), etc. The common property of all L_i-phase inducers is the membrane-solution interface expansion in the gel phase to the point at which chain tilt would exceed approximately 50° (Vierl et al., 1994).

C. Nonlamellar Phases

Nonlamellar membrane phases are of the inverse hexagonal ($H_{II}=H_\alpha$; see Fig. 12) or cubic (Q_α) type, by and large (Seddon, 1990; Lindblom and Rilfors, 1992); the latter phase in the living world is most often bicontinuous and thus provides two three-dimensional isotropic matrices, with symmetries easily obtained by X-ray diffraction.

The inverse hexagonal phase consists of parallel, water-filled tubes separated by the lipids forming a hexagonal lattice perpendicular to the tubes' long axis (Seddon, 1990). Such phases were inferred to exist, for example, in the region of tight junctions between the cells and as intermediates in membrane fusion. To explain their existence physicochemically, an inherent tendency of the involved lipids to form a curved surface (the negative membrane spontaneous curvature) and immanent membrane frustrations were invoked (Tate et al., 1991). The observation that lipid extracts from the membranes grown with the different fatty acids have not only a similar lamellar-to-nonlamellar phase transition temperature (Lindblom and Rilfors, 1992) but also a nearly constant spontaneous curvature (Osterberg et al., 1995) circumstantially supports such an interpretation.

Three evolutionarily conserved families of cubic membranes (Fig. 18) were unequivocally identified in cells through analysis of electron micrographs (Landh, 1995). In each of these families, one or more (parallel) periodically arranged membranes (Engblom, 1996) partition the subcellular space into two or more independent, albeit convoluted, subspaces of membrane potential-determined dimensions. Such special membrane organization gives rise to periodic cubic surfaces, the details of which seem to be related to the system activity (Landh, 1995).

Bicontinuous membrane organization is very attractive biologically as it provides the greatest possible surface contact with the surrounding or, better, incorporated solvent. From the physicochemical point of view, membrane elastomechanics provides good arguments for the emergence of cubic phases: to minimize average curvature and maintain a high negative local curvature, the membrane should generate saddles; when the latter are arranged in a periodic lattice they give rise to cubic phases (Seddon, 1990). Alternatively, the periodicity of lipid bilayers in a cubic phase was speculatively associated with the standing-wave character of vibrational motions: the periodic curvature was proposed to reflect a dominating mode of standing-wave oscillations, with the different cubic bilayer structures representing alternative standing-wave conformations of the bilayer (Larsson, 1997).

Nonbilayer structures typically emerge when the lateral repulsion between the hydrocarbon chains starts to exceed the corresponding lateral pressure in the solvated interfacial region (Cevc, 1993a). The former pressure increases with chain disorder and with the amount of material in the hydrocarbon membrane interior. The latter pressure chiefly depends on the membrane hydration and therefore is proportional to interfacial polarity and thickness. Consequently, any increase in the lipid headgroup hydrophilicity lowers the system's tendency to form a nonlamellar phase. On the other hand, any change that enhances interchain repulsion increases the probability of nonbilayer structure generation. During the course of a monotonic system variation, which affects the lateral pressure in the interfacial or hydrocarbon region, a lower limit for the bilayer-to-nonbilayer phase transition temperature is normally reached. At this limiting temperature the bilayer-to-nonbilayer phase transition coincides with the bilayer chain-melting phase transition. Bilayers kept below this characteristic temperature are stable in the lamellar phase, whereas above such a temperature nonlamellar phases prevail. Relatively nonpolar lipids, such as phosphatidylethanolamines at neutral or low pH, phosphatidylserines and phosphatidic acids in acidic buffers, or phosphatidylglycerol and cardiolipin (diphosphatidylglycerol) in the presence of sufficiently high concentrations of protons or other membrane-binding ions (lithium, calcium, magnesium, etc.), spontaneously

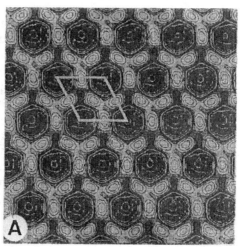

Fig. 18   Cubic phase (Fd3m). Freeze fracture image of one of the main fracture planes (111) observed in a hydrated dioleoylglycerol-dioleoylphosphatidylcholine mixture (left) and the corresponding domain in electron density distribution (right). Bicontinuous cubic structures abound in Golgi system, thylakoids, etc. (From Delacroix et al., 1996.)

form inverted nonbilayer phases. Mixtures of glycerophospholipids with comparably long fatty acids, fatty alcohols, and fatty amines have a propensity to form nonlamellar phases as well. Sometimes these are of (inverted) cubic type. The addition of suitable alkanes also relieves the "strain" of the tentative voids in the hydrocarbon region that would arise if the pure system formed a nonbilayer phase and thus promotes H_{II} phase creation. Low amounts of PEG-lipid induce the cubic phase in phosphatidylethanolamine dispersions (Koynova et al., 1997). A transition of phosphatidylethanolamine from an expanded lamellar gel phase to a cubic phase was also observed in the presence of a concentrated sucrose, and at least some other sugar, solution (Tenchov et al., 1996).

Maintenance of sufficient membrane fluidity and, in most cases, of lamellarity is crucial for the living system's functioning: organisms try to keep their membranes close to a fluid-phase optimum (Neidleman, 1987; de Mendoza and Cronan, 1983). Often this includes some distance to the bilayer-to-nonbilayer phase transition temperature (Lindblom and Rilfors, 1989); this is mainly true for membranes that lack cholesterol. Said self-regulation may be related to the cell requirement for optimum (local) membrane curvature, especially in the vicinity of certain proteins (see also Section XIII for more discussion).

D. Phase Transition Shifts

1. Hydrocarbon Chain Effects

On the absolute temperature scale, the phase behavior of lipid bilayers is determined mainly by the hydrocarbon chains. The chain-melting phase transition temperature of an arbitrary phospholipid, as a function of the number of carbon atoms per chain, n_c, is

$$T_m(n_c) = T_m(\infty)1 + n_m/n_c + n_h/n_c^2 + \cdots \qquad (2)$$

The factor n_m in Eq. (2) corresponds approximately to the length of the shortest segment for which a first-order chain-melting phase transition is possible; the n_h term allows a phenomenological description of headgroup and other end effects. Parameter $T_m(\infty) \sim 414$ K refers to the chain-melting transition temperature of a hypothetical lipid with infinitely long chains. An alternative formula was proposed by C. Huang (1991).

Application of Eq. (2) to the chain-melting phase transition data (Cevc, 1991b) of fully saturated and hydrated diacyl-phosphatidylcholines suggests

$$T_m(n_c) = 414(1 - 3.20/n_c - 10.64/n_c^2) \text{ K} \qquad (3)$$

This reproduces available experimental data for $T_m(n_c)$ to within 2 degrees or 0.5%, as illustrated in Fig. 19.

According to Eq. (3), the membrane fluidization temperature (T_m) increases with effective chain length. The latter, as a rule, increases with the length of the longest ordered and aligned segment on each chain. This conclusion is independent of the cause of the reduced chain packing in the membrane interior: chain unsaturation (which effectively decouples the two hydrocarbon segments disjoined by a double bond) and chain asymmetry (which causes the terminal hydrocarbon segments to lose close contact) affect the bilayer chain-melting phase transition temperature comparably on the effective chain-length scale. Thermodynamic consequences of *trans* unsaturation are approximately 50% smaller than the effects of double bonds in the *cis* conformation because of the smaller membrane perturbations by the former double bonds. Double bonds in the middle of the chain are more influential than chain unsaturation near the chain ends (Cevc, 1991b).

A simple model was proposed to calculate and describe these effects on the lipid chain-melting transition temperature quantitatively, starting with the known lipid chemical composition. In short, if the length of first hydrocarbon segment exceeds the extension of the second part of the chain by more than three to four carbon atoms, the lipid chain-melting polymorphism is governed by the length of the former. The chain-melting phase transition temperature in such a situation is determined chiefly by the longer hydrocarbon part. From the thermodynamic point of view, the effective chain length of a lipid with asymmetric chains is always approximately equal to the length of the shorter chain plus one (see Fig. 19 and Cevc, 1991b).

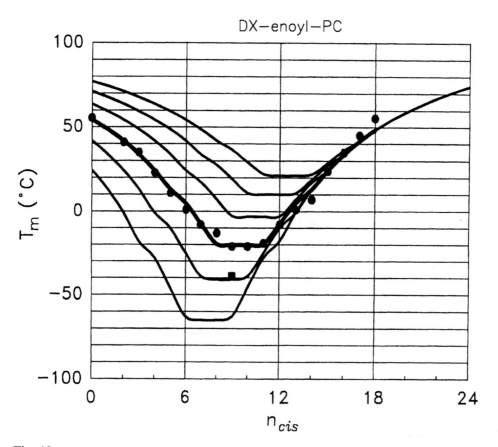

Fig. 19 Effect of lipid chain length and unsaturation on chain-melting phase transition in bilayer membranes. n_{cis} denotes the position of a single double bond in the chains of diacylphosphatidylcholines with $n_c = 24$ to 14, from top to bottom, respectively. (From Cevc, 1991b.)

The same model also explains the lipid sensitivity to hydrocarbon type and attachment, usually with an accuracy close to 100%. For lipid mixtures with identical headgroups and diverse hydrocarbon chains, the temperature of the gel-to-fluid phase transition, in the simplest approximation, is given by the arithmetic mean of the individual transition temperatures, except when the chain lengths differ by more than three or four methylene groups.

2. Headgroup Effects

The lowest enthalpy transitions in lipid membranes are most sensitive to perturbation in the interfacial region. Measuring the shift in the temperature of such transitions induced by various agents therefore offers a means of studying the interactions between such agents and a membrane.

Thermodynamic consequences of chemical or physicochemical lipid headgroup variations, consequently, can be as important as chain modifications, on the physiological temperature scale. One must keep in mind, however, that chain unsaturation renders membranes more sensitive to the phase modulation by lipid headgroups and ionic solution. This may be one of the reasons why living organisms maintain their membranes fluid by using unsaturated rather than short-chain lipids.

3. Solution Effects

Solution effects can play an important role in controlling membrane properties.

Solvent (water): The initial effect of changing lipid hydration on the chain-melting phase transition temperature is relatively small. A moderate decrease of total water content or a major rise in the bulk salt concentration changes the chain-melting phase transition temperature by less than 2%, or less than 5°C, as long as the headgroup protonation remains the same. This notwithstanding, hydration generally plays an important role in controlling the membrane properties. Extensive dehydration can shift the phase transition temperature by 50°C and more upward (Cevc and Marsh, 1985).

The lipid chain-melting phase transition ($L_\beta \to L_\alpha$ or $P_{\beta'} \to L_\alpha$ at $T = T_m$) therefore acts as a thermodynamic osmometer: the chain-melting phase transition shift is directly proportional to the change of the logarithm of the bulk water activity coefficient (Cevc, 1988). The sensitivity of the osmometer decreases with the effective hydrocarbon chain length, however.

The lipid pretransition ($L_{\beta'} \to P_{\beta'}$ at $T = T_p$) has different features. This transition also reacts to the changes in membrane hydration but behaves as a strongly interface-biased osmometer. The sensitivity of the "pretransition osmometer" can be quite high: for a sufficiently hydrated surface, even a small decrease of the interfacial water activity, or a moderate increase of the interfacial solute concentration, will strongly shift the pretransition temperature, most frequently upward (Cevc, 1991a).

The magnitude of the solvent-dependent pretransition temperature shift (ΔT_p) decreases with the lipid chain length but is always much greater than the shift of the chain-melting transition temperature (ΔT_m) owing to the smaller enthalpy of the pretransition compared with the chain-melting enthalpy, as one would expect on the basis of Eq. (4).

Being determined largely by the chains, the lipid crystal-to-gel transition ($L_c \to L_\beta$ at $T = T_s$) is only slightly sensitive to the membrane hydration (Cevc, 1991a). The corresponding subtransition temperature also depends very little on the concentration of solutes in the aqueous subphase (Cevc, 1991d; Tenchov et al., 1996). However, the likelihood of a subtransition and the rate of crystalline phase formation both increase with decreasing water activity in the system (Vierl et al., 1994).

Noncharged solutes, including many sugars and other nonionic molecules that do not interact with the lipid bilayer directly and strongly, affect lipid polymorphism mainly by changing the bulk water activity and thus the membrane hydration. Additional membrane perturbations and extra transition temperature shifts may arise from the direct lipid-solute interactions, such as intermolecular hydrogen bonds. Some solutes, such as trehalose, can replace the lipid-bound water altogether and then efficiently compensate the thermodynamic consequences of lipid dehydration (Crowe et al., 1984).

Ions in the bilayer bathing solution strongly influence membrane properties. This offers a means for partial control of the membrane phase behavior and structure. The mechanism is much faster and energetically less expensive than that based on the adaptation of the lipid chain composition. Solution-dominated lipid bilayer control is therefore believed to be most relevant for short-term transient modifications of membrane properties or for localized membrane adaptation, including membrane contact points, membrane complexes, or lipid-protein aggregates (Cevc, 1987; 1988; 1993e).

Chain-melting phase transition shifts obtained with the most ubiquitous noncharged phospholipids (phosphatidylcholine and phosphatidylethanolamine) and with certain derivatives of such lipids are given in Table 2; corresponding information pertaining to normally charged lipids is summarized in Table 3. These data indicate, first, that acidification of the bulk solution, that leads to at least partial phosphate group protonation, increases the temperature at which lipid chains melt. Alkalization of the aqueous subphase or lipid deprotonation by chemical means has the opposite effect. Consequently, pH changes may induce isothermal melting or freezing of lipid chain within a very short period of time (≤ 1 s). Other electrolytes, such as NaCl, that contain no ions that bind directly to the membrane chiefly affect the bilayer transition temperatures via osmotic (hydration) effects.

This lipid phase behavior can be rationalized quantitatively in terms of the various contributions, $\Delta T_{m,j}$, from the polar membrane region to the corresponding total shift of the lipid chain-melting phase transition temperature:

$$\Delta T_{m,p} = \Delta G_{p,m}/\Delta S_{ref,m} = \sum_i \Delta G_{i,m}/\Delta S_{ref,m} \tag{4}$$

where $\Delta S_{ref,m}$ is the entropy change at the phase transition of the lipid in the reference state; a reasonable choice of such a state is the anhydrous lipid, for which the value of $\Delta S_{ref,m}$ is approximately proportional to the chain length, n_c (Cevc and Marsh, 1985). $G_{p,m}$ is the change in free energy of the polar membrane region at the phase transition, which chiefly stems from the change in membrane hydration. Using Eq. (4), the phase behavior of fully hydrated lipids is modeled rather reliably by using the following individual shifts:

$$\Delta T_{m,bond} \sim 1.5 \pm 1\,K \leq \Delta T_{m,el} \sim 5.5 \pm 0.5\,K \leq \Delta T_{m,h}^{H} \sim 7 \pm 1\,K \ll \Delta T_{m,h}^{PO4}$$
$$\sim 13 \pm 1.5\,K$$

Table 2 Effect of Lipid Chain Length, Headgroup Protonation, or Methylation on the Chain-Melting Phase Transition Temperature (°C) of Lipid Bilayers or Its Shifts (in Italics)

Lipid[a]	0		8		13	Lipid[a]	0		8		13
DMPE[b]	54	*4.5*	49.5	*25.5*	24	DTPC	59	*4*	55	*26*	29
		6		*7.5*	*−1*			*6*		*8*	*−1*
DMPE(CH₃)	48	*5.5*	42.5	*17.5*	25	DTPE(CH₃)	53	*6*	47	*17*	30
		6		*11.5*	*0*			*6*		*11*	*0*
DMPE(CH₃)₂	42	*11*	31	*6*	25	DTPE(CH₃)₂	46	*10*	36	*6*	30
		6		*8*	*2*			*7*		*7.5*	*1.5*
DMPC	36	*13*	23	*0*	23	DTPC	39	*10.5*	28.5	*0*	28.5
DPPE	67	*3.5*	63.5	*21.5*	42	DHPE	71.5	*3*	68.5	*24.5*	44
		5.5		*5.5*	*−1*			*5.5*		*7*	*−1*
DPPE(CH₃)	61.5	*3.5*	58	*15*	43	DHPE(CH₃)	66	*4.5*	61.5	*16.5*	45
		5.5		*10*	*0*			*5.5*		*11*	*0*
DPPE(CH₃)₂	56	*8*	48	*5*	43	DHPE(CH₃)₂	60.5	*10*	50.5	*5.5*	45
		6		*6*	*1*			*6.5*		*7*	*1.5*
DPPC	50	*8*	42	*0*	42	DHPC	54	*10.5*	43.5	*0*	43.5

[a] DM=1,2-dimyristoyl-*sn*-glycero; DT=1,2-ditetradecyl-*rac*-glycero; DP=1,2-dipalmitoyl-*sn*-glycero; DH=1,2-dihexadecyl-*rac*-glycero; PE ≡P(CH₂)₂NH₃ ≡ phosphatidylethanolamine; PE(CH₃) ≡ phosphoryl-*N*-methylethanolamine; PE(CH₃)₂ ≡ phosphoryl-*N,N*-dimethylethanolamine; PE(CH₃)₃ ≡ P(CH₂)₂N(CH₃)₃ ≡ PC ≡ phosphatidylcholine.

Table 3 Effect of Lipid Ionization[a] and Chain Length on the Bilayer Chain-Melting Phase Transition Temperature (°C) and Its Shifts (in Italics)

Lipid[b]	0[d]		~3		8		13	Lipid[b]	0[d]		~3		8		13
DMPG	42		*18*		24	*0*	24	DTPG	46		*18*		28	*0*	28
DMPS	52	*8*	44	*8*	36	*21*	(15)	DTPS	56	*8*	48	*7*	41	*21.5*	19.5
DMPA[c]	45	*−10*	55	*5*	50	*22*	28	DTPA	48	*−11*	59	*4*	55	*21*	34
		−3		*8*		*18*	*−4*			*−4*		*8*		*18*	*−3*
DMPA(CH₃)	48	*1*	47	*15*	32	*0*	32	DTPA(CH₃)	52	*1*	51	*14*	37	*0*	37
DPPG	58		*16*		42	*0*	42	DHPG	62		*16*		46	*0*	46
DPPS	68.5	*7*	61.5	*7.5*	54	*22*	32	DHPS	72	*8*	64	*8*	56	*21*	34
DPPA[c]	62	*−12*	74	*6*	68	*23*	45	DHPA	62	*−14*	76	*5*	71	*21*	50
		1		*12*		*20*	*−3*			*−4*		*11*		*19*	*−2*
DPPA(CH₃)	63	*1*	62	*16*	48	*0*	48	DHPA(CH₃)	66	*1*	65	*13*	52	*0*	52

[a] Ionization states are at pH 0 for PG, PA(CH₃), PA, and PS⁺, at pH 3–4 for PA(CH₃), PA$^{0.5-}$, and PS⁻, at pH 8 for PG⁻, PS⁻, PA(CH₃)⁻, and PA⁻, and at pH 13 for PG⁻, PS^{2-}, PA(CH₃)⁻, and PA^{2-}.
[b] DM=1,2-dimyristoyl-*sn*-glycero; DT=1,2-ditetradecyl-*rac*-glycero; DP=1,2-dipalmitoyl-*rac*-glycero; DH=1,2-dihexadecyl-*rac*-glycero; PG=phosphatidylglycerol; PA=phosphatidic acid; PA(CH₃)=phosphatidic acid methyl ester; PS=phosphatidylserine.

(The precise values may vary somewhat with the detailed lipid type and state.) The electrostatic shift arising from the second charge on a lipid is relatively small, $\Delta T_{m,el}^{(++,--)} - \Delta T_{m,el}^{(+-)} \leq 3$ K. The corresponding values for lipids with ether bonds in the backbone region are higher by a small amount (≤ 1 K) than for the diacyl lipids.

A similar approach can also be used to explain the effects of a polar membrane region on the other phase transitions, such as the bilayer pretransition or membrane conversion from a lamellar to a nonlamellar phase. For example, the maximally protonated phosphatidylethanolamine and its monomethylated derivative, with two alkyl or alkenoyl chains comprising more than 16 carbons per chain, undergo a chain-melting phase transition at nearly the same temperature with the formation of a nonlamellar phase(s). Conversely, at pH 7 the chain-melting phase transition temperature and the temperature of the transition to a nonbilayer phase differ substantially, by 27, 25, and 16 degrees for the distearoyl-, dieladoyl-, and dioleoyl-phosphatidylethanolamine, respectively. However, the alkylated versions of the former two lipid types no longer form a nonlamemellar phase in excess buffer at 100 °C.

4. Effect of Proteins

The effect of proteins is typically to lower the chain-melting phase transition temperature when they decrease the chain order in the fluid phase more than in the ordered phase; in the opposite situation the reverse effect is observed. No protein effect on the lipid phase transition(s) typically signals the lack to identify protein interactions with lipids in either phase. Most proteins that adsorb to the lipid bilayer shift transition temperatures downward by at least a few degrees.

VIII. PHASE TRANSITION KINETICS

The observed thermal relaxation behavior of all multilamellar vesicles is quantitatively similar (Laggner and Kriechbaum, 1991). The relaxation times vary from approximately 50 ms to 4 s, with a pronounced maximum just above the phase transition temperature. Large unilamellar vesicles exhibit a single relaxation process without a large maximum in the relaxation time. The relaxation time of such vesicles is approximately 80 ms over most of the transition range (van Osdol et al., 1991). At all temperatures, the thermal response data are consistent with a single primary relaxation process of the lipid. The less accurate bulk modulus data exhibit two relaxation times; one of which may be an experimental artifact (van Osdol et al., 1991).

Phase transitions involving better packed membrane states, such as the lipid pretransition or subtransition, normally exhibit strong hysteresis. This seems to be a consequence of kinetic trapping rather than of an inherently slow process.

IX. LATERAL DIFFUSION

The lateral diffusion constant of common phospholipids is typically of the order of 10^{-9} cm^2 s^{-1}, the precise value depending on the lipid characteristics and on the degree of membrane hydration. Other factors affecting the lipid (headgroup) packing are influential as well. This is due to the proportionality between the lipid lateral diffusion constant and the excess lipid molecular area, as long as molecular motion in the plane of the bilayer is based on the migration of lipids through the interstitial sites or on lipid exchange. Diffusion via vacancies, on the other hand, makes the diffusion constant proportional to the vacant site or molecular area (Cevc and Marsh, 1987).

For example, sphingosylphosphocholine, phosphatidylcholine, and phosphatidylserine all diffuse with a similar lateral diffusion coefficient, 3×10^{-10} cm^{2}s^{-1}, but the latter phospholipid also has a much faster lateral diffusion component, 2×10^{-9} cm^2 s^{-1}, according to measurements done with the fluorescently labeled lipids. It seems that this fast component measured with phosphatidylserine exists only in ATP-containing cells and is limited to the inner membrane leaflet. This indicates that the two leaflets act as separate, uncoupled membrane domains (el Hage-Chahine et al., 1993).

Lateral diffusion is one to two orders of magnitude more rapid in the fluid than in the ordered lamellar lipid phase, the specific values for phosphatidylcholine being 4×10^{-9} and 10^{-10} cm^2 s^{-1}, respectively. Lateral mobility of the lipid molecules moving along the defect lines in a bilayer is much higher, however.

The lateral transport of water-soluble molecules inside a membrane is of no practical importance owing to the low probability of finding such molecules in the core of the lipid bilayer, but domain boundaries can change this. The transport rate of such molecules in the interfacial region was shown to be anomalously high, however, at least for the water molecules (Heim et al., 1995).

X. COMPOSITION-STRUCTURE RELATIONSHIP

For illustration, we will compare the properties of bilayers made of phosphatidylcholine (PC) and of sphingomyelin (SPM) to relate membrane composition, structure, and dynamics. Subsequently, the interactions of both kinds of phospholipids with cholesterol will be described.

Lipids containing the choline group reside in the outer leaflet of plasma membranes, by and large (Fig. 6). However, sphingolipids are synthesized in the Golgi complex and then enriched on the cell surface and in endocytotic organelles. This probably involves intracellular machinery that preferentially shuttles the lipids in natural vesicles to the cell surface. The mechanism appears to involve the formation of domains of sphingolipid and cholesterol in the luminal leaflet of Golgi membranes. For example, approximately 50% of the plasma membrane sphingomyelin is recycled in less than 30 min in Chinese hamster fibroblasts by an ATP- and microtubule-dependent process (el Hage-Chahine et al., 1993). Lipid transfer between the different membranes involves lipid transfer protein(s) (van Wijk et al., 1992).

The resemblance of PC and SPM, the two most common choline-based phospholipids, is limited to the headgroups. Other regions on these kinds of molecules differ in structure substantially. The PC chains are esterified to the glycerol and, for most naturally occurring PCs, are almost identical in length; the acyl chain attached to carbon C-1 of the glycerol (*sn* nomenclature) is normally saturated, whereas the other aliphatic residue, in the *sn*-2 position, is unsaturated. The latter residue typica contains between one and six double bonds, all in the *cis* configuration. The hydrophobic part of SPM is composed of one acyl chain bound in an amide linkage to the primary amino group at C-2 of the sphingosine (see Fig. 6); the second chain is a part of the sphingosine base (C-6 to C-18), which does not belong to the polar or interfacial region. The interphase contains the components of the amide linkage (between the acyl chain and the primary amino group at C-2 of the sphingosine) as well as the free hydroxyl group attached to C-3; the *trans* double bond between C-4 and C-5 atoms is probably also part of the interphase. This allows hydrogen bonding in the interfacial region of SPM, owing to the presence of free hydroxyl, amide, and carbonyl groups on SPM, which may serve as donors as well as acceptors of hydrogens. In comparison, PC through its two carbonyl esters can accept only hydrogens. The *trans* double bond between the C-4 and C-5 atoms of the sphingosine moiety (4-sphingenine) has the ability to induce dipoles in the interfacial region (Barenholz and Thompson, 1980). Better chain stacking and closer lipid packing may result from this; membrane stabilization and lower membrane permeability are further consequences (Abrahamsson et al., 1977).

Physical studies of SPM are fewer than those of PC (Barenholz and Thompson, 1980; Barenholz, 1984). Whereas single-crystal x-ray diffraction data are available for dimyristoyl PC (DMPC) (Hauser et al., 1981; Pearson and Pascher, 1979), the corresponding information pertaining to SPM can only be deduced from the data available for ceramides, cerebrosides, and SPM model compounds (Sundaralingam, 1972; Pascher, 1976; Lofgren and Pascher, 1977; Kang et al., 1976).

Phase behavior, lateral diffusion, and percolation in the range of coexisting gel and fluid phases in binary mixtures of SPM and PC are known. Mixtures of dimyristoyl PC and D-erythro-*N*-lignoceryl SPM or D-erythro-*N*-palmitoyl SPM have similar chain-melting phase transition temperatures. Chain mismatch is considerably different, however, and much larger for the former than for the latter, almost symmetric, SPM variant. This mismatch may be involved in signal transduction via the sphingomyelin cycle (Riboni et al., 1997).

N-lignoceryl SPM exhibits a complex thermotropic behavior. This relates to the two forms of the gel phase of the lipid (gel I and gel II), which differ in the extent of opposing monolayer interdigitation (Fig. 20) (Levin et al., 1985). Sphingomyelin in biological membranes is therefore expected to contribute to the membrane complexity (Barenholz and Thompson, 1980; Barenholz, 1984; Thompson et al., 1977; Calhorn and Shipley, 1979).

Dimyristoylphosphatidylcholine and palmitoyl SPM ($n_c = 16$) mix nearly ideally; their percolation threshold locus is close to the liquidus on the phase diagram. In contrast, behenoyl SPM ($n_c = 24$) and dimyristoyl PC ($n_c = 14$) mix nonideally, with the percolation threshold locus close to the solidus. The latter mixture also forms particles, at least some of which are not multilamellar vesicles.

It is clear that cholesterol greatly affects membrane properties (Yeagle, 1993), but the differential interaction of cholesterol with various lipids is not yet fully understood. For example, the sensitivity of different phospholipids to cholesterol oxidase is not identical, and the "protection" of cholesterol against oxidation by various lipids decreases in the order SPM > PC ≫ PS (Patzer et al., 1978b). The rate of cholesterol desorption from highly curved membranes containing PC and SPM is not the same either. This is true for spontaneous transfer of cholesterol from one population of vesicles to another, for cyclodextrin-catalyzed desorption of cholesterol from mixed monolayers, and for cyclodextrin-catalyzed desorption of dehydroergosterol from unilamellar vesicles. The existence of two kinetically distinct cholesterol pools seems to be responsible for this (Bar et al., 1986, 1987, 1989). One pool transfers with a first-order rate and relatively short half-life of a few seconds; the nonexchangeable

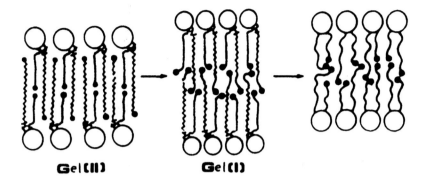

Fig. 20 Molecular packing model including interdigitation of fully hydrated C24:0 sphingomyelin. The schemes from left to right represent gel II, gel I, and liquid crystalline phase. Gel I is a fully interdigitated system and gel II is a mixed interdigitated system. The gel II-gel I transformation should increase bilayer width and reduce the other at the bilayer center. (For more details, see Shmeeda et al., 1994.)

pool does not transfer within a time scale of 12 h. The size of the two pools and, to a lesser degree, the rate of transfer from the first pool depend on the exact lipid composition and temperature. At 37 °C, 90% of the cholesterol in bovine brain SPM bilayers appears to reside in the nonexchangeable pool, compared with 20% in 1-palmitoyl-2-oleyl PC vesicles. This may be reflected in different cholesterol transfer kinetics from rat microsomes to liposomes of SPM or PC. Both liposomes accept cholesterol at a similar rate, once the nonexchangeable cholesterol pool has formed, but SPM enriched bilayers lose cholesterol by desorption less readily. The overall result is a positive correlation between SPM and cholesterol levels (Shmeeda et al., 1994). However, cholesterol always partitions nearly exclusively into the saturated phospholipid phase (Gawrisch and Huster in Barenholz, 1998a).

These studies indicate that it may be possible to manipulate the cholesterol steady-state level in the membranes of living cells by regulating the balance of SPM to PC using liposomes or modified lipoproteins as acceptors or donors for the purpose (Shmeeda et al., 1994).

XI. FORCES AND FACTORS INVOLVED IN MEMBRANE DOMAIN FORMATION

A. Modeling Lipid-Protein Interactions

The interaction between lipid bilayers and macromolecules, such as DNA or proteins, involves several free energy contributions. When a membrane is in the "fluid state," and hence behaves as a flexible, two-dimensional liquid crystal, the incorporation of macromolecules can induce bending and stretching deformations in the lipid bilayer. In a multicomponent membrane, the presence of macromolecules can also induce local demixing and phase separation of different lipid components. If some of the lipids are charged, electrostatic interactions often dominate and can induce compositional inhomogeneities in the lipid membrane. The structure, stability, and phase behavior of the resulting bilayer-macromolecule system depends on the delicate interplay of all strongly coupled contributions to the system free energy.

Incorporation of an integral hydrophobic protein into a lipid membrane may enforce elastic deformations of its immediate lipid environment. For instance, if the thickness of a protein (d_P) is greater than that of the unperturbed lipid bilayer (d_B), the surrounding lipid chains may stretch to minimize the exposure of hydrophobic protein regions to the aqueous solvent. For a negative "hydrophobic mismatch" ($d_P < d_B$), the lipid chains around the protein may be compressed. Furthermore, even for $d_P = d_B$, the presence of a rigid protein boundary can decrease the conformational entropy of surrounding lipid chains. This was described quantitatively within the framework of a molecular model (Fattal and Ben-Shaul, 1993; 1996).

Unspecific lipid-mediated interactions induced by the hydrophobic mismatch between the transmembrane proteins and their neighboring lipids were proposed to drive protein aggregation as well as lipid domain and ion channel formation. Theoretical investigation allowed for protein and lipid movement, as well as for chain, interfacial, and headgroup repulsion, expressed in terms of the mean-field chain statistics and the opposing-forces model, in a three-dimensional approximation with a nonparametric surface (Fattal and Ben-Shaul, 1993; Aranda-Espinoza et al., 1996; Kralchevsky et al., 1995, Harries et al., 1998).

B. Lateral Domain Formation Induced by Short Basic Peptides

The myristoylated alanine-rich C-kinase substrate, MARCKS, exists in a punctate distribution in the plasma membrane of macrophages and other cells. This substrate aggregates on the membrane surface and induces migration of the negatively charged lipids into the regions below it, that is, formation of membrane lateral domains important in signal transduction (Yang and Glaser, 1995). For example, the multivalent acidic lipid phosphatidylinositol 4,5-bisphosphate (PIP$_2$) is sequestered into domains with the monovalent acidic lipid phosphatidylserine (PS), whereas the phosphoinositide-specific phospholipase C (PLC), which hydrolyzes PIP$_2$, is excluded from such domains (Glaser et al., 1996). Therefore, PIP$_2$ in domains is protected from the activity of the phospholipase. Phosphorylation of MARCKS by protein kinase C causes MARCKS to dissociate from the lipid bilayer and thus exposes PIP$_2$ to PLC activity. Yang and Glaser (1995) demonstrated that the peptide corresponding to the basic effector region, MARCKS(151–175), spontaneously forms lateral domains enriched in the acidic phosphatidylserine in phospholipid vesicles.

Pentalysine and heptalysine also form domains enriched in the acidic lipid phosphatidylserine and inhibit hydrolysis of PIP$_2$ by PLC. This suggests that electrostatic interactions are involved in domain formation (Glaser et al., 1996; Glaser in Barenholz, 1998a; McLaughlin et al., in Barenholz, 1998a), the domains formed in charged phospholipid vesicles by basic peptides representing a minimum free energy state. Specifically, the energy gain due to the formation of the domain has to exceed the energy losses in three different processes: (a) the electrostatic repulsion between the negatively charged lipids that have migrated into the domain, (b) the electrostatic repulsion between the positively charged peptides (for example, pentalysine) forming the domain, and finally (c) the entropy of mixing. Denisov et al. (in press) and McLaughlin et al., in Barenholz (1998a) presented a semiquantitative   description   of   such   domain   formation   based   on   the

Gouy-Chapman-Stern theory. They assumed the charges on the lipids and the adsorbed peptides to be smeared uniformly over the membrane surface. In a more realistic approach based on the Monte Carlo-like sampling procedure and Poisson-Boltzmann equation, combined with a rather detailed picture of basic peptides and the phospholipid membrane, Ben-Tal and coworkers have calculated the electrostatic free energy required to concentrate phosphatidylserine in a lateral domain with a similar final result for the free energy cost of migration of the phospholipid into domains. Lateral domains seem to form at low, but not at high, ionic strength, in accord with the experimental data (Ben-Tal et al., 1996).

C. Influence of Chain Unsaturation and Cholesterol

Investigation of lateral organization in biological membranes, especially in the submicrometer range, is difficult. New methods operating on the short time and distance scales pertinent for lipid-lipid interactions are needed. It is already possible, for example, to study lateral lipid organization by deuterium nuclear magnetic resonance (NMR) as well as by two-dimensional NOESY proton NMR spectroscopy with magic angle spinning (MAS). The advantage of both methods is that they require only isotopic labeling, which, in contrast to spin and fluorescent labeling, does not perturb the membrane packing.

In one specific application of this approach, the lateral lipid organization in membranes resembling neural and retinal membranes in lipid composition (PC/PE/PS/cholesterol, 4:4:1:1 mole ratio) was investigated. The mixed-chain phospholipids contained saturated stearic acid (S) in the *sn*-1 position and monounsaturated oleic acid (O) or polyunsaturated docosahexaenoic acid (D) in the *sn*-2 position. The chain order for the latter was evaluated from proton NMR chain signal MAS sideband intensities. Averaged over all lipids, cholesterol increased the order in *sn*-1 chains. The effect was twice as large for the monounsaturated as for the polyunsaturated lipids. In the absence of cholesterol, chain order decreased in the sequence SOPE > SOPC > SOPS for the monounsaturated and SDPC ≫ SDPE > SDPS for the polyunsaturated mixtures. This suggests that cholesterol induces the formation of lipid microdomains in a headgroup- and chain unsaturation-dependent fashion. In contrast, addition of 5 mM calcium ions removed order parameter differences between the polyunsaturated lipids, indicative of a more homogeneous lipid distribution. In the monounsaturated mixture, calcium ions increased lipid order in the sequence SOPE > SOPS > SOPC.

The preferential interaction between cholesterol and polyunsaturated SDPC, followed by SDPE and SDPS, was confirmed by proton MAS NOESY cross-peak intensity differences. Cholesterol was thus found to associate preferentially with the lipids with saturated or mixed chains. Cholesterol therefore seems to form PC-enriched microdomains in mixtures of polyunsaturated SDPC-SDPE-SDPS membranes in which the saturated *sn*-1 chains are preferentially oriented toward cholesterol (Gawrisch and Huster, in Barenholz, 1998a).

D. Phospholipid Domains and Protein Function

Composition-induced lateral domains are functional in biological membranes (Thompson et al., 1995) and might be involved in signal transduction.

It is probably not a coincidence, for example, that membranes from the retinal rod outer segment, sperm cells, postsynaptic neurons, and other excitable cells contain up to 50% highly unsaturated phospholipids with arachidonoyl (20:4n6) and docosahexaenoyl (22:6n3) chains. In addition, n-3 (ω-3) fatty acid deficiency is also associated with the visual and cognitive deficits. Alkenoyl chains of the latter type therefore modulate cell membrane functions associated with neuronal signaling and illustrate the role played by lipid chain unsaturation.

The extent of formation of metarhodopsin II (MII), the G protein-activating form of photoactivated rhodopsin, was measured in various fluid-phase bilayers. MII formation was increasingly likely with increasingly acyl chain polyunsaturation, which potentiates the formation of the active receptor conformation. Addition of cholesterol decreased MII generation, however, as a function of phospholipid acyl chain composition. A chain packing model for bilayers containing mixed-chain phospholipids (*sn*-1 saturated, *sn*-2 unsaturated, and bi-polyunsaturated compounds) and cholesterol was described. Such a model was able to account for the formation of clusters or microdomains with increasing *sn*-2 chain unsaturation. In the model, lateral interactions between the saturated chains were maximized and cholesterol was placed at the center of the domains. Here, the molecules interacted almost exclusively with the saturated acyl chains (Litman, in Barenholz, 1998a).

E. Peptide Orientation in Microdomains and Pore Formation

The effect of cholesterol on the partitioning of peptide GALA (WEAALAEALAEA-LAEHLAEALAEALEALAA) into a bilayer and on the peptide assembly into pores has been studied experimentally as well. GALA was thus found to undergo a conformational change from a random coil to an amphipathic α-helix, when the pH of the surrounding medium was reduced from 7.0 to 5.0. Simultaneously, the vesicles begin to leak. The rate of transport for neutral and negatively charged membrane vesicles is similar at pH 5.0. A mathematical model (Parente et al., 1990) that assumes GALA incorporation into the vesicle bilayer and its irreversible aggregation into pores containing 10±2 peptides was proposed. Cholesterol in the membranes counteracts this effect, partly by minimizing the peptide binding to the membrane. Cholesterol also makes the peptide aggregation in the membrane more reversible (Nicol et al., 1996). Similar reversibility of aggregation was observed with other peptides, such as pardaxin (Rapaport et al., 1996).

Orientation of GALA in the fluid-phase phosphatidylcholine bilayers was determined by a BODIPY-avidin/biotin binding assay. In brief, the peptide and its variants were labeled at the N- or C-terminus. BODIPY-avidin molecules were then added to solution or were encapsulated in the lipid vesicles. It was concluded that in a pore, the peptides arrange themselves perpendicular to the membrane such that, on the average, two-thirds of the molecules orient themselves with the N-terminus on the internal side of the membrane. The pore is stable for at least 10 min. When the peptides bind or form an aggregate with less than 10 participants, the peptides are mostly oriented parallel to the surface. However, when the aggregate attains some critical size, the peptides rearrange and rapidly form a pore in which the hydrophilic sides of the helices are present (Nir et al., in Barenholz, 1998a).

F. Membrane Skeleton Effects

Movements of transferrin and α_2-macroglobulin receptor molecules in the plasma membrane of cultured NRK cells and of band 3 in human erythrocyte ghosts were investigated by single-particle tracking. This was done with spatial precision of a nanometer and temporal resolution of 0.2 ms. For the purpose, the proteins were labeled with colloidal gold particles. The movement of the latter suggested that plasma membrane is compartmentalized into many small domains, approximately 0.25 μm^2 large (0.01 μm^2 for band 3). The proteins move between individual compartments every 25 s (350 ms for band 3), on the average. Within one compartment, membrane proteins diffuse freely. Long-range diffusion is a series of successive inter-compartmental hops.

The boundaries between individual compartments are made of the membrane-associated part of the cytoskeleton or of the membrane skeleton ("membrane-skeleton fence" model).

To investigate mechanically the interactions between integral membrane proteins and intercompartmental boundaries, transferrin receptor (TR)-particle complexes were dragged along the plasma membrane with laser tweezers. The majority (90%) of the TR-particle complexes, which exhibited hop diffusion, were pulled past the intercompartmental boundary by a trapping force of 0.25 pN. A dragging force of 0.05 pN left half of the TR molecules at such boundaries. This escape occurred in the forward as well as in the backward direction.

The domain boundaries have an effective elastic constant of 1 to 10 mN/m. Some 10% of transferrin receptors in the investigated membranes were bound to elastic structures (Kusumi, in Barenholz, 1998a). All this is consistent with the view that the compartment boundaries are defined by the membrane skeleton.

G. Lipid DNA Interaction

DNA does not interact with zwitterionic or acidic lipids unless di- or multivalent ions are added. The polymer has an affinity for cationic species, however. An aqueous suspension of vesicles comprising cationic lipids or a mixture of the latter and neutral, "helper" lipids, such as phosphatidylethanolamines or cholesterol, is therefore sensitive to addition of DNA. This can, but need not, lead to liposome disruption and to stable lipid-DNA complexes. These can be of three basic types: (a) a lamellar (L_α^c or "sandwich") complex, comprising alternating lipid bilayers with an intercalated monolayer of DNA, with parallel and equidistant polyelectrolyte chains (Rädler et al., 1996; Battersby et al., 1998); (b) a "spaghetti-like" complex, comprising DNA strands coated by a lipid bilayer (Lasic, 1997); or (c) a "honeycomb-like" (or H_c^{ll}) complex, consisting of a bundle of hexagonally packed monolayer-coated DNA units (Lasic, 1997; Koltower et al., 1998).

The propensity for the formation of one or the other of these lipid-DNA complexes is affected by the choice of the phospholipids and their phase preference: H_{II}-forming phospholipids (typically phosphatidylethanolamines) prefer type (c) and phospholipids forming mainly lamellar phases (typically phosphatidylcholines) favor type (a) lipid-DNA complex structures; cationic lipids are also influential. To explain the stability and phase behavior of some of these structures, a free energy model has been proposed. This model takes into account the electrostatic interactions, the elastic membrane energy, and the mixing entropy contributions to relate the variation in DNA packing with the lipid composition and lipid/DNA ratio (Ben-Shaul, in Barenholz, 1998a). Incubation of cells with DNA-cationic liposome complexes (lipoplexes) is a common means of cell transfection (referred to as lipofection). The process efficiency is affected by the lipoplex DNA^-/L^+ ratio, but other factors also play a role. The precise relationship between the lipoplex structure, stability, or electrostatics and lipofection optimum remains unclear (Lasic and Templeton, 1996) despite extensive physicochemical characterization (Rädler et al., 1996; Zuidam and Barenholz, 1997, 1998; Battersby et al., 1998; Hirsch-Lerner and Barenholz, 1998; Hübner et al., 1999). Some results suggest that DNA-lipid interactions are reversible and that one may treat the system as a phase diagram (Zuidam and Barenholz, 1997, 1998). Other changes, such as size increase, are permanent, however. In either case, the kinetics of mixing is of paramount importance (Hübner et al., 1999) but more so for the irreversible system reorganization (Battersby et al., 1998), such as vesicle fusion and bilayer sandwiching.

XII. RECOGNITION

Macromolecular interactions at and with an interface are of paramount importance in life. One of the chief reasons for this is the role that they play in intermolecular and cell-cell recognition. It is therefore crucial to understand and to quantify the forces near surfaces of biological interest.

Molecular recognition can be interpreted at different levels of complexity. The simplest is the "key and lock" model; the most advanced are sophisticated computer simulations of protein-ligand binding with atomic detail. The former approach is easy to grasp but offers no means of quantitative evaluation or predictions. Computer simulation is extremely informative, especially when it can be correlated with the results for single molecules, but has the problem of long computation times and, very often of unclear starting assumptions. The gap between the worlds can be closed by using suitable semiphenomenological models. Such models must pay attention to the molecular detail and contain sufficient, but not too many, adjustable parameters to ensure sufficient agreement between the experimental and calculated data. The mean-field theory of surface electrostatics and hydration is an example of this (Cevc, 1990).

Nonspecific adsorption of a protein to a membrane is affected little by the overall repulsion between the macromolecule and the bilayer surface. Instead, such adsorption is governed by the number of defects and/or by the availability of hydrophobic binding sites at the interphase. Even simple, artificial lipid membranes offer a sufficient number of hydrophobic binding sites to an approaching macromolecule to result in protein adsorption to a nonstabilized lipid bilayer within a few seconds. The lipid phase may play a role in this, as illustrated by the fact that certain antibodies preferentially bind to the lipids in a nonbilayer phase (Janoff and Rauch, 1986).

In order to suppress nonspecific protein binding to a bilayer, an optimum number of water-soluble and sufficiently mobile polymers can be attached to the lipids by the cell or a synthetic chemist. Such polymers increase the repulsion between the membrane surface and macromolecules because of the generation of a thick, mobile, but also strongly hydrated interface. Protein affinity to bind to or to be inserted into such a modified surface is thus lowered and the surface-induced protein denaturation or complement insertion is hampered (Blume and Cevc, 1992). This makes polymer-coated liposomes much less attractive for phagocytosis (Blume and Cevc,

1993; Emanuel et al., 1996). Such liposomes, consequently, remain in the blood circulation much longer than simple lipid vesicles and accumulate in tumors spontaneously. Pathogens use a similar trick to protect themselves from the immune system or from digestion in a surrounding fluid. The role of lipid proper in the process may be significant as well, as implied by studies of glycoconjugate function in the cell membrane. One possible reason for this is that the lipid moiety promotes or reduces carbohydrate exposure of (glycolipid) headgroups and even restricts the number of permitted conformations for a given glycolipid oligosaccharide.

Evidence for the regulation of glycolipid function by a ceramide moiety was found, for example, in studies of enzyme substrate specificity, antiglycolipid recognition, and bacterial-host cell interactions. Characteristics of verotoxin binding to its glycolipid receptor globotriaosyl ceramide, moreover, revealed the modulation of receptor function by the glycolipid fatty acid content and its role in in vitro binding assays, cell cytotoxicity, and intracellular routing (Lingwood, 1996).

Control of intermolecular associations in the membrane and at the membrane-cytoskeleton interface involves specific phosphorylation and dephosphorylation cascades. These involves proteins as well as phospholipids and are regulated by the extracellular matrix. The binding of growth factors and hormones to their specific receptor, tyrosine kinase, occurs in such fashion (Kinnunen, 1991).

A. Molecular Basis of Specific (Protein) Binding

The efficacy of specific association between a receptor and its ligand is sensitive to the proximity of the polar surface. As a rule, the apparent binding constant is lower for surface-attached ligands than for the same freely accessible, dissolved molecule. The magnitude of such binding suppression increases with increasing lipid (and thus membrane surface) polarity (Cevc, 1995).

This suggests that the proximity of a highly hydrophilic surface will hamper the binding of a large molecule to its specific receptor in biological membranes. Immunological interactions near the surface of lipid bilayers, consequently, are promoted and enhanced by hapten incorporation into a relatively apolar bilayer (Yasuda et al., 1977; Dancey et al., 1978). Alternatively, and more elegantly, similar gains can be achieved by introducing a sufficiently long spacer between the presenting surface and the actual binding entity (Dancey et al., 1979). This is most likely the reason why phosphatidylethanolamine-caproyl derivatives are better immunogens than the corresponding simple phosphatidylethanolamine derivatives (Dancey et al., 1978; 1979) and also why phosphatidylethanolamine-rich vesicles present antigens better to the immune system than phosphatidylcholine membranes (Yasuda et al., 1977).

B. Molecular Basis of Nonspecific (Protein) Adsorption

Nonspecific protein adsorption to a model membrane surface involves phenomena similar to those in specific binding.

An arbitrary macromolecule that approaches a lipid bilayer surface must penetrate through the repulsive, interfacial barrier before it can adsorb nonspecifically to the membrane. Such binding normally involves only a few amino acids at a time, to minimize the energetic cost of macromolecular penetration. Once the repulsive barrier is overcome, the initial energy loss is compensated by the subsequent energy gain from the protein accommodation in the interfacial region.

Nonspecific macromolecular adsorption to an interface differs in several aspects from specific molecular binding. In the former case, participating molecules are subject to only a few structural constraints. (Denaturation of a protein molecule and substantial structural lipid reorganization at or near the adsorption site are very common after adsorption.) In the latter case, the bound entities are confined to an energetically unfavorable region, from the point of view of interactions with lipid. Membrane lipid composition and protein biochemistry therefore both play a role, as seen in Table 4.

Free energy released by the protein after its adsorption to a lipid bilayer normally stems from the insertion of the protein's hydrophobic residues into the hydrocarbon membrane core. Imperfections in this core, therefore, increase the probability and stability of nonspecific membrane-protein association.

Protein adsorption to lipid bilayers is particularly strong in the phase transition or the phase separation region. To minimize protein loss on a lipid bilayer, proximity to the phase transition temperature of a mixed lipid system should be avoided. Ordered phases, moreover, should be given preference over fluid phases. Instead of lipids with a low chain-packing density (such as charged or strongly hydrophilic lipids), moderately polar and densely packed lipids should be used for practical applications.

Cellular recognition involves a plethora of molecules, most of which belong to the class of proteins or carbohydrates. By virtue of C-type lectin domains, these receptors mediate adhesion, recognizing specific carbohydrate-based ligands on the partner cell (Rosen, 1993).

Selectins are receptors that initiate rolling of leukocytes on activated platelets or endothelium through Ca^{2+}-dependent recognition of cell surface carbohydrates (McEver, 1994). Selectins are also found on endothelial cells. Inappropriate surface display of selectins contributes to inflammation (McEver, 1994).

Exogenous carbohydrates alter the cytolytic capabilities of natural killer (NK) cells, as does cell treatment with glycosidases or inhibitors of glycosylation. Target recognition, tissue distribution, and postbinding events in the lytic cascade are carbohydrate dependent as well (McCoy et al., 1991).

Recognition plays a pivotal role in elimination of aging and senescent cells. For example, apoptotic cells are marked for disposal by mechanisms that are poorly understood but appear to involve surface sugar changes and exposure of phosphatidylserine; both are recognized by as yet unidentified phagocyte receptors (Savill, 1997). It is conceivable that there is a hierarchy of recognition steps that

Table 4 Selectivity of Interaction of Spin-Labeled Phospholipids with Peripheral Proteins Bound to Dimyristoyl-Phosphatidylglycerol Bilayers[a]

Peripheral Protein	Interaction Hierarchy
Myelin basic protein	$PS^- > CL^- > PA^{2-} > PG^- > PI^- > PA^- > PE^{\pm} > PC^{\pm b}$
Apocytochrome c	$PI^- > CL^- > PS^- > PC^{\pm} > PG^- > PE^{\pm}$
Cytochrome c	$PI^- > PG^- > CL^- > PS^- \approx PC^{\pm} > PE^{\pm}$
Lysozyme	$CL^- > PG^- \gg PE^{\pm} > PC^{\pm} > PS^- > PI^-$
Polylysine	$CL^- > PS^- \geq PG^- > PI^- > PC^{\pm} > PE^{\pm}$

[a] PS, phosphatidylserine; CL, cardiolipin; PA, phosphatidic acid; PG, phosphatidylglycerol; PI, phosphatidylinositol; PE, phosphatidylethanolamine; PC, phosphatidylcholine.

are involved at different stages of the death program. In erythrocytes, reactions of reactive oxygen species with the membrane constituents and subsequent proteolysis are important. So is immunoglobulin G binding, for the triggering of cellular removal (Bartosz, 1991). Furthermore, in aging erythrocytes, cell deformability decreases while fragility to hemolytic factors increases; surface charge density remains the same, however. A different situation is encountered in lymphocytes, where the interaction mediated by CD2 with LFA-3 adhesion receptors is dramatically altered by surface charge and adhesion receptor density. This occurs in such a way that this pathway is latent in resting T lymphocytes but becomes active over a period of hours following T-cell activation. Receptor lateral diffusion in the membrane is part of the mechanism (Dustin et al., 1991).

Involvement of a multitude of cell surface molecules, in addition to T cell receptor (TCR)–major histocompatibility complex (MHC)–peptide complexes, in the binding and signaling for lymphocyte-mediated lysis has been demonstrated. Two proposed mechanisms of lymphotoxicity currently appear to be valid: (a) the formation of pores in target cell membranes by secreted molecules of lymphocyte origin, such as perforin and granzymes, and (b) a nonsecretory process initiated by receptor-mediated triggering of apoptosis-inducing target cell surface molecules, but not involving the secretion of pore-forming agents and granzymes (Berke, 1994). These two mechanisms are not mutually exclusive and are probably used by different types of effector cells or by the same effector cells at different stages of differentiation.

Adhesive interactions between cells and extracellular matrices profoundly affect spatiotemporal positioning, site-specific gene expression, and proliferation rate; aberrant adhesion also often contributes to disease formation. At the molecular level, these phenomena rely on the recognition of adhesive components of the extracellular matrix by membrane-bound receptor molecules, typically via formation of three-dimensional aggregates of glycoproteins and proteoglycans, and ultimately by the transduction of chemical and physical signals to the cell interior (Tuckwell et al., 1993).

Various blood group antigen (BGA)–related glycodeterminants are expressed on the cell surface at definite stages of cell differentiation during embryogenesis, organogenesis, tissue repair, regeneration, remodeling, and maturation (Glinsky, 1992). The appearance of the BGA-related glycoepitopes on the cell membrane is a consequence of the association of MHC and peptides, with subsequent elimination of cells having a high density of BGA-related glycoepitopes on their surface. In cancer it has been considered a key mechanism of phenotypic divergence of tumor cells immunoselection, tumor progression, and metastasis (Glinsky, 1992).

XIII. LIPID-PROTEIN INTERACTIONS

Many integral membrane proteins require specific lipids for optimal activity and are inhabited by other lipid species (Sandermann, 1978). Membrane protein activity is also sensitive to the lipid bilayer dynamics and physicochemical state. From the biological (Veld et al., 1993), physical (McElhaney, 1993; Mouritsen and Bloom, 1993), physicochemical (Quinn, 1989; Kinnunen, 1991) Electron Spin Resonance (ESR) and NMR spectroscopic (Marsh (1995a; 1995b) and Watts (1991), respectively), and functional points of view, this phenomenon has been reviewed repeatedly.

The molecular basis of lipid-protein interactions remains obscure. In general, these interactions can be divided into the highly specific and the obligatory. The former are required for protein activity (e.g., with phospholipases and other lipid-metabolizing enzymes or with protein kinase C and β-hydroxybutyrate kinase or pyruvate oxidase, the latter after substrate binding); the obligatory interactions mediate protein binding (in)to a membrane and maintain protein structure. It is widely accepted, for example, that "annual" lipids stabilize integral membrane proteins (Veld et al., 1993). The reasons for this are, among other, the hydrophobic mismatch between integral proteins and membrane lipids (including steric elastic strain and van der Waals forces), the interactions in the interfacial region (charge-charge, hydrogen bonds, solvation, etc.), and, occasionally, small-scale 'phase' transitions.

The specificity of lipid interactions with both integral and peripheral proteins is reflected in the lateral inhomogeneity of membrane lipid distribution (Marsh, 1987, 1995a,b, 1998 and Marsh in Barenholz, 1998a).

To give but a few specific examples: Bovine spinal cord myelin basic protein (MBP) has a preference for phosphatidylglycerol over phosphatidylcholine, but the two lipids phase separate incompletely even when the mole fraction of dimyristoyl-phosphatidylcholine is below 0.25 (Sankaram et al., 1989). Phosphatidylglycerol interacts more strongly with photosystem 1 or photosystem 2 in (sub)thylakoid membranes than the corresponding monogalactosyldiacylglycerol and phosphatidylcholine analogues (Li et al., 1989). A much weaker preferential ganglioside–Na^+,K^+)-ATPase interaction was observed by means of spin-label ESR in the order GDlb greater than GM1 approximately equal to GM2 approximately equal to GM3 (Esmann et al., 1988).

Specifically, protein kinase C (PKC) requires one molecule of diacylglycerol and four molecules of phosphatidylserine (PS) for proper action; the optimum being reached with approximately 10 molecules of PS. Other anionic lipids do not fully activate the enzyme, despite the fact that they can bind to the protein in Ca^{2+}-dependent manner. The activation by PS is relatively insensitive to membrane electrostatics, probably because the intimate protein-lipid interaction is stereospecific for the L-form and involves both the carboxyl and amino groups on PS. The interfacial conformation of the polar headgroup is critical, and neither lyso-PS nor the 2-acetyl derivative supports activity (Marsh, 1995a).

β-Hydroxybutyrate dehydrogenase is activated by phosphatidylcholine (PC) in a cooperative allosteric fashion, nearly insensitive to the precise (fluid) chain composition and glycerol backbone (D- and L-stereoisomers activate similarly well). Whereas PC headgroup elongation has little effect, replacement of the phosphate and trimethylammonium groups by an isopropyl link abolishes activation; the phosphinate analogue of PC is also inactive, in contrast to the phosphono derivative, which works well (Marsh, 1995b).

Pyruvate kinase activation occurs primarily by lipid chains binding to hydrophobic regions that are exposed on the protein following the substrate-induced conformation change. This explains the strong, linear chain length dependence, which resembles that observed in studies of lipid-lipid interactions (Cevc and Marsh, 1987). Indeed, for many proteins of the integral (intramembrane) type, the effects of lipid chain length come mostly from a limited modulation of the overall interaction. Sensitivity to chain composition for such proteins is thus only modest (Marsh, 1995b).

A further example of this is rhodopsin, which also has a linear chain length dependence but a much smaller incremental change ($\sim 0.02\ kT$ per CH_2) than pyruvate kinase (0.155 and 0.12 kT per CH_2 on single- and double-chain lipids, respectively); the value for Ca^{2+}-ATPase is even lower (0.01 kT/CH_2). These differences suggest a much smaller free energy of association with lipids for the latter two lipids (Marsh, 1995b).

However, for rhodopsin interaction with PC molecules with less than 13 or more than 17 carbon atoms per chain, the chain length effects are much more dramatic. In these more extreme cases, the lipid chain is incapable of matching the apolar stretch of the protein and aggregation or phase separation of the protein takes place (Ryba and Marsh, 1992; Piknova et al., 1997). For the Ca^{2+}-ATPase from sarcoplasmic reticulum (Johannsson et al., 1981) and for adenosinetriphosphatase (Caffrey and Feigenson, 1981) an optimal chain length of C(18:1) to C(20:1) has been found. The overall lipid-protein interactions exhibit a less sharp maximum in the chain length dependence than enzyme activity, especially in the case of Ca^{2+}-ATPase, suggesting that the latter reflects more subtle variations (Marsh, 1995b).

The relative association constants for different lipid classes interacting with various integral proteins are given in Table 4. [Of the lipids not included in this table, gangliosides show relatively little selectivity over that for PC in interaction with the Na^+,K^+-ATPase or with the acetylcholine receptor (Marsh, 1995b).] For thylakoid membranes, monogalactosyl diglyceride appears to take the part played by PC in mammalian membranes. Phosphatidylglycerol, which occurs frequently in nonmammalian cells, than takes the role in preferential association with the proteins. Interestingly, chilling-resistant and -sensitive plants are quantitatively different in this respect (Li et al., 1990).

Selectivity of interactions is found with most integral proteins, rhodopsin and Ca^{2+}-ATPase being exceptions (Marsh, 1987). In general, phosphatidylcholine, which is the most widely observed phospholipid of mammalian membranes, exhibits one of the lowest selectivities, often followed by phosphatidylethanolamine. Higher affinities are normally found for negatively charged lipids; an exception is phosphatidylglycerol, which is the most polar and has a relatively low propensity for H-bond formation. It is noteworthy that the presence of other membrane proteins can affect the selectively of lipid-protein interactions (Marsh, 1995b), either because of their effect on the state of lipids or by competition.

Electrostatics does not necessarily dominate the interactions between proteins and charged lipids, as demonstrated by the fact that lipids with the same nominal charge often have different protein affinities (Marsh, 1998). In particular, negatively charged phosphatidylglycerol often resembles in this respect the zwitterionic phosphatidylcholine. The screening of electrostatic contributions by high-ionic-strength electrolytes also more often than not leaves the protein selectivity little affected. This notwithstanding, the Debye-Hückel theory of electrolytes, in some cases at least (Cevc, 1990), can reasonably describe the observed salt dependence (Esmann and Marsh, 1985). The proviso for this is that additional adjustable parameters are introduced into the model in order to extend the range of applicability empirically.

The lipid selectivity is additionally affected by protonation (pH titration), if the pK_a values of the groups involved are in the studied range. However, some of the observed changes may be due to nonelectrostatic effects, as discussed previously. Most important among such nontrivial effects are the changes in lipid (and protein) surface polarity and hydration (Cevc, 1987; 1988; 1989; 1990) but also of hydrogen bonding capability (see Sec. VII for comparison).

The intrinsic, spontaneous off-rates for lipid exchange are inversely proportional to the relative association rates. When the lipid selectivity is reduced, either by pH titration or by electrostatic screening by the salt, the on-rates do not change, indicating that they are diffusion controlled, whereas the off-rates increase, reflecting the reduced specificity.

In general, the off-rates for phosphatidylcholine are in the region of 10^7 s^{-1}, which is of the same order as, but significantly slower than, the intrinsic lipid-lipid exchange rates (10^8 s^{-1}) (Sachse et al., 1987), arising from lateral diffusion in fluid lipid bilayers. In the gel phase, the exchange of protein-associated lipids is slow ($\ll 10^6$ s^{-1}), but the number of such lipids is not affected markedly by the phase transition (Marsh, 1995a). Specifically, the exchange off-rates for spin-labeled stearic acid and phosphatidylcholine in complexes of fixed lipid/myelin proteolipid protein (MPP) ratio are in the range between 1 and 9×10^6 s^{-1} in a fluid phase bilayer at 30 °C, respectively, in the gel phase, no exchange was observed in such a time domain (Horvath et al., 1993a). Approximately 11 phosphatidylcholine molecules were concluded to be associated with the MPP protein monomer in the gel phase (Horvath et al., 1993b); this value might change with the lipid chain length, however, as found with rhodospin and Ca^{2+}-ATPase (Marsh, 1998).

For many proteins, the number of perturbed lipids was found to scale with the number of α-helices (Fig. 21) but the selectivity of interaction depends on other factors as well (Fig. 22).

As a result of lipid-protein interactions, biomembranes may contain laterally separated domains that, at their interfaces, provide mismatched regions that facilitate passage of polar components through the bilayer. Individual domains may have different "phase" transition temperatures (Quinn, 1989) and divergent affinities for dissolved molecules. Local variations in ion (especially protons, Ca^{2+}, and some other divalent species) concentration, regional interactions between the cell and its surroundings (extracellular matrix, neighboring cells), and coupling with the cytoskeleton are among the consequences of this. Proximity and binding of hormones, metabolites, drugs, and other bioactive substances, especially when these are amphiphilic and hence relatively potent, are important as well. In the following section, material flux across the membrane is discussed in greater detail.

XIV. BARRIER PROPERTIES

Membranes are an obstacle to the motion of hydrophilic molecules through the bilayer but may facilitate the corresponding molecular motion along the bilayer surface. Conversely, fatty substances move relatively easily in the plane of the noncompartmentalized lipid bilayer but are strongly restricted in mobility perpendicular to the bilayer surface. For water-soluble substances, such as inorganic ions, the membrane interior is a permeability barrier maximum, for fat-soluble substances, such as organic ions, the same region acts as a trap. In either case, exchange of material between the aqueous compartments separated by a membrane is hindered. A substantial transbilayer concentration or electrostatic potential difference is therefore required to drive the solute or solvent flow across a membrane (Cevc, 1993a).

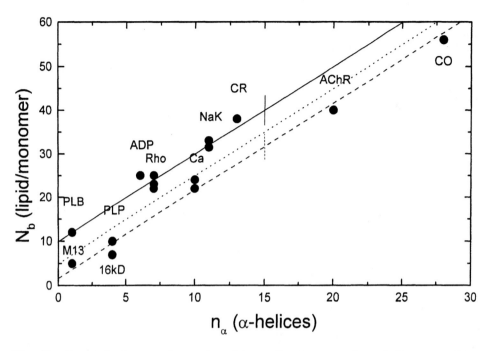

Fig. 21 Correlation between the number of the predicted transmembrane helices, n_α, and the number of first-shell motionally restricted lipids, N_b, per monomer of different integral membrane proteins. Lines were calculated assuming monomeric helical sandwiches (full), protein dimers (dotted), or hexamers (dashed). (From Marsh, 1998.)

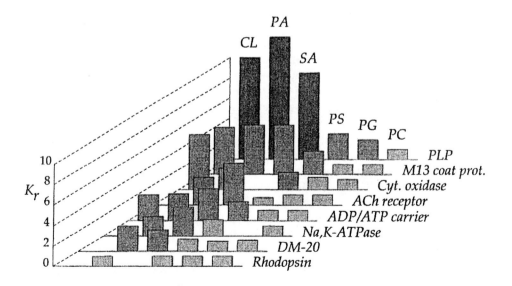

Fig. 22 Lipid headgroup selectivity of various integral membrane proteins. (From Marsh, 1998.)

However, membranes are complex permeability barriers. One of the reasons for this is the membrane sensitivity to ambient changes: a membrane often responds to the solute-membrane interactions or to the variable transbilayer gradient. Alcohols, for example, not only flow across a membrane but also increase the permeability of the lipid bilayer by increasing the polarity of membrane interior (Orme et al., 1988); this may be due to a concurrent decrease in aliphatic chain packing density (Vierl et al., 1994). The permeability of biological membranes often decreases with increasing transbilayer solute concentration gradient as well (Sten-Knudsen, 1978), the transport versus concentration difference curve then showing an apparent saturation. (To minimize the danger of data misinterpretation, it is thus advisable to measure the rate of transport for different solute concentrations and then extrapolate the corresponding permeability value toward zero solute concentration.)

Experiments suggest that the resistance of protein or carrier-free lipid bilayers to solute permeation is inhomogeneously distributed throughout a membrane; the highest and rate-limiting barrier is often located in the interfacial region. For example, the interfacial barrier for the small water-soluble molecules was measured and found to be $10-10^3$ times higher than the remaining transmembrane resistance (Dix et al., 1978).

The rate of simple amphiphile transport across a protein-free lipid bilayer is proportional to the length of the permeant's apolar side chains; the constancy of the incremental free energy of transfer between the aqueous and hydrocarbon compartment (approximately -1.3 kT per molecule) explains this. Increasing permeant polarity diminishes transmembrane flux. Glycerol, for example, is transported across the lipid bilayers > 10 times less rapidly than ethylene glycol (ethanediol) and approximately three orders of magnitude less efficiently than ethanol (see Table 5). This is chiefly due to decreased partitioning of the more polar compounds in the phospholipid bilayer. Likewise, the rate of transport decreases with the solute capacity for hydrogen bond formation and, in the first approximation, is commensurate with the number of OH groups on the permeant.

The permeability of mixed phospholipid bilayers is dominated by the short-chain components, especially by phosphatidylcholines (Singer, 1982), or by the charged lipid species. Addition of cholesterol diminishes the membrane permeability to ionic solutes, by and large (Bittman et al., 1984). Addition of cholesterol always increases the permeability of gel-phase lipid bilayers to water; in the fluid membranes cholesterol has the opposite effect. Moreover, a new liquid ordered (LO) phase is formed (Mouritsen and Jorgenssen, 1995).

The permeability of a simple lipid bilayer to ions is very low; it is higher for organic anions than for organic cations, however. (The dipolar membrane potential was suggested to be responsible for this but the contribution of interfacial water is probably comparably important.) Inorganic ions diffuse across the membrane even less readily than most organic ions. Small charged solutes therefore rely on channels to cross biological membranes in appreciable quantity.

More generally speaking, the transport across cell and organelle membranes rarely, if ever, reflects simple lipid bilayer properties. Instead, it is dominated by the low permeation resistance of defects, channels (pores), and various carriers (for a series of reviews see Bronner, 1984). Even protein adsorption to or insertion of proteins into the phospholipid bilayer tends to facilitate transmembrane transport. This is due to either the generation of membrane defects and/or pores or the carrier or channel activity of the proteins. Simple, but efficient, channels are created in the lipid bilayers by polyene macrolide antibiotics (amphotericin, filipin, nystatin, etc.) or by some polypeptides (such as gramicidins and mellitin).

Channel properties are determined by the protein structure by and large (Wu, 1991). This makes the solute selectivity of different channels sensitive to the primary protein sequence. But it is also possible to affect the channel properties and conductivity by changing the host lipid matrix and its characteristics. Chain ordering and the resulting increase in lipid packing density, for example, suppress sugar translocation through the lactose permease channel when this protein is incorporated the ordered-phase dimyristoylphosphatidylcholine bilayers (Dornmair et al., 1989).

Pores may arise spontaneously in the defect-rich bilayer regions as well. Lipid membranes undergoing a phase transition and bilayers under mechanical or electrical stress, consequently, are very leaky; so are lipid vesicles that have adsorbed proteins. Small sonicated vesicles or lipid vesicles in the biological fluid consequently lose the water-soluble contents rapidly. Mixed lipid vesicles also contain their contents poorly (Hertz and Barenholz, 1975; Matsuzaki et al., 1996).

The mode of protein insertion, folding, and translocation may be determined directly by the surface properties of a biomembrane. Although any accurate theoretical model of the underlying processes will inevitably, have to be complex, owing to the

Table 5 Permeabilities P_m (cm s^{-1}) and Activation Energies E_a (kJ mol^{-1}) for Transport Across Fluid Phosphatidylcholine Bilayers[a]

Cationic	H$^+$	Na$^+$	K$^+$	Rb$^+$	Tl$^+$	Cd^{2+}
P_m	5×10^{-12}–3×10^{-9}	10^{-12}	3×10^{-12}	3.3×10^{-12}	(4×10^{-12})	(3×10^{-12})
E_a	71–83	70–80	63			25

Alkylamines	Methylamine	Ethylamine	Propylamine	Butylamine	Pentylamine
P_m	1×10^{-6}	7×10^{-7}	4.3×10^{-6}	5.2×10^{-5}	4×10^{-4}

Anionic	Cl$^-$	Br$^-$	J$^-$	SCN$^-$	NO$_3^-$
P_m	7×10^{-12}	1.8×10^{-11}	3×10^{-11}	(10^{-11})	(5×10^{-11})
E_a	79				

"Silent"	H$_2$O	H$^+$/OH$^-$	HCl	TlCl	CdCl$_2$	HgCl$_2$
P_m	4×10^{-3}	10^{-9} 10^{-4}–10^{-3}	1.5×10^{-11} 7×10^{-8}	1.1×10^{-7}	4×10^{-9}	(10^{-2})

Nonionic					
Alkanols	Methanol	Ethanol	Butanol	Propanol	Pentanol
P_m	3.3×10^{-3}	2×10^{-3}	6×10^{-3}	5.7×10^{-2}	(10^{-2})
Polyalcohols	Glucose $C_6O_6H_{12}$	Fructose $C_6O_6H_{12}$	Sucrose $C_{12}O_{11}H_{22}$	Propantriol (Glycerol = $C_3O_3H_7$)	Ethandiol (Ethyl glycol = $C_2O_2H_5$)
P_m	3×10^{-12}–3×10^{-11}	4×10^{-10}	8×10^{-14}	5×10^{-6}	2.5×10^{-5}

[a] When two values differing by several orders of magnitude are given, the permeation process is biphasic.

multiplicity of interactions involved, it is clear that the spatial anisotropy and the unidirectional nature of the interface will take a dominant part in any description.

Transport of smaller solutes across membranes is catalyzed by the specific membrane proteins. The activity of most of them is modulated by the bulk physical characteristics of the lipid bilayer; specific lipid requirements are rare (Veld et al., 1993).

XV. INTERFACIAL CATALYSIS AND MODULATION

Interfacial complexity is of high biological relevance (Watts, 1995). For example, charged lipids can interact in a stoichiometric way with charged protein residues and attract the oppositely charged molecules from the solution. Proteins in turn change conformation and often form denatured "molten globules" or else become activated on interaction with the membrane surface. Moreover, the pH, salt concentration, and water activity gradient are normally nonuniform near the membrane surface.

Phospholipase A_2, for example, requires a lipid-water interface for complete activation and calcium ions as cofactor. Diffusion of the phospholipid into the active site is facilitated by the essentially hydrophobic and dehydrated protein interior, which must be in close contact at its mouth with the membrane surface (Marsh, 1997).

Increased conductivity of the surface-associated water was postulated to be involved in many important phenomena, such as surface catalysis or lateral communication along cellular membranes (Heberle et al., 1994; Heim et al., 1995). The lateral electrical conductivity of the adsorbed water film is up to 10^4 times higher than that of the bulk water, only quantitatively affected by the detailed surface properties. Fast proton transport seems to be responsible for this, as seen from the fact that the conductivity in surface-adsorbed D_2O is approximately 40% higher than that in surface-adsorbed H_2O. A contribution from electronic conductivity cannot be ruled out entirely, however. In contrast, the involvement of classical ion transport appears to be small, if it exists at all (Heim, M., R. Guckenberger and G. Cevc, unpublished).

XVI. STRUCTURE-FUNCTION RELATIONSHIP

Lipid bilayers are the backbone of all biological membranes. Lipids act as a "solvent" for the integral membrane proteins; lipids, such as phosphatidylinositol, play a role as second messengers; lipids serve as anchors for many receptors and act as catalytic factors; and lipids provide an energy storage system (Small, 1986), can trigger gene expression (Vigh, 1998), etc. Lipids, moreover, progressively gain importance as the basic component of innovative pharmacological and biotechnological applications (Gregoriadis and McCormack, 1998). It is, therefore, important to understand how the physicochemical properties of lipid bilayers, such as the phase state, are controlled by the structural variations at the molecular level.

At least primitive cells maintain structural homeostasis. This has often been explained in terms of an optimum spontaneous membrane curvature (Tate et al., 1991). The phenomenon is more complex, however, and must be considered locally rather than globally. Membrane homeostasis involves proteins as well as lipids, some of which are crucial for the functioning of the former. For example, in the membrane of *Acholeplasma laidlawii* a constant surface charge density, similar phase equilibria, and nearly constant spontaneous curvature are maintained for the polar lipids (Dahlqvist et al., 1995). Specifically, by studying the effect of various lipids on protein activity, it was concluded that regulation of packing conditions in *A. laidlawii* membranes by the DGlcDAG synthase is governed not by the absolute chain order but more by the "spontaneous curvature," that is, by the packing properties in the membrane within a certain range of conditions (Karlsson et al., 1996). The nonbilayer phase propensity of the membrane lipids is also important for the side chain cleavage activity of cytochrome P450SCC (Schwarz et al., 1997). Furthermore, the crucial role of maintaining the right membrane polymorphism was confirmed for an *Escherichia coli* strain (AD93) unable to synthesize the nonbilayer lipid phosphatidylethanolamine (Rietveld et al., 1994). This may have to do with the fact that nonbilayer lipids are required for efficient protein transport across the plasma membrane of *E. coli* (Rietveld et al., 1995).

The binding of protein kinase C and of the insect protein apolipophorin III into lipid membranes is affected by the presence of lipids that favor positive membrane curvature (Escriba et al., 1997). Phospholipase C activity is also sensitive to lipid packing but prefers negative membrane curvature and therefore is highest in sonicated lipid mixtures (Rao et al., 1993). Membrane stress and defects resulting from membrane sonication are also likely to be involved.

It has been speculated that the activity of some antitumor agents may involve destabilization of nonlamellar membrane structures, probably via interference with cellular signaling pathways (Escriba et al., 1995). The change in the content of non-bilayer phase preferring phosphatidylethanolamine in sarcolemma was also shown to play a crucial role in the sarcolemmal disruption during prolonged ischemia and/or reperfusion (Post et al., 1995).

Heterotrimeric G proteins (peripheral proteins) conduct signals from membrane receptors (integral proteins) to regulatory proteins localized to various cellular compartments (Escriba et al., 1997). They are in excess over any G protein–coupled receptor type on the cell membrane, which is necessary for signal amplification. Thus, the protein-lipid interactions are crucial for the cellular localization of receptors and, consequently, for signal transduction. G-protein α-subunits are able to bind to various lipid bilayers, but the presence of phospholipids that prefer nonlamellar structures, e.g., phosphatidylethanolamines, is very helpful for this (Escriba et al., 1997). Conversely, the binding of G protein–coupled receptor kinase 2 and the G-protein β-subunit is not increased by lipids prone to form hexagonal phases. Cholesterol and membrane surface charge are influential as well.

Anatomical structures involved in the generation and maintenance of the difference between apical and basolateral domains include the tight junction, which serves to restrict lateral diffusion within the membrane, and the cortical cytoskeleton, which can selectively bind and retain transmembrane proteins at a particular surface. The difference in protein and lipid composition of the apical and basolateral domains allows the cell to perform a variety of vectorial functions (Le Gall et al., 1995).

Membrane phospholipid generation is co-ordinated with the cell cycle, at least in a colony-stimulating factor 1–dependent macrophage cell line: net phospholipid accumulation in S phase results from an interaction between cell cycle-dependent oscillations in the rates of phosphatidylcholine (PC) biosynthesis and degradation; CTP: phosphocholine cytidylyltransferase is the rate-controlling factor in the former process (Jackowski, 1994). The selective anterograde transport of plasma membrane proteins may be mechanistically related to the sphingolipid domains. Some proteins, such as pp60src kinase, can induce specific changes in the cytoskeleton-membrane interface and thus provoke specific configurational changes in the membrane as well as cytoskeleton architecture. The resulting membrane architecture was postulated to correspond to distinct metabolic-differentiation states of the cell and the formation and maintenance of proper three-dimensional membrane structure, exemplified in neurites and pseudopods (Kinnunen, 1991).

Gramicidin aggregation and function in the membrane were reported to be sensitive to the changes in phosphatidylserine headgroup properties. This could be due to a modified electrostatic energy and curvature stress (Lundbaek et al., 1997), but effects of hydration and hydrogen bonding cannot be excluded.

XVII. CONCLUDING REMARKS: THE LIPID CONNECTION

Some of the differences in thermodynamic behavior of lipid and biological membranes probably result from the polar headgroup rather than from the bilayer core effects. This was demonstrated for the lipid chain-melting phase transition and implied for the bilayer-to-nonbilayer phase transitions in lipid membranes; in either case, the effect of bilayer hydration seems to exceed that of membrane electrostatics. It is therefore possible to control the bilayer state easily by modulating the interfacial polarity via protonation or chemical modification (e.g., alkylation) of the surface polar residues. Alternatively, the interactions between the bilayer surface and solutes can play a similar role, as can conformational changes involving nonlipid membrane components.

Specific lipid-lipid and lipid-protein interactions govern the well-controlled, but highly dynamic, membrane architecture; their understanding is therefore important in understanding the selective modulation of membrane properties by the cell and its environment. Moreover, the specific binding of large molecules, such as antibodies, to the membrane is affected by the proximity of its polar surface. Further still, the presence of net surface charges may raise or lower local macromolecular concentration depending on the relative sign of the charges involved, equal signs lowering and opposite signs increasing such concentration. On the other hand, a receptor attached to the strongly polar surface will always attract and bind its ligand less effectively than a similar receptor in a solution. (The reason for this is the non-Coulombic repulsion between the ligand-presenting polar surface and an approaching macromolecule, such force being promoted by the surface hydrophilicity and by the finite width of the interfacial region.)

Bacteria adapt to changes in the environment (temperature, hydrostatic pressure, pH, etc.) by altering the lipid composition of their membranes. Such organisms maintain their membranes in the lipid-crystalline state by adapting lipid headgroup composition, by altering the degree of acyl chain saturation and branching, as well as by variations of the acyl chain length and the sterol content of the membrane. A different adaptation prevents the formation of nonbilayer structures in the membrane, which would disrupt the closed membrane organization and thus catastrophically increase the membrane permeability. A balance between lipids that prefer lamellar phases and those that tend to adopt nonbilayer structures keeps the integral membrane proteins optimally active under physiologic conditions (Veld et al., 1993).

Significant changes in the lipid membrane composition should affect many, if not all, membrane-associated processes; they can also affect the expression of various genes. The lipid composition of cells therefore changes permanently. The magnitude of the resulting variability is commensurate with the cell's sensitivity to external and internal factors, including experimental boundary conditions. It must therefore be kept in mind that, while great biological diversity in the membrane lipid composition is not unexpected, the observed differences may vary from one experimental setup to another. Various cells also have a different "buffering" or "resistance" capacity to the induced change in the membrane lipid composition.

Normally, a cell will change its lipid metabolism and membrane composition depending on the requirement for specific membrane organization and dynamism; proteins associated with or incorporated in the lipid membrane environment will participate in the process. Current evidence suggests that genetic as well as environmental factors play a role in this, in as yet unknown proportions, and that cell-to-cell variability is appreciable. The ability to manipulate the lipid membrane composition and thereby to modify the cell function offers a means of studying the role lipids play in nature. This capability also opens ways for medical approaches that can be referred to as membrane-lipid replacement therapy. In every instance, such manipulations have a chance for success only if the structure and the properties of the membrane as a whole are properly understood.

REFERENCES

Abrahamsson S, Dahlen B, Lofgren H, Pascher I, Sundell S. In: Abrahamsson S, and Pascher I, eds. Structure of Biological Membranes: New York: Plenum, 1977, p 1.

Ansell GB, Hawthorne JN, Dawson RMC. Historical introduction. In: Ansell GB, Hawthorne JW, Dawson RMC, eds. Form and Function of Phospholipids. Amsterdam: Elsevier, 1973, p 1–8.

Aranda-Espinoza H, Berman A, Dan N, Pincus P, Safran S. Interaction between inclusions embedded in membranes. Biophys J 71:648–656, 1996.

Bangham AD. Liposomes—The Babraham connection. Chem Phys Lipids 64:275–285, 1993.

Bangham AD, Hill MW, Miller WGA Preparation and use of liposomes as models of biological membranes. Methods Membr Biol 1:1–68, 1974.

Bar LK, Barenholz Y, Thompson TE. The fraction of cholesterol undergoing spontaneous exchange between small unilamellar phosphatidylcholine vesicles. Biochemistry 25:6701–6705, 1986.

Bar LK, Barenholz Y, Thompson TE. Dependence of the fraction of cholesterol undergoing spontaneous exchange between small unilamellar vesicles and their phospholipid composition. Biochemistry 26:5460–5465, 1987.

Bar LK, Chong PC, Barenholz Y, Thompson TE. Spontaneous transfer between phospholipid bilayers of dehydroergosterol, a fluorescent cholesterol analog. Biochim Biophys Acta 983:109–112, 1989.

Bar LK, Barenholz Y, Thompson TE. The effect of sphingomyelin composition on the phase structure of phosphatidylcholine-sphingomyelin bilayers. Biochemistry 36:2507–2516, 1997.

Barenholz Y. Sphingomyelin-lecithin balance in membranes, composition, structure and function relationships. In: Shinitzky M, ed. Physiology of Membrane Fluidity. Vol. I. Boca Raton, FL: CRC Press, 1984, pp 131–173.

Barenholz Y. Design of liposome-based drug carriers, from basic research to application as approved drugs. In: Lasic DD, Papahadjopoulos D, eds. Medical Application of Liposomes. Amsterdam: Elsevier, 1998, Chap. 7.1 pp. 545–565.

Barenholz Y, ed. Abstract Book of Domain Organization in Membranes, Biological Implications, Research Workshop of the Israel Science Foundation, Jerusalem, Israel March 1998a.

Barenholz Y, Gatt S. In: Hawthorne JN, Ansell, GB, eds. Phospholipids. New York: Elsevier, 1982, p 129–177.

Barenholz Y, Lasic DD, eds. Liposomes: Handbook of Nonmedical Applications of Liposomes. Vols I–IV, Boca Raton, FL: CRC Press, 1996.

Barenholz Y, Thompson TE. Sphingomyelins in bilayers and biological membranes. Biochim Biophys Acta 604:129–158, 1980.

Barenholz Y, Suurkuusk J, Mountcastle D, Thompson TE, Biltonen RL. (1976) A calorimetric study of the thermotropic behaviour of aqueous dispersions of natural and synthetic sphingomyelins. Biochemistry 15:2441–2447, 1976.

Barenholz Y, Ceastaro B, Lichtenberg D, Freire E, Thompson TE, Gatt S. Characterization of micellar and liposomal dispersions of gangliosides and phospholipids. In: Svennerholm L, Mandel P, Dreyfus H, Urban PF, eds. Structure and Function of Gangliosides. New York: Plenum, 1980, pp 105–123.

Barenholz Y, Cohen T, Haas E, Ottolenghi M. Lateral organization of pyrene-labeled lipids in bilayers as determined from the deviation from equilibrium between pyrene monomers and excimers. J Biol Chem 271:3085–3090, 1996.

Bartosz G. Erythrocyte aging: Physical and chemical membrane changes. Gerontology 37:33–67, 1991.

Battersby BJ, Grimm R, Hübner S, Cevc G. Evidence for three-dimensional interlayer correlations in cationic lipid-DNA complexes as observed by cryo-electron microscopy. Biochim Biophys Acta 1372:379–383, 1998.

Behr JP. Gene transfer with synthetic cationic amphiphiles, prospects for gene therapy. Bioconj Chem 5:382–389, 1994.

Ben-Tal N, Honig B, Peitzsch M, Denisov G, McLaughlin S. Binding of small basic peptides to membranes containing acidic lipids: Theoretical models and experimental results. Biophys J 71:561–575, 1996.

Berke G. The binding and lysis of target cells by cytotoxic lymphocytes: Molecular and cellular aspects. Annu Rev Immunol 12:735–773, 1994.

Bittman R, Clejan S, Lund-Katz S, Phillips MC. Influence of cholesterol on bilayers of ester and ether-linked phospholipids. Permeability and 13-C-nuclear magnetic resonance measurements. Biochim Biophys Acta 772:117–126, 1984.

Blume G, Cevc G. Liposomes for the sustained drug release in vivo. Biochim Biophys Acta 1029:91–97, 1990.

Blume G, Cevc G. Drug-carrier and stability properties of the long-lived lipid vesicles, cryptosomes, in vitro and in vivo. J Liposome Res 2:355–368, 1992.

Blume G, Cevc G. Molecular mechanism of lipid vesicle longevity in vivo. Biochim Biophys Acta 1146:157–168, 1993.

Borenstain V, Barenholz Y. Characterization of liposomes and other lipid assemblies by multiprobe fluorescence polarization. Chem Phys Lipids 64:117–127, 1993.

Borenstain V, Barenholz Y, Rachmilewitz EA, Eldor A. Phosphatidylserine in outer leaflet of β-thalassemic patients may explain the chronic hypercoagulable state and thrombotic episodes. Am J Hematol 44:63–65, 1993.

Braganza LF, Worcester DL. Hydrostatic pressure induces hydrocarbon chain interdigitation in single-component phospholipid bilayers. Biochemistry 25:2591–2596, 1986.

Bronner F, ed. Current Topics in Membrane Transport. Ion Channels: Molecular and Physiological Aspects. Orlando, FL: Academic Press, 1984.

Brown RE. Sphingolipid organization in biomembranes, what physical studies of model membranes reveal. J Cell Sci 111:1–9, 1998.

Caffrey M, Feigenson GW. Fluorescence quenching in model membranes. 3. Relationship between calcium adenosinetriphosphatase enzyme activity and the affinity of the protein for phosphatidylcholines with different acyl chain characteristics. Biochemistry 20:1949–1961, 1981.

Caffrey M, Koynova R, Hogan J, Moynihan D. Lipidat, a database of lipid phase transition temperature enthalpy changes and associated information. In: Lasic DD, Barenholz Y, eds. Handbook of Nonmedical Applications of Liposomes. Vol II. Chap 6. Boca Raton, FL: CRC Press, 1996, pp 85–104.

Calhoun WI, Shipley GG. Fatty-acid composition and thermal behaviour of natural sphingomyelins. Biochim Biophys Acta 555:436–441, 1979.

Cantu L, Corti M, Degiorgio V. Mixed micelles of gangliosides. J Phys Chem 94:793–795, 1990.

Cevc G. How membrane chain melting properties are controlled by the polar surface of the lipid bilayers. Biochemistry 26:6305–6310, 1987.

Cevc G. Effects of lipid headgroups and (nonelectrolyte) solution on the structural and phase properties of bilayer membranes. Ber Bunsenges 92:953–961, 1988.

Cevc G. Colloidal and phase behaviour of biomacromolecules interdepend and are regulated by the supramolecular surface polarity. Examples with lipid bilayer membranes. J Phys 50:1117–1134, 1989.

Cevc G. Membrane electrostatics. Biochim Biophys Acta Rev Membr 1031–3:311–382, 1990.

Cevc G. Isothermal lipid phase transitions. Chem Phys Lipids 57:293–307, 1991a.

Cevc G. How membrane chain melting phase transition temperature is affected by the lipid chain asymmetry and degree of unsaturation: Analysis and predictions based on the effective chain length model. Biochemistry 30/29:7186–7193, 1991b.

Cevc G. Hydration force depends on the interfacial structure of the polar surface. J Chem Soc Faraday II 87:2733–2739, 1991c.

Cevc G. Polymorphism of bilayer membranes in the ordered phase and the molecular origin of lipid pretransition and rippled lamellae. Biochim Biophys Acta 1062:59–69, 1991d.

Cevc G. Lipid properties as a basis for the modelling and design of liposome membranes. In: Gregoriadis G, ed. Liposome Technology. 2nd ed. Boca Raton, FL: CRC Press, 1992, pp 1–36.

Cevc G. Solute transport across lipid bilayers. In: Cevc G, ed. Phospholipids Handbook. New York: Marcel Dekker, 1993a, pp 639–661.

Cevc G. Electrostatic characterization of liposomes. Chem Phys Lipids 64:63–187, 1993b.

Cevc G. Lipid hydration. In: Westhof E, ed. Hydration of Biological Macromolecules. New York: Macmillian, 1993c, pp 338–390.

Cevc G. Lipid hydration. In: Westhof E, ed. Hydration of Biological Macromolecules. New York: Macmillan, 1993d, pp. 338–390.

Cevc G, ed. Phospholipids Handbook. New York: Marcel Dekker, 1993e.

Cevc G. Biophysical view of the role of interfaces in biomolecular recognition. Biophys Chem 55:43–53, 1995.

Cevc G. Lipid suspensions on the skin. Permeation enhancement, vesicle penetration and transdermal drug delivery. Crit Rev Ther Drug Carrier Syst 13:257–388, 1996.

Cevc G, Marsh D. Hydration of noncharged lipid bilayer membranes theory and experiments with phosphatidylethanolamines. Biophys J 47:21–32, 1985.

Cevc G, Marsh D. Phospholipid Bilayers. Physical Principles and Models. New York: Wiley-Interscience, 1987.

Cevc G, Richardsen H. Lipid vesicles and membrane fusion. Adv Drug Delivery Rev, in press.

Cevc G, Watts A, Marsh D. Nonelectrostatic contribution to the titration of the ordered-fluid phase transition of phosphatidylglycerol bilayers. FEBS Lett 120:267–271, 1980.

Cevc G, Strohmaier L, Berkholz J, Blume G. Molecular mechanism of protein interactions with the lipid bilayer membrane. Stud Biophys 138:57–70, 1990.

Cevc G, Hauser M, Kornyshev AA. Effects of the interfacial structure on the hydration force between laterally uniform surfaces. Langmuir 11:3103–3110, 1995.

Chapman D. In: Ansell GB, Hawthorne JN, Dawson RMC, eds. Form and Function of Phospholipids. Amsterdam: Elsevier, 1973, p 117–169.

Clerc S, Barenholz Y. A quantitative model for using acridine orange as a transmembrane pH gradient probe. Anal Biochem 259:104–111, 1988.

Comfurius P, Williamson P, Smeets EF, Schlegel RA, Bevers EM, Zwaal RF. Reconstitution of phospholipid scramblase activity from human blood platelets. Biochemistry 35:7631–7634, 1996.

Cooper VG, Yedgar S, Barenholz Y. Diffusion coefficients of mixed micelles of Triton X-100 and sphingomyelin and of sonicated sphingomyelin liposomes, measured by autocorrelation spectroscopy of Rayleigh scattered light. Biochim Biophys Acta 363:86–97, 1974.

Correa-Freire M, Barenholz Y, Thompson TE. Glucocerebroside transfer between phosphatidylcholine bilayers. Biochemistry 21:1244–1248, 1982.

Cossins AR. Adaptation of biological membranes to temperature, the effect of temperature activation of goldfish upon the viscosity of synaptosomal membranes. Biochim Biophys Acta 470:395–411, 1977.

Cossins AR, Prosser CL. Evolutionary adaptation of membranes to temperature. Proc Natl Acad Sci USA 75:2040–2044, 1978.

Crain RC, Zilversmit DB. Two nonspecific phospholipid exchange proteins from beef liver 2. Use in studying the asymmetry and transbilayer movement of phosphatidylcholine, phosphatidylethanolamine, and sphingomyelin in intact rat erythrocytes. Biochemistry 19:1440–1447, 1980.

Crowe JH, Crowe LM, Chapman D. Preservation of membranes in anhydrobiotic organisms: The role of trehalose. Science 223:701–703, 1984.

Cullis PR, Hope NJ, de Kruijff B, Verkleij AJ, Tilcock CPS. Structural properties and functional roles of phospholipids in biological membranes. In: Kuo JF, ed. Phospholipids and Cellular Regulation. Boca Raton, FL: CRC Press, 1985, pp 1–59.

Dahlqvist A, Nordstrom S, Karlsson OP, Mannock DA, McElhaney RN, Wieslander A. Efficient modulation of glucolipid enzyme activities in membranes of *Acholeplasma laidlawii* by the type of lipids in the bilayer matrix. Biochemistry 34:13381–13389, 1995.

Dancey GF, Yasuda T, Kinsky SC. Effect of liposomal model membrane composition on immunogenicity. J Immunol 120:1109–1114, 1978.

Dancey GF, Isakson PC, Kinsky SC. Immunogenicity of liposomal model membranes sensitized with dinitrophenylated phosphatidylethanolamine derivatives containing different length spacers. J Immunol 122:638–643, 1979.

Danielli JF. The bilayer hypothesis of membrane structure. In: Wissman G, Claiborne R, eds. Cell Membranes. Biochemistry, Cell Biology and Pathology. HP Pub. New York, 1975, pp 3–11.

Deamer DW. The first living systems, a bioenergetic perspective. Microbiol Mol Biol Rev 61:239–261, 1997.

de Gier J. Osmotic behavior and permeability properties of liposomes. Chem Phys Lipids 64:187–196, 1993.

Delacroix H, Gulik-Krzywicki T, Seddon JM. Freeze fracture electron microscopy of lyotropic lipid systems. Quantitative analysis of the inverse micellar cubic phase of space group Fd3m. J Mol Biol 258:88–103, 1996.

de Mendoza D, Cronan JE Jr. Thermal regulation of membrane lipid fluidity in bacteria. TIBS February:49–52, 1983.

Denisov G, Wanaski S, Luan P, Glaser M, McLaughlin S. Biophys J, in press.

Deuticke B. Properties and structural basis of simple diffusion pathways in the erythrocyte membrane. Rev Physiol Biochem Pharmacol 78:1–97, 1977.

Devaux PF. Static and dynamic lipid asymmetry in cell membranes. Biochemistry 30:1163–1173, 1991.

Devaux PF. Protein involvement in transmembrane lipid asymmetry. Annu Rev Biophys Biomol Struct 21:417–439, 1992.

Devaux PF. Lipid transmembrane asymmetry and flip-flop in biological membranes and in lipid bilayers. Curr Opin Struct Biol 3:489–494, 1993.

Dix JA, Kivelson D, Diamond JM. Molecular motion of small nonelectrolyte molecules in lecithin bilayers. J Membr Biol 40:315–342, 1978.

Dornmair K, Overath P, Jahnig F. Fast measurement of galactoside transport by lactose permease. J Biol Chem 264:342–346, 1989.

Dustin ML, Springer TA. Role of lymphocyte adhesion receptors in transient interactions and cell locomotion. Annu Rev Immunol 9:27–66, 1991.

Edidin M. Lipid microdomains in cell surface membranes, Curr Opin Struct Biol 7:528–532, 1997.

Egberts E, Marrink SJ, Berendsen HJ. Molecular dynamics simulation of a phospholipid membrane. Eur Biophys J 22:423–436, 1994.

el Hage Chahine JM, Cribier S, Devaux PF. Phospholipid transmembrane domains and lateral diffusion in fibroblasts. Proc Natl Acad Sci USA 90:447–451, 1993.

Emanuel N, Kedar E, Bolotin EM, Smorodinsky NI, Barenholz Y. Preparation and characterization of doxorubicin-loaded sterically stabilized immunoliposomes. Pharm Res 13:352–359, 1996.

Engblom J. Bicontinuous cubic phase: A model for investigating the effects on a lipid bilayer due to a foreign substance. Chem Phys Lipids 84:155–164, 1996.

Escriba PV, Sastre M, Garcia-Sevilla JA. Disruption of cellular signaling pathways by daunomycin through destabilization of nonlamellar membrane structures. Proc Natl Acad Sci USA 92:7595–7599, 1995.

Escriba PV, Ozaita A, Ribas C, Miralles A, Fodor E, Farkas T, Garcia-Sevilla JA. Role of lipid polymorphism in G protein–membrane interactions: Nonlamellar-prone phospholipids and peripheral protein binding to membranes. Proc Natl Acad Sci USA 94:11375–11380, 1997.

Esmann M, Marsh D. Spin label studies on the origin of the specificity of lipid-protein interactions in Na^+/K^+-ATPase membranes from *Squalus acanthias.* Biochemistry 24:3572–3578, 1985.

Esmann M, Marsh D, Schwarzmann G, Sandhoff K. Ganglioside-protein interactions: Spin-label electron spin resonance studies with (Na^+,K^+)-ATPase membranes. Biochemistry 27:2398–2403, 1988.

Etemadi A. Membrane asymmetry; a survey and critical appraisal of the methodology. II. Methods for assessing the unequal distribution of lipid. Biochim Biophys Acta 604:347–422, 1980a.

Etemadi A. Methods for assessing the asymmetric orientation and distribution of proteins. I. Biochim Biophys Acta 604:423–475, 1980b.

Fattal DR, Ben-Shaul A. A molecular model for lipid-protein interaction in membranes: The role of hydrophobic mismatch. Biophys J 65:1795–1809, 1993.

Fattal DR, Ben-Shaul A. Molecular theory of acyl chain packing in lipid and lipid proteomembranes. In: Lasic DD, Barenholz Y, eds. Handbook of Non-Medical Application of Liposomes. Vol I. Boca Raton, FL: CRC Press, 1996, pp 125–151.

Felgner PO, Freire E, Barenholz Y, Thompson TE. Asymmetric incorporation of trisialoganglioside into dipalmitoylphosphatidylcholine vesicles. Biochemistry 20:2168–2172, 1981.

Felgner PL, Thompson TE, Barenholz Y, Lichtenberg D. Kinetics of transfer of gangliosides from their micelles to dipalmitoyl phosphatidylcholine vesicles. Biochemistry 22:1670–1674, 1983.

Friedrichson T, Kurzchalla TV. Microdomains of GPI-anchored proteins in living cells revealed by crosslinking. Nature 394:802–805, 1998.

Fromherz P, Rüppel D. Lipid vesicle formation: The transition from open disks to closed shells. FEBS Lett 179:155–158, 1985.

Füldner HH. Characterization of a third phase transition in multilamellar dipalmitoyllecithin liposomes. Biochemistry 20:5707–5710, 1981.

Gershon H, Ghirlando R, Guttman SB, Minsky A. Mode of formation and structural features of DNA-cationic liposome complexes used for transfection. Biochemistry 32:7143–7151, 1993.

Glaser M, Wanaski S, Buser CA, Boguslavsky V, Rashizada W, Morris A, Rebecchi M, Scarlata SF, Runnels LW, Prestwich GD, Chen J, Aderem A, Ahn J, McLaughlin S. Myristoylated alanine-rich C kinase substrate (MARCKS) produces reversible inhibition of phospholipase C by sequestering phosphatidylinositol 4,5-bisphosphate in lateral domains. J Biol Chem 271:26187–26193, 1996.

Glinsky GV. The blood group antigens (BGA)–related glycoepitopes. A key structural determinant in immunogenesis and cancer pathogenesis. Crit Rev Oncol Hematol 12:151–166, 1992.

Gliozzi A, Relini A. Lipid vesicles as model system for Archea membranes. In: Lasic DD, Barenholz Y, eds. Handbook of Nonmedical Applications of Liposomes. Vol. II. Boca Raton, FL: CRC Press, 1996, pp 329–348 (and references listed therein).

Goll J, Carlson FD, Barenholz Y, Litman BJ, Thompson TE. Photon correlation spectroscopic study of the size distribution of phospholipid vesicles. Biophys J 38:7–13, 1982.

Gomez-Munoz A. Modulation of cell signaling by ceramides. Biochim Biophys Acta 1391:92–109, 1998.

Gregoriadis G, McCormack B, eds. Targeting of Drugs, Strategies for Stealth Therapeutic Systems. NATO ASI Series, Life Sciences, Vol 300. New York: Plenum, 1998.

Gudgin EF, Cameron DG, Mantsch HH. Dependence of acyl chain packing of phospholipids on the head group and acyl chain length. Biochemistry 20:4496–4500, 1981.

Hannun YA. The sphingomyelin cycle and the second messenger function of ceramide. J Biol Chem 269:3125–3128, 1994.

Hannun YA. Functions of ceramide in coordinating cellular responses to stress. Science 274:1855–1859, 1996.

Harder T, Simons K. Caveolae DIG and dynamics of sphingolipid-cholesterol microdomains. Curr Opin Cell Biol 9:534–542, 1997.

Harlos K. Pretransitions in the hydrocarbon chains of phosphatidylethanolamines. A wide angle X-ray diffraction study. Biochim Biophys Acta 511:348–353, 1978.

Harries D, Mag S, Gelbart W, Ben-Shaul A. Structure, stability and thermodynamics of lamellar DNA-lipid complexes. Biophys J 75:159–173, 1998.

Hauser H. In: Frank F, ed. Water. Vol. 4. New York: Plenum, 1975, Chap. 4, pp 209–303.

Hauser H, Pascher I, Pearson RH, Sundell S. Preferred conformation and molecular packing of phosphatidylethanolamine and phosphatidylcholine. Biochim Biophys Acta 650:21–51, 1981.

Heberle J, Riesle J, Thiedemann G, Oesterhelt D, Dencher NA. Proton migration along the membrane surface and retarded surface to bulk transfer. Nature 370:379–382, 1994.

Heim M, Cevc G, Guckenberger R, Knapp H, Wiegräbe W. Lateral electrical conductivity of mica-supported lipid bilayer membranes measured by scanning tunnelling microscopy. Biophys J 69:489–497, 1995.

Helenius A. Properties of detergents. Methods Enzymol 56:734–756, 1979.

Helfrich W. Elastic properties of lipid bilayers: Theory and possible experiments. Z Naturforsch 28c:693–703, 1973.

Helfrich W. Size distributions of vesicles: The role of the effective rigidity of membranes. J Phys 47:321–329, 1986.

Hertz R, Barenholz Y. Permeability and integrity properties of lecithin-sphingomyelin liposomes. Chem Phys Lipids 15:138–156, 1975.

Hirsch-Lerner D, Barenholz Y. Probing DNA–cationic lipid interactions with the flurophore trimethylammonium diphenyl-hexatriene (TMADPH). Biochim Biophys Acta 1370:17–30, 1998.

Hodgkin MN, Pettitt TR, Martin A, Michell RH, Pemberton AJ, Walkelam MJO. Diacylglycerols and phosphatidates, which molecular species are intracellular messengers? Trends Biochem Sci 23:200–204, 1998.

Hooper NM, Turner AJ. Lipid Modification of Proteins. The Practical Approach Series. 1992, IRL Press.

Horrocks LA, Sharma M. In: Hawthorne JN, Ansell GB, eds. Phospholipids. New York: Elsevier, 1982, pp 51–93.

Horvath LI, Brophy PJ, Marsh D. Exchange rates at the lipid-protein interface of the myelin proteolipid protein determined by saturation transfer electron spin resonance and continuous wave saturation studies. Biophys J 64:622–631, 1993a.

Horvath LI, Brophy PJ, Marsh D. Spin label saturation transfer EPR determinations of the stoichiometry and selectivity of lipid-protein interactions in the gel phase. Biochim Biophys Acta 1147:277–280, 1993b.

Hosemann M, Hentschel R, Helfrich W. Direct x-ray study of the molecular tilt in dipalmitoyl lecithin bilayers. Z Naturforsch 35a:643–644, 1980.

Huang C. Studies of phosphatidylcholine vesicles. Formation and physical characterization. Biochemistry 8:344–352, 1969.

Huang C. Empirical estimation of the gel to liquid crystalline phase transition temperatures for fully hydrated saturated phosphatidylcholines. Biochemistry 30:26–30, 1991.

Hübner S, Battersby BJ, Grimm R, Cevc G. Lipid-DNA complex formation: Reorganization and rupture of lipid vesicles in the presence of DNA as observed by cryo-electron microscopy. Biophys J 76:3158–3166, 1999.

Hui SW, Mason JT, Huang C-H. Acyl chain interdigitation in saturated mixed-chain phosphatidylcholine bilayer dispersions. Biochemistry 23:5570–5577, 1984.

Ipsen JH, Mouritsen OG, Bloom M. Relationship between lipid membrane area, hydrophobic thickness, and acyl-chain orientational order. The effects of cholesterol. Biophys J 57:405–412, 1990.

Israelachvili J. Intermolecular and Surface Forces. 2nd ed. London: Academic Press, 1992.

Israelachvili JN, Marcelja S, Horn RG. Physical principles of membrane organization. Q Rev Biophys 13:121–200, 1980.

Jackowski S. Coordination of membrane phospholipid synthesis with the cell cycle. J Biol Chem 269:3858–3867, 1994.

Jain MK, Wagner RC. Introduction to Biological Membrane. New York: Wiley, 1980.

Janoff AS, Rauch J. The structural specificity of anti-phospholipid antibodies in autoimmune disease. Chem Phys Lipids 40:315–332, 1986.

Johannsson A, Keightley CA, Smith GA, Richards CD, Hesketh TR, Metcalfe JC. The effect of bilayer thickness and n-alkanes on the activity of the (Ca^{2+} and Mg^{2+})-dependent ATPase of sarcoplasmic reticulum. J Biol Chem 256:1643–1650, 1981.

Jülicher F, Lipowsky R. Domain-induced budding of vesicles. Phys Rev Lett 70:2964–2967, 1993.

Kader JC, Douady D, Mazliak I. In: Hawthorne JN, Ansell GB, eds. Phospholipids. New York: Elsevier, 1982, p 279–311.

Kang S, Bergamini VW, Hasen MJ. Conformation of N-formyl-1,3-dihydroxy-delta31-pentene, a model compound of sphingomyelin. J Theor Biol 63:117–124, 1976.

Kapitulnick J, Tschershedsky M, Barenholz Y. Fluidity of rat liver microsomal membrane, dramatic increase in birth. Science 206:843–844, 1979.

Kapitulnick J, Tschershedsky M, Barenholz Y. Fluidization of the rat liver microsomal membrane at birth. In: Microsomes, Drug Oxidation and Chemical Carcinogenesis. New York: Academic Press, 1980, pp 549–552.

Karlsson OP, Rytomaa M, Dahlqvist A, Kinnunen PK, Wieslander A. Correlation between bilayer lipid dynamics and activity of the diglucosyldiacylglycerol synthase from *Acholeplasma laidlawii* membranes. Biochemistry 35:10094–10102, 1996.

Kinnunen PK. On the principles of functional ordering in biological membranes. Chem Phys Lipids 57:375–399, 1991.

Kinnunen PKJ. On the mechanisms of the lamellar hexagonal H_{II} phase transition and the biological significance of H_{II} propensity. In: Lasic DD, Barenholz Y, eds. Handbook of Nonmedical Applications of Liposomes (Theory and Basic Sciences). Vol I. Boca Raton, FL: CRC Press, 1996, Chap 6, pp 153–172.

Kirchner S, Cevc G. Calculation of the model independent surface polarity profiles from the experimental hydration force data. Langmuir 10:1934–1947, 1994.

Kolesnick R, Golde DW. The sphingomyelin pathway in tumor necrosis factor and interleukin-1 signaling. Cell 77:325–328, 1994.

Koltover I, Salditt T, Rädler JO, Safinya CR. An inverted hexagonal phase of cationic liposome–DNA complexes related to DNA release and delivery. Science 281:78–81, 1998.

Koynova R, Tenchov B, Rapp G. Low amounts of PEG-lipid induce cubic phase in phosphatidylethanolamine dispersions. Biochim Biophys Acta 1326:167–170, 1997.

Kralchevsky PA, Paunov VN, Denkov ND, Nagayama K. Stresses in lipid membranes and interaction between inclusions. J Chem Soc Faraday Trans 91:3415–3432, 1995.

Kutchai H, Barenholz Y, Ross TF, Wermer DE. Developmental changes in plasma membrane fluidity in chick embryo heart. Biochim Biophys Acta 436:101–112, 1976.

Kuypers FA, Yuan J, Lewis RA, Snyder M, Kiefer CR, Bunyaratvej A, Fucharoen S, Ma Li, Lori S, de Jong K, Schrier SL. Membrane phospholipid asymmetry in human thalassemia. Blood 91:3044–3051, 1998.

Laggner P, Kriechbaum M. Phospholipid phase transitions: Kinetics and structural mechanisms. Chem Phys Lipids 57:121–145, 1991.

Landh T. From entangled membranes to eclectic morphologies: cubic membranes as subcellular space organizers. FEBS Lett 369:13–17, 1995.

Larsson K. On periodic curvature and standing wave motions in cell membranes. Chem Phys Lipids 88:15–20, 1997.

Lasic DD. On the thermodynamic stability of liposomes. J Colloid Interface Sci 140:302–304, 1990.

Lasic DD. Liposomes in Gene Delivery. Boca Raton, FL: CRC Press, 1997.

Lasic DD, Barenholz Y. Liposomes, past, present and future. In: Lasic DD, Barenholz Y, eds. Handbook of Nonmedical Applications of Liposomes. Vol IV. Boca Raton, FL: CRC Press, 1996, Chap 21, pp 299–315.

Lasic DD, Martin F, eds. Stealth Liposomes. Boca Raton, FL: CRC Press, 1995.

Lasic DD, Templeton NS. Liposomes in gene therapy. Adv Drug Delivery Rev 20:221–266, 1996.

Le Gall AH, Yeaman C, Muesch A, Rodriguez-Boulan E. Epithelial cell polarity: New perspectives. Semin Nephrol 15:272–284, 1995.

Levin IW, Thompson TE, Barenholz Y, Huang C. Two types of hydrocarbon chain. Interdigitation in sphingomyelin bilayers. Biochemistry 24:6282–6286, 1985.

Li G, Knowles PF, Murphy DJ, Nishida I, Marsh D. Spin-label ESR studies of lipid-protein interactions in thylakoid membranes. Biochemistry 28:7446–7452, 1989.

Li G, Knowles PF, Murphy DJ, Marsh D. Lipid-protein interactions in thylakoid membranes of chilling-resistant and -sensitive plants studied by spin label electron spin resonance spectroscopy. J Biol Chem 265:16867–16872, 1990.

Lichtenberg D, Barenholz Y. Liposomes, preparation, characterization and preservation. In: Glick D, ed. Methods of Biochemical Analysis. Vol. 33. New York: Wiley, 1988, pp 337–462.

Lichtenberg D, Freire E, Schmidt CF, Barenholz Y, Felger PL, Thompson TE. Effect of surface curvature on stability, thermodynamic behavior and osmotic activity of dipalmitoylphosphatidylcholine single lamellar vesicles. Biochemistry 20:3462–3467, 1981.

Lin H, Wang Z, Huang C. The influence of acyl chain-length asymmetry on the phase transition parameters of phosphatidylcholine dispersions. Biochim Biophys Acta 1967:17–28, 1991.

Lindblom G, Rilfors L. Cubic phases and isotropic structures formed by membrane lipids—possible biological relevance. Biochim Biophys Acta 988:221–256, 1989.

Lindblom G, Rilfors L. Nonlamellar phases formed by membrane lipids. Adv Colloid Interface Sci 41:101–125, 1992.

Lingwood CA. Aglycone modulation of glycolipid receptor function. Glycoconj J 13:495–503, 1996.

Lipka G, Op den Kamp JA, Hauser H. Lipid asymmetry in rabbit small intestinal brush border membrane as probed by an intrinsic phospholipid exchange protein. Biochemistry 30:11828–11836, 1991.

Lipowsky R. Budding of membranes induced by membrane domains. J Phys II 2:1825–1840, 1992.

Lipowsky R. Domain-induced budding of fluid membranes. Biophys J 64:1133–1138, 1993.

Löbbecke L, Cevc G. Effect of short-chain alcohols on the phase behaviour and interdigitation of phosphatidylcholine bilayer membranes. Biochim Biophys Acta 1237:59–69, 1995.

Lofgren H, Pascher I. Molecular arrangements of sphingolipids. The monolayer behaviour of ceramides. Chem Phys Lipids 20:263–284, 1977.

Lundbaek JA, Maer AM, Andersen OS. Lipid bilayer electrostatic energy, curvature stress, and assembly of gramicidin channels. Biochemistry 36:5695–5701, 1997.

Marai L, Kuksis A. Molecular species of lecithins from erythrocytes and plasma of man. J Lipid Res 10:141–145, 1969.

Marsh D. Selectivity of lipid-protein interactions. J Bioenerg Biomembr 19:677–689, 1987.

Marsh D. Handbook of Lipid Bilayers. Boca Raton, FL: CRC Press, 1990.

Marsh D. Specificity of lipid-protein interactions. Biomembranes 1:59–64, 1995a.

Marsh D. Lipid-protein interactions and heterogeneous lipid distribution in membranes. Mol Membr Biol 12:59–64, 1995b.

Marsh D, Horvath LI. Stucture, dynamics and composition of the lipid-protein interface. Perspectives from spin-labelling. Biochimica et Biophysica Acta 1376:267–296, 1998.

Matsuzaki K, Murase O, Fujii N, Miyajima K. An antimicrobial peptide, magainin 2, induced rapid flip-flop of phospholipids coupled with pore formation and peptide translocation. Biochemistry 35:11361–11368, 1996.

McCoy JP Jr, Chambers WH. Carbohydrates in the functions of natural killer cells. Glycobiology 1:321–328, 1991.

McElhaney RN. Physical studies of lipid organization and dynamics in mycoplasma membranes. Subcell Biochem 20:53–108, 1993.

McEver RP. Selectins. Curr Opin Immunol 6:75–84, 1994.

McLaughlin S, Aderem A. The myristoyl-electrostatic switch: A modulator of reversible protein-membrane interactions. Trends Biochem Sci 20:272–276, 1995.

Mouritsen OG, Andersen OJ, eds. In: Search of a New Biomembrane Model. Copenhagen: Munksgaard, 1998.

Mouritsen OG, Bloom M. Models of lipid-protein interactions in membranes. Annu Rev Biophys Biomol Struct 22:145–171, 1993.

Mouritsen OG, Jørgensen K. Micro, nano- and meso-scale heterogeneity of lipid bilayer and its influence on macroscopic membrane properties. Mol Membr Biol 12:15–20, 1995.

Mulukutla S, Shipley G. Structure and thermotropic properties of phosphatidylethanolamine and its N-methyl derivatives. Biochemistry 23:2514–2519, 1984.

Neidleman SL. Effects of temperature on lipid unsaturation. Biotechn Gen Eng Rev 5:245–268, 1987.

Nicol F, Nir S, Szoka FC Jr. Effect of cholesterol and charge on pore formation in bilayer vesicles by a pH-sensitive peptide. Biophys J 71:3288–3301, 1996.

Nishizuka Y, Tanaka C, Endo M. eds. The biology and medicine of signal transduction. Adv Second Messenger Phosphorylation Res 24, 1990.

Orme FW, Moronne MM, Macey RI. Modification of the erythrocyte membrane dielectric constant by alcohols. J Membr Biol 104:57–68, 1988.

Osterberg F, Rilfors L, Wieslander A, Lindblom G, Gruner SM. Lipid extracts from membranes of *Acholeplasma laidlawii* A grown with different fatty acids have a nearly constant spontaneous curvature. Biochim Biophys Acta 1257:18–24, 1995.

Pagano RE, Schroit AJ, Struck DK. In: From Physical Structure to Therapeutic Applications. New York: Elsevier, 1981, p 323.

Pagano RE, Watanabe R, Wheatley C, Chen C. Use of BODIPY sphingomyelin to study membrane traffic along endocytic pathway. Chem Phys Lipids, in press.

Pal R, Barenholz Y, Wagner RR. Depletion and change of cholesterol from the membrane of vesicular stomatitis virus by interaction with serum lipoproteins or poly(vinylpyrrolidone) complexed with bovine serum albumin. Biochemistry 20:530–539, 1981.

Pal R, Petri WA Jr, Barenholz Y, Wagner RR. Lipid and protein contributions to membrane surface potential of vesicular stomatitis virus probed by a fluorescent pH indicator, 4-heptadecyl-7-hydroxycoumarin. Biochim Biophys Acta 729:185–192, 1983.

Pal R, Petri WA, Ben Yashar V, Wagner RR, Barenholz Y. Characterization of fluorophore 4-heptadecyl-7-hydroxy coumarin: A probe for the head group region of lipid bilayers and biological membranes. Biochemistry 24:573–581, 1985.

Parente RA, Nir S, Szoka FC Jr. Mechanism of leakage of phospholipid vesicle contents induced by the peptide GALA. Biochemistry 29:8720–8728, 1990.

Pascher I. Molecular arrangements in sphingolipids conformation and hydrogen bonding of ceramide and their implication on membrane stability and permeability. Biochim Biophys Acta 455:433–451, 1976.

Patzer EJ, Moore NF, Barenholz Y, Shaw JM, Wagner RR. Lipid organization of the membrane of vesicular stomatitis virus. J Biol Chem 253:4544–4550, 1978a.

Patzer EJ, Wagner RR, Barenholz Y. Cholesterol oxidase as a probe for studying membrane organization. Nature 274:394–395, 1978b.

Pearson RH, Pascher I. The molecular structure of lecithin dihydrate. Nature 281:499–501, 1979.

Peters R. Translational diffusion in the plasma membrane of single cells as studied by fluorescence microphotolysis. Cell Biol Int Rep 5:733–760, 1981.

Pike LJ, Casey L. Localization and turnover of phosphatidylinositol 4,5-bisphosphate in caveolin-enriched membrane domains. J Biol Chem 271:26453–26456, 1996.

Piknova B, Marsh D, Thompson TE. Fluorescence quenching study of percolation and compartmentalization in two-phase lipid bilayers. Biophys J 71:892–897, 1996.

Piknova B, Marsh D, Thompson TE. Fluorescence quenching and ESR study of percolation in a two-phase lipid bilayer containing bacteriorhodopsin. Biophys J 72:2660–2668, 1997.

Pomorski T, Muller P, Zimmermann B, Burger K, Devaux PF, Herrmann A. Transbilayer movement of fluorescent and spin-labeled phospholipids in the plasma membrane of human fibroblasts: A Quantitative approach. J Cell Sci. 109:687–689, 1996.

Post JA, Bijvelt JJ, Verkleij AJ. Phosphatidylethanolamine and sarcolemmal damage during ischemia or metabolic inhibition of heart myocytes. Am J Physiol 268:H773–H780, 1995.

Priev A, Samuni AM, Tirosh O, Barenholz Y. The role of hydration in stabilization of liposomes; resistance to oxidative damage of PEG-grafted liposomes. In: Targeting of Drugs: Strategies for Stealth Therapeutic Systems. G. Gregoriadis and B. McCormack, eds., New York: Plenum Publishing Corporation, 147–162, 1998.

Quinn PJ. Membrane lipid phase behavior and lipid-protein interactions. Subcell Biochem 14:25–95, 1989.

Rädler JO, Koltover I, Salditt T, Safinya CR. Structure of DNA–cationic liposome complexes: DNA intercalation in multilamellar membranes in distinct interhelical packing regimes. Science 275:810–814, 1997.

Rao NM, Sundaram CS. Sensitivity of phospholipase C (*Bacillus cereus*) activity to lipid packing in sonicated lipid mixtures. Biochemistry 32:8547–8552, 1993.

Rapaport D, Peled R, Nir S, Shai Y. Reversible surface aggregation in pore formation by pardaxin. Biophys J 70:2503–2512, 1996.

Reizer J, Grossowicz N, Barenholz Y. The effects of growth temperature on the thermotropic behavior of the membranes of a thermophylic bacillus, composition-structure-function relationships. Biochim Biophys Acta 815:268–280, 1985.

Riboni L, Viani P, Rosseria B, Rinetti A, Tettamenti G. The role of sphingolipids in the process of signal transduction. Prog Lipid Res 36:153–195, 1997.

Rietveld AG, Chupin VV, Koorengevel MC, Wienk HL, Dowhan W, de Kruijff B. Regulation of lipid polymorphism is essential for the viability of phosphatidylethanolamine-deficient *Escherichia coli* cells. J Biol Chem 269:28670–28675, 1994.

Rietveld AG, Koorengevel MC, de Kruijff B. Non-bilayer lipids are required for efficient protein transport across the plasma membrane of *Escherichia coli*. EMBO J 14:5506–5513, 1995.

Rintoul DA, Chous S, Silbert DF. Physical characterization of sterol-depleted LM-cell plasma membrane. J Biol Chem 254:10070–10077, 1979.

Roseman M, Thompson TE. Mechanism of the spontaneous transfer of phospholipids between bilayers. Biochemistry 19:439–444, 1980.

Rosen SD. Cell surface lectins in the immune system. Semin Immunol 5:237–247, 1993.

Rosenberg H. Phospholipids In: Ansell GB, Hawthorne JN, Dawson RMC, eds. Form and Function of Phospholipids. Amsterdam: Elsevier, 1975, pp 333–344.

Rothman JE, Lenard J. Membrane asymmetry. Science 195:743–753, 1977.

Rouser G, Nelson CJ, Fleischer S, Simon G. In: Chapman D, ed. Biological Membrane. Vol 1. New York: Academic Press, 1968, p 5.

Rouser G, Kritchevsky G, Yamamoto A, Baxter CF. Lipids in the nervous system of different species as a function of age brain, spinal cord peripheral nerves, purified whole cell preparations and subcellular particulates. Adv Lipid Res 10:261–360, 1972.

Rowe ES. Induction of lateral phase separations in binary lipid mixtures by alcohol. Biochemistry 26:46–51, 1987.

Ruocco MJ, Shipley GG. Characterization of the sub-transition of hydrated dipalmitoylphosphatidylcholine bilayers Kinetic hydration and structural study. Biochim Biophys Acta 691:309–320, 1982.

Russell NJ. Mechanisms of thermal adaptation in bacteria, blueprints for survival. Trends Biochem Sci 9:108–112, 1984.

Ryba NJ, Marsh D. Protein rotational diffusion and lipid/protein interactions in recombinants of bovine rhodopsin with saturated diacylphosphatidylcholines of different chain length studied by conventional and saturation-transfer electron spin resonance. Biochemistry 31:7511–7518, 1992.

Sachse J-H, King MD, Marsh D. ESR determination of lipid translational diffusion coefficients at low spin label concentrations in biological membranes, using exchange-broadening, exchange narrowing, and dipole-dipole interactions. J Magn Reson 71:385–404, 1987.

Sandermann H Jr. Regulation of membrane enzyme by lipids. Biochim Biophys Acta 515:209–237, 1978.

Sankaram MB, Brophy PJ, Marsh D. Selectivity of interaction of phospholipids with bovine spinal cord myelin basic protein studied by spin-label electron spin resonance. Biochemistry 28:9699–9707, 1989.

Sankaram MB, Marsh D, Thompson TE. Determination of fluid and gel domain sizes in two-phase lipid bilayers. An ESR spin label study. Biophys J 63:340–349, 1992.

Sankaram MB, Marsh D, Gierasch LM, Thompson TE. Reorganization of lipid domain structure in membranes by a transmembrane peptide. Biophys J 66:1959–1968, 1994.

Savill J. Recognition and phagocytosis of cells undergoing apoptosis. Br Med Bull 53:491–508, 1997.

Schroit AJ, Madsen JW, Tanaka Y. In vivo recognition and clearance of red blood cells containing phosphatidylserine in their plasma membrane. J Biol Chem 260:5131–5138, 1985.

Schwarz D, Kisselev P, Pfeil W, Pisch S, Bornscheuer U, Schmid RD. Evidence that nonbilayer phase propensity of the membrane is important for the side chain cleavage activity of cytochrome P450SCC. Biochemistry 36:14262–14270, 1997.

Seddon JM. Structure of the inverted hexagonal (Hii) phase, and nonlamellar phase transitions of lipids. Biochim Biophys Acta 1031:1–69, 1990.

Seddon JM, Hogan JL, Warrender NA, Pebay-Peyroula E. Structural studies of phospholipid cubic phases. Prog Colloid Polymer Sci 81:189–197, 1990.

Seddon JM, Templer RH, Warrender NA, Huang Z, Cevc G, Marsh D. Phosphatidylcholine-fatty acid membranes: effects of headgroup hydration on the phase behaviour and structural parameters of the gel and inverse hexagonal (H_{II}) phases. Biochim Biophys Acta 1327:131–147, 1997.

Shepherd JCW, Büldt G. Zwitterionic dipoles as a dielectric probe for investigating head group mobility in phospholipid membranes. Biochim Biophys Acta 514:83–94, 1978.

Shinitzky M, Barenholz Y. Fluidity parameters of lipid regions determined by fluorescence polarization. Biochim Biophys Acta 515:367–394, 1978.

Shinitzky M, Yuli I. Lipid fluidity at the submacroscopic level, determination by fluorescence polarization. Chem Phys Lipids 30:261–282, 1982.

Shmeeda HR, Golden EB, Barenholz Y. Membrane lipids and aging. In: Shinitzky M, ed. Handbook of Biomembranes. Mammalian Membrane, Structure and Function. Vol 2. Weinheim: Balaban VCH, 1994, pp 1–82.

Simons K, Ikonen E. Functional rafts in cell membranes. Nature 387:569–572, 1997.

Sinensky M. Homeoviscous adaptation, a homeostatic process that regulates the viscosity of membrane lipids in *E. coli*. Proc Natl Acad Sci USA 71:522–525, 1974.

Singer M. Permeability of bilayers composed of mixtures of saturated phospholipids. Chem Phys Lipids 31:145–159, 1982.

Singer SJ. Architecture and topography of biological membranes In: Weissmann G, Claiborne R, ed. Cell Membranes, Biochemistry, Cell Biology and Pathology. New York: HP Publishing, 1975, pp 35–44.

Singer SJ, Nicolson GL. The fluid mosaic model of the structure of cell membranes. Science 175:720–731, 1972.

Slater JL, Huang C-H. Interdigitated bilayer membranes. Prog Lipid Res 27:325–359, 1988.

Small DM. Surface and bulk interactions of lipids and water with a classification of biologically active lipids based on these interactions. Fed Proc Fed Am Soc Exp Biol 29:1320–1326, 1970.

Small DM. Lipids. New York: Plenum, 1986.

Spiegel S, Milstien S. Sphingolipid metabolites, members of a new class of lipid second messengers. J Membr Biol 146:225–237, 1995.

Spiegel S, Foster D, Kolesnick R. Signal transduction through lipid second messengers. Curr Opin Cell Biol 8:159–167, 1996.

Sten-Knudsen O. Passive transport processes. In: Giebisch G, Tosteson DC, Ussing HH, eds. Membrane Transport in Biology. I. Concepts and Models. New York: Springer, 1978, pp 1–113.

Straume M, Litman BJ. Influence of cholesterol on equilibrium and dynamic bilayer structure of unsaturated acyl chain phosphatidylcholine vesicles as determined from higher order analysis of fluorescence anisotropy decay Biochemistry 26:5121–5126, 1987.

Strickland KP. The chemistry of phospholipids. In: Ansell GB, Hawthorne J, Dawson RMC, eds. Form and Function of Phospholipids. Amsterdam: Elsevier, 1975, pp 9–42.

Sundaralingam M. Discussion paper: Molecular structures and conformations of the phospholipids and sphingomyelins. Ann N Y Acad Sci 195:324–355, 1972.

Tanford S. The Hydrophobic Effect. New York: Wiley, 1980.

Tardieu A, Luzzati V, Reman FC. Structure and polymorphism of the hydrocarbon chains of lipids: A study of lecithin-water phases. J Mol Biol 75:711–733, 1973.

Tate MW, Eikenberry EF, Turner DC, Shyamsunder E, Gruner SM. Nonbilayer phases of membrane lipids. Chem Phys Lipids 57:147–164, 1991.

Tenchov B, Rappolt M, Koynova R, Rapp G. New phases induced by sucrose in saturated phosphatidylethanolamines: An expanded lamellar gel phase and a cubic phase. Biochim Biophys Acta 1285:109–122, 1996.

Thompson TE, Lentz B, Barenholz Y. In: Smenza G, Caraafoli E, eds. Biochemistry of Membrane Transport. New York: Springer-Verlag, 1977, p 47.

Thompson TE, Sankaram MB, Biltonen RL, Marsh D, Vaz WLC. Effects of domain structure on in-plane reactions and interactions. Mol Membr Biol. 12:157–162, 1995.

Tirosh O, Kohen R, Katzhendler J, Gorodetsky R, Barenholz Y. Novel synthetic phospholipid protects lipid bilayers against oxidative damage role of hydration layer and bound water. J Chem Soc Perkin Trans 2:383–389, 1997.

Tirosh O, Barenholz Y, Katzhendler Y, Priev A. Hydration of polyethelene glycol–grafted liposomes. Biophys J 74:1371–1379, 1998.

Torchillin VP. Effect of polymers attached to lipid headgroups on properties of liposomes. In: Lasic DD, Barenholz Y, eds. Handbook of Non-Medical Applications of Liposomes. Vol I. Boca Raton, FL: CRC Press, Chap 13, pp 263–284.

Trahms LW, Klabe D, Boroske E. H-NMR study of the three low temperature phases of DPPC-water systems. Biophys J 42:285–293, 1983.

Tuckwell DS, Weston SA, Humphries MJ. Integrins: A review of their structure and mechanisms of ligand binding. Symp Soc Exp Biol 47:107–136, 1993.

Ulrich AS, Watts A. Molecular response of the lipid headgroup to bilayer hydration monitored by ^{2}H-NMR. Biophys J 66:1441–1449, 1994.

Ulrich AS, Sami M, Watts A. Hydration of DOPC bilayers by differential scanning calorimetry. Biochim Biophys Acta 1191:225–230, 1994.

Van Deenen PWM, de Gier J, Houtsmuller VMTM, Montpoort A, Mulder U. In: Frazer AC, ed. Biochemical Problems of Lipids. Amsterdam: Elsevier, 404.

van Helvoort A, van Meer G. Intracellular lipid heterogeneity caused by topology of synthesis and specificity in transport. Example: sphingolipids. FEBS Lett 369:18–21, 1995.

van Helvoort A, Smith AJ, Sprong H, Fritzsche I, Schinkel AH, Borst P, van Meer G. MDR1 P-glycoprotein is a lipid translocase of broad specificity, while MDR3 P-glycoprotein specifically translocates phosphatidylcholine. Cell 87:507–517, 1996.

van Osdol WW, Johnson ML, Ye Q, Biltonen RL. Relaxation dynamics of the gel to liquid-crystalline transition of phosphatidylcholine bilayers. Effects of chainlength and vesicle size. Biophys J 59:775—785, 1991.

van Wijk GM, Gadella TW, Wirtz KW, Hostetler KY, van den Bosch H. Spontaneous and protein-mediated intermembrane transfer of the antiretroviral liponucleotide 3'-deoxythymidine diphosphate diglyceride. Biochemistry 31:5912–5917, 1992.

Veld GI, Driessen AJ, Konings WN. Bacterial solute transport proteins in their lipid environment. FEMS Microbiol Rev 1:293–314, 1993.

Vierl U, Cevc G. Time-resolved X-ray reflectivity measurements of protein adsorption onto model lipid membranes. Biochim Biophys Acta 1325:165–177, 1997.

Vierl U, Nagel N, Löbbecke L, Cevc G. Solute effects on the colloidal and phase behavior of lipid bilayer membranes. Ethanol-dipalmitoyl-phosphatidylcholine mixtures. Biophys J 67:1067–1079, 1994.

Vigh L, Maresca B, Harwood JL. Does the membrane's physical state control the expression of heat shock and other genes? Trends Biochem Sci 23:369–374, 1998.

Wallach DFH. Membrane Molecular Biology of Neoplastic Cells. New York: Elsevier, 1976.

Watts A. Magnetic resonance studies of lipid-protein interface and lipophilic Molecule partitioning. Ann N Y Acad Sci 625:653–667, 1991.

Watts A. Biophysics of the membrane interface. Biochem Soc Trans 23:959–965, 1995.

Weissmann G, Clailborne R. Cell Membranes, Biochemistry, Cell Biology and Pathology. New York: HP Pub., 1975.

White D. In: Ansell GB, Hawthorn JN, Dawson RM, eds. Form and Function of Phospholipid. London: Elsevier, 1973, pp. 441–482.

Wieslander A, Nordstrom S, Dahlqvist A, Rilfors L, Lindblom G. Membrane lipid composition and cell size of *Acholeplasma laidlawii* strain A are strongly influenced by lipid acyl chain length. Eur J Biochem 227:734–744, 1995.

Williamson P, Shlegel RA. Back and forth: The regulation and function of transbilayer phospholipid movement in eukaryotic cells. Mol Membr Biol 11:199–216, 1994.

Williamson P, Kulick A, Zachowski A, Schlegel RA, Devaux PF. Ca^{2+} induces transbilayer redistribution of all major phospholipids in human erythrocytes. Biochemistry 31:6355–6360, 1992.

Wimley WC, Thompson TE. Transbilayer and interbilayer phospholipid exchange in dimyristoylphosphatidylcholine/dimyristoylphosphatidylethanolamine large unilamellar vesicle. Biochemistry 30:1702–1709, 1991.

Wirtz KWA. In: Jost PC, Griffith OH, ed. Lipid-Protein Interactions. Vol 1. New York: Wiley, 1982, p 151.

Wong PTT, Mantsch HH. A low-temperature structural phase transition in aqueous dimyristoyl phosphatidylcholine bilayers observed by Raman scattering. Can J Chem 60:2137–2140, 1982.

Wood BL, Gibson DF, Tait JF. Increased erythrocyte phosphatidylserine exposure sickle cell disease, flow cytometric measurement and clinical associations. Blood 88:1873–1880, 1996.

Wood R. Tumor lipids. J Am Oil Chem Soc Res, Champaign IL, 1973.

Wu J. Microscopic model for selective permeation in ion channels. Biophys J 60:238–251, 1991.

Yang L, Glaser M. Membrane domains containing phosphatidylserine substrate can be important for the activation of protein kinase C. Biochemistry 34:1500–1506, 1995.

Yang L, Glaser M. Formation of membrane domains during the activation of protein kinase C. Biochemistry 35:13966–13974, 1996.

Yasuda T, Dancey GF, Kinsky SC. Immunogenicity of liposomal model membranes in mice. Dependence on phospholipid composition. Proc Natl Acad Sci USA 74:1234–1236, 1977.

Yeagle LP. The Membranes of Cells. 2nd ed. San Diego: Academic Press, 1993.

Yechiel E, Barenholz Y, Henis YL. Lateral mobility and organization of phospholipids and proteins in rat myocyte membrane. Effects of aging and manipulation of lipid composition. J Biol Chem 260:9132–9136, 1984.

Zachowski A. Phospholipids in animal eukaryotic membranes: Transverse asymmetry and movement. Biochem J 294:1–14, 1993.

Zasadzinski JAN, Schnier J, Gurley V, Elings V, Hansma PK. Scanning tunneling microscopy of freeze-fracture replicas of biomembranes. Science 239:1013–1015, 1988.

Zuidam NJ, Barenholz Y. Electrostatic parameters of cationic liposomes commonly used for gene delivery as determined by 4-heptadecyl-7-hydroxy-coumarin. Biochim Biophys Acta 1329:211–222, 1997,

Zuidam NJ, Barenholz Y. Electrostatic and structural properties of plasmid DNA-lipid complexes commonly used for gene delivery. Biochim Biophys Acta 1368:115–128, 1998.

Zuidam NJ, Minsky A, Barenholz Y. Modulation of secondary and tertiary DNA structure in lipoplexes commonly used for gene delivery. Submitted.

Zwaal RF, Schroit AJ. Pathophysiologic implications of membrane phospholipid asymmetry in blood cells. Blood 89:1121–1132, 1997.

Zwaal RF, Confurius P, Bevers EM. Mechanism and function of changes in membrane-phospholipid asymmetry in platelets and erythrocytes. Biochem Soc Trans 21:248–253, 1993.

8

Attempts to Mimic Biomembrane Processes: Recognition- and Organization-Induced Functions in Biological and Synthetic Supramolecular Systems

H. Ringsdorf
University of Mainz, Mainz, Germany

I. INTRODUCTION: FUNCTIONAL SUPRAMOLECULAR SYSTEMS TODAY

A. Tradition and Innovation in Science

Two decisive aspects of the field of science are tradition and innovation. Tradition is the basis, for it is the cumulation of wisdom in the body of knowledge. To know what a subject is all about and to control it creates self-confidence, thus paving the way for innovations. Innovation is the adventure, because with the challenge comes the risk of calling into question (or even losing) one's own scientific identity, gained through tradition.

Persisting in tradition without innovation, however, soon leads to tiresome routine, to the science of yesterday: the longing for new adventures withers and dies (1). On the other hand, pure innovation harbors the danger of superficiality. The sum of knowledge is immense and growing! Tradition and solid, successful work are honored and admired. Nevertheless, science can be justified only by challenge and demands the willingness to give up long-held classical or traditional views (2) in the attempt to discover new horizons.

B. Molecular Architecture and Molecular Engineering of Self-Assembled Systems

In recent years "supramolecular systems" has become the cumulative title to describe the rapidly emerging achievements at the interfaces between chemistry, physics, and biology. As classical organic chemistry is now able to construct highly complex molecules and provide biologists and biochemists with detailed structural insights into biological processes, chemists are becoming increasingly interested in investigating organic chemistry beyond the covalent bond, as coined by J.-M. Lehn (3–6). The function of supramolecular systems is achieved by the non–covalent-bonded interplay between different functional units. This type of interaction was first postulated to be a possible source of molecular functions as early as 1894, when Emil Fischer casually introduced his metaphor for selective interaction of molecules: the "lock-and-key" principle (7). This idea has been a guiding light for one of the most stimulating fields in modern science.

In nature, a unit that demonstrates this perfectly is the cell membrane, where the interplay between molecular self-organization and molecular recognition of the individual constituents (phospholipids, glycolipids, glycoproteins, membrane-spanning peptide helices, the cytoskeleton, etc.) leads to the construction of this natural supramolecular system. It combines order and mobility, and its function is based on its self-organization. The scientific effort to understand, construct, and mimic natural molecular assemblies can no longer focus on only single molecular performances but has to address self-organized molecular aggregates—the whole being more than the sum of its parts. This principle of nature was adopted as a basic idea of the ancient philosophies of Asia and Europe: only the mutuality of the parts creates the whole and its ability to function.

The common perspectives of life science and materials science are illustrated in Fig. 1 by showing examples of functional supramolecular systems, many of whose properties, structures, and functions are determined by their supramolecular order (8,9). The cell membrane with its carbohydrate recognition structures on the surface (glycocalyx) is structured by different phospholipids and stabilized by functional proteins and lipids, which extend through the membrane, as well as by polymer chains and nets spread out below the cell membrane, for example, the cytoskeleton in red blood cells. Descending in complexity from this perfect natural system, one finds liposomes with their relatively simple spherical bilayer structure and, finally, micelles with their various lyotropic structures. These suprastructures have long been important in studies of synthetic and natural systems and, indeed, colloid science determined the course of physical chemistry in the 1920s. Developments in monolayers and multilayers (Langmuir and Langmuir-Blodgett systems) (10) have resulted in ultrathin films in which order and mobility are combined. Of more interest to materials science are liquid crystals (11), which are already of industrial importance, both as low-molar-mass and as macromolecular products. Their order is based on their form anisotropy (molecular shape), and it is possible for rods, disks, and boards (12,13) to self-organize. Too often, it is forgotten that nature abounds with liquid crystalline materials. Many of the natural polymers and cell membranes themselves can also be regarded as lyotropic liquid crystalline systems. Besides molecular self-organization, another crucial property of supramolecular systems, as shown in Fig. 1, is their capability of specific molecular interactions. Many processes occurring at natural or synthetic membrane surfaces start with a specific molecular recognition event leading to enzymatic reactions, such as the cleavage of phospholipids by various phospholipases (9) or to protein crystallization, as in the interaction of biotinylated membranes with streptavidin (9,14). The study of molecular recognition processes also includes the alteration

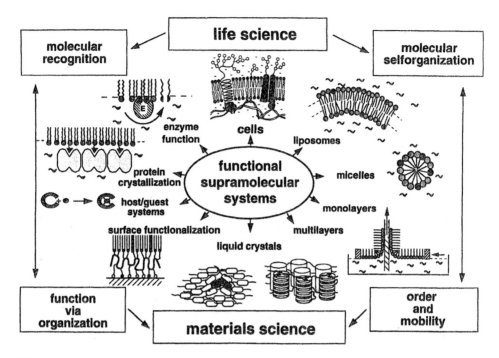

Fig. 1 Functional supramolecular systems—connecting links between life science and materials science. (From Ref. 9.)

of surfaces and the structuring of ultrathin layers, which are of particular interest in catalysis research, sensor technology, and tribology.

II. FUNCTION BASED ON ORGANIZATION AND MOLECULAR RECOGNITION

The broad interest in supramolecular science also comes from the perspectives this field appears to offer. The molecular evolution of life—as yet an unsolved problem—can be addressed in a new light. Along the way to this ultimate intellectual goal, significant progress is expected, e.g., in molecular biology, drug delivery systems, and in biophysics with a strong impact on pharmaceutics and medicine. Furthermore, the next decades will plunge us much deeper into what we already call the age of information. The growing need to increase information storage capacities and processing speed will finally force us to use to perfection the principles nature developed. We will learn to use nature's minute building blocks and molecules as a source and carrier of information and to exploit their cooperativity and self-assembly to transmit information.

In this chapter, two examples designed to contribute to a better understanding of function based on molecular recognition and organization will be discussed. The first one is the function of phospholipase A_2 at phospholipid monolayers. It is a purely biological example in which nature's skills in inducing functions by organization and recognition are taken advantage of (9,15). The other example—based on the interaction of barbituric acid amphiphile monolayers with triaminopyrimidine—is a

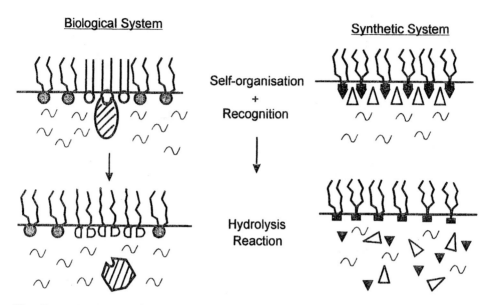

Fig. 2 Molecular organization and molecular recognition as a basis for function in biological and synthetic systems: two examples of hydrolytic cleavage reactions at monolayers.

purely "synthetic" approach to this problem (15,16). Figure 2 elucidates the principle for both examples discussed in this chapter.

A. An Enzyme That Likes Organization: Interaction of Phospholipase A_2 with Phospholipid Monolayers

Phospholipases are an important class of enzymes (17). They catalyze the hydrolysis of ester bonds in membrane-forming phospholipids. The enzyme of this class that has been investigated most thoroughly is phospholipase A_2 (9). It is a relatively small molecule (molecular mass 12–14 kDa), soluble in water, and remarkably stable. It catalyzes stereospecifically the ester cleavage at the C-2 position in naturally occurring glycero-phospholipids to yield the corresponding lysolecithin and the free fatty acid (Fig. 3).

One of its most important functions is to open, modify, and rebuild cell membranes. It cannot be excluded that the death of tumor cells—as described in Sec. IV—is caused by membrane-bound phospholipases (see Fig. 25). Most unusual and very interesting is the fact that the hydrolytic activity of phospholipase A_2 depends very strongly on the physical state of the substrate: It increases many times in going from a homogeneous solution to organized substrates such as micelles, liposomes, or monolayers. The highest activity in liposomes is found when the lipid is in the phase transition region. This property of phospholipase A_2 makes it appropriate for investigations in monolayers because it is easy to bring about large changes in both the composition and the physical state of phospholipid monolayers.

1. Hydrolysis of Substrate Lipids of Phospholipase A_2 at Monolayers followed by Fluorescence Spectroscopy

In order to observe the orientation-dependent hydrolytic activity of phospholipase A_2 (PLA-2) at lecithin (L-α-DPPC) monolayers, a Langmuir trough equipped with a flu-

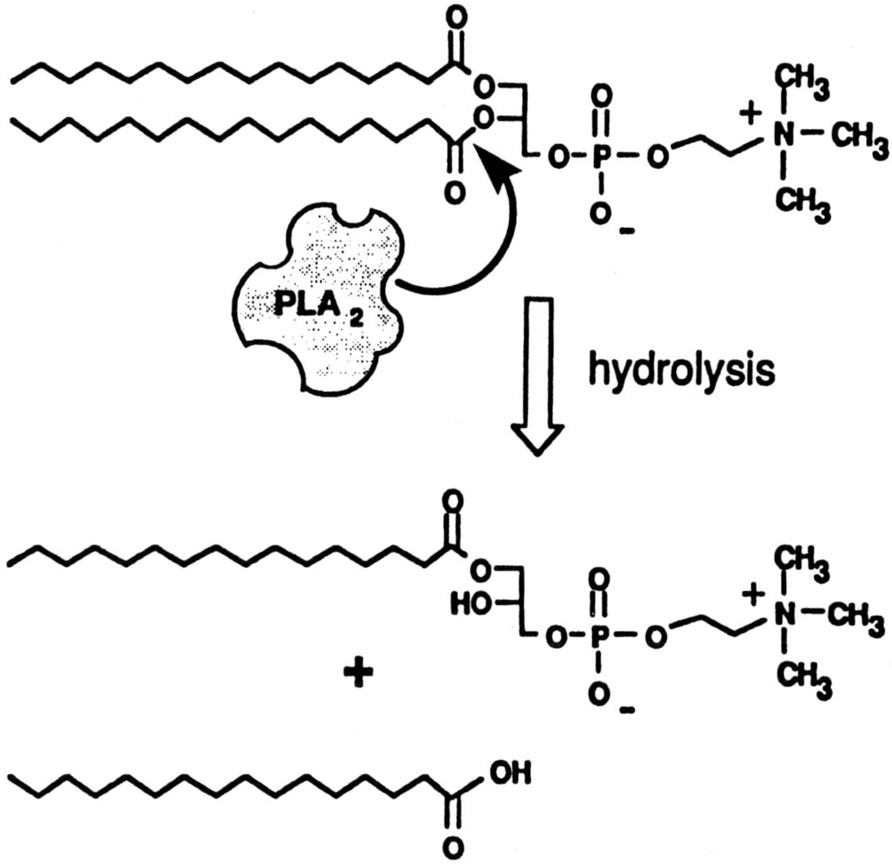

Fig. 3 Cleavage reaction of phospholipids as catalyzed by phospholipase A_2 (schematic). The enzyme hydrolyzes the ester bond in the C-2 position of a lecithin, resulting in the corresponding lysolecithin and the free fatty acid.

orescence microscope was used. The monolayer was doped with a sulforhodamine-lipid [SR-DPPE=N-(Texas-red-sulfonyl)-dipalmitoyl-L-phosphatidyl-ethanolamine], and the phospholipase was labeled with fluorescein isothiocyanate (FITC). In combination with a fluorescence microscope, this double-labeling technique makes it possible to observe processes in the lipid monolayer and the behavior of the enzyme separately and simultaneously (9). Using a rhodamine filter, one observes only fluorescence from the monolayer doped with the sulforhodamine lipid, which reflects changes in the morphology of the lipid layer. On the other hand, using a fluorescein filter, the lipid layer becomes virtually transparent, and fluorescence is observed only from the fluorescein attached to the protein. This as well as the whole process to be discussed is schematically demonstrated in Fig. 4.

The lipid monolayer is compressed into the phase transition region. This leads to the formation of solid-analogous lipid domains in a liquid-analogous matrix (Fig. 4A). Injection of the labeled phospholipase A_2 into the subphase (Fig. 4B) is followed by specific recognition between the enzyme and its substrate. The attack by the protein

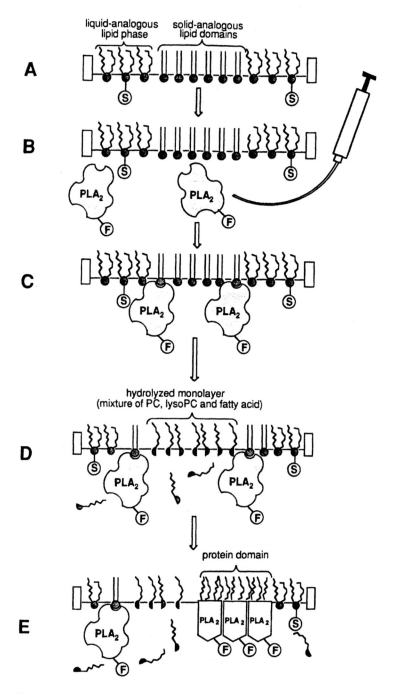

Fig. 4   Hydrolysis of a lipid monolayer by phospholipase A_2 (schematic). (A) Monolayer in the phase transition region with solid-analogous lipid domains in a liquid-analogous lipid matrix mixed with a sulforhodamine marker. (B) Injection of the FITC-labeled phospholipase A_2. (C) Specific recognition of the substrate lipids by the enzyme and preferential attack at the domain boundaries. (D) Hydrolysis of the solid-analogous lipid domains and accumulation of the hydrolysis products in the monolayer. (E) Aggregation of the enzyme.

takes place preferentially at the boundary between the solid-analogous and the liquid-analogous phase (Fig. 4C). As hydrolysis progresses, the solid-analogous lipid domains disappear due to the hydrolytic cleavage of the phospholipid, and the hydrolysis products dissolve or accumulate in the monolayer (Fig. 4D). After a certain time, the phospholipase A_2 starts to aggregate, forming its own domains (Fig. 4E). This may indicate a deactivation of the enzyme caused by a change of its conformation.

2. Hydrolysis of Solid-Analogous L-α-DPPC Domains

The process described schematically in Fig. 4 can be observed directly under the fluorescence microscope (9), as illustrated by using L-α -dipalmitoylphosphatidylcholine. The enzyme is injected under a monolayer of DPPC in the phase transition region. Viewed through the sulforhodamine filter, dark solid-analogous lipid domains—which are typical of L-α-DPPC—are seen in the bright matrix of the liquid-analogous lipid and its sulforhodamine marker (see Fig. 5A and the scheme in Fig. 4A and B).

Via the fluorescein filter, only a homogeneous fluorescence of the protein is visible below the DPPC domains (see Fig. 5B). The first signs that the enzyme recognizes its substrate in the monolayer and starts to hydrolyze DPPC appear after 10 min. First, areas of erosion are formed specifically at the concavities previously present in the lipid domains (see Fig. 5C and Fig. 4C). The enzyme then starts to degrade the domains inward from these points on (Fig. 5C and E). In consequence, there is local accumulation of the acid and lysolecithin, which visibly leads to autocatalytic activation of the phospholipase A_2 at this point (18). This would explain why further hydrolysis from the initial point of attack is preferred. After about 60 min only a few remnants of the lipid domains are left (Fig. 5G, I, and L). After 30 min, observation through the fluorescein filter reveals small bright dots (Fig. 5F), which appear dark through the sulforhodamine filter: the protein is starting to aggregate at the monolayer. These bright dots increase in number and size as the reaction progresses (Fig. 5F, H, and K) until at its end (after about 60 min), they have a typical kidney shape (see Fig. 5L). The monolayer now consists only of almost completely destroyed solid-analogous lipid domains and the new protein domains in a liquid matrix of lysolipid and palmitic acid (see Fig. 5L and M). It is noticeable that the domains of the phospholipase A_2 are similar in shape to some of the lipid domains present at the start of the reaction (compare Fig. 5L and M, and Fig. 5A).

To examine whether there is a template effect, the hydrolysis experiment was repeated with L-α-dimyristoyl phosphatidylcholine (L-α-DMPC), a phospholipid that forms solid domains of very different morphology (very small and star-shaped domains) at the gas-water interphase (9,19). After injection of the enzyme PLA-2, a completely analogous hydrolysis process takes place. The small, star-shaped lipid domains disappear after 50–60 min and the protein aggregates grow as the reaction progresses.

The fact that they show the same shape as in the DPPC experiments (Fig. 5) demonstrates that the shape of the protein domains does not depend on the shape of the original lipid domains (19).

3. Compression of a Hydrolyzed L-α-DPPC Monolayer

Considering the fact that in living systems the enzyme PLA-2 constantly interacts with cell membranes without completely destroying them, one has to assume that an inhibitor effect is in place in living systems. It is known that PLA-2 has an autocatalytic start

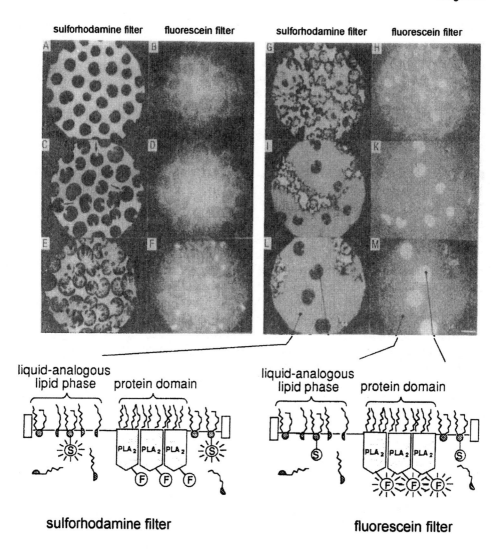

Fig. 5 Time dependence of the hydrolysis of an L-α-DPPC monolayer by phospholipase A_2 observed by fluorescence microscopy using a double-labeling technique; the monolayer is seen through the sulforhodamine filter (left) and the enzyme is seen through the fluorescein filter (right). (A) Dark solid-analogous DPPC domains in a bright lipid-analogous matrix; (B, D) homogeneous fluorescence of the enzyme in the subphase; (C) first signs of hydrolysis of the lipid domains (arrow); (E) progressive degradation of the lipid domains and first signs of the formation of enzyme aggregates at the monolayer [bright spots through the fluorescein filter (E), dark spots through the sulforhodamine filter (F)]; (G)–(K) further hydrolysis of the lipid domains (G,I) with simultaneous growth of the protein domains (H,K); (L, M) after almost complete degradation of the lipid domains, there are now only protein domains in a fluid matrix composed of the hydrolysis products (lysolecithin and the long-chain acid) and DPPC; these enzyme domains have a specific shape and appear dark through the sulforhodamine filter (L) and bright through the fluorescein filter (M). (A, B) $t = 0$ min (immediately after enzyme injection); (C, D) $t = 10$ min; (E, F) $t = 30$ min; (G, H) $t = 40$ min; (I, K) $t = 50$ min; (L, M) $t = 60$ min; $\pi = 22$ mN m^{-1}; $T = 30°$C; 5 mm = 20 μm.

phase—as mentioned earlier—but also an autocatalytic termination mechanism, as demonstrated in these DPPC and DMPC monolayer experiments. What further role do the PLA-2 domains play and are they still enzymatically active? More information about the nature of the protein in the domains was obtained by changing the surface pressure in a hydrolyzed L-α-DPPC monolayer. Expansion to the gas-analogous state ($\pi = 0$ mN/m) has no effect on the shape of the protein domains: they are very stable and do not disappear. The processes observed during recompression of the expanded monolayer are demonstrated both schematically and as they appear under the fluorescence microscope.

New solid-analogous lipid domains are formed in the fluid matrix (Fig. 6A). This shows that even after complete hydrolysis of the original lipid domains, there is still uncleaved DPPC present in the monolayer. The most important finding is that new lipid domains are produced not only in the fluid phase but also directly at the edges of the protein domains (bright through the fluorescein filter, Fig. 7B). This demonstrates that lipids crystallized onto the protein domains are not attacked and thus not degraded by the organized enzyme. This points to the fact that the aggregated phospholipase A_2 is no longer active at all.

What causes the enzyme domain formation? The first evidence that the lysolecithin produced by the enzymatic lipid cleavage plays an important role was gained by experiments with mixed lyso-DPPC–DPPC monolayers (9). Mixed domains are formed. They are not only *not* attacked by the phospholipase A_2 but are just used to dock the enzyme, leading to spontaneous formation of PLA-2 domains already after 2 min. The same process may take place in or at cell surfaces. Lysolecithin formed during the enzyme reaction may act as an "inhibitor" and induces the domain formation of now inactivated PLA-2. This might help to explain why the phospholipases—widespread in nature—do not destroy cell membranes in an uncontrolled fashion.

B. Hydrogen Bond–Induced Interaction and Function of Barbiturate-Lipid Monolayers with Triaminopyrimidine (TAP)

In order to examine the influence of preorientation and molecular recognition on reactions in nonbiological systems, a purely synthetic lipid model system was synthesized for use at the air-water interface. The barbituric acid lipids used were designed to enable them to undergo a substrate-catalyzed hydrolysis reaction, as proposed in Fig. 2. The system used here is based on the interaction of barbituric acid lipids with substrates such as 2,4,6-triaminopyrimidine (TAP) (see Fig. 7), or melamine, and urea (see Fig. 11).

The strand formation of these H-bond donors and H-bond acceptors has already been carefully investigated in nonaqueous solvents (20). The hydrolysis to be discussed is the retro-Knoevenagel reaction of lipid **1** yielding free barbituric acid and the analogue benzaldehyde derivative. The reaction can be easily followed via ultraviolet-visible (UV-Vis) spectroscopy as shown in Fig. 8.

The barbituric acid lipid **1** forms a stable solid analogously packed monolayer at 25°C. It is interesting to note that cospreading of **1** with TAP leads basically to a monolayer showing the same surface pressure–area diagram. Changes start only at the onset of the retro-Knoevenagel reaction leading to free barbituric acid and the benzaldehyde lipid (see Fig. 8). This points to a perfect fit of one TAP molecule

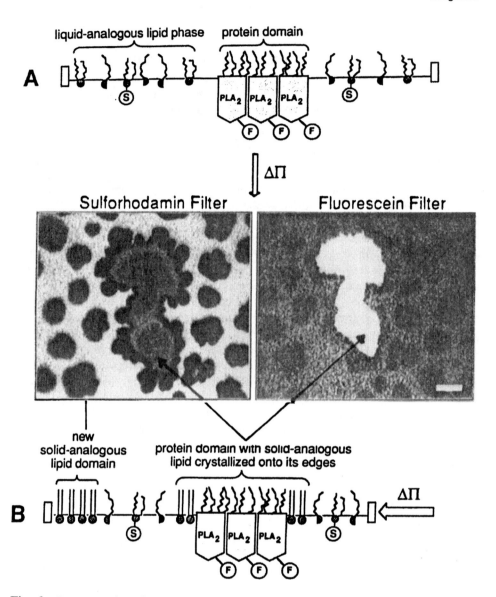

Fig. 6 Recompression of an L-α-DPPC monolayer after hydrolysis by phospholipase A$_2$. (A) Schematic representation of a protein domain in a fluid matrix composed of hydrolysis products and DPPC. (B) Compression leads to formation of new solid-analogous DPPC domains both in the fluid matrix and directly on the edges of the protein domains (appearance under the fluorescence microscope and schematic representation). 5 mm = 20 μm.

in between two lipids leading to H-bonded strands, as known for barbituric acid derivatives and TAP in organic solvents (20).

   The recognition, insertion, and C=C bond cleavage reaction of **1** and TAP in monolayers was studied by reflectance spectroscopy on a Langmuir trough equipped with a UV-vis spectrometer (15,16). The results of this organization and recognition induced hydrolysis experiment are summarized in Fig. 9.

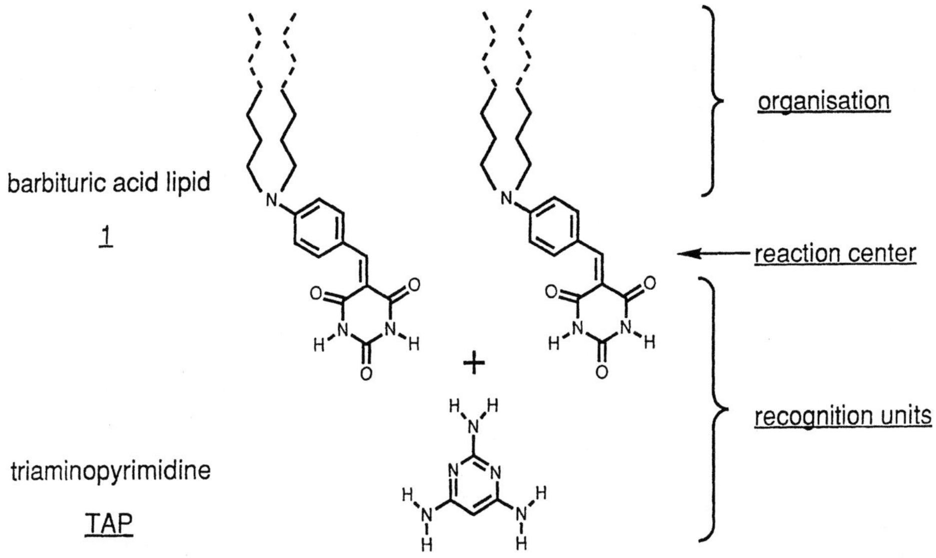

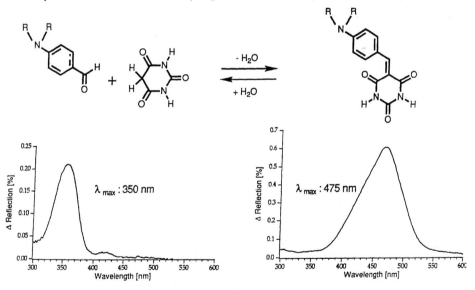

Fig. 7 Concept of the interaction of barbituric acid lipid **1** with triaminopyrimidine (TAP): model system for reaction induced by organization and recognition at monolayers.

Fig. 8 Formation (Knoevenagel condensation) and hydrolysis (Retro-Knoevenagel reaction) of the benzyliden-barbituric acid lipid **1** in solution.

The UV-vis spectrum of a lipid **1** monolayer on a 10^{-4} M HCl but TAP-free subphase (Fig. 9) is characterized not only by the same adsorption maximum as in solution (475 nm; see Fig. 8) but also by an aggregation band around 430 nm. This points to stacking of the aromatic chromophore in the oriented system. Under these—that is, TAP-free—conditions, the monolayer is stable over a period of at least 17 h and the reflectance spectra do not change at all.

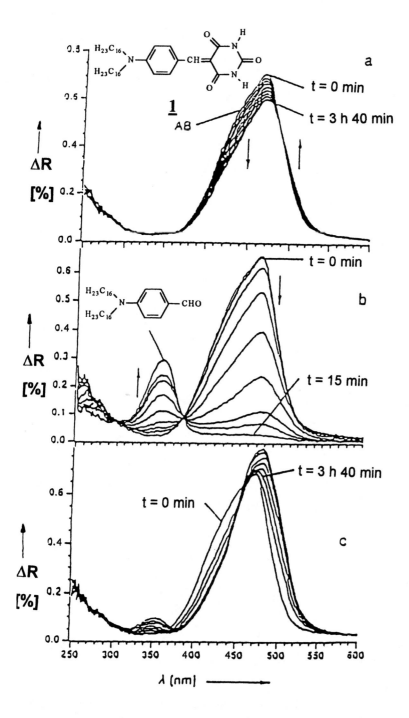

Fig. 9 UV-visible reflectance spectra of the barbituric acid lipids **1** at the air-water interface: (a) On an HCl containing 10^{-4} M TAP subphase (pH 3); disappearance of the aggregation band at 430 nm due to TAP insertion; no hydrolysis. (b) On a 10^{-4} M TAP subphase (pH 6.5); fast recognition-induced hydrolysis of barbituric acid lipid **1**. (c) On a 10^{-4} M NaOH subphase (pH 10); very slow basic hydrolysis of barbituric acid lipid **1**.

In contrast, the aggregation band at 430 nm disappears within 4 h if the lipid **1** is spread on TAP-containing 10^{-4} M HCl subphase at pH 3 (Fig. 9a). This can be explained by insertion of TAP between the lipid molecules disturbing the dye stacking. This is a uniform and defined process of insertion of TAP into the monolayer, as can be concluded from the three isosbestic points at 360, 486, and 544 nm. While following the insertion of TAP into a monolayer of **1** at pH 6.5—the normal pH of a 10^{-4} molar TAP subphase—one can see a second process: a cleavage of the C=C double bond of the chromophore (Fig. 9b). This leads to formation of the *p*-aminobenzaldehyde-lipid ($\lambda_{max.} = 350$ nm) and barbituric acid. Again, two isosbestic points can be detected, indicating a uniform and defined hydrolysis process. The barbituric acid headgroup of the lipid is cleaved quantitatively after 15 min. Particularly astonishing is the fact that the hydrolysis on a 10^{-4} M NaOH subphase (pH 10) is just starting at that time (Fig. 9c). The steps of a possible mechanism of this TAP-induced C=C bond cleavage are shown schematically in Fig. 10, speculating about similarities with comparable enzymatic reactions.

In this mechanism, water is perfectly situated to attack the double bond. The high activity of these water molecules can be explained in two ways. On the one hand, there is activation of the C=C double bond due to intermolecular hydrogen bonds from the neighboring carbonyl groups to TAP. On the other hand, hydrophobic pockets at the air-water interface are formed, helping to orient the water molecules as discussed for many enzymatic reactions: The hydrogens orient toward the hydrophobic alkyl chains of the monolayer. This mechanism of the recognition and orientation-induced C=C cleavage (retro-Knoevenagel) reaction is supported by several other experiments. When melamine ($pK_B = 8.9$) is used as a substrate instead of TAP ($pK_B = 7.3$), the speed of the insertion process is similar, but the hydrolysis process is approximately nine times slower. This can be attributed to the weaker hydrogen bonds that are formed

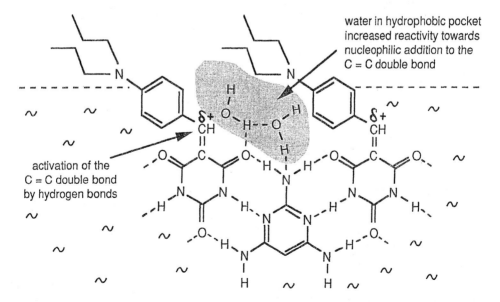

Fig. 10 Hydrogen bond–activated C=C double bond of lipid **1** attacked by oriented H_2O molecules in hydrophobic pockets at the air-water interphase (schematic).

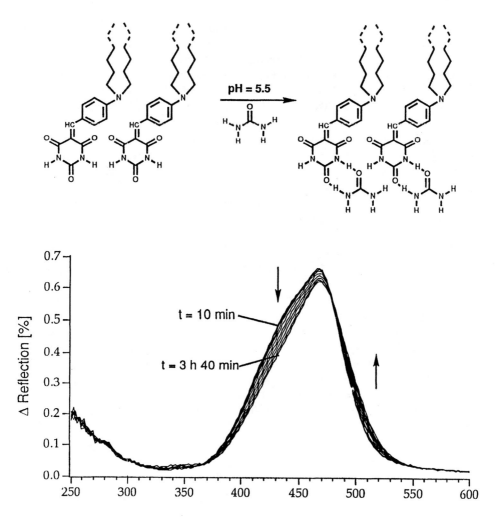

Fig. 11 UV-visible reflectance spectrum and schematic presentation of the insertion of urea into a monolayer of lipid **1** formation of a strand structure.

with the less basic melamine, so that the activation of the C=C double bond is decreased. The importance of the hydrogen bond–induced activation of the double bonds can also be shown by the use of urea as a substrate in the subphase.

Figure 11 demonstrates the insertion of urea into the lipid **1** monolayer, forming strands via hydrogen bonds. The reflectance spectra show a process similar to that shown in Fig. 8A. The aggregation band disappears (again, one can detect three isosbestic points). As there is no activation of the carbonyl groups conjugated to the C=C double bond, hydrolysis is not induced. In order to show that the enhanced concentration of hydroxyl ions at the monolayer due to the insertion of the base TAP is not the reason—or at least not the only reason—for this remarkable cleavage speed, monolayers of lipid **1** on subphases of water-soluble organic bases, such as dimethylamine and trimethylamine, were examined. These organic bases are enriched at the surface, and therefore there is an enhanced OH concentration at the surface.

They are orders of magnitude more basic than TAP ($pK_{b(dimethylamine)} = 3.29$; $pK_{b(trimethylamine)} = 4.26$; $pK_{b(TAP)} = 7.2$). It was found that for monolayers of lipid **1** on these surface-active substrates the hydrolysis is faster than with the even stronger, but not surface active, base NaOH. Nevertheless, even these faster hydrolysis processes are still more than 20 times slower than with the much weaker base TAP. This shows that there must be other factors in the mechanism that influence the reaction speed.

To investigate whether the proposed hydrophobic pocket (Fig. 10) plays a role in this reaction, the pockets were blocked by introducing one or two methyl groups in *ortho* positions of the chromophore (**1**) (16c). The methyl groups are about the same size as the water molecules in such a pocket. The UV-vis reflectance spectra of the TAP interaction with the methyl-substituted lipid **1** show that even after 9 h only a small amount of lipid is hydrolyzed under the same conditions as in the experiment represented in Fig. 9 for the unsubstituted lipid **1**. This point to the fact that the hydrophobic pocket plays a role in the reaction mechanism.

The principle of molecular recognition is underlined when one examines the interaction of the following pyrazolin-3,5-dion lipid with TAP at the air-water interface:

$$H_3C(H_2C)_{14}H_2C \diagdown$$
$$H_3C(H_2C)_{14}H_2C \diagup N - \bigcirc - CH = \diagup\diagdown \; \substack{NH \\ NH} \; O, O$$

The aggregation band of this chromophore does not disappear after injection of TAP. There is no fit for a strand structure, as the angles of the five-membered ring structure are not compatible with the angles of the recognition unit, which is a six-membered ring. For this reason, no strand is formed and, as the proposed mechanism suggests, the monolayer is not cleaved at all.

III. INTERACTION OF ENZYMES AND MULTIENZYME COMPLEXES WITH FUNCTIONALIZED MODEL MEMBRANES

Spheres, disks, cylinders, and columns are often the characteristic structural building blocks of both natural and synthetic supramolecular systems. Spheres form rings, discs form columns, and rings and disks organize themselves into cylinders and hexagonal superstructures. The complex-looking molecular architecture of such supramolecular systems nevertheless leads to structures whose function is based on their organization. In the field of materials science this is demonstrated by the high photoconductivity of diskotic triphenylene derivatives found only in the liquid crystalline D_h mesophase of these column-forming molecules (13). Perfect examples from life science (and to be discussed here) are proteasomes (Fig. 12), multienzyme complexes that show their highest enzymatic activity only if all protein subunits are correctly assembled to form a cylindrical or barrel-shaped multienzyme complex.

A. Properties and Structure of Proteasomes

When two or more enzymes catalyze two or more steps in a metabolic cascade, they often form noncovalently associated multienzyme complexes. The formation of these

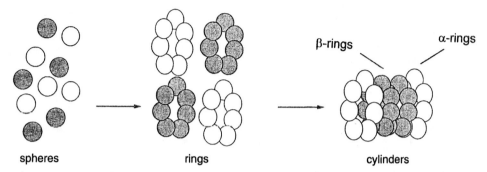

Fig. 12 Representation of barrel- or cylinder-shaped proteasomes, self-organized from four rings (α-rings on the outside, β-rings on the inside). The rings in turn have been self-organized from seven spherical protein subunits each. High catalytic activity exists only in the completely assembled form.

multienzyme complexes results in greater catalytic efficiency. The organized complexes allow much more efficient catalytic turnover than the nonassociated enzymes. In addition, it cannot be excluded that multicatalytic enzyme complexes play a more important role. Many proteins without known functions may be, in fact, "building blocks" for assembling these protein complexes, which then disassemble when the specific function is finished. An example of such a process is the immune cascade.

Examples of multienzyme complexes that have attracted increasing interest during the past few years are the so-called proteasomes. Proteasomes are high-molecular-weight multisubunit enzyme complexes (approximately 700 kDa) with at least three distinct proteolytic activities (trypsin-like, chymotrypsin-like, and peptidyl-glutamyl-peptide hydrolyzing) (21, 22). They are highly conserved ubiquitous in eukaryotic cells from yeast to human. Proteasomes have also been found in the archaebacterium *Thermoplasma acidophilum* (23). Eukaryotic proteasomes have a rather complex subunit composition; they typically contain 10–20 distinctly different but related subunits, apparently encoded by one gene family. Proteasomes from the archaebacterium *T. acidophilum* have the same quaternary structure, in which seven of the protein subunits assemble into one ring and four of these rings collectively form the cylindrical or barrel-shaped complex (Fig. 12). However, in contrast to the eukaryotic proteasomes, the *T. acidophilum* proteasome consists of only two different protein subunits, α and β. Therefore each proteasome particle contains 14 α-subunits and 14 β-subunits. It could be shown by immunoelectron microscopy that the 14 α-subunits are located in the two outer rings (seven subunits per ring) of the proteasome, and the 14 β-subunits constitute the two juxtaposed inner rings (see Fig. 12) (24).

An important issue is the identification of the functional roles of the α- and β-subunits in the *T. acidophilum* proteasome. Although an unambiguous assignment of functions is not yet possible, it is proposed that the α-subunits serve as regulatory and targeting function while the β-subunits, alone or in combination with the α-subunits, carry the active site responsible for the catalytic function. Dissociation and reassociation experiments have shown that neither single α-subunits nor single β-subunits are able to catalyze the hydrolysis of a polypeptide significantly (25). Even pure α-subunits organized into double rings do not show any catalytic activity. These experiments show clearly that only the completely self-organized cylindrical proteasome particle has full catalytic activity. Therefore, it can be speculated that the active site is located between one α- and one β-ring or between the two β-rings and that both neighboring rings are necessary to form the binding pocket for the enzyme substrate.

To further locate the catalytic site in the proteasome particle, the interaction of proteasomes with specifically functionalized model membranes was studied. Interaction of the proteasomes with the substrate derivatized membranes should give information about their ability to recognize and hydrolyze the substrate lipid headgroup at membrane surfaces. In addition, interaction of the proteasomes with membrane-incorporated inhibitor lipid headgroups may result in binding and orientation of the enzyme complexes relative to the membrane. Proteasomes from *T. acidophilum* show a chymotrypsin-like activity (26). Therefore, a substrate and an inhibitor of chymotrypsin have been used as headgroups in the synthesis of functionalized ligand lipids (see Fig. 14). With these kinds of lipids, optimal conditions for the specific interaction of the proteasomes with the membrane-incorporated ligand lipids could be tested by using chymotrypsin as a model receptor protein. These conditions could then be used for experiments with the much more sophisticated multienzyme complex (27).

B. Functionalized Ligand Lipids with Substrate and Inhibitor Headgroups for Chymotrypsin and Proteasomes

Oligopeptides with the Ala-Ala-Phe sequence act as substrates for chymotrypsin. An easy way to measure the enzymatic activity in solution is summarized in Fig. 13. Substrate **2** (succinyl-alanine-alanine-phenylalanine-7-amido-4-methyl-coumarin; Suc-Ala-Ala-Phe-AMC) is an oligopeptide linked to a fluorescent dye. Chymotrypsin hydrolyzes the amide bond on the carboxyl side of aromatic amino acids. This hydrolysis means that the enzyme cleaves the amide bond between the phenylalanine and the fluorescent dye to release chromophore **4**, 7-amino-4-methyl-coumarin (Fig. 13A). Hydrolysis of the substrate can easily be detected by fluorescence spectroscopy, because the amide-linked chromophore shows maximum fluorescence at 395 nm whereas the free chromophore has its maximum fluorescence at 435 nm. Inhibitor **5** contains the same oligopeptide sequence as the substrate but has the reactive α-chloromethylketone unit as the end group instead of the fluorescent dye. Inhibition of chymotrypsin is based on the irreversible alkylation of histidine 57, which is part of the active site of the enzyme (Fig. 13B).

For the monolayer studies, a substrate lipid **6** and an inhibitor lipid **7** have been synthesized (Fig. 14). Their spreading behavior on a Langmuir trough was characterized by pressure-area isotherms (27). At 20C both collapse at approximately 40 mN/m and show the existence of liquid-expanded phases, phase transition plateaus, and solid analogous phases. In lipids **6** (dioctadecylamine-diethylenoxide-substrate, DODA-EO$_2$-S) and **7** (dioctadecylamine-diethylenoxide-inhibitor, DODA-EO$_2$-I), the lipid headgroup is decoupled from the membrane forming alkyl chains by a long, flexible, hydrophilic spacer. In this case, the accessibility of the headgroups for proteins is greatly enhanced because of the distance between the membrane surface and the ligand at the monolayer. This is especially important for the proteasomes because binding of the barrel-shaped protein to monolayers and liposomes requires a long spacer to allow interaction of the headgroups with the, presumably, deep binding pocket of the multienzyme complexes.

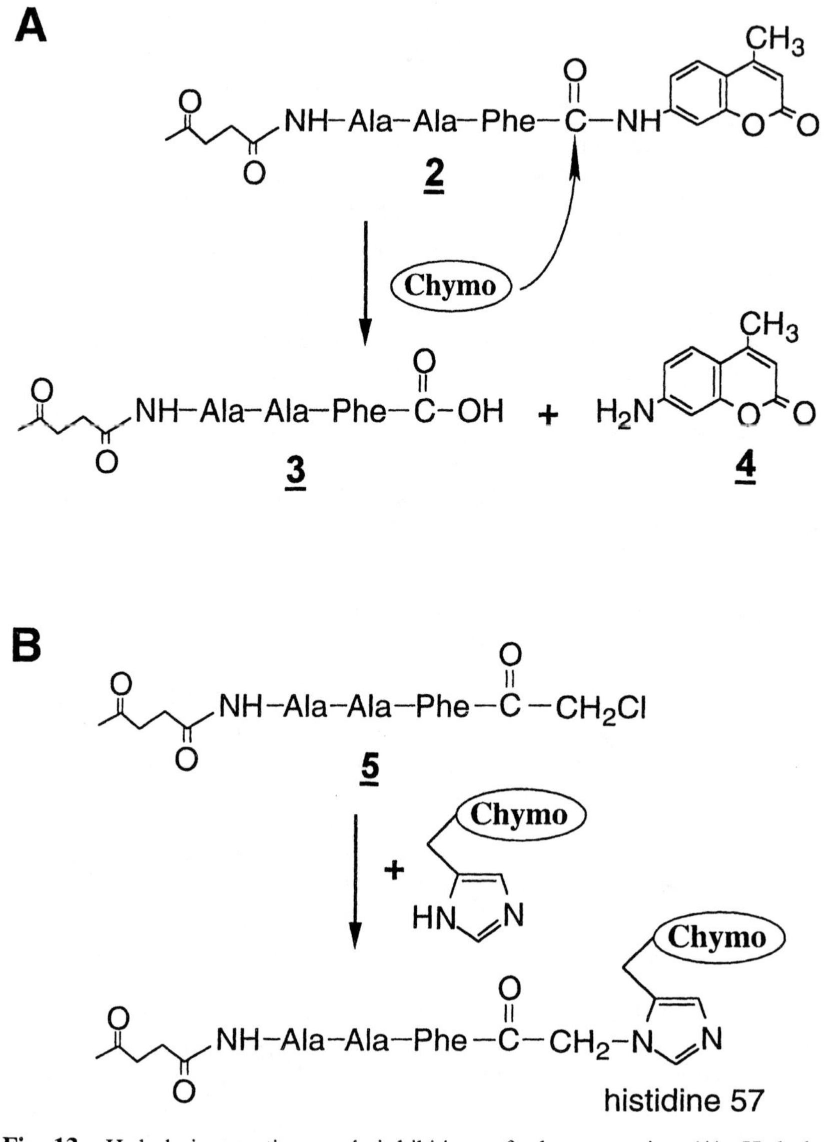

A

B

Fig. 13 Hydrolysis reaction and inhibition of chymotrypsin. (A) Hydrolysis of the fluorescence-labeled oligopeptide Suc-Ala-Ala-Phe-AMC (**2**) ($\lambda_n = 395$ nm) by chymotrypsin leads to release of the fluorescent dye 7-amino-4-methyl-coumarin (**4**) ($\lambda_n = 435$ nm). (B) The inhibition of chymotrypsin by the inhibitor Suc-Ala-Ala-Phe-chloromethylketone (**5**) is due to alkylation of histidine 57 in the catalytic center of chymotrypsin.

C. Interaction of Chymotrypsin and Proteasomes with Functionalized Model Membranes

1. Hydrolysis of Substrate Lipids with Chymotrypsin and Proteasomes

The accessibility of the membrane-linked substrate lipid headgroups for chymotrypsin and proteasomes was tested by measuring the hydrolysis of the substrate headgroup by

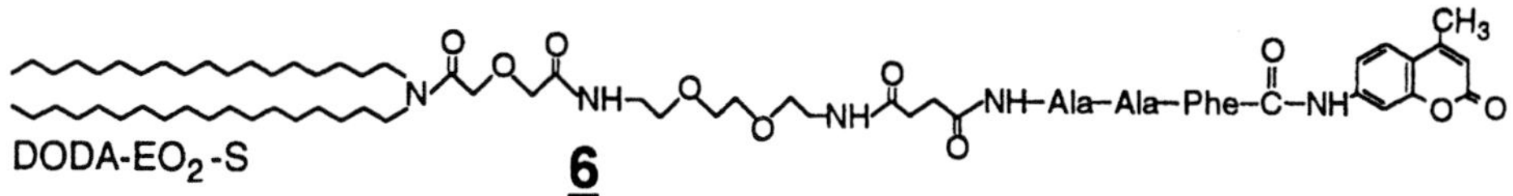

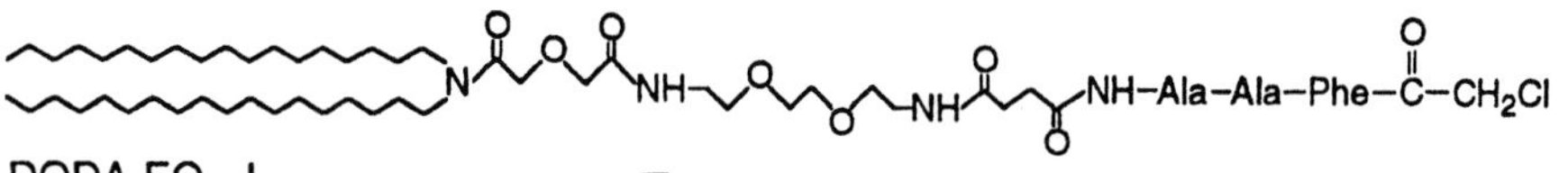

Fig. 14 Lipids carrying a substrate (**6**) and an inhibitor (**7**) headgroup for the interaction with chymotrypsin and proteasomes at a gas-water interface.

the proteins. Hydrolysis of the substrate can be monitored in homogeneous solution, in micelles, in liposomes, and in monolayers as shown schematically in Fig. 15.

a. Hydrolysis in Homogeneous Solution

The hydrolysis of the amide bond in the water-soluble oligopeptide Suc-Ala-Ala-Phe-AMC **2** (reaction shown in Fig. 13A) by chymotrypsin was monitored in aqueous phosphate buffer. The fluorescence spectrum of the covalently bound chromophore in **2** shows its fluorescence maximum at 395 nm (Fig. 16a). After addition of the enzyme, the fluorescence intensity at 395 nm decreases and a new fluorescence band with a maximum at 435 nm appears (Fig. 16b–j), demonstrating the release of the chromophore **4** (7-amino-4-methylcoumarin). The fluorescence spectra show that the hydrolysis reaction in isotropic solution is fast and the presence of an isosbestic point indicates a homogeneous process.

Hydrolysis of the water-soluble substrate **2** by proteasomes was monitored in the same way. The resulting spectra show the same characteristics; the only difference is the time scale. The hydrolytic activity of the proteasomes is much lower than that of chymotrypsin—hours instead of minutes.

b. Hydrolysis in Mixed Liposomes

Hydrolysis of the substrate lipid DODA-EO$_2$-S **6** by chymotrypsin and proteasomes was monitored in mixed liposomes consisting of 1–2% of the ligand lipid **6** mixed with 98–99% of dimyristoylphosphatidylcholine (DMPC). These functionalized mixed liposomes were prepared by sonication in phosphate buffer. The enzymatic hydrolysis of the ligand lipid **6** was initiated by injection of the proteins into the liposomes solutions. Figure 17 shows the fluorescence spectra obtained during hydrolysis of DODA-EO$_2$-S **6** incorporated in liposomes by chymotrypsin. Again, the fluorescence spectrum of the bound chromophore shows its fluorescence maximum at 395 nm (Fig. 17a).

During the hydrolysis of the substrate headgroup of lipid **6**, the intensity of this band decreases, commensurate with an intensity increase of the fluorescence band for the released chromophore at 435 nm (Fig. 17b–j). In comparison with the hydrolysis of the water-soluble substrate **2** in solution, the time necessary for the hydrolysis of

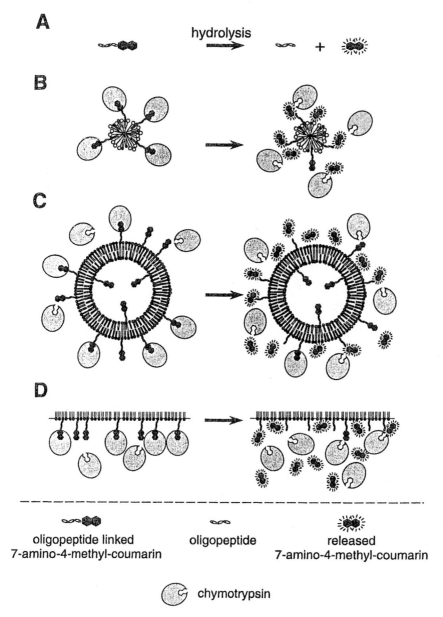

Fig. 15 Schematic representation of the hydrolysis of the soluble and lipid-linked substrate Suc-Ala-Ala-Phe-AMC by chymotrypsin in four different model systems: (A) water-soluble substrate; (B) DODA-EO$_2$-S (**6**) in micelles; (C) DODA-EO$_2$-S (**6**) in liposomes; (D) DODA-EO$_2$-S (**6**) in monolayers.

the substrate lipid **6** in liposomes is much longer. Presumably, the slower hydrolysis is a result of the hindered accessibility of the membrane-bound substrate headgroup. This slow hydrolysis leads to a slow photobleaching of the chromophore. Therefore no isosbestic point is visible in Fig. 17. In addition, it is noteworthy that full hydrolysis of all liposome-linked substrate headgroups is not possible, as shown by the remaining

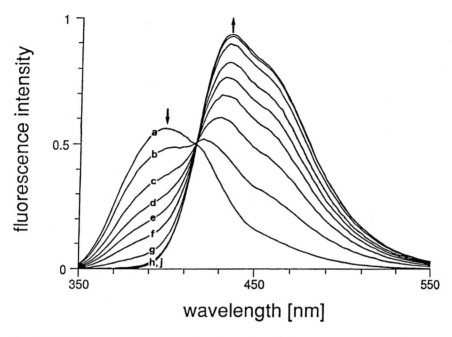

Fig. 16 Fluorescence spectra as a function of time during the hydrolysis of the water-soluble substrate Suc-Ala-Ala-Phe-AMC (**2**) by chymotrypsin: (a) $t = 0$ s; (b) $t = 10$ s; (c) $t = 90$ s; (d) $t = 170$ s; (e) $t = 250$ s; (f) $t = 360$ s; (g) $t = 600$ s; (h) $t = 1\,200$ s; (j) $t = 2\,400$ s; $T = 23°$C; phosphate buffer, pH 7.5; enzyme concentration $= 5$ μg/mL.

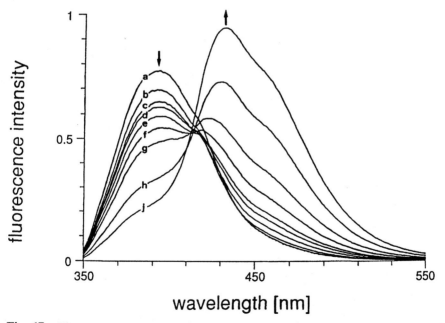

Fig. 17 Fluorescence spectra as a function of time during the hydrolysis of the substrate lipid DODA-EO$_2$-S (**6**) by chymotrypsin in mixed liposomes (DMPC/substrate lipid (**6**) $= 98:2$): (a) $t = 0$ min; (b) $t = 10$ min; (c) $t = 1$ h; (d) $t = 2$ h; (e) $t = 3$ h; (f) $t = 6$ h; (g) $t = 10$ h; (h) $t = 23$ h; (j) $t = 47$ h; $T = 23°$C; phosphate buffer, pH 7.5; enzyme concentration $= 100$ μg/mL.

fluorescence intensity at 395 nm. This partial hydrolysis is a result of the system used, because some of the substrate headgroups are trapped inside the liposomes (approximately 40%) and, therefore, are not accessible to the enzyme. But all the peptide headgroups at the outer surface of the liposomes are recognized and cleaved by chymotrypsin, which proves that the enzyme can specifically interact with its substrate in organized systems.

The hydrolysis experiments with both enzymes in homogeneous solution showed that the chymotryptic activity of proteasomes is much lower than the activity of chymotrypsin itself. Increasing the temperature can significantly increase the hydrolytic activity of the multienzyme complexes, which in nature show their highest catalytic turnover at approximately 90°C. Therefore hydrolysis of DODA-EO$_2$-S **6** in liposomes by proteasomes was performed at 70°C, where the liposomes are stable but the hydrolysis occurs at a rate that can be monitored. Figure 18a shows the fluorescence spectrum of the liposomes before addition of the proteasomes (covalently bound chromophore).

After incubation of these liposomes with proteasomes for 3 days, about 60% of the substrate headgroups were cleaved by the enzyme (Fig. 18b). This result clearly demonstrates that the proteasomes, as highly organized cylindrical multienzyme complexes, can also specifically recognize and hydrolyze membrane-bound substrates linked to the membrane with a long flexible spacer. It has to be pointed out that the length and flexibility of the spacer are essential: in comparable experiments with spacer-free substrate lipids, only chymotrypsin was able to interact. In the case of the complex barrel-shaped proteasomes, the membrane-bound functional headgroups could not reach the binding pockets.

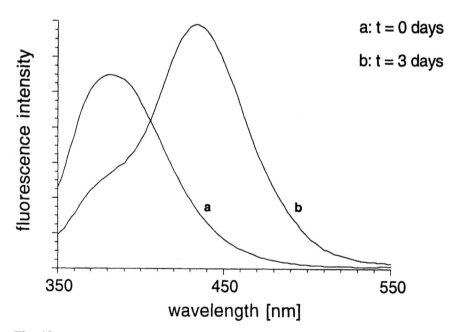

Fig. 18 Fluorescence spectra as a function of time during the hydrolysis of the substrate lipid DODA-EO$_2$-S (**6**) by proteasomes in mixed liposomes (DMPC/substrate lipid (**6**) = 99:1); MOPS (morpholinopropansulton) buffer, $T = 70$°C, pH 7.5; proteasome concentration = 300 μg/mL.

2. Fixation of Chymotrypsin at Inhibitor Lipid—Containing Liposomes

The interaction of chymotrypsin with the inhibitor lipid headgroup of DODA-EO$_2$-I **7** was investigated in monolayers and giant vesicles. The chymotrypsin was labeled with fluorescein isothiocyanate (FITC) so that it could be detected upon binding to the inhibitor lipid **7**–containing model membranes. The specific binding of FITC-labeled chymotrypsin to inhibitor lipid–containing liposomes was performed with giant vesicles. These liposomes (diameter 10–40 μm) are visible with a light microscope. Therefore, interaction of the labeled protein with the liposomes could easily be detected with a normal fluorescence microscope. Giant vesicles made of 10% DODA-EO$_2$-I **7** and 90% DMPC were prepared in a sealed Teflon chamber with glass windows directly under the microscope. After a 30-min incubation of these liposomes with FITC-labeled chymotrypsin, the excess protein was removed and the liposomes were imaged with the florescence microscope. A schematic representation of the molecular recognition process between the fluorescein-labeled chymotrypsin and the inhibitor lipid–containing liposomes is shown in Fig. 19 together with real images of the giant vesicles as observed in the light microscope in the phase-contrast and the fluorescein filter modes.

The outer membrane of the liposomes, visible in the phase-contrast mode (Fig. 19A), shows very intense fluorescence in the fluorescein filter (Fig. 19B). This

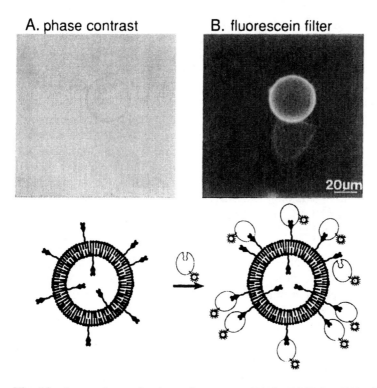

A. phase contrast   **B. fluorescein filter**

20µm

Fig. 19 Interaction of giant liposomes (90% DMPC, 10% DODA-EO$_2$-I) with fluorescein-labeled chymotrypsin: (A, left) image of a giant liposome in the phase contrast mode; (B, right) image of a giant liposome with fluorescein-labeled chymotrypsin as seen with the fluorescein filter after 30 min of incubation with the fluorescence-labeled enzyme.

demonstrates that the FITC-labeled chymotrypsin is linked to the membrane surface. In contrast, no vesicles with fluorescent membranes could be detected after incubation of pure DMPC liposomes with FITC-labeled chymotrypsin. Also, incubation of the normal DODA-EO$_2$-I $\underline{7}$–containing liposomes but with inhibited FITC-labeled chymotrypsin (presaturated with the soluble inhibitor alanine-alaninephenylalanine-chloromethylketone) does not yield fluorescent membranes. As expected, the enzyme cannot bind to the surface of the vesicles under these conditions. These control experiments demonstrate that the intense membrane fluorescence, visible in Fig. 19B, is due only to the specific interaction of chymotrypsin with the inhibitor headgroups of DODA-EO$_2$-I exposed at the surface of the liposome-membranes. Similar experiments were conducted with FITC-labeled proteasomes. After labeling of these multienzyme complexes with about 100 FITC molecules per enzyme complex, the proteasomes were fluorescent, but electron microscopy revealed that the cylindrical particle structure was no longer intact. Proteasomes with a smaller number of FITC molecules per particle (10) retained their structural integrity, but they did not show enough fluorescence to detect their surface binding with the fluorescence microscope.

D. Fixation and Orientation of Proteasomes from *Thermoplasma acidophilum* at Inhibitor Lipid–Containing Monolayers

In the preceding sections, it was demonstrated that chymotrypsin, a small "single-molecule" enzyme, and proteasomes, sophisticated barrel-shaped multienzyme complexes, are both able to interact specifically with functionalized liposomes and hydrolzye the substrate. To investigate the ability of the complex proteasomes to interact specifically with inhibitor lipid–containing monolayers, high-resolution electron microscopy (EM) was used (27). Binding of these multienzyme complexes to the functionalized model membranes, containing inhibitor lipid DODA-EO$_2$-I $\underline{7}$ was supposed to induce a preferred orientation of the proteasomes relative to the membrane. This orientation may indicate the position of the active site in the proteasome particles. Several methods have already been used to obtain structural information about the proteasomes. The best results (see Fig. 20) so far have been obtained by image analysis of single proteasome particles on carbon-coated grids via electron micrographs (23). The proteins are shown either side on, the barrel with its four rings visible (indicated in Fig. 20 by an arrow), or end on, looking into the barrel, with just the outer ring visible, indicated by a ring in Fig. 20.

Several investigations (28) demonstrated that proteasomes show their highest enzymatic activity only in the completely assemble state. They indicated that the active site for the chymotryptic activity is probably located either between one α-ring and one β-ring or between the two inner β-rings of the proteasomes. Assuming that this is correct, there are two possible locations for the catalytic center. The active site could be (1) inside the particle, accessible only through the top of the proteasome, or (2) at the outside surface of the protein between the different rings. Depending on the location of the binding site, specific interaction of the proteasomes with inhibitor lipid–containing monolayers could induce a preferred orientation of the particles relative to the membrane. Two orientations are possible: location of the active site inside the particle might induce orientation of the proteasomes perpendicular to the membrane surface (Fig. 21A); parallel orientation of the protein relative to the membrane would suggest binding of the inhibitor headgroups to the outside of the

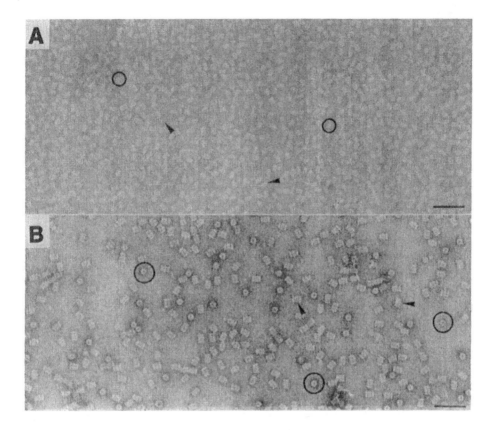

Fig. 20 Two electron micrographs at different magnifications of negatively stained (uranyl acetate) proteasome particles on carbon-coated grids, seen side on (A, rectangular shape; arrows) and end on (B, ring shape; circles).

proteasomes (Fig. 21B). Thus, experiments showing the interaction of proteasomes with inhibitor lipids were devised in order to help elucidate the position of the active site in the enzyme particle.

Monolayer-proteasome complexes for the EM investigations were prepared using a Langmuir film balance with a subphase volume of only 5 mL. After spreading of the appropriate amount of the inhibitor lipid DODA-EO$_2$-I **7**, pure or in mixture with stearyl-oleyl-phosphatidyl-choline (SOPC) on a buffer subphase, the monolayer was compressed to a surface pressure of 20 mN/m. The proteasomes were injected into the subphase underneath the inhibitor-functionalized monolayer and incubated for approximately 24 h. The resulting monolayer-proteasome complexes were lifted onto carbon-coated, hydrophobic electron microscopy grids using the Langmuir-Schäfer technique (Fig. 22). The samples were negatively stained with uranyl acetate, air dried, and subsequently imaged with an electron microscope.

Figure 23 shows electron micrographs of two different areas of transferred lipid-proteasome layers (90% SOPC and 10% DODA-EO$_2$-I **7**). Densely packed proteasomes, which are all oriented perpendicular to the membrane as schematically represented in Fig. 21A, are visible in the electron micrographs in Fig. 23A. This was the

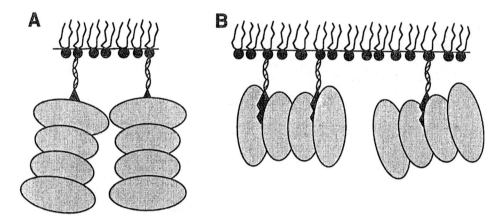

Fig. 21 Cartoon representations of the possible orientation of proteasomes relative to the membrane surface after specific interaction of the membrane-incorporated inhibitor lipid headgroups with the active site of the proteasomes: (A) location of the active site inside the particles induces a preferred orientation of the proteasomes perpendicular to the membrane and (B) binding of the membrane-in-corporated inhibitor headgroups to an active site on the outside of the particles induces a preferred orientation of the proteasomes parallel to the membrane (27).

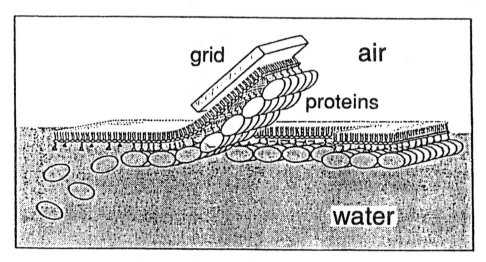

Fig. 22 Schematic representation of a Langmuir-Schäfer transfer of monolayer-protein complexes onto electron microscopy grids. After placing a hydrophobized electron microscopy grid onto the lipid layer, the grid is incubated for about 5 min and subsequently lifted. The transferred protein layer can be negatively stained with uranyl acetate.

orientation found in nearly all of the experiments. Figure 23B shows one of the fewer examples in which within the aggregates the protesomes are oriented parallel to the membrane.

But in contrast to "hanging" and not crystallizing edge-on proteasomes (Fig. 23A), these side-on aggregates of proteasomes are regulary ordered in a two-dimensional lattice. With electron microscopic (EM) image analysis, it could be demonstrated that these proteasome domains are indeed two-dimensional (2-D)

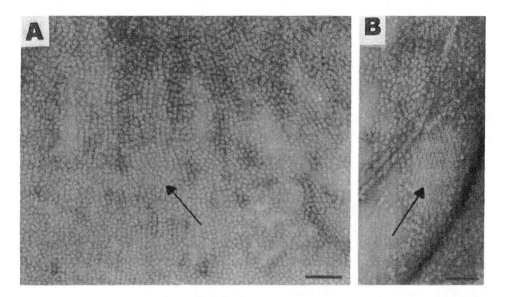

Fig. 23 Electron micrographs of proteasome domains oriented at a monolayer of SOPC/DODA-EO$_2$-I (**7**) (9:1). (A) In this image, the hexagonally packed proteasomes within the domains are all oriented perpendicular to the membrane. (B) In this image, all aggregated proteasomes are oriented parallel to the membrane. The regular order within these protein aggregates indicates that these aggregates are two-dimensional protein crystals.

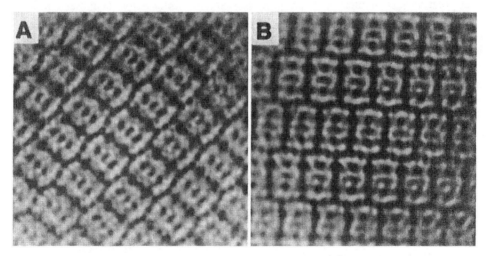

Fig. 24 Side view of two proteasome domains after averaging and image analysis: (A) average based on 22 particles; (B) average based on 67 particles (27).

protein crystals (Fig. 24A and B). All four rings in one particle can be clearly identified, but the positions of the single subunits in one ring (the mass distribution) are not exactly defined. The resolution that could be obtained with these images was limited to 16 Å because the number of particles in one domain was too small to obtain higher resolution.

The preceding results show that proteasome aggregates with only one orientation of the enzyme barrels could not be obtained. Both types of orientations could be observed in the preparation. The fact that hanging proteasomes were found indicated overwhelmingly that the binding sides sit inside the barrels. Their crystallization in this form is prevented by the easy rotation of the proteasomes around the binding tether. On the other hand, the long hydrophilic spacer of **7** allows a side-on aggregation too. Now rotation of the barrel is prevented, or at least hindered, thus allowing the 2D crystallization of the multienzyme complex. In the meantime, highly diffracting three-dimensional crystals of proteasome-inhibitor complexes from *T. acidophilum* could be prepared. With these crystals, the structure of these multienzyme complexes was determined with a resolution of 3.4 Å (29). Thus, the position of the active sites could be exactly defined. The position agrees with the EM results: the catalytic centers are located in the interior of the particle and can be unequivocally assigned to the 14 β-subunits. This points to the fact that the active sites for the chymotryptic activity are accessible only from the ends of the cylinders. This is also an agreement with all investigations with space-free substrates and inhibitors. They are able to interact with chymotrypsin as a small protein but did not show any interaction with the barrel-shaped structure of the multienzyme complex.

IV. SELECTIVE OPENING OF PHASE-SEPARATED LIPOSOMES

A. Death of a Tumor Cell—Is Simulation Possible?

The great variety of biomembrane processes and the interaction of molecular and cellular effects are reflected in a rather fascinating event, namely the death of a tumor cell (8). Figure 25 illustrates such an event: a cell that cannot escape from an active immune system.

The scanning electron micrograph shows the attack of an activated macrophage on a tumor cell (Fig. 25A). On a molecular level, via antigen-antibody interaction, the macrophage recognizes the target cell and establishes close membrane contact. Then, on a cellular level, patching of the antigen-antibody complexes presumably takes place. In the subsequent membrane processes, which are not yet completely understood, the defensive cell is able to drill holes in the originally stable membrane of the tumor cell. As a result, the cytoplasmic interior leaks out, which causes the death of the cell. What remains is a ghost cell (Fig. 25B). Can we mimic the process?

Looking for possibilities to simulate such a "hole drilling" attack of an activated macrophage on cells, one can consider liposomes as their simplest spherical model. Using vesicles, it is important not to destroy the whole liposome by membrane-destabilizing processes. Only labile domains within a stable lipid matrix should be selectively pulled out. One way to prevent spontaneous and complete destabilization of the membrane is to use phase-separated liposomes in which the matrix of these vesicles is stabilized by polymerization.

B. Ordered Membranes from Monomeric and Prepolymerized Amphiphiles

In the past few years, numerous methods for stabilizing model membranes have been developed, mainly by using polymeric systems (8). On the one hand, polymerizable amphiphiles can be used to self-organize in monomeric model membranes that can be polymerized in the oriented state (Fig. 26A). On the other hand, the use

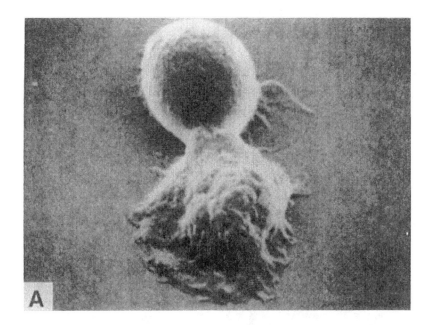

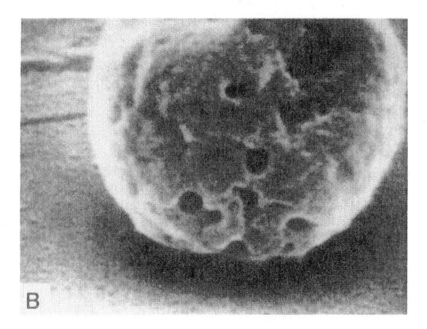

Fig. 25 Scanning electron micrograph of a tumor cell attacked by an activated macrophage. (A) Close contact between tumor cell (upper part) and activated macrophage via antibody-antigen binding (lower part). (B) Holes in the membrane of the tumor cell after the attack, probably caused by phospholipase-induced cleavage of membrane phospholipids (see Sec. II.A, Fig. 3).

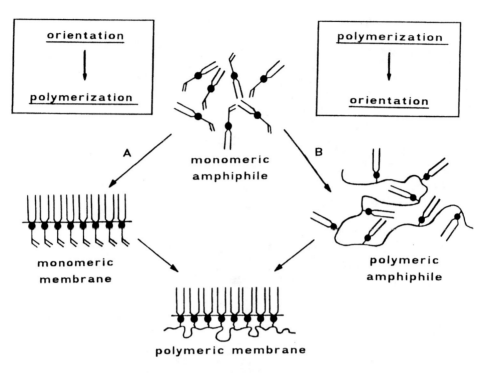

Fig. 26 Strategies for the preparation of the polymeric model membranes. (A) Orientation of the monomers in model membranes with subsequent polymerization. (B) Polymerization of the monomers in isotropic solution and subsequent orientation of the polymeric amphiphiles in model membranes (30b).

of oligomeric amphiphiles or the incorporation of spacer groups into the polymeric amphiphile to decouple the motions of the polymer chains from the motions of the ordered membrane allows the formation of membrane from prepolymerized amphiphiles (Fig. 26B) (30). The polyreaction behaviour of amphiphiles with polymerizable units was first studied in monomolecular films (31). In 1979, the formation of polymerized vesicles by using a diacetylene group containing lipid was mentioned for the first time (32). In the meantime, many polymerizable groups have been incorporated in various lipid structures to prepare polymerized liposomes. The development of structural variations of polymerizable amphiphiles (lipids, surfactants) and their use in monolayers, liposomes, multilayers, and liquid crystals have enriched the monomeric and polymeric landscape and have helped to bridge the (still too wide) gap between polymer science and life science (8, 33).

A survey of possible molecular architectures used to stabilize liposomes is given in Fig. 27. Vesicles can be stabilized not only by polymerization of polymerizable lipids but also through membrane-spanning lipids (34) or polycondensation reactions. Another concept for the preparation of polymeric membranes is based on the use of amphiphiles that have already been polymerized, as shown in Fig. 26. For this, one can use either macromolecules with low degrees of polymerization or polymers containing flexible spacers between the hydrophobic units and the polymer chain, decoupling the different motion and organization pro-

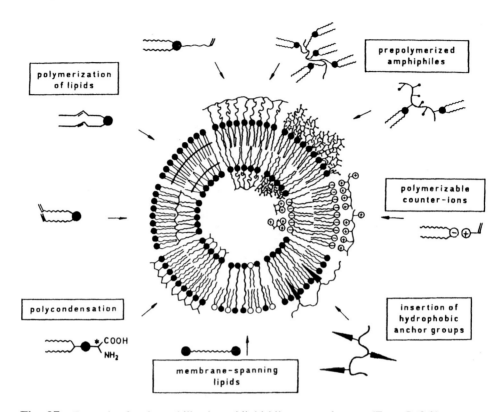

Fig. 27 Strategies for the stabilization of lipid bilayer membranes. (From Ref. 8.)

cesses taking place between the oriented alkyl groups and the much more flexible polymer chain. Nature commonly uses this spacer concept, especially when several functional groups jointly introducing a reaction have to be kept close together. Already used for polymer therapeutics (35), a similar concept has also been applied for the synthesis of liquid crystalline polymers (36). Nature uses polymers too to stabilize cell membranes. Polymer chains and polymer nets provide them with flexibility and stability (cytoskeleton). Different proteins, such as acetin, tubulin, and ankryn, form fiberlike microfilaments and microtubules are used and anchored at the inside of the plasma membrane. A simulation with liposomes can be achieved by the interaction of water-soluble polymers via hydrophilic or ionic processes (8). Hydrophobized water-soluble polymers (polymeric micelles) were used especially by Okumura and Sunamoto to prepare so-called artificial cell walls or wigs for liposomes (8, 33).

C. Formation of Domains in Vesicle Membranes

Phase-separated liposomes can be used to simulate the hole formation process in cells as shown in Fig. 25. In order to avoid complete destruction of these liposomes by phase separation leading to fusion, it is important that they are composed of a polymerized lipid matrix surrounding the phase-separated domains of cleavable and nonpolymerizable lipids, as shown schematically in Fig. 28.

polymeric lipid matrix

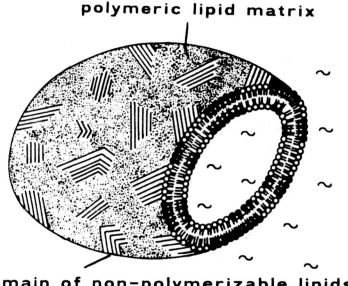

domain of non-polymerizable lipids

Fig. 28   Scheme of a phase-separated liposome. The polymerized lipids form the stable matrix, surrounding the domains of nonpolymerizable, cleavable lipids.

The formation of domains can be predetermined by choosing suitable incompatible lipid mixtures. Examples are mixtures of hydrocarbon and fluorocarbon lipids (37). Mixed liposomes made from DMPC and the fluorocarbon amine **8** (ratio 5:1) show two phase transitions, e.g., in differential scanning calorimetry (DSC) measurements. This observation clearly demonstrates that these two lipids are immiscible because these phase transition temperatures are identical to those of the pure respective lipids (37a). Freeze-fracture electron micrographs reveal that phase-separated liposomes with domains of the individual lipids are formed from the lipid mixture instead of two different liposomes populations (37a).

Some of the functional lipids needed for the decorking experiments are shown in Fig. 29. To obtain a stable polymeric matrix, the polymerizable lipids **9** and **10** with hydrocarbon chains, respectively, have been used. Small unilamellar liposomes can be prepared from both lipids by means of sonication. The tightness of the polymeric matrix of vesicles formed from the lipid **9** was demonstrated by time-dependent release measurements using eosin as a fluorescent marker (8).

D.   Corkscrews for Corked Liposomes

In order to simulate the attack on a tumor cell by a macrophage, as shown in Fig. 25, it is important not to destroy the whole liposome by membrane-destabilizing processes. Only labile domains within a stable lipid matrix should be selectively dissolved (8). Figure 30 shows different processes for selectively opening such stabilized phase-separated polymeric liposomes.

The polymeric components of these mixed liposomes form the stabilizing, shape-maintaining matrix that cannot be attacked by the various "opening processes." In order to destroy the labile domains, at least two methods are conceivable: dissolving

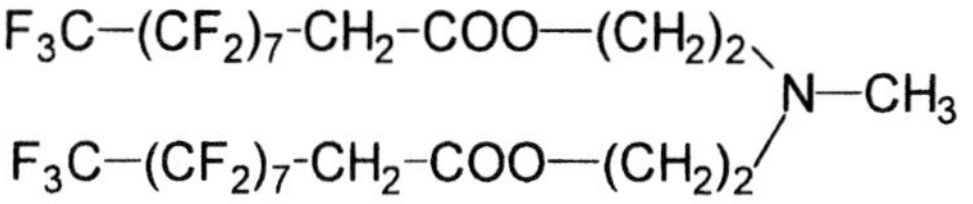

F₃C—(CF₂)₇-CH₂-COO—(CH₂)₂

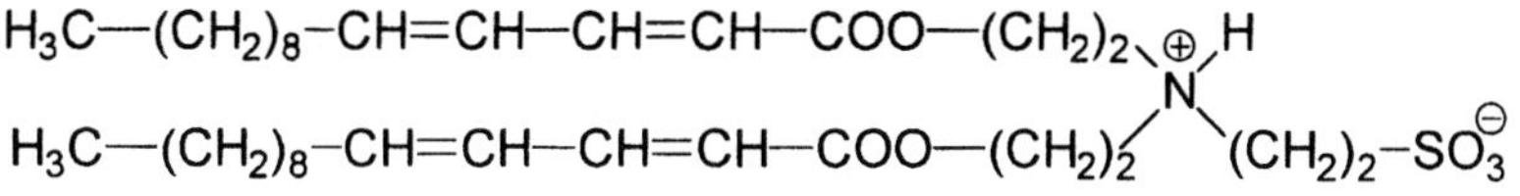

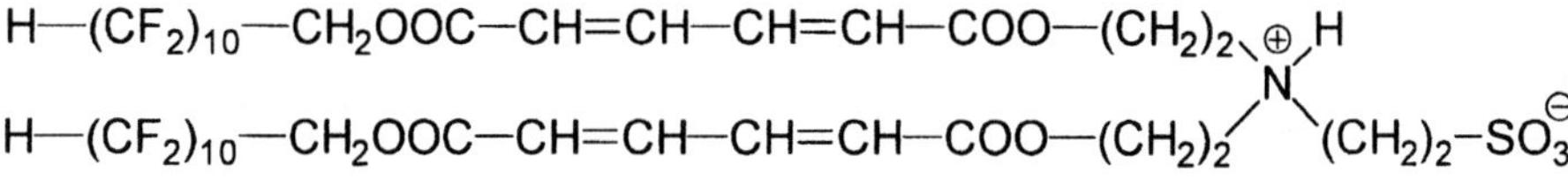

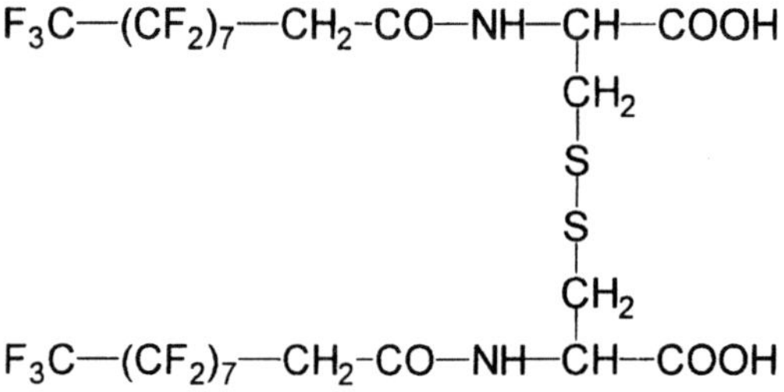

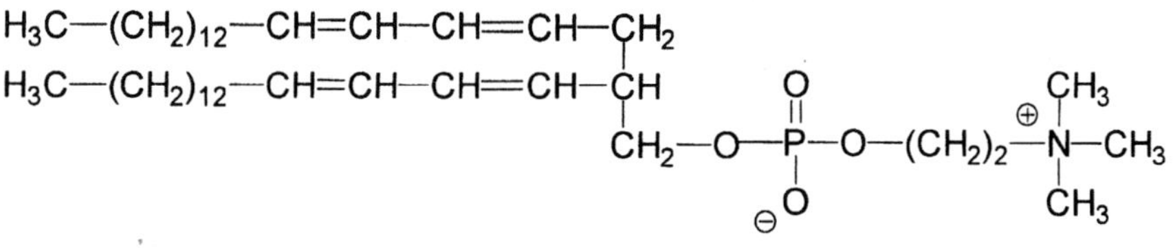

F₃C—(CF₂)₇-CH₂-COO—(CH₂)₂ $\rangle$ N—CH₃   **8**

H₃C—(CH₂)₈—CH=CH—CH=CH—COO—(CH₂)₂

H₃C—(CH₂)₈—CH=CH—CH=CH—COO—(CH₂)₂ $\overset{\oplus}{N}$ (CH₂)₂—SO₃$^{\ominus}$ **9**

H—(CF₂)₁₀—CH₂OOC—CH=CH—CH=CH—COO—(CH₂)₂   **10**

H—(CF₂)₁₀—CH₂OOC—CH=CH—CH=CH—COO—(CH₂)₂ $\overset{\oplus}{N}$ (CH₂)₂—SO₃$^{\ominus}$

F₃C—(CF₂)₇—CH₂-CO—NH—CH—COOH
　　　　　　　　　　　　|
　　　　　　　　　　　CH₂
　　　　　　　　　　　|
　　　　　　　　　　　S
　　　　　　　　　　　| **11**
　　　　　　　　　　　S
　　　　　　　　　　　|
　　　　　　　　　　　CH₂
　　　　　　　　　　　|
F₃C—(CF₂)₇—CH₂-CO—NH—CH—COOH

H₃C—(CH₂)₁₂—CH=CH—CH=CH—CH₂

H₃C—(CH₂)₁₂—CH=CH—CH=CH—CH   **12**

　　　　　　　　　　　　　　CH₂—O—P—O—(CH₂)₂—$\overset{\oplus}{N}$—CH₃

with $\overset{O}{\underset{\ominus O}{\|}}$ P and CH₃ groups

Fig. 29 Lipids used for the preparation of phase-separated and polymerized vesicles.

or cleaving the lipids. On the one hand, it is possible to dissolve the lipids of those domains with surfactants (detergents), thus forming mixed micelles, or with organic solvents (e.g., acetone or ethanol). On the other hand, lipid cleavage through enzymatic or chemical reactions can be carried out. This offers a great number of synthetic ways to open the labile lipid domains in polymerized liposomes locally. Reductive cleavage of the water-insoluble disulfides **11**, for example, with dithiothreitol (DTT), leads to the formation of the corresponding partially water-soluble, single-chain thiols.

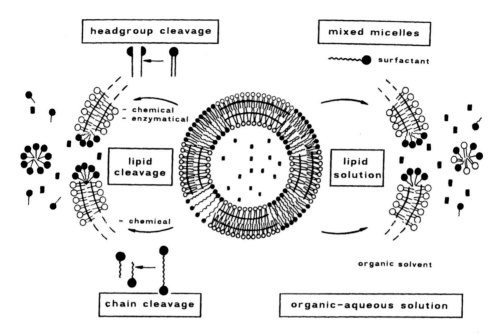

Fig. 30 Methods for uncorking phase-separated liposomes composed of a stable polymeric matrix with labile lipid domains. (From Ref. 8.)

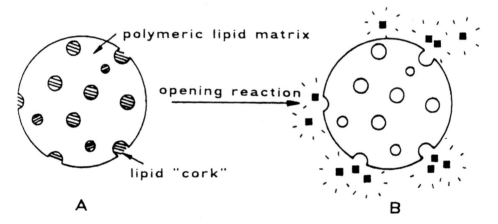

Fig. 31 Hole formation in polymerized vesicles containing labile cleavable lipid domains (schematic). (A) No fluorescence of the entrapped marker owing to self-quenching. (B) Fluorescence of the released and therefore diluted marker after hole formation in the membrane.

The combination of the polymerizable taurine lipids **9** with the cork-forming amphiphile **11** results in stable polymeric liposomes with cleavable "cork" domains (8). Evidence for the opening of the labile domains in mixed monomeric and polymeric liposomes can be obtained by electron microscopy as well as by the release of entrapped eosin. Figure 31 illustrates the opening process schematically.

A drastic change in release behavior occurs if phase-separated liposomes from the lipid **9** and the domain-forming fluorocarbon "cork" **11** (9:1) are compared before and after UV polymerization. In the monomeric case, the domains of **11** are rapidly destroyed by DTT during the disulfide cleavage, but because of the lateral diffusion of the monomeric lipids, the membrane defects are healed rather quickly. After cleavage of the whole cork component (about 1 h), the release rate decreases to that without DTT (8). After addition of DTT to the analogous polymerized liposomes, the stiff polymeric matrix with extremely low lateral lipid mobility is decorked and eosin is rapidly and almost completely released. Does this uncorking experiment really lead to hole formation comparable to the process shown in Fig. 25 for cells? Or does it just destroy the lipid membrane of the vesicles?

The hole formation in phase-separated liposomes was confirmed by electron microscopy. Figure 32 shows a scanning electron micrograph of a phase-separated polymeric liposome made from the taurine lipid **9** and the cleavable cystine lipid **11** before and after the destruction of the labile domains (8). As can be seen in Fig. 32B, the lipid bilayers do not collapse. The function of the cytoskeleton of the cell,

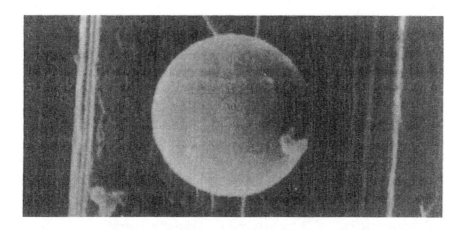

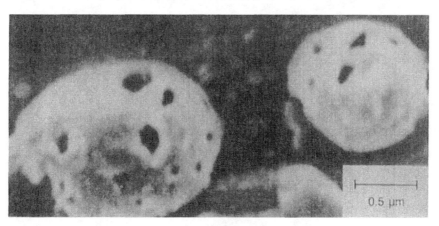

Fig. 32 Scanning electron micrograph of phase-separated, polymerized liposomes (90 mol% of polymerizable lipid **9** and 10 mol% cleavable - S- S- lipid **11**). (Top) Liposome before the cleavage reaction. (Bottom) Perforated membrane after the cleavage reaction (uncorked liposome!).

i.e., stabilization of the membrane, is thus fulfilled by the polymeric matrix of the liposomes. The pore formation observed is not an artifact of the preparation technique for the scanning electron micrograph. This could be proven by means of freeze-fracture electron microscopy, again showing the holes.

The opening reactions discussed so far proceed under nonphysiological conditions and with a relatively high concentration of cleavage reagent. The biochemical opening, in contrast, uses phospholipases and allows the process to take place under physiological conditions. As already discussed in Sec. II.A, phospholipase A_2 specifically catalyzes the hydrolysis of the ester bond at the C-2 position of the L-α-glycerophospholipids. This enzymatic reaction can be transferred to model membranes composed of polymerized and natural lipids (38). To investigate this cleavage reaction in liposomal membranes, the polymerizable lecithin **12** and DPPC (1:1) were used. Here, DPPC acts as the cork, cleavable by enzymes. The liposomes formed from these two lipids do not show domains because they are homogeneously miscible. Polymerization, however, induces a phase separation. It has to be pointed out that the highly specific phospholipase A_2 attacks the ester group of the dienoly lecithin neither in the monomeric nor in the polymeric form. Indeed, liposomes prepared only from the dienoly lecithin **12** remain tight and do not release eosin after addition of the enzyme. Thus, the release of 6-carboxyfluorescein from mixed liposomes is clearly due only to the cleavage of the saturated lecithin (DPPC). After the addition of phospholipase A_2, a dramatic change in the release behavior occurs. The polymerized phase-separated vesicles rapidly lose the entrapped marker; after 10 min, 90% of the vesicle contents are released. Again, the hole formation could be confirmed by EM measurements (8).

1. Synthetic Corkscrews for Cells?

The experiments discussed so far show that it is indeed possible to simulate the biological process of local cell membrane opening by using phase-separated, partially polymerized liposomes. Figure 33 leads back to a biological system and does not

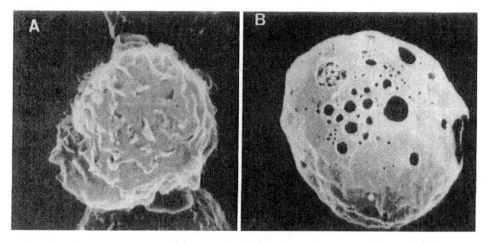

Fig. 33 Scanning electron micrograph of the hole formation in leukemia cells (39a, b). (A) Untreated leukemia cell. (B) Leukemia cell after treatment with O-methyllysolecithin.

merely show an additional polymerized liposome before and after it has been uncorked. This scanning electron micrograph shows leukemia cells of a cell culture before (A) and after (B) treatment with special surfactants, namely methyl ether derivatives of lysophospholipids, which are being discussed as antitumor agents (39). The result of the uncorking of the cell by the synthetic lysolipid can clearly be seen in Fig. 33B. It is impressive to see the similarity between such a "synthetic" uncorking process of cells and the uncorking process of partially polymerized vesicles shown in Fig. 32. Even the sizes of the two very different uncorked "vesicles" in Figs. 32 and 33B are comparable, namely about 3–5 μm.

As impressive as the synthetic uncorking process of cells and polymerized liposomes may appear, it is still miles away from the natural mechanisms. The hole-drilling process as shown in Fig. 25 for a tumor cell using an activated macrophage as a "corkscrew" is very likely caused by the enzyme phospholipase A_2 (as discussed in Sec. II.A). Bound to the membrane of the activated macrophage, PLA-2 may hydrolyze DPPC domains, leading to the cell-destroying lysolecithins. In whatever form the natural process of uncorking cells by activated macrophages may proceed, it can at least be simulated using liposomes and cells.

REFERENCES

1. Ronald Searle, the scalpel-sharp British cartoonist, illustrates that even for the old gold diggers it was an absurd adventure to pan for gold by the light of the silvery neon in the main street of downtown Klondike. (R Searle. From Frozen North to Filthy Lucre. New York: Viking Press, 1964.) Already in 1788 (!) Georg Christoph Lichtenberg, the great critical philosopher and physicist from Göttingen, was clear and outspoken in this respect when he characterized one of his colleagues: "he was still hanging at his University like a wonderful chandelier without a tiny candle lit for years" (8).
2. Erwin Chargaff, biochemist and critical essayist, despicts this as follows: "In Science today smaller and smaller rooms are furnished more and more luxuriously and completely." (E Chargaff. Bemerkungen. Stuttgart: Klett-Cotta, 1981, p 33.)
3. JM Lehn. Supramolecular Chemistry. Weinheim: VCH, 1995.
4. Supramolecules ("Übermolekeln") were defined by Wolf in 1937 to describe associations of molecules whose function was based on their organization: (a) KL Wolf, F Frahm, H Harms. Über den Ordnungszustand der Moleküle in Elüssigkeiten. Z Phys Chem B 36:17, 1937. (b) KL Wolf, R Wolff. Übermolekeln. Angew Chem 61:191, 1949. In this contribution the term supramolecular systems was used not only for small molecular complexes of single molecules, such as substrate-receptor complexes. It was also applied to large molecular aggregates (high aggregation numbers) whose function is again based on their organization. An ideal system to exemplify this definition is the biomembrane (see Fig. 1).
5. F Vögtle. In: C Elschenbroich, F Hensel, H Hopf, eds. Supramolekulare Chemie. Stuttgart: Teuber, 1989.
6. JL Atwood, JE Davies, DD MacNicol, F Vögtle, eds. Comprehensive Supramolecular Chemistry Pergamon, in preparation. Several volumes have already been edited, e.g. JP Sauvage. Templating. In MW Hosseini, ed. Self-Assembly and Self-Organization. New York: Plenum, New York, 1996.
7. E Fischer. Einfluß der Configuration auf die Wirkung der Enzyme. Ber Dtsch Chem Ges 27:2985, 1994.
8. H Ringsdorf, B Schlarb, J Venzmer. Molecular architecture and function of polymeric oriented system: models for the study of organization, surface recognition and dynamics of biomembranes. Angew Chem Int Ed Engl 27:113–158, 1988.

9. M Ahlers,W Müller, A Reichert, H Ringsdorf, J Venzmer. Specific interactions of proteins with functional lipid monolayers—Ways of simulating biomembrane processes. Angew Chem Int Ed Engl 29:1269–1285, 1990.

10. A Ullman. In: G Roberts, ed. Langmuir Blodgett Films. New York: Plenum, 1990.

11. D Demus, J Goodby, GW Gray, H Spiess,V Vill, eds. Handbook of Liquid Crystals.Vols 1–3. Weinheim: Wiley-VCH, 1998.

12. M Ebert, OH Schoenherr, JH Wendorff, H Ringsdorf, P Tschirner. Sandics: A new class of mesophases, displayed by highly substituted rigid rod polyesters and polyamides. Liq Cryst 7:63, 1990.

13. D Adam, P Schuhmacher, J Simmerer, L Häußling, K Siemensmeyer, KH Elzbach, H Ringsdorf, D Harrer. Fast photoconduction in the highly ordered columnar phase of a discotic liquid crystal. Nature 371:141, 1994.

14. W Müller, H Ringsdorf, E Rump, G Wildburg, X Zhang, L Angermeier,W Knoll, M Liley, J Spinke. Attempts to mimick docking processes of the immune system: Recognition-induced formation of protein multilayers. Science 262:1706, 1993.

15. A Müller, A Dress, F Vögtle. From simplicity to complexity in chemistry. In: S Denzinger, A Dittrich,W Paulus, H Ringsdorf, eds. Function Based on Organization and Recognition. Weinheim: VCI, 1996, p 63.

16. (a) R Ahuja, PL Carnso, D Möbius,W Paulus, H Ringsdorf, G Wildburg. Formation of molecular strands by hydrogen bonds at the gas-water interface. Molecular recognition and quantitative hydrolysis of barbituric acid lipids. Angew Chem Int Ed Engl 32:1033, 1993. (b) TM Bohanon, S Denzinger, R Fink,W Paulus, H Ringsdorf, M Weck. Barbituric acid/2,4,6-triaminopyrimidine aggregates in water and their competitive interaction with a monolayer of barbituric acid lipids at the gas-water interface. Angew Chem Int Ed Engl 34:58, 1995. (c) TM Bohanon, P-L Caruso, S Denzinger, R Fink, D Möbius,W Paulus, JA Preece, H Ringsdorf, D Schollmeyer. Molecular Recognition-Induced Function and Competitive Replacement by Hydrogen-Bonding Interactions: Amphiphilic Barbituric Acid Derivatives, 2,4,6-Triaminopyrimidine, and Related Structures at the Air-Water Interface. Langmuir 15:174, 1999.

17. M Waite. The Phospholipases. New York: Plenum, 1987.

18. (a) A Plückthun, EA Dennis. Activation, aggregation and product inhibition of cobra venom phospholipase A_2 and comparison with other phospholipases. J Biol Chem 260:11099, 1985. (b) MK Jain, DV Jahagirdar. Action of phospholipase A_2 on bilayers. Effect of fatty acid and lysophospholipid additives on the kinetic parameters. Biochim Biophys Acta 814:313, 1985. (c) MK Jain, MR Egmont, HM Verheij, R Apitz-Castro, R Dijkman, GH De Haas. Interaction of phospholipase A_2 and phospholipid bilayers. Biochim Biophys Acta 688:341, 1982.

19. (a) DW Grainger, A Reichert, H Ringsdorf, C Salesse. An enzyme caught in action: Direct imaging of hydolytic function and domain formation of phospholipase A_2 in phosphatidylcholine monolayer. FEBS Lett 252:74, 1989. (b) DW Grainger, A Reichert, H Ringsdorf, C Salesse. Hydrolytic action of phospholipase A_2 in monolayers in the phase transition region: Direct observation of enzyme domain formation using fluorescein microscopy. Biochim Biophys Acta 1023:365, 1990.

20. (a) J-M Lehn, M Mascal, A Decian, J Fischer. Molecular recognition directed self-assembly of ordered supramolecular strands by cocrystallization of complementary molecular components. J Chem Soc Chem Commun 479–491, 1990. (b) JA Zerkowsky, CT Seto, DA Wierda, GM Whitesides. Design of organic structures in the solid state: Hydrogen-bonded molecular "Tapes." J Am Chem Soc 112:9025–9026, 1990.

21. (a) AP Arrigo, K Tanaka, AL Goldberg,WJ Welch. Identity of the 19 S prosome particle with the large multifunctional protease complex of mammalian cells (the proteasome). Nature 331:192, 1988. (b) B Dahlmann, M Rutschmann, H Reinauer. Activation of the multicatalytic proteinase from rat skeletal muscle by fatty acids or sodium dodecyl sulphate. Biochem J 228:171, 1985. (c) PE Falkenburg, C Haass, PM Kloetzel, B Niedel, F Kopp, L

Kuehn, B Dahlmann. *Drosophila* small cytoplasmic 19 S ribonucleoprotein is homologous to the rat multicatalytic proteinase. Nature 331:190, 1988. (d) JR Harris. Release of macromolecular protein component from human erythrocyte ghosts. Biochim Biophys Acta 150:534, 1968.

22. JA Rivett. Purification of a liver alkaline protease which degrades oxidatively modified glutamine synthetase. J Biol Chem 260:12600, 1985.

23. B Dahlmann, F Kopp, L Kühn, B Niedel, G Pfeifer, R Hegerl, W Baumeister. The multicatalytic proteinase (prosome) is ubiquitous from eukaryotes to archaebactaria. FEBS Lett 251:125, 1989.

24. A Grizwa, B Dahlmann, Z Cejka, U Santarius, W Baumeister. Localization of a sequence motif complementary to the nuclear localization signal in proteasomes from *Thermoplasma acidophilum* by immunoelectron microscopy. J Struct Biol 109:168, 1992.

25. P Zwickl, J Kleinz, W Baumeister. Critical elements in proteasome assembly. Nat Struct Biol 1:765, 1994.

26. B Dahlmann, L Kuehn, A Grizwa, P Zwickl, W Baumeister. Biochemical properties of the proteasome from *Thermoplasma acidophilum*. Eur J Biochem 208:789, 1992.

27. A Reichert, H Ringsdorf, P Schuhmacher, W Baumeister, T Scheybani. In: JP Sauvage, MW Hosseini, eds. Comprehensive Supramolecular Chemistry. Vol. 9. 1996, pp 313–350.

28. P Zwickl, A Grziwa, G Pühler, B Dahlmann, F Lottspeich, W Baumeister. Primary structure of the *Thermoplasma* proteasome and its implications for structure, function, and evolution of the multicatalytic proteinase. Biochemistry 31:964, 1992.

29. J Löwe, D Stock, B Jap, P Zwickl, W Baumeister, R Huber. Crystal structure of the 20 S proteasome from the archaeon *T. acidophilum* at 3.4 Å resolution. Science 268:533, 1995.

30. (a) T Kunitake, N Nakashima, K Takarabe, M Nagai, A Tsuge, H Yanagi. Vesicles of the polymeric and monolayer membranes. J Am Chem Soc 103:5945, 1981. (b) R Elbert, A Laschewsky, H Ringsdorf. Hydrophilic spacer groups in polymerizable lipids: Formation of biomembrane models from bulk polymerized lipids. J Am Chem Soc 107:4134, 1985.

31. (a) A Cemel, T Fort, J Lando. Polymerization of vinyl stearate multilayers. J Polym Sci Part A-1 10:2061, 1972. (b) R Ackermann, O Inacker, H Ringsdorf. Polymerisation von Acryl- und Methacrylverbindungen in monomolekularen Schichten. Kolloid Z Z Polym 249:1118, 1971.

32. D Day, H Hub, H Ringsdorf. Polymerization of mono- and bi-functional diacetylene derivatives in monolayers at the gas-water interface. Isr J Chem 18:325, 1979.

33. (a) Y Okumura, J Sunamoto. Supramolecular assembly of functionalized lipids. Supramol Sci 3:171, 1996. (b) E Sackmann. Supported membranes: Scientific and practical applications. Science 271:43, 1996. (c) T Kunitake. Ultrathin films as biomimetic films. Polym J 23:613, 1991. (d) T Kunitake. Compr Supramol Chem 9:351, 1996.

34. (a) J Fuhrhop, R Bach. Monolayer lipid membranes (MLMs) from bolaamphiphiles. Adv Supramol Chem 2:25, 1992. (b) SC Kushwaha, M Kates, G Sprott, I Smith. Novel complex polar lipids from the methanagenic archaebacterium *Methanospirillum hungatei*. Science 211:1163, 1981. (c) J Fuhrhop, K Ellermann, H David, J Mathieu. Monomolecular membranes from synthetic macrotetrolides. Angew Chem Int Ed Engl 21:440, 1982. (d) H Bader, H Ringsdorf. Liposomes from α,ω-dipolar amphiphiles with a polymerizable diyne moiety in the hydrophobic chain. J Polym Sci Polym Chem Ed 20:1623, 1982.

35. (a) L Gros, H Ringsdorf, H Schupp. Polymere Antitumormittel auf molekularer und zellulärer Basis? Angew Chem Int Ed 20:305–325, 1981. (b) R Duncan. Drug-polymer conjugates: Potential for improved chemotherapy. Anticancer Drugs 3:175, 1992. (c) R Duncan. Polymer therapeutics for tumour specific delivery. Chem and Jnd 7:237, 1997. (d) P Vasey, SB Kaye, R Duncan, LS Murray, S Burtles, J Cassidy. Phase I Clinical

and Pharmacokinetic Study of PK1 [N-(2-Hydroxypropyl)methacrylamide Copolymer Doxorubicin]: First member of a New Class of chemotherapeutic agents – Drug Polymer Conjugates. Clinical Cancer Res 5: 83, 1999.

36. (a) A Roviello, A Sirigu. Mesophasic structures in polymers. A preliminary account on the mesophases of some poly-alkanoates of p,p''-di-hydroxy-α,α'-di-methyl benzalazine. J Polym Sci Polym Lett Ed 13:455, 1975. (b) H Finkelmann, H Ringsdorf, J Wendorff. Model consideration and example of enantiotropic liquid crystalline polymers. Makromol Chem 179:273, 1978.

37. (a) R Elbert, T Folda, H Ringsdorf. Saturated and polymerizable amphiphiles with fluorocarbon chains. Investigation on monolayers and liposomes. J Am Chem Soc 106:7687, 1984. (b) T Kunitake, S Tawaki, N Nakashima. Excimer formation and phase separation of hydrocarbon and fluorocarbon bilayer membranes. Bull Chem Soc Jpn 56:3235, 1983.

38. R Büschl, B Hupfer, H Ringsdorf. Polyreaction in oriented systems. Mixed monolayers and liposomes from natural and polymerizable lipids. Makromol Chem Rapid Commun 3:589, 1982.

39. (a) P Munder, M Modolell, W Bausert, H Oettgen, O Westphal. Alkyllysophospholipids in cancer therapy. Prog Cancer Res Ther 16:411, 1981. (b) R Andreesen, A Schulz, U Costabel, P Munder. Tumor cytotoxicity of human macrophages after incubation with synthetic analogs of 2-lysophospatidylcholine. Immunobiology 163:335, 1982. (c) H Eibl. Phospholipids as functional building blocks of biological membranes. Angew Chem Int Ed Engl 23:257, 1987.

9

Self-Assemblies of Biomembrane Mimics

T. Kunitake
Kyushu University, Fukuoka, Japan

I. BIOMEMBRANE MIMICS BASED ON SYNTHETIC BILAYERS

Current attempts to mimic the structure and functions of biomembranes can be traced back to 1964, when Bangham and Horne (1) demonstrated that the bilayer structure is formed spontaneously from aqueous dispersions of egg yolk lecithin. It was shown by this finding that, among the constituents of biomembranes, the lipid component by itself could readily produce cell-like organization upon simple dispersion in water. Their initial publication was followed by extensive efforts to develop artificial cells and biomembrane mimics. Typical examples include that of Gebicki and Hicks (2), who observed formation of globular aggregates from dispersed thin films of oleic acid and linoleic acid. Hargreaves and Deamer (3) described vesicular aggregates from 1:1 mixtures of saturated fatty acids of C_{12} to C_{20} chains and single-chain lysolecithin. These aggregates were, however, not sufficiently stable and devoid of definite evidence for the existence of the isolated bilayer structure. Since that time, there have been numerous investigations of how to produce cell-like aggregates from synthetic and semiartificial analogues.

At this point, we must make a clear distinction between liquid crystalline dispersion and bilayer membrane. As pointed out earlier by Gray and Winsor (4) physical properties (such as viscosity and stability) of the mesophases are determined by the intermicellar forces, i.e., the lattice forces rather than forces from jointing or close packing. The intermicellar forces are indispensable for maintaining molecular organization of liquid crystalline dispersions. On the other hand, a bilayer membrane should be able to exist as an isolated entity without relying on the lattice forces for keeping its structural integrity. Thus, the formation of bilayer membranes requires a self-assembling capacity greater than that of liquid crystalline dispersions.

In this section, we discuss how we can design totally synthetic amphiphiles that show self-assembling properties similar to those of biolipid molecules. It will be seen that the formation of a stable bilayer membrane is achieved by a large variety of synthetic compounds. Their structural variation far exceeds that of biological origin. We can conclude that spontaneous bilayer assembly is a widely observable physicochemical phenomenon that covers biotic and abiotic worlds.

Subsequently, we describe examples of functional conjugation of such synthetic bilayer membranes with biological macromolecules. Biomembranes are composed of lipid bilayers, proteins, and polysaccharides. Interaction of biomembranes and DNA is deeply entwined in some cellular processes. Thus, it is important to see how synthetic bilayer systems interact with proteins and DNA.

II. CHEMICAL CONSTITUTIONS OF REPRESENTATIVE SYNTHETIC BILAYER MEMBRANES

When we started investigations of biomembrane mimics, we made an assumption that the unique structure of the polar headgroup of biolipid molecules was determined by the biosynthetic and physiological requirements rather than by the physical chemistry of membrane formation. A readily available compound that satisfies the latter requirement was didodecyldimethylammonium bromide $\underline{1}$.

$$CH_3(CH_2)_{11} \diagdown \overset{+}{\underset{N}{}} \diagup CH_3$$
$$CH_3(CH_2)_{11} \diagup \diagdown CH_3 \qquad Br^-$$

1

This compound and related double-chain ammonium halides have been known to show exceptional lyotropic liquid crystalline behavior in water (5) that is quite different from those of other micelle-forming surfactants. The didodecyl derivative 1 gives transparent aqueous dispersions on sonication or on heating. When this dispersion was negatively stained with uranyl acetate and observed by electron microscopy, single-walled vesicles and multiwalled vesicles with a layer thickness of 30–50 Å were found (Fig. 1) (6). Their aggregation characteristics such as critical aggregate concentration and molecular weight and other physicochemical measurements were consistent with the existence of such large, ordered aggregates. This finding was the first example of a totally synthetic bilayer membrane, and it was clear that unique two-dimensional self-assembly was not a monopoly of biolipid molecules.

Subsequent investigations revealed that bilayer formation is observable for a variety of synthetic amphiphiles. Their (now conventional) molecular structures are summarized in Fig. 2 as combinations of molecular modules. Bilayer-forming double-chain amphiphiles are close analogues of natural lecithin molecules and are composed of molecular modules of tail, connector, space, and head. The tail may consist of normal alkyl chains of C_{10} to C_{20} and the polar headgroup can be cationic, anionic, nonionic, or zwitterionic. The connector portion that links the hydrophobic alkyl chains and the hydrophilic headgroup helps promote alignment of the alkyl chains, and a spacer unit that intervenes between the headgroup and

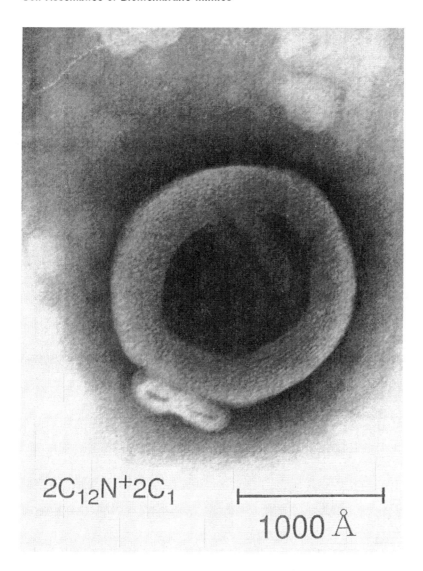

$2C_{12}N^+2C_1$

1000 Å

Fig. 1 Electron micrograph of a bilayer structure formed by didodecyldimethylammonium bromide (negatively stained with UO_2^{2+}).

the connector unit exerts a significant influence on the molecular orientation within the bilayer.

The design principle of bilayer-forming single-chain amphiphiles is illustrated in Fig. 3. Characteristic molecular structures of smectic liquid crystalline materials and micelle-forming surfactants are combined to give novel amphiphiles that form aqueous bilayer aggregates. The development of the bilayer structure is improved with increasing chain length of the flexible tail (C_n portion). The kind of the headgroup could be varied as much as those of the double-chain counterpart, and the structure of the rigid segment is extensively variable. An extensive list of this class of single-chain compounds has been compiled (7). As a somewhat different design, Cho and

(a)

Tail | Rigid segment | Spacer | Head

(b)

Tail | Connector | Head | Spacer

(c)

Tail | Connector | Spacer | Head

(d)

Tail | Connector | Spacer | Head

Fig. 2 Structural elements (modules) of bilayer-forming amphiphiles: (a) single-chain amphiphile, (b) double-chain amphiphile, (c) triple-chain amphiphile, and (d) quadruple-chain amphiphile.

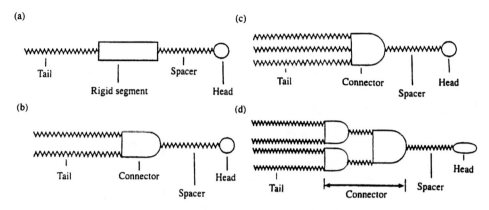

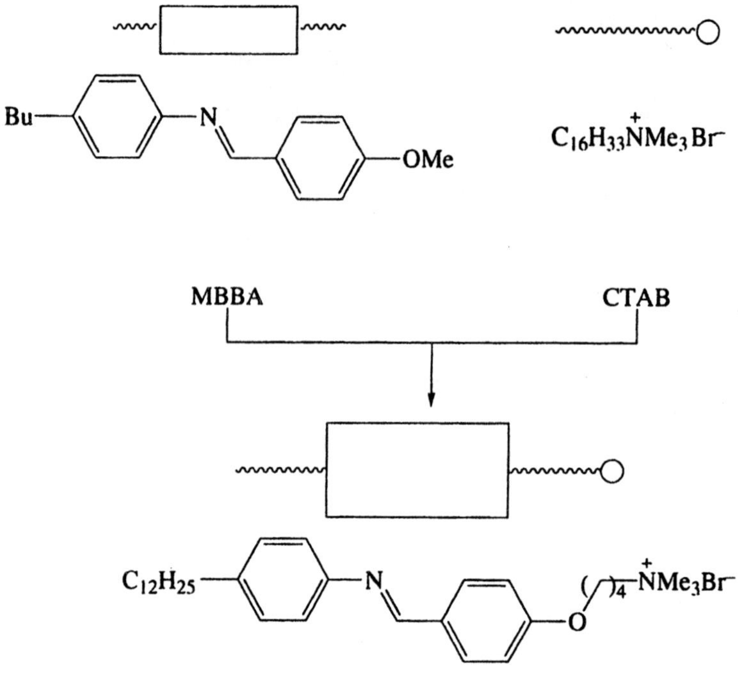

Bu—⬡—N=CH—⬡—OMe

$C_{16}H_{33}\overset{+}{N}Me_3\,Br^-$

MBBA   CTAB

$C_{12}H_{25}$—⬡—N=CH—⬡—O—$(CH_2)_4$—$\overset{+}{N}Me_3Br^-$

Fig. 3 Design principle of bilayer-forming, single-chain amphiphile.

coworkers (7a) reported that an ammonium derivative of cholesterol formed a bilayer vesicle. It has to be noted that the "rigid segment" implies the structural unit that promotes molecular alignment by aromatic stacking and other strong intermolecular interactions. Menger and Yamasaki (8) reported that a combination of a very long alkyl chain (C_{24}–C_{35}) and a multicharged oligoethylenimine moiety gave bilayer aggregates.

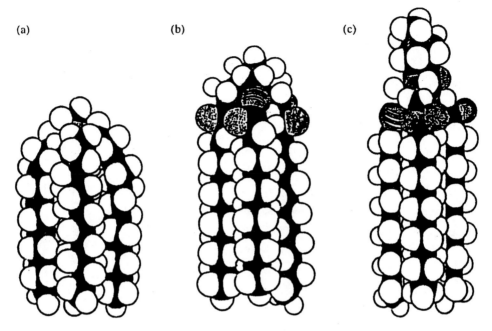

Fig. 4 CPK model of triple-chain amphiphiles with different connectors. The parallel alignment of the alkyl chains in (a) is inferior to that in (b) and (c).

The hydrophile-lipophile balance is not the most crucial factor that determines bilayer formation. For example, triple-chain ammonium amphiphiles that would form reversed micelles according to the balance give well-developed bilayer structures if the molecular structure allows superior alignment of alkyl chains; see Fig. 4. Certain degrees of hydrophile-lipophile imbalance are accommodated by the stabilization gained by molecular alignment. Even some quadruple-chain ammonium amphiphiles, **2**, are capable of forming stable bilayer membranes (9). The native counterpart of these amphiphiles is four-chained cardiolipin, which, however, cannot form a stable bilayer assembly by itself.

$$CH_3(CH_2)_{13}OC(CH_2)_2$$
$$CH_3(CH_2)_{13}OCCHNHC(CH_2)_2CNH(CH_2)_4$$
$$CH_3(CH_2)_{13}OCCHNHC(CH_2)_2CNHCHCO(CH_2)_3\overset{+}{N}(CH_3)_3 \quad Br^-$$
$$CH_3(CH_2)_{13}OC(CH_2)_2$$

2

Some archaeobacteria contain bipolar macrocyclic membrane lipids, **3**, that span the bilayer thickness. Their synthetic counterparts include ionene-type amphiphiles and bolaamphiphiles, **4**.

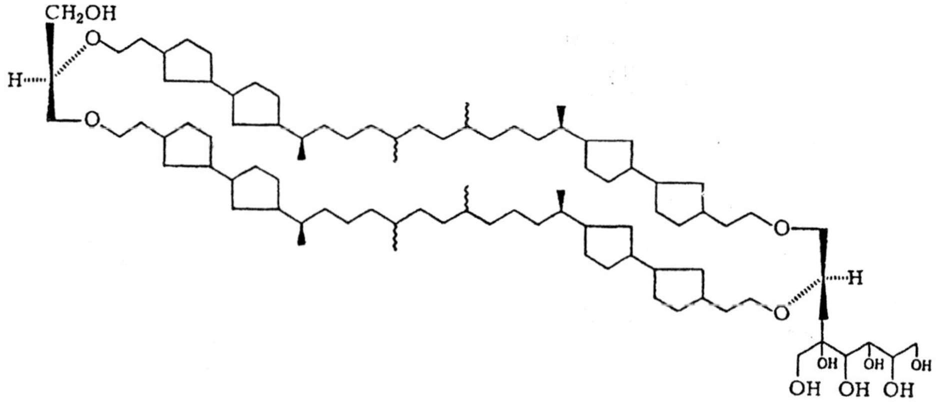

3

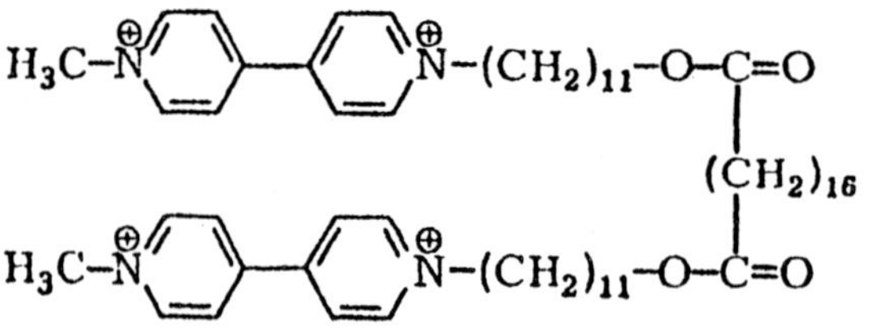

4

III. PHYSICOCHEMICAL CHARACTERISTICS OF SYNTHETIC BILAYER MEMBRANES

Aggregate morphologies and molecular packing of synthetic bilayer membranes are very similar to those of biolipid bilayer membranes. This is true because of the essentially identical physicochemical characteristics of biological and artificial bilayers. In the following we describe phase transition and phase separation of double-chain synthetic amphiphiles.

The phase transition from gel to liquid crystal is commonly observed for aqueous dispersions of synthetic bilayer membranes. Differential scanning calorimetry (DSC) gives transition enthalpy changes characteristic of melting of crystalline alkyl chains. The transition temperature, T_c, is consistent with the phase transition ranges estimated by other methods such as reaction rates in bilayer matrices, nuclear magnetic resonance (NMR) line broadening of the crystalline bilayer, fluorescence spectral changes of probe molecules, turbidity changes, and the extent of positron annihilation. DSC data (T_c, enthalpy change, and entropy change) for a whole family of bilayer-forming, double-chain amphiphiles were compiled (10,11).

Phase separation in synthetic bilayer membranes has been detected by the presence of separate component peaks in DSC, by λ_{max} shifts in absorption spectroscopy, and by excimer emission in fluorescence spectroscopy.

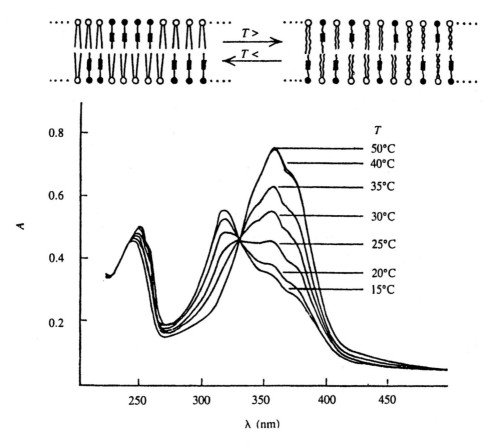

Fig. 5 Absorption spectral changes as a result of phase separation in a 1:10 mixed bilayer of an azobenzene amphiphile and dioctadecyldimethylammonium bromide.

Azobenzene-containing bilayers give large peak shifts caused by relative orientation and aggregation of the azobenzene units in bilayers. This is applied to the detection of phase separation (Fig. 5) (12). A 10:1 mixed dispersion of azobenzene-containing amphiphile and dialkyldimethylammonium amphiphile (n = 18) shows λ_{max} at 355 nm in the high-temperature region, indicating the presence of the isolated azobenzene species. On lowering the temperature, a new peak due to formation of the clustered azobenzene component appears at 315 nm. The spectral change corresponds to the phase-separated azobenzene component.

IV. LESS CONVENTIONAL BILAYER MEMBRANES

A. Fluorocarbon Bilayer Membrane

A variety of less conventional bilayer membranes have been developed on the basis of the preceding design principle. A first example is fluorocarbon amphiphiles with ammonium headgroups (13). Replacement of hydrocarbon chains with fluorocarbon chains in the tail module of Fig. 2 provides fluorocarbon amphiphiles, eg., **5** and

6, that show aggregation morphologies that are not much changed from the hydrocarbon counterpart. Phospholipid analogues with fluorocarbon chains were synthesized by Riess and coworkers (14) and were shown to form highly stable vesicles (longer shelf life and higher thermal stability). Liang and Hui (15) prepared a monolayer vesicle from a bolaamphiphile with a semifluorocarbon chain. Immiscibility of hydrocarbon and fluorocarbon chains leads to an interesting molecular organization. Mixed vesicles of hydrocarbon and fluorocarbon components possess unique phase-separated domains, and the latter monolayer vesicle may contain all the fluorocarbon halves on the outer side of the vesicle membrane.

$$CF_3(CF_2)_7C_2H_4\overset{O}{\overset{\|}{C}}OCH_2CH_2$$

$$CF_3(CF_2)_7C_2H_4\underset{O}{\overset{\|}{C}}OCH_2CH_2 \overset{N^+}{\underset{CH_3}{\diagdown}} \quad Br^-$$

$$2C_{11}^{F} - de - N^+$$

5

$$CF_3(CF_2)_7C_2H_4\overset{O}{\overset{\|}{C}}OCH_2CH_2$$

$$CH_3(CH_2)_{10}\underset{O}{\overset{\|}{C}}OCH_2CH_2 \overset{N^+}{\underset{CH_3}{\diagdown}} \quad Br^-$$

$$C_{11}^{F} , C_{12} - de - N^+$$

6

Figure 6 shows controlled phase separation in mixed hydrocarbon-fluorocarbon bilayers in the case of triple-chain amphiphiles. Pure hydrocarbon bilayers give totally phase-separated domains with pure fluorocarbon bilayers. These two components become at least partially miscible upon addition of a third amphiphile that is composed of hydrocarbon and fluorocarbon chains.

B. Bilayer Formation Assisted by Complementary Hydrogen Bonds

The molecular modules of bilayer-forming amphiphiles need not all be connected by covalent bonding. Hydrogen bonding in artificial molecular systems is presumably most effective in the solid state or in noncompetitive organic media, and its use in aqueous media appears not profitable. Contrary to this supposition, we found that hydrogen bonding acted efficiently in aqueous dispersion when combined with hydrophobic association (16).

An equimolar mixture of **7** and **8** gave a stable, transparent dispersion in water upon ultrasonication. Transmission electron microscopy showed the presence of disklike aggregates with diameters of several 100 Å and a thickness of about 100 Å.

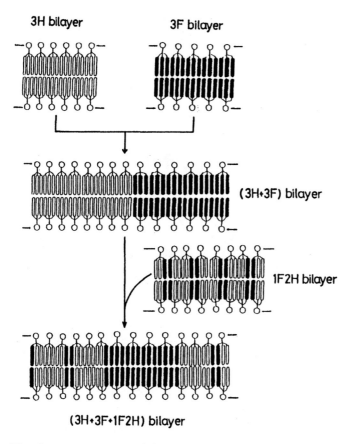

Fig. 6 Schematic illustrations of the mixing behavior of fluorocarbon (3F) and hydrocarbon (3H) bilayers in multicomponent bilayers.

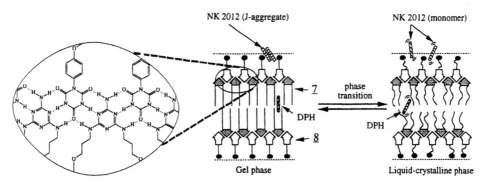

Fig. 7 Schematic illustration of the physical states of bilayer **7**·**8** and membrane-bound dyes.

DSC measurements and spectral properties of membrane probes clearly showed the presence of a gel-to-liquid crystal phase transition similar to those of the conventional bilayer membrane (Fig. 7). Such self-assembly assisted by the interplay of hydrogen bonding and hydrophobic force produces many exciting possibilities. For instance, the hydrophilic head unit involved in the original bilayer assembly is replaced by a second headgroup that gives a more stabilized assembly; one part of the headgroups may be reversibly modified.

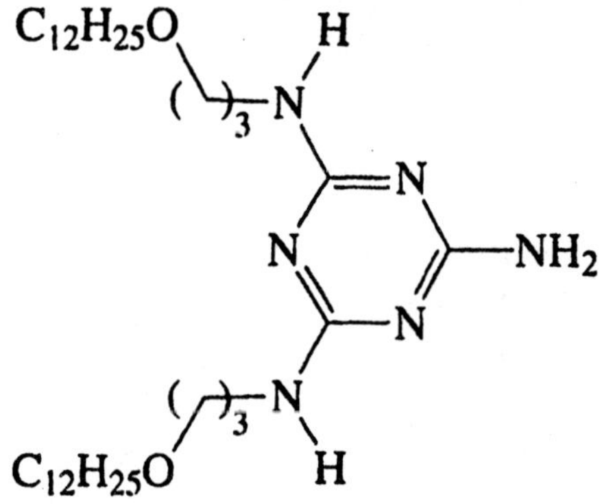

7

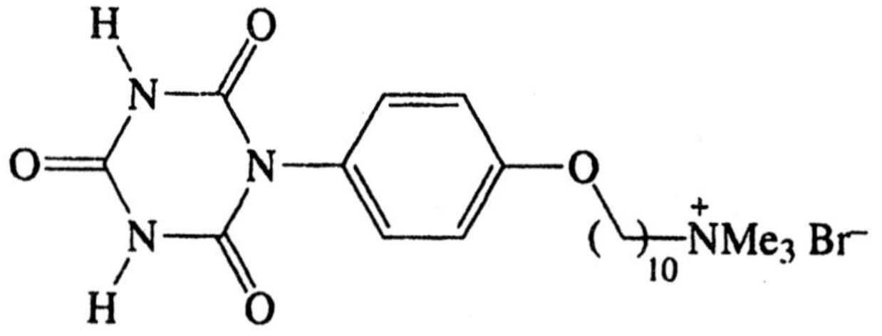

8

C. Bilayer Membrane of Ion-Paired Amphiphiles

Amphiphiles composed of single alkyl chains usually cannot form stable bilayer assemblies, unlike their double alkyl chain counterparts. This situation may be remedied by ion pairing with oppositely charged components. Jokela et al. (17) showed that the combination of oppositely charged surfactants gave rise to lamellar phases, and Kaler and others (18) claimed spontaneous formation of vesicles on mixing of cationic and anionic single-tail surfactants. As another example, cetyltrimethylammonium bromide, a common micelle-forming surfactant, was converted to a bilayer-forming polyelectrolyte when the bromide ion was replaced with polyacrylate (19).

D. Bilayer Assembly Assisted by Metal Complexation

Metal complexation improves self-assembly of amphiphilic molecules, and weakly associating molecules can be transformed to stable bilayer aggregates. For example, dithiooxamide derivative **9** cannot form stable monolayer membranes in water; however, its 1:1 complex with Cu^{2+} gives a coordination polymer that produces a vesicle-like structure (20). Suh and coworkers (21) prepared a sulfonate amphiphile that contained two ligand sites. This compound formed bilayer membranes by itself and by complexation with metal ions (Fig. 8). The aggregate morphology varied

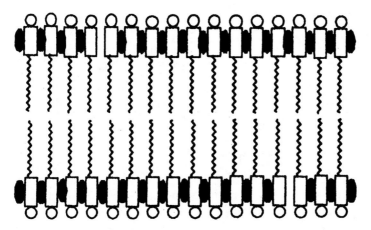

Fig. 8 A coordinately polymerized bilayer membrane.

depending on which metal ion (Fe^{2+}, Fe^{3+}, Co^{2+} or Co^{3+}) coordinated with the bilayer membrane.

9

Yonezawa et al (22) prepared an ammonium amphiphile with the cyanobenzene unit at the other molecular end, which formed a 2:1 complex with $PtCl_2$. Electron microscopy of an aqueous dispersion of the bis-ammonium complex showed formation of rigid rods with a diameter of 5 nm without staining. The rod is probably made of bundles of linearly extended atomic Pt chains as illustrated in Fig. 9 (22).

V. INCORPORATION OF ENZYMES INTO SYNTHETIC BILAYER MEMBRANES

Ringsdorf discussed in the previous chapter the interaction of chymotrypsin, a hydrolytic enzyme, and proteasomes with functionalyzed liposomes. Chymotrypsin can readily cleave a peptide moiety that is covalently attached to a lipid component in a lecithin liposome. Earlier, O'Brien and coworkers (23) demonstrated successful incorporation of purified, delipidated rhodopsin (Rh) into membrane vesicles composed of photopolymerized dienoylphosphatidylcholine and dioleoyl-phosphatidylcholine. It was shown that approximately 50% of the Rh molecules functions in a natural manner. This result was an indication that sensitive vertebrate proteins could be usefully incorporated into polymerized membrane bilayers.

Synthetic bilayer membranes can provide unique chemical organizations that are different from those of the biolipid. This feature is not fully exploited in the preceding examples. Flexibility in the molecular design of synthetic bilayer membranes should be useful in constructing new supramolecular conjugates of bilayers and proteins.

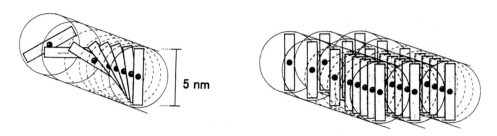

5 nm

Fig. 9   Rodlike morphologies of Pt-containing molecular membrane.

In the following, we discuss unique features arising from incorporation of enzymes into synthetic bilayer systems.

Immobilization of enzymes has been extensively investigated in order to take advantage of their specific catalytic activities for material conversion and biosensors. Organization of proteins at the air-water interface and on solid substrates attracted much fundamental interest, as discussed by Ringsdorf. Bilayer membranes dispersed in water are readily transformed to regular molecular films by simple casting (24). The interbilayer space of cast multibilayer films constitutes a hydrophilic, two-dimensional cavity that can incorporate a variety of water-soluble molecules. Because the bilayer components are regularly aligned in the cast film with closely packed headgroups, guest molecules may be incorporated in ordered arrangements.

As a typical example, we carried out a systematic study to place biochemically important porphyrin derivatives in specific spatial arrangements in matrices of ammonium bilayer membranes (25). Aqueous mixtures of anionic Cu(II) porphyrins and bilayer dispersions of double-chain ammonium amphiphiles were cast on Teflon sheets to produce regular multilayer films. The orientation of doped Cu(II) porphyrins is determined by anisotropy of the electron spin resonance (ESR) spectral patterns. It was found that the mode of porphyrin orientation was determined by the distribution of anionic substituents on guest porphyrins and the supramolecular structure of host bilayers. As shown in Fig. 10, type III porphyrins that possess evenly distributed anionic substituents are incorporated horizontally on the ammonium bilayer surface. Type I porphyrins, in which anionic substituents are localized on one side of the porphyrin ring, are incorporated in the spacer portion of the bilayer parallel to the molecular axis. A type II porphyrin with three sulfonate substituents gives a random orientation.

We extended this approach to anisotropic arrangements of globular proteins such as myoglobin, hemoglobin, cytochrome *c*, and horseradish peroxidase. Positively charged metmyoglobin is incorporated without denaturation (26) into a cast film of a phosphate bilayer membrane. The heme plane of metmyoglobin is oriented at an angle of 15–20° against the bilayer surface, as shown in Fig. 11. The observed orientation appears to optimize the electrostatic attraction between the positive protein surface and the negative membrane surface. The molar ratio of 1:40 (protein/amphiphile) corresponds to a situation in which regularly oriented metmyoglobin molecules almost completely cover the polar bilayer surface. The protein intercalation is also achieved by immersing a preformed film in an aqueous solution of myoglobin (27).

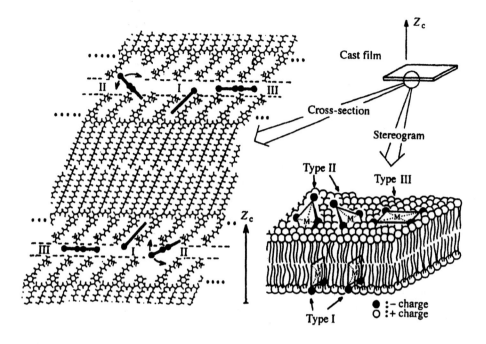

Fig. 10 Schematic representation of orientation of three types of anionic porphyrins in a cast multibilayer film.

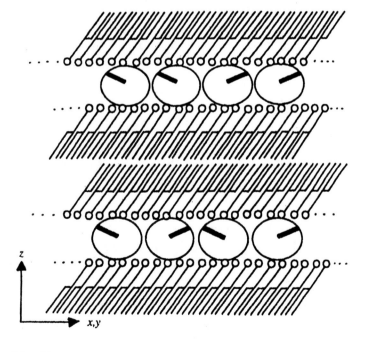

Fig. 11 Schematic illustration of met-Mb molecules immobilized in cast multilayer films. The protein molecules show one-dimensional ordering along the z axis. The heme plane is represented by filled rectangles.

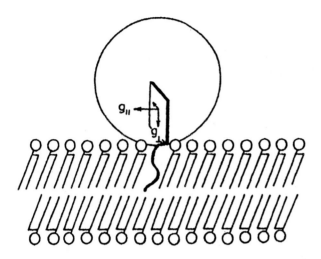

Fig. 12 Schematic illustration of anisotropic incorporation of lipid-anchored Mb into the DPCC bilayer membrane.

Orientational control is possible by using hydrophobic interaction. Hamachi and coworkers (28,29) synthesized myoglobin with a long alkyl chain attached to the heme cofactor. The binding assay based on gel filtration and ultrafiltration clearly indicated that the engineered myoglobin was efficiently bound to lecithin bilayer vesicles, unlike natural myoglobin. In this case, the modified myoglobin molecule is placed on the lipid bilayer surface in a fixed orientation by inserting the long alkyl chain into the lipid membrane interior as an anchor (Fig. 12), as demonstrated by anisotropic ESR patterns of cast lipid films.

VI. FUNCTIONAL CONVERSION OF PROTEINS INCORPORATED IN SYNTHETIC BILAYER MEMBRANES

Myoglobin is a dioxygen binding protein and does not display catalytic activity in the natural environment. It can be embedded in the membrane surface of mixed (ammonium and phosphate) amphiphiles together with dihydronicotinamide adenine dinucleotide (NADH) and flavin mononucleotide (FMN) (30). In this unique microenvironment, a myoglobin molecule can efficiently accept an electron from NADH via FMN and then release the electron catalytically to dioxygen or 1,2-naphthoquinone-4-sulfonate. Myoglobin is converted from an oxygen storage protein to a redox enzyme in the electron transport scheme of Fig. 13.

Subsequently, this unique assembly was employed to induce the aniline-hydroxylase activity of myoglobin by organizing NADH, FMN, and myoglobin on a cationic bilayer. Electron transfer from NADH to myoglobin is greatly accelerated by an ammonium bilayer membrane, whereas a phosphate membrane completely suppresses this process. Aniline is effectively converted to hydroxylamine with dioxygen as oxidant in the presence of a cationic bilayer. The membrane-bound FMN acted as an effective electron mediator from NADH to myoglobin, inducing, as a result, the hydroxylase activity of myoglobin (31).

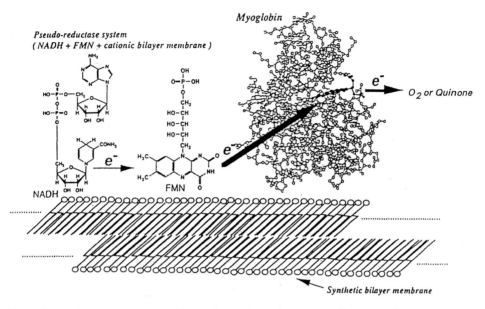

Fig. 13 Cofactor-protein assembly on the surface of synthetic bilayer membranes.

Cytochrome c (Cyt-c) is an electron-transporting protein in the mitochondrial respiratory chain. Peroxydase activity (peroxydation of o-methoxyphenol with hydrogen peroxide, Eq. 1) of Cyt-c is greatly enhanced when Cyt-c is electrostatically bound to a phosphate bilayer membrane (32). Oxidative N-demethylation of N,N-dimethylaniline Eq. 2 is another case of enhanced activity of membrane-bound Cyt-c. A lysine cluster on the surface of the Cyt-c molecule appears to be bound to phosphate bilayer membranes, causing a spin-state change of the iron heme in the active site. This change leads to enhanced demethylase activity. It is presumed from the kinetic analysis that the membrane-bound Cyt-c is activated either by enhanced affinity of Cyt-c for hydrogen peroxide or by additional enhancement of the catalytic activity. Synthetic bilayer membranes operate as active effectors through noncovalent interactions in addition to their ability to assemble multiple bioactive components (33,34).

VII. SYNTHETIC BILAYERS AND DNA

A. DNA Conjugates

There are two major reasons for studying the conjugate structure and interactions of DNA and cationic bilayers (35). The first is that conjugates are known to mimic certain characteristics of viruses by being efficient chemical carriers of genes (DNA sections) for delivery in cells. The second is that conjugates are models of aggregated DNA in two dimensions. DNA-membrane interactions should also provide clues to relevant molecular forces in the condensation of DNA in chromosomes and viral capsids.

The interaction of DNA and bilayer-forming amphiphiles has been investigated by Okahata and coworkers (36). They prepared DNA-lipid complexes by simply mixing aqueous solutions of DNA and a cationic amphiphile. The complex can be transformed to a self-standing cast film, which contains typical double-strand DNA structures and shows characteristic dye intercalation (36).

They subsequently prepared a related complex of DNA and a surface monolayer of cationic amphiphile, **10**, by using the Langmuir-Blodgett technique (Fig. 14). It was shown from the measurement of transferred weight that 95% of the monolayer was covered by DNA molecules (37).

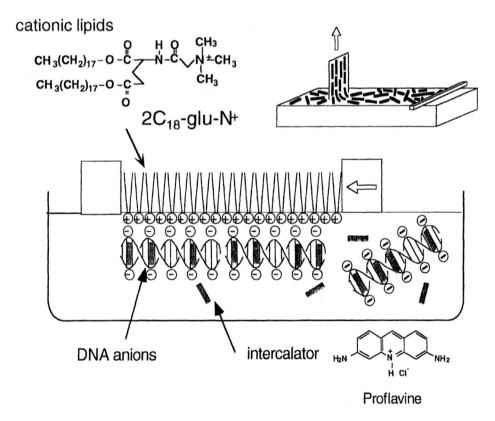

Fig. 14 Schematic illustration of formation of a DNA-oriented LB film by using a polyion complex of DNA-intercalator and cationic lipid monolayers.

cationic lipids

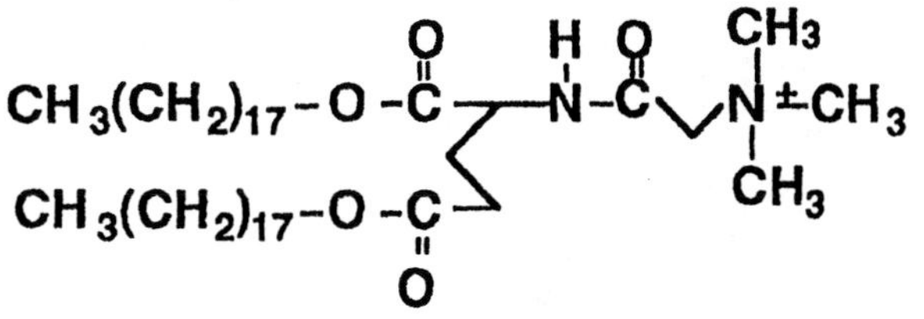

$$CH_3(CH_2)_{17}-O-\overset{O}{\overset{\|}{C}}$$
$$CH_3(CH_2)_{17}-O-\overset{}{\underset{O}{\overset{\|}{C}}}$$

with $-N\overset{H}{\underset{}{}}-\overset{O}{\overset{\|}{C}}\overset{}{\underset{}{}}-N^{\pm}CH_3$ and CH_3 groups

$$2C_{18}\text{-glu-N}^+$$

10

A similar complex of a cationic lipid and DNA was found to exhibit a multilamellar structure composed of alternating lipid bilayer and DNA monolayer with distinct interhelical DNA spacings (38). A detailed study was carried out on aggregation behavior, morphology, and interfacial properties as related to the solution structure of **11**/**12**-DNA complexes, and a structural model was presented as shown in Fig. 15. (39).

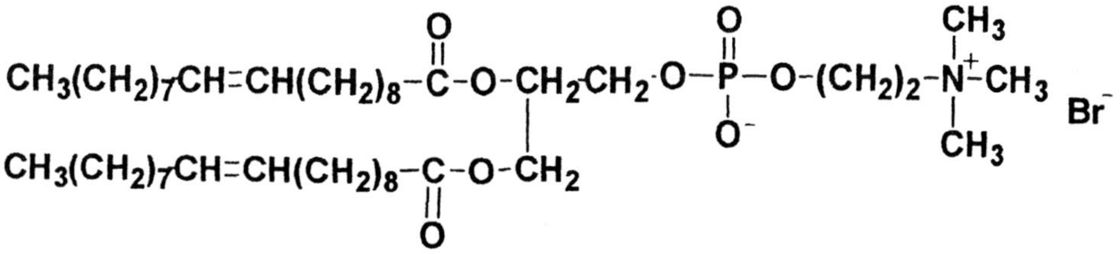

$$CH_3(CH_2)_7CH{=}CH(CH_2)_8{-}\overset{O}{\overset{\|}{C}}{-}O{-}CH_2CH_2{-}O{-}\overset{O}{\overset{\|}{P}}{-}O{-}(CH_2)_2{-}\overset{CH_3}{\underset{CH_3}{N^+}}CH_3 \quad Br^-$$
$$CH_3(CH_2)_7CH{=}CH(CH_2)_8{-}\overset{}{\underset{O}{\overset{\|}{C}}}{-}O{-}CH_2$$

11

$$CH_3(CH_2)_7CH{=}CH(CH_2)_8{-}O{-}\overset{}{\underset{|}{CH}}{-}CH_2{-}N^+(CH_3)_3 \quad Cl^-$$
$$CH_3(CH_2)_7CH{=}CH(CH_2)_8{-}O{-}CH_2$$

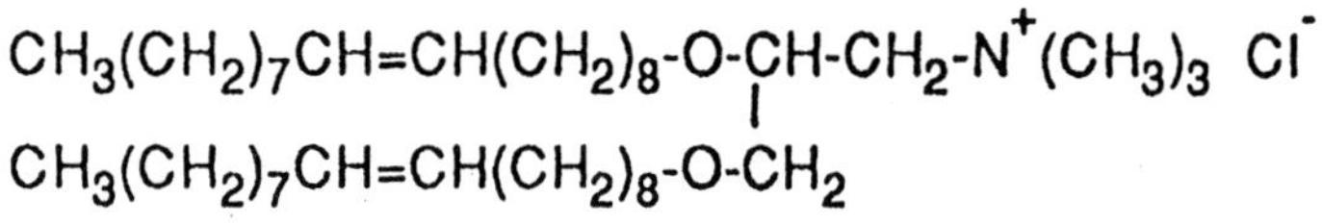

12

B. Transfection

The transfer of extracellular genetic materials into different types of eukaryotic cells has been a major goal of gene therapy. Felgner et al (40) reported in 1987 that the uptake of exogenous DNA by eukaryotic cells is facilitated if the anionic polynucleic acid is previously complexed with cationic liposomes. This approach of gene transfection is often called lipofection and was proved, over the past 10 years, to be mediated by a wide class of cationic lipids and cationic polymers, yielding no less transfection efficiency than previously established synthetic nonviral gene delivery methods (Fig. 16). Synthetic ammonium amphiphiles as described in the previous section thus became a powerful tool for this purpose. Because the repulsive force among the phosphate groups of the DNA backbone is suppressed by charge neutralization by cationic vesicles, a highly compact structure of the DNA, chromatin, is formed and is entrapped in the aqueous compartment of the vesicle.

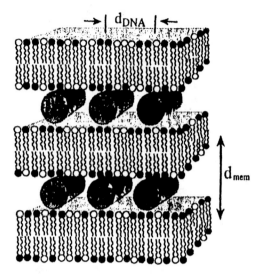

Fig. 15 The structure of self-assembled DNA-lipid complexes.

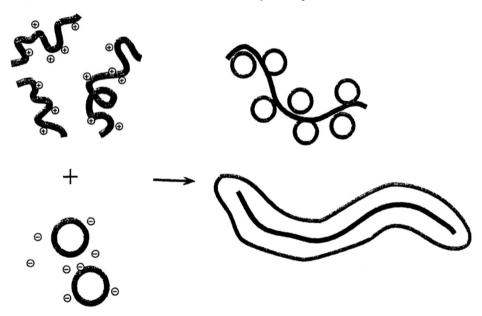

Fig. 16 Interaction models of DNA and vesicles of cationic amphiphiles in the DNA transfection process.

We searched for compounds that would efficiently transfect DNA into eukaryotic cells from among synthetic amphiphiles with cationic heads and long hydrocarbon tails (41). The gene transfer efficiency was examined by the transient expression of cytochrome b_5 from its complementary DNA (cDNA) in COS cells. Among various synthetic amphiphiles often used, **12**, **13** was highest in efficiency. Amphiphiles with a phase transition temperature (T_c) lower than 37°C, as measured by differential scanning calorimetry, could introduce DNA into the cells. Electron microscopic observation indicated that amphiphiles possessing DNA transfection ability formed vesicular structures in aqueous solution. Fluid and vesicular bilayer structures were much more effective than rigid and helical bilayer structures (42).

$$2C_{12}\text{-L-Glu-ph-}C_n\text{-N}^+ \quad (n = 2, 4, 6, 8)$$

$$CH_3(CH_2)_{11}OCO\text{-}\underset{|}{CH}\text{-NH-CO-}C_6H_4\text{-O-}(CH_2)_n\text{-}N^+(CH_3)_3 \ Br^-$$

$$CH_3(CH_2)_{11}OCO\text{-}(CH_2)_2$$

13

Van der Woude et al. (43) made a detailed study of the transfection efficiency of **12**. Optimal transfection was dependent on the amphiphile concentration, which determined the efficiency of vesicle interaction with the target cell membrane as well as the toxicity of the amphiphiles toward the cell. A low lipid/DNA ratio prevented the complex from interacting with the cell surface, whereas the complex became toxic at a larger ratio. Translocation efficiency was independent of the initial vesicle size but was affected by the size of the DNA. The mechanism of DNA entry appears to involve translocation of the nucleic acids through pores across the membrane rather than delivery via fusion or endocytosis (43).

More recently, this group developed novel double-chained pyridinium compounds that displayed highly efficient transfection properties. The transfection efficiency of **14** was enhanced by an order of magnitude relative to that of lipofectin (**12**), and they were essentially nontoxic toward cells. It was suggested that the enhancement is related to delivery rather than the packaging of the amphiphile-DNA complex (44).

14

C. Functional Control and Reaction

We pursued more active roles of artificial bilayer membranes in functional control and reaction of DNA. To the best of our knowledge, this important theme has not been investigated by other groups. Dna protein is the initiator of chromosomal DNA replication in *Escherichia coli*. The lipid interaction of DnaA protein is accompanied by a decrease in the affinity of ATP in the natural biological process. An ammonium bilayer membrane, but not a phosphate membrane, inhibited the ATP binding to DnaA protein and stimulated the release of ATP from the ATP-DnaA complex. A phosphate bilayer component dispersed in the ammonium matrix membrane had little effect on binding and release of ATP, but clustered phosphate components in mixed artificial membranes inhibited the ATP binding and stimulated the ATP release. Cluster formation is an important parameter in decreasing the affinity of DnaA protein for ATP (45).

$$\text{COO}^{-} \; \left(\text{Ca}^{2+} \right) \; {}^{-}\text{OOC}$$

Nitr-5 = Ca^{2+} complex

15

In another example, cleavage of λ-DNA by a restriction enzyme, *Hind*III, is suppressed when the latter is bound to a phosphate bilayer membrane, **14**, at 37°C by Coulombic attraction. The membrane-bound enzyme is not denatured, because its cleavage activity is recovered by addition of Ca^{2+}. At elevated temperatures beyond T_c of the bilayer, Ca^{2+} addition is not effective, and *Hind*III is probably denatured. This result was applied to photocontrol of the enzyme activity. Photoresponsive ligand **15** forms a stable Ca^{2+} complex but releases Ca^{2+} ions on illumination with 360-nm light. Thus, in the presence of an Nitr-5–Ca^{2+} complex, the cleavage of λ-DNA by *Hind*III is suppressed by phosphate bilayer, **14** and the enzyme activity is restored by Ca^{2+} release on illumination (46).

Lanthanide ions are reported to hydrolyze phosphodiester linkages in nucleic acids (47). Because these ions are strongly bound to the phosphate bilayer membranes and alter the physicochemical properties of the membrane, (48) combined use of

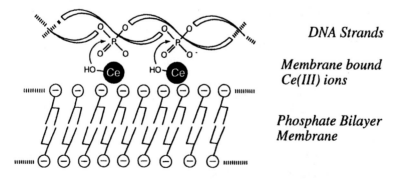

DNA Strands

Membrane bound Ce(III) ions

Phosphate Bilayer Membrane

Fig. 17   Schematic illustration of DNA-Ce^{3+}-1 complex and concomitant hydrolysis of DNA.

lanthanide ions and phosphate membranes leads to controlled cleavage of DNA. The scission of DNA by Ce^{3+} ions was examined in the presence of phosphate bilayer membranes by the plasmid relaxation assay using pBR322. The original supercoiled form of this plasmid DNA is transformed to a nicked form by hydrolytic cleavage and to a linear form by another cleavage at a nearby site, as can be distinguished by gel electrophoresis. The hydrolytic cleavage by Ce^{3+} is enhanced in the presence of a liquid-crystalline bilayer of **16**. In contrast, the influence of bilayer addition is not recognized clearly at lower temperatures where the phosphate bilayer is in the crystalline state. It is likely, as illustrated in Fig. 17, that Ce^{3+} ions are concentrated and activated at the surface of a phosphate bilayer membrane. Maintenance of the fluid bilayer organization appears essential for the activity. Other phosphate bilayers that are disordered considerably by Ce^{3+} ion binding suppress the hydrolytic activity of Ce^{3+} (49).

$$CH_3(CH_2)_{13}\ O\overset{O}{\overset{\|}{C}}-CH-N\overset{H}{\underset{}{|}}-\overset{}{\underset{O}{\overset{\|}{C}}}-\langle\!\!\!\!\bigcirc\!\!\!\!\rangle-O\ (CH_2)_6\ O-\overset{O}{\underset{OH}{\overset{\|}{P}}}-OH$$

$$CH_3(CH_2)_{13}\ O\overset{}{\underset{O}{\overset{\|}{C}}}-CH$$

16

REFERENCES

1. AD Bangham, RW Horne. Negative staining of phospholipids and their structural modification by surface-active agents as observed in the electron microscope. J Mol Biol 8:660, 1964.
2. JM Gebicki, M Hicks. Ufasomes are stable particles surrounded by unsaturated fatty acid membranes. Nature 243:232, 1973.
3. WR Hargreaves, DW Deamer. Liposomes from ionic, single-chain amphiphiles. Biochemistry 17:3759, 1978.
4. WG Gray, PA Winsor. Generic relationships between non-amphiphilic and amphiphilic mesophases of the "fused" type. (In: S Friberg, ed., Lyotropic Liquid Crystals. Washington, DC: American Chemical Society, 1976, pp 1–12.
5. AW Ralston, DN Eggenberger, PL DuBrow. Conductivities of quaternary ammonium chlorides containing two long-chain alkyl groups. J Am Chem Soc 70:977, 1948.
6. T Kajiyama, A Kumano, M Takayanagi, Y Okahata, T Kunitake. Crystal–liquid crystal phase transformation of synthetic bimolecular membranes. Contemp Top Polym Sci 4:829, 1984.
7. T Kunitake, R Ando, Y Ishikawa. DSC studies of the phase transition behavior of synthetic bilayer membranes. Part II. Bilayer membranes of single-chain and triple-chain amphiphiles. Mem Fac Eng Kyushu Univ 46(2):245, 1986.
7a. I Cho, C.-S. Kim. New synthetic vesicles formed by polymeric quaternary ammonium bromide with double alkyl chains. Chem Lett (10):1543, 1985.
8. FM Menger, Y Yamasaki. Hyperextended amphiphiles. Bilayer formation from single-tailed compounds. J Am Chem Soc 115:3840, 1993.
9. N Kimizuka, H Ohira, M Tanaka, T Kunitake. Bilayer membranes of four-chained ammonium amphiphiles. Chem Lett 1990:29, 1990.

10. T Kunitake, R Ando, Y Ishikawa. DSC studies of the phase transition behavior of synthetic bilayer membranes. Part I. Bilayer membranes of double-chain amphiphiles. Mem Fac Eng Kyushu Univ 46(2):221, 1986.

11. JM Kim, T Kunitake. DSC studies of the phase transition behavior of synthetic bilayer membranes. Part III. Bilayer membranes of phosphate amphiphiles. Mem Fac Eng Kyushu Univ 49(2):93, 1989.

12. M Shimomura, T Kunitake. Fusion and phase separation of ammonium bilayer membranes. Chem Lett 1981:1001, 1981.

13. T Kunitake, Y Okahata, S Yasunami. Formation and enhanced stability of fluoroalkyl bilayer membranes. J Am Chem Soc 104:5547, 1982.

14. C Santaella, P Vierling, JG Riess. Highly stable liposomes from fluoroalkylated glycerophosphoquinolines. Angew Chem Int Ed Engl 30:567, 1991.

15. K Liang, Y Hui, Vesicle of a hybrid bolaamphiphile: Flip-flop behavior of spin labels. J Am Chem Soc 114:6588, 1992.

16. N Kimizuka, T Kawasaki, T Kunitake. Self-organization of bilayer membranes from amphiphilic networks of complementary hydrogen bonds. J Am Chem Soc 115:4387, 1993.

17. P Jokela, B Jönsson, A. Khan. Phase equilibria of catanionic surfactant—water systems. J Phys Chem 91:3291, 1987.

18. EW Kaler, AK Murthy, BE Rodriguez, JAN Zasadzinski. Spontaneous vesicle formation in aqueous mixtures of single-tailed surfactants. Science 245:1371, 1989.

19. M Wakita, KA Edwards, SL Regen, L Steven, D Turner, SM Gruner. Use of a polymeric counter ion to induce bilayer formation from a single-chain surfactant. J Am Chem Soc 110:5221, 1982.

20. I Ichinose, Y Ishikawa, T Kunitake. Aggregation and complexation of a dithiooxamide derivative with a long alkyl chain. Complex Salt Chem Prepr Jpn 43:293, 1989.

21. J Suh, K-J Lee, G Bae, O-B Kwon, S Oh. Coordinatively polymerized bilayer membranes prepared by metal complexation of an amphiphilic o,o'-dihydroxyazobenzene derivative. Langmuir 11:2626, 1995.

22. Y Iwasaki, T Yonezawa, T Kunitake. 1D and 2D organization of Pt(II) based on ammonium derivatized coordination compounds. 35th Joint Kyushu Meeting of Chemistry-Related Societies, Fukuoka, Japan, July 1998.

23. P Tyminski, LH Latimer, DF O'Brien. Rhodopsin in polymerized bilayer membranes. J Am Chem Soc 107:7769, 1985.

24. N Nakashima, R Ando, T Kunitake. Casting of synthetic bilayer membranes on glass and spectral variation of membrane-bound cyanine and merocyanine dyes. Chem Lett 10:1577, 1983.

25. Y Ishikawa, T Kunitake. Design of spatial disposition of anionic porphyrines in matrices of ammonium bilayer membranes. J Am Chem Soc 113:621, 1991.

26. I Hamachi, S Noda, T Kunitake. Layered arrangement of oriented myoglobins in cast films of a phosphate bilayer membrane. J Am Chem Soc 112:6744, 1990.

27. I Hamachi, T Honda, S Noda, T Kunitake. Oriented intercalation of myoglobin into multilayered films of synthetic bilayer membranes. Chem Lett 7:1121, 1991.

28. I Hamachi, S Higuchi, K Nakamura, H Fujimura, T Kunitake. Lipid anchored myoglobin. 2. Effect of the anchor structure on membrane binding. Chem Lett 7:1175, 1993.

29. I Hamachi, K Nakamura, A Fujita, T Kunitake. Anisotropic incorporation of lipid-anchored myoglobin into a phospholipid bilayer membrane. J Am Chem Soc 115:4966, 1993.

30. I Hamachi, S Noda, T Kunitake. Functional conversion of myoglobin bound to synthetic bilayer membranes: From dioxygen storage protein to redox enzyme. J Am Chem Soc 113:9625, 1991.

31. I Hamachi, A Fujita, T Kunitake. Aniline-hydroxylase activity of myoglobin by coupling with a membrane-bound electron transport system. Chem Lett 8:657, 1995.

32. A Fujita, H Senzu, T Kunitake, I Hamachi. Enhanced peroxidase activity of cytochrome *c* by phosphate bilayer membrane. Chem Lett 7:1219, 1994.
33. I Hamachi, A Fujita, T Kunitake. Enhanced *N*-demethylase activity of cytochrome *c* bound to a phosphate-bearing synthetic bilayer membrane. J Am Chem Soc 116:8811, 1994.
34. I Hamachi, A Fujita, T Kunitake. Protein engineering using molecular assembly: Functional conversion of cytochrome *c* via noncovalent interactions. J Am Chem Soc 119:9096, 1997.
35. CR Safinya, I Koltover, J Raedler. DNA at membrane surfaces: An experimental overview. Curr Opin Colloid Interface Sci 3:67, 1998.
36. Y Okahata, K Ijiro, Y Matsuzaki. A DNA-lipid cast film on a quartz-crystal microbalance and detection of intercalation behaviors of dye molecules into DNAs in an aqueous solution. Langmuir 9:19, 1993.
37. K Ariga, Y Okahata. In situ characterization of Langmuir-Blodgett films during a transfer process. Evaluation of transfer ratio and water incorporation by using a quartz crystal microbalance. Langmuir 10:3255, 1994.
38. JO Raedler, L Koltover, T Salditt, CR Safinya. Structure of DNA-cationic liposome complexes: DNA intercalation in multilamellar membranes in distinct interhelical packing regimes. Science 275:810, 1997.
39. JO Raedler, I Koltover, A Jamieson, T Salditt, CR Safinya. Structure and interfacial aspects of self-assembled cationic lipid-DNA gene carrier complexes. Langmuir 14:4272, 1998.
40. PL Felgner, TR Gadek, R Raman, HW Chan, M Wenz, JP Northrop, GM Ringold, M Danielson. Lipofection: A highly efficient, lipid-mediated DNA-transfection procedure. Proc Natl Acad Sci USA 84:7413, 1987.
41. A Ito, R Miyazoe, J Mitoma, T Akao, T Osaki, T Kunitake. Synthetic cationic amphiphiles for liposome-mediated DNA transfection. Biochem Int 22:235, 1990.
42. T Akao, T Osaki, J Mitoma, A Ito, T Kunitake. Correlation between physicochemical characteristics of synthetic cationic amphiphiles and their DNA transfection ability. Bull Chem Soc Jpn 64:3677, 1991.
43. I van der Woude, HW Visser, MBA ter Beest, A Wagenaar, MHJ Ruiters, JBFN Engberts, D Hoekstra. Parameters influencing the introduction of plasmid DNA into cells by the use of synthetic amphiphiles as a carrier system. Biochim Biophys Acta 1240:34, 1995.
44. I van der Woude, A Wagenaar, AAP Meekel, MBA ter Beest, MHJ Ruiters, JBFN Engerts, D Hoekstra. Novel pyridinium surfactants for efficient nontoxic in vitro gene delivery. Proc Natl Acad Sci USA 94:1160, 1997.
45. T Mizushima, Y Ishikawa, E Obana, M Hase, T Kubota, T Katayama, T Kunitake, E Watanabe, K Sekimizu. Influence of cluster formation of acidic phospholipids on decrease in the affinity for ATP of DnaA protein. J Biol Chem 271:3633, 1996.
46. A Baba, N Kimizukka, T Kunitake. Controlled cleaveage of DNA by the use of bilayer-bound restriction enzyme. 43th SPSJ Annual Meeting, Nagoya, May 1994.
47. M Komiyama. Sequence-specific and hydrolytic scission of DNA and RNA by lanthanide complex–oligo DNA hybrids. J Biochem 118:665, 1995.
48. N Nakashima, R Ando, H Fukushima, T Kunitake. Drastic changes of circular dichroism of synthetic phosphate bilayers due to interaction with metal ions. Chem Lett 10:1503, 1985.
49. N Kimizuka, E Watanabe, T Kunitake. Lanthanide Ion–mediated hydrolysis of DNA on phosphate bilayer membrane. Chem Lett, 29, 1999.

10

Biosimulation with Liposomes and Lipid Monolayers

J. Sunamoto
Niihama National College of Technology, Niihama, Japan

I. INTRODUCTION

Liposomes are basically spherical vesicles that have an interior water pool enclosed by lipid bilayers. They have been studied from various points of view as they represent the simplest structural model for cells and may be considered an important tool in various research fields such as supermolecular chemistry, physics, biophysics, chemistry, biochemistry, biology, biotechnology, and medicine.

After a long-term study of self-aggregates of lipids in aqueous media by many pioneers since the 1800s, Alec D. Bangham in England, in 1965, first showed the structural characteristics of liposomes (1). I believe that this was really the opening of a window in liposome sciences. Therefore, at an early stage after the discovery of liposomes, they were called "banghasomes." In the late 1960s, Weismann proposed the name "liposome", from lipid+soma (fat body). In 1911, before the discovery of liposomes, Lehmann (2) had shown an optical micrograph of self-aggregates of lipids (most probably large multilamellar vesicles) and called them "Künstlische Zellen (artificial cells)." In 1996, Lasic and Barenholz (3) coedited four excellent volumes of the *Handbook of Nonmedical Applications of Liposomes, Theory and Basic Sciences.* In the first and last chapters of this handbook, they briefly describe the history of liposomes and their perspectives as well. This handbook is rather comprehensive and covers well most of the important and interesting topics, including both basic sciences and applications of liposomes. Nevertheless, several topics of interest in biosimulation or biofunctionalization of liposomes are still missing in this handbook.

In this chapter, therefore, I intend to introduce several topics related to biosimulation using lipid assemblies of liposomes and monolayers, not discussed much in previous major books on liposome sciences. However, I would like to ask readers to consider it as a sort of monograph because most of the topics introduced · were developed in my laboratory.

Biosimulation, or mimicking of nature, refers to the simulation of the structure and function of any component of living systems composed of small molecules such as amino acids, lipids, nucleotides, saccharides, coenzymes, or hormones; biomacromolecules such as peptides, proteins, enzymes, DNAs, RNAs, or polysaccharides; and even tissues, organs, or cells. Biosimulation involves astute use of the chemistry and physicochemistry of supramolecular assemblies of lipids, proteins, or polysaccharides. Although the primary purpose of biosimulation is, of course, to understand detailed mechanisms of living systems at the molecular level, the information and knowledge obtained through these basic investigations are beneficial in developing new biomaterials and biosystems designed for biotechnology and medicine (Fig. 1).

II. BIOSIMULATION USING LIPOSOMES

In 1964, Bangham and Horne (1) first demonstrated that dispersion of naturally occurring phospholipids in an aqueous medium results in a vesicle (liposome) that has an

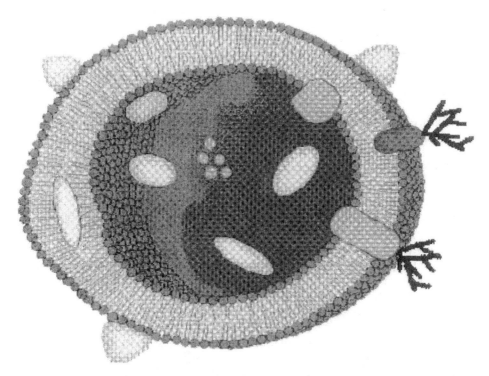

Fig. 1 Liposomes as molecular cargo. Hydrophobic or lipophilic materials such as membrane proteins or hydrophobic drugs can be embedded in a lipid bilayer. Hydrophilic or water-soluble materials are encapsulated in the inner water pool.

interior water phase surrounded by a lipid bilayer membrane. Since then, liposomes have been considered an excellent model of cells and employed in basic studies to better understand structural and functional characteristics of cells (3). They are also potent carriers for drugs, toxins, proteins, enzymes, antigens, antibodies, and nucleotides (4–10). As a liposome is a sort of supermolecule of naturally occurring lipids, it is a very convenient biodegradable material. At the same time, this feature also becomes disadvantageous because of its physicochemical and biochemical instability. In addition, conventional liposomes are not cell specific, fusogenic, immunogenic, or effective in reconstitution of membrane proteins. To overcome these disadvantages of liposomes, several methodologies to make them more stable and more functional have been envisioned by mimicking intact cells.

A. Structural Simulation of Plant Cell Walls (11,12)

For example, gram-positive bacterial cell walls are chiefly composed of peptidoglycan, whereas plant cell walls are usually composed of cellulose. Cell walls are necessary to maintain the shape and stiffness of cells and to protect the cell membranes against chemical and physical stimuli such as osmotic pressure and temperature changes (13–15). Naturally occurring polysaccharides are considered a good candidate for coating of the liposomal surface to attain better stability of the liposome. This is a supramolecular assembly of an artificial cell wall of the naturally occurring polysaccharide derivatives at the outermost surface of an artificial cell liposome.

Interactions with simple polysaccharides and liposomal membranes have been investigated by Brooks and Seaman (16), Minetti et al. (17), and Sunamoto and Iwamoto (18). These studies show that dextran, pullulan, and amylopectin strongly adhere to the liposomal surface, mostly by hydrophobicity, and sometimes induce aggregation of liposomes (18). Conventional polysaccharides adsorbed on the liposomal surface, however, easily desorb upon dilution. Therefore, these conventional, water-soluble polysaccharides were hydrophobized by chemically conjugating cholesterol or fatty acid in order to achieve a strong interaction with liposomal membranes.

1. Preparation of Hydrophobized Polysaccharides

Cholesteryl group–conjugated polysaccharides were first synthesized using an aminoethylcarboxymethyl derivative of pullulan followed by condensation with cholesteryl chloroformate (19). In this procedure, however, trace amounts of carboxylic acid and aminoethylcarboxymethyl groups remained unreacted in the final product. In order to synthesize entirely nonionic hydrophobized polysaccharides, the reaction between polysaccharide and cholesteryl N-(6-isocyanatohoxyl)carbamate (1) was proposed (20) (Fig. 2).

No degradation of the polymer during the modification was ascertained by size exclusion column chromatography (SEC). The degree of substitution of the cholesteryl group was determined by ^{1}H nuclear magnetic resonance (NMR). When pullulan [molecular weight (MW) 55,000] is substituted by 2.0 cholesterol groups per 100 anhydroglucoside units, it is denoted by CHP-55-2.0. Similarly, monoalkyl or dialkyl fatty acids are also conjugated: hexadecyl (C_{16}) and α,α'-dodecyl diglyceryl diether ($2C_{12}$) groups. Various hydrophobized polysaccharides such as pullulan (P), dextran (D), mannan (M), phosphorylated mannan (M-P), inulin (I), levan (L), amylopectin (Ap), and amylose (As) are synthesized by the same procedure.

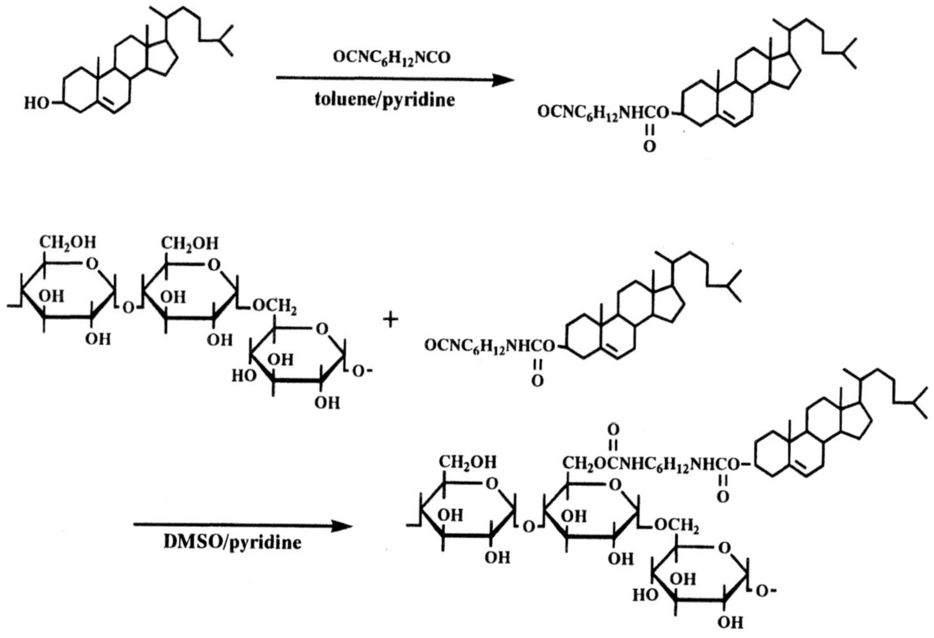

Fig. 2 Synthetic route of cholesteryl group-conjugated pullulan (CHP).

2. Coating of the Outermost Surface of Liposomes

In an attempt to mimic the morphology and function of cell walls of plant and bacterial cell membranes, several approaches have been used. Tirrell and coworkers assembled synthesized polyelectrolytes to phospholipid bilayer membranes (20) and studied the pH-induced permeability change of the lipid membrane (21). Ringsdorf and coworkers coated lipid bilayer membranes with hydrophobized polymers (22) and studied the thermoreversible expansion and construction of the polymer on the liposomal surface (23).

To investigate the coating efficiency of hydrophobized polysaccharides on liposomes, the fluorescence depolarization method has been employed. The measurement is based on the decrease in mobility of the fluorescein isothiocyanate (FITC)-labeled polysaccharide upon binding to the liposomal surface. The p-value change of the fluorescent probe is monitored after the addition of FITC-pullulan or FITC-CHP to the liposomal suspension as a function of time. The original p-values were 0.099 for CHP and 0.097 for pullulan in the absence of liposome. The change in the p-value of the FITC-labeled polysaccharide was biphasic. The first process was completed within 10 min (24), whereas the subsequent slow process required several tens of hours to reach equilibrium (25). FITC-pullulan also bound to liposomes even though the initial p-value change of pullulan was smaller than that of CHP. In contrast to the case of CHP, with pullulan, a further increase in the p-value was not observed after the initial fast increase.

The coating efficiency of the polysaccharide on the liposomal membrane has also been investigated by the gel chromatography method using the same FITC-labeled polymers. The amount of polysaccharide bound to liposome is determined from

Table 1 Coating Efficiency of Liposome with FITC-Polysaccharides at 25°C

	Coating efficiency (%)	
Polysaccharide derivatives[a]	1 h	12 h
FITC-0.55-pullulan-50	2	2
FITC-0.49-CHP-50-1.6	70	80

[a] (FITC-polysaccharide)/(lipid)=0.1 by weight.

the absorbance of FITC at 495 nm. The percent coating efficiency is defined as 100 times the ratio of the amount of polysaccharide bound to the liposome to the amount of polysaccharide initially employed (Table 1).

For pullulan, for example, the coating efficiency is only 2% after 1 h of incubation, and no increase in the coating efficiency was observed even after 12 h. The adsorbed pullulan is detached from the liposome during gel chromatographic separation. The binding of pullulan to the liposomal surface is not strong enough to isolate pullulan-coated liposomes. With CHP, however, this value becomes approximately 70% at 1 h and 80% after 12 h. The binding of CHP to the liposome is biphasic, as is the *p*-value change (see earlier). These results reveal that the hydrophobic group conjugated in part to polysaccharide greatly contributes to the effective coating of the liposomal surface.

The difference in coating efficiency among three hydrophobic anchors (CH, C_{16}, and $2C_{12}$) and various polysaccharides such as pullulan, dextran, and mannan has been investigated. The concentration of bound polysaccharide was determined by the phenol–sulfuric acid method. Shown in Fig. 3 is an example of a gel chromatographic separation of CHP-55-1.7–coated liposomes on a Sepharose 4B column. At an early stage of the coating, the amount of polysaccharides bound to the liposomal surface increased with increasing amount of the polysaccharide initially added and then gradually leveled off.

Except for the $C_{16}P$-coated liposomes, the binding isotherms of CHP-, CHD-, CHM-, and $2C_{12}P$-coated liposomes fit a Langmuir-type adsorption (26), and the binding constant, K, is calculated from Eq. (1):

$$\frac{1}{q} = \frac{1}{q_s K C} + \frac{1}{q_s} \tag{1}$$

where q is the amount of hydrophobized polysaccharides bound to the liposome, C is the amount of free (uncoated) polysaccharide, and q_s is the saturation capacity. A plot of $1/q$ against $1/C$ gives a straight line (Fig. 4).

From the slope and intercept of the straight line, the binding constant (K) and the saturation capacity (q_s) are obtained (Table 2). In this study, the liposomes were coincubated with hydrophobized polysaccharide long enough for the coating to reach equilibrium (12 h at 50°C).

The binding constant apparently increased with an increase in both the DS (degree of substitution) of the cholesteryl moiety and the molecular weight of the polysaccharide. The chemical structure of the parent polysaccharide slightly affected

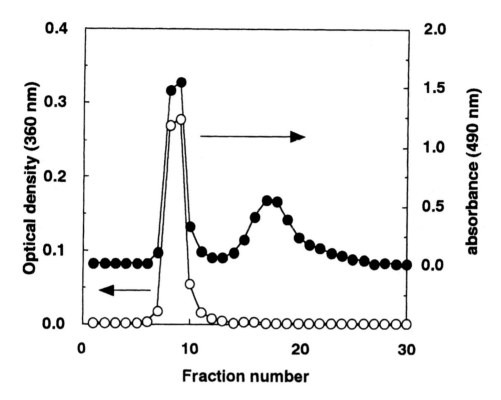

Fig. 3 Gel chromatographic separation of the CHP-55-1.7–coated liposomes from free CHP-55-1.7. The polysaccharide fractions were monitored at 490 nm by the phenol–sulfuric acid method, and liposome fractions were detected by turbidity at 360 nm. R value (CHP)/(lipid) was 1.0 (by weight).

the binding constant. Regarding the q_s value, there is no significant difference between CHP and CHD irrespective of their molecular weight and DS. The q_s value of CHM is somewhat larger than those of CHP and CHD. This may be due to more branching of mannan (the branching ratio is around 35%) compared with the case of pullulan (almost no branching) or dextran (branching is approximately 5%). In the case of $2C_{12}P$-55-2.3, both K and q_s are smaller than those of other cholesteryl group–conjugated polysaccharides. These results also reveal that the cholesteryl group is a more efficient hydrophobic anchor for coating the liposomal surface than a simple fatty acid.

The changes in the vesicle size before and after polysaccharide binding are shown in Table 3. In the case of CHP-55-1.7, CHP-55-2.5, CHP-108-1.3, and CHD-70-1.7, the size of the liposome increased by 20–30 nm upon polymer binding. Both the molecular weight and the DS of hydrophobized pullulan slightly affected the size of the liposome upon polysaccharide binding. The diameter of the CHM-85-2.3–coated liposome (approximately 180–200 nm) is larger than those of the CHP- or CHD-coated liposome (approximately 150–170 nm). The hydrophobized polysaccharides, CHP, $C_{16}P$, $2C_{12}P$, and CHD, all form self-aggregates with a similar size in water (27,28). The diameter of the self-aggregates is approximately 20–30 nm.

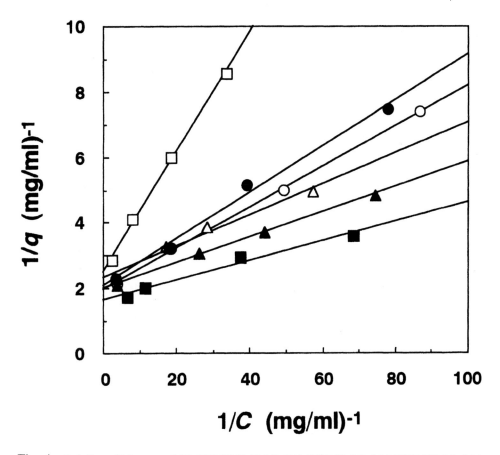

Fig. 4 Relation of $1/q$ versus $1/C$: (●) CHP-55-1.7; (○) CHP-55-2.5; (▲) CHP-108-1.3; (△) CHD-70-1.7; (■) CHM-85-2.3; and (□) $2C_{12}P$-55-2.3.

Table 2 Parameter of Langmuir-Type Adsorption

Hydrophobized polysaccharide	K (mg/mL)$^{-1}$	q_s (mg/mL)
CHP-55-1.7	2.5 ($\pm$0.2)$\times$10	5.2 ($\pm$0.5)$\times$10^{-1}
CHP-55-2.5	3.2 ($\pm$0.2)$\times$10	5.1 ($\pm$0.3)$\times$10^{-1}
CHP-108-1.3	5.2 ($\pm$0.2)$\times$10	5.0 ($\pm$0.1)$\times$10^{-1}
CHD-70-1.7	4.5 ($\pm$0.4)$\times$10	4.5 ($\pm$0.4)$\times$10^{-1}
CHM-85-2.3	4.0 ($\pm$0.3)$\times$10	6.5 ($\pm$0.4)$\times$10^{-1}
$2C_{12}P$-55-2.3	1.5 ($\pm$0.1)$\times$10	3.8 ($\pm$0.2)$\times$10^{-1}

However, CHM forms larger self-aggregates (approximately 50 nm) than the other hydrophobized polysaccharides (28). The particle size of hydrophobized polysaccharide–coated liposomes seems to correlate with the size of the polysaccharide self-aggregates.

The diameter of the $2C_{12}P$-55-2.3–coated liposome does not change much with an increase in the polysaccharide/lipid ratio (R value) and is almost comparable

Table 3 Average Hydrodynamic Diameters of
Polysaccharide-Coated Liposomes[a]

Polysaccharide	R	Diameter (± 2 nm)
None		137
CHP-55-1.7	0.1	159
	0.3	155
	0.5	157
	1.0	156
CHP-55-2.5	0.1	170
	0.3	168
	0.5	172
	1.0	166
CHP-108-1.3	0.1	171
	0.3	169
	0.5	168
	1.0	166
CHD-70-1.7	0.1	150
	0.3	156
	0.5	157
	1.0	164
CHM-85-2.3	0.1	189
	0.3	199
	0.5	188
	1.0	181
$2C_{12}P$-55-2.3	0.1	147
	0.3	156
	0.5	157
	1.0	154
$C_{16}P$-55-2.4	0.05	201
	0.1	262 (± 7)
	0.2	252 (± 3)
	0.3	232
	0.5	185 (± 6)
	1.0	161
	0.1+0.9	227 (± 4)
	0.2+0.8	237

[a] [Liposomal lipid] $= 4.0 \times 10^{-4}$ M, (DPPC)/(chol.) $=$
3/1 by mol. $R =$ (hydrophobized polysaccharide)/(lipid)
(by weight).

to that of the CHP-coated liposome. On the other hand, for the $C_{16}P$-coated liposome, both the vesicle size and the polydispersity in size distribution become maximum at $R = 0.1$. Although there is no direct evidence, this may reflect aggregation of liposomes by intervesicular bridging with the polysaccharide (19). The vesicle size significantly decreases by 20–40 nm when more $C_{16}P$ is added to the liposome.

  If the binding of polysaccharide to liposome is not strong enough and the amount of polymer is insufficient to coat the whole liposomal surface, then bridging of the polymers occurs between neighboring liposomes. The aggregation induced by

$C_{16}P$ at the initial stage is probably due to the colloidal instability of the self-aggregates of $C_{16}P$ compared with that of cholesteryl group–bearing polysaccharides. This would result from the bridging among liposomes.

Interestingly, this clearly resembles structural characteristics of biological membranes, in which cholesterol plays an important role in packing of lipid membranes.

3. Physicochemical Stability of Polysaccharide-Coated Liposomes

The membrane permeability or leakiness of liposomes is usually studied by the fluorescent probe release technique (30) using, for example, 6-carboxyfluorescein (CF) as the probe. Structural stabilization of the polysaccharide-coated liposomes is also studied by this technique (Fig. 5).

If a liposome is coated with a cholesteryl group–conjugated polysaccharide, the membrane permeability of the liposome decreases even in the presence of serum or plasma compared with that of a conventional liposome without a polysaccharide coat. The polysaccharide coating of the liposome certainly depresses the lysis of the liposome by serum or plasma proteins and also by enzymes such as lipases or lipid peroxidases. These results suggest that these polysaccharide-coated liposomes would be promising as stable carriers of water-soluble drugs for systemic administration to animals (see later).

B. Simulation of Molecular Recognition on the Cellular Surface

Saccharide determinants associated with mammalian cell surfaces, such as glycolipids and glycoproteins, generally play an important role in various types of biological recognition, as in antigen-antibody interactions, immunogenicity, and cell-cell adhesion.

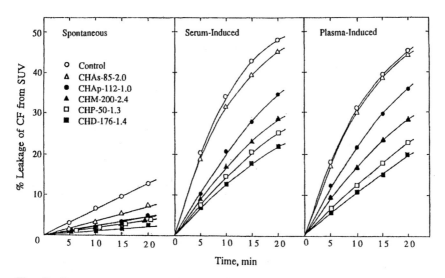

Fig. 5 Spontaneous, serum-induced, and plasma-induced leakages of CF from SUV at 37°C. [Egg PC]=1.0×10^{-4} M. [Serum]=[plasma]=18% (v/v); [polysaccharide]/[egg PC]=1.0 by weight. Buffer, 20 mM Tris-HCl containing 200 mM NaCl (pH 7.4). (○) control; (△) CHAs-85-2.0; (●) CHAp-112-1.0 (▲) CHM-200-2.4; (□) CHP-50-1.3; (■) CHD-176-1.4.

Several sugar moieties specific to receptors have been found (31). For example, liver parenchymal cells have a receptor specific to galactose (32), phagocytic cells have a receptor for mannose (33), and fibroblasts have a mannose-6-phosphate–specific receptor (34). In this sense, polysaccharide-coated liposomes may be used as cell recognition devices enabling simulation of molecular recognition on the cellular surface.

1. Quantitative Estimation of the Interaction Between Lectin and Galactose-Bearing CHP (35)

In many reports related to the clustering effect, it was shown that the affinity of saccharide determinants for the lectins (36,37) or receptors on the viral surface (38) drastically increases with their conjugation to the polymer. We can estimate binding constants between lectin and each saccharide attached to the polymer by competitive titration of the fluorescent ligand using the fluorescence polarization technique (39,40).

Values of the fluorescence polarization (p-value) of the fluorescent ligand MUG increase on binding of MUG to the RCA_{120} lectin. However, the p-value decreases when galactosides are added to the system. The depolarization is due to the release of MUG from the lectin by competitive binding of the galactosides added to the system.

Figure 6 shows the change in the p-value of MUG as a function of the concentration of Lac-CHP (lactose moiety–conjugated CHP). The following scheme is based

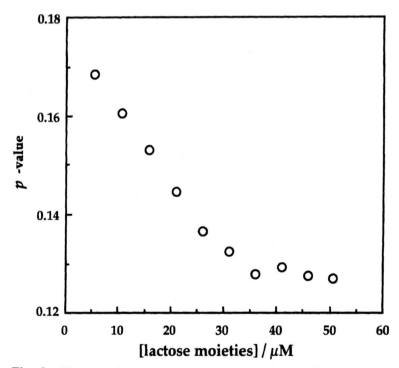

Fig. 6 Changes in fluorescence polarization of MUG at different concentrations of Lac(2)-22-CHP. A 1.5-mL sample of solution mixture, 2.4×10^{-1} mM MUG and 1.1×10^{-1} mM RCA_{120} lectin, was titrated with portions of 1.0 mg/mL Lac(2)-22-CHP suspension in 150 mM PBS at 25°C and pH 7.4.

on the facts that (a) RCA_{120} lectin has two carbohydrate recognition sites, (b) one galactoside can occupy one binding site, and (c) those sites are completely independent of each other during the sugar binding.

$$RCA \cdot MUG + S \overset{1/K_S}{\rightleftharpoons} RCA + MUG + S \overset{K_C}{\rightleftharpoons} RCA \cdot S + MUG \tag{2}$$

According to Eq. (2), the binding constants for the substrate, S, and fluorescent ligand, MUG, are defined as $K_C = [RCA \cdot S]/[RCA][S]$ and $K_S = [RCA \cdot MUG]/[RCA][MUG]$. [RCA] is comparable to the number of binding site of the lectin as represented by the concentration. When the molar fractions of bound and free MUG are given by f_B and f_F, the following equation is valid:

$$\frac{f_F}{f_B}[RCA]_T - f_F[MUG]_T = \frac{1}{K_S} + \frac{K_C}{K_S}[S] \tag{3}$$

where $[RCA]_T$ and $[MUG]_T$ are the total concentrations of the respective binding sites of RCA_{120} lectin and MUG in solution. If p_B ($= 0.290$) and p_F ($= 0.070$) represent the polarization of bound and free MUG, the respective molar fractions are

$$f_F = \frac{p_B - p}{p_B - p_F}, \qquad f_B = \frac{p - p_F}{p_B - p_F} \tag{4}$$

and

$$\frac{f_F}{f_B} = \frac{q_B}{q_F}\frac{p_B - p}{p - p_F} \tag{5}$$

where q_B and q_F are the quantum yields of bound and free MUG, respectively. The term $q_F/q_B = 1$ signifies that no changes in fluorescence intensity occur when MUG is bound to RCA_{120} lectin (40). Equation (3) is then converted to Eq. (6):

$$p_{app} = \frac{p_B - p}{p - p_F}[RCA]_T - \frac{p_B - p}{p_B - p_F}[MUG]_T = \frac{1}{K_S} + \frac{K_C}{K_S}[S] \tag{6}$$

When [S] is much higher than [RCA], [S] is considered to be equal to $[S]_T$, which is the total concentration of the substrate in the solution. The binding constant for the galactoside, K_C, is obtained from the p-value change of the reaction mixture as a function of the galactoside concentration. The results obtained are summarized in Table 4.

Table 4 Binding Constants Between RCA_{120} Lectin and Galactosides of Synthesized CHP Derivatives

Sample	Kc (mM^{-1})
Gal(2)-23-CHP	47.5±4.7
Gal(6)-24-CHP	53.5±0.1
Lac(2)-8-CHP	67.5±5.2
Lac(2)-17-CHP	80.7±7.7
Lac(2)-22-CHP	114.4±18.7
Galactose	1.0±0.1
Lactose	1.5

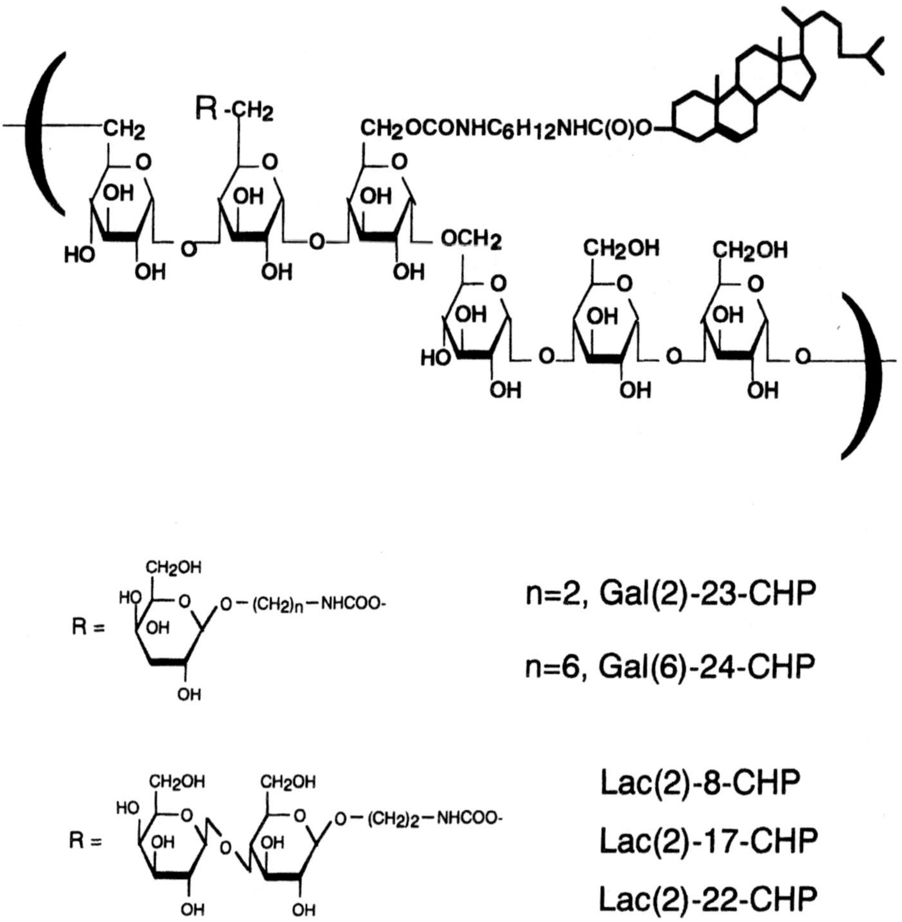

Fig. 7 Structures of Gal-CHPs and Lac-CHPs.

Clearly, the binding constant of the galactoside group conjugated to CHP is much higher than that of the corresponding free small molecules. Polyvalent conjugation of the galactosides to the polymer results in an increase in the local concentration, which would enhance the affinity for the lectin. In fact, the binding affinity increased with an increase in the substitution degree of lactose moieties of Lac-CHP (lactose-conjugated CHP). The larger binding constant of Lac-CHP compared with Gal-CHP shows that RCA_{120} lectin binds to lactose more tightly than galactose. Gal(6)-24-CHP, which has a longer alkyl spacer, had a lager binding constant than Gal(2)-23-CHP. The increase in mobility of the galactose group with Gal(6)-24-CHP (Fig. 7) results in a higher binding constant than for the Gal(2)- or Lac (2) derivative.

C. Biosimulation of Carbohydrate-Carbohydrate Interaction

Aqueous two-phase systems have been widely used for separation and purification of proteins, enzymes, and cells (41,42). The partition of liposomes in an aqueous

PEO-dextran two-phase system was investigated in cell separation (43,44). The partition is affected by the size and the lipid composition of the liposome employed. Surface properties of the liposome are, of course, predominant in affecting the partition.

Sunamoto and coworkers (45) investigated the effect of a polysaccharide coating on the partitioning of liposomes in an aqueous two-phase system (PEO-pullulan or PEO-dextran) in order to obtain information about the carbohydrate-carbohydrate interaction in water. The partitioning of the liposomes is drastically changed by coating the surface with the hydrophobized polysaccharides (Table 5).

Partitioning to the bottom phase increases with increasing density of the polysaccharide on the liposomal surface, but partitioning to the PEO-rich top phase does not change at all. The partitioning efficiency levels off above the point where the weight ratio of polysaccharide to lipid becomes approximately 0.5.

Sharpe and Warren (44) also investigated the partitioning of glycolipid-containing liposomes in the PEO-dextran system. In their case, surface modification of the liposomes by carbohydrates markedly increased the partitioning of the liposomes to the PEO-rich top phase. Considering the higher hydrophobicity of the PEO-rich top phase compared with the polysaccharide-rich bottom phase, they ascribed the stronger partitioning of the glycolipid-containing liposomes to the top phase to the hydrophobicity of the carbohydrates (44).

In contrast, with polysaccharide-coated liposomes partitioning to the bottom polysaccharide-rich phase occurred (Table 5). This suggests that a specific interaction takes place between the liposomal polysaccharide and the polysaccharide in the bulk bottom phase. This is a more important factor than the hydrophobicity of the polysaccharide. No significant differences were observed in the partitioning behavior among the liposomes studied, even though the liposomes were coated with pullulans with different molecular weights and DS values of the cholesteryl moiety. This result was the same in the two different polysaccharide systems, pullulan-50 and dextran-40. Only the density of the polysaccharide on the liposomal surface seems to be important. However, a significant difference was observed when the liposome was coated by structurally different polysaccharides. In order to assess this partitioning quantitatively, the results were analyzed as follows. The partitioning of the polysaccharide-coated liposome from the interface to the bottom polysaccharide-rich phase can be described by specific binding between the two carbohydrates:

$$PS_{1(liposome)} + PS_{2(bulk)} \overset{K}{\rightleftharpoons} (PS_{1(liposome)} \cdot PS_{2(bulk)})$$

The binding constant, K, is calculated on the basis of a Langmuir-type adsorption equation:

$$K = [PS_{1(liposome)} \cdot PS_{2(bulk)}]_b / [PS_{1(liposome)}]_f [PS_{2(bulk)}]_f \qquad (7)$$

where $[PS_{1(liposome)} \cdot PS_{2(bulk)}]_b$ is the amount of complex produced between the liposomal polysaccharide and the polysaccharide in the bulk bottom phase, $[PS_{1(liposome)}]_f$ is the amount of liposomal polysaccharide remaining at the interface, and $[PS_{2(bulk)}]_f$ is the amount of free polysaccharide in the bulk bottom phase. In addition, the following assumptions are made: $[PS_{1(liposome)}]_f = [PS_{1(liposome)}]_t - [PS_{1(liposome)}]_b$, $[PS_{1(liposome)} \cdot PS_{2(bulk)}]_b = [PS_{1(liposome)}]_b$, and $[PS_{2(bulk)}]_f = [PS_{2(bulk)}]_t - [PS_{2(bulk)}]_b$. However, as $[PS_{2(bulk)}]_t$ is much larger than

Table 5 Partition Efficiency (%) of Hydrophobized Polysaccharide-Coated Liposomes in Aqueous 6% (w/w) PEO-20/8% (w/w) Pullulan-50 System[a]

Liposomal polysaccharide R^b	Partition (%)		
	Top phase	Interface	Bottom phase
None	0.5 (0.1)	97.7 (1.1)	1.8 (0.9)
CHP-55-1.7			
05	0.4 (0.1)	85.7 (4.8)	13.9 (4.7)
0.1	0.4 (0.1)	80.6 (2.8)	19.0 (2.8)
0.2	0.4 (0.1)	69.9 (1.0)	29.7 (1.1)
0.3	0.4 (0.1)	55.8 (5.1)	43.8 (5.1)
0.5	0.4 (0.1)	43.8 (4.0)	55.8 (4.0)
1	0.4 (0.1)	49.4 (6.4)	50.2 (6.4)
CHP-55-2.5			
05	0.4 (0.1)	86.7 (3.4)	12.9 (3.4)
0.1	0.5 (0.1)	81.3 (2.1)	18.2 (2.0)
0.2	0.4 (0.1)	70.1 (3.8)	29.5 (3.8)
0.3	0.4 (0.1)	56.6 (6.3)	40.0 (6.3)
0.5	0.4 (0.1)	53.2 (4.6)	46.4 (4.7)
1	0.3 (0.1)	48.3 (4.6)	51.4 (4.4)
CHP-108-1.3			
05	0.4 (0.1)	87.5 (2.2)	12.1 (2.1)
0.1	0.4 (0.1)	84.0 (1.7)	15.6 (1.7)
0.2	0.3 (0.1)	70.9 (3.8)	28.8 (3.8)
0.3	0.4 (0.1)	52.2 (8.8)	47.4 (8.8)
0.5	0.4 (0.1)	44.6 (7.3)	55.0 (7.3)
1	0.4 (0.1)	43.1 (6.6)	56.5 (6.6)
CHD-70-1.7			
05	0.3 (0.1)	88.0 (2.0)	11.7 (2.0)
0.1	0.3 (0.1)	83.6 (1.6)	16.1 (1.4)
0.2	0.3 (0.1)	76.2 (0.7)	23.5 (0.7)
0.3	0.3 (0.1)	72.4 (7.3)	27.3 (7.3)
0.5	0.3 (0.1)	69.9 (2.8)	29.8 (2.8)
1	0.3 (0.1)	72.4 (3.4)	27.3 (3.4)
CHM-33-2.3			
05	0.7 (0.1)	90.3 (1.1)	9.0 (1.1)
0.1	0.8 (0.1)	86.5 (2.8)	12.7 (2.8)
0.2	0.7 (0.1)	82.1(1.7)	17.2 (1.7)
0.3	0.7 (0.1)	72.8 (1.4)	26.5 (1.4)
0.5	0.8 (0.1)	72.0 (3.3)	27.2 (3.4)
1	0.7 (0.1)	72.2 (1.9)	27.1 (1.9)

[a] All the experiments were performed in triplicate. Values in parentheses are standard deviations.
[b] R = Weight ratio of hydrophobized polysaccharide to lipid (mg/mg).

$[PS_{2(bulk)}]_b$, $[PS_{2(bulk)}]_f$ is almost equal to $[PS_{2(bulk)}]_t$. Equation (7) can be converted to Eq. (8):

$$[PS_{1(liposome)}]_b = [PS_{1(liposome)}]_t \cdot K \cdot [PS_{2(bulk)}]_t/(1 + K[PS_{2(bulk)}]_t) \qquad (8)$$

Table 6 Binding Constant, K, for the Interaction Between the Liposomal Polysaccharides and the Bulk Polysaccharide (Pullulan-50) in the Aqueous 6% (w/w) PEO-20/8% (w/w) Pullulan-50 System

Liposomal polysaccharide	K (mg^{-1})
CHP-55-1.7	1.7 $(\pm 0.3) \times 10^{-2}$
CHP-55-2.5	1.8 $(\pm 0.2) \times 10^{-2}$
CHP-108-1.3	2.2 $(\pm 0.3) \times 10^{-2}$
CHD-70-1.7	5.5 $(\pm 0.3) \times 10^{-3}$
CHM-33-2.3	5.3 $(\pm 0.3) \times 10^{-3}$

where $[PS_{1(liposome)}]_b$ is the amount of liposomal polysaccharide partitioned to the bottom phase, $[PS_{1(liposome)}]_t$ the initial amount of liposomal polysaccharide added, $[PS_{2(bulk)}]_t$ the total amount of the polysaccharide in the bottom phase, and $[PS_{2(bulk)}]_b$ the amount of the polysaccharide interacting with the liposomal polysaccharide in the bottom phase. A plot of $[PS_{1(liposome)}]_b$ against $[PS_{1(liposome)}]_t$ yields a straight line. The binding constant K, is obtained from the slope of this straight line (Table 6).

The affinity between the liposomal polysaccharide (PS_1) and the polysaccharide in the bulk bottom phase (PS_2) controls the partitioning efficiency. The sequence of the strength of interaction between the two carbohydrates is as follows: for the PEO-dextran two-phase system, dextran$_{(liposome)}$/dextran$_{(bulk)}$ > mannan$_{(liposome)}$/dextran$_{(bulk)}$ > pullulan$_{(liposome)}$/dextran$_{(bulk)}$, and for the PEO-pullulan system, pullulan$_{(liposome)}$/pullulan$_{(bulk)}$ > dextran$_{(liposome)}$/pullulan$_{(bulk)}$ $\approx$ mannan$_{(liposome)}$/pullulan$_{(bulk)}$. Similarly, ganglioside (GM3, GD1a, GD1b, or GT1b)–reconstituted liposomes are also partitioned using the same aqueous two-phase systems (46).

D. Applications of Polysaccharide-Coated Liposomes in Biotechnology and Medicine

1. Cell Specificity of Polysaccharide-Coated Liposomes

The in vitro efficiency of internalization of the polysaccharide-coated liposomes by phagocytic cells was investigated by the radioisotope technique (47,48).

Endocytosis efficiencies for polysaccharide-coated liposomes are usually greater than those for conventional liposomes. It is considered that the internalization of polysaccharide-coated liposomes by phagocytic cells is realized through receptor-mediated endocytosis, because macrophages provide receptors specific for galactose, *N*-acetylgalactosamine, mannose, fucose and *N*-acetylglucosamine (31). Specificity of the polysaccharide-coated liposomes for human blood neutrophils and guinea pig alveolar macrophages was also investigated using CHM-coated and CHAp-coated liposomes. The uptake by these white cells is also higher than that of conventional liposomes without polysaccharide coatings. Such cell specificity for the polysaccharide-coated liposomes can be utilized in preparing drug carriers for active drug targeting in drug delivery systems (DDSs).

2. Chemotherapy of Infectious Diseases Using Macrophage-Specific Liposomes

In order to modulate more effectively the in vivo (adjuvant) activity of an immunopotentiator, Sunamoto and coworkers (31,47,49–53) attempted to encapsulate several immunomodulators in macrophage-specific liposomes coated with CHM and obtained several successful results.

The best treatment for bacterial infectious diseases involves the use of antibiotics. Therefore, it seems logical to employ macrophage-specific liposomes as carriers of antibiotics to phagocytic cells, which infectious harbors for microorganisms to survive intracellularly. When antibiotics are encapsulated in polysaccharide-coated liposomes, their bactericidal activities are greatly enhanced (54,55).

3. Antibody-Conjugated Liposomes (Immunoliposomes)

In the case of certain cancer cells, one can also use an antigen expressed on the surface of the target cancer cell. A promising technique that is used to safely bind a sensor device to the target cell consists in attaching a monoclonal antibody or its fragment to the liposomal surface. This method involves coating the outer surface of liposome with a polysaccharide derivative that bears an antibody or its fragment.

Instead of antibody, of course, we can conjugate antigen to the liposomal surface. The most important process for the enhancement of immunogenicity by endogeneous antigen is effective internalization of the antigen-loading liposomes into antigen-presenting cells (APCs). The adjuvant activity of liposomes for inducing effective humoral and cellular immune responses has been demonstrated in several systems (56–60). On the basis of this concept, in vivo immunization of WKA/H rats was also tried with the use of polysaccharide-coated liposomes bearing a human T-lymphotropic virus type 1 (HTLV-1)–related protein, *gag-env* hybrid protein (61).

4. Serum-Free Cell Culture Using Liposomes

Cell-specific liposomes can be employed as carriers for water-insoluble materials such as fatty acids, cholesterol, and lipophilic vitamins in a serum-free medium. To prepare liposomes specific for fibroblasts, CHM was further and partly phosphorylated, and egg phosphatidylcholine liposomes were coated with CHMP. The interaction between the CHMP-coated liposomes and mouse fibroblasts (L-cells) was investigated. Using such modified liposomes, in which several water-insoluble nutrients were encapsulated, serum-free culture of L cells was performed successfully (62).

E. Simulation of Function of Boundary Lipids in Cell Membranes

Extraction of intrinsic membrane proteins or enzymes from intact tissue epithelia or cells without any denaturation or deactivation is a basic requirement in the study and engineering of membrane proteins. In 1976, Kriss and coworkers (63) first demonstrated that a variety of mammalian cell components are extracted from liposomes to cells and from cells to liposomes during coincubation of the two. Huestis and coworkers (64–71) also found that membrane proteins are readily extracted from red blood cells to phospholipid liposomes.

In discussing the higher order structure, location, and function of membrane proteins in cell membranes, we cannot disregard the important role of boundary lipids (72). A boundary lipid is defined as a motion-restricted lipid that is next to a membrane protein with a specific interaction and is distinguished from other bulk matrix lipids. The boundary lipid is, therefore, considered to be strongly related to the function of biological membranes. Some still do not trust the concept of the boundary lipid. The lipid-protein interaction has been extensively investigated in various membranes since Jost et al. (72) first suggested the importance of the boundary lipids around membranous cytochrome oxidase on the basis of studies using the electron spin resonance (ESR) technique. In mammalian cell membranes, sphingomyelin has been speculated to be a candidate for a boundary lipid. The possibility of intra- and/or intermolecular hydrogen bonding in sphingomyelin bilayers was also proposed by Schmidt et al. (73). It has also been reported that most membrane-binding enzymes show no activity where no boundary lipids exist (74). Therefore, the investigation of lipid-protein interactions is very important in understanding the roles of both lipids and proteins in membranes.

Sunamoto and coworkers (75) first synthesized a lipid, 1,2-dialkylamido-1,2-deoxyphosphatidylcholine (D_nDPC, $n = 12$, 14, or 16) (Fig. 8). Membrane characteristics of D_{14}DPC-containing liposomes have been extensively studied in terms of various physicochemical aspects by fluorescence depolarization and differential scanning calorimetry (DSC) (75), ^{2}H-NMR (76), ^{31}P-NMR (77), Fourier transform infrared (FTIR) (78,79), and ESR (80) investigations. These investigations showed that D_{14}DPC behaves as a boundary lipid in lipid bilayer membranes and makes the liposomal membrane more physicochemically and biochemically stable (81). This was the first direct experimental evidence for the existence of boundary lipids.

Employment of this artificial boundary lipid improved the reconstitution efficiency of cell membrane proteins in artificial cell membranes. Moreover, it is an approach of biosimulations to understand the function of biomembranes. This would make the complex system of intact cells simpler and more convenient for investigation.

1. Reconstitution of Human Platelet Membrane Proteins (82)

The process of platelet activation and aggregation is one of the crucial steps in the blood clotting process, in which numerous proteins on the platelet membrane are involved. The structure and function of these platelet membrane proteins are the

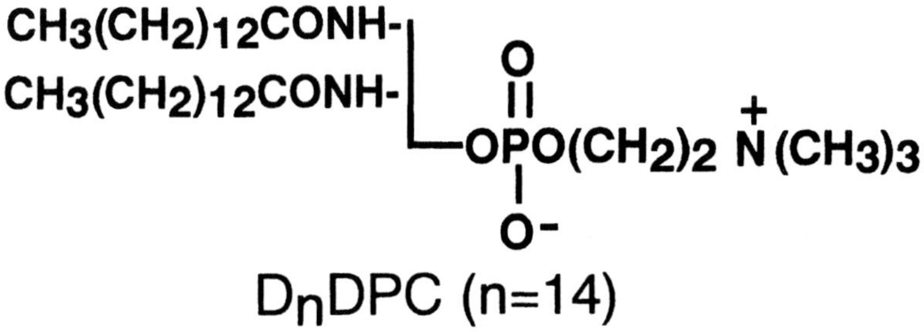

$$CH_3(CH_2)_{12}CONH-$$
$$CH_3(CH_2)_{12}CONH-$$
$$-OPO(CH_2)_2\ \overset{+}{N}(CH_3)_3$$

$$D_nDPC\ (n=14)$$

Fig. 8 Structure of D_nDPC ($n=14$).

key to understanding the mechanism of the platelet coagulation process. Some of these proteins have been identified, although the detailed process is yet to be elucidated. In studies of platelet membrane proteins, such as the search for the heparin binding protein, it is often desired to use the platelet membrane proteins in an isolated plasma membrane or in an artificially reconstituted system. This is to avoid possible interference from or complication by other platelet cell components or platelet activities, particularly activation. Platelets are easily activated by chemical, mechanical, or physiological stimulation. The activation triggers various responses of platelets such as morphological transformation, aggregation, and secretion of chemical substances including serotonin. Different techniques are employed for the isolation of the platelet plasma membrane without any activation, although these techniques generally require complex and careful processing because of the susceptibility of platelets to exogenous stimuli. Kobayashi et al. (83) reported that coincubation of rabbit platelets with an aqueous suspension of dilauroylphosphatidylcholine (DLPC) produces vesicles containing the platelet membrane proteins.

In our quest for the heparin binding proteins, we need a method for obtaining a platelet membrane protein sample that is simpler and more convenient than the conventional procedures. Direct protein extraction with $D_{14}DPC$-containing liposomes seems to be an ideal option for obtaining a membrane protein sample directly from platelets.

The extraction of platelet membrane proteins by $D_{14}DPC$-containing liposomes starts without a noticeable lag period and reaches completion after 1 h. In the case of 60 mol % $D_{14}DPC$-containing DMPC liposomes, the half-time for the extraction is less than 10 min. Table 7 summarizes the amounts of proteins and lipids found in the liposomal fraction after 1 h of incubation of platelets with liposomes of various compositions. The amount of extracted proteins increases with increasing $D_{14}DPC$ content in the DMPC liposomes.

Extraction with 80 mol % $D_{14}DPC$-containing liposomes yields approximately four times more proteins than extraction with pure DMPC liposomes. The amount

Table 7 Amounts of Proteins and Lipids in the Liposome Fractions After the Interaction with Human Platelets[a]

$D_{14}DPC$ (mol %)	Proteins extracted (mg)	Phospholipids recovered[b] (mmol)
0	28.2 (4.8)[c]	6.24 (0.52)
40	59.2 (7.4)	5.76 (0.43)
80	108.2 (18.1)	5.41 (1.47)

[a] Human platelets (3.0×10^9 cells) were incubated with liposome suspensions [final volume, 3.0 mL; total liposomal lipid concentration ($D_{14}DPC+DMPC$), 3.0 mM] at 37°C for 1 h. After the incubation, the liposomal fraction was separated by centrifugation and the protein and the lipid contents were determined.
[b] Amount of phospholipids in the liposome before the coincubation with platelet was 9 mmol.
[c] Mean of three experiments with the standard deviation in parentheses.

of the membrane protein extracted by 40 mol % $D_{14}DPC$-containing liposomes was 4.6% of the total membrane proteins. This was comparable to the value (5%) reported by Kobayashi et al. (83) for a DLPC suspension. The extraction is occasionally accompanied by secretion of a small amount of cytosolic enzyme lactate dehydrogenase (LDH) and serotonin to the exterior of the platelet. Leakage of LDH from the cell reveals loss of integrity of the platelet plasma membrane, and the amount of extracellular serotonin indicates the extent of possible activation of the platelet. After 1 h of incubation with the liposomes, 4–8% of whole platelet LDH and less than 8% of whole platelet serotonin are found at the exterior of the platelet cell. On the other hand, their secretion without the liposomal treatment is 3% and 5%, respectively. These results indicate that the extraction of the platelet membrane proteins causes neither significant lysis nor significant activation of the platelets.

No significant difference is observed in the protein band patterns, and more than 20 proteins of the parent platelet membrane are also seen in the liposomal fraction, including four densely stained bands for actin (40 kDa), GPIIIa (90 kDa), GPIIb (120 kDa), and myosin (200 kDa). No extra bands are detected in the liposomal fraction. This indicates that the direct extraction process is free from noticeable contamination by soluble cytosolic proteins or decomposition of the platelet proteins. The extraction process has some selectivity in the kinds of proteins extracted. There is no noticeable difference in the composition of the proteins extracted from the nonactivated platelets and from the preactivated platelets. An ideal method for obtaining platelet membrane proteins should meet the following criteria: should be (a) free from lysis or uncontrolled activation of platelets, (b) nondestructive to the proteins, and (c) simple. By using the $D_{14}DPC$-containing liposomes, one can effectively extract platelet membrane proteins with controlled platelet activation. Control of activation is one of the most important points in dealing with the platelet membrane because of possible modification of the membrane status by activation.

Direct protein extraction with $D_{14}DPC$-containing liposomes is a one-pot process, is clearly much more convenient than the conventional "extraction and reconstitution" procedures, and satisfies most of the requirements. Although the conventional methods may yield more proteins per platelet cell, the convenience and fidelity of the direct extraction procedure are more significant in most cases of platelet membrane protein preparation.

2. Reconstitution of the Human Erythrocyte and Its Ghost Membrane Proteins

In 1976, Dunnik et al. (63) found that during exposure of DPPC or egg PC liposomes to red blood cells (RBCs), spleen cells, or tumor cells, cell components such as phospholipids and membrane proteins transferred from these cells to the liposomes. Bouma et al. (84) found that at least four membrane proteins [band 4.2, band 7, acetylcholinesterase (AChE), and band 3] were transferred from human erythrocytes to DMPC liposomes. They also found that transferred band 3 still kept its anion transport activity and native orientation (66,85) and that the transfer of AChE was affected by the fluidity of the erythrocyte and liposome membranes (67). For proteins in liposomal membrane to be more stable, the $D_{14}DPC$-containing liposome is also adopted to erythrocytes.

a. Proteins Extracted from Human Erythrocyte Ghosts

After coincubation of erythrocyte ghosts with DMPC (40 mol %)–$D_{14}DPC$ (60 mol %) liposomes, gel filtration of the liposomal supernatant was carried out. All the proteins bound to the liposomes, and no soluble proteins were contained in this fraction. Further analysis by sucrose density gradient centrifugation showed that the liposomal supernatant was totally free of cell fragments. The size of liposomes remained unchanged after exposure to the ghost cells, indicating that the liposomes were stable enough during the exposure. Even 3 weeks after the exposure to erythrocyte ghosts, the liposomes were quite stable. An increase in the molar ratio of $D_{14}DPC$ to DMPC increased the total amount of membrane proteins extracted (Table 8). Compared with conventional DMPC liposomes, the DMPC(40)-$D_{14}DPC$(60) liposomes could extract four times more membrane proteins.

Sodium dodecyl sulfate–polyacrylamide gel electrophoresis (SDS-PAGE) analysis of the extracted erythrocyte ghost membrane proteins showed that the band at 88 kDa was stained more significantly in liposomes containing more $D_{14}DPC$. The band at 74 kDa (band 4.2) and the band around 55 kDa (band 4.5) were also extracted. The SDS-PAGE analysis revealed that proteins bigger than band 3, for example, spectrin, are not extracted under these conditions.

Cook et al. (67) reported that there was an induction period before AChE extraction from erythrocytes to DMPC liposomes. Also, in the extraction with $D_{14}DPC$-containing liposomes, an induction period was observed in the band 3 extraction, and the extraction of AChE and band 3 reached a plateau at the same time. Therefore, there is no big difference in the induction period for AChE, total proteins, and band 3 extractions (Fig. 9). Considering the large difference in structure between AChE and band 3, it is unlikely that the initiation and the termination of the extraction are due to the proteins themselves. Therefore, it is assumed that the initiation and termination of the membrane protein extraction would be regulated by other factors such as a change in cell membrane dynamics. The significant increase in total proteins extracted 100 min after the incubation may be due mostly to the

Table 8 Proteins, Phospholipids, AChE Activity, Band 3, and Cholesterol Found in Liposome Fractions After Interaction with Erythrocyte Ghosts[a]

	Liposomes (DMPC:$D_{14}DPC$ by mol)			
	100:0	80:20	60:40	40:60
Proteins (mg)	0.89	1.73	2.64	3.55
	(4.8)[b]	(9.3)	(14.3)	(19.2)
Phospholipids (mg)	10.9	10.0	10.5	9.7
	(80.3)	(74.0)	(77.4)	(72.0)
AChE activity	(9.6)	(36.5)	(51.2)	(58.7)
Band 3	(3.8)	(11.6)	(17.8)	(28.0)
Cholesterol	(8.8)	(12.8)	(18.8)	(22.2)

[a] Human erythrocyte ghosts (protein concentration, 9.24 mg/mL) were incubated with liposome suspensions (final volume, 4.0 mL; total liposomal lipid concentration, 10 mM) in 10 mM HEPES buffer containing 150 mM NaCl (pH 7.4) at 37.0°C for 2 h.
[b] Values in parentheses are relative ratios (%) of items found in the liposome to those found in the erythrocyte ghosts.

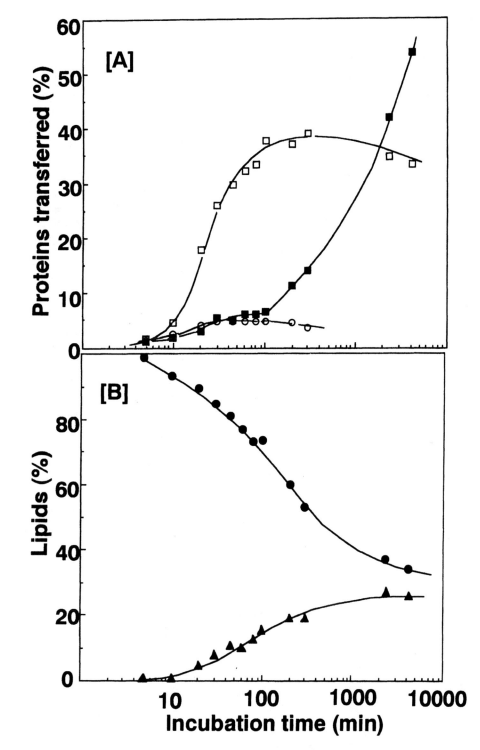

Fig. 9 Time course of protein transfer [A] (○, band 3; ■, total proteins; and □, AChE), and the amount of lipids in the liposomal supernatant [B] (▲, cholesterol; ●, total phospholipids).

further extraction of peripheral proteins, because the amount of integral proteins, other than AChE and band 3, is too small to account for the phenomenon.

b. Transfer of Blood Type Antigens from Human Erythrocytes (86)

Blood type antigens of human erythrocytes are also extracted by $D_{14}DPC$-containing liposomes. Some of the major blood group antigens are on lipids and proteins of the RBC membrane. After incubation of RBCs with $D_{14}DPC$-containing liposomes, transfer of the antigens for the blood groups ABO, Rh D, MN, and P_1 was examined by agglutination of the liposomes by respective antisera. Approximately 80% of the ABO blood group antigens are associated with carbohydrates on band 3 (87). The ABO-blood group antigens can be successfully transferred and reconstituted into the liposomes membrane without losing their antigenicity. This indicates that the antigenicity of the blood group antigens is kept even after transfer to the liposome. The presence of $D_{14}DPC$ in the liposome is crucial to the efficient transfer of the ABO antigen. On the other hand, the transfer of the Rh D, MN, and P_1 antigens is not constantly detectable by agglutination.

3. Reconstitution of α-Melanoma-Stimulating Hormone (α-MSH) Receptor (88)

Fortunately, with B16 melanoma cells, protein extraction is possible using two different cell culture systems (Fig. 10). Amounts of proteins extracted by liposomes after coincubation with mouse B16 melanoma cells are shown in Table 9.

In both culture systems, the addition of $D_{14}DPC$ to DMPC liposomes increased the total amount of proteins extracted. The amount of proteins extracted is larger in the case of suspension culture than in monolayer culture. Lipids are almost quantitatively recovered from the supernatant after sedimentation of cell pellets in both culture systems. When the liposomal supernatant is submitted to gel filtration to eliminate soluble proteins, however, some lipids are lost during the gel filtration. The amount of lipids lost is greatest in the case of the DMPC (20 mol %)–$D_{14}DPC$ (80 mol %) liposomes and is closely related to the efficiency of the protein extraction or the amount of proteins extracted (Table 9). This suggests that some of the proteins extracted to and reconstituted in the liposomes dissociate again from the liposomes accompanied by some amounts of lipids during the gel filtration procedure. This is also revealed by a change in size of liposomes after coincubation with the cells. When the $D_{14}DPC$ content is 80 mol %, the size of the liposomes increases two-

Table 9 Amounts of Proteins Extracted and Lipids Recovered After Exposure to Liposomes (1.0×10^{-4} mol) with B16 Melanoma (1.0×10^7 Cells) at 37.0°C for 60 min in Different Culture Systems

Liposome DMPC:$D_{14}DPC$ (mol %)	In suspension culture			In monolayer culture		
	Protein extracted (mg)	Lipid recovered		Protein extracted (mg)	Lipid recovered	
		mg	%		mg	%
100:0	479.0	60.3	89.0	29.5	54.1	79.8
60:40	1529.6	57.1	84.0	420.7	50.9	75.2
20:80	1891.4	24.2	36.0	1245.2	38.5	56.9

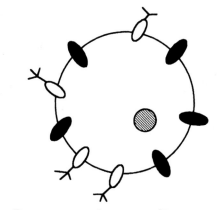

Suspension culture

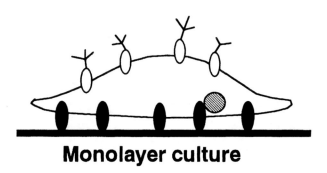

Monolayer culture

 : Proteins which concern cell attachment

 : Other receptor proteins

Fig. 10   Cell morphology in two different cell culture systems.

to threefold after the protein extraction procedure. With other liposomes, the particle size of the liposomes does not change within experimental error before and after exposure to the cells. This suggests that the extraction of greater amounts of proteins, occasionally accompanied by cellular lipids, might lead to an increase in the size of the liposome.

The most interesting finding is that the proteins extracted are quite different in both amount and kind between the two cell culture systems. When DMPC(20)-D$_{14}$DPC(80) liposomes are used with the suspension culture system, three kinds of proteins (112, 101, and 51 kDa) are mainly detected. The proteins of 112 and

101 kDa seem to be vitronectin-binding proteins responsible for cell adhesion (89). In the monolayer culture, on the other hand, proteins of 95, 77, 68, 48, and 35 kDa are detected. The protein of 48 kDa is considered to be the α-MSH (α-melanoma-stimulating hormone) receptor (90). This result suggests that the location of membrane proteins depends on whether the cells are suspended in culture medium or adhere to a substrate. In monolayer culture, most probably, proteins concerned with cell attachment to a substrate may be located on the basal side of the cell, whereas other receptor proteins are mostly located on the top of the cell facing the bulk medium phase (bottom of Fig. 10). This is analogous to the clustering that is known as "capping." In suspension culture, on the other hand, all the membrane proteins are considered to distribute rather homogeneously in the cytoplasm membrane (top of Fig. 10).

When DMPC(20)-D_{14}DPC(80) liposomes were adopted to the suspension culture system, a tumor antigen ganglioside, GM_3, was detected in the proteoliposome. This is detected by an enzyme immunostaining method using anti-GM_3 monoclonal

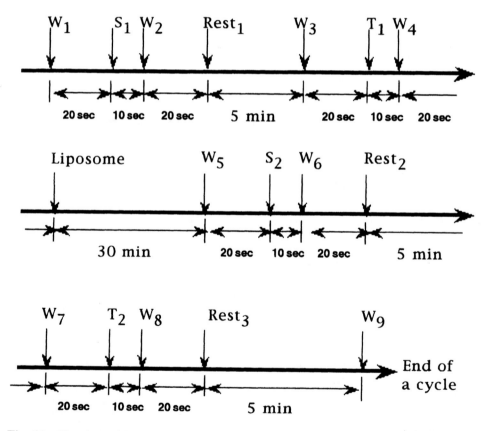

Fig. 11 Flowchart of the liposomal treatment of the frog tongue and the measurement of nerve response of the tongue against taste stimuli; W stands for washing with pure water, S for stimulation with an aqueous NH$_4$Cl solution as the standard, and T for stimulation with an aqueous solution of a taste stimulus such as L-alanine, sucrose, L-leucine, or quinine hydrochloride.

antibody (M2590). In liposomes with a $D_{14}DPC$ content less than 80 mol %, additional cellular lipids except the liposomal lipids originally employed are not detected at all by thin-layer chromatography (TLC). Judging from these results, it seems reasonable to consider that lipid exchange or extraction between the liposome and the B16 melanoma cell hardly occurs during the process of protein extraction under relatively mild conditions. Only when a relatively tough and stable liposome of high $D_{14}DPC$ content is employed are some proteins extracted together with some cellular lipids.

4. Direct Extraction of Insulin Receptor from Bovine Placenta Membrane (91)

The insulin receptor is an integral membrane glycoprotein composed of two β-subunits (MW 135,000) and two α-subunits (MW 95,000) linked together by disulfide bonds. However, the relationship between the two subunits and function is less well understood. The tyrosine kinase activity domain of the receptor is located in the cytoplasmic region of the β-subunit and is activated when insulin binds to the α-subunit. Insulin-mediated receptor kinase activation leads to receptor autophosphorylation, which enhances the ability of the receptor to catalyze the phosphorylation of exogenous substrates. The α-subunit also contains a single membrane-spanning region with a consensus ATP binding site sequence in the cytoplasmic region. The kinase activity leads to the generation of an intracellular message that carries the insulin signal to metabolic targets. However, the detailed mechanism is still controversial.

The direct protein extraction of whole insulin receptor from tissue epithelium of bovine placental amnion by $D_{14}DPC$-containing liposomes was examined with attention to the subunit construction of the extracted receptor. Most of proteins recovered in the buffer are cytosolic soluble proteins secreted from the tissue, even after extensive prewashing of the amnion surface with buffer. In contrast, the proteins extracted from the plasma membrane by the liposomes contain predominantly insulin receptor subunits. In addition, a high-molecular-weight protein corresponding to the $\alpha\beta$ subunit complex which keeps the insulin binding activity is found on the proteoliposome. The protein also carries its N-acetylglucosamine moiety, indicating that the protein holds the $\alpha\beta$ subunit complex of the insulin receptor. A competitive binding assay revealed that the insulin receptor protein maintained its full insulin binding activity even after extraction by the liposomes. This is analyzed as the bi-binding site model. The binding property was compared with that of the protein purified using an aqueous micelle. The specific binding strength significantly improved with the liposome-extracted receptors. In any event, the extracted insulin receptor on the liposome still maintained its original biochemical functions, namely specific insulin binding on the α-subunit and autophosphorylation on the β-subunit.

5. Reconstitution of Taste Receptor Proteins in Animal Tongue Epithelium to Liposomes (92,93)

Membrane proteins in tongue epithelium are of special interest in relation to taste transduction. Despite many years of research, molecular biological attempts to purify taste receptors were unsuccessful, and the receptor mechanisms involved in the sense of taste are still unclear. To investigate taste transduction, a neurophysiological study of the nerve response to taste stimuli as expressed on the surface of an animal tongue is another possible approach. To understand the mechanisms of taste transduction, therefore, it is expected to combine the two methodologies, molecular biological investigation of taste receptor proteins and neurophysiological study of taste stimuli (92). The two methods have been used independently so far.

Figure 11 shows a flowchart of the protein extraction with liposomes and subsequent measurement of the nerve response of the tongue. A healthy adult bullfrog (*Rana catesbeiana*) was employed.

A given concentration of an aqueous solution of taste stimulus (100 mM NH_4Cl at S_1 and S_2 stages, and 250 mM L-alanine, 1.0 M sucrose, 100 mM L-leucine, and 0.1 mM quinine hydrochloride, at T_1 and T_2 stages) was applied at the T_1 stage for 10 s, and the nerve response was measured. Before and after each stimulation, the tongue was thoroughly washed again with deionized water for 20 s (W_{1-8} in Fig. 11). In addition, a suitable rest was interposed for 5 min between taste stimulations, and Ringer's solution was applied to the tongue during this period. The cycle between W_5 and W_9 was repeated four or five times to ascertain the recovery of the nerve response after the liposomal treatment and the reproducibility of the measurement. To evaluate the nerve responses of the bullfrog, the initial peak (the phasic wave) height was integrated during 10 s after each taste stimulation. The response to an aqueous NH_4Cl solution was always measured before and after each taste stimulus as a control.

When the tongue is exposed to the liposomes for the period shorter than 30 min, no significant change is observed in the nerve response to NH_4Cl even 90 min after the liposomal treatment. Because it is generally accepted that salty stimuli are transduced by a stimulus influx, the result suggests that no serious damage to the ion channel occurs with salty stimuli under the conditions used. Figure 12 shows the responses of the frog gustatory nerve to four taste stimuli before and after the treatment with $D_{14}DPC$-DMPC liposomes.

To understand the results of these neurophysiological studies, gel electrophoretic analysis was also carried out for the proteins extracted by the liposomes. With soluble proteins of water washings and liposomal extracts, the major proteins were almost identical to each other. With the proteoliposome, the major proteins found were 147 kDa (12.8%), 95 kDa (6.5%), 79 kDa (4.5%), 57 kDa (2.3%), 28 kDa (3.4%), and 15 kDa (10.4%). Judging from the obvious reduction of the gustatory nerve responses to taste stimuli upon treatment with the $D_{14}DPC$-DMPC liposomes, it seems reasonable to assume that a sort of taste receptor protein of the tongue is surely lost or damaged.

This effective reconstitution of the taste receptor protein(s) on the liposome is also confirmed by both the quartz-crystal microbalance (QCM) method and affinity chromatography (93). Specific binding of the proteoliposome to an L-alanine–conjugated QCM surface is monitored by changes in the vibrational frequency of the QCM upon adsorption of the proteoliposome. The vibrational frequency change with the $D_{14}DPC$-DMPC proteoliosome is larger than those observed with other liposomes. Furthermore, the rate constant of specific binding of the liposome to the QCM surface (k_1) is estimated from the binding isotherm. The k_1 value of the $D_{14}DPC$-DMPC proteoliposome is the largest among them. Affinity chromatography study also reveals that the $D_{14}DPC$-DMPC proteoliposome shows specific binding to an L-alanine– or L-leucine–conjugated gel column. All the results strongly support the conclusion that a taste receptor protein(s) is certainly extracted with the $D_{14}DPC$-DMPC liposomes directly from the frog tongue epithelium. This method was also applied to the extraction of rat taste receptor proteins (93).

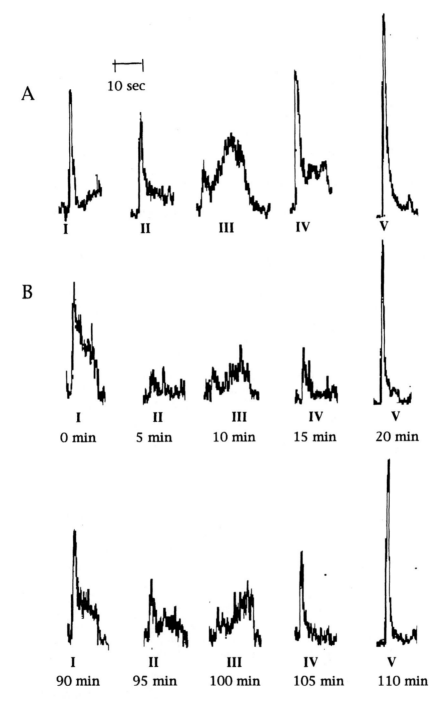

Fig. 12 Integrated gustatory nerve response of a bullfrog tongue to different taste stimuli without liposomal treatment (A) and after liposomal treatment (B). Time given below each nerve response refers the time after the liposomal treatment. I, NH_4Cl; II, L-alanine; III, sucrose; IV, L-leucine; and V, quinine hydrochloride. Amplified responses of the whole nerve were rectified and digitally integrated into 0.3 s and plotted versus time.

6. Direct Extraction of Tumor Anitigen Proteins from Tumor Cells BALB RVD (94)

Such a convenient technique was further used to produce an improved liposomal vaccine by direct extraction of a tumor surface antigenic protein (TSAP) from BALB RVD leukemia cells to the $D_{14}DPC$-containing liposomes. Tumor rejection antigens have been found on a variety of tumors. Analysis of cellular immune responses in syngeneic or semisyngeneic mice revealed that $CD8^+$ cytotoxic T lymphocytes (CTLs) are efficiently generated in vitro against leukemia in which tumor rejection antigens are demonstrated. T-cell immune responses to antigens are potentiated when protein antigens are associated with liposomes (95). This adjuvant effect of liposomes appears to be dependent on macrophages (96). To prepare a more potent liposomal vaccine, these findings required us to overcome the following problems: (a) better isolation of TSAP, (b) more effective reconstitution of TSAP to liposomes, (c) enhanced uptake of the TSAP-containing liposomes by antigen-presenting cells (APCs), and subsequently (d) more effective generation of CTLs.

Thus, membrane proteins were directly extracted from intact BALBRVD cells to the $DMPC(20)$-$D_{14}DPC(80)$ liposomes. The total amount of membrane proteins extracted from the cell to the liposomes is largely affected by the amount of $D_{14}DPC$ added. When 80 mol % $D_{14}DPC$ was added to DMPC, the highest efficiency of extraction was observed. Mixing of $D_{14}DPC$ with DMPC or egg PC liposomes led to a decrease in membrane fluidity (75). The amount of proteins extracted from the cell to the liposomes is largely related to the fluidity of the liposomal membrane.

Because most of the membrane proteins of BALB RVD are sialoproteins, sialic acid can be chosen as the marker for protein extraction. By determining the ratio of the amount of sialic acid to that of total proteins, the difference in the retention efficiency for the saccharide moieties of the glycoproteins can be investigated for two different protein extraction methods, direct liposomal extraction and n-butanol extraction. Clearly, from Table 10, the total amount of protein-bound sialic acid directly extracted from cell to the $DMPC(20)$-$D_{14}DPC(80)$ liposomes is approximately seven times greater than that extracted from the cells by n-butanol, despite the fact that the total amount of proteins is 5.5-fold less. More surprisingly, the total amount of sialic acid residues on the intact BALB RVD cells determined was 25.6 μg per 1.0×10^8 cells. This value is almost comparable to that obtained for the membrane proteins-extracted liposome (27.8 μg, see Table 10) within experimental error. These results indicate that glycoproteins on the tumor cell membrane are nicely extracted by the

Table 10 Sialic Acid Content and Total Amount of Membrane Protein as Extracted from 1.0×10^8 BALB RVD Cells by Two Different Extraction Methods: Direct Liposomal Extraction and n-Butanol Extraction

Extraction method	Sialic acid content (mg)	Total amount of proteins (mg)	Amount of sialic acid (mg) per 1.0 mg of proteins
Liposomal extraction	27.8	524	53×10^{-3}
2.5% n-butanol extraction	3.9	2860	1.4×10^{-3}

liposomes without serious damage and loss of the glyco part under the conditions employed.

To understand the cell specificity of the proteoliposome, the phagocytic behavior of mouse splenic cells with the liposomes was investigated. Compared with conventional liposomes and n-butanol extracts with reconstituted liposomes, the proteoliposomes showed a drastic increase in cell uptake. Judging from the results for specific lectin-induced aggregation of proteoliposomes, it seems reasonable to consider that mannose and/or galactose terminals of membrane glycoproteins are well represented on the surface of the liposomes and subsequently effective receptor-mediated uptake of the liposomes takes place. With the n-butanol extracts and reconstituted liposomes, the splenic cell uptake significantly decreases.

After mice were immunized with the TSAP-containing liposomes twice at an interval of 2 weeks, BALB RVD leukemia cells were challenged by intradermal injection (97). When mice are immunized with the proteoliposomes, CTLs are largely produced and the tumor growth is effectively prevented. With the n-butanol extracts and reconstituted liposomes, on the other hand, neither CTL induction nor an immunization effect is significantly observed. This was the first success in prevention of cancer in experimental animals using a TSAP-reconstituted liposomal vaccine.

a. Human Lung Cancer Cell (98)

Expecting clinical application in practice, a model was prepared using human lung cancer xenograft nude mice, and the liposomal TSAP extraction technique was applied to both the suspension and the homogenate of the xenograft. The nude mice were sensitized with the $D_{14}DPC$ proteoliposomes, the xenograft was transplanted to the sensitized mice, and then the xenograft growth was monitored for the no-treatment group and the tumor homogenate–injected group.

The transplantable human squamous cell carcinoma cells were suspended in phosphate-buffered saline (PBS) and coincubated with the $D_{14}DPC$-containing liposomes. The proteoliposomes thus obtained and the disrupted tumor cell suspension were subcutaneously injected in nude mice. The proteoliposome treatment was promising for practical clinical applications.

In conclusion of this section, the functions of membrane proteins are intrinsic in versatile cell functions. Nevertheless, the understanding of the membrane proteins or enzymes lags behind that of soluble proteins or enzymes. As already mentioned, one of the reasons is the difficulty of effective isolation from complex cell membranes, purification, and correct reconstitution into a simpler system that makes the investigation more convenient. On the basis of this fundamental idea, Sunamoto and his coworkers established an effective, simple, and reproducible methodology for reconstitution of cell membrane proteins into artificial cell membranes without loss of structure and function. Before accomplishing this, of course, one had to understand the microenvironment around membrane proteins in biomembranes to provide the most adequate situation for them even in artificial circumstances. This required us to understand the importance of the boundary lipids surrounding membrane proteins in biomembranes and led to the development of an artificial boundary lipid. As introduced briefly in this chapter, the synthesis of an artificial boundary lipid (D_nDPC) and the direct extraction of membrane enzymes or receptors from intact cells and tissue epithelium to liposomes were successful.

F. Simulation of Cell Fusion

The liposome is composed of a lipid bilayer membrane (99). The same bilayer membrane structure also constitutes the fundamental part of the cytoplasmic membrane (100). The bilayer membrane is a sort of supramolecular assembly of lipids and therefore can self-organize. As a consequence, a unique mode of interaction, membrane fusion, is observed between liposome and intact cells.

The cell-liposome fusion has potential application in medicine or cell engineering. The restricted passage of water-soluble substances through the bilayer membrane allows the liposome to hold these substances in its interior water pool. On fusion between the intact cell and liposome, the liposomal contents are directly introduced into the cytosol. This feature can be useful, for example, in obtaining local physicochemical information about the cytosol such as pH, visocity, and polarity by site-specific introduction of probe molecules. On the other hand, the fusion can be particularly valuable for introduction of substances that are susceptible to degradation with lysosomal enzymes. The introduction of such substances—genes, antisense oligonucleotides, enzymes, or peptides—is often hampered by lysosomal degradation if they are internalized via endocytosis (Fig. 13).

Several methods other than endocytosis have been proposed for transportation of various substances into the cell, e.g., microinjection (101), electroporation (102), utilization of erythrocyte ghosts (103), lipofection (104), and use of diethylaminoethyl (DEAE) dextran (105). However, some of these techniques require conditions rather harsh to the target cells. Direct fusion between liposomes and intact cells would be more moderate and convenient.

Although there are many possible applications of cell-liposome fusion, the technique has one major problem: fusion of conventional liposomes with intact cells hardly occurs. Therefore, an effective fusogen is required. Cell-liposome fusion using

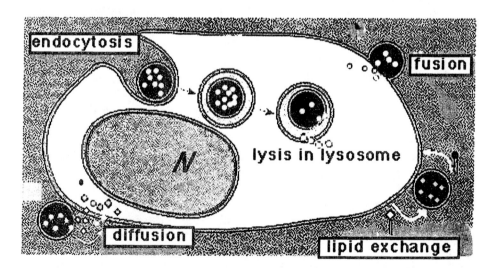

Fig. 13   Modes of cell-liposome interaction; endocytosis and lysis by lisosomal enzymes, diffusion across lipid bilayer, lipid exchange, and membrane fusion.

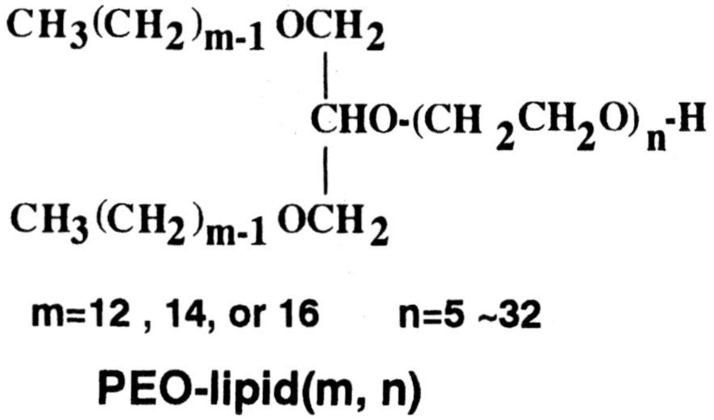

$$CH_3(CH_2)_{m-1} OCH_2$$
$$|$$
$$CHO-(CH_2CH_2O)_n-H$$
$$|$$
$$CH_3(CH_2)_{m-1} OCH_2$$

m=12 , 14, or 16   n=5 ~32

PEO-lipid(m, n)

Fig. 14   The structure of PEO-lipids.

lysozyme (106,107) and, most notably, using virus proteins was first reported. Okada and coworkers developed a liposome that is modified with a fusogenic protein of Sendai virus. Using this liposome, they demonstrated successful introduction of biologically active materials such as diphtheria toxin A fragment (DTA) and a plasmid encoding DTA into a target cell (108–111) although careful handling of the virus is required.

Poly(ethylene oxide) (PEO) has been commonly used for induction of cell-cell fusion of protoplasts (112) and several mammalian cells (113). PEO has also been applied to fusion between monolayer cultures of eukaryotic cells and glycolipid-containing liposomes that were adhered to the cell by lectin (114). However, the fusion with PEO is accompanied by problems of cytotoxicity or undesirable homofusion between cells or between liposomes.

In a quest for simple and well-regulated cell-liposome fusion under conditions that are mild to cells, an artificial lipid that bears a PEO moiety as the hydrophilic headgroup (PEO-lipids, Fig. 14) was proposed (115–120). With the hydrophobic anchor, the PEO moiety is to be localized near the outer surface of liposome, where the fusion is supposed to take place. In this section, the fusion of the PEO-lipid–modified liposomes will be introduced, paying attention to the physicochemical characteristics and cell fusogenicity.

1.   Synthesis and Physicochemical Characterization of PEO-Lipids

Membrane fusion involves reorganization of membrane lipids. Therefore, the liposomal membrane more prone to reorganization should be in favor of fusion. Presumably, this could be achieved by loosening the packing of the lipid molecules in the bilayer membrane. Considering these factors, the synthesis of an artificial lipid, PEO-lipid, and subsequent reconstitution of the lipid to liposomes were proposed (Fig. 14).

In addition, after reconstitution of PEO-lipid into liposomes, the liposomal membrane still has to provide a barrier to the leakage of encapsulated substances. This, however, favors tight lipid packing of the liposomal membrane. As a compromise, the PEO-lipids have two C_n alkyl moieties. A double alkyl chain anchor provides better barrier function than a single alkyl one, which significantly disturbs the integrity of bilayer membranes and often results in loss of barrier function. Meanwhile, C_{12} as the hydrophobic leg (shorter than liposomal phosphatidylcholine) was selected for PEO-lipid liposomes to cause as little as possible perturbation of the lipid packing.

The PEO-lipids were first synthesized by Kuwamura in 1961 (121,122) and later by Okahata et al. (123). In the procedure, the PEO moiety was constructed by polymerization of ethylene oxide in the presence of a base starting at the hydroxy group of 1,3-dialkyloxy-2-propanol. The successive addition of ethylene oxide leaves ambiguity in the number of oxyethylene units is the PEO-lipids produced. The average length of the PEO moiety is estimated by ^{1}H-NMR analysis.

An early study of Kuwamura (121) concerns the solution property of the PEO-lipids as surfactants. His study determined some of the basic properties of the PEO-lipids in aqueous media, including the clouding point and surface tension–concentration relationship. Also, Okahata et al. (123) investigated self-aggregation of several different PEO-lipids(m,n) (m=8–18 and n=6–30) in water. All of the PEO-lipids they examined, except PEO-lipids($8,n$), yielded aggregates. Different morphologies among those aggregates were shown by either static light scattering or electron microscopy with negative staining. The critical aggregate concentration (CAC) of PEO-lipids determined from surface tension measurements is in the range 10^{-5}–10^{-6} M and decreases with an increase in the alkyl chain length. In any event, the PEO-lipids themselves undergo self-aggregation in an aqueous medium (124).

A laser dynamic light scattering study of the aggregate indicated that PEO-lipid(12,a5) forms large aggregates, probably of lamellar structure, and PEO-lipid(12,a13) forms vesicles with a diameter of approximately 100 nm. PEO-lipid(12,a31) showed relatively weaker light scattering in an aqueous medium than the other two. Dynamic light scattering measurements using a powerful argon laser (488 nm, 1 W) revealed the formation of aggregates smaller than 10 nm in diameter. The aggregate of PEO-lipid(12,a31) is thus likely to be an aqueous micelle. This result is conceivable from the proposed relationship (125) between the morphology of the aggregate and the relative area occupied by the head and tail of the surfactant. With a longer PEO chain, the PEO moiety occupies more space or surface area (see later) (126,127). This makes PEO-lipid(12,a31) preferred to form micelles, which allows large headgroups. In a previous report (123), Okahata et al. stated that PEO-lipid(12,a10) forms irregular vesicles and PEO-lipid(12,a15) forms vesicles and a lamellar phase. For PEO-lipid(12,a28), Okahata et al. found a lamellar structure by electron microscopic observation.

PEO-lipids are spontaneously reconstituted into liposomes by injecting an ethanol solution of the PEO-lipids into a preformed liposome suspension. With PEO-lipid(12,a31), the PEO-lipid–reconstituted liposome can be separated from free PEO-lipid by ultracentrifugation. Hence, for example, PEO-lipid(12,a31) could be stably reconstituted into the liposome up to 30 mol % to liposomal egg PC (119).

Even after the incorporation of PEO-lipids, the liposome keeps encapsulated materials for a reasonable period of time. For the reconstitution of 30 mol % PEO-lipids, the leakage of water soluble FITC-dextran (molecular weight 19,600) from the liposome is less than 20% for 1 h. However, in the reconstitution of 60 mol % PEO-lipid(12,a13) or PEO-lipid(12,a31), drastic leakage of FITC-dextran is observed, suggesting extensive damage of the liposomal membrane.

Kodama et al. (128) examined the behavior of PEO-lipid(12,a11.5)–reconstituted DMPC liposomes by high-sensitivity differential scanning calorimetry (DSC). Incorporation of PEO-lipid into the liposomes up to 20 mol % causes a decrease in the gel–liquid crystalline phase transition temperature from 24.5°C for simple DMPC liposomes to 21.4°C. At the same time, the cooperativity in the transition decreases and the transition enthalpy becomes larger. A sharp transition peak is observed for the incorporation of up to 10 mol % PEO-lipid. At PEO lipid contents higher than 10 mol %, the DSC peak becomes broader and asymmetric, indicating that partial phase separation occurs in the liposomal membrane.

2. Liposome Fusion with Plant Protoplast

The interaction between the PEO-lipid–functionalized liposome and carrot protoplasts was also studied (115,116). Generally, plant cells are well protected by cellulose or the cell wall. To remove this protection, plant cells are first treated with cellulase. The carrot (*Daucus carota* L.) protoplasts must be used within 2 h after cellulase treatment.

PEO-lipid–functionalized liposomes containing both hydrophobic phospholipid probes (NBD-PE and Rh-PE) in the membrane and a hydrophilic probe (FITC-dextran) in the interior of the same liposome were coincubated with the protoplasts. After washing, the protoplasts were observed under a fluorescence microscope. The hydrophobic probes were localized only on the plasma membrane of the protoplasts and the hydrophilic probes were distributed in the cytoplasm. This observation strongly suggests direct fusion between the protoplast membrane and the PEO-lipid–reconstituted liposome.

The interaction was further examined by the resonance energy transfer (RET) technique. The RET donor and acceptor, N-(7-nitrobenz-2-oxa-1,3-diazol-4-yl) (NBD)-PE (1.75 mol %) and N-(Lissamine rhodamine B sulfonyl) (Rh)-PE (0.25 mol %), were embedded in the egg PC liposomal membrane, and the labeled liposomes were incubated with carrot protoplasts for 30 min at 30 min at 30 ãC. From this RET effect (Table 11), it was inferred that there are two factors for optimal cell-liposome fusion. First, the fusion depends on the PEO-lipid content in the liposome. The fusion was at its optimum with 23 mol % PEO-lipid. Second, the length of the PEO-moiety affected the fusion efficiency. The PEO-lipid with the shorter PEO moiety [PEO-lipid(12,a11.5)] showed the more fusion. However, it should be noticed again that the RET % does not correspond to the actual fusion efficiency. The result shows only the tendency of the fusion efficiency.

Table 11 Interaction of PEO-Lipid-Reconstituted Egg PC Liposome with Protoplast

PEO-lipid (mol %)	RET%	
	PEO-	PEO-
0	0	1.0
5	1.2	2.0
9	2.7	2.7
23	18.1	5.1
33	1.0	6.1
50	0	4.0

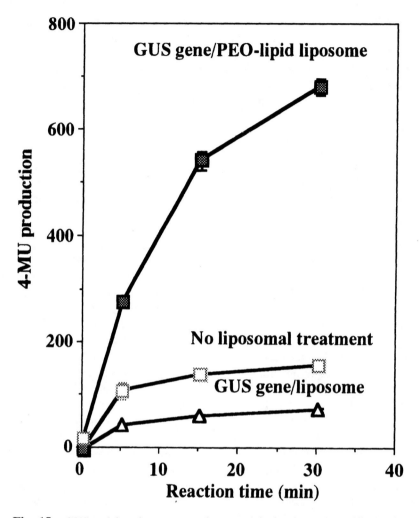

Fig. 15   GUS activity of carrot protplast cytosol after treatment with the liposomes.

The fusion between liposomes and protoplasts can be proved by *GUS* gene introduction (129). Carrot protoplasts were treated with the liposomes encapsulating the
GUS (β-glucuronidase) gene, cultured for several days, and then homogenized. Production of 4-methylumbelliferone (4-MU) from 4-methylumbelliferyl glucuronide
was monitored using fluorescence of 4-MU at 455 nm. Figure 15 clearly showed that
only the PEO-lipid–reconstituted liposomes can express the *GUS* gene.

3.   Liposome Fusion with HeLa Cells (130)

HeLa cells, cervical carcinoma, adherent cells, were chosen to investigate the
liposome-cell fusion (117). HeLa cells were separately cultured as single clones.
The cells were coincubated with PEO-lipid(12,a13)–reconstituted liposomes
encapsulating a plasmid encoding DTA (diphtheria toxin fragment A). DTA blocks
cell protein synthesis upon introduction into the cytosol. After complete removal
of free liposomes the cells were cultured for a week and the colony-forming efficiency

Table 12 Colony Formation of HeLa Cells

Treatment	PEO-lipid(12,a13)	Number of colonies[a]
Balanced salt solution	None	360.8±16.9
Liposomes+PEO-lipid	None	362.8±14.6
	30	346.5±15.8
	60	307.5±4.8
Plasmid only	None	356.3±19.0
Plasmid in liposome+PEO-lipid	None	352.5±13.7
	30	223.2±59.3
	60	244.0±10.5

[a] Mean±SE from four measurements.

was determined. As shown in Table 12, significantly retarded colony formation was observed for cells treated with the PEO-lipid–reconstituted liposomes, suggesting that the liposomes certainly fused with the HeLa cells.

4. Liposome Fusion with Jurkat Cells (131)

Higashi et al. (131) examined the interaction of PEO-lipid–reconstituted liposomes with Jurkat cells, human lymphoblastoma cells cultured in suspension using DTA. DTA-dependent toxicity to Jurkat cells was observed only when DTA was encapsulated in PEO-lipid(12,31)–reconstituted liposomes. Cytochalasin B, an inhibitor of endocytosis, did not suppress the toxicity. On the basis of these results, Higashi et al. concluded that DTA was introduced into the cytosol by cell-liposome fusion. The efficiency of introduction of DTA was estimated at 30–50%. PEO-lipids with a shorter PEO moiety [PEO-lipid(12,5) or PEO-lipid(12,13)] did not show fusogenecity to Jurkat cells. Interestingly, reconstitution with PEO-lipid(12,13) resulted in enhanced uptake of the liposome by Jurkat cells (131).

5. Liposome Fusion with B16 Melanoma Cells (132)

The effect of the cell cycle on the induction of cell-liposome fusion by PEO-lipid was investigated using B16 melanoma cells. No membrane fusion was observed between B16 melanoma and liposomes as reconstituted with 20 mol % PEO-lipid(12,31) for either an asynchronously growing or a trypsin-harvested B16 cell population. Only B16 cells in the mitotic stage underwent fusion. In contrast, conventional liposomes were taken up by B16 cells in asynchronous growth but not by cells in mitosis. Introduction of encapsulated DTA into the cells was seen for liposomes reconstituted with 20 mol % of PEO-lipid(12,31) but not with 60 mol % PEO-lipid(12,15). Therefore, in the case of B16 melanoma, the fusion requires very particular conditions: cells in the mitotic stage and liposomes functionalized with PEO-lipid(12,31).

G. Simulation of Immunogenicity

1. Direct Stimulation of T Cells by Gangliosides

Gangliosides are present on the surface of melanoma cells, and these gangliosides have been considered possible antigens in immunization against the melanoma. In general, presentation of an antigen to immunocompetent cells is affected not only by the chemical structure of the antigen but also by such factors as its dose, density, and/or mobility. Therefore, a systematic study of immunological stimulation by gangliosides requires precise control of these factors. Reconstitution of gangliosides in liposomes can provide a microsurface functionalized with the oligosaccharide moieties of the gangliosides controllable at the molecular level.

Immunization of C57BL/6 mice with ganglioside-modified egg PC liposomes revealed that GT- or GQ-modified liposomes completely suppress in vivo growth of B16 melanoma, whereas the more abundant GM- or GD-modified liposomes are not effective (133). This suppression was not accompanied by a significant increase in antibody titer and was not observed in T-lymphocyte–deficient nude mice. These observations suggest that the tumor rejection should be based on direct stimulation of T cells by gangliosides.

In vitro direct stimulation of rat T cells by ganglioside liposomes was monitored as the change in the intracellular calcium ion concentration of individual T cells, which was visualized by fluorescence from Fluo-3 incorporated in the cell cytosol and laser confocal fluorescence microscopy (134). The internal Ca^{2+} concentration increased after T cells were mixed with ganglioside-containing liposomes. This was the first finding of direct stimulation of T cells by ganglioside-reconstituted liposomes. The number of T cells responsive to the liposomes depends on the chemical structure and the surface density of the gangliosides reconstituted. Among the gangliosides examined GD_{1a}, GT_{1b}, and GQ_{1b} effectively stimulated T cells, whereas GM_3, GD_3, or GD_{1b} did not (Fig. 16).

GT_{1b} causes stimulation only when incorporated in liposomes. Interestingly, the effects of both the ganglioside structure and the dose on the T-cell stimulation

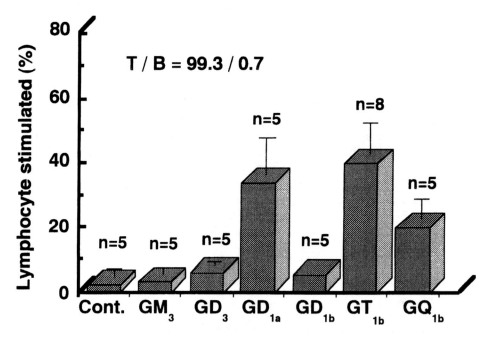

Fig. 16 Stimulation of rat T lymphocytes with liposomes containing various sialolipids at 37°C. The T/B value is the ratio of T cells to B cells.

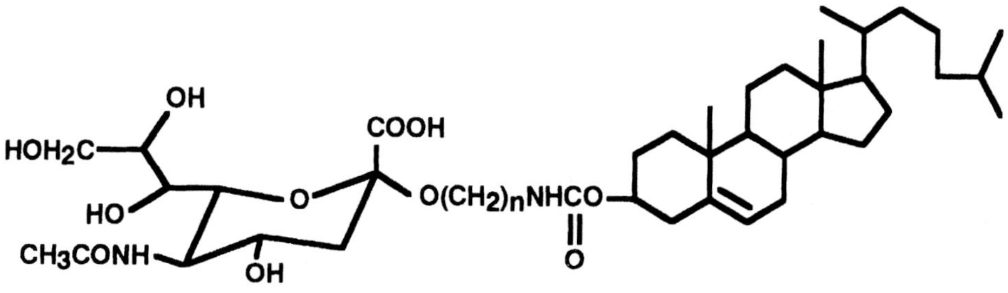

SC(n), n=2, 4

Fig. 17 Structure of sialyl cholesterol, SC(n).

coincided with those on the in vivo growth suppression of B16 melanoma. The common saccharide sequence in GD_{1a}, and GT_{1b}, NeuAc-Gal-GalNAc-, seems to be essential for the T-cell stimulation.

When the surface density of gangliosides on the liposomes was modified by changing the molar ratio of the ganglioside to the matrix lipids, the stimulation by GT_{1b} was significant only at ganglioside contents higher than 3 mol %. Above this threshold, the number of stimulated T-cells remained at a similar level up to 9 mol %.

2. Direct Stimulation of T-cell by Sialyl Cholesterols as Ganglioside Model Compounds

More systematic investigation of the direct T-cell stimulation is needed with respect to the effect of the structure, the density, and the distance from the liposomal surface of the terminal sialic acid of gangliosides. Hence, artificial sialolipids were developed by attaching a sialic acid moiety to a cholesterol anchor via a linker of different lengths (Fig. 17) (135).

Liposomes as reconstituted with these model sialolipids, sialyl cholesterols, successfully induce T-cell stimulation comparable to that by GT_{1b} and GD_{1a} (Fig. 15). This confirms that the sialic acid moiety is crucial to the T-cell stimulation. Sialyl cholesterols with the longer linker (SC6) were more effective in T-cell stimulation that those with the shorter linker (SC2), suggesting that the distance of the sialic acid moiety from the liposomal surface should be important in the recognition by T-cell receptors. The T-cell stimulation also has an apparent threshold in the amount of sialyl cholesterols incorporated in liposomes. Surprisingly, sialyl cholesterols are more effective for stimulation even at a lower content in liposomes than native gangliosides (1% for SC6 and 2% for SC2). Furthermore, sialyl cholesterols are able to induce T-cell proliferation.

III. BIOSIMULATION USING LIPID MONOLAYERS

A. Vertical Sectioning of Lipid Monolayer Using Laser Scanning Confocal Microscopic (LSCM) Technique (136)

Fluorescence microscopy has become more versatile with laser scanning confocal microscopy (LSCM). Confocal imaging (a) reduces interference from out-of-focus structures, (b) is capable of electronic contrast enhancement, and furthermore (c)

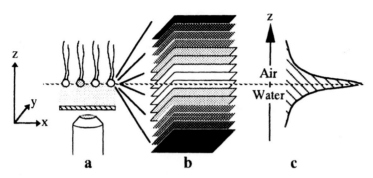

Fig. 18 General procedure for obtaining a vertical fluorescence intensity profile: (a) scanning with LSCM; (b) stack of 2-D images; (c) processing of stack of 2-D images into a fluorescence intensity profile $I(z)$.

offers the possibility of surface profiling (137). This methodology was introduced to investigate the vertical distribution of fluorescent probes at the air-water interface, and its validity was demonstrated by an example of well-investigated interactions between biotin or its derivatives and streptavidin (138,139).

The LSCM is a BioRad MRC-600 in combination with a Nikon Diaphot inverted microscope equipped with a long-distance objective (Nikon CF Plan Achromat 10×, numerical aperture 0.3). A five-phase stepper motor drive controlled by the MRC-600 software is connected to the fine focus control of the microscope. An argon ion laser operating at 488 nm and a krypton-argon mixed-gas laser (both lasers are from Ion Laser Technology, Salt Lake City, UT) operating at 488 and 568 nm are used as excitation sources. Automatic gain and black level settings of the MRC 600 control unit are switched off. For one vertical scan, a stack of two dimensional (2-D) images in x-y planes is recorded with increasing z values (Fig. 18).

The stack of intensity integrals of the 2-D images yields a vertical (z-direction) fluorescence intensity profile of the scanned volume element. With the current setup, the x-y plane can be as small as a single line scan (1416 μm) or up to 1416×944 μm^2. Therefore, the recording of a stack of 100 2-D images consumes from several minutes down to less than a second for a single line scan. For scanning the air-water interface, a miniaturized Langmuir trough with a quartz window at the bottom (USI-System, Fukuoka, Japan) is mounted on the microscopic stage.

Streptavidin can bind to biotin-conjugated lipids and spontaneously self-organizes to give a thin two-dimensional crystalline layer of the protein at the lipid-water interface. Specific binding of streptavidin to the biotin-lipid can be achieved by injecting the protein underneath the biotin-lipid monolayer. In a desthiobiotin-lipid system, the protein can be replaced by addition of free biotin to the subphase, because the affinity of streptavidin for desthiobiotin is lower than that for biotin (140). This process is followed by mixing a monolayer of DMPE-desthiobiotin-lipid with the fluorescent lipid Rh-PE. FITC-streptavidin is injected into the subphase. Finally, the streptavidin bound to the lipid layer is replaced by the addition of free biotin. The resulting fluorescence intensity profiles of rhodamine Rh-PE and FITC-streptavidin could be independently detected by two photomultipliers with one vertical scan using the 488- and 568-nm lines of the krypton-argon mixed-gas laser. The ternary interaction of the

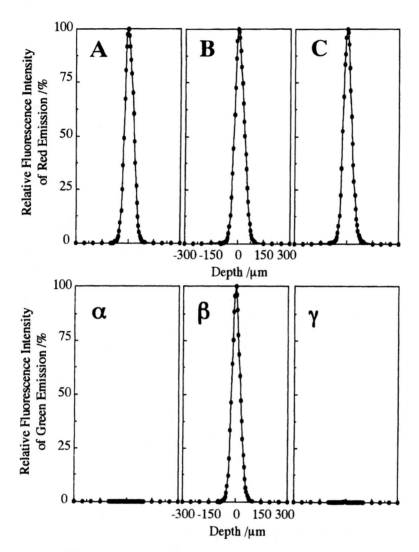

Fig. 19 Fluorescence intensity profiles of DMPE-desthiobiotin-lipid monolayer containing 1% (by weight) of rhodamine Rh-PE (A, B, C) and FITC-streptavidin (α, β, γ). Profiles A and α before the FITC-streptavidin injection; profiles B and β after the injection; and profiles C and γ after the addition of free biotin (scanned area, 1416×944 mm²).

desthiobiotin-lipid/streptavidin/biotin system is carefully documented by the LSCM vertical scans as seen in Fig. 19.

Figure 18 represents an example of intensity profiles obtained in any scan of fluorescent probe–containing monolayers that can be formed by basically any kind of preparation technique. Fluorescence is detected in the range of a couple of tens of micrometers along the monolayer's vertical section. The point with the highest fluorescence intensity is almost coincident with the air-water interface as it can be seen by focusing onto the water surface. Therefore, the peak position is assigned as the zero point of the vertical scan depths. The optical section thickness (defined as the distance $Z_{1/2}$ between the focus positions at which the collected intensity is 50% of the peak value) was found to be between 20 and 25 μm with the present system. This section thickness is well above the confocal effect that is theoretically expected (141) to be about 5 μm for our experimental setup.

This discrepancy is not surprising because the theory assumes a perfectly transparent sample, an ideal lens with a perfectly flat field, and no spherical or chromatic aberration. To increase the signal intensity, the confocal aperture in the system is usually set rather large, further reducing the confocal effect. This is, however, unavoidable because caution has to be used about photobleaching of the fluorescent probe by the scanning laser beam. Therefore, neutral density filters are used to cut the laser light to only 1% transmission, leading to weaker emission signals. Photobleaching can also be reduced by limiting both the number of single scans per entire vertical scan and the scanned area. In principle, higher concentrations of fluorescent dye in the monolayer lead to higher emission signals as long as no fluorescence-reducing processes such as quenching occurred. Therefore, it can be seen that a decrease in surface area of a given monolayer leads to consecutively stronger emission signals in the vertical scan with the LSCM. In this sense, this new methodology applies directly to what is found for conventional epifluorescence microscopy.

To demonstrate the versatility of the new methodology, another experiment, concerning the initial interaction between a CHP self-aggregate and a lipid monolayer, was performed. CHP-50-1.6 was further labeled with fluorescein (DS 0.49 per 100 glucose units, FITC-0.49-CHP-50-1.6) and injected into the subphase of a mixed D_{14}DPC-DMPC lipid monolayer. An increase in the fluorescence intensity at the air-water interface certainly indicates an interaction between the CHP aggregates and the monolayer. When the same polysaccharide but without hydrophobic modification was used (FITC-pullulan), no increase in the fluorescence intensity at the air-water interface was observed (Fig. 20).

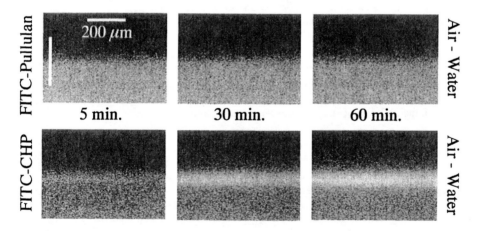

Fig. 20 Interaction of FITC-pullulan (top) and FITC-CHP-0.49-50-1.6 (bottom) (subphase concentration, 5×10^{-8} M) with D_{14}DPC-DMPC (60/40 mol %) monolayer (surface pressure, 30 mN/m). Each picture is a reconstruction of a stack of single line scans with a z-step width of 2.5 n/mm.

As can be seen in Fig. 19 for FITC-CHP, there is an increase in the fluorescence intensity at the air-water interface with time. This experiment reflects the capability of the proposed methodology to monitor (time resolved) the interaction of fluorescent substrates with the air-water interface.

In another investigation of the same system using surface pressure and surface potential measurements (142,143), a slow penetration of CHP into the DMPC lipid monolayer was observed. The proposed new methodology could reveal a rather rapid preadsorption of the FITC-0.49-CHP-50-1.6 to the lipid monolayer interface.

Fluorescence from a lipid monolayer can be vertically sectioned by scanning laser confocal microscopy over a range of several tens of micrometers. Of course, this method cannot directly compare with other analytical tools with resolution in the molecular range. However, this is a simple method for gaining information about the vertical distribution of fluorescent molecules at the air-water interface. The existence of a vertical concentration gradient of fluorescent molecules can be monitored, and quantitative comparisons of different interfacial situations can also be directly studied by means of fluorescence intensity at the air-water interface. Furthermore, this method can easily be combined with conventional epifluorescence studies, and any fluorescent molecules are applicable.

B. Simulation of Biological Phenomena at the Air-Water Interface (144)

Artificial bioactive surfaces have great potential with respect to mimicking of biomembrane processes and the creation of new materials with powerful characteristics (145). Still, the handling of bioactive materials, with a boundless number of interactions, is as complex as their inherent properties (146). This is one main reason why investigations and applications of membrane proteins are well behind those of soluble proteins. If one wants to succeed in reconstitution of membrane proteins, it is crucial to avoid any denaturation or deactivation. It has been known for several years that liposomes and biomembranes transform at the air-water interface (147). However, the systems so far described either are very complex, particularly if biomembranes were transformed (148), or require isolation and solubilization of the membrane proteins prior to the reconstitution (149) so that it is uncertain whether their natural activity and/or structure can be successfully preserved.

As previously shown (94,95,150), membrane proteins can be efficiently transferred to liposomes containing an artificial boundary lipid, $D_{14}DPC$. Membrane proteins keep their native orientation when reconstituted into the liposomal membrane. Furthermore, it is believed that $D_{14}DPC$ stabilizes the reconstituted membrane proteins in the lipid bilayer membranes (94,95,150).

The following method allows the formation of a stable protein-containing lipid monolayer from membrane protein–containing liposomes by direct protein extraction from the intact cell membrane (Fig. 21).

Membrane proteins from human erythrocyte ghosts and intact bovine erythrocytes were transferred to large unilamellar vesicles containing 60 mol % $D_{14}DPC$ and 40 mol % DMPC as already described. When a suspension of proteoliposomes so obtained is poured into a trough, a spontaneous and fast increase in surface pressure is observed. In any case, the surface pressure increase continues until the collapse pressure of the corresponding $D_{14}DPC$-DMPC monolayer spread from an organic solution is reached. It can also be shown by transforming liposomes without proteins reconstituted that, with

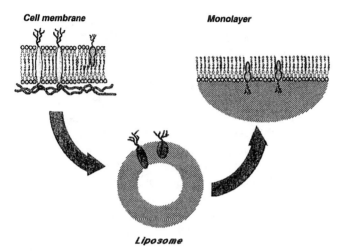

Fig. 21 Two-step direct protein transfer from intact cell membrane to lipid monolayer via liposome.

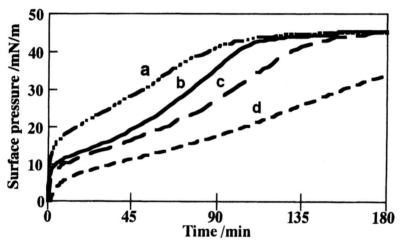

Fig. 22 Spontaneous increase in surface pressure of $D_{14}DPC$-DMPC liposomal suspensions at 37°C with (a, protein concentration 40 mg/mL, $D_{14}DPC$-DMPC molar ratio 3:2) and without ($D_{14}DPC$-DMPC molar ratios; b, 4:1; c, 3:2; d, 0:1) proteins reconstituted (lipid content 1 mmol).

increasing content of $D_{14}DPC$ in the liposomal membrane, the surface pressure increases more rapidly (Fig. 22).

For separation of the resulting surface-active film from the liposomes remaining in the subphase, the so called wet-bridge method is applicable (151). A commercially available miniaturized film balance (USI Systems, Fukuoka, Japan) is equipped with a customized trough made of polytetrafluoroethylene (PTFE) (Fig. 23).

A corrugated thin strip of wet filter paper is used to bridge between the spreading well and the trough. The whole device is placed into a temperature-controlled cabinet to keep the temperature constant at 37°C. The atmosphere inside the box is saturated

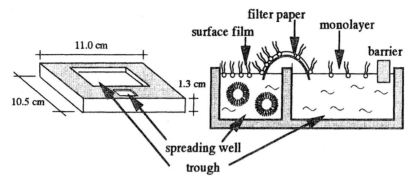

Fig. 23 Sketch of trough and "wet bridge" (top view and cross section).

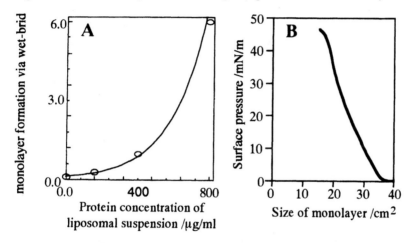

Fig. 24 (A) Velocity of monolayer formation (size of monolayer formed via wet bridge per time) at 15 mN/m from proteoliposomes as a function of protein concentration of liposomal suspension. (B) Isotherm of D_{14}DPC-DMPC monolayer with membrane proteins at 37°C.

with moisture to prevent the filter paper bridge from drying. After the barrier is moved close to the end of the trough where the filter paper dips into the subphase, the liposomal suspension (lipid content 1–2 mmol) is poured into the spreading well. An increase in the surface pressure of the trough compartment indicates the transformation of liposomes in the spreading well and the expansion of the surface film over the paper bridge to the trough due to the lateral gradient in surface pressure. The barrier is gradually moved so that the surface film gradually expands at a constant surface pressure onto the surface of the trough. It is necessary to form the surface-active film at a constant surface pressure of 10–15 mN/m. It is known that proteins undergo conformational changes under zero or very low surface pressure conditions at the air-water interface (152). Conditions are usually chosen so that the trough is fully covered with the surface-active film within a few minutes. At lower surface pressures (i.e., at 10 mN/m rather than 15 mN/m), the velocity of monolayer formation via the wet bridge is significantly higher. Also, at higher protein concentrations (153) of the liposomal suspension, the transformation process is almost exponentially accelerated (Fig. 24).

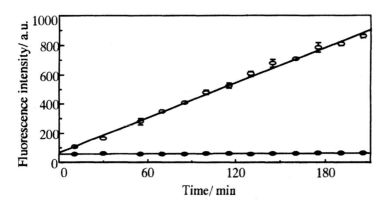

Fig. 25 Increase of thiocholine concentration in the subphase due to AChE hydrolysis of the monolayer. Fluorescence intensity corresponds directly to thicholine concentration. (○) For transformation of proteoliposomes. (●) For conventional liposomes without proteins.

The wet bridge is removed, and the surface-active film is allowed to stand for at least 30 min at 15 mN/m. After this procedure, the surface pressure (adjusted to a certain value between 15 and 40 mN/m) remains stable for at least several hours, and surface pressure–area isotherms show only slightly visible hysteresis. Furthermore, at about 45 mN/m, collapse of the surface layer is observed (Fig. 23). Because of this typical behavior, the surface-active film is referred as a monolayer. If the monolayer is further expanded and the surface pressure drops to 0 mN/m, a slow time-dependent increase in the surface pressure subsequently occurs. Such behavior is interpreted in terms of unfolding of proteins at the air-water interface (152).

To demonstrate that membrane proteins are certainly transferred from erythrocytes to the monolayer via liposomes (153), the enzymatic activity of acetylcholinesterase (AchE) transferred was investigated in the monolayer. This enzyme is anchored via a phosphatidylinositol moiety in the membrane of erythrocytes and the active site is not located in the membrane (154). Therefore, this enzyme should retain its activity even if transferred to a monolayer at the air-water interface. Acetylthiocholine was injected as a substrate into the subphase and the thiocholine produced was quantitatively determined by a picomole fluorescence assay (155). Enzymatic activity was found over at least 3 h (Fig. 25). In addition, after cleaning the water surface by aspiration, no further increase in enzymatic activity was observed, ensuring the reconstitution of enzymatic activity at the interface. Of course, it is proved that the subphase is not contaminated with liposomes.

REFERENCES

1. AD Bangham, W Horne. Negative staining of phospholipids and their structural modification by surface active agents as observed in the electron microscope. J Mol Biol 8:660, 1964.
2. O Lehmann. Die Flussige Kristalle. Leipzig: Akdemische Verlagsgesell-schaft, 1911.
3. DD Lasic, Y Barenholz, eds. Handbook of Nonmedical Applications of Liposomes. Theory and Basic Sciences. Vols I–IV. Boca Raton, FL: CRC Press, 1996.

4. D Papahadjopoulos, ed. Liposomes and Their Use in Biology and Medicine. Vol 308. Ann NY Acad Sci 38:1978.

5. CG Knight, ed. Liposomes: From Physical Structure to Therapeutic Application. Amsterdam: Elsevier, North Holland Biomedical Press, 1981.

6. K Yagi, ed. Medical Applications of Liposomes, Special Issue. Tokyo: Scientific Societies Press. 1986.

7. KH Schmist, ed. Liposomes as Drug Carriers. Stuttgart: George Theime, 1986.

8. G Gregoriadis, ed. Liposomes as Drug Carriers. New York: Wiley, 1988.

9. G Gregoriadis, ed. Liposome Technology. 2nd ed. Vols 1, 2, 3. Boca Raton, FL: CRC Press, 1993.

10. JR Phillippot, F Schuber, eds. Liposomes as Tools in Basic Research and Industry. Boca Raton, FL: CRC Press, 1995.

11. T Sato, J Sunamoto. Recent aspects in the use of liposomes in biotechnology and medicine. Prog Lipid Res 31:345, 1992.

12. E-C Kang, K Akiyoshi, J Sunamoto. Surface coating of liposome with hydrophobized polysaccharide. J Bioactive Compatible Polym 12:14, 1997.

13. HJ Rogers, HR Perkins, JB Ward, eds. Microbial Cell Walls and Membranes. London: Chapman & Hall, 1980.

14. BD Davis, R Dulbecco, HN Eisen, HS Ginsberg, eds. Microbiology. New York: Harper & Row, 1980.

15. AL Lehninger, ed. Biochemistry. New York: Worth, 1975.

16. DE Brooks, GVF Seaman. Electroviscous effect in dextran-erythrocyte suspensions. Nature 238:251, 1972.

17. M Minetti, P Aducci, V Viti. Interaction of neutral polysaccharides with phosphatidylcholine multilamellar liposomes. Phase transitions studied by the binding of fluorescein-conjugated dextrans. Biochemistry 18:2541, 1979.

18. J Sunamoto, K Iwamoto. Liposomal membranes. XII. Adsorption of polysaccharides on liposomal membranes as monitored by fluorescence depolarization. J Biochem 91:975, 1982.

19. J Sunamoto, T Sato, T Taguchi, H Hamazaki. Naturally occurring polysaccharide derivatives which behave as an artificial cell wall on artificial cell liposomes. Macromolecules 25:5665, 1992.

20. K Seki, DA Tirrell. pH-dependent complexation of poly(acrylic acid) derivatives with phospholipid vesicle membranes. Macromolecules 17:1692, 1984.

21. M Maeda, A Kumano, DA Tirrell. H^+-induced release of contents of phosphatidylcholine vesicles bearing surface-bound polyelectrolyte chains. J Am Chem Soc 110:7455, 1988.

22. G Decher, E Kuchinka, H Ringsdorf, J Venzmer, D Bitter-Suermann, C Weisgerber. Interaction of amphiphilic polymers with model membranes. Angew Makromol Chem 166/167:71, 1989.

23. H Ringsdorf, J Venzmer, FM Winnik. Interaction of hydrophobically-modified poly-N-isopropylacrylamides with model membranes—or playing a molecular accordion. Angew Chem Int Ed Engl 30:315, 1991.

24. G Gluck, H Ringsdorf, Y Okumura, J Sunamoto. Vertical sectioning of molecular assemblies at air/water interface using laser scanning confocal fluorescence microscopy. Chem Lett 209, 1996.

25. B Demé, V Rosilio, A Baszkin. Polysaccharides at interfaces. 1. Adsorption of cholesteryl-pullulan derivatives at the solution-air interface. Kinetic study. Colloids Surf B 4:357, 1995.

26. Y Baba, Y Kawano. Selective adsorption of hemoglobin with copper-phthalocyanine trisulfonate derivative immobilized on gel beads. Chem Lett 181, 1994.

27. K Akiyoshi, J Sunamoto. Supramolecular assembly of hydrophobized polysaccharide. Supramol Sci 3:157, 1996.

28. EC Kang, J Scheurer, F Mellinger, C Imiya, K Akiyoshi, J Sunamoto, unpublished results.

29. J Sunamoto, K Iwamoto, H Kondo, S Shinkai. Liposomal membranes. VI. Polysaccharide-induced aggregation of multilamellar liposomes of egg lecithin. J Biochem 88:1219, 1980.

30. FC Szoka Jr, K Jacobson, D Papahadjopoulos. The use of aqueous space markers to determine the mechanism of interaction between phospholipid vesicles and cells. Biochim Biophys Acta 551:295, 1979.

31. M Monsigny, C Kieda, A-C Roche. Membrane lectins. Biol Cell 36:289, 1979.

32. RL Hudgin, WE Pricer, G Ashwell, RJ Stockert, AG Morrell. The isolation and properties of a rabbit liver binding protein specific for asialoglycoproteins. J Biol Chem 249:5536, 1974.

33. T Wileman, R Boshaus, P Stahl. Uptake and transport of mannosylated ligands by alveolar macrophages. Studies on ATP-dependent receptor-ligand dissociation. J Biol Chem 260:7387, 1985.

34. HD Fisher, A Gauzalez-Noriega, WS Sly, DJ More. Phosphomannosyl-enzyme receptors in rat liver. Subcellular distribution and role in intracellular transport of lysosomal enzymes. J Biol Chem 255:9608, 1981.

35. I Taniguchi, K Akiyoshi, J Sunamoto. Self-Aggregate Nanoparticles of Cholesteryl and Galactoside Groups-Substituted Pullulan and Their Specific Binding to Galactose Specific Lectin, RCA_{120}. Macromol Chem Phys. 200:1554, 1999.

36. K Matsuoka, SI Nishimura. Synthetic glycoconjugates. 5. Polymeric sugar ligands available for determining the binding specificity of lectins. Macromolecules 28:2961, 1995.

37. K Kobayashi, A Tsuchida, T Ushui, T Akaike. A new type of artificial glycoconjugate polymer: A convenient synthesis and its interaction with lectins. Macromolecules 30:2016, 1997.

38. GB Sigal, M Mammen, G Dahman, GM Whitesides. Polyacrylamides bearing pendant a-sialoside groups strongly inhibit agglutination of erythrocytes by influenza virus: The strong inhibition reflects enhanced binding through cooperative polyvalent interactions. J Am Chem Soc 118:3789, 1996.

39. MI Khan, MK Mathew, P Balaram, A Surolia. Fluorescence-polarization studies on binding of 4-methylumbelliferyl beta-D-galactopyranoside to *Ricinus communis* (castor-bean) agglutinin. Biochem J 191:395, 1980.

40. GS Jacob, C Kirmaier, SZ Abbas, SC Howard, CN Steininger, JK Welply, P Scudder. Binding of sialyl Lewis x to E-selectin as measured by fluorescence polarization. Biochemistry 34:1210, 1995.

41. PA Albertsson. Partition of Cell Particles and Macromolecules. New York: Wiley, 1986.

42. H Walter, DE Brooks, D Fisher. Partitioning in Aqueous Two-Phase Systems. Theory, Methods, Uses and Applications to Biotechnology. New York: Academic Press, 1985, p. 41.

43. E Eriksson, PA Albertsson. The effect of lipid composition on the partition of liposomes in aqueous two-phase systems. Biochim Biophys Acta 507:425, 1978.

44. PT Sharpe, GS Warren. The incorporation of glycolipids with defined carbohydrate sequence into liposomes and the effects on partition in aqueous two-phase systems, Biochim Biophys Acta 772:176, 1984.

45. E-C Kang, K Akiyoshi, J Sunamoto. Partition of polysaccharide-coated liposome in an aqueous two-phase system. J Biol Macromol 16:348–353, 1994.

46. EC Kang, K Akiyoshi, J Sunamoto. Partition of ganglioside-reconstituted liposomes in aqueous two-phase systems. J Bioactive and Compatible Polym. in press.

47. J Sunamoto, T Sato. Improved drug delivery directed to specific tissue using polysaccharide-coated liposomes. In: T Tsuruta, A Nakajima, eds. Multiphase Biomedical Materials. Amsterdam: VSP, 1989, p 167.

48. J Sunamoto, T Sato, RM Ottenbrite. Enhanced biological activity of polymeric drugs by encapsulating in liposomes. In: RM Ottenbrite, LA Utracki, S Inoue, eds. Current Topics in Polymer Science. Vol 1. New York: Hanser, 1987, p 101.

49. J Sunamoto, T Sato. Development of cell-specific liposomes and its application in biotechnology. J Chem Soc Jpn Chem Ind Chem 1989:161, 1989.

50. T Sato, K Kojima, T Iida, J Sunamoto, RM Ottenbrite. Macrophage activation by poly(maleic acid-alt-2-cyclohexyl-1,3-doxap-5-ene) encapsulated in polysaccharide-coated liposomes. J Bioactive Compatible Polym 1:448, 1986.

51. RM Ottenbrite, J Sunamoto, T Sato, K Kojima, K Sahara, K Hara, M. Oka. Improvement of immunopotentiator activity of polyanionic polymers by encapsulation into polysaccharide-coated liposome. J Bioactive Compatible Polym 3:184, 1988.

52. M Akashi, H Iwasaki, N Miyauchi, T Sato, J Sunamoto, K Takemoto. Potent immunomodulating activity of (polyvinyladenine and vinyladenine-alt-maleic acid)copolymer. J Bioactive Compatible Polym 4:124, 1989.

53. M Oka. Enhancement of antitumor activity in mouse alveolar macrophages by immunoactivators encapsulated within polysaccharide-coated liposomes. Acta Med Nagasaki 34:88, 1989.

54. S Kohno, T Miyazaki, K Yamaguchi, H Tanaka, T Hayashi, M Hirota, A Saito, K Hara, T Sato, J Sunamoto. Polysaccharide-coated liposome with antimicrobial agents against intracytoplasmic pathogen and fungus. J Bioactive Compatible Polym 3:137, 1988.

55. S Kohno, T Miyazaki, K Yamaguchi, H Tanaka, T Hayashi, M Hirota, A Saito, K Hara, T Sato, J Sunamoto. Polysaccharide-coated liposome with antimicrobial agents against intracytoplasmic pathogen and fungus. J Bioactive Compatible Polym 3:137, 1988.

56. G Gregoriadis. Liposomes for drugs and vaccines. Trends Biotechnol 3:235, 1985.

57. G Steele Jr, T Ravikumar, D Ross, V King, RE Wilson, T Dodson. Specific active immunotherapy with butanol-extracted, tumor-associated antigens incorporated into liposomes. Surgery 96:352, 1984.

58. SJ LeGrue. Carrier and adjuvant properties of liposome-borne tumor-specific antigens. Cancer Immunol Immunother 17:135, 1984.

59. O Bakouche, D Gerlier. Enhancement of immunogenicity of tumour virus antigen by liposomes: The effect of lipid composition. Immunology 58:507, 1986.

60. LP Kahl, CA Scott, R Lelchuk, G Gregoriadis, FY Liew. Vaccination against murine cutaneous leishmaniasis by using *Leishmania major* antigen/liposome. J Immunol 142:4441, 1989.

61. Y Noguchi, T Noguchi, T Sato, Y Yokoo, S Itoh, M Yoshida, T Yoshiki, K Akiyoshi, J Sunamoto, E Nakayama, H Shiku. Priming for in vitro and in vivo anti HTLV-1 cellular immunity by virus related protein reconstituted into liposome. J Immunol 146:3599, 1991.

62. T Sato. Characterization of polysaccharide-coated liposomes and their applications. Ph.D. thesis, Kyoto University, 1990.

63. JK Dunnik, JD Rooke, S Aragon, JP Kriss. Alteration of mammalian cells by interaction with artificial lipid vesicles. Cancer Res 36:2385, 1976.

64. SR Bouma, FW Drislane, WH Huestis. Selective extraction of membrane-bound proteins by phospholipid vesicles. J Biol Chem 252:6759, 1977.

65. AC Newton, SL Cook, WH Huestis. Transfer of band 3, the erythrocyte anion transporter, between phospholipid vesicles and cells. Biochemistry 22:6110, 1983.

66. WH Huestis, AC Newton. Intermembrane protein transfer. Band 3, the erythrocyte anion transporter, transfers in native orientation from human red blood cells into the bilayer of phospholipid vesicles. J Biol Chem 261:16274, 1986.

67. SL Cook, SR Bouma, WH Huestis. Cell to vesicle transfer of intrinsic membrane proteins: Effect of membrane fluidity. Biochemistry 19:4601, 1980.

68. AC Newton, WH Huestis. Biochemistry 27:4655, 1988.

69. E Ferrell, WH Huestis. Phosphoinositide metabolism and the morphology of human erythrocytes. J Cell Biol 98:1992, 1984.

70. E Ferrell Jr, K-J Lee, WH Huestis. Membrane bilayer balance and erythrocyte shape: A quantitative assessment. Biochemistry 24:2849, 1985.

71. E Yang, WH Huestis. Mechanism of intermembrane phosphatidylcholine transfer: Effects of pH and membrane configuration. Biochemistry 32:12218, 1993.

72. PC Jost, OH Griffith, RA Capaldi, G Vandertkooi. Evidence for boundary lipid in membranes. Proc Natl Acad Sci USA 70:480, 1973.

73. CF Schmidt, Y Barenholz, TE Thompson. A nuclear magnetic resonance study of sphingomyelin in bilayer systems. Biochemistry 16:2649, 1977.

74. H Saderman Jr. Biochim Biophys Acta 515:209, 1978.

75. J Sunamoto, M Goto, K Iwamoto, H Kondo, T Sato. Synthesis and characterization of 1,2-dimyristoylamido-1,2-deoxyphosphatidylcholine as a boundary lipid model. Biochim Biophys Acta 1024:209, 1990.

76. J Sunamoto, K Nagai, M Goto, B Lindman. Deuterium nuclear magnetic resonance studies on the interaction of glycophorin with 1,2-dimyristoylamido-1,2deoxyphosphatidylcholine. Biochim Biophys Acta 1024:220, 1990.

77. Z Zhou, Y Okumura, J Sunamoto. NMR study of choline methyl group of phospholipids. Proc Jpn Acad 72(B):23, 1996.

78. T Kawai, J Umemura, T Takenaka, M Goto, J Sunamoto. Fourier transform infrared study on the phase transitions of a 1,2-dimyristoylamido-1,2-deoxyphosphatidylcholine–water system. Langmuir 4:449, 1988.

79. DW Grainger, J Sunamoto, K Akiyoshi, M Goto, K Knutson. Mixed monolayers and cast films of acylester- and acylamido- phospholipids. Langmuir 8:2479, 1992.

80. M Goto, J Sunamoto. Effect of artificial boundary lipid on the membrane dynamics of human glycophorin-containing liposome. Bull Chem Soc Jpn 65:3331, 1992.

81. J Sunamoto, M Goto, K Akiyoshi. Structural stability of lecithin liposomes as improved by adding an artificial boundary lipid. Chem Lett 2141, 1990.

82. Y Okumura, M Ishitobi, M Sobel, K Akiyoshi, J Sunamoto. Transfer of membrane proteins from human platelets to liposomal fraction by interaction with liposomes containing an artificial boundary lipid. Biochim Biophys Acta 1194:335, 1994.

83. T Kobayashi, H Okamoto, J Yamada, M Setaka, T Kwan. Biochim Biophys Acta 778:210, 1984.

84. SR Bouma, FW Drislane, WH Huestis. Selective extraction of membrane-bound proteins by phospholipid vesicles. J Biol Chem 252:6759, 1977.

85. AC Newton, SL Cook, WH Huestis. Transfer of band 3, the erythrocyte anion transporter, between phospholipid vesicles and cells. Biochemistry 22:6110, 1983.

86. K Suzuki, Y Okumura, T Sato, T Yasuda, A Oki, M Oki, J Sunamoto. Direct extraction of A and B blood group antigens from human red cells by liposomes. Transfusion 36:966, 1996.

87. WM Watkins, P Greenwell, AD Yates, PH Johnson. Regulation of expression of carbohydrate blood group antigens. Biochimie 70:1579, 1988.

88. J Sunamoto, Y Mori, T Sato. Direct transfer of membrane proteins form B16 melanoma cell to artificial cell liposome. Proc Jpn Acad 68(B):69, 1992.

89. R Pytela, MD Pierschbacher, E Ruoslahti. A 125/115-kDa cell surface receptor specific for vitronectin interacts with the arginine-glycine-aspartic acid adhesion sequence derived from fibronectin. Proc Natl Acad Sci USA 82:5766, 1985.

90. T Scimonelli, AN Eberle. Photoaffinity labelling of melanoma cell MSH receptors. FEBS Lett 226:134, 1987.

91. T Ueda, J Sunamoto. In Preparation; preliminary findings presented at the 4th Pacific Polymer Conference, Kauai, Hawaii, December 1995.

92. M Nakamura, K Tsujii, Y Katsuragi, K Kurihara, J Sunamoto. Biochem Biophys Res Commun 201:415, 1994.

93. M Nakamura, K Tsujii J Sunamoto. Liposome-induced release of cell membrane proteins from intact tissue epithelium. J Med Biol Eng Comp 36:645, 1998.

94. J Sunamoto, T Noguchi, T Sato, K Akiyoshi, R Shibata, E Nakayama, H Shiku. Direct transfer of tumor surface antigen–presenting protein (TSAP) from tumor cell to liposome for making liposomal vaccine. J Controlled Release 20:143, 1992.

95. O Bakouche, D Gerlier. Presentation of an MuLV-related tumour antigen in liposomes as a potent tertiary immunogen after adoptive transfer. Immunology 57:219, 1986.

96. PD Monte, FC Szoka Jr. Effect of liposome encapsulation on antigen presentation in vitro. Comparison of presentation by peritoneal macrophages and B cell tumors. J Immunol 142:1437, 1989.

97. R Shibata, T Noguchi, T Sato, K Akiyoshi, J Sunamoto, H Shiku, E Nakayama. Induction of in vitro and in vivo anti-tumor responses by sensitization of mice with liposomes containing a crude butanol extract of leukemia cells and transferred intermembranously with cell surface proteins. Int J Cancer 48:434, 1991.

98. T Ariyasu, O Ike, S Hitomi, H Wada, Y Okumura, J Sunamoto. Sensitization of nude mice using direct liposome transfer of tumor cell antigens. J Bioactive Compatible Polym 11:191–202, 1996.

99. DD Lasic. Liposomes, from Physics to Applications. Amsterdam: Elsevier; 1993.

100. RB Gennis. Biomembranes, Molecular Structure and Function. New York: Springer-Verlag, 1989.

101. JB Gurdon, HR Lana, G Marbaix. Use of frog eggs and oocytes for the study of messenger RNA and its translation in living cells. Nature 233:177, 1971.

102. U Zimmermann. Electric field–mediated fusion and related electrical phenomena. Biochim Biophys Acta 694:227, 1982.

103. M Furusawa, T Nishimura, M Yamaizumi, Y Okada. Injection of foreign substances into single cells by cell fusion. Nature 249:449, 1974.

104. PL Felgner, TR Gadek, M Holm, R Roman, HW Chan, M Wenz, JP Northrop, GM Ringold, M Danielsen. Lipofection: A highly efficient, lipid-mediated DNA-transfection procedure. Proc Natl Acad Sci USA 84:7413–7417, 1987.

105. A Vaheri, JS Pagano. Virology 27:434, 1965.

106. T Arvinte, K Hildenbrand, P Wahl, C Nicolau. Lysozyme-induced fusion of liposomes with erythrocyte ghosts at acidic pH. Proc Natl Acad Sci USA 83:962–966, 1986.

107. T Arvinte, P Wahl, C Nicolau. Low pH fusion of mouse liver nuclei with liposomes bearing covalently bound lysozyme. Biochim Biophys Acta 899:143, 1987.

108. T Uchida, J Kim, M Yamaizumi, Y Miyake, Y Okada. Reconstitution of lipid vesicles associated with HVJ (Sendai virus). Purification and some properties of vesicles containing nontoxic fragment A of diphtheria toxin. J Cell Biol 80:10, 1979.

109. M Nakanishi, T Uchida, H Sugawa, M Ishiura, Y. Okada. Efficient introduction of contents of liposomes into cells using HVJ (Sendai virus). Exp Cell Res 159:399, 1985.

110. K Kato, M Nakanishi, Y Kaneda, T Uchida, Y Okada. Expression of hepatitis B virus surface antigen in adult rat liver. Co-introduction of DNA and nuclear protein by a simplified liposome method. J Biol Chem 266:3361, 1991.

111. M Nakanishi, K Ashihara, T Senda, T Kondo, K Kato, T Mayumi, In: VHL Lee, M Hashida, Y Mizushima, eds. Trends and Future Perspectives in Peptide and Protein Drug Delivery. Amsterdam: Harwood Academic Publishers, 1995.

112. KN Kao, MR Michayluk, Method for high-frequency intergeneric fusion of plant protoplasts. Planta 115:355, 1974.

113. OF Ahkong, JI Howell, JA Lucy, F Safwat, MR Davey, EC Cocking. Fusion of hen erythrocytes with yeast protoplasts induced by polyethylene glycol. Nature 255:66, 1975.

114. F Szoka, K-E Magnisson, J Wojcieszyn, Y Hou, A Derzko, K Jacobson. Use of lectins and polyethylene glycol for fusion of glycolipid-containing liposomes with eukaryotic cells. Proc Natl Acad Sci USA 78:1981. 1685–1689.

115. J Sunamoto, K Tanaka, K Akiyoshi, T Sato. Development of fusogenic liposome. Polym Prepr 31:155, 1990.

116. T Sato, J. Sunamoto. Recent aspects in the use of liposomes in biotechnology and medicine. Prog Lipid Res 31:345, 1992.

117. Y Okumura, M Yamauchi, M Yamamoto, J Sunamoto. Interaction of a fusogenic liposome with HeLa cell. Proc Jpn Acad 69(B):45, 1993.

118. N Higashi, J Sunamoto. Endocytosis of poly(ethylene oxide)–modified liposome by human lymphoblastoid cells. Biochim Biophys Acta 1243:386, 1995.

119. N Higashi, Y Okumura, M Yamauchi, M Nakanishi, J Sunamoto, Fusion between Jurkat cell and PEO-lipid modified liposome, Biochim Biophys Acta 1285:183, 1996.

120. M Haratake, J Sunamoto, Interaction of poly(ethylene oxide)–bearing lipid reconstituted liposomes with murine B16 melanoma cells. Abstract of 5th Ikctani Conference–International Symposium on Biomedical Polymers, Kagoshima, 1995, p 210.

121. T Kuwamura. Preparation and properties of higher glyseryl a,a'-diethers. Kogyo Kagaku Zasshi 64:1958, 1961.

122. T Kuwamura. Nonionic surfactants derived from glyseryl higher diethers. Kogy Kagaku Zasshi 64:1965, 1961.

123. Y Okahata, S Tanamochi, M Nagai, T Kunitake. Synthetic bilayer membranes prepared from dialkyl amphiphiles with nonionic and zwitterionic head groups. J Colloid Interface Sci 82:401, 1981.

124. Y Okumura, N Morone, J Sunamoto. Unpublished results.

125. JN Israelachvili, S Marcelja, RG Horn. Physical principles of membrane organization. Q Rev Biophys 13:121, 1980.

126. V Rosilio, G Albrecht, A Baszkin, Y Okumura, J Sunamoto. Monolayers of poly(ethylene oxide)–bearing lipids at air-water interface. Chem Lett 1996:657, 1996.

127. V Rosilio, G Albrecht, Y Okumura, J Sunamoto, A Baszkin. Surface properties and miscibility of monolayers of dimyristoylphosphatidylcholine and poly(ethylene oxide) lipids at the water/air interface. Langmuir 12:2544, 1996.

128. M Kodama, S Tsuchiya, K Nakayama, Y Takaichi, M Sakiyama, K Akiyoshi, K Tanaka, J Sunamoto. Thermal characterization of a polyethyleneglycol (PEG)-derivative induced vesicle fusion as revealed by high sensitivity differential scanning calorimetry. Thermochim Acta 163:81, 1990.

129. J Zheng, K Tanaka, J Sunamoto. Unpublished results.

130. Y Okumura, M Yamauchi, M Yamamoto, J Sunamoto. Interaction of a fusogenic liposome with HeLa cell. Proc Jpn Acad 69(B):45, 1993.

131. N Higashi, M Yamauchi, Y Okumura, M Nakanishi, J Sunamoto. Fusion between Jurkat cell and PEO-lipid modified liposome. Biochim Biophys Acta 1285:183, 1996.

132. M Haratake, J Sunamoto. Abstracts of Papers, 5th Iketani Conference–International Symposium on Biomedical Polymers, Kagoshima, Japan, 1995, p 210.

133. J Sunamoto, H Shiku. Effective transfer of membrane proteins from intact cells to liposomes and preparation of liposomal vaccine. Ann NY Acad Sci 613:116, 1990.

134. E Kato, K Akiyoshi, T Furuno, M Nakanishi, A Kikuchi, K Kataoka, J Sunamoto. Interaction between ganglioside-containing liposome and rat T-lymphocyte: Confocal fluorescence microscopic study. Biochem Biophys Res Commun 203:1750, 1994.

135. E Kato, K Akiyoshi, J Sunamoto. Presented at 1995 International Chemical Congress of Pacific Basin Societies, Honolulu, 1995.

136. G Glück, H Ringsdorf, Y Okumura, J Sunamoto. Vertical sectioning of molecular assemblies at air/water interface using laser scanning confocal fluorescence microscopy. Chem Lett 1996:209, 1996.

137. CJR Sheppard, A Choudhury. Opt Acta 24:1051, 1977; T Wilson. Scanning optical microscopy. Scanning 7:79, 1985; JG White, WB Amos, M Fordham. An evaluation of confocal versus conventional imaging of biological structures by fluorescence light microscopy. J Cell Biol 105:41, 1987.

138. W Müller, H Ringsdorf, E Rump, G Wildburg, X Zhang, L Angermaier, W Knoll, M Liley, J Spinke. Attempts to mimic docking processes of the immune system: Recognition-induced formation of protein multilayers. Science 262:1706, 1993.

139. M Ahlers, R Blankenburg, DW Grainger, P Meller, H Ringsdorf, C Salesse. Specific recognition and formation of two-dimension streptavidin domain in monolayers: Applications to molecular devices. Thin Solid Films 180:93, 1989.

140. M Ahlers, M Hoffmann, H Ringsdorf, AM Rourke, E Rump. Specific interaction of desthiobiotin lipids and water-soluble biotin compounds with streptavidin. Makromol Chem Macromol Symp 46:307, 1991.

141. GQ Xiao, GS Kino. A real-time confocal scanning optical microscope. Scanning Imaging Technol SPIE 809:107, 1987.

142. A Baszkin, V Rosilio, F Puisieux, G Albrecht, J Sunamoto. The effect of polysaccharide adsorption on surface potential of phospholipid monolayers spread at water-air interface. Chem Lett 299, 1990.

143. A Baszkin, V Rosilio, G Albrecht, J Sunamoto. Cholesteryl-pullulan and cholesteryl-amylopectin interactions with egg phosphatidylcholine monolayers. J Colloid Interface Sci 145:502, 1991.

144. G Glück, Y Okumura, J Sunamoto. Reconstitution of membrane proteins into lipid monolayer. Two-step transfer technique: From cell to liposome, from liposome to lipid monolayer. Chem Lett 1995:1031, 1995.

145. H Ringsdorf, B Schlarp, J Venzmer. Molecular architecture and function in polymeric oriented systems. Models for the study of organization, surface recognition, and dynamics in biomembranes. Angew Chem Int Ed Engl 27:113, 1988.

146. H Sandermann. Regulation of membrane enzymes by lipids. Biochim Biophys Acta 515:209, 1978.

147. F Pattus, P Desnuelle, R Verger. Spreading of liposomes at the air/water interface. Biochim Biophys Acta 507:62, 1978.

148. F Pattus, C Rothen, M Streit, P Zahler. The spreading of biomembranes at the air-water interface: Structure, composition, enzymatic activities of human erythrocyte and sacroplasmic reticulum membrane films. Biochim Biophys Acta 647:29, 1981.

149. T Wiedmer, U Brodbeck, P Zahler, BW Fulpius. Interactions of acetylcholine receptor and acetylcholinesterase with lipid monolayers. Biochim Biophys Acta 506:161, 1978.

150. K Suzuki, Y Okumura, T Sato, J Sunamoto. Membrane protein transfer from human erythrocyte ghosts to liposomes containing an artificial boundary lipid. Proc Jpn Acad 71B:93, 1995.

151. S-P Heyn, RW Tillmann, M Egger, HE Gaub. A miniaturized micro-fluorescence film balance for protein-containing lipid monolayers spread from a vesicle suspension. J Biochem Biophys Methods 22:145, 1991.

152. JC Skou. Influence of the degree of unfolding and the orientation of the side chains on the activity of a surface-spread enzyme. Biochim Biophys Acta 31:1, 1959.

153. P Böhlen, S Stein, W Dairman, S Udenfrien. Fluorometric assay of proteins in the nanogram range. Arch Biochem Biophys 155:213, 1973.

154. I Silman, AH Futerman. Modes of attachment of acetylcholinesterase to the surface membrane. Eur J Biochem 170:11, 1987.

155. R Parvari, I Pecht, H Soreq. A microfluorometric assay for cholinesterases, suitable for multiple kinetic determinations of picomoles of released thiocholine. Anal Biochem 133:450, 1983.

11

Enzymatic Reactions at Interfaces: Interfacial and Temporal Organization of Enzymatic Lipolysis

I. Panaiotov
University of Sofia, Sofia, Bulgaria

R. Verger
Laboratoire de Lipolyse Enzymatique, CNRS, Marseilles, France

I. INTRODUCTION

Lipases (1–5) and phospholipases (6,7) are water-soluble enzymes that can catalyze the hydrolysis of the ester bonds of triglycerides and phospholipids, respectively. They play an important role in lipid metabolism and have been found in most organisms from the microbial (8–10), plant (11,12), and animal kingdoms (13,14).

The enzymatic lipolytic reaction is an important example of heterogeneous catalysis. The water-soluble lipolytic enzymes act at the interfaces of insoluble lipid substrates, where the catalytic reactions are coupled with various interfacial phenomena. The mechanisms of the enzymatic lipolysis depend strongly on the mode of organization of the lipid substrate in interfacial structures such as monolayers, micelles, liposomal dispersions, or oil-in-water emulsions.

In the past two decades, various kinetic models have been proposed to describe the interfacial mechanisms of enzymatic lipolysis. In the simplest one, proposed by Verger et al. (15), the coupling between the classical Michaelis-Menten chemical step and the process of penetration and interfacial activation of the enzyme is described. Instantaneous solubilization of the reaction products is assumed. By using this simple approach, many kinetic experiments on the hydrolysis of synthetic medium-chain lipids, generating water-soluble lipolytic products, have been analyzed.

However, the natural substrates of lipolytic enzymes are, in fact, long-chain lipids generating water-insoluble lipolytic products. The accumulation of the insoluble products at the interface can lead either to surface substrate dilution and inhibition or to autoacceleration of the enzyme binding, enhancing the hydrolysis process. In general, the kinetic models must take into account the processes of interfacial molecular reorganization and segregation of the insoluble lipolytic products, modifying the "interfacial quality" of the lipid structures. In the presence of water-soluble acceptors, a process of complexation and solubilization of the lipolytic products into the aqueous subphase takes place. In the presence of inhibitors that interact either directly with the enzyme or indirectly by affecting the "interfacial quality" of the substrate, a process of interfacial "competitive inhibition" occurs.

When the lipid substrate is organized in micelles, liposomal dispersions, or emulsions, the special colloid chemical features of these systems play a role in the hydrolysis kinetics. In such dispersed systems, the interface/volume ratio is very large and the contribution of the phenomena taking place at the interface is amplified. One must also take into account the following processes: the possible exchange of enzyme molecules between lipid particles, the lipid exchange and the substrate replenishment in the micelles when individual molecules are hydrolyzed to form water-soluble products, the reorganization of the reaction products in the lipid bilayers leading to alteration and destabilization of the spherical liposomal structures, etc.

Various kinetic models describing the role of the interfacial processes just mentioned stem from the work of Verger et al. (15). One of these models is the "surface dilution model" describing catalysis with mixed micelles (16,17). Another important adaptation to liposomal dispersions is the "scooting and hopping modes" of enzyme action (18,19). From these models, information about the mechanism of enzyme action and some kinetic parameters have been obtained. However, it is still not easy to interpret the values of the kinetic constants deduced from these kinetic schemes. Some confusion exists in the literature because many investigators have considered that the Verger (15), Dennis (16,17), and Jain (18,19) models, as well as their extensions, are restricted exclusively to monolayer, micellar, and vesicular systems, respectively. We would like to illustrate in this chapter that these models are, in fact, interconnected and to what extent they sometimes overlap. Our objective is to unify and generalize the ideas developed to describe the catalytic steps of the lipolytic enzymes acting upon various interfacial structures.

The recent progress in our understanding of the three-dimensional (3D) structures of the lipolytic enzymes as well as the molecular events of interfacial catalysis will be briefly reviewed. Some attractive applications of lipases will also be listed.

II. STRUCTURAL ASPECTS OF SOME LIPOLYTIC ENZYMES

The elementary catalytic steps of the action of lipases are based on a triad, composed of a nucleophilic serine residue activated by an hydrogen bond network in relay with hystidine and aspartate or glutamate (Fig. 1).

A unique property of the lipases is the phenomenon of "interfacial activation" (20,21). The activity of lipases is low on monomeric solutions of lipid substrates but strongly enhanced once organized lipid structures are formed. This property is quite different from that of the usual esterases acting on water-soluble carboxylic ester

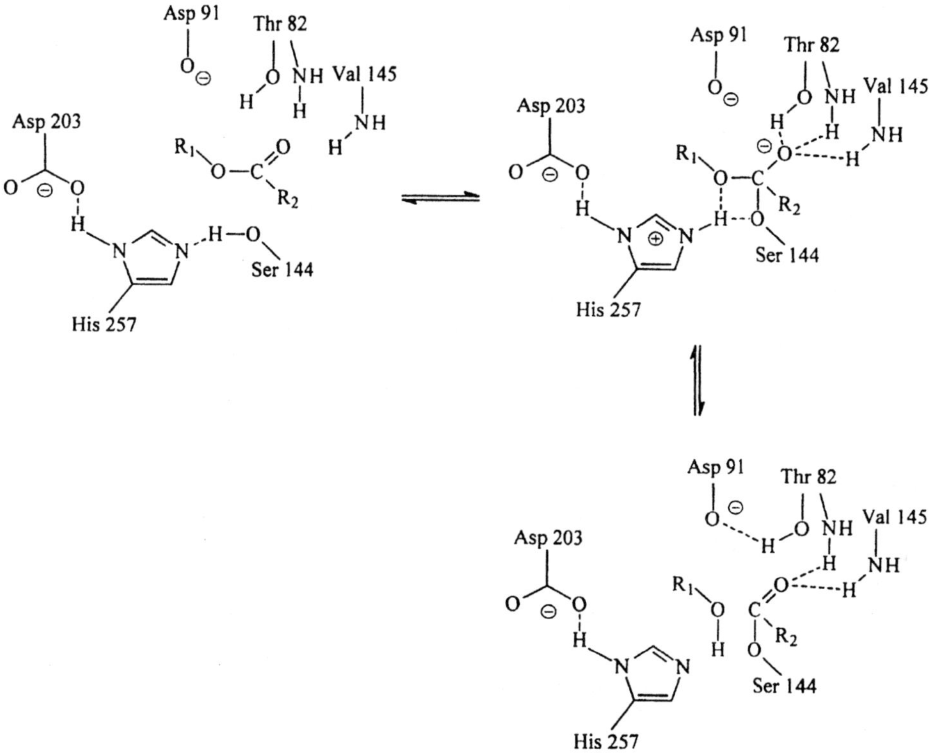

Fig. 1 Catalytic mechanism of lipases based on a "catalytic triad" of serine (nucleophile), histidine, and aspartate or glutamate (hydrogen bond relay). The tetrahedral intermediate is stabilized by an "oxyanion hole." Numbering of amino acids refers to lipase from *Rhizopus oryzae*.

molecules, and for a long time lipases were defined as a special class of esterases acting specifically at the interfaces of insoluble lipid substrates.

The existence of a conformation change of the enzyme at the water-lipid interface was postulated very early by Desnuelle and coworkers (20,21). However, our knowledge of the molecular aspects of some catalytic steps is essentially based on structural information obtained more recently by X-ray crystallography, site-directed mutagenesis, and classical biochemical methods. In 1990, the first two lipase structures were solved by X-ray crystallography and revealed a unique mechanism, unlike that of any other enzyme: their three-dimensional structures suggested that the "interfacial activation" phenomenon might be due to the presence of an amphiphilic peptidic loop covering the active site of the enzyme in solution, just like a lid or flap (22,23). From the X-ray structure of cocrystals between lipases and substrate analogues or inhibitors, there is strong indirect evidence that, when contact occurs with a lipid-water interface, this lid undergoes a conformational rearrangement, rendering the active site accessible to the substrate (24,25). More structural and numerous biochemical data on highly purified lipases provided evidence that not all lipases participate in the phenomenon of interfacial activation. Thus, the lipases with known tertiary structure originating from *Pseudomonas glumae* (26) and *Candida antarctica* B (27) have an amphiphilic lid covering the active site but do not show interfacial activation. Among the pancreatic lipases with known tertiary structures, human pancreatic lipase contains a 23-amino-acid lid and shows interfacial activation (25), whereas guinea pig pancreatic lipase does not, probably due to the presence of a minilid composed of 5 amino acid residues (28).

Human pancreatic lipase (HPL) is the major lipolytic enzyme involved in the digestion of dietary triglycerides (29). Pancreatic lipase hydrolyzes primary ester bonds of tri- and diglycerides, thus generating 2-monoglycerides and fatty acids, which are adsorbed through the intestinal barrier, in the form of mixed micelles with bile salts (5).

A specific lipase-anchoring protein, colipase, is present in the exocrine pancreatic juice. It forms a 1:1 complex with the lipase that facilitates its adsorption to bile salt–covered lipid-water interfaces (5,25). The general features of interfacial catalysis by the pancreatic lipase–colipase complex have been particularily well investigated and refined at a molecular level. First, X-ray studies of the 3D structure of HPL by Winkler et al. (22) confirmed the existence of two distinct domains in pancreatic lipase: a larger N-terminal domain comprising residues 1–335 and a smaller C-terminal domain made up of residues 336–449 (Fig. 2). The large N-terminal domain is a typical α/β hydrolase fold dominated by a central parallel β-sheet. It contains the active site with a catalytic triad formed by Ser 152, Asp 176, and His 263, all of which are conserved in other members of the mammalian lipase family, e.g., in lipoprotein lipase and hepatic lipase. The active site is covered by a surface loop between the disulfide-bridge Cys 237 and 261. This surface loop includes a short one-turn α-helix with a tryptophan residue (Trp 252) completely buried and sitting directly on top of the active site Ser 152. Under this closed conformation, the "lid" prevents the substrate from having access to the active site.

The open structure of the lipase-procolipase complex illustrates how colipase might anchor the lipase at the interface in the presence of the bile salts: colipase binds to the noncatalytic β-sheet of the C-terminal domain of HPL and exposes the hydrophobic tips of its fingers at the opposite side of its lipase-binding domain. This hydrophobic surface, in addition to the hydrophobic back side of the lid as well as the β9-loop, helps to bring the catalytic N-terminal domain of HPL into close contact with the lipid-water interface.

Apart from the apolipoprotein C_{II} activating lipoprotein lipase, no colipase has been found in other organs or organisms. Thus, the mechanism discussed seems to be specific to the case of the pancreatic lipase-colipase system.

During the past few years, new results on the mechanisms of the interfacial activation of lipases were obtained (30–38) and have been reviewed (39,40).

III. KINETIC MODELS OF INTERFACIAL ENZYMATIC LIPOLYSIS

A. Basic Kinetic Model

The simplest adaptation of the Michaelis-Menten kinetic model to the interfacial hydrolysis of synthetic short- and medium-chain lipids generating soluble products is illustrated in Fig. 3. The first step is the fixation of a water-soluble enzyme (E) to the lipid-water interface by means of reversible adsorption-desorption mechanisms. The penetration (adsorption) step, leading to a more favorable energetic state of the enzyme (E*), is followed by a two-dimensional Michaelis-Menten catalytic step. The enzyme in the interface (E*) binds a substrate molecule (S) to form the (E*S)

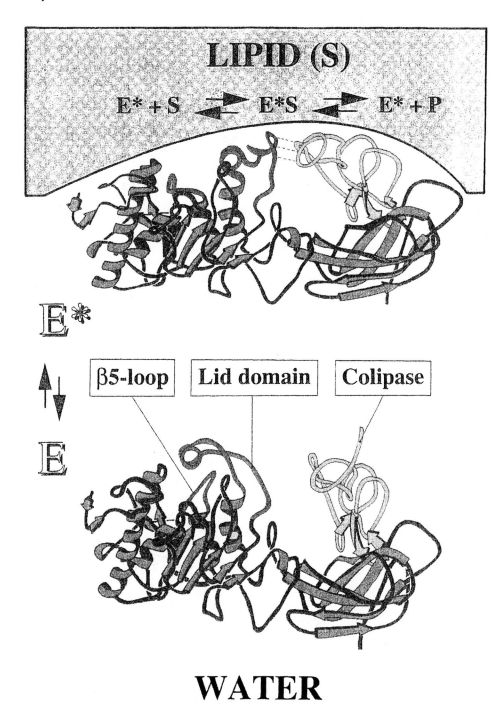

Fig. 2 Structure of the HPL-procolipase complex in the closed conformation (E) and structure of the HPL-procolipase complex in the open conformation (E*S). These two structures show the conformational changes in the lid, the β5-loop, and the colipase during "interfacial activation." (Adapted from Refs. 24 and 25.)

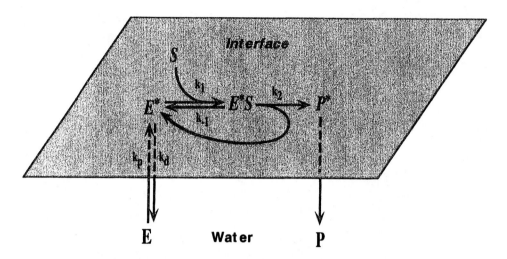

Fig. 3   Adaptation of the Michaelis-Menten kinetic scheme describing interfacial catalysis of short- and medium-chain lipids with soluble reaction products. (From Ref. 39.)

complex, followed by its decomposition. The reaction products P* are soluble in the aqueous phase and diffuse instantaneously away (Fig. 3).

Various intermediate states between E and E*, for instance, an adsorbed but nonactivated form E_a, have been suggested (41–43). The actual experimental evidence does not allow us to distinguish such intermediate states of the enzyme. For this reason, we prefer to use the simplest model of Fig. 3.

To be consistent with the fact that the enzyme-catalyzed reaction occurs at the interface, the concentrations of E, E*, S, E*S, P*, and P must be expressed as bulk concentrations C_E and C_P and surface concentrations Γ_{E*}, Γ_S, Γ_{E*S}, and Γ_{P*} (molecules cm^{-2}), respectively. The following kinetic and mass-conservation equations correspond to this simplified model:

$$\frac{d\Gamma_{E*S}}{dt} = k_1\Gamma_{E*}\Gamma_S - (k_2 + k_{-1})\Gamma_{E*S} \tag{1}$$

$$\frac{d\Gamma_{E*}}{dt} = k_p C_E + (k_2 + k_{-1})\Gamma_{E*S} - (k_d + k_1\Gamma_S)\Gamma_{E*} \tag{2}$$

$$\frac{d\Gamma_{P*}}{dt} = k_2\Gamma_{E*S} \tag{3}$$

$$C_{E0}V = C_E V + \Gamma_{E*}A + \Gamma_{E*S}A \tag{4}$$

The mass conservation equation describes the partitioning of the enzyme at an initial bulk concentration C_{E0}, between the aqueous bulk phase (volume V) and the lipid interface (area A).

By solving Eqs. (1)–(4) with the experimental initial condition $t = 0$, $\Gamma_{P* \, \text{st}} = 0$, assuming that the surface substrate concentration Γ_S remains constant during the hydrolysis, the following expression for the product surface concentration Γ_{P*} can be obtained (15):

$$\Gamma_{P*} = \frac{k_2 \Gamma_S C_{E0}}{K_m^* \dfrac{k_d}{k_p} + (K_m^* + \Gamma_s)\dfrac{A}{V}}\left[\tau + \frac{\tau_1^2}{\tau_1 - \tau_2}\left(e^{-t/\tau_1} - 1\right) - \frac{\tau_2^2}{\tau_1 - \tau_2}\left(e^{-t/\tau_2} - 1\right)\right] \quad (5)$$

where τ_1 is the induction time, describing the establishment of the penetration-desorption steady state, τ_2 is the characteristic time of establishment of the interfacial catalytic steady state, and $K_m^* = (k_2 + k_{-1})/k_1$ (molecules cm^{-2}) is the interfacial Michaelis-Menten constant.

At steady state, from Eq. (5), the following expression for the hydrolysis rate v is obtained (15):

$$v \equiv \frac{d\Gamma_{p*}}{dt} = \frac{k_2 \Gamma_s}{K_m^* \dfrac{k_d}{k_p} + (K_m^* + \Gamma_s)\dfrac{A}{V}} C_{E0} \quad (6)$$

The coefficient of proportionality between v and C_{E0} describes the coupling between the enzymatic catalytic action at the lipid interface and the enzyme partitioning between the aqueous bulk phase and the surface. This coefficient contains the ratio between the specificity constant k_2/K_m^* and the volume-surface partitioning of the enzyme k_d/k_p at steady state. When the characteristic time of enzyme desorption is larger than the characteristic time of the catalytic steps ($k_d \ll k_2/K_m^*$), the enzyme molecules are confined at the interface during many catalytic turnover cycles. This kinetic situation, corresponding to better catalytic efficacy (larger enzymatic rate v) has been called the "scooting mode" (18,19). On the contrary, when the enzyme molecules desorb after few catalytic cycles, the kinetic situation corresponds to lower catalytic efficacy (smaller enzymatic rate v) and has been called the "hopping mode" (18,19).

Formula (6) can, in principle, be adapted to various interfacial structures: monolayers at the air-water and oil-water interfaces, dispersions of bilayer liposomes; micelles, and oil-in water emulsions.

A more general scheme is illustrated in Fig. 4. The two-dimensional Michaelis-Menten catalytic steps ($E^* + S \rightleftarrows E^* S \rightarrow P^*$) are coupled with the following

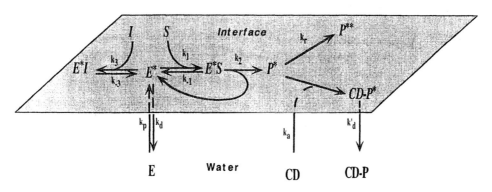

Fig. 4 General model of interfacial catalysis. (From Ref. 39.)

processes: the enzyme partitioning between the bulk phase and the surface ($E \rightleftarrows E^*$), the accumulation and molecular reorganization of the insoluble reaction products ($P^* \rightarrow P^{**}$), the competitive inhibition of the interfacial reaction ($E^*+I \rightleftarrows E^*I$), and the solubilization of the insoluble lipolytic products in the presence of acceptors ($CD+P^* \rightarrow CD-P$). The general scheme must also take into account the possible exchange of enzyme, substrate, and product molecules between the dispersed particles leading to alteration and destabilization of the colloid chemical system.

There exists an impressively large body of experimental data on the kinetics of hydrolysis. As an introduction to this field, we have first given a list of the definitions of rates and related quantities most commonly used in this context. In order to simplify the reader's difficult task, we have used the same units throughout this chapter as those listed in Table 1 (for example, time is expressed throughout in minutes and not in seconds).

B. Lipid Monolayers as Substrates for Lipolytic Enzymes

1. Soluble Reaction Products

The lipid monolayer at the air-water interface is one of the simplest traditional well-defined model systems for studying lipolysis. In this case, an important simplification of Eq. (6) is possible. It is experimentally established that in the monolayer system a limited number of enzyme molecules initially present in aqueous phase has penetrated in the forms E^* and E^*S. A typical order of magnitude of the volume-interface enzyme partitioning at steady state is $k_d/k_p = C_E/\Gamma_{E^*} \approx 10^2$ cm^{-1}. The typical value of the area-volume ratio is $A/V \approx 1$ cm^{-1} and the following inequality is fulfilled (15,44):

$$K_m^* \frac{k_d}{k_p} \gg \left(K_m^* + \Gamma_s \right) \frac{A}{V} \tag{7}$$

From Eqs. (6) and (7), the following simplified equation for the enzymatic rate in the monolayer, v_m, at steady state is obtained:

$$v_m \equiv \frac{d\Gamma_{P^*}}{dt} = \frac{k_2}{\dfrac{k_d}{k_p} K_m^*} \Gamma_s C_{E0} \equiv Q_m \Gamma_s C_{E0} \tag{8}$$

where $Q_m \equiv (1/C_{E0}\Gamma_s)(d\Gamma_{P^*}/dt) = k_2/(k_d/k_p).K_m^*$ is the overall kinetic constant called the interfacial quality (15), taking into account the influence of the physicochemical properties of the interface on the enzyme activity.

Under these conditions $C_E \approx C_{E0}$ and

$$\frac{k_d}{k_p} = \frac{C_{E0}}{\Gamma_{E^*}} \tag{9}$$

Table 1 Definition of Rates and Related Quantities

Quantities	Dimensions (units)	Comments
Hydrolysis rate $\dfrac{dC_p}{dt} = -\dfrac{dC_S}{dt}$	Bulk product concentration (time)$^{-1}$ (μmol L^{-1} min^{-1})	As usual, the hydrolysis rate can be expressed in terms of either product appearance or substrate disappearance.
$\dfrac{d\Gamma_{P*}}{dt} = -\dfrac{d\Gamma_S}{dt}$	Interfacial product concentration (time)$^{-1}$ (molecule cm^{-2} min^{-1})	The product or substrate concentration could be expressed either in volume or in surface units.
International units (IU) $\dfrac{dm_P}{dt}$	Product mass time^{-1} (μmol min^{-1})	Universal expression of the amount of enzyme based on its catalytic activity.
Specific hydrolysis rate Turnover (k_2) $\dfrac{1}{C_{E0}}\dfrac{dC_P}{dt}$	time^{-1} (min^{-1})	The classical way to express the specific hydrolysis rates
Specific activity (SA) $\dfrac{1}{C_{E0}}\dfrac{dC_P}{dt}$	Substrate mass time^{-1} (enzyme mass)$^{-1}$ (μmol min^{-1} mg^{-1}) $\equiv$ (IU mg^{-1})	The usual empirical way to express the same quantity in other units
Interfacial specific activity (SA*) $\dfrac{1}{\Gamma_{E*}}\dfrac{d\Gamma_{P*}}{dt}$	Substrate mass time^{-1} (enzyme mass)$^{-1}$ (μmol min^{-1} mg^{-1})	An adaptation to heterogeneous catalysis
Specificity constant $\dfrac{k_2}{K_m}$	(time)$^{-1}$ (bulk substrate concentration)$^{-1}$ (L min^{-1} mol^{-1})	The classical Michaelis-Menten expression of the enzyme efficacy in homogeneous systems
$\dfrac{k_2}{K_m^*}$	(time)$^{-1}$ (interfacial substrate concentration)$^{-1}$ (cm^2 min^{-1} molecule^{-1})	An adaptation in heterogeneous medium
Interfacial quality (Q) $\dfrac{1}{C_{E0}\Gamma_S}\dfrac{d\Gamma_{P*}}{dt}$	Volume time^{-1} (enzyme mass)$^{-1}$ (cm^3 min^{-1} molecule^{-1})	This global kinetic constant of the hydrolysis takes into account the influence of the various physicochemical parameters of the interface on the enzyme activity.

From Eqs. (8) and (9), the following expressions for the interfacial specific activity SA*
and the specificity constant k_2/K_m^* can be obtained:

$$SA^* = \frac{1}{\Gamma_{E^*}} \frac{d\Gamma_{P^*}}{dt} = \frac{k_2}{K_m^*} \Gamma_S \tag{10a}$$

$$\frac{d(SA^*)}{d\Gamma_S} = \frac{k_2}{K_m^*} \tag{10b}$$

The hydrolysis course of short- or medium-chain lipids, generating soluble products, is
easily followed by measuring the decrease of the film area required to maintain the
monolayer at constant surface pressure and constant substrate concentration in
the so-called zero-order trough (Fig. 5) (15,45–48). During hydrolysis under the
barostatic conditions, each desorbed molecule in the left reaction compartment is supplied instantaneously by the right reservoir. The surface substrate concentration Γ_s is
maintained constant and the decrease of the surface area ΔA reflects directly the
kinetics of product formation.

Phospholipase A_2 (PLA$_2$) from different sources was injected into the stirred
subphase in the reaction compartment. Two typical curves of the relative decrease
in surface area $\Delta A / A_r$ with time t, recorded with dinonanoylphosphatidylcholine
(DNPC) monolayers at $\pi = 5$ mN m^{-1} are presented in Fig. 6 (15). The relative
decrease in area is expressed as a fraction of the film area A_r of the reactional

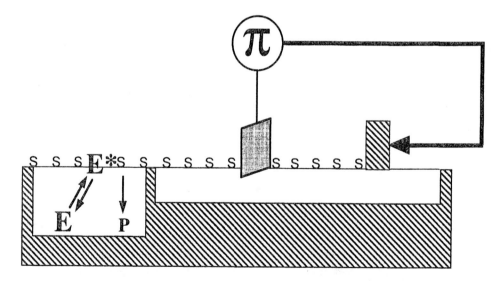

Fig. 5 Principle of the surface barostat balance consisting of a substrate (S) reservoir and a
reaction compartment with area A_r containing the enzyme (E) solution. The two compartments
are connected by a narrow surface canal. The surface pressure (π) and the surface substrate
concentration are maintained constant during the hydrolysis and solubilization of the reaction
product (P). (Adapted from Ref. 45.)

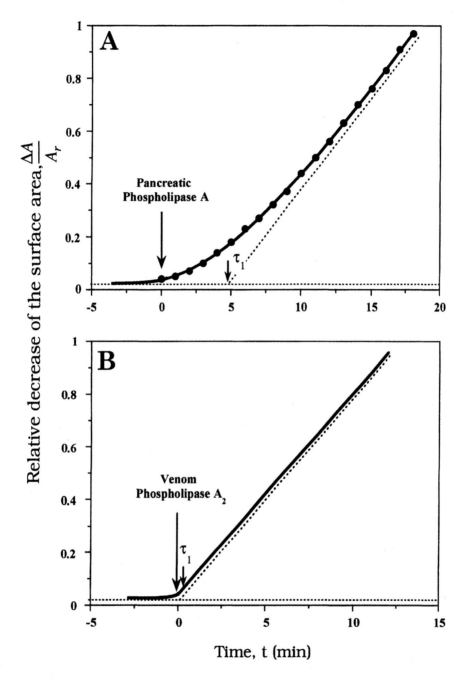

Fig. 6 Kinetics of the hydrolysis of a DNPC monolayer upon injection of PLA$_2$ from different sources (relative decrease of surface area $\Delta A / A_r$ with time t at constant surface pressure $\pi = 5$ mN m^{-1}); pH 8, Tris-HCl, 10 mM; NaCl, 0.1 M; CaCl$_2$, 20 mM. The film area of the reaction compartment $A_r = 93.7$ cm^2. (A) Injection of 5.9 μg of pancreatic PLA$_2$ (molecular weight 14,000) into 230 mL of stirred subphase onto the left reaction compartment ($C_{E0} = 2.57 \times 10^{-5}$ mg cm$^{-3} \equiv 1.1 \times 10^{12}$ molecules cm^{-3}). (B) Injection of 14 μg of bee venom PLA$_2$ into the left reaction compartment. (Adapted from Ref. 15.)

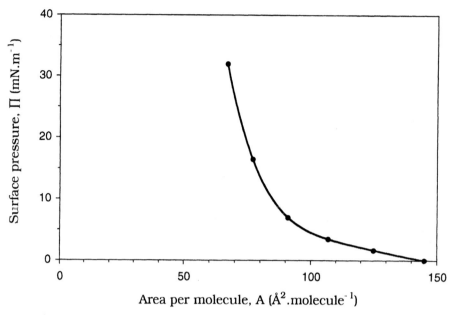

Fig. 7 Equilibrium surface pressure (π) versus area (A) isotherm of DNPC monolayer. (Adapted from Ref. 49.)

compartment. As usually, an initial latency period is observed until a steady state is established in the product formation.

From the slope of the kinetic curve for pancreatic PLA$_2$ (Fig. 6A) during the steady-state period and the value of $\Gamma_S = 1.12 \times 10^{14}$ molecules cm^{-2} at $\pi = 5$ mN m^{-1} from the equilibrium surface pressure–area isotherm of the DNPC monolayer (Fig. 7) (49), the hydrolysis rate $v_m = d\Gamma_{P*}/dt = (1/A_r)(d\Delta A/dt)\Gamma_S = 0.84 \times 10^{13}$ molecules cm^{-2} min^{-1} was obtained. By then using Eq. (8) and $C_{E0} = 1.1 \times 10^{12}$ molecules cm^{-3} the value of $Q_m = 6.8 \times 10^{-14}$ cm^3 min^{-1} molecule^{-1} was deduced.

The experimental observations confirm the proportionality between the enzymatic velocity v_m and the enzyme concentration in the subphase C_{E0}, as predicted by Eq. (8) (15). However, the proportionality between v_m and Γ_S predicted by Eq. (8) is only apparent. In general, the enzyme partitioning and the catalytic properties of the penetrated enzyme depend on the state of the lipid monolayer. The enzymatic activity often shows a bell-shaped dependence of velocity (v_m) versus surface pressure (π). An example of the bell-shaped surface pressure dependence of the hydrolysis rate $v_m = d\Gamma_{P*}/dt$ of the DNPC monolayer by pancreatic PLA$_2$ with maximal enzyme activity at $\pi = 12$ mN m^{-1} is presented in Fig. 8 (47). An independent measurement of the amount of enzyme that is really involved in interfacial catalysis would be useful for analyzing the role of the surface pressure of the lipid monolayer in the amount and catalytic properties of the penetrated enzyme. The interfacial concentration Γ_{E*} of radioactively labeled enzyme has been determined in the steady state after 15 min of hydrolysis and is presented in the same figure. The amount of radioactive enzyme present at or close to the interface is only a few percent of the entire amount of the enzyme initially present in the water phase ($C_E \approx 10^{-5}$ mg cm^{-3}; $\Gamma_E \approx 10^{-7}$ mg cm^{-2}) and decreases linearly with increasing surface pressure. Using the data from Figs. 7 and 8 and Eqs. (10a) and (10b), the surface

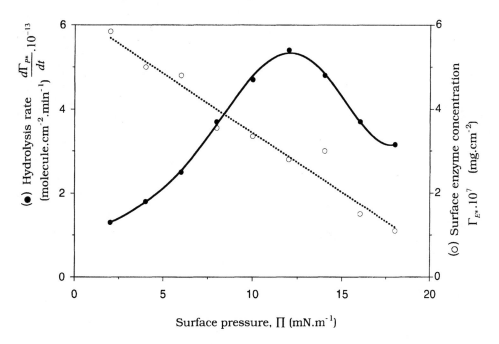

Fig. 8 Surface pressure (π) dependence of the hydrolysis rate ($v_{\mathrm{m}} = d\Gamma_{\mathrm{P*}}/dt$) in DNPC monolayer by pancreatic PLA$_2$ (●). The interfacial excess of radioactively labeled enzyme (○) was determined in the steady state of hydrolysis. (Adapted from Ref. 47.)

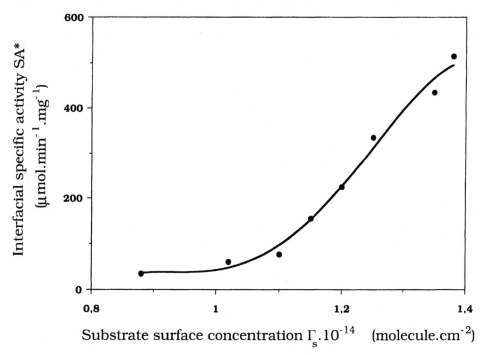

Fig. 9 Substrate surface concentration (Γ_S) dependence of the interfacial specific activity (SA*), calculated by using Eq. (10) and data from Figs. 7 and 8.

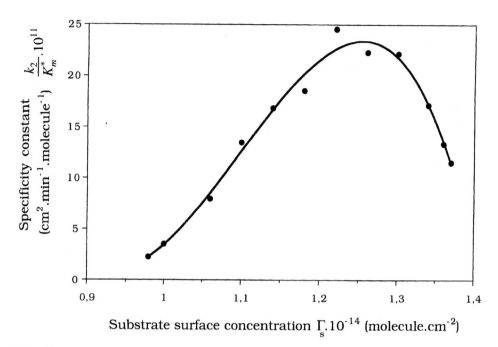

Fig. 10   Substrate surface concentration (Γ_S) dependence of the specificity constant (k_2/K_m^*), calculated by using Eq. (11) and data from Fig. 9.

concentration dependence of the interfacial specific activity SA* and the specificity constant k_2/K_m^* were calculated and are represented in Figs. 9 and 10, respectively. The interfacial specific activity SA* (Fig. 9), accounting for the enzyme amount involved in the catalysis, increases continuously with π without reaching an optimum value in the range of surface pressures investigated. However, the specificity constant k_2/K_m^* (Fig. 10) confirms the existence of a surface state optimum, leading to an optimal fit between the active site of the enzyme and the substrate.

Two non–mutually exclusive interpretations of the optimal surface pressure, taking into account the packing of the substrate monolayer (50) or(and) conformational changes of the enzyme at the interface (44,47), have been proposed. Peters et al. (51) argue in favor of the first interpretation: a reorganization of the headgroups of dipalmitoylglycerol monolayers undergoing the phase transition, causing a change in the hydrophobicity of the interface. The exact values of the optimal pressures for various enzyme-substrate systems have been determined (44,46–48,52,53). For example, maximal activity of the venom PLA$_2$ from *Vipera berus* against a dilauroylphosphatidylcholine (DLPC) monolayer was observed at $\pi = 22$ mN m^{-1} with $Q_m = 3.1 \times 10^{-13}$ cm^3 min^{-1} molecule^{-1} (54). A short induction time ($\tau \ll 1$ min) for the venom and a longer one ($\tau \approx 5$ min) for pancreatic PLA$_2$ are observed in Fig. 6. The simplest idea that the lag time is related to slow mixing of the injected enzyme below the surface and diffusion-controlled enzyme adsorption was tested and rejected (47,55). The origin of the slow hydrolysis phase can be also related to slow enzyme partitioning E $\rightleftarrows$ E* or slow catalytic steps E*+S $\rightleftarrows$ E*S$\rightarrow$P.

It has been found experimentally that the formation and decomposition of the interfacial Michaelis-Menten complex is not rate limiting, i.e., $\tau_1 \gg \tau_2$ (15). Then, the induction time τ_1, reflecting the establishment of the enzyme penetration-desorption steady state during hydrolysis, can be obtained from Eq. (5) as the intercept of the asymptote with the time axis. As seen in Fig. 6A and B, the pancreatic enzyme penetrates more slowly into the lipid monolayer than the venom PLA$_2$.

2. Inhibition

The next development of the simplest kinetic model of hydrolysis of lipids generating water-soluble products takes into account the presence of insoluble inhibitors. A step for the competitive inhibition of E* was added (Fig. 4) (47) and theoretically analyzed in detail (56,57). By solving the corresponding kinetic equations, the following expression for the hydrolysis rate v at steady state is obtained (56):

$$v = \frac{d\Gamma_{P*}}{dt} = \frac{k_2 \Gamma_S}{K_m^* \dfrac{k_d}{k_p} + \left(K_m^* + \dfrac{K_m^*}{K_I^*} \Gamma_I + \Gamma_S \right) \dfrac{A}{V}} \tag{11}$$

where Γ_I is the interfacial inhibitor concentration and $K_I^* = k_{-3}/k_3$ (molecules cm^{-2}) is the interfacial dissociation constant for the enzyme-inhibitor complex (E*I).

Numerous compounds may interfere with phospholipase activity by interacting either directly with the enzyme or indirectly by affecting the interfacial state of the lipid substrate. As an example, the inhibition by gangliosides of the PLC from *Bacillus cereus* acting upon DLPC monolayers is mainly due to the modification of the physicochemical state of the lipid monolayer (58).

As far as the inhibitory effect of soluble proteins is concerned, a competition between the enzyme and the protein during the adsorption kinetics at the air-water interface seems to be the rate-determining step (59,60). For a general review of the covalent inhibition of digestive lipases, the reader is referred to Ref. 61.

3. Insoluble Reaction Products

For a natural long-chain lipid substrate such as dipalmitoylphosphatidylcholine (DPPC), the reaction products are insoluble and remain at the interface in the absence of product acceptors. Two opposed effects are observed (62). First, the accumulation of the insoluble lipolytic products, making the substrate inaccessible to the enzyme, leads to inhibition by the reaction products.

The processes of interfacial molecular reorganization and segregation of the insoluble products, represented schematically by the step P*→P** (Fig.4), often take place. Such processes modify the microheterogeneous structure of the lipid monolayer during the lipolysis and often lead to increased enzyme binding and lipolytic autoactivation (62). As a result of the increase of the surface enzyme concentration, biphasic kinetics are often observed. They are characterized by an unusual lag time, which depends on the products accumulating at the interface.

The morphology of lipid monolayers can be visualized by using either epifluorescence microscopy, atomic force microscopy (AFM), or Brewster angle microscopy (BAM) (63–65). Visual observation by epifluorescence microscopy of various phospholipid monolayers could distinguish several mechanisms of action of PLA$_2$ on lipid domains. After reaching a substantial level of hydrolysis, highly organized two-dimensional domains of enzyme of regular sizes and morphologies were clearly seen (Fig. 11) (66). Morphological differences in the assemblies of the products during

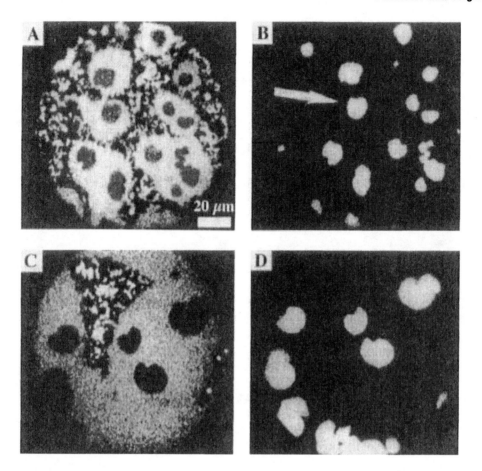

Fig. 11 Formation of enzyme domains of fluorescein-labeled phospholipase A_2 in an L-α-DPPC monolayer after extended hydrolysis times. Photographs corresponds to the same view seen through two different filters. In (A) and (C) (after approximately 60 and 90 min hydrolysis time, respectively), the monolayer is imaged through a sulforhodamine filter (monolayer); the bright signal represents the fluid phase of the lipid monolayer and the dark areas solid-phase lipid and enzyme domains. In (B) and (D), the corresponding views at the respective time points are seen through a fluorescein filter; the bright areas represent specifically the enzyme domains (example: see arrow) from (A) and (C). Scale bar in (A) in 20 μm. (From Ref. 66.)

the hydrolysis of phospholipid monolayers by PLD from *Streptomyces* were also reported (67).

Theoretical analysis of the enzyme activity modified by the insoluble lipolytic products is a difficult task. The enzyme lipolysis cannot be treated by means of Eq. (8), based on the assumption that during hydrolysis the surface enzyme Γ_{E^*} and substrate Γ_S concentrations are constant (the substrate being in large excess or Γ_S being kept constant by regulatory processes as in the barostat shown in Fig. 5). The theoretical description of the interfacial morphology also presents inherent difficulties. The classical thermodynamic Goodrich treatment cannot be applied to microheterogeneous two-dimensional systems in which the dependence of the

Table 2 Global Kinetic Constants of Hydrolysis of Monolayers of DPPC, DOPC, and DLPC by *Vipera berus* PLA$_2$ ($C_E = 1.03$ nM) and Monolayers of DO, TO, DC, and DL by HPL ($C_E = 0.38$ nM) in the Absence or Presence of β-CD ($C_{CD} = 0.8$ mg cm^{-3}) in the Aqueous Subphase (pH 8.0) of the Reaction Compartment

	Q_m(cm^3 min^{-1} molecule^{-1})	
Substrate	Without β-CD	With β-CD
Phospholipids, PL		
DPPC	4.2×10^{-15}	2.3×10^{-13}
DOPC	3.7×10^{-15}	3.5×10^{-13}
DLPC	3.1×10^{-13}	—
Glycerides, G		
DO	—	1.9×10^{-12}
TO	—	1.9×10^{-12}
DC	2.8×10^{-12}	—
DL	2.4×10^{-12}	—

thermodynamic parameters on the dimensions of small interfacial phases should be taken into account. Some information about the degree of interfacial heterogeneity can be obtained from computer simulation of the gel-to-fluid phase transition in a lipid bilayer (68). Thus, the complete kinetic treatment of the interfacial hydrolysis of long-chain lipids, generating water-insoluble products, requires additional kinetic equations describing their reorganization and inhibition or acceleration of the interfacial hydrolysis. This objective seems, however, illusory at present due to the complexity of the system. This is the reason why by using Eq. (8) an apparent kinetic constant Q_{mapp} can be estimated. For the maximal activity of the venom PLA$_2$ from *V. berus* against a long-chain DPPC monolayer $Q_{mapp} = 1.1 \times 10^{-15}$ (cm^3 min^{-1} molecule^{-1}), which is 280 times smaller than the true kinetic constant $Q_m = 3.1 \times 10^{-13}$ (cm^3 min^{-1} molecule^{-1}) for medium-chain DLPC (54).

In order to solubilize the water-insoluble lipolytic products of hydrolysis by various kinds of PLA$_2$, a large excess of serum albumin was used originally (55). However, in order to overcome the drawbacks associated with the presence in the bulk phase of a high concentration of a surface-active protein SA, the use of the non–surface-active α- and β-cyclodextrin (CD) as selective lipolytic product acceptors has been proposed. A theoretical approach describing the kinetics of formation and solubilization of the CD-P* complex (Fig. 4) was developed (69,70).

In the presence of β-CD in the subphase, the kinetic constants of enzymatic hydrolysis of monomolecular films of long-chain PC (69) and glycerides G (70) were found to be comparable to those measured with medium-chain substrates. As an llustration, the values of the global kinetic constant of hydrolysis $Q_m \equiv (1/C_{E0}\Gamma_S)(d\Gamma_{P*}/dt)$ at the optimal surface pressures are given in Table 2.

C. Bilayer Liposomes as a Substrate for Lipolytic Enzymes

When the lipid substrate is organized as liposomal dispersions, the description of the hydrolysis kinetics must take into account the colloid chemical features of the system. First, as a result of the large ratio A/V (which is measure of the degree of dispersion), of the order of magnitude of 10^4–10^5 cm^{-1}, the following inequality is fulfilled:

$$K_m^* \frac{k_d}{k_p} \ll (K_m^* + \Gamma_s) \frac{A}{V} \tag{12}$$

From Eq. (6) a simplified expression for the enzymatic rate of hydrolysis (v_1) of the dispersion liposomal system, generating soluble products, is obtained:

$$v_1 = \frac{d\Gamma_{p^*}}{dt} = \frac{k_2}{\dfrac{A}{V}(\Gamma_s + K_m^*)} \Gamma_s C_{E0} \equiv Q_1 \Gamma_s C_{E0} \tag{13}$$

where $Q_1 \equiv k_2/[(A/V)(\Gamma_s + K_m^*)]$ is an overall kinetic constant taking into account the influence of the liposomal interface on the enzyme activity.

It is experimentally established that in a dispersion system with a large A/V ratio, practically all enzyme molecules initially present in water are fixed at the interface and

$$C_{E0} V \cong \Gamma_{E^*} A \tag{14}$$

From Eqs. (13) and (14), the following expression for the interfacial specific activity SA* can be obtained:

$$SA^* = \frac{1}{\Gamma_{E^*}} \frac{d\Gamma_{P^*}}{dt} = \frac{k_2}{K_m^* + \Gamma_S} \Gamma_S \tag{15}$$

It is easy to show that for dispersion systems with large A/V ratios SA* is identical to the usual definition of specific activity SA

$$SA = \frac{1}{C_{E0}} \frac{dC_{P^*}}{dt} = \frac{1}{C_{E0}} \frac{A}{V} \frac{d\Gamma_{P^*}}{dt} = \frac{k_2}{K_m^* + \Gamma_S} \Gamma_S \tag{16}$$

The surface concentrations of the substrate (Γ_S) and the product (Γ_{P^*}) from Eq. (13) can be expressed in a more usual way as the number of molecules (or moles) per liposome, using the following relationships, according to Jain and Berg's nomenclature (18,19):

$$a\Gamma_{p^*}(t) = N(t)$$
$$a\Gamma_S(t) = N_T - N(t) \tag{17}$$

where a is the surface area of one liposome, $N(t)$ is the number of substrate molecules per one enzyme-containing vesicle hydrolyzed after a time t, and N_T is the total number of phospholipid molecules initially present in the outer monolayer of one vesicle. For example, at early stages of hydrolysis $N(t) \ll N_T$ and $a\Gamma_{S0} \approx N_T$ and from

Eqs. (13) and (17) one obtains:

$$\frac{dN(t)}{dt} = \frac{k_2}{\frac{A}{V}(\Gamma_{S0} + K_m^*)} N_T C_{E0} \equiv Q_i N_T C_{E0} \tag{18}$$

Let us now briefly discuss the role of the exchange of enzyme molecules between vesicles in the effectiveness of the catalytic reaction.

Two extreme kinetic situations are considered in Jain's model of enzyme action (18,19). In the "scooting mode" (Fig. 12A), the enzyme molecules are irreversibly bound to the vesicles. A bound enzyme E^* remains at the interface between many catalytic turnover cycles and will hydrolyze all substrate molecules in the outer monolayer of the vesicle. In the pure "hopping mode" (Fig. 12B) the enzyme molecules can exchange between vesicles. The binding $(E \rightarrow E^*)$ and the desorption $(E^* \rightarrow E)$ of one enzyme molecule occur between catalytic turnover cycles. Hopping from one vesicle to another, the available substrate molecules in the outer monolayers of all vesicles will be finally hydrolyzed.

The description of the hydrolysis kinetics in the scooting mode takes into account the catalytic activity of the enzyme confined at the interface, assuming a Poissonian random distribution of all enzyme molecules between the available vesicles (18,19).

$$P_j = \frac{(C_E/C_V)^j}{j!} e^{-C_E/C_V} \tag{19}$$

where P_j is the probability that a vesicle has j bound enzyme molecules; C_V is the vesicle concentration, and C_E/C_V is an average number of enzyme molecules per vesicle.

The probability for vesicles without any bound enzyme is

$$P_{j=0} = e^{-C_E/C_V} \tag{20}$$

At small enzyme-vesicle ratios $C_E/C_V \ll 1$, $P_{j=0} = 1 - C_E/C_V$ and $P_{j=1} = C_E/C_V$; that is, the number of vesicles containing only one enzyme molecule is equal to the number of enzyme molecules in the dispersion system. For large enzyme/vesicle ratios

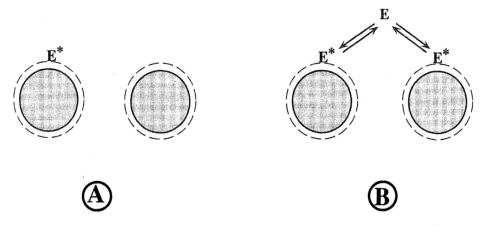

Fig. 12 Diagram of two extreme modes of liposome interfacial catalysis: (A) scooting mode; (B) hopping mode. (Adapted from Ref. 18.)

$C_E/C_v \gg 1$ $P_{j=0} \approx 0$ and $P_{j\neq 0} = 1-e^{-C_E/C_v} \approx 1$; that is, there are practically no vesicles without any bound enzyme. An attempt to describe the more complex kinetic situation in the hopping mode is presented in Ref. 18. For small enzyme/vesicle ratios, the two limiting kinetic mechanisms are easily distinguishable. In the scooting mode, $P_{j=0} = 1-C_E/C_v$. In the hopping mode, $P_{j=0}$ depends on the competition between the catalytic step (k_2/K_m^*) and the enzyme desorption (k_d) and in the limit situation of very rapid exchange, "intensive hopping mode," $P_{j=0} \rightarrow 0$.

It should be noted, however, that in addition to the enzyme exchange the other parallel processes of intervesicle exchange of substrates and products can also interfere and must be taken into account in a general pattern of interfacial catalysis.

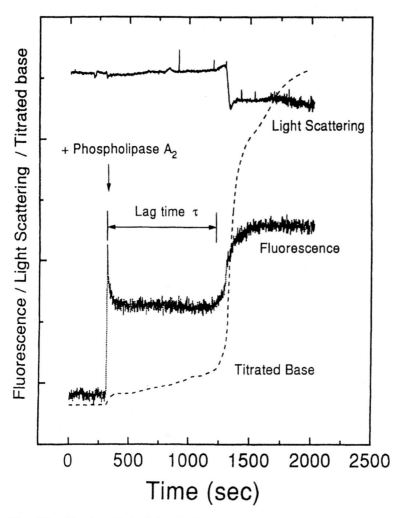

Fig. 13 Kinetics of hydrolysis of DPPC, organized in large unilamellar liposomes, by venom PLA_2 at 39°C. The hydrolysis reaction is monitored by pH-stat titration, intrinsic fluorescence from PLA_2 emitted at 340 nm upon excitation at 285 nm, and 90° light scattering from the suspension at 285 nm (all data are given in arbitrary units). The reaction is initiated at time 300 s by adding PLA_2 to the equilibrated vesicle suspension. (From Ref. 68.)

The common experimental methods for following the course of enzymatic hydrolysis in liposomal dispersions are based on measurements of the concentration of the products released with time (classical pH-stat titration with NaOH, spectroscopic methods, etc.). The binding of phospholipases to small unilamellar liposomes and the hydrolysis of various anionic or zwitterionic phospholipids are reviewed in Refs. 18 and 19. A typical reaction time course of venom PLA_2–catalyzed hydrolysis of DPPC organized in large unilamellar vesicles is presented in Fig. 13 (68). The PLA_2 is added at time 300 s to the equilibrated vesicle suspension and the product release is monitored by classical pH-stat titration. Biphasic kinetics with a drastic change in the reaction course after a characteristic lag time τ is observed. The catalytic reaction is also monitored by intrinsic fluorescence from PLA_2, emitted at 340 nm upon exitation at 285 nm.

Another possibility for following the course of the enzymatic reaction is based on the analysis of the colloid chemical characteristics of the dispersion during lipolysis. A simple method is based on the decrease with time of the turbidity of the dispersion upon lipase action monitored by 90° light scattering at 285 nm (Fig. 13).

The formation and reorganization of the lipolytic products may induce structural rearrangements in the liposomal bilayers. It has been proposed that the so-called lipid packing defects in the liposomal bilayer may promote high PLA_2 activity (71–78). Some information about the structural microheterogeneity of the nonhydrolyzed lipid bilayer near the main phase transition was obtained from computer simulations on a microscopic molecular model (68,79,80). The dynamic heterogeneity is described in terms of the formation of lipid domains, characterized by a length scale in the nanometer range, as illustrated in Fig. 14. A correlation between the lag times and the degree of dynamic bilayer heterogeneity with respect to both temperature and lipid chain length variations was found (Fig. 15).

Another approach, based on the analysis of the structural rearrangements in the liposomal dispersion, is the study of the acceleration of the surface transformation of perfectly closed vesicles into open interfacial structures as a result of the activity of venom PLA_2 against bilayer PC liposomes (54). A theoretical analysis of the coupling between the interfacial catalysis, represented by the kinetic scheme in Fig. 4, and the process of destabilization and reorganization of vesicles spread at the air-water interface was developed.

The slow transformation of spread intact liposomes into the surface film at the air-water interface was observed for the first time by Verger and coworkers (81) and treated theoretically by Schindler (82), who proposed a detailed kinetic scheme with five parameters. A simplified kinetic scheme with only two parameters was subsequently used by Panaiotov and coworkers (83,84) as a first approximation. This rough theoretical scheme distinguishes a diffusion process with diffusion coefficient D from the other processes summarized in a single transformation of perfectly closed vesicles into surface-active structures of both aggregate and individual lipid molecules (Fig. 16, left), and a global kinetic constant K of the transformation can be obtained.

After a liposomal suspension has been spread at the air-water interface, the kinetics of surface film formation may be described by two simultaneous processes: (a) irreversible diffusion of liposomes to the liquid bulk phase and (b) irreversible transformation of closed vesicles into surface-active structures of aggregate and individual lipid molecules. In the general case, in which the rates of two simultaneous processes are of the same order, both processes influence the phenomenon; otherwise,

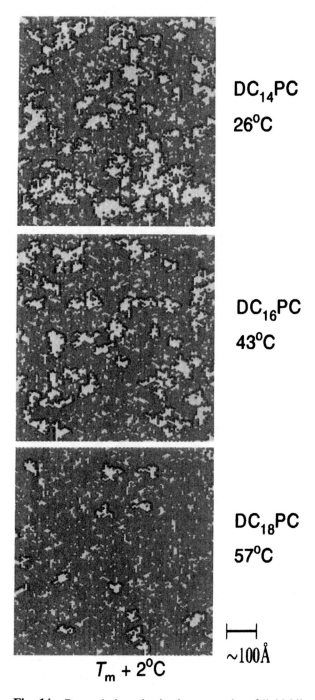

DC$_{14}$PC
26°C

DC$_{16}$PC
43°C

DC$_{18}$PC
57°C

T_m + 2°C

$\sim$100Å

Fig. 14 Dynamic lateral microheterogeneity of lipid bilayers as obtained by Monte Carlo computer simulations. The photographs show typical schematic top-view bilayer configurations (550 Å × 550 Å) in the fluid phase 2°C above the respective gel-to-fluid phase transition temperature T_m, for a lipid bilayer composed of DC$_{14}$PC, DC$_{16}$PC, and DC$_{18}$PC. Gel and fluid regions of the bilayer are also denoted. The interfacial regions between the dynamic coexisting gel and fluid region are highlighted in green. (From Ref. 68.)

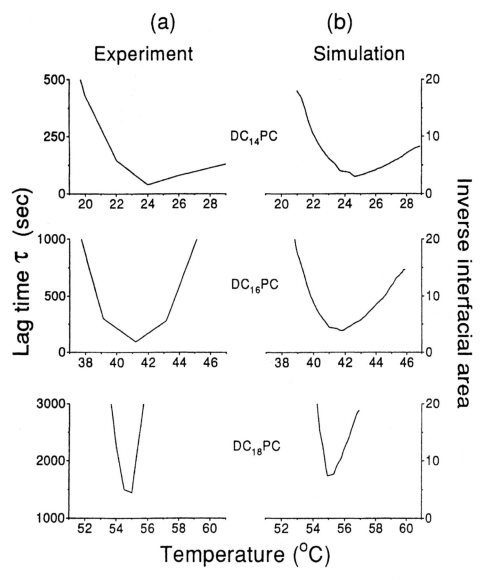

Fig. 15 Correlation between the lag time (τ) and the microheterogeneity of the lipid bilayer: (a) experimental results for the PLA$_2$-catalyzed hydrolysis of unilamellar vesicles in the temperature regions of the gel-to-fluid phase transition for DC$_{14}$PC, DC$_{16}$PC, and DC$_{18}$PC lipid bilayers; (b) Monte Carlo computer simulation of the temperature dependence of the inverse interfacial area of lipid bilayers composed of DC$_{14}$PC, DC$_{16}$PC, and DC$_{18}$PC in their respective gel-to-fluid phase transition regions. (From Ref. 68.)

the phenomenon is controlled by the faster process. Process (a) was described by Fick's diffusion equations and process (b) by an appropriate irreversible adsorption kinetic equation.

It has been established that the liposomal bilayer structure can be altered by various chemical reactions, affecting the hydrophilic-hydrophobic balance of the

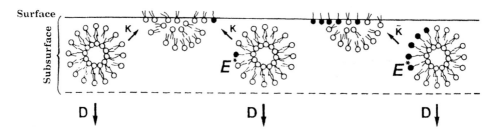

Fig. 16 Diagram of the surface transformation of the three populations of liposomes: without any enzyme (left) and containing at least one enzyme molecule before (center) and after (right) formation by products of a domain with a critical size. Double-tailed white symbols represent phospholipids composing the liposomes. Single-tailed dark symbols represent lipolytic products. (From Ref. 54.)

organized lipid monolayer (85). The resulting structural alterations in the case of lipolysis are extensive enough to destabilize the liposome structure. An accelerated transformation of the closed spherical structure into a surface film at the air-water interface was observed. Figure 16 oversimplifies the transformation at the air-water interface of three populations of liposomes: without any enzyme (left) and containing at least one enzyme molecule before (center) and after (right) formation of a defect with a critical size by the lipolytic products.

The surface transformation of intact and defective liposomes can be described by the model taking account the liposome diffusion from the interface into the bulk phase and the surface transformation. The rate constant $\tilde{K}$ of surface transformation of defective liposomes is assumed to be larger than that of intact spherical bilayer structures. The overall interfacial transformation rate dn^*/dt is therefore given by the sum of the transformations of intact (with or without any bound enzyme) and defective vesicles (54):

$$
\begin{aligned}
\frac{dn^*}{dt} = {} & KP_{j=0}C_0 d \operatorname{erf} \frac{L}{2\sqrt{Dt}}\left(1 - \frac{n^*}{C_0 d}\right) \\
& + K(1 - P_{j=0})C_0 d \operatorname{erf} \frac{L}{2\sqrt{Dt}}\left(1 - \frac{n^*}{C_0 d}\right)e^{-t/\tau} \\
& + \tilde{K}(1 - P_{j=0})C_0 d \operatorname{erf} \frac{L}{2\sqrt{Dt}}\left(1 - \frac{n^*}{C_0 d}\right)\left(1 - e^{-t/\tau}\right)
\end{aligned}
\tag{21}
$$

where $n^*(t)$ is the surface concentration of the adsorbed destructed liposomes, C_0 is the initial liposomal concentration, d is the liposome diameter, D is the diffusion coefficient, L is the thickness of the layer of spread liposomal suspension, and τ is the time of formation of a defect of a critical size.

The transformation of the non–surface-active closed vesicles into a surface film was monitored by measuring the evolution of the surface pressure π and surface potential ΔV with time t. Figure 17 shows the effects of phospholipase A_2 hydrolysis on the spreading kinetics $\pi(t)$ and $\Delta V(t)$ of DOPC (panels A and B) and DPPC (panels C and D) liposomal suspensions: either free of enzyme (0) or preincubated for 1 min with various enzyme concentrations C_E (various enzyme/vesicle ratios C_E/C_V). The acceleration of the spreading kinetics observed and the increase in the final π

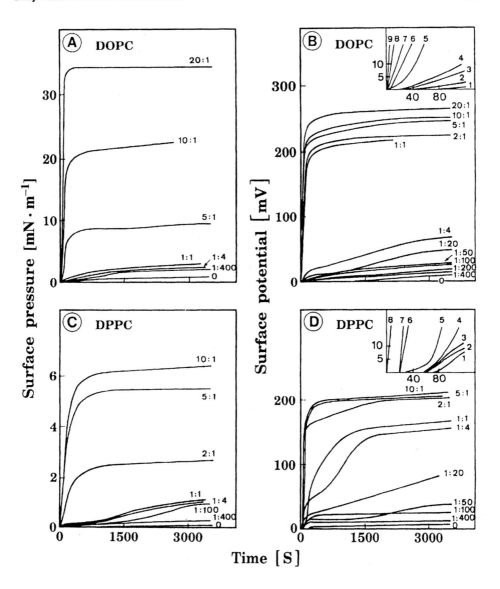

Fig. 17 Variation with time of surface pressure (π) (A, DOPC; C, DPPC) and surface potential (ΔV) (B, DOPC; D, DPPC) after spreading of 100 μL (DOPC) and 50 μL (DPPC) of enzyme-free liposomal suspension (0) or the same suspension preincubated for 1 min with enzyme at various enzyme/vesicle (C_E / C_V) ratios (1:400, 1:200, 1:100, 1:50, 1:20, 1:4, 1:1, 2:1, 5:1, 10:1, and 20:1). Inserts: the beginning of the kinetics $\Delta V(t)$. Panel B (curve 1, 1:400; curve 2, 1:200; curve 3, 1:100; curve 4, 1:20; curve 5, 1:4; curve 6, 1:1; curve 7, 2:1; curve 8, 5:1; curve 9, 10:1). Panel D (curve 1, 1:400; curve 2, 1:100; curve 3, 1:50; curve 4, 1:20; curve 5, 1:4; curve 6, 2:1; curve 7, 5:1; curve 8, 10:1). (From Ref. 54.)

and ΔV values recorded as the enzyme concentration increased may be attributable to competition among the processes involved in the mechanism described in Fig. 16. The increase in the rate and the efficiency of spreading with the enzyme concentration may be due to the fact that the surface transformation of hydrolyzed vesicles plays a more important role than the vesicle diffusion into the bulk phase. The quantitative interpretation is based on Eq. (21).

The values of the rate constants of surface transformation of intact $(K = 0.9 \times 10^{-1}$ min$^{-1})$ and defective $(\tilde{K} = 2.5 \times 10^{-1}$ min$^{-1})$ DPPC liposomes were obtained, as well as the value of the characteristic time τ of defect formation with a critical size $(P = 100$ molecules) at the early stage of hydrolysis. The specific area of this experimental system is $A/V = C_V a = 8.9 \times 10^{13}$ (liposomes cm^{-3}) $\times 2 \times 10^{-10}$ (cm^2 liposome^{-1}) $= 1.8 \times 10^4$ cm^{-1}. Then, by using Eq. (18) for hydrolysis, generating soluble products, a value of the apparent global kinetic constant in the absence of acceptor $Q_1 = 2.3 \times 10^{-16}$ cm^3 min^{-1} molecule^{-1} was estimated.

In spite of the fact that the processes of enzyme exchange in the hopping mode are not adequatelly described, the preceding calculations were performed in the framework of the two limiting kinetic mechanisms, scooting mode $(P_{j=0} = e^{-C_E/C_V})$ and the intensive hopping mode $(P_{j=0} \rightarrow 0)$ (54). This estimation leads to the conclusion that the hopping mode seems to be more suitable for describing the hydrolysis of DOPC and DPPC liposomes. A comparison of the apparent hydrolysis rate obtained in the absence of acceptor for DPPC liposomes $Q_1 = 2.3 \times 10^{-16}$ cm^3 min^{-1} molecule^{-1} and that obtained for a DPPC monolayer at optimal surface pressure in the absence of acceptor $Q_m = 4.2 \times 10^{-15}$ cm^3 min^{-1} molecule^{-1} shows the better efficacy of the hydrolysis process at the monolayer surface state optimum.

In conclusion, the interfacial and temporal organization of enzymatic lipolysis (ITOEL) in liposomal dispersions can be adequately described in terms of the interfacial adaptation of the Michaelis-Menten scheme, as long as the enzyme is confined to the interface in the framework of the so-called scooting mode. The interpretation of the role of various processes of intervesicle exchange of enzyme, substrate, and product in hopping mode, as well as the role of the size repartition and the destabilization of the dispersion in the lipolysis, needs a more adequate description.

D. Micelles as a Substrate for Lipolytic Enzymes

The action of water-soluble lipases and phospholipases against lipids organized in pure or mixed (with detergent) micelles has been intensively studied. The micellar system presents a high interfacial area with an A/V ratio of typical values of 10^5 or 10^6 cm^{-1} up to 10^8 cm^{-1} for concentrated reverse micelles.

The first attempt to describe the ITOEL in mixed micelles is the "surface dilution model" (16,17). The proposed kinetic model, presented in Fig. 18, is similar to the basic model in Fig. 3. A soluble enzyme (E) first associates nonspecifically or specifically with a detergent-lipid micelle (A) to form an enzyme–mixed micelle complex (EA). Once the enzyme is fixed on the micelle, it binds an individual lipid molecule (B) in the catalytic site, forming the Michaelis-Menten complex (EAB). As catalysis occurs, the products (P) are formed and (EA) is regenerated.

Generally speaking, it is difficult to determine experimentally the exact value of the water-micelle interfacial area. For this reason, the first step of association of the enzyme with the micellar interface is considered a bulk step. [A] is the bulk concentration of micellar sites capable of associating with one enzyme molecule. The concentration of bound enzyme [EA] is then expressed as a volumic concentration. In the framework

$$E + A \underset{k_{-1}}{\overset{k_1}{\rightleftharpoons}} EA \qquad\qquad EA + B \underset{k_{-2}}{\overset{k_2}{\rightleftharpoons}} EAB \xrightarrow{k_3} EA + P$$

bulk step   **surface step**

Fig. 18 Surface dilution kinetic scheme describing the enzymatic activity toward lipid substrate organized in mixed micelles. (From Ref. 39.)

of nonspecific fixation, assuming that all lipid and detergent molecules are at the micellar surface, [A] is related to the sum of concentrations of lipid and detergent.

$$[A] = \frac{xC_L + yC_T}{n} \ (\text{mol cm}^{-3}) \tag{22}$$

where C_L and C_T are the bulk concentrations of lipid and detergent Triton X-100, respectively; x and y are the average surface areas per molecule of lipid and detergent, respectively; n is the micellar surface area per binding site for the enzyme; and $n[A]$ corresponds to the entiere interface capable of associating with enzyme per unit volume. For a specific fixation on substrate molecules, [A] is proportional only to the molar concentration of lipid. The concentration of the lipid molecules assembled at the micellar interface [B] is expressed as surface concentration in the second catalytic step.

$$[B] = \frac{C_L}{xC_L + yC_T} \ (\text{mol cm}^{-2}) \tag{23}$$

Assuming that x, y, and n are constant as [A] and [B] are varied and that $x \approx y$, the definitions (22) and (23) were used in the form (16):

$$[A] = \frac{x}{n}[A'], \qquad [A'] = C_L + C_T \tag{24}$$

$$[B] = \frac{1}{x}[B'], \qquad [B'] = \frac{C_L}{C_L + C_T} \tag{25}$$

The following equation for the rate of hydrolysis $v \equiv dC_P/dt$ at steady state was obtained (16):

$$v = \frac{V[A][B]}{K_S^A K_m^B + K_m^B[A] + [A][B]} \tag{26}$$

where $K_S^A = k^{-1}/k_1$ (mol cm^{-3}), $K_m^B = (k_{-2} + k_3)/k_2$ (mol cm^{-2}), $V = k_3 E_T$ (mol cm^{-3} min^{-1}), and E_T (mol cm^{-3}) is the total enzyme concentration.

It is easy to demonstrate that using the notations from the general kinetic scheme of Fig. 3 and appropriate dimensions for the kinetic constants, one can convert expression (26) in terms of Eq. (6). In fact,

$$C_P \equiv \Gamma_P \frac{A}{V}, \qquad n[A] = xC_L + yC_T \equiv \frac{A}{V}, \qquad k_3 \equiv k_2, \qquad k_3 E_T \equiv k_2 C_{E0}$$

$$[B] \equiv \Gamma_S \qquad K_m^B = \frac{k_{-2} + k_3}{k_2} \equiv K_m^* = \frac{k_2 + k_{-1}}{k_1}$$

$$(27)$$

At steady state

$$K_S^A = \frac{k_{-1}}{k_1} = \frac{[E][A]}{[EA]} (\text{mol cm}^{-3}) \quad \text{and} \quad \frac{k_d}{k_p} = \frac{C_E}{\Gamma_{E^*}} (\text{cm}^{-1})$$

thus

$$\frac{K_S^A}{[A]} = \frac{[E]}{[EA]} = \frac{C_E}{C_{E^*}} = \frac{C_E}{\Gamma_{E^*}(A/V)} = \frac{k_d}{k_p(A/V)}$$

The meaning of the dimensionless volume-interface enzyme partitioning $k_d/k_p(A/V)$ at steady state is a ratio between the number of enzyme molecules in the volume $C_E V$ and that at the interface $\Gamma_{E^*} A$. By using the expressions (27), one can obtain Eq. (6) from Eq. (26).

As already discussed, it is not easy to distinguish the enzyme interfacial binding from the subsequent catalytic steps. The original idea underlying the simplified kinetic treatment by Dennis et al. was to try to separate these two steps experimentally by measuring the hydrolysis rate (v) when the lipid surface concentration [B] was decreased (surface dilution is realized by using various substrate/detergent ratios) at a constant concentration of enzyme binding sites [A], or inversely by measuring v as a function of [A] at constant [B]. For that purpose the following useful form of Eq. (26) is deduced (16):

$$\frac{1}{v} = \frac{nK_S^A K_m^B}{V[B']} \frac{1}{A'} + \frac{1}{V}\left(1 + \frac{xK_m^B}{[B']}\right)$$

$$(28)$$

If the hydrolysis rate is expressed as specific activity $SA = v/E_T$, Eq. (28) can be rewritten $(V/E_T = k_3)$:

$$\frac{1}{SA} = \frac{nK_S^A K_m^B}{k_3[B']} \frac{1}{[A']} + \frac{1}{k_3}\left(1 + \frac{xK_m^B}{[B']}\right)$$

$$(29)$$

Kinetic studies of the PLA$_2$ action toward DPPC in mixed spherical micelles with Triton X-100 (16) are shown in Fig. 19. The reciprocal values of the specific activity SA are represented in accordance with Eq. (29) as a function of the reciprocal values of [A'].

The intercepts $(1/k_3)[1 + (xK_m^B/[B'])]$ and the slopes $nK_S^A K_m^B/k_3[B']$ obtained from Fig. 19 are represented as a function of $1/[B']$ in Fig. 20A and B, respectively. From the data in Fig. 20 and the values $x = 5.1 \times 10^9$ cm^2 mol^{-1}, $n = 3.6 \times 10^{10}$ cm^2 mol^{-1}, and $M_E = 13,400$ (for more details see Ref. 16), the following

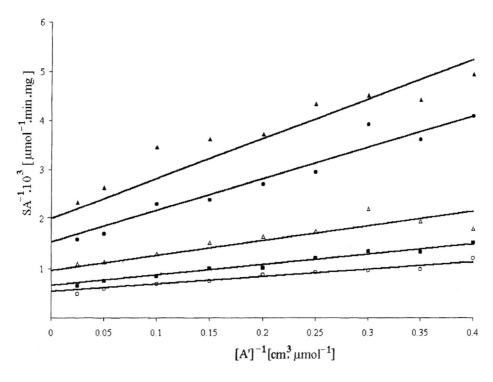

Fig. 19 Activity of PLA$_2$ toward DPPC in mixed micelles with Triton X-100: Plot of 1/SA versus 1/[A'] in accordance with Eq. (29) at several values of [B']: (○) 0.333; (■) 0.200; (△) 0.143; (●) 0.111; (▲) 0.077. (Adapted from Ref. 16.)

constants were obtained:

$$k_3 = 7 \times 10^3 \mu\text{mol S min}^{-1} \text{ mgE}^{-1} = 9.4 \times 10^4 \, (\text{min}^{-1})$$

$$K_S^A = 7 \times 10^{-7} \text{mol cm}^{-3}$$

$$K_m^B = 1.8 \times 10^{-10} \text{mol cm}^{-2} \equiv 1.1 \times 10^{14} \text{ molecules cm}^{-2}$$

At maximal bulk concentration of micellar sites for enzyme adsorption $[A] = (x/n)$ $[A'] = \frac{1}{7}4 \times 10^{-5} = 5.7 \times 10^{-6}$ mol cm^{-3}, the A/V ratio is equal to $[A/V] = [A]$ $n = 2.1 \times 10^5$ cm^{-1}.

The dimensionless volume-micellar surface enzyme partitioning $K_S^A/[A] = 7 \times 10^{-7}/7 \times 10^{-6} = 10^{-1}$; that is, approximately 90% of all enzyme molecules are bound at the micellar interface.

The value of the interfacial quality Q can also be calculated. At maximal $[A] = 5.7 \times 10^{-6}$ mol cm^{-3} and surface concentration of lipid molecules $\Gamma_S = [B] = [B']/x = 0.077/5.1 \times 10^9 = 1.5 \times 10^{-11}$ mol cm^{-2}, the corresponding specific activity is

$$\frac{1}{C_{E0}} = \frac{dC_P}{dt} = \frac{1}{E_T} \frac{dC_P}{dt} = 4.5 \times 10^{-4} \text{molS min}^{-1} \text{ mgE}^{-1}$$

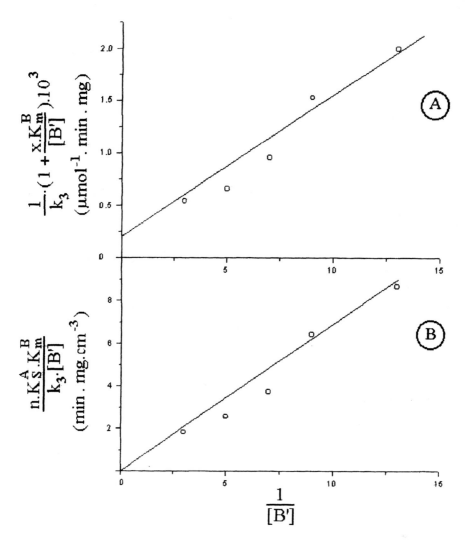

Fig. 20 Replots of the data from Fig. 19. Intercepts (A) and slopes (B) as a function of $1/[B']$. (Adapted from Ref. 16.)

With these values and $C_P \equiv \Gamma_P A/V$ from Eqs. (27), one can obtain

$$Q_{mic} = \frac{1}{\Gamma_S C_{E0}} \frac{d\Gamma_{P^*}}{dt} = \frac{dC_{P^*}/dt}{[B]\frac{A}{V} E_T} = \frac{4.5 \times 10^{-4}}{(1.5 \times 10^{-11})(2.1 \times 10^5)}$$

$$= 1.4 \times 10^2 \mathrm{cm}^3 \, \mathrm{min}^{-1} \, \mathrm{mgE}^{-1} = 3.2 \times 10^{-15} \mathrm{cm}^3 \, \mathrm{min}^{-1} \, \mathrm{molecule}^{-1}$$

It should be noted that K_S^A and K_m^B in Eq. (26) are constant, but in an experimental mixed micelle system, it is possible for variations of x and y with [B] to cause changes in the physical state of the interface and consequently affect these constants.

In Refs. 86 and 87, the process of micelle binding was separated experimentally from the catalytic step by using two independent methods. It should also be noted that the "surface dilution" approach is based on some hypothesis related to the colloid chemical features of the system, assuming that the size and aggregation number of mixed micelles remain constant over the whole concentration range as well as during lipolysis. Indeed, the mean size of spherical micelles is, above the critical micelle concentration (CMC), relatively insensitive to the surfactant concentration. For other kinds of micelles, the system is highly polydisperse, and the mean aggregation number is concentration dependent and is very sensitive to small changes in the molecular interactions (88). An implicit assumption of the surface dilution model is rapid reorganization of the lipid components when the micelle concentration is decreased or when other amphiphilic molecules are added. It has been suggested that in some cases this assumption is not valid and the substrate replenishment in the micelles may limit the observed activity (89,90). To evaluate whether or not substrate replenishment and product exchange have an effect on the enzyme kinetics, a direct measurement of phospholipid exchange by using fluorescent probes, has been performed. The rate constants for lipid exchange were compared with turnover rates of several phospholipases (91).

The role of amphiphilic molecules in enzyme activity was recently reviewed (92). Surfactants can influence both the conformational state of the enzyme and the lipid interfacial organization.

In some systems, the surfactant behaves as a neutral diluent of the lipid substrate. For example, such behavior was observed with mixed micelles of phosphatidylinositol and dodecyl maltoside, used to assay PLC activity (86).

In other systems, the presence of surfactant strongly affects the lipolysis. An inhibition effect by the anionic surfactant sodium bis-(2-ethylhexyl)-sulfosuccinate (AOT) on the hydrolysis of reverse micelles of *p*-nitrophenyl butyrate by lipases from *Candida rugosa* has been observed (93). The penetration degree into the micellar interface, associated with the amphipathic lid region of these lipases, is probably affected by the presence of AOT and consequently the interfacial enzyme activation is unfavored. An inhibition effect of the activity of *B. cereus* PLC against mixed micelles PC of gangliosides has been observed (58). The effect of gangliosides is mainly due to an alteration of the substrate lipid organization.

An interesting kinetic system consisting of mixed micelles of substrate and strong competitive inhibitors, possessing very similar molecular structure, has been analyzed (94) on the basis of the kinetic model for water-insoluble inhibitors developed by Ransac et al. (56).

E. Lipolysis at the Oil-Water Interface

The studies of lipase hydrolysis of natural long-chain glycerides at the oil-water interface seem more representative of what is occurring in vivo. This is why oil-in-water emulsions are often used as lipase substrates (20,21,46,48,53,95). The kinetic data are treated on the basis of the model of Fig. 3 and the role of the colloid chemical organization of such disperse systems is often underestimated. The calculation of the overall kinetic constant Q_e is a difficult task because the emulsion systems have a complex composition (various surface-active components are added) and the A/V ratio is generally unknown. It is not always easy to correlate the lipase activity with the specific area A/V of the emulsified system used, as

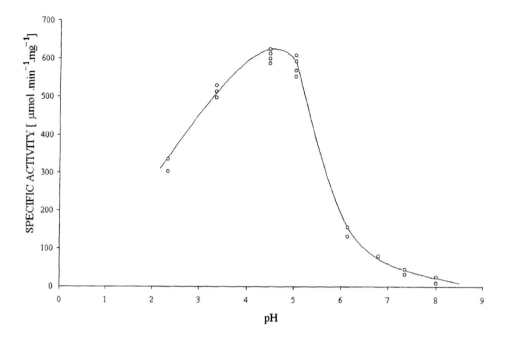

Fig. 21 Influence of pH on HGL specific activity on Intralipid measured after 1 min of incubation of the enzyme (15 nM) with the emulsion at constant pH and a temperature of 37°C. Five milliliters of 30% Intralipid was added to 10 mL of 150 mM NaCl, 3.5 mM CaCl$_2$, and 30 μM bovine serum albumin (BSA). (Adapted from Ref. 101.)

Benzonana and Desnuelle did in their pioneering work (95). This is why commercially available soybean oil emulsions, stabilized with egg phosphatidylcholine, are often used as lipase substrates. These commercial emulsions are known as Intralipid (10% or 20%) and are used for human parenteral nutrition. Furthermore, they are stable over periods of months and their particle size distribution has been determined by several methods, allowing the measurement of their specific surface (96–99). The lipids are aggregated into two main types of particles coexisting in the dispersion, namely (a) emulsion particles with a mean diameter of 360 nm containing a soybean oil core emulsified by an egg phosphatidylcholine monolayer and (b) egg phosphatidylcholine unilamellar liposomes 85 nm in diameter. For $C_e = 9 \times 10^{12}$ emulsion particles per cm^3, the ratio $A/V = 3.7 \times 10^4$ cm^{-1} is obtained. Chemical analysis of such emulsions (100) shows that the composition of the monolayer corresponds to 97% phosphatidylcholine and 3% soybean oil.

Unfortunately, not all lipases are able to hydrolyze Intralipid [such an pancreatic lipase (101,102)]. Thus, in order to calculate some kinetic parameters using calibrated emulsions, we have selected the experimental data obtained with gastric lipase, a more penetrating enzyme than pancreatic lipase (see Fig. 21). For maximal activity SA = 600 μmolS min^{-1} mgE^{-1} at pH 5 with $A/V = 3.7 \times 10^4$ cm^{-1} and $\Gamma_s = 0.03 \times 1.67 \times 10^{-4} = 5 \times 10^{-6}$ μmol cm^{-2}, an experimental value for the overall kinetic

Fig. 22 Typical evolution in shape and morphology of the oil drop resulting from the action of PL.

constant $Q_e = 3.2 \times 10^3$ cm^3 min^{-1} mgE$^{-1} = 2.7 \times 10^{-13}$ cm^3 min^{-1} molecule^{-1} was obtained. It is worth noticing that both lipases act synergistically on Intralipid as a substrate (102).

We will now briefly discuss the results obtained with long-chain triacylglycerides at an oil-water (O/W) interface by means of the oil drop method (103–106). The enzymatic reaction was monitored by recording the interfacial tension changes resulting from the accumulation of a monolayer of products at the interface during the early stages of hydrolysis. The general kinetic model of Fig. 3 and the approximation for the monolayer $A/V \approx 1$ were used to calculate the overall kinetic constant $Q_{\mathrm{O/W}}$. After 1 min of hydrolysis by HPL (molecular weight 50,000) at a soybean oil-water interface, a maximum interfacial specific activity SA* $\equiv (1/\Gamma_{\mathrm{E^*}})d\Gamma_{\mathrm{P^*}}/dt = 41$ μmolS min^{-1} mgE^{-1} was found. The maximum amount of adsorbed enzyme of 1% of the total HPL injected was independently determined by ELIZA test. Then, the dimensionless volume-interface enzyme partitioning $k_d/k_p(A/V) = C_E V/\Gamma_{\mathrm{E^*}}A = 99/1 \approx 10^2$ (106). The typical specific area is $A/V \approx 1$ cm^{-1}. The interfacial concentration of the soybean oil is $\Gamma_S = 1.67 \times 10^{-4}$ μmol cm^{-2}. From these data, the overall kinetic constant can easily be obtained:

$$Q_{\mathrm{O/W}} = \frac{1}{\Gamma_S C_E}\frac{d\Gamma_{\mathrm{P^*}}}{dt} = \frac{10^{-2}}{\Gamma_S \Gamma_{\mathrm{E^*}}(A/V)}\frac{d\Gamma_{\mathrm{P^*}}}{dt} = \frac{41 \times 10^{-2}}{1.67 \times 10^{-4}}$$
$$= 2.43 \times 10^3 \, \text{cm}^3 \, \text{min}^{-1} \, \text{mgE}^{-1} = 2.0 \times 10^{-13} \, \text{cm}^3 \, \text{min}^{-1} \, \text{molecule}^{-1}$$

The effects of HPL and colipase (Co) on monomolecular films at air-water and oil-water interfaces were compared—$Q_e(+\mathrm{Co}) = 8.4 \times 10^{-13}$ cm^3 min^{-1} molecule^{-1}. Colipase behaves like a true cofactor of HPL (106).

One has to emphasize that the applicability of this approach is limited to the early stages of lipolysis. With advancement of the reaction, the accumulation of lipolytic products leads to the formation of more complex multilayered structures. A typical evolution in shape and morphology of the oil drop resulting from the action of PL is shown in Fig. 22.

Table 3 Specific Area (A/V), Volume-Surface Enzyme Partitioning $[k_d/k_p(A/V)]$, and Interfacial Quantity (Q) for Phospholipolysis of DPPC by PLA$_2$ and Lipolysis of TO by HPL and HGL

Parameter	DPPC-PLA$_2$			TO-HPL		
	Monolayer A/W	Liposomal bilayer	Mixed micelles	Monolayer A/W	Monolayer O/W	Emulsion
$\frac{A}{V}$ (cm^{-1})	≈ 1	1.8×10^4	2.1×10^5	≈ 1	≈ 1	3.7×10^4
$\frac{k_d}{k_p(A/V)}$ (dimensionless)	$24\left(=\frac{96}{4}\right)$	<0.1	$\approx 0.1\left(=\frac{9.09}{90.0}\right)$		$99\left(=\frac{99}{1}\right)$	<0.1
Q (cm^3 min^{-1} molecule^{-1})	(+CD) 2.3×10^{-13}			(+CD) 1.9×10^{-12}	(+Co) 8.4×10^{-13}	
	(−CD) 4.2×10^{-15}	0.23×10^{-15}	3.2×10^{-15}		(−Co) 2.0×10^{-13}	2.7×10^{-13}

F. Comparative Kinetics of Enzyme Action on Variously Organized Lipid Substrates

The mechanism of the enzymatic hydrolysis of phospholipids has been extensively studied by using monolayers at the air-water (A/W) interface, liposomal bilayers, or micelles as phospholipase substrates. The catalytic action of lipases on the tri- and diglyceride substrates organized as monolayers at A/W and O/W interfaces or emulsions has also been investigated. Typical results obtained with two systems that are representative of both classes of enzymatic reactions (phospholipolysis of DPPC by PLA$_2$ and lipolysis of triolein (TO) by HPL) were analyzed on the basis of the simplified model in Fig. 3 in the framework of two approximations: (a) $A/V \approx 1$ for monolayers at A/W and O/W interfaces [Eq. (8)] and (b) $A/V \gg 1$ for dispersion systems [liposomes, micelles, and emulsions, Eq. (13)]. The kinetic parameters obtained are summarized in Table 3 and Fig. 23.

The A/V ratio is a measure of the degree of dispersion and varies from 1 for the monolayer systems to many orders of magnitude for the dispersion systems used. The dimensionless volume-surface enzyme partitioning $k_d/k_p(A/V)$ at steady state represents the ratio between the enzyme molecules in the bulk $C_E V$ and those at the interface $\Gamma_{E^*} A$. As previously discussed, in a monolayer system only few percent of the initially present enzyme molecules have bound at the interface (4% at A/W; 1% at O/W), whereas in the dispersion systems practically all enzyme molecules are at the interface (90% in the example of mixed micelles discussed earlier but practically 100% in the most studied dispersed systems). As illustrated schematically in Table 3 and Fig. 23, one can see that using either the phospholipid–phospholipase A$_2$ or the triglyceride-lipase system, there is no

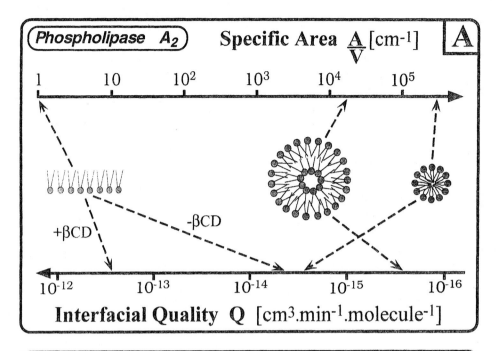

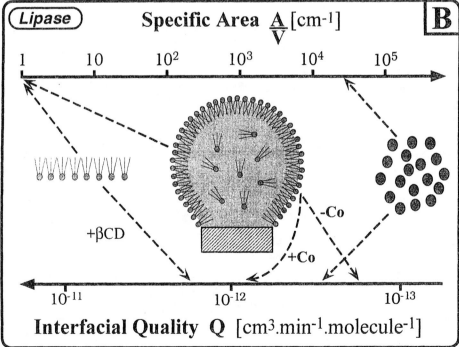

Fig. 23 Specific area A/V and interfacial quality Q for variously organized lipid substrates. (A) Phospholipolysis by PLA$_2$ of DPPC organized as a monolayer at the A/W interface, liposomal bilayers, or micelles. (B) Lipolysis by HPL of TO, organized as monolayers at A/W and O/W interfaces or emulsions.

obvious relationship between the specific area (A/V) and the interfacial quality (Q). This lack of direct correlation is expected because both parameters are not theoretically correlated when using the kinetic models previously described. However, it is remarkable that the monolayer system having the smallest specific area $(\approx 1 \text{ cm}^{-1})$ reveals the highest interfacial quality with both pancreatic lipase and pancreatic phospholipase A_2. One has to take into account the fact that the monolayer system is experimentally the most versatile one and was thus optimized in terms of lipid packing, chain length specificity, and product solubilization, which is not the case for the other dispersed systems such as liposomes, micelles, and emulsions. As a consequence, when using long-chain DPPC monolayers as substrates for PLA$_2$ in the absence of product acceptor (β-CD), the interfacial quality decreases by two orders of magnitude, reaching approximately the value characteristic of mixed Triton-DPPC micelles. In the case of the soybean oil drop and the monolayer system, both specific area and interfacial quality are of the some order of magnitude when the pancreatic lipase–colipase system is used. In contrast, when the specific protein cofactor of pancreatic lipase (colipase) is omitted, the interfacial quality of the oil drop system reaches a value close to that observed with gastric lipase acting on the Intralipid emulsion. It should also be noted that only the

Table 4 Applications of Lipases

Field of application	Main action	Use
Foodprocessing		
Dairy products and bakery	Lipid hydrolysis	Improving flavor and preserving
Brewery	Lipid hydrolysis	Improving aroma and speeding up fermentation process
Meat and fish products	Lipid hydrolysis	Enhancing flavor and removing excess fat
Butter, oil, and margarine processing	Interesterification of vegetable oils and animal fats	Bioprocessing procedures
Fine chemistry		
Lipids	Hydrolysis and interesterification	Production of fatty acids, mono- and diglycerides
Organic synthesis	Esterification and hydrolysis	Synthesis of chiral esters and resolution of racemic mixtures
Analytical chemistry		Glyceride determination and fatty acid distribution analysis
Detergents, cleaning, pollution agents	Lipid hydrolysis	Stain removal, greasy water and antiprocessing, leather cleaning
Cosmetics	Lipid hydrolysis	Grease removal and special formulas
Paper manufacturing	Wax hydrolysis	Paper pulp processing
Medical and pharmaceutical	Enzyme therapy	Treatment for pancreatic insufficiency (defective fat absorption processes)
	Inhibition	Treatment for overweight using specific digestive lipase inhibitors

monolayer system allows an accurate kinetic comparison between phospholipolysis and glycerolipolysis.

IV. SOME PRESENT-DAY APPLICATIONS OF LIPASES

See Table 4.

V. CONCLUSIONS AND PROSPECTIVES

In the past, a large number of kinetic data dealing with the enzymatic action against variously organized lipid structures have been accumulated. The interpretation of these data is implicitly or explicitly based on the usual kinetic models of interfacial lipolysis developed in the 1970s by Verger et al. (15), Denis et al. (16), and Jain et al. (18). These classical kinetic analyses were further developed to describe competitive inhibition, reorganization, and solubilization of lipolytic products. Some fruitful attempts to account for the colloid-chemical features of the systems used have also been made. However, the role of the colloid chemical organization is still poorly understood.

In the future, the kinetic analysis will be completed and generalized in order to take into account more adequately some colloid chemical properties such as aggregation processes, reorganization and solubilization of the reaction products, autoacceleration of the hydrolysis as a result of the accumulation of insoluble products, formation of more complex and multilayered structures at the oil-water interface.

Efforts to better understand the molecular events of interfacial catalysis will also continue. In order to bridge the gap between the various model systems used and the in vivo situation, analysis of steady states close to equilibrium, with minimum entropy production, or of dissipative structures far from equilibrium, based on the principles of irreversible thermodynamic processes, is certainly worthwhile.

ACKNOWLEDGMENTS

The authors thank A. Tiss and K. Balashev for their help in drawing the figures. Part of this work and the travel expenses of I.P. and R.V. to write this review were supported by the E.U. research program BIOTECH No. BIO4-CT97-2365 and the Bulgarian National Foundation for Scientific Research Projects.

REFERENCES

1. P Wooley, SB Petersen. Lipases: Their Structure, Biochemistry and Application. Cambridge: Cambridge University Press, 1994.
2. B Rubin, EA Dennis, eds. Lipases, Methods in Enzymology. Vols 284 and 286. New York: Academic Press, 1997.
3. L Alberghina, RD Schmid, R Verger, eds. Lipases: Structure, Mechanism and Genetic Engineering. GBF Monographs. Vol 16. Weinheim: VCH, 1991.
4. H Brockerhoff, RG Jensen. Lipolytic Enzymes. New York: Academic Press, 1974.
5. B Borgström, HL Brockman. Lipases. Amsterdam: Elsevier, 1984.
6. M Waite. In DJ Hanahn, ed. The Phospholipases. New York: Plenum, 1987.
7. EA Dennis, ed. Phospholipases, Methods in Enzymology. Vol 197. New York: Academic Press, 1991.

8. KE Jaeger, S Ransac, BW Dijkstra, C Colson, M van Heuvel, O Misset. FEMS Microbiol Rev 15:29, 1994.

9. EJ Gilbert. *Pseudomonas* lipases: Biochemical properties and molecular cloning. Enzyme Microb Technol 15:634, 1993.

10. KE Jaeger, S Wohlfahrt. Bacterial lipases: Biochemistry, molecular genetics and applications in biotechnology. Bioengineering 9:39, 1993.

11. AHC Huang. In Ref. 5, p 419.

12. KD Mukherjee, MJ Hills. In Ref. 1, p 49.

13. F Carriere, Y Gargouri, H Moreau, S Ransac, E Rogalska, R Verger. in Ref. 1, p 181.

14. F Carriere, S. Bezzine, R Verger. Molecular evolution of the pancreatic lipase and two related enzymes towards different substrate selectivities. J Mol Catal B Enzymatic 3:55, 1997.

15. R Verger, MCE Mieras, GH de Haas. Action of phospholipase A at interface J Biol Chem 248:4023, 1973.

16. RA Deems, BR Eaton, EA Dennis. Kinetic of phospholipase A_2 activity toward mixed micelles and its implications for study of lipolytic enzymes. J Biol Chem 250:9013, 1975.

17. GM Carman, RA Deems, EA Dennis. Lipid signaling enzymes and surface dilution kinetics. J Biol Chem 270:18711, 1995.

18. MK Jain, OG Berg. The kinetics of interfacial catalysis by phospholipase A_2 and regulation of interfacial activation: Hopping versus scooting. Biochim Biophys Acta 1002:127, 1989.

19. MK Jain, MH Gelb, J Rogers, OG Berg. Kinetic basis for interfacial catalysis by phospholipase A_2. Methods Enzymol 249:567, 1995.

20. L Sarda, P Desnuelle. Action de la lipase pancréatique sur les esters en emulsion. Biochim Biophys Acta 30:513, 1958.

21. P Desnuelle, L Sarda, G Ailhaud. Inhibition de la lipase pancréatique par diéthyl-*p*-nitrophényl phosphate en émulsion. Biochim Biophys Acta 37:570, 1960.

22. FK Winkler, A D'Arcy, W Hunziker. Structure of human pancreatic lipase. Nature 343:771, 1990.

23. L Brady, AM Brzozowski, ZS Derewenda, E Dodson, G Dodson, S Tolley, JP Turkenburg, L Christiansen, Huge-Jensen, L Norskov, L Thim, U Menge. A serine protease triad forms the catalytic centre of a triacylglycerol lipase. Nature 343:767, 1990.

24. AM Brzozowski, U Derewenda, ZS Derewenda, GG Dodson, DM Lawson, JP Turkenburg, F Bjorkling, B Huge-Jensen, SA Patkar, L Thim. A model for interfacial activation in lipases from the structure of a fungal lipase-inhibitor complex. Nature 351:491, 1991.

25. H van Tilbeurgh, M-P Egloff, C Martinez, N Rugani, R Verger, C Cambillau. Interfacial activation of the lipase-procolipase complex by mixed micelles revealed by X-ray crystallography. Nature 362:184, 1993.

26. MEM Noble, A Cleasby, LN Johnson, LGJ Frenken, MR Egmond. The crystal structure of triacylglycerol lipase from *Pseudomonas glumae* reveals a partially redundant catalytic aspartate. FEBS Lett 331:123, 1993.

27. J Uppenberg, MT Patkar, S Hansen, A Jones. The sequence, crystal structure determination and refinement of two crystal forms of lipase-B from *Candida antarctica*. Structure 2:293, 1994.

28. A Hjorth, F Carrière, C Cudrey, H Wöldike, E Boel, DM Lawson, F Ferrato, C Cambillau, GG Dodson, L Thim, R Verger. A structural domain (the lid) found in pancreatic lipases is absent in the guinea pig (phospho) lipase. Biochemistry 32:4702, 1993.

29. F Carrière, JA Barrowan, R Verger, R Laugier. Secretion and contribution to lipolysis of gastric and pancreatic lipases during a test meal in humans. Gastroenterology 105:876, 1993.

30. T Giller, P Buchwald, DB Kaelin, W Hunziker. Two novel human pancreatic lipase related proteins, HPLRP1 and HPLRP2. Differences in colipase dependence and in lipase activity. J Biol Chem 267:16509, 1992.

31. K Thirstrup, R Verger, F Carrière. Evidence for a pancreatic lipase subfamily with new kinetic properties. Biochemistry 33:2748, 1994.

32. MP Egloff, F Marguet, G Buono, R Verger, C Cambillau, H van Tilbeurgh. The 2.46 Å resolution structure of the pancreatic lipase-colipase complex inhibited by a C11 alkyl phosphonate. Biochemistry 34:2751, 1995.

33. ML Jennens, ME Lowe. C-terminal domain of HPL is required for stability and maximal activity but not colipase reactivation. J Lipid Res 36:1029, 1995.

34. F Carrière, K Thirstrup, S Hjorth, F Ferrato, PF Nielsen, C Withers-Martinez, C Cambillau, E Boel, L Thim, R Verger. Pancreatic lipase structure-function relationships by domain exchange. Biochemistry 36:239, 1997.

35. C Withers-Martinez, F Carrière, R Verger, D Bourgeois, C Cambillau. A pancreatic lipase with a phospholipase A_1 activity: Crystal structure of a chimeric pancreatic lipase–related protein 2 from guinea pig. Structure 4:1363, 1996.

36. M Martinelle, M Holmquist, IG Clausen, S Patkar, A Svendsen, K Hult. The role of Glu87 and Trp89 in the lid of *Humicola lanuginosa* lipase. Protein Eng 9:519, 1996.

37. GH Peters, OH Olsen, A Svendsen, RC Wade. Theoretical investigation of the dynamics of the active site lid in *Rhizomucor miehei* lipase. Biophys J 71:119, 1996.

38. L Byuing-In, ET Yoon, W Cho. Roles of surface hydrophobic residues in the interfacial catalysis of bovine pancreatic phospholipase A_2. Biochemistry 35:4231, 1996.

39. I Panaiotov, M Ivanova, R Verger. Interfacial organization of enzymatic lipolysis. Curr Opin Colloid Interface Sci 2:517, 1997.

40. R Verger. Interfacial activation of lipases: Fact and artifacts. Rev TIBTECH January 15:32, 1997.

41. HM Verheij, MR Egmond, GH de Haas. Chemical modification of the α-amino group in snake venom phospholipase A_2. A comparison of the interaction of pancreatic and venom phospholipase with lipid-water interface. Biochemistry 20:94, 1981.

42. M Martinelle, K Hult. Kinetics of triglyceride lipase. In P Woolley, SB Petersen, eds. Lipases. Cambridge: Cambridge University Press, 1994, p 159.

43. K Hult, M Holmquist. Methods Enzymol 286:386, 1997.

44. F Pattus, AJ Slotboom, GH de Haas. Regulation of phospholipase A_2 activity by the lipid-water interface: A monolayer approach. Biochemistry 13:2691, 1979.

45. R Verger, GA de Haas. Enzyme reactions in a membrane model—A new technique to steady enzyme reactions in monolayers. Chem Phys Lipids 10:127, 1973.

46. R Verger. Enzyme kinetics of lipolysis. Methods Enzymol 64B:340, 1980.

47. R Verger, GH de Haas. Interfacial enzyme kinetics of lipolysis. Annu Rev Biophys Bioeng 5:77, 1976.

48. S Ransac, F Carrière, E Rogalska, R Verger, F Marguet, G Buono, EP Melo, JMS Cabral, MPE Egloff, H van Tilbeurgh, G Cambillau. The kinetics, specificities and structural features of lipases. In: AF Jos, Op den Kamp, eds. NATO ASI Series H96. Berlin: Springer-Verlag, 1996, p 265.

49. P Joos, RA Demel. The interaction energies of cholesterol and lecithin in spread mixed monolayers at the air-water interface. Biochim Biophys Acta 189:447, 1969.

50. JM Muderhwa, HL Brockman. Lateral lipid distribution is a major regulator of lipase activity. Implications for lipid-mediated signal transduction. J Biol Chem 267:24184, 1992.

51. GH Peters, S Toxvaerd, NB Larsen, T Bjørnholm, K Schaumburg, K Kjaer. Structure and dynamics of lipid monolayers: Implications for enzyme catalysed lipolysis. Struct Biol 2:395, 1995.

52. H Verheij, M Boffa, C Rothen, M Bryckaert, R Verger, GH de Haas. Correlation of enzymatic activity and anticoagulant properties of phospholipase A_2. Eur J Biochem 112:25, 1980.

53. E Cernia, L Battinelli, S Soro. Biocatalysed hydrolysis of triglycerides in emulsion and as monolayers. Thin Solid Films 284–285:727, 1996.

54. V Raneva, Tz Ivanova, R Verger, I Panaiotov. Comparative kinetics of phospholipase A_2 action on liposomes and monolayers of phosphatidylcholine spread at the air-water interface. Colloids Surf B Biointerfaces 3:357, 1995.

55. R Verger, F Pattus. Lipid-protein interactions in monolayers. Chem Phys Lipids 30:189, 1982.

56. S Ransac, C Rivière, JM Soulié, C Gancet, R Verger, GH de Haas. Competitive inhibition of lipolytic enzymes. A kinetic model applicable to water insoluble competitive inhibitors. Biochim Biophys Acta 1043:57, 1990.

57. BH Havsteen, VR Castellanos, GMJ Meseguer, E Valero, GM Moreno. Kinetic theory of the action of lipases. J Theor Biol 157:523, 1992.

58. JJ Danielle, B Maggio, ID Bianco, FM Goni, A Alonso, G Fidelio. Inhibition by gangliosides of *Bacillus cereus* phospholipase C activity against monolayers, micelles and bilayer vesicles. Eur J Biochem 239:105, 1996.

59. M Ivanova, I Panaiotov, A Bois, Y Gargouri, R Verger. Inhibition of pancreatic lipase by ovalbumin and β-lactoglobulin A at the air-water interface. J Colloid Interface Sci 136:363, 1990.

60. M Ivanova, R Verger, A Bois, I Panaiotov. Proteins at the air-water interface and their inhibitory effects on enzyme lipolysis. Colloids Surf 54:279, 1991.

61. Y Gargouri, S Ransac, R Verger. Covalent inhibition of digestive lipases: An in vitro study. Biochim Biophys Acta 1344:6, 1997.

62. T Wieloch, B Borgström, G Pieroni, F Pattus, R Verger. Product activation of pancreatic lipase. J Biol Chem 257:11523, 1982.

63. H Möhwald. Phospholipid and phospholipid-protein monolayers at the air-water interface. Annu Rev Phys Chem 41:441, 1990.

64. D Hönig, D Möbius. Direct visualization of monolayers at the air-water interface by Brewster angle microscopy. J Phys Chem 95:4590, 1991.

65. D Vollhard. Morphology and phase behaviour of monolayers. Adv Colloid Interface Sci 64:143, 1996.

66. DW Grainger, A Reichert, H Ringsdorf, C Salesse. Hydrolytic action of phospholipase A_2 in monolayers in the phase transition region: Direct observation of enzyme domain formation using fluorescence microscopy. Biochim Biophys Acta 1023:365, 1990.

67. T Kondo, T Kakiuchi, M Shimomura. Fluorescence microscopic imaging of hydrolysis of phospholipid monolayers by phospholipase D at the air-water interface. Thin Solid Films 244:887, 1994.

68. T Hønger, K Jørgensen, RL Biltonen, OG Mouristen. Systematic relationship between PLA_2 activity and dynamic lipid microheterogeneity. Biochemistry. 35:9003, 1996.

69. M Ivanova, Tz Ivanova, R Verger, I Panaiotov. Hydrolysis of monomolecular films of long chain phosphatidylcholine by phospholipase A_2 in the presence of β-cyclodextrin. Colloids Surf B Biointerfaces 6:9, 1996.

70. M Ivanova, R Verger, I Panaiotov. Mechanisms of the desorption of monolayers of long chain lipolytic products by cyclodextrins: Application to lipase kinetics. Colloids Surf B Biointerfaces 10:1, 1997.

71. R Apitz-Castro, MK Jain, GH de Haas. Origin of the latency phase during the action of phospholipase A_2 on unmodified phosphatidylcholine vesicles. Biochim Biophys Acta 688:349, 1982.

72. JAF Op den Kamp, J De Gier, LLM Deenen. Hydrolysis of phosphatidylcholine liposomes by pancreatic phospholipase A_2 at the transition temperature. Biochim Biophys Acta 345:253, 1974.

73. CR Kensil, EA Dennis. Action of cobra venom phospholipase A_2 on the gel and liquid crystalline states of dimyristoyl phosphatidylcholine vesicles. J Biol Chem 254:5843, 1979.

74. E Goormaghtigh, M Van Campenhoud, JM Ruysschaert. Lipid phase separation mediates binding of porcine pancreatic phospholipase A_2 to its substrate. Biochem Biophys Res Commun 101:1410, 1981.

75. D Lichtenberg, G Romero, M Menashe, RL Biltonen. Hydrolysis of dipalmitoylphosphatidylcholine large unilamellar vesicles by porcine pancreatic phospholipase A_2. J Biol Chem 261:5334, 1986.

76. NE Gabriel, NV Agman, MF Roberts. Enzymatic hydrolysis of short-chain lecithin/long-chain phospholipid unilamellar vesicles: Sensitivity of phospholipases to matrix phase state. Biochemistry 26:7409, 1987.

77. N Cheriani-Gruszka, S Almong, RL Biltonen, D Lichtenberg. Hydrolysis of phosphatidylcholine in phosphatidylcholine-cholate mixtures by porcine pancreatic phospholipase A_2. J Biol Chem 263:11808, 1988.

78. JYA Lehtonen, PKJ Kinnunen. Phospholipase A_2 as a mechanosensor. Biophys J 68:1888, 1995.

79. OG Mouristen, K Jørgensen. Dynamic lipid-bilayer heterogeneity: A mesoscopic vehicle for membrane function. Bioessays 14:129, 1992.

80. OG Mouristen, K Jørgensen. Dynamical order and disorder in lipid bilayers. Chem Phys Lipids 73:3, 1994.

81. F Pattus, P Desnuelle, R Verger. Spreading of liposomes at the air-water interface. Biochim Biophys Acta 507:62, 1978.

82. H Schindler. Exchange and interactions between lipid layers at the surface of a liposome solution. Biochim Biophys Acta 555:316, 1979.

83. Tz Ivanova, G Georgiev, I Panaiotov, MA Surpas, JE Proust, F Puisieux. Behaviour of liposomes prepared from lung surfactant analogues and spread at the air-water interface. Prog Colloid Polym Sci 79:24, 1989.

84. I Panaiotov, JE Proust, V Raneva, Tz Ivanova. Kinetics of spreading at the air-water interface of DOPC liposomes influenced by photodynamic lipid peroxidation. Thin Solid Films 244:845, 1994.

85. M Haubs, H Ringsdorf. Photosensitive monolayers, bilayer membranes and polymers. New J Chem 11(2):151, 1987.

86. SR James, A Paterson, KT Harden, PC Downes. Kinetic analysis of phospholipase $C\beta$ isoforms using phospholipid-detergent mixed micelles. J Biol Chem 270:11872, 1995.

87. A Abousalham, J Nari, M Teissère, N Ferté, G Noat, R Verger. Study on sunflower–phospholipase D: Fatty-acid specificity using detergent/phospholipids mixed micelles. Eur J Biochem 248:374, 1997.

88. JN Israelachvili. In: Intermolecular and Surface Forces. 2nd ed. London: Academic Press, 1992, p 359.

89. MJ Jain, J Rogers, O Berg, MH Gelb. Interfacial catalysis by phospholipase A_2: Activation by substrate replenishment. Biochemistry 30:7340, 1991.

90. J Rogers, BZ Yu, MK Jain. Basis for the anomalous effect of competitive inhibitors on the kinetic hydrolysis of short-chain PC by PLA_2. Biochemistry 31:6056, 1992.

91. CE Soltys, MF Roberts. Fluorescence studies of phosphatidylcholine micelle mixing: Relevance to phospholipase kinetics. Biochemistry 33:11608, 1994.

92. DN Rubingh. The influence of surfactants on enzyme activity. Curr Opin Colloid Interface Sci 1:598, 1996.

93. C Otero, L Robledo. Lipase activity in anionic reverse micelles. Inhibition effect of the system. Prog Colloid Polym Sci 98:219, 1995.

94. GH de Haas, R Dijkman, JWP Boots, HM Verheij. Competitive inhibition of lipolytic enzymes XI. Biochim Biophys Acta 1257:87, 1995.
95. G Benzonana, P Desnuelle. Étude cinétique de l'action de la lipase pancréatique sur des triglycerides en emulsion. Biochim Biophys Acta 105:121, 1965.
96. M Rotenberg, M Rubin, A Bor, D Meyuhas, Y Talmon, D Lichtenberg. Physicochemical characterization of Intralipid® emulsion. Biochim Biophys Acta 1086:265, 1991.
97. MJ Groves, M Vinderberg, APR Brain. The presence of liposomal material in phosphatide stabilized emulsion. J Dispersion Sci Technol 6:237, 1985.
98. E Granot, RJ Deckelbaum, S Eisenberg, Y Oschry, G Bengtsson-Olivecrona. Core modification of human low-density lipoprotein by artificial triacylglycerol emulsion. Biochim Biophys Acta 833:308, 1985.
99. G Nalbon, D Lairon, M Charbonnier-Augeire, JL Vigne, J Leonardi, C Chabert, JC Hauton, R Verger. Pancreatic phospholipase A_2 hydrolysis of phosphatidylcholines in various physicochemical states. Biochim Biophys Acta 620:612, 1980.
100. K Miller, D Small. Structure of triglyceride-rich lipoproteins: an analysis of core and surface phases. In: A Neuberger, LLM van Deenen, eds. New Comprehensive Biochemistry. Vol 14. Amsterdam: Elsevier, 1987, p 1.
101. Y Gargouri, G Piéroni, C Rivière, JF Sauniere, P Lowe, L Sarda, R Verger. Kinetic assay of human gastric lipase on short- and long chain triacylglycerol emulsions. Gastroenterology 91:919, 1986.
102. Y Gargouri, G Piéroni, C Rivière, P Lowe, JF Saunier, L Sarda, R Verger. Importance of HGL for intestinal lipolysis: An in vitro study. Biochim Biophys Acta 879:419, 1986.
103. S Nury, G Piéroni, C Rivière, Y Gargouri, A Bois, R Verger. Lipase kinetics at the triacylglycerol-water interface using surface tension measurements. Chem Phys Lipids 45:27, 1987.
104. S Labourdenne, N Gaudry-Rolland, S Letellier, M Lin, A Cagna, G Esposito, R Verger, C Rivière. The oil-drop tensiometer: Potential applications for studying the kinetics of (phospho)lipase action. Chem Phys Lipids 71:163, 1994.
105. S Labourdenne, MG Ivanova, O Brass, A Cagna, R Verger. Surface behaviour of long-chain lipolytic products (a 1-to-1 mixture of oleic acid and diolein) spread as monomolecular films in the presence of long-chain triglycerides. Colloids Surf B Biointerfaces 6:173, 1996.
106. S Labourdenne, O Bras, M Ivanova, A Cagna, R Verger. Effects of colipase and bile salts on the catalytic activity of human pancreatic lipase. A study by the oil-drop tensiometer. Biochemistry 36:3423, 1997.

12

Mimicking Physics of Cell Adhesion

R. Simson
Software Design and Management AG, Munich, Germany

E. Sackmann
The Technical University of Munich, Munich, Germany

I. INTRODUCTION

Cell adhesion is a most fascinating and complex process. It is regulated by a subtle interplay of specific receptor-mediated lock and key forces, a phalanx of interfacial forces, and, most important, by the membrane elasticity. This competition of forces and elasticity is illustrated in Fig. 1. Adhesion is further complicated by the fact that, in general, it is accompanied by lateral phase separation, which results in a clustering of receptors. An example of this effect is the formation of focal adhesion plaques during adhesion of fibroblasts. It appears obvious that such a complex process requires control by simple principles of regulation and that their understanding demands detailed experimental and theoretical studies.

The main purpose of the present contribution is to summarize physical principles that regulate adhesion of soft elastic shells on surfaces, stressing the ubiquitous role of entropy-controlled membrane elasticity. In this context, the analogies to the physics of wetting will also be pointed out. A second purpose is to show how the physical principles of adhesion can be elucidated by studying the interaction of model membranes (giant vesicles with reconstituted receptors) with functionalized solid supports, employing surface-sensitive techniques. The consequences of the interplay of specific and universal forces are illustrated for two situations:

The formation of adhesion plaques caused by the competition between strong receptor-mediated attraction and weak repulsion (e.g., caused by undulation forces)

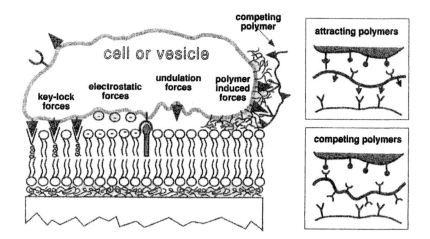

Fig. 1 Cartoon illustrating cell adhesion as the interplay of receptor-mediated specific and universal forces as well as membrane elasticity. Universal forces include electrostatic interaction, undulation forces, van der Waals interaction, hydration forces, and manifold forces mediated by macromolecules. The latter may consist of a repulsion between glycocalyces or competition between cells and extrinsic proteins (e.g., hyaluronic acid, fibrinogen) for binding sites on target cells or tissues.

> The electrostatically induced adhesion and its control by the osmotic repulsion generated by the counterions.

These examples demonstrate the close correlation between adhesion and phase separation, resulting in the formation of pinning domains. They show how important entropic effects are for regulating the adhesion strength. The entropic (undulatory) repulsion forces control the nucleation of pinning centers and subsequently impede their fusion, resulting in the long-time stabilization of adhesion plaques. The translational entropy associated with adhesion-induced phase separation can reduce the work of separation by several orders of magnitude. Physically, this is a consequence of the mechanical equilibrium condition at the membrane-substrate interface, which is governed by the Young equation. An intriguing aspect of this control mechanism is that it could be exploited by cells in order to avoid a strong attraction to solids, which may otherwise lead to cell death (1,2).

In the first part of this chapter, we show that the control of cell adhesion by membrane elasticity can be well described in terms of the classical theory of elasticity of soft shells if one considers the limit of weak deformations. This leads to a powerful, yet simple method for local measurements of the work of adhesion (or unbinding) and of adhesion-induced lateral tensions. This method is based on the analysis of the contour of the adhering shells near the substrate in terms of the elastic boundary conditions, where the contour is observed by employing a microinterferometric technique. In this context we also introduce the concept of undulation forces, which was first postulated by Helfrich (3,4).

In the second part we present a model system demonstrating that the interplay of strong receptor-mediated attraction forces and weak long-range repulsion forces (such as undulation forces) naturally leads to the formation of adhesion plaques, a well-known phenomenon in cell adhesion. In a third part we study membrane adhesion enforced through electrostatic attraction of oppositely charged surfaces. Owing to the control through the cushion of counterions (separating the membranes in order to ensure charge neutrality), their adhering membrane decays into regions of strong attraction and decoupled regions. The latter form blisters that are pressurized by the osmotic pressure generated by the counterions. A major new feature is adhesion-induced lateral phase separation. A most important consequence of this is the breakdown of the classical Young law due to the generation of a lateral osmotic pressure difference between the strongly adhering and the quasi-free membrane regions. This leads to a dramatic relaxation of the work of adhesion, as shown in the fourth part. In the appendix we describe methods for the engineering of models of cell or tissue surfaces on solid supports.

II. ARCHITECTURE AND ELASTICITY OF COMPOSITE MEMBRANES AND BILAYERS

The elasticity of cell plasma membranes is controlled by the combined action of two coupled shells: the lipid protein bilayer and the associated macromolecular network of the cytoskeleton. The latter may consist of a quasi-two dimensional, fishnet-like meshwork as in the case of erythrocytes. Or it may be composed of partially cross-linked actin filaments, forming a cortex that is roughly 1 μm thick. In the case of erythrocytes the network is made up of

> Tetramers of spectrin (a highly flexible molecule with a persistence length of ~ 10 nm and an end-to-end distance of ~ 80 nm) and of
> Oligomers of actin (~ 35 nm in length), which can act as multifunctional cross-linkers of spectrin tetramers that exhibit actin binding sites on both ends.

Spectrin tetramers form a surprisingly well-defined triangular network, with actin forming the interstices. However, as indicated in Fig. 2a, the network exhibits a high number of defects consisting of adjacent pairs of pentagons and heptagons.

The actin cortex is composed of an ill-defined network of partially cross-linked and partially entangled actin filaments. Its structure and elasticity are regulated by a manifold of actin binding proteins (ABPs). In highly mobile cells such as the amoeba *Dictyostelium discoideum*, the actin cortex forms an active contractile shell due to the homogeneous distribution of the motor protein myosin II. The viscoelastic module of the cortex can be controlled through the mesh size of the network, its degree of cross-linkage, and, most important, the coupling of the cortex to the lipid-protein bilayer. This coupling can be mediated in various ways as shown in Fig. 2.

The spectrin-actin network of erythrocytes is coupled to integral membrane proteins by two types of anchoring proteins: ankyrin and band IV.1. The key aspect is that the coupling strength can be regulated by phosphorylation of the coupling proteins, which decreases their binding constants by a factor (5) of 10. This suggests that the bilayer cytoskeleton coupling strength is regulated in a dynamic way by phosphorylation and dephosphorylation of the anchoring proteins (Fig. 2b).

The coupling of the actin cortex to the bilayer in eukaryotic cells is much more complex. Of the three types of coupling mechanisms that have been identified, two are presented in Fig. 2b. An interesting case is the coupling through

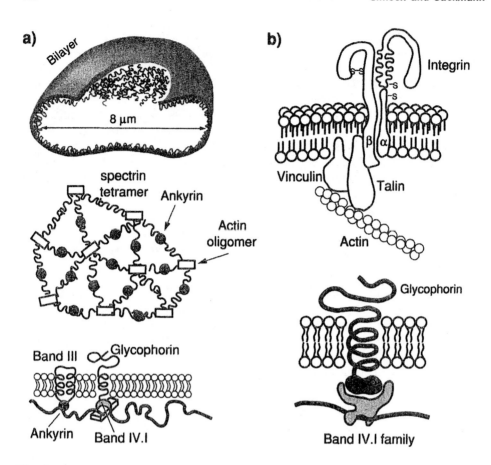

Fig. 2 (a) Coupling of the bilayer to the fishnet-like spectrin-actin network in the erythrocyte membrane. The cytoskeleton is composed of spectrin tetramers forming the sides and of actin oligomers forming the corners of a triangular network. The network structure is reminiscent of hexagonal phases (6). Coupling is mediated by ankyrin, which couples the centers of spectrin tetramers to the anion exchange protein band III, and band IV, which couples the actin cross-links to glycophorin C. The latter protein is a highly glycosylated receptor important for determining the blood group. Note that the degree of coupling is regulated by ATP. (b) Coupling of actin cortex to the plasma membrane is mediated by talin and vinculin, which in turn bind to receptors of the integrin family, and by the anchoring proteins from the band IV.1 family, which bind to receptors of the glycophorin type, as shown in (a) for erythrocytes.

hisactophilin, a small cylindrically shaped protein exhibiting several histidine-rich loops on one side and a myristic acid chain at the opposite face. It reconstitutes readily in membranes containing negatively charged lipids through combined electrostatic and hydrophobic forces. The binding of actin can be controlled by variation of the pH value, because the histidine charges are neutralized below ~pH 6.5. Finally, many of the actin-bilayer coupling proteins, notably talin, can directly couple to negatively charged membranes through electrostatic forces and thus mediate the membrane binding of actin filaments (7).

III. MEMBRANES AS SOFT ELASTIC SHELLS

The elasticity of biomembranes can be astonishingly well described by the classical theory of elasticity of shells (8). Essential new factors, however, are the entropic tension generated by thermal membrane undulations and the specific surface energy (or tension) due to the coupling of the bilayer with the surrounding aqueous phase. These tensions result in partial penetration of water into the semipolar region of the bilayer.

The general theory of the elasticity of adhering shells is extremely complex and leads to nonlinear equations. However, it can be strongly simplified if we consider only situations of weak bending corresponding to weak adhesion. In-plane deformations can be neglected because they contribute only in fourth order to the elastic energy of adhesion (9). A further simplification is possible because the regulation of adhesion by membrane elasticity is determined by the elastic boundary conditions at the shell-substrate interface, as will be shown in the following.

Membrane deformations can be described by the three well-known modes of deformation: lateral extensions, bending, and shearing. Because the shear deformation is determined by the cytoskeleton, it is customary to describe pure shear by the traditional energy-density g of the theory of rubber elasticity (10).

$$g = \frac{1}{2}\mu(\lambda^2 + \lambda^{-2} - 2) \tag{E.1}$$

Here, μ is the two-dimensional shear elastic constant (in N/m^2) and λ is the ratio of the extended to the original length $[\lambda = (L_0 + \delta L)/L_0]$. Pure shear is generated, for instance, by a lateral tension $-\tau$, acting in one direction and a tension of opposite sign, $+\tau$, acting in a direction perpendicular to the first one. Note that shear elasticity does play a role for composite membranes but vanishes in fluid bilayers.

The two-dimensional energy density associated with an isotropic lateral extension g_{ex} is expressed in terms of the area change δA:

$$\delta A = (L_0 + \delta L)^2 - L_0^2$$
$$g_{ex} = \frac{1}{2}K(\delta A/A)^2 \tag{E.2}$$

where K is the area compressibility modulus (in J/m^2).

The strain resulting from a bending deformation is expressed in terms of the local curvature of the membrane (bilayer or composite shell), which is characterized by the two principal radii of curvature R_x and R_y. These are defined as

$$R_x^{-1} = C_x = \partial^2 u/\partial x^2 \qquad R_y^{-1} = C_y = \partial^2 u/\partial y^2 \tag{E.3}$$

where $u(\mathbf{r})$ is the deflection of the membrane from its equilibrium shape and $\mathbf{r} = (x, y)$ is the local coordinate system (see Fig. 3). The bending energy of the closed shell is then expressed as (3,11)

$$G_{bend} = \frac{1}{2}\kappa \oiint_{surface} dA(C_X + C_y - C_0)^2 \tag{E.4}$$

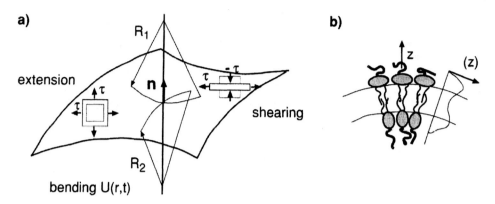

Fig. 3   (a) Schematic view of radii of main curvature and modes of deformation of soft elastic shells. (b) Generation of spontaneous curvature by intrinsic gradient of lateral tension across the membrane.

where κ is the bending modulus (exhibiting the dimension of an energy), and the integration extends over the surface area A. The mean curvature $C_x + C_y$ could be expressed equivalently as $\Delta u = \partial^2 u/\partial x^2 + \partial^2 u/\partial y^2$. C_0 is the spontaneous curvature, which accounts for possible membrane asymmetries. C_0 can be best visualized in terms of a gradient of lateral pressure $\nabla \pi(z)$ across the membrane (z direction), generating a bending moment of

$$M = \int z \nabla \pi(z) dz \tag{E.5a}$$

and a spontaneous curvature

$$C_0 = M/\kappa \tag{E.5b}$$

Spontaneous curvatures are essential in order to generate small invaginations in cell membranes such as coated pits or caveoli. They can be most easily generated by adsorption of proteins such as clathrin in the case of coated pits or by electrical charging of one of the monolayers (12).

It should be noted that the proceeding energy functional accounts only for open shells or for local deformations of closed shells. For closed stratified shells such as bilayers or composite membranes, one also has to consider a global contribution of the bending energy which accounts for the fact that the two opposing shells undergo a net extensional deformation (11,13). Because this aspect is not relevant for cell adhesion, we ignore this subtlety.

It is often helpful to relate the three elastic constants to the three-dimensional bulk Young modulus E and to the Poisson ratio v of the membrane.

$$K = \frac{ED_m}{3(1-v^2)}, \qquad \mu = \frac{ED_m}{3(1-2v)}, \qquad \kappa = \frac{Ed_m}{12(1-v^2)} \tag{E.6}$$

where d_m is the membrane thickness. For macromolecular networks, which are practically incompressible, it is $v = 1/2$ and $K \gg \mu$.

Table 1 Typical Elastic Constants of Lipid Bilayer Membrane and Biological Membrane and Comparison with Scaled Model of Erythrocyte Membrane Made up of Polyethylene

Membrane	Area compression modulus K (mN/m)	Bending modulus κ ($k_B T$)	Shear elasticity μ (J/m^2)
DMPC-L$_\alpha$	140	25	3
1:1 DMPC-cholesterol	690	100	—
DMPC-L$_\beta$	900	—	—
DMPC+1% SDS[a]	—	5	—
Erythrocyte	1000	5	0.006
Polyethylene	5×10^3	5×10^3	300
Dictyostelium cells[b]	—	500	6

[a] SDS, sodium dodecyl sulfate.
[b] Note that removal of talin reduces κ to a value characteristic for a 1:1 mixture of DMPC-cholesterol.

In Table 1 we summarize characteristic values of the elastic constants for bilayers composite membranes of erythrocytes, and *Dictyostelium* cells and compare them, with those of a hypothetical shell made up of synthetic material such as polyethylene. Please note that:

> Erythrocytes and lipid bilayers are orders of magnitude softer with respect to bending and shearing than the synthetic material, but they are equally resistant to lateral extensions.
> Bilayers rupture (lyse) at lateral extensions exceeding $\delta A/A = 3\%$. Their tensile strength is Γ_{max} 2–20 mN/m (14).
> The bending stiffness of lipid bilayers can be reduced drastically by small amounts of amphiphilic solutes, such as the surfactant sodium dodecyl sulfate (SDS) or bipolar lipids (15).

IV. ADHESION OF VESICLES AND COMPOSITE MEMBRANES

A. Adhesion of Vesicles as Wetting of Soft Elastic Fluid

The adhesion of vesicles with fluid membranes can be considered as partial wetting of solids by fluid droplets with curvature elasticity. The total free energy of adhesion therefore comprises contributions accounting for wetting as well as the energy costs associated with the adhesion-induced shape changes Eq.(E.4):

$$G_{adv} = p \oiiint dV + \frac{1}{2}\sigma \oiint dA + WA_c + \frac{1}{2}\kappa \oiint dA(\Delta u - C_0)^2 \tag{E.7}$$

The first term accounts for the osmotic pressure p of the shell (the integration extends over the volume V), the second for the change of the surface tension σ due to adhesion-induced area changes, and the third for the gain in adhesion energy (W is the gain in adhesion energy per unit area and A_c the area of contact). The last term accounts for the bending elastic contribution. Note that the first three terms of Eq. (E.7) account for the free energy of a partially wetting fluid droplet (17), whereas the last is unique for soft elastic shells.

Minimizing the preceding energy functional [Eq. (E.7)], Seifert and Lipowsky (11) calculated the shapes of adhering vesicles and their state of adhesion as a function of the adhesion energy and of the degree of deflation. The latter is expressed as

$$v = V/V_0 = 3V/4\pi R_0^3 = 6\pi^{1/3} V/A^{2/3} \tag{E.8}$$

where V is the actual volume of the vesicle and V_0 is the volume a spherical vesicle of the same area A_0 and radius R_0 would have. Thus, v varies between $v = 0$ (complete deflation) and $v = 1$ (sphere). R_0 is often called the sphere equivalent radius. The excess area can be generated by osmotic deflation or by thermal expansion of the bilayer according to $A = A_0(1 + \alpha T)$. Due to the large area expansivity ($\alpha \sim 3 \times 10^{-3}$), the membrane expands much faster with temperature than the volume.

The behavior of the vesicle is controlled by three dimensionless parameters

$$W^* = WR_0^2/\kappa, \qquad p^* = pR_0^2/\kappa, \qquad \sigma^* = \sigma R_0^2/\kappa \tag{E.9}$$

where W^*, p^*, and σ^* are the reduced (dimensionless) adhesion energy, osmotic pressure, and surface tension, respectively. This reduction is the consequence of the scale invariance of the bending energy of spherical shells.

Figure 4 shows a scenario of vesicle shapes and states of adhesion in terms of a phase diagram in (v, W^*) space. The remarkable results are the following:

> The state of adhesion may be controlled equally well by changes in the adhesion energy and in the excess area (= degree of deflation).
> Transitions between the free and the bound state may be discontinuous or continuous.

The transition from the free to the bound state depends critically on the deflation parameter v. Thus, for a given value of W^* one can go from a free state to a bound state and back to a free state by continuous increase of the excess area. Correspondingly, the vesicle shape changes from a prolate to a inside budded state and the transition indeed appears to be discontinuous (18).

B. Adhesion of Composite Membranes Is Determined by Elastic Boundary Conditions

The situation is much more complex for composite membranes, such as those of *Dictyostelium* cells. The elasticity is determined by the coupling between the two shells constituting the membrane and the bilayer shell is in general an open system (Fig. 5a). The shear elastic modulus is high and the cortical layer is in general heterogeneous.

Fortunately, the situation may be greatly simplified in the case of weak adhesion. The elastic control of adhesion is dominated by the elastic boundary condition at the cell-substrate interface and may be evaluated quantitatively by analysis of the topology of the shells near the substrate. To see this, we consider Fig. 5. The global deformation of the shell of radius R is determined by the deflection $u(\mathbf{r})$ of the spherical cap at the interface, which is related to the radius a of the contact disk by $u \approx a^2/2R$ (Fig 5b).

a)

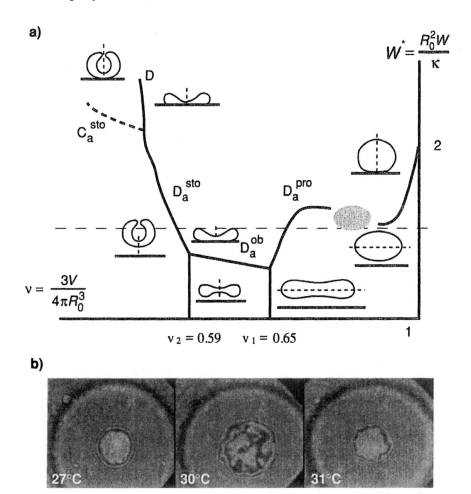

$$W^* = \frac{R_0^2 W}{\kappa}$$

$$v = \frac{3V}{4\pi R_0^3}$$

$v_2 = 0.59 \qquad v_1 = 0.65$

b)

Fig. 4 (a) Phase diagram of vesicle shapes and states of adhesion (bound-unbound) in a phase space spanned by the reduced adhesion energy W^* and the degree of deflation v. Shown are cross sections of typical vesicle shapes, such as stomatocyte (sto), oblate (ob), and prolate (pro). Note that the transitions between an adhering and a free state may be discontinuous (D_a) or continuous (C_a). The so-called reentrant transition from a free through a bound to a free state by continuous increase of the excess area and thus decrease of v is marked by a dashed horizontal line. (b) The sequence of images show a vesicle simultaneously observed with RICM and bright-field microscopy. With increasing temperature (decreasing v) the adhesion disk exhibits area changes reminiscent of a reentrant transition. After several minutes at 31°C, the vesicle suddenly loses contact and swims away.

In all practical circumstances the shells exhibit excess area. Therefore one has to consider only lateral extensions associated with the compression and dilatation of the two leaflets of the shells during bending (no change of area of neutral plane). The relative change in area is thus

$$\frac{\delta A}{A} \approx \frac{1}{2}(\nabla u)^2 \tag{E.10}$$

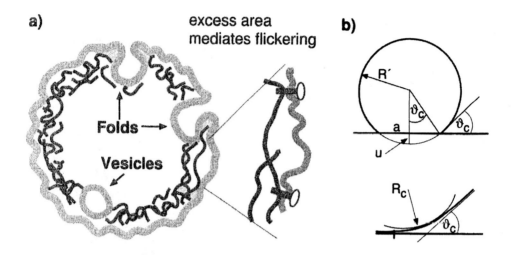

Fig. 5 (a) Cartoon of the composite cell membrane, consisting of cortical layer and protein-lipid bilayer. Note that the bilayer, which is locally pinned to the actin cortex, exhibits local invaginations or folds and may exchange vesicles with the environment. It is therefore an open system. The enlargement stresses the fact that the bilayer may undulate between pinning centers. The mean squared amplitudes ($\langle u^2 \rangle$) depend on the distance Δ between the pinning points as $\langle u^2 \rangle = (k_B T/\kappa)\Delta^2$. For erythrocytes it is $\Delta \sim 100\text{Å}$, $\kappa \approx 10 k_B T$, and therefore $\langle u^2 \rangle^{1/2} \approx 30$ nm. (b) Geometric parameters characterizing adhesion-induced deformation for the case of weak adhesion ($a \ll R$). We assume that the shell is composed of two leaflets separated by a neutral surface. The contact angle is $\theta_c \approx a/R$. The deflection of the adhesion disk (height of the spherical cap) is $u \approx a^2/2R$. The area of the spherical cap is $A \approx \pi(a^2 + u^2)$. The area change due to cap formation is $\Delta A \approx \pi u^2$. If the total volume were kept constant, the radius of the shell would change to $R' \approx R + u^2/4R$. Note that the shell must be bent at the contact line (forming a radius of curvature R_C), allowing a smooth transition into the contact zone, in order to avoid infinitely large bending energies associated with a sharp bend. The curvature energy can be accounted for by introducing a line tension τ.

According to Fig. 5, the strains associated with the deformations of the spherical cap (at the adhesion zone) can be approximated (8) as $\partial u/\partial x \approx u/R$ and $\partial^2 u/\partial x^2 \approx u/R^2$. Because $u \approx a^2/2R$, the bending and extension elastic energy densities are thus on the order of

$$g_{\text{bend}} \approx E d_m^3 a^4/R^6, \qquad g_{\text{extension}} \approx E d_m a^8/R^8 \qquad (E.11)$$

where d_m is the bilayer thickness. The elastic energy density associated with the change in surface tension during adhesion is on the order of

$$g_{\text{tension}} = \sigma \delta A/A \approx \sigma a^2/R^2 \qquad (E.12)$$

[since $\delta A \approx \pi u^2$ and $A \approx \pi(a^2 + u^2)$; see Fig. 5]. For an enlightening discussion of the deformation of closed shells the reader should consult the textbook of Landau and Lifshitz (8).

The ratio u/R is typically on the order of $a/R \approx 10^{-1}$ (see Fig. 5b) and the extensional deformation is thus to first order approximately determined (19), by the energy associated with the change in surface tension given by Eq. (E.12).

The contribution of the bending energy to the total adhesion energy is dominated by the high local curvature R_c of the membrane at the contact line (see Fig. 5). There, the membrane has to go smoothly over into the flat contact area in order to avoid sharp edges. Following an idea of Bruinsma (19) this bending energy can be accounted for by introducing a line tension τ. It accounts for the loss of adhesion energy due to the bending-induced reduction of the contact area along the rim of the contact disk.

In summary, the total free energy (per unit area) of the adhering shell can thus be approximated by

$$\Delta g_{adh} \approx 2\tau/a + \sigma a^2/R^2 - W_{ad} \tag{E.13}$$

Here, W_{ad} is the gain in free surface energy (or the work of separation) per unit area due to adhesion. By minimizing Eq. (E.13), one can establish relationships between the geometric parameters (the ratio a/R, the contact angle θ_c, and the contact curvature R_c) and the material properties (the tension σ, the bending modulus κ, and the adhesion energy W_{ad}). However, these correlations can be more easily established by considering the elastic boundary conditions at the contact zone. These are determined by

The equilibrium of the lateral membrane tensions, which must be the same in the adhering and the decoupled regions of the membrane. This is expressed by Young's law (13)

$$W = \sigma(1 - \cos\theta_c) \tag{E.14}$$

W is the so-called spreading pressure. It is only equal to the free surface energy W_{ad} for one-component membranes. The equilibrium of the bending moments, which provide the following relationship between the work of separation and the local curvature $\partial^2 u/\partial n^2|_c$ of the membrane at the contact line (8, 11):

$$W = \frac{\kappa}{2}\frac{\partial^2 u(\mathbf{r})}{\partial n^2} \tag{E.15}$$

where $\mathbf{n}$ is the normal to the contact line (8,11).

According to Eqs. (E.14) and (E.15), measurements of the contact angle and contact curvature allow one to determine the membrane tension and the work of unbinding, provided the bending stiffness is measured separately. This is possible for vesicles but exceedingly difficult for cells. In the following we show how all three parameters can be determined by analysis of the deformation of the shells in a hydrodynamic shear field. We will also show that such measurements are important to quantify the often small effects of mutations of the cytoskeleton on membrane elasticity and adhesion strength.

Direct and precise measurements of the contact curvature by micro-interferometry is difficult. It can, however, be determined by analysis of the surface profile $S(x)$ along sections perpendicular to the contact line. Following Bruinsma (19),

$S(x)$ is obtained by minimizing the elastic energy per unit length of the contact line (L). This energy may be expressed as

$$\Delta G_{\text{profile}} = \int_0^\infty dx \left\{ \frac{1}{2} \kappa (\Delta S(x))^2 + \frac{\sigma}{2} (\nabla S)^2 \right\} \tag{E.16}$$

where x is the local axis perpendicular to the contact line. Minimizing $\Delta G_{\text{profile}}$ with respect to x yields the following differential equation for $S(x)$:

$$\sigma \frac{\partial^2 S}{\partial x^2} - \kappa \frac{\partial^4 S}{\partial x^4} = 0 \tag{E.17}$$

In order to account for the straight profile (see Fig. 5b), $S(x)$ must be linear for large x and zero for $x = 0$. These boundary conditions are fulfilled by

$$S(x) = \theta_c x - \theta_c \lambda [1 - \exp(-x/\lambda)] \quad \text{for } x > 0 \tag{E.18a}$$

$$S(x) = 0 \qquad\qquad\qquad\qquad \text{for } x \leq 0 \tag{E.18b}$$

The characteristic length scale embedded in this equation,

$$\lambda = \sqrt{\kappa/\sigma} \tag{E.19}$$

is equal to the distance between the contact line and the intersection of the tangent to the profile with the horizontal axis. The λ has the following meaning: For distances large compared with λ the profile is determined by surface tension and for distances smaller than λ by bending rigidity. Analysis of the profile in terms of Eq. (E.18) yields rather reliable values of σ/κ and W/κ.

A great advantage of the present procedure is that it enables local measurements of the relative work of separation W in cases in which the contact zone of the adhering shells is nonisotropic, exhibiting regions of various adhesion strengths W. An example will be shown in Fig. 8.

V. DYNAMIC SURFACE ROUGHNESS OF FLUID MEMBRANES AND UNDULATION FORCES

Because of the effect of pronounced thermally excited bending undulations, fluid membranes are dynamically rough surfaces. This is a consequence of the low bending modulus of fluid bilayers, which is of the order of 50 $k_B T$ for lipid bilayers containing 50% cholesterol, corresponding to the situation of the cell plasma membrane (see Table 1). The bending modulus may be further reduced by a factor of 10 in the presence of 1 mol % of small amphiphilic solutes. Examples are sodium dodecyl sulfate, the ion carrier valinomycin, and short bipolar lipids (bola lipids) (16).

Bending undulations are exceptionally large for erythrocytes despite the coupling between bilayer and cytoskeleton (20). It is still a matter of debate whether this is due to a very small bending modulus of the membrane or to biochemically induced fluctuating forces. A highly intriguing possibility suggested by Prost et al. [personal communication (21)] is the amplification of bending

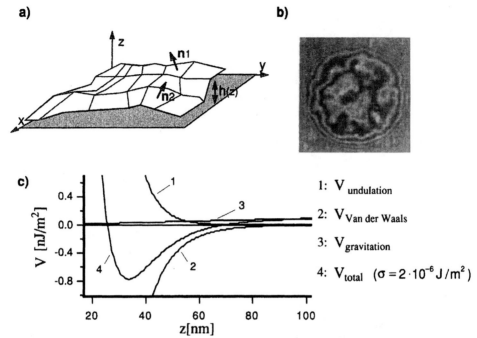

Fig. 6 (a) Dynamically rough membrane close to a wall (average distance $\langle h \rangle$) subjected to an interfacial potential $V(h)$. The random roughness is characterized by the local normal vector $\mathbf{n}(\mathbf{r})$ and a correlation length ζ corresponding to the distance over which orientations are correlated: $\langle \mathbf{n}(0)\mathbf{n}(\mathbf{r}) \rangle \propto \exp\{-r/\zeta\}$. (b) RICM image, showing the dynamic roughness of a vesicle close to a substrate. Dark areas correspond to small distances and bright areas to large distances between bilayer and substrate. (c) Superposition of van der Waals, gravitation, and undulation potentials for a lateral membrane tension of $\sigma 2 \times 10^{-6}$ J/m^2 (V_{total}). Note that tension may be induced by adhesion. The van der Waals potential was calculated, using a Hamaker constant $H \approx 10^{-20}$ J, which is characteristic of a bilayer close to a glass wall (22).

undulations by nonequilibrium processes such as ion pumping, if membranes are close to walls.

The basic idea of undulation forces can be best understood by considering a flat piece of membrane of dimension $L \times L$ close to a wall (at an average distance $\langle h \rangle$; see Fig. 6). The influence of the wall can be represented by an interfacial potential $V(h)$. In order to avoid nonlinear behavior, $V(h)$ is approximated by a harmonic potential with the force constant $k = \frac{1}{2}\partial^2 V/\partial h^2$. The total free energy may then be expressed as

$$G(h) = \oint dA \left[\frac{1}{2}\kappa(\Delta h)^2 + \frac{1}{2}\sigma(\nabla h)^2 + \frac{1}{2}k(h - \langle h \rangle)^2 \right] \tag{E.20}$$

provided spontaneous curvature effects can be neglected.

Membrane undulations are statistically excited and completely overdamped due to the coupling of the undulations to the hydrodynamic flow of the environment. The local displacement $u(\mathbf{r})$ of the membrane from the equilibrium height is expanded in plane wave Fourier modes

$$u(\mathbf{r}, t) = \sum_q u_q(t) \exp(i\mathbf{q}\mathbf{r}) \tag{E.21}$$

where $\mathbf{q}$ is the two-dimensional wave vector and $u_q(t) = u_q(0) \exp\{-t/T_q\}$. The mean square amplitudes $\langle u_q(0)^2 \rangle$ of each mode are easily calculated by inserting Eq. (E.21) into Eq. (E.20) and applying the equipartition theorem.

$$L^2 \langle \bar{u}_q^2 \rangle = \frac{k_B T / \kappa}{q^4 + \sigma' q^2 + \zeta_p^{-4}} \tag{E.22a}$$

$$\zeta_p = (\kappa/k)^{1/4} \tag{E.22b}$$

Note that ζ_p has the dimensions of a length. It is called "correlation length". It increases with the rigidity of the membrane and decreases with the force constant. A physical interpretation of ζ_p will be given in the following (Eq.(24b)).

The key parameters characterizing the random surface are, of course, the mean square amplitude $\langle u^2 \rangle$ and the correlation function $\langle u(0)u(\mathbf{r}) \rangle$:

The mean square amplitude, which is also called the roughness (23), is obtained by integrating the right side of Eq. (E.22a) over all modes. For this purpose the sum in Eq. (E.21) is replaced by an integration and the discrete amplitudes u_q by a spectrum $u(q)$.

$$\langle u^2 \rangle = (2\pi)^{-2} \int_{q_{min}}^{q_{max}} u(q)^2 q \, dq \tag{E.23}$$

where $q_{min} = \pi/L$ is the maximum value and q_{max} the maximum value of the wave number. $q_{max} = \pi/d_m$ is determined by the membrane thickness d_m.

The spatial correlation function is a measure of the spatial range of randomly induced local excitations. It can also be visualized as a measure of the range over which two local normal vectors to the surface are correlated (see Fig. 6). In this image the membrane is a two-dimensional analogue of a wormlike macromolecule. Following the Wiener-Khichin theorem, the spatial correlation function is obtained by inverse Fourier transformation of the frequency spectrum $u(q)$.

The integration leads to a complex integral that can be solved only numerically, for instance, by Monte Carlo techniques (24). However, asymptotic solutions have been derived for the two limiting cases: $\sigma = 0$ (bending-dominated regime) and $\sigma \to \infty$ (tension-dominated regime).

A. The Bending-Dominated Regime $\sigma = 0$

The asymptotic solutions are

$$\langle u_s^2(\sigma = 0) \rangle \approx \frac{k_B T}{8\kappa} \zeta_p^2 \tag{E.24a}$$

and

$$\langle u_{\rm s}(\mathbf{r}) u_{\rm s}(0) \rangle \approx \frac{k_{\rm B}T}{8\pi\kappa} \zeta_{\rm p}^2 \exp\left\{\frac{-r}{\sqrt{2}\zeta_{\rm p}}\right\} \tag{E.24b}$$

These equations have the following simple interpretation, which leads directly to the concept of undulation forces as first introduced by Helfrich (3): The local fluctuations decay exponentially with a correlation length $\sqrt{2}\zeta_{\rm p}$. This length corresponds to the persistence length of wormlike polymer chains. According to reflection interference contrast microscopy (RICM) studies (21), $\zeta_{\rm p} \sim 1\,\mu$m. Because $\zeta_{\rm p}$ is small compared with the dimension of vesicles or cells (e.g., erythrocytes), we can consider the rough membrane as being made up of small segments (or cushions) of dimension $\zeta_{\rm p} \times \zeta_{\rm p}$. These segments exhibit independent Brownian motions in the normal direction and thus perform random kicks against the wall. These kicks give rise to a disjoining pressure very similar to the pressure generated by an ensemble of gas molecules confined to a finite volume. As each segment has an energy content of $k_{\rm B}T$, they will exhibit a pressure

$$p \approx \frac{k_{\rm B}T}{2\zeta_{\rm p}^2 \langle h \rangle} \tag{E.25}$$

Because the average distance between the membrane and the wall is determined by the minimum of the interaction potential $V(h)$, we can assume $\zeta_{\rm p} \approx \langle h \rangle$. Together with Eq. (E.24), this yields

$$p \approx c \frac{(k_{\rm B}T)^2}{16\kappa\langle h \rangle^3} \tag{E.26}$$

where c is a numerical factor of the order of one. Thus, one obtains an interfacial potential

$$V_{\rm und}(\sigma - 0) \approx c \frac{(k_{\rm B}T)^2}{48\kappa h^2} \tag{E.27}$$

The prefactor c has been calculated (23) using renormalization group theory as $c/40 \approx 0.06$. A remarkable result is that the undulation potential exhibits the same distance dependence as the van der Waals potential.

B. The Tension-Dominated Regime

This case holds for $\sigma \gg \kappa/\zeta_{\rm p}^2$. The calculation, which is much more complicated than in the bending-dominated case, has been performed with the help of Monte Carlo studies (24). The mean square amplitudes are strongly reduced and $\zeta_{\rm p}$ and $\zeta_{\rm s}$ are related by a logarithmic law instead of Eq.(E.24a)

$$\langle u_{\rm s}^2 \rangle \approx c \frac{k_{\rm B}T}{2\pi\sigma} \ln\left\{\frac{\sigma\zeta_{\rm p}^2}{\kappa}\right\} \tag{E.28}$$

Therefore the undulation potential obeys an exponential law. Monte Carlo calculations yield

$$V_{und}(\sigma \neq 0) \approx 0.185 \frac{k_B T \sigma}{\kappa} \left(\frac{h_\sigma}{h}\right)^{1/4} \exp\left\{-\frac{h}{h_\sigma}\right\} \qquad (E.29)$$

where $h_\sigma = (k_B T / 2\pi\sigma)^{1/2}$

The preceding model has been extended to the case of closed vesicles (25). In the limit of small wavelengths, a relationship similar to Eq. (E.29) is obtained. An important result of this study is that increasing the tension of the membrane leads to unbinding of adhering vesicles.

The competition between van der Waals attraction and undulation forces has been studied by analysis of membrane undulations using microinterferometry (RICM). The force constants, the average distances $\langle h \rangle$, the mean squared amplitudes, and the persistence lengths have been measured for lipid vesicles (22,26). Figure 6 shows superpositions of van der Waals and undulation potentials for various membrane tensions. It is seen that for tension-free membranes the undulation forces may overcompensate the van der Waals attraction. Second minima of the potential are obtained at tensions on the order of 10^{-5} J/m^2. The minima shift to larger distances with decreasing tensions, and the equilibrium distances of the second minima are on the order of 30 nm.

C. Effect of Shear Elasticity on Biological Membranes

Like tension, shear elasticity is expected to suppress membrane undulations. This effect has been studies by Monte Carlo techniques and has been shown to contribute cubic terms to the elastic energy (24). The mean squared amplitudes are (for tension-free states)

$$\langle u_q^2 \rangle \sim \frac{k_B T L^2}{2c(k_B T \mu)^{1/2} q^3 + \kappa q^4} \qquad (E.30)$$

where μ is the shear elastic modulus and c is a numerical constant ($c \approx 1.3$). In normal cell membranes, where $\mu \geq 1$ J/m^2, the membrane undulations are completely suppressed. However, in erythrocytes we have $\mu \sim 5 \times 10^{-6}$ J/m^2 and the shear elasticity of the spectrin-actin network does not affect the undulations at wavelengths $q_c^{-1} < 500$ nm. Excitation of the shear modes may suppress the long-wavelength undulations at wavelengths on the order of the cell dimension or larger.

VI. INTERPLAY OF SPECIFIC ATTRACTIVE FORCES AND WEAK (FOR EXAMPLE, UNDULATORY) REPULSIVE FORCES LEADS TO FORMATION OF ADHESION PLAQUES

The control of adhesion through competition between strong (receptor-mediated) attraction and weak repulsive interactions such as undulation forces can be studied in the model system shown in Fig. 7. Specific forces are, for instance, easily modelled through biotin-streptavidin-biotin linkers. For that purpose, lipids carrying biotin headgroups are reconstituted into the bilayers of both a vesicle (acting as model cell) and a supported membrane (acting as target cell). Another more biologically relevant

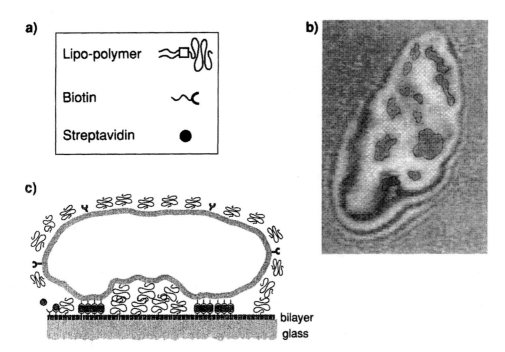

Fig. 7 (a) Model system for studying the interplay of specific and universal forces. Lipopolymers have been reconstituted in order to mimic the glycocalyx of cells. (b) RICM micrograph of a vesicle adhering to a supported membrane, mediated by biotin-streptavidin-biotin linkages. Lipids with biotinylated headgroups were incorporated at a concentration of 0.1 mol % into both the vesicle and the supported membranes. Lipopolymers were reconstituted to mimic the glycocalyx. The adhesion zone decays into domains of tight adhesion (encircled dark areas) and regions of weak adhesion (bright areas). The bright regions are dominated by undulation forces and exhibit pronounced flickering. (c) Schematic view of formation of adhesion plaques by adhesion-induced phase separation.

possibility is to reconstitute monopolar adhesion proteins such as contact site A proteins of *Dictyostelium* cells into both membranes. These receptors form strong homophilic bonds.

In order to mimic the effects of the glycocalyx, lipopolymers (lipids with hydrophilic macromolecular headgroups such as polyethyleneoxide) are reconstituted into both membranes. An additional advantage of this approach is that it facilitates the formation of unilamellar giant vesicles. Observation of the adhesion of these vesicles by RICM shows a surprising result (Fig. 7b). At small receptor concentrations, $c < 0.1\%$, the contact zone is not homogeneous but decays into small domains of relatively tight adhesion that are separated by regions of very weak adhesion. The latter exhibit lateral distances of the order of 100 nm and show pronounced flickering.

A. Measurement of Local Adhesion Strength

The adhesion strength (spreading pressure) and the membrane tension can be measured locally by analysis of the surface profile in terms of the elastic boundary conditions, employing the procedure described in the discussion of Eq. (E.18). Because

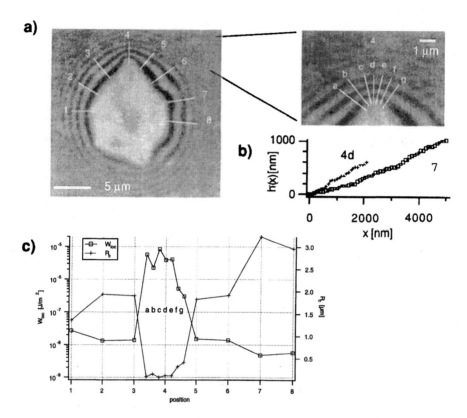

Fig. 8   Measurement of spreading pressure W of a vesicle near pinning centers by analysis of the vesicles contour perpendicular to the contact line. Vesicle and supported membrane contain 5 mol % lipopolymer (molecular weight 2000), to mimic the glycocalyx, and 0.1 mol % of lipid exhibiting biotinylated headgroups. The linkage is formed by streptavidin. (a) RICM image of vesicles (left) and enlarged view of a pinning center. The white bars indicate directions along which the contour has been analyzed. (b) Two examples of contours: one in the region of the pinning center and one in the region of weak adhesion. (c) Plot of local adhesion energy (solid line) and contact curvature (dashed line) for different positions (1–8) as denoted by white lines.

the bending modulus κ can be measured by flicker spectroscopy (15), both the membrane tension σ and the spreading pressure W can be obtained from two geometric parameters, θ_c and λ. Figure 8 shows such a measurement. The spreading pressure is $W_w \approx 10^{-8}$ J/m² for the weakly adhering and $W_s \approx 10^{-5}$ J/m² for the strongly adhering domains. The former value is determined by the competition between van der Waals and undulation forces (22).

The binding energy of the streptavidin-biotin bond (27) is about $W_0 \approx 35 k_B T$. The value of $W_s \approx 10^{-5}$ J/m² would thus correspond to a receptor density of $n_R \sim 10^{-4}$ receptors per nm² or to a receptor-to-lipid molar ratio of $r_{RL} \sim 10^{-4}$. This is much smaller than the molar fraction of biotinylated lipid incorporated in the membranes. In the following we will explain this discrepancy in terms of the osmotic pressure difference between the strongly adhering domains and the decoupled regions of the vesicle. This pressure difference is generated by the adhesion-induced lateral segregation of the receptors. It reduces the work of unbinding (or the free energy of binding) of the adhesion plaques, which is measured by the procedure of contour analysis mentioned earlier.

How can weak entropic undulation forces control strong attraction forces? The answer is, by nucleation of attraction sites (pinning centers). In order to form a tight local bond (of interfacial distance ~ 1 nm), work has to be performed against the repulsive undulation forces. Statistically formed single linkages facilitate the formulation of adjacent bonds and thus act as nucleation sites of tight adhesion. The domain structure is intrinsically unstable and the pinning centers should merge in order to reduce the energy associated with the line tension τ, which is generated by the bending of the membrane along the rim of the adhesion plaques (see Fig. 7c). However, it is well known from the theory of nucleation and growth that such coarsening processes are very slow because the average size of the domains grows roughly with the cubic root of time. For receptor concentrations on the order of 1 mol%, the adhesion plaques merge on a time scale of 10 min, forming a continuous tight adhesion region (18). In contrast, it was found that the undulation forces are strong enough to break bonds on a time scale of seconds instead of weeks.* Again, this discrepancy could be explained in terms of a reduction of the adhesion strength due to the gain in translational entropy of the unbound receptors (see later).

The formulation of local pinning centers is a familiar phenomenon observed during cell adhesion. In the case of fibroblasts, adhesion plaques are formed by clustering of integrin receptors, a process that may be initiated by the binding of integrins of the extracellular matrix. The cytoplasmic domains of clustered integrins form sites of attachment to intracellular stress fibers, composed of actin bundles. The origin of receptor clustering is not known yet. One possibility is the mechanism demonstrated in Fig. 7c. However, undulatory forces could not contribute remarkably to long-range repulsion forces because membrane flickering is surpressed due to the coupling of the bilayer to the cytoskeleton. Long-range forces could be mediated by one of the polymer-induced forces shown in Fig. 1 or by electrostatic forces, which are discussed in the next section. On the other hand, receptor clustering could also be caused by formation of receptor complexes mediated by cell signaling processes (28). Evidence in favor of a clustering due to adhesion-induced phase separation is provided by the finding that mutants of fibroblasts lacking vinculin also form adhesion domains although the actin binding protein is considered essential for the formation of adhesion plaques (29).

VII. ELECTRICALLY INDUCED ADHESION AND PHASE SEPARATION

Adhesion is also controlled by electrostatic forces. A model system for studying electrostatically mediated adhesion is shown in Fig. 9. A giant vesicle containing

---

* The undulation force acting on the pinning centers is of the order of $f \sim (k_B T/\kappa)(D^2/d^3)$, where D is the mean distance between the pinning centers and d is the distance between the membrane and the substrate. D is on the order of 1 μm and $\kappa \approx 30 k_B T$ for lipid bilayers. Therefore it is $f \sim 10^{-11}$ N for $d = 30$ nm. This is smaller by an order of magnitude than the streptavidin-biotin binding force measured by AFM (26). $f_{max} \sim 10^{-10}$ N.

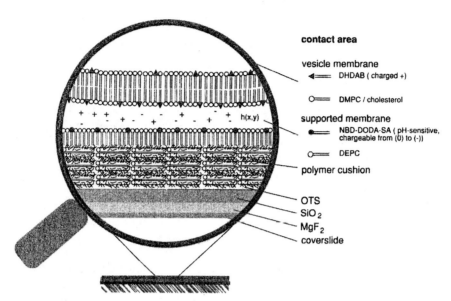

Fig. 9 Model system for studying electrostatic control of adhesion of soft shells. Monolayers of isopentyl cellulose (called hairy rods) form a soft cushion between solid and supported membrane, with the uppermost monolayer serving as the inner leaflet of the membrane. The top monolayer contains about 1 mol % dissociable acidic lipids and the giant vesicle about 1–10% positive lipids. The membranes are separated by cushions of counterions ensuring electron neutrality.

positively charged lipids interacts with a supported membrane containing acidic lipids as "receptors." The supported membrane consists of a multilayer of rodlike molecules with substituted alkyl chains (so-called hairy rods), which mimic the bottom monolayer, and of a superimposed lipid monolayer. Simultaneously, the hairy rod multilayer forms a soft cushion, and artifacts due to surface roughness and surface charges of the solid support are avoided (see Appendix).

Because the electrical charge carried by the acidic membrane can be varied through pH changes, the adhesion strength can be controlled in a reversible way. When the adhesion is switched on by increasing the pH from 2.8 to 4.5, a surprising behavior is found: Initially, the vesicle exhibits a homogeneous adhesion disk. With a further increase in pH and thus in charge of the supported membrane, the adhesion zone decays into regions of tight contact (gray areas) and blisters (light areas), as shown in Fig. 10. Reconstruction of the surface profile of the blisters (by analysis of the interference pattern) shows that they form spherical caps, indicating that they are pressurized. As demonstrated in Fig. 11b, the blisters form both within the adhesion disk and at the advancing front of the adhesion disk. Owing to the osmotic pressure, the blisters tend to merge as demonstrated in Fig. 11c.

Blisters form as a consequence of an intrinsic instability of two oppositely charged membranes exhibiting different charge densities (σ_1 and σ_2). Following Gingell and Parsegian (30), this is a consequence of the competition between the attraction of the membrane surfaces and the electrostatic repulsion generated by the cushion of counterions that is needed to establish charge neutrality. A homogeneous state of tight adhesion is possible only if the absolute charge densities σ_1 and σ_2 of the two interacting

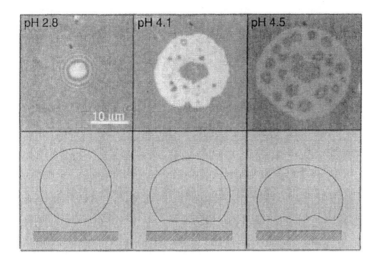

Fig. 10 RICM images of a giant vesicle demonstrating adhesion-induced phase separation of charged lipids (mimicking receptors), followed by blister formation (31). The vesicle bilayer contains 1 mol % positively charged lipids and the supported membrane 10% dissociable acidic lipid (with succinic acid headgroups). At pH 2.8 the bottom monolayer is uncharged. After increasing the surface charge of the supported membrane, one observes the formation of blisters of the adhering bilayer (nucleating at pH 4.1), which grow in number with increasing surface charge. The large blister in the center of the adhesion disk is due to a small vesicle intercalated between adhering membranes (32).

membranes are identical ($\sigma_1 = -\sigma_2$). If $|\sigma_1| \neq |\sigma_2|$, the surfaces are separated by the osmotic pressure of the counterions. For the situation $|\sigma_1| \gg |\sigma_2|$ the electrostatic potential can be approximately expressed as

$$V(h) = \frac{8\pi}{\varepsilon\kappa}\left(\frac{1}{2}\sigma_1^2 \exp\{-2\kappa h\} - |\sigma_1|\,|\sigma_2| \exp\{-\kappa h\}\right) \qquad (E.31a)$$

where κ is the reciprocal of the Debye screening length and $\varepsilon = 80\varepsilon_0$ (for water). The potential $V(h)$ exhibits a minimum

$$V^*(h^*, \sigma_2) = -\frac{4\pi}{\varepsilon\kappa}\{\sigma_2^2 + O|\sigma_2^4|/|\sigma_1^4|\} \qquad (E.31b)$$

at an equilibrium distance

$$h^* = \kappa^{-1}\ln(|\sigma_1|/|\sigma_2|) \qquad (E.32)$$

Remarkably, V^* does not depend remarkably on the majority charge density σ_1.

This situation holds for fixed surface charges on solids. For fluid membranes, however, it is completely different. Charges may segregate and the membranes can easily deform. The free energy expression has to be augmented by the translational entropy $(-\sigma_i \ln\{\sigma_i/\sigma_0\})$ of each species of charges. Because $|\sigma_1 \gg |\sigma_2|;$, one can write

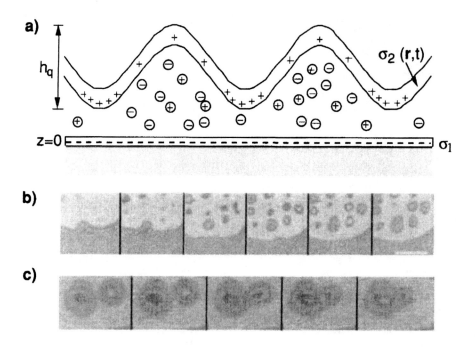

Fig. 11 Dynamics of blister formation. (a) Illustration of decomposition of surface charge of minority charged membrane into patches of enriched charge density (with $\sigma_1 = -\sigma_2$), which may form close contact with the support. Note that the charge-depleted areas eventually form spherical caps. (b) Growth of number of blisters with increasing contact area. Note that blisters from both within the already formed contact area and at the leading front during spreading (see first three images). (c) Merging of blisters demonstrating that they are pressurized. Note the similarity to fusion of soap bubbles.

for the free energy of interaction.

$$G_{el}(\sigma_1\sigma_2) = V^*(h^*, \sigma_2) + (k_B T/e)\sigma_2 \ln\{\sigma_2/\sigma_0\} \tag{E.33}$$

The theory of thermodynamic stability predicts that the state of homogeneous lateral charge distribution is stable only if $\partial^2 G_{el}/\partial\sigma^2 > 0$. It thus follows that the minority charge distribution σ_2 decays laterally above a threshold concentration

$$\sigma_2 > \sigma_2^* = \frac{\varepsilon\kappa k_B T}{8\pi e} \tag{E.34}$$

Such a threshold behavior has indeed been verified experimentally, as shown in Figs. 10 and 11. It is important to note that the charges in the supported membrane with the higher charge density are not redistributed because the free energy G does not depend remarkably on σ_1. The minority charged membrane is thus "slaved" by the majority charged membrane.

Blistering is a typical dynamic instability phenomenon (32). It is accompanied by a lateral transport of counterions from regions of strong adhesion to decoupled areas of the membrane-substrate contact zone or to regions outside the adhesion disk. Because the escape of ions through the bilayer of the vesicle is suppressed, an osmotic pressure results that leads to pressurization of the blisters, which thus form spherical caps. The major reason for the adhesion-induced blistering is the competition between strong attraction and long-range osmotic repulsion (generated by the cushion of counterions). Similar dynamic instabilities would be expected if the osmotic pressure was generated by dissolved or membrane-anchored macromolecules intercalated between the two membranes. Strong attractive forces could be generated by salt bridges or receptors. Indeed, similar blistering was observed for erythrocytes that are strongly attracted to a polylysine-coated surface (33,34). Another biological example is encountered during the attack on virus-infected cells by cytotoxic killer cells. The killer cells adhere tightly to the target cell. Blisters are formed within the contact area by fusion of vesicles with the membrane of the cytotoxic cells. This process is accompanied by the secretion of perforin, which forms large pores in the infected cell, causing its death (35).

VIII. ADHESION OF CELLS AND MODIFICATIONS BY CYTOSKELETAL MUTATIONS

An intriguing question is whether cells make use of the possibility of controlling adhesion by the interplay of interfacial forces and membrane elasticity. In particular, the coupling between the actin-based cytoskeleton and the bilayer membrane could affect the adhesion strength in two ways: (a) by control of the lateral mobility of the receptors, such as integrins or other cell adhesion molecules with intracellular actin binding sites, and (b) by modification of the membrane elasticity. As discussed below, the mobility of receptors controls the adhesion-induced lateral segregation of the receptors and thus the work of adhesion (see Sec. IX).

Unfortunately, the profile analysis method described above allows a direct measurement of only two of the three interesting parameters (tension, bending modulus, and adhesion energy). This problem can be overcome by analyzing perturbations of the membrane surface profile that are induced by weak hydrodynamic flow fields (36). The flow field generates an additional membrane tension $\Delta\sigma$ at the side of the cell facing flow (Fig. 12). This additional tension changes the characteristic length λ according to

$$\lambda - \Delta\lambda = \sqrt{\kappa/(\sigma + \Delta\sigma)} \tag{E.35}$$

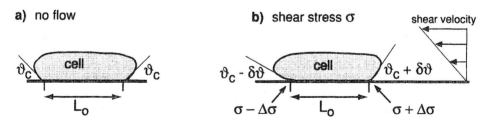

Fig. 12 Illustration of the changes in shape and contact angle of an adhering soft shell, induced by a hydrodynamic flow field. The load on the cell has to be compensated by an additional lateral tension in the direction of the load. Both contact angle and lateral membrane tension therefore increase at the edge of the cell facing the flow and decrease at the opposite side.

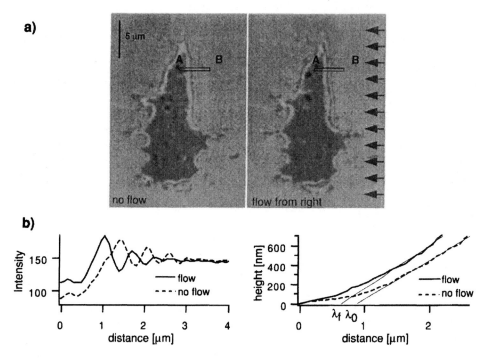

Fig. 13 Measurement of membrane tension, bending elastic modulus, and work of adhesion (in terms of spreading pressure) of ameboid cells of *Dictyostelium discoideum*. Analysis of the surface topology—with and without the influence of a weak hydrodynamic flow field—in terms of elastic boundary conditions allows one to determine all three parameters simultaneously. (a) RICM image showing the contact area of a wild-type cell adhering to bovine serum albumin–covered glass, without (left) and under the influence of hydrodynamic flow (right). (b) Interferogram (left) of the cell surface profile along the marked line (A–B) perpendicular to the contact line, and reconstructed cell contour (right). Note that λ decreases from the undisturbed value λ_0 to λ_f under the influence of the flow (36).

By measuring λ in the absence and in the presence of flow, one obtains two equations from which both the membrane tension σ and the bending modulus can be determined. A typical application of this technique to wild-type *Dictyostelium* is shown in Fig. 13.

This method provides new possibilities for characterizing changes in cellular properties caused by mutations, toxic substances, or drugs in a quantitative way. Well-aimed mutations of the actin-based cytoskeleton are of particular interest because they yield insight into the control of cell adhesion or cell locomotion by the actin-myosin cortex and the plasma membrane. Two interesting examples are summarized in Table 2. Talin is a ubiquitous protein in eukaryotic cells that mediates the coupling of actin filaments to membrane receptors. Cortexillin I is a cross-linking protein that favors the formation of actin bundles. Its removal leads to a strongly impaired cell division. An important finding of these experiments is that the undisturbed contact angle formed by wild-type cells on bovine serum albumin (BSA)–coated glass does not differ significantly from the angles observed for the mutants. However, remarkable differences between wild-type and mutant cells become evident under the influence of an external force, such as a shear flow. Employing this

Table 2 Modifications of Contact Angle, Membrane Bending Modulus, Membrane Tension, and Work of Adhesion by Mutagenic Removal of Cortexillin I (an Actin Bundling Protein) and Talin (an Actin-Membrane Coupling Protein)

Parameter	AX2 (19)	Talin-null (9)	Cortexillin I-null (14)
Contact angle[a] (deg)	21.5±1.8	21.8±1.9	22.3±1.8
Bending modulus ($k_B T$)	391±156	71±21	94±18
Membrane tension (μN/m)	3.1±1.4	0.8±0.3	2.2±0.8
Work of adhesion (10^{-6} J/m^2)	22.0±12.2	5.9±2.4	15.2±4.7

[a] The numbers in parentheses denote the number of cells examined for each wild-type or mutant strain.
[b] The effective contact angle was measured at zero shear stress. Means are shown with their standard deviations.
From Ref. 35.

method, we could show that the membrane tension, the bending modulus, and the adhesion strength are reduced drastically for the talin-null mutant, but the reductions are less pronounced if cortexillin I is removed (Table 2). Most remarkable is the reduction of the bending modulus by a factor of 5 to 6 if talin is removed. Indeed, this reduced value of $\kappa = 70 k_B T$ corresponds very well to the bending rigidity of a lipid bilayer containing 50% cholesterol. The cholesterol content of the *Dictyostelium* plasma membrane is 20 mol%. This result therefore provides very strong evidence that talin is indeed a major cytoskeleton-membrane coupling protein.

Another interesting result is the strong reduction of the adhesion energy by removal of talin. This remarkable reduction may be attributed to the effect of undulation forces. These forces are expected to be larger for the talin-deficient cells, which exhibit both a reduced lateral tension and a lower bending energy of the membrane. Moreover, experiments on model membrane show that lipid bilayers that are only weakly coupled to actin shells exhibit a pronounced flickering.

Although these experiments are still preliminary, they show that concerted studies of model systems and cells provide helpful tools for gaining insight into the structure of complex biological machines such as the composite plasma membrane. These studies are of high practical interest, because they allow us to characterize changes of the structure and behavior of cells by mutations in a quantitative way.

IX. REGULATION OF ADHESION STRENGTH THROUGH PHASE SEPARATION

The adhesion-induced segregation of the receptors and their accumulation in adhesion domains (pinning centers) provide a most attractive and effective mechanism for the regulation of the adhesion strength of cells. This becomes evident if one compares the measured value of the spreading pressure W obtained by local surface profile analysis (as described earlier) with the total binding energy E_B of the receptors. The latter can be calculated in the case of the electrostatic attraction between oppositely charged lipid molecules. The contour analysis yields $W \approx 10^{-7}$ J/m^2, and the electrostatic binding energy (31) is $E_B \approx 10^{-3}$ J/m^2.

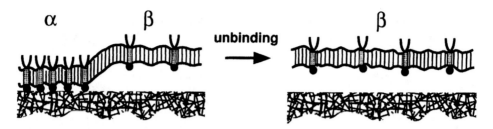

Fig. 14 Schematic view of relaxation of adhesion strength by segregation of receptors, resulting in a reduction of the Gibbs free energy of unbinding.

This discrepancy by a factor of 10^4 can be explained as follows: According to the classical Young equation [Eq. (E.14)], the gain in Gibbs free energy W associated with adhesion is related to the membrane tension σ. Thus, W is equal to the work of unbinding (per unit area), which is called the spreading coefficient in the literature on wetting (17). However, the Young equation breaks down if the adhesion is associated with lateral phase separation. In this case, the total work of adhesion is determined not only by the gain in surface energy of the adhesion disks but also by the free energy associated with adhesion-induced phase separation and thus the difference in osmotic pressure between the adhesion disks and the decoupled regions of the membrane. Provided the membrane within the adhesion domains does not undergo a phase transition, the work of unbinding W and the binding energy per unit area E_B differ essentially by the lateral osmotic pressure difference between areas exhibiting strong (α) or weak (β) adhesion (Fig. 14):

$$W = E_B + \Pi_\alpha^{osm} - \Pi_\beta^{osm} \tag{E.36}$$

Here Π^{osm} is the two-dimensional osmotic pressure. The osmotic pressure difference is roughly determined by the translational entropy of the receptors and can be expressed as

$$\Pi_\alpha^{osm} - \Pi_\beta^{osm} \approx k_B T \ln c_\alpha/c_\beta \tag{E.37}$$

where c_α, c_β are the receptor concentrations in the two respective membrane regions. Note that all these considerations hold only in thermodynamic equilibrium.

This is a remarkable result, which shows that adhesion-induced receptor segregation can reduce the adhesion strength (work of unbinding) by several orders of magnitude. The work required to unbind strongly adhering domains is reduced dramatically by the gain in translational entropy due to the receptor redistribution taking place after unbinding. This gain in entropic free energy is particularly large for small receptor concentrations. Cells could thus drastically reduce the work of unbinding of adhesion plaques in an interactive way, for instance, by reducing the receptor concentration in the decoupled membrane regions. This could be achieved, for example, by vesicle-mediated internalization of receptors.

The entropic relaxation of the work of unbinding would also drastically alter the kinetics of cell adhesion and unbinding. The effects that have been discussed are generally ignored in present kinetic models of adhesion, which do not consider the adhesion-induced formation of domains (37,38).

X. APPENDIX: SUPPORTED MEMBRANES AS PHANTOM CELLS

One purpose of the present chapter is to demonstrate the advantages offered by studying cell adhesion on solid supports. Powerful surface-sensitive techniques can be applied, such as microinterferometry or fluorescence microscopy based on excitation of fluorescence probes by total internal reflection (39) or standing waves (40). A second advantage is that target cells or tissue surfaces may be mimicked by biofunctionalization of the solid supports, which may thus act as phantom cells. These can be prepared either by deposition of ultrathin polymer films or by using supported membranes with reconstituted receptors and glycocalyces of defined lateral density and mobility.

Polymer films can consist of natural constituents of the extracellular matrix (e.g., collagen and hyaluronic acid) or of hydrophilic synthetic polymers carrying functional groups (oligosaccharides of glycoproteins). In order to mimic the softness of cell surfaces, supported membranes are separated from the solid by soft polymer cushions.

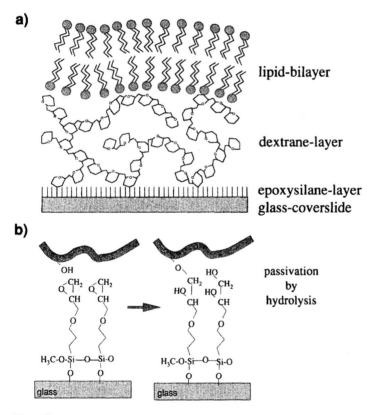

Fig. 15 (a) Self-assembly of supported membrane on a soft polymer cushion (e.g., dextran) that is anchored to the substrate as shown in (b). The pure lipid bilayer can be first deposited onto the polymer film by single bilayer spreading (43). The lipid-coupled receptors as well as ion channels can be reconstituted afterward by transfer of vesicles, into which they have been reconstituted (44). (b) Method for grafting hydrophilic polymers to Si-SiO$_2$ or indium–tin oxide surfaces. A monolayer of alkyl silane carries functional groups (e.g., epoxy groups, photoaffinity labels, or amines) to which polymer is coupled as indicated. The fraction of functional groups can be controlled, for instance, by hydrolysis of some of the groups. The hydrolyzed groups serve for passivation of the surface (43).

The preparation of such composite films on solids provides a challenging and complex problem of wetting. The stability of each film—the polymer cushion on the solid and the membrane on the polymer film—requires that the corresponding spreading coefficients are positive. Moreover, the interfacial interaction potential between the solid and the membrane must be repulsive or exhibit a second shallow minimum (41).

Various successful strategies for modeling cell surfaces or tissues have been developed. One possibility is partial anchoring of the polymer on the substrate as shown in Fig. 15a. Dextran is well suited for this approach: it swells readily, can be easily functionalized, and prevents nonspecific binding of proteins. Single bilayers spread spontaneously on such films, forming self-healing membranes with a high electrical resistance.

A second approach is to separate the membrane from the solid by stilts. For that purpose small quantities ($\sim$1 mol %) of lipopolymers (lipids carrying polymeric headgroups such as polyethylene oxide) are reconstituted into the membranes (42). A third attractive method is the formation of cushions from hairy rod molecules. These are rodlike macromolecules (such as cellulose) that carry alkyl chains of variable length. In analogy to lipid monolayers, hairy rods form amphiphilic monolayers on the air-water interface, with the alkyl chains forming a hydrophobic brush facing the air. Multilayers of hairy rods, deposited in such a way that the alkyl chains of the top layer are exposed, can therefore mimic the inner leaflet of a supported membrane onto which the top monolayer can be deposited. These composite films are well suited for reconstituting amphiphilic receptors such as gangliosides or lipid-anchored cell adhesion molecules.

Using hairy rods, where the alkyl chains are silane coupled to the rodlike backbone, the deposited multilayers can be rendered hydrophilic by treatment with hydrochloric acid (HCl). Thus modified, the multilayers can serve as a soft cushion

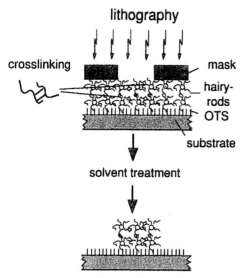

Fig. 16 Formation of patterned models of cell surfaces. Hairy rod multilayers are structured laterally by photolithography in such a way that a checkerboard-like arrangement of hydrophilic and hydrophobic domains is formed. This forms the basis for the deposition of lipid membranes on one or both types of pattern by self-organization (45).

for the spreading of lipid bilayer membranes. A great advantage of the hairy rod layers is that they can easily be stabilized by slight photochemical cross-linkage of the hairy rods (45). Hairy rod films are also well suited for the formation of functionalized patterns on solids. For this purpose the film is first laterally structured by photolithography and the lipid protein membrane (e.g., a monolayer) is deposited subsequently as shown in Fig. 16.

REFERENCES

1. D Hanein, B Geiger, L Addadi. Differential adhesion of cells to enantiomorphous crystal surfaces. Science 263:1413–1416, 1994.
2. CS Chen, M Mrksich, S Huang, GM Whitesides, DE Ingber. Geometric control of cell life and death. Science 276:1425–1428, 1997.
3. W Helfrich. Steric interactions of fluid membranes in multilayer systems. Z Naturforsch 33a:305–315, 1978.
4. W Helfrich, RM Servus. Undulations, steric interactin and cohesion of fluid membranes. Nuovo Cimento 3D:137–151, 1984.
5. V Bennet. Spectrin-based membrane skeleton: A multipotential adaptor between plasma membrane and cytoplasm. Phys Rev 70:1029–1065, 1990.
6. DR Nelson, L Peliti. Fluctuations on membranes with crystalline and hexatic order. J Phys Fr 48:1085–1092, 1987.
7. G Isenberg, WH Goldmann. Actin-membrane coupling: A role for talin. J Musc Res Cell Motil 13:587–589, 1992.
8. LD Landau, EM Lifschitz. Elastizitätstheorie. Lehrbuch der theoretischen Physik VII. Berlin: Akademie Verlag, 1989.
9. F Brochard, JF Lennon. Frequency spectrum of the flicker phenomenon in erythrocytes. J Phys Fr 11:1035–1047, 1975.
10. EA Evans, R Skalak. Mechanics and Thermodynamics of Biomembranes. Boca Raton, FL: CRC Press, 1980.
11. U Seifert, R Lipowsky. Adhesion of vesicles. Phys Rev A 42:4768, 1990.
12. E Sackmann. Physics of vesicles. In: R Lipowsky, E Sackmann, eds. Handbook of Biological Physics. Vol 1, Chap 5. Amsterdam: Elsevier, 1995.
13. EA Evans. Bending resistance and chemically induced moments in membrane bilayers. Biophys J 14:923–931, 1974.
14. D Needham, E Evans. Giant vesicle bilayers composed of mixtures of lipids, cholesterol and polypeptides. Faraday Discuss Chem Soc 81:267–280, 1986.
15. W Häckl, U Seifert, E Sackmann. Effects of fully and partially solubilized amphiphiles on bilayer bending stiffness and temperature dependence of the effective tension of giant vesicles. J Phys II 7:1141–1157, 1997.
16. E Sackmann. Membrane bending energy concept of vesicle shapes and shape changes. FEBS Lett 346:3–16, 1996.
17. P deGennes. Wetting: statics and dynamics. Rev Mod Phys 57:827–863, 1985.
18. A Albersdörfer, T Feder, E Sackmann. Adhesion-induced domain formation by interplay of long-range repulsion and short-range attraction force: A model membrane study. Biophys J 73:245–257, 1997.
19. R Bruinsma. Adhesion and rolling of leucocytes: A physical model. In: T Riste, D Sherrington, eds. Proceedings of NATO, Advanced Institute on Physics of Biomaterials, Geilo. NATO ASI Series. Vol 322. Dordrecht: Kluwer, 1996, p 61.
20. A Zilker, M Ziegler, F Sackmann. Spectral analysis of erythrocyte flickering in the 0.3–4 μm-1 regime by microinterferometry combined with fast image processing. Phys Rev A 46:–7998, 1992.

21. J Prost, J-B Manneville, R Bruinsma. Fluctuation-magnification of nonequilibrium membranes near a wall. Eur Phys J B1:465–480, 1998.

22. JO Rädler, TJ Feder, HH Strey, E Sackmann. Fluctuation analysis of tension-controlled undulation forces between giant vesicles and solid substrates. Phys Rev E 51:4526–4536, 1995.

23. R Lipowsky. Genetic interaction of flexible membranes. In: R Lipowsky, E Sackmann, eds. Handbook of Biological Physics. Vol 1. Chap 11. Amsterdam: Elsevier, 1995.

24. R Lipowsky, M Giradet. Shape fluctuations of polymerized solid-like membranes. Phys Rev Lett 65:2893–2897, 1990.

25. U Seifert. Self-consistent theory of bound vesicles. Phys Rev Lett 74:5060–5063, 1995.

26. J Rädler, E Sackmann. Imaging optical thicknesses and separation distances of phospholipid vesicles at solid surfaces. J Phys II 3:727, 1993.

27. E-L Florin, VT Moy, HE Gaub. Adhesion forces between individual ligand-receptor pairs. Science 264:415–417, 1994.

28. A Ullrich, J Schlessinger. Signal transduction by receptors with tyrosine kinase activity. Cell 61:203–212, 1990.

29. H Goldmann, RM Ezzell, ED Adamson, V Niggli, G Isenberg. Vinculin, talin and focal adhesions. J Musc Res Cell Motil 17:1–5, 1996.

30. A Gingell, D Parsegian. On the electrostatic interaction across a salt solution between two bodies bearing unequal charges. Biophys J 12:1192, 1972.

31. J Nardi, T Feder, R Bruinsma, E Sackmann. Electrostatic adhesion between fluid membranes: Phase separation and blistering. Europhys Lett 37:371–376, 1997.

32. J Nardi, R Bruinsma, E Sackmann. Adhesion induced reorganisation of charged fluid membranes. Phys Rev E, 58:6340–6345, 1998.

33. WT Coakley, LA Hewison, T Tilley. Interfacial instability and the agglutination of erythrocytes by polylysine. Eur Biophys J 13:123–130, 1985.

34. D Gallez, WT Coakley. Far-from-equilibrium phenomena in bioadhesion processes. Heterog Chem Rev 3:443–475, 1996.

35. JD Young, ZA Cohn. How killer cells kill. Sci Am 258(1):38, 1988.

36. R Simson, E Wallraff, J Faix, J Niewöhner, G Gerisch, E Sackmann. Membrane bending modulus and adhesion energy of wild-type and mutant cells of *Dictyostelium* lacking talin or cortexillins. Biophys J 74:514–522, 1998.

37. GJ Bell. Models of the specific adhesion of cells to cells. Science 200:618–627, 1978.

38. EA Evans. Detailed mechanics of membrane adhesion and separation. I. Continuum of molecular cross bridges. Biophys J 48:175–183, 1985.

39. TH Watts, HE Gaub, HM McConnell. T-cell mediated association of peptide antigen and major histocompatibility complex protein detected by energy transfer in an evanescent wave-field. Nature 320:179–181, 1986.

40. D Braun, P Fromherz. Cell adhesion studied by fluorescence interference on silicon (abstr). Eur Biophys J 26:3–5, 1997.

41. M Kühner, E Sackmann. Ultrathin hydrated dextran films grafted on glass: Preparation and characterization of structural, viscous, and elastic properties by quantitative microinterferometry. Langmuir 12:4866–4876, 1996.

42. E Sackmann. Supported membranes: Scientific and practical applications. Science 271:43–48, 1996.

43. J Rädler, E Sackmann. Functionalization of solids by ultrathin soft polymer-lipid film composites: Modeling of cell surface and cell recognition processes. Curr Opin Solid State Materi Sci 2:330–336, 1997.

44. S Gritsch, F Jähnig, X Nollert, E Sackmann. Langmuir, 14:3118–3125, 1998.

45. G Wiegand, T Jaworek, G Wegener, E Sackmann. Heterogeneous surface of structured hairy-rod polymer films: preparation and method of functionalization. Langmuir 13:3563–3569, 1997.

13

Physicochemistry of Microbial Adhesion From an Overall Approach to the Limits

H. J. Busscher, R. Bos, and H. C. van der Mei
University of Groningen, Groningen, The Netherlands

P. S. Handley
University of Manchester, Manchester, United Kingdom

I. INTRODUCTION

Microbial adhesion to surfaces is ubiquitous, simply because it represents a mode of survival for the organisms (1). For example, clearance of microorganisms from the oral cavity by salivary flow leads to cell death in the gastrointestinal tract, which can be avoided by adhesion to teeth or oral mucosa. Adhesion and biofilm formation on biomedical implants protect the organisms against the host immune system and environmental attacks such as by antimicrobials (2). Often, as on marine and other aquatic surfaces, nutrients accumulate at surfaces (3), constituting another reason for microorganisms to adhere. In bioreactors, microbial adhesion is stimulated to maintain a biofilm for optimal production conditions (4). In contrast, in many biomedical applications or in drinking water systems, microbial adhesion is undesirable (5).

The study of microbial adhesion encompasses a broad range of scientific disciplines, ranging from medicine, dentistry, and microbiology to colloid and surface science. Initially, the involvement of colloid and surface scientists originated from the simple realization that microorganisms are, with respect to their dimensions, colloidal particles and that their adhesions should be predictable by surface thermodynamic (6,7) or DLVO-theory (Derjaguin-Landau-Verweij-Overbeek) like approaches (8,9). Considering the ubiquitous nature of microbial adhesion, such generalized predictive models would be extremely valuable.

The aim of this chapter is to point out the merits of a generalized physicochemical approach to microbial adhesion phenomena, while emphasizing the limitations of physicochemistry due to the extremely complex structural and chemical features of microbial cell surfaces.

II. STRUCTURAL COMPLEXITIES OF MICROBIAL CELL SURFACES

Microorganisms are not smooth particles, although from a physicochemical point of view they have generally been considered similar to inert polystyrene particles. Even polystyrene particles are now known not to be smooth and are chemically heterogeneous, with the possibility that they carry 10-nm-long "appendages" (10,11). Even so, microorganisms, in contrast to polystyrene particles, may have much longer, usually very thin surface structures protruding from the cell surface, radiating outward into the surrounding liquid, and being responsible for adhesion to a variety of surfaces. Bacteria in particular have a wide range of surface structures that have been described on the basis of their ultrastructure and distribution on the cell surface. There are many morphologically distinct types of surface structure, and almost every bacterial species has its own distinct type of surface appendage. In order to illustrate the principle of surface structure diversity, the major types of structures found on bacteria are described in this section with particular emphasis on oral bacteria.

A. Definitions of Surface Structures

So far two major, morphologically distinct, types of surface structure have been found on bacteria that are responsible for adhesion to a wide range of different substrata. These two types of structure are called fimbriae and fibrils, and their incidence and distribution on the surfaces of bacteria are outlined here. Our aim is to give an indication of the diversity of surface appendages that have been found on bacteria and that in many cases are responsible for attachment.

B. Fimbriae

Fimbriae (Latin: thread fiber, or fringe) are proteinaceous appendages 0.2–2.0 μm in length, with a high content of hydrophobic amino acid residues, and they may be rigid rodlike structures about 7–10 nm in diameter (Fig. 1a). The degree of fimbriation on a cell can vary from a few fimbriae to several hundred, peritrichously distributed all over the cell surface. Fimbriae are thought to be the most common type of surface structure and have been detected on many different genera of gram-negative bacteria (bacteria that have an outer membrane on the cell wall), such as members of the family Enterobacteriaceae. These organisms live mostly in the human gut. Fimbriae, however, are much less common on gram-positive bacteria (bacteria that have no outer membrane on the cell wall). There are many ultrastructurally distinct types of fimbriae, which are distinguished from each other by their width and flexibility. The type 1, 987P, CFA1, CS1, CS2, and P fimbriae found on different strains of *Escherichia coli* are rigid structures with a diameter of approximately 7.0 nm and with an apparent axial hole running down the center [reviewed by Paranchych and Frost (12)]. Flexible very thin fimbriae 2–5 nm in diameter have also been detected and are represented by the K88, K99, F41, and CS3 fimbriae, also found on different *E. coli* strains.

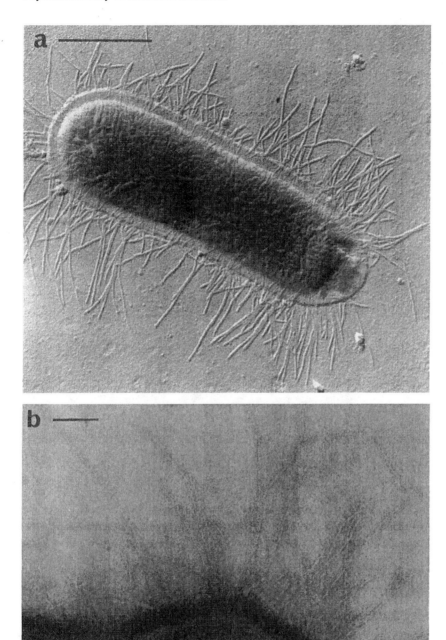

Fig. 1 (a) *Escherichia coli* cell with rigid, peritrichous type 1 fimbriae, approximately 7 nm in diameter and up to 1.0 μm long. Shadowed with palladium. Bar marker 0.5 μm. (b) *Actinomyces naeslundii* PK2407 cell with peritrichous, flexible, 3–4 nm fimbriae, up to 1.0 μm long. Negatively stained with 1% methylamine tungstate. Bar marker 0.1 μm.

These flexible fimbriae are also known to exist on some species of oral bacteria such as *Streptococcus salivarius* (13) and *Porphyromonas gingivalis* [formerly known as *Bacteroides gingivalis* (14,15)]. *Actinomyces naesludii*, which attach to teeth, the oral mucosa, and other oral bacteria, also express a dense array of the thinner type of flexible peritrichous fimbriae (Fig. 1b).

C. Fibrils

The term "fibril" has been proposed to describe a second type of surface structure, which is ultrastructurally distinct from fimbriae [reviewed by Handley (16)]. Fibrils are much shorter than fimbriae and are usually less than 0.2 μm in length, although they may be up to 0.4 μm on some strains.

Fibrils are usually peritrichous and often have no measurable width as they clump together, and they may be densely or sparsely distributed on the cell surface. Fibrils are commonly found on oral bacteria and in some organisms they are localized in lateral or polar tufts, consisting of one or two lengths of fibrils. The distribution and lengths of fibrils found on freshly isolated strains of oral streptococci from dental plaque are diagrammatically represented in Fig. 2. Streptococci are organisms commonly found in the human mouth, and most freshly isolated strains of streptococci carry some surface structures. Completely smooth-walled, bald, strains are comparitively rare in freshly isolated oral organisms. Where the appropriate studies

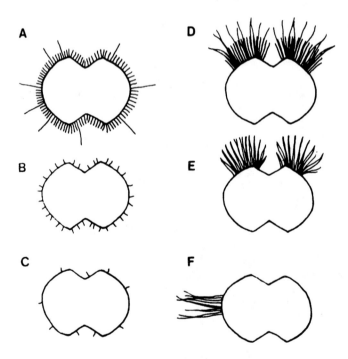

Fig. 2   Diagrammatic representation of the surface structure found on a range of oral streptococcal species. (A) Two lengths of peritrichous fibrils found on *S. salivarius* and *S. sanguis* strains. (B) Fibrils of one length with an intermediate density, found on *S. sanguis* strains. (C) Fibrils of one length, sparsely distributed, found on *S. sanguis* strains. (D) Lateral tufts of two lengths of fibrils, found on *Streptococcus oralis* and *Streptococcus crista* strains. (E) Lateral tufts of one length of fibril, found on *S. crista* and *S. sanguis* strains. (F) Polar tufts of one length of fibril as found on *S. sanguis*. (Adapted from Ref. 17.)

have been done, fibrils and fimbriae are always associated with adhesion. Therefore, there must be a very strong selection pressure for surface structures in an environment where organisms are subjected to high shear forces and where they will be swallowed unless they attach readily to one of the many surfaces in the oral cavity.

The structural diversity among surface appendages on oral bacteria is illustrated in Fig. 3a and b. Some *S. salivarius* strain from the tongue and some *Streptococcus sanguis* strain from dental plaque carry densely packed peritrichous fibrils (Fig. 3a) that clearly clump together in methylamine tungstate, the stain used to detected them in the electron microscope. Other strains of *S. sanguis* from dental plaque carry much more sparsely distributed fibrils, also with no clear width (Fig. 3b). This obvious difference in density of fibrils found on different strains from the same habitat (the teeth) indicates that it is probably not the number of fibrils that controls continued survival and therefore adhesion in a given habitat.

Some oral streptococci carry tufts of fibrils in a lateral (17) (Fig. 3c) or polar position (Fig. 3d) on the cell surface. Lateral tufts may carry one length of fibrils or two lengths of fibrils in the tuft (18). Sometimes it is possible to detect these two distinct "morphotypes" in the same culture (Fig. 3c). The two cell types can be separated and the physicochemical properties of the two cell populations can be compared. In this way, using a range of physicochemical tests on both cell types, it has been deduced that the long fibrils in the tuft have hydrophobic properties (19). The function of the tuft structures in adhesion is known in some cases, and some of the tufted streptococci adhere specifically by their tufts of fibrils to other plaque bacteria (20), showing that cell surface adhesion properties are located in the tuft fibrils only and not on the bare cell wall of the rest of the cell surface. It is therefore probable that the tufts adhere to their partner bacteria because only the tufts exhibit hydrophobic properties. This asymmetry of adhesion properties due to the presence of asymmetrically distributed surface structures will be discussed later in the chapter.

D. Cell Surface Properties That Change on Subculturing

Some organisms do not have detectable surface appendages; however, they may have cell wall polymers that are lost on repeated subculture in the laboratory, showing how quickly cell surface properties can be lost when the environmental selective pressures are no longer present. This loss of polymers leads to a change of surface physicochemical properties. For example, all cells in a population of a freshly isolated strain of *Lactobacillus acidophilus* (an organism found in the gastrointestinal tract and vagina of humans) have a thick ruthenium red (RR)–stainable layer outside the cell wall (21). The ability to stain with RR indicates the presence of polyanionic polymers such as polysaccharides (22,23). The function of this RR staining layer on *L. acidophilus* is not known, but after subculturing in the laboratory, cultures arose in which only 50% of the cells still carried this thick RR stainable layer (24) (see Fig. 4). This structural loss in half the cells was also detected by a physicochemical method. Particulate microelectrophoresis detected two charged subpopulations with zeta potentials of -2.0 and -19.7 mV, of which the former was identified as the initially isolated cells, all of which had an RR-stainable layer on the cell wall (24). Microbial cell surfaces can therefore change very quickly on continued subculture.

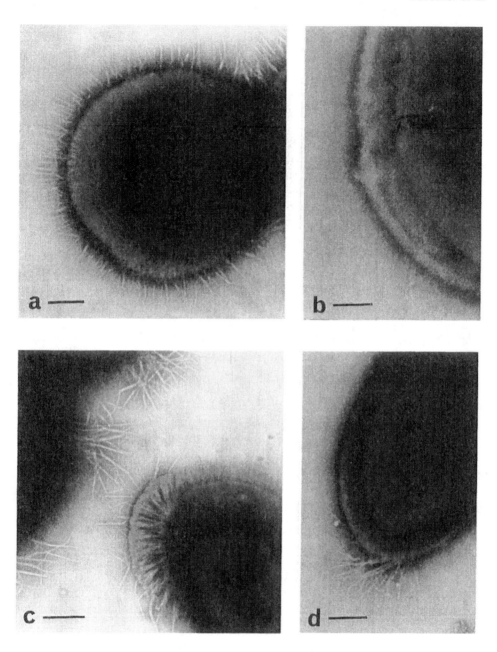

Fig. 3 (a) *S. salivarius* CN3928 showing peritrichous clumped fibrils, 191 nm long. Bar marker 100 nm. (b) *S. sanguis* LMH with sparsely distributed peritrichous fibrils, 71 nm long. Bar marker 100 nm. (c) *S. oralis* CN3410 showing lateral fibrillar tufts of both lengths, long (289 nm) and short (159 nm) fibrils, as well as a further peritrichous "fuzz" all over the cell surface. One tuft shows two lengths of fibrils and the others show one length of fibril, both in the same population of cells. Bar marker 300 nm. (d) *S. sanguis* PSH2 carries tufts of fibrils (374 nm) on the polar cap of the end cell in a chain. Bar marker 200 nm. All cells were negatively stained with 1% methylamine tungstate.

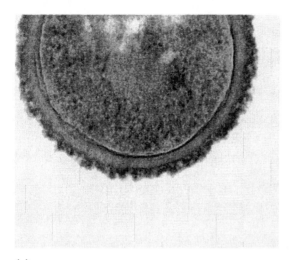

(a)

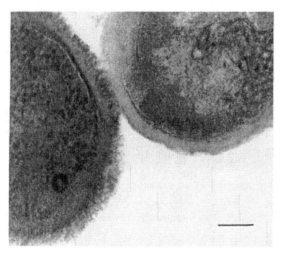

(b)

Fig. 4 *Lactobacillus acidophilus* RC14 is surrounded by a thick, ruthenium red/uranyl acetate-stainable layer with no obvious surface appendages (a) that disappears on a proportion of the cells upon repeated subculturing (b). The bar denotes 0.15 nm. (From Ref. 24.)

III. PHYSICOCHEMICAL MEASUREMENTS ON MICROBIAL CELL SURFACES

From the previous section it can be concluded that relevant physicochemical measurements on microbial cell surfaces require a microscopic resolution that cannot be accomplished with most currently employed methods. Yet, contact angles and zeta potentials of microbial cell surfaces are frequently measured to be used as input data for predictive, physicochemical models, and overall chemical composition data for cell surfaces are being unexpectedly obtained by techniques such as X-ray photoelectron spectroscopy (XPS) (25–27) and Fourier transform infrared spectroscopy (FTIR) (28,29).

A. Contact Angle Measurements

Contact angles with liquids can be measured on microbial cell surfaces essentially by the same methods as applied on solid surfaces. However, in order to obtain a sufficiently large surface for positioning a microliter-sized liquid droplet, microbial lawns are prepared from microbial suspensions on membrane filters (7,30). The only nontrivial step in contact angle measurements on microbial lawns involves the degree of drying of the lawns. Immediately after preparation of a lawn, water droplets spread completely, but during air drying water contact angles gradually increase until a so-called plateau contact angle is measured. It is generally advocated that these plateau contact angles be presented.

   Van der Mei and Busscher (31) compared plateau water contact angles with contact angles on pressed tablets of freeze-dried microorganisms for a collection of hydrophilic and hydrophobic strains (see Fig. 5). As can be seen, little difference exists in both types of contact angles when hydrophilic strains are considered, but on the hydrophobic strains plateau contact angles were higher than the contact angles on freeze-dried tablets. It was

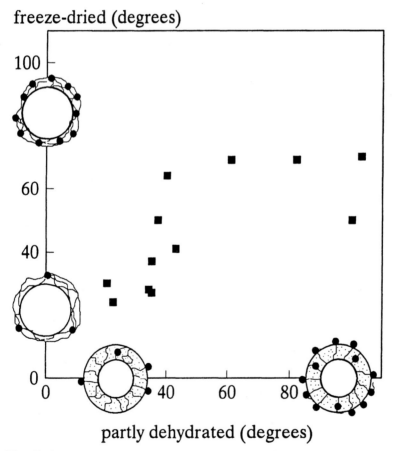

Fig. 5 Hypothesis for the behavior of surface appendages on microbial cell surfaces during contact angle measurements, based on the relationship between water contact angles on freeze-dried microbial strains and "plateau contact angles" on partly dehydrated organisms.

Table 1 A Compilation of Contact Angles (Degrees) on Various Microbial Strains for Water (W), Formamide (F), Methyleneiodide (M), and α-Bromonaphthalene (α-Br)

Strain	θ_W	θ_F	θ_M	$\theta_{\alpha-br}$	Growth medium[a]	Reference
Streptococci						
S. salivarius HB	42	48	56	33	THB	32
S. salivarius HBC12	21	28	55	32	THB	
S. sanguis PSH 1b	60	60	53	36	BHI	19
S. sanguis CR311	74	67	55	47	BHI	
S. sanguis CR311 VAR3	31	39	55	46	BHI	
S. mitis BA	103	55	49	35	THB	32
S. mitis ATCC9811	68	53	48	31	THB	
Peptostreptococci						
P. micros HG1108	66	69	51	34	THB	33
P. micros HG1109	33	49	58	43	THB	
Staphylococci						
S. epidermidis HBH276	29	3	51	34	BHI	34
S. epidermidis 236	32	31	52	35	BHI	
S. epidermidis 3294	107	44	53	35	BHI	
Escherichia coli						
E. coli O2K2	57	30	59	35	BHI	35
E. coli O83K?	54	36	57	39	BHI	
E. coli Hu734	17	18	48	33	BHI	36
Lactobacilli						
L. plantarum RC6	25	31	49	24	MRS	37
L. acidophilus RC14	91	47	55	38	MRS	
Serratia marcescens						
S. marcescens RZ30	54	55	88	91	BHI	38
S. marcescens 3164	21	18	58	49	BHI	

[a] THB, Todd-Hewitt broth; BHI, brain-heart infusion broth; MRS, de Mann, Rogosa, and Sharpe broth.

concluded that the water content of the cell surfaces after drying to the plateau state was high enough to prevent collapse of microbial cell surface structures and that plateau contact angles were therefore a realistic reflection of the overall, intrinsic cell surface hydrophobicity. Presumably, in a freeze-dried state, all surface structures collapse, and the cell surface is in a state remote from its physiologic one.

Despite being an overall cell surface property, microbial contact angles vary greatly between different strains and even the presence of tufts of fibrils as on *S. Sanguis* strains is reflected in measured contact angles (19) (see Table 1). Both *S. salivarius* HBC12 and *S. sanguis* CR311 VAR3 have bald cell surfaces and are hydrophilic (water contact angles 21° and 31°, respectively) compared with their peritrichously fibrillated (*S. salivarius* HB) and tufted (*S. sanguis* PSH 1b and CR311) parent strains. Hence, hydrophobicity is conveyed to the cell surface by fibrils and fimbriae and unlikely for the bald cell surface.

We have developed a routine for measuring contact angles with water, formamide, methyleneiodide, and α-bromonaphthalene. The use of these four liquids allows us to calculate the microbial cell surface free energy according to either:

1. The equation of state (39,40)
2. The concept of dispersion and polar surface free energy components (41,42)
3. The Lifshitz–van der Waals/acid-base approach (43–45)

In combination with the Young equation

$$\gamma_{lv} \cos \theta = \gamma_{sl} - \gamma_{sv} \tag{1}$$

in which θ is the contact angle and γ_{ij} is the surface free energy of the liquid (l), solid (s) or vapor (v) interface, each of these three approaches enables the calculation of the interfacial free energy of adhesion ΔG_{adh}, defined as

$$\Delta G_{adh} = \gamma_{sm} - \gamma_{sl} - \gamma_{bl} \tag{2}$$

in which (m) denotes the microbial cell surface (6,7).

In the equation of state approach, ΔG_{adh} is not divided into any components, whereas the other two approaches distinguish a dispersion and a polar component according to

$$\Delta G_{adh} = \Delta G_{adh}^{d} + \Delta G_{adh}^{p} \tag{3}$$

or a Lifshitz–van der Waals/acid-base component

$$\Delta G_{adh} = \Delta G_{adh}^{LW} + \Delta G_{adh}^{AB} \tag{4}$$

in which

$$\Delta G_{adh}^{AB} = +2\left[\left(\sqrt{\gamma_{mv}^{\oplus}} - \sqrt{\gamma_{sv}^{\oplus}}\right)\left(\sqrt{\gamma_{mv}^{\ominus}} - \sqrt{\gamma_{sv}^{\ominus}}\right)\right.$$
$$- \left(\sqrt{\gamma_{mv}^{\oplus}} - \sqrt{\gamma_{lv}^{\oplus}}\right)\left(\sqrt{\gamma_{mv}^{\ominus}} - \sqrt{\gamma_{lv}^{\ominus}}\right)$$
$$\left. - \left(\sqrt{\gamma_{sv}^{\oplus}} - \sqrt{\gamma_{lv}^{\oplus}}\right)\left(\sqrt{\gamma_{sv}^{\ominus}} - \sqrt{\gamma_{lv}^{\ominus}}\right)\right] \tag{5}$$

where $\oplus$ and $\ominus$ indicate the hydrogen-donating and electron-donating surface free energy parameters, which together yield the acid-base surface free energy component, i.e.,

$$\gamma_{i}^{AB} = 2\sqrt{\gamma_{i}^{\oplus}\gamma_{i}^{\ominus}} \tag{6}$$

Application Eq. (2) is the ultimate goal of many contact angle measurements on solid substrata and on microbial lawns.

B. Particulate Microelectrophoresis

The measurement of microbial electrophoretic mobilities and the derivation therefrom of zeta potentials by particulate microelectrophoresis proceeds according to the standard methodologies in physicochemistry (46), although the calculation of zeta potentials from electrophoretic mobilities is not always straightforward (47). Furthermore, it is important to realize that the plane of shear may be removed far away from the microbial cell surface if long fibrils or fimbriae are present that may collapse onto the cell surface upon increasing the ionic strength, as demonstrated, e.g., by dynamic light scattering (48). The ionic composition of microbial suspensions influences the results not only from a physicochemical perspective but also from a biological one. Organisms that are accustomed to high ionic strengths in their natural environments may lyse when suspended in water or other nonphysiological suspending fluids.

In our experience, the pH dependence of microbial zeta potentials in 10 mM potassium phosphate is highly characteristic for a given isolate, as can be seen from the compilation of results given in Fig. 6. Consequently, isoelectric points can be

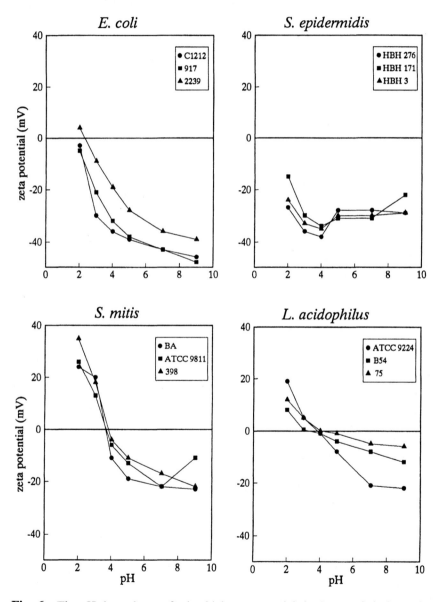

Fig. 6 The pH dependence of microbial zeta potentials is characteristic for a given isolate. (All zeta potentials shown were measured in 10 mM potassium phosphate, with pH adjusted by addition of HCl or KOH.)

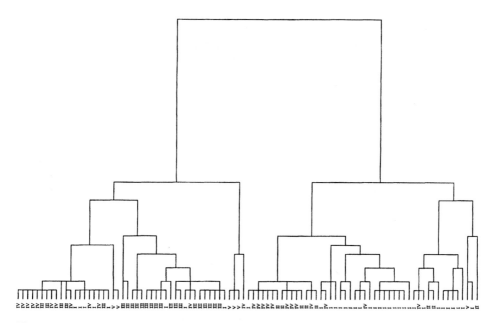

Fig. 7 Dendogram of the results of a hierarchical cluster analysis of the pH-dependent zeta potentials of 96 different microbial strains, including I, 35 oral streptococci; II, 8 thermophilic streptococci; III, 23 staphylococci; IV, 24 lactobacilli; V, 6 *E. coli*.

measured that vary from extremely low to above pH 4. Interestingly, tufts of fibrils on a strain such as *S. sanguis* CR311 increase the isoelectric point from below pH 2 for the bald variant to pH 2.8 for the tufted one, despite the fact that the tufts cover only a portion of the cell surface. Although the pH dependence of microbial zeta potentials is characteristic for a given isolate, it is generally not at the strain or species level. A hierarchical cluster analysis, with the pH-dependent zeta potentials ($2 \leq \mathrm{pH} \leq 9$) of 96 isolates as input, yields a dendogram with two clear clusters (see Fig. 7). The distinction between the two clusters is determined by whether or not an isolate has an isoelectric point. Although not all strains and species are grouped together, nearly all streptococci possess an isoelectric point and are consequently grouped together, whereas staphylococci are consistently placed in the cluster of isolates lacking an isoelectric point. Lactobacilli and *E. coli* are divided over both clusters.

Most naturally occurring surfaces are negatively charged (49), and repulsive EL (electrostatic) interactions interfere with the attractive LW (Lifshitz-Van der Waals) forces that mediate microbial adhesion to surfaces. Microbial zeta potentials are most often employed to calculate the EL interaction energy, which reads for the sphere (microorganism)–plate configuration

$$\Delta G^{\mathrm{EL}}(d) = \pi^{\varepsilon}\varepsilon_0 a(\zeta_{\mathrm{m}}^2 + \zeta_{\mathrm{s}}^2)\left\{\frac{2\zeta_{\mathrm{m}}\zeta_{\mathrm{s}}}{\zeta_{\mathrm{m}}^2 + \zeta_{\mathrm{s}}^2}\ln\frac{1 + \exp(-\kappa d)}{1 - \exp(-\kappa d)} + \ln(1 - \exp(-2\kappa d))\right\} \quad (7)$$

where ε is the dielectric constant of the suspending fluid, ε_0 is the permittivity of free space a is the microbial cell radius, $\zeta_{\mathrm{m,s}}$ are the microbial and substratum zeta potentials, κ is the Debye-Hückel length, and d is the separation distance between the interacting surfaces (9).

C. X-Ray Photoelectron Spectroscopy

X-ray photoelectron spectroscopy (XPS) provides a means of obtaining the chemical composition of microbial cell surfaces (26) at an overall level similar to that of intrinsic cell surface hydrophobicity by contact angles and charge properties by micro-electrophoresis. As XPS is a high-vacuum technique, extensive sample preparation, including washing, centrifuging, and freeze-drying, is involved before microbial cell surfaces can be studied by XPS (50). These steps obviously bring the cell surface to a state that is far remote from its physiological one, and some authors believe that the integrity of the vulnerable cell surface of especially gram-negative bacteria, as compared with gram-positive bacteria, is disrupted by this extensive preparation (51), with a potential impact on the results.

XPS spectra of microbial cell surfaces are fairly similar, with carbon, nitrogen, oxygen, and phosphorus the main elements detected, albeit in different amounts on different isolates (see Fig. 8). Decomposition of C_{1s} and O_{1s} electron binding energies has furthermore indicated the presence of lipids, proteins, and polysaccharides. Tufts of fibrils on *S. sanguis* CR311 increased the N/C ratio from 0.066 for the bald variant to 0.085 for the parent strain, indicating a nitrogen (protein)–rich composition of the tufts (19).

D. Fourier Transform Infrared Spectroscopy

Fourier transform infrared spectroscopy (FTIR) is definitely not a trivial method for analyzing microbial cell surfaces, especially not when used in a transmission mode. Yet, quantitative analysis of transmission IR spectra of freeze-dried microorganisms in KBr pellets (see Fig. 9) demonstrated that the ratios of the absorption band areas for amide (AmI and AmII), polysaccharide (PII), and phosphate (PI) versus carbo-hydrate (CH) correlated well with XPS data. These observations have led us to conclude that FTIR of freeze-dried microorganisms yields spectra that are determined largely by the composition of the cell surface (28,29), or stated in another way: the microbial cell surface is, from an overall compositional point of view, the most variable within different strains.

E. Relationships

In the literature there is an ongoing search for relationships between overall, physicochemical cell surface properties, and for several collections of strains, including some tufted strains, comprehensive relationships have been described (see Fig. 10), whereas for others such relationships were completely lacking.

In terms of relationships, the following seem to have been established for gram-positive microorganisms (19,21,27,31–33,37,38,52–58):

> Intrinsic microbial cell surface hydrophobicity increases with increasing number of poteinaceous groups on the cell surface, whereas polysaccharides convey hydrophilicity to the cell surface.
> Microorganisms with a high intrinsic cell surface hydrophobicity have elevated isoelectric points (pH $\geq$ 3).
> Microbial cell surfaces with a low isoelectric point are rich in polysaccharides and phosphates.

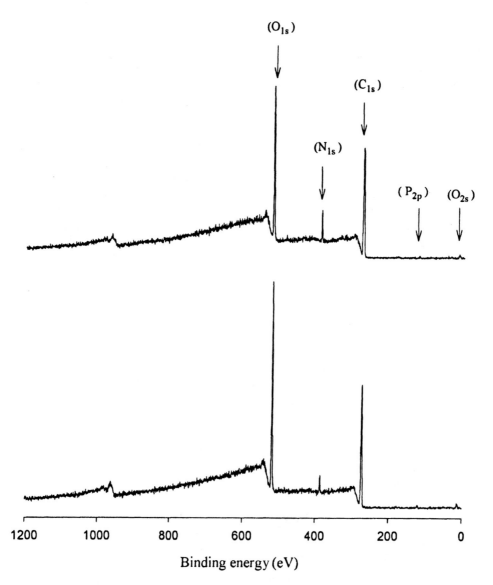

Fig. 8 XPS spectra of *S. salivarius* HB (upper line) and *S. salivarius* HBC12 (lower line) with elemental surface concentration O/C ratios of 0.432 and 0.495 and N/C ratios of 0.104 and 0.053, respectively.

The existence of these relationships for gram-positive strains, especially because they include isoelectric points that are measured under physiological conditions of the organisms, indicates the virtue of an overall physicochemical characterization of microbial cell surfaces. Application of these characteristics in an adhesion model will be relatively straightforward for bald and peritrichous strains, but for strains with tufts of fibrils or long surface appendages with different compositions and lengths the development of a physicochemical model will be more difficult and possibly subject to arbitrary decisions.

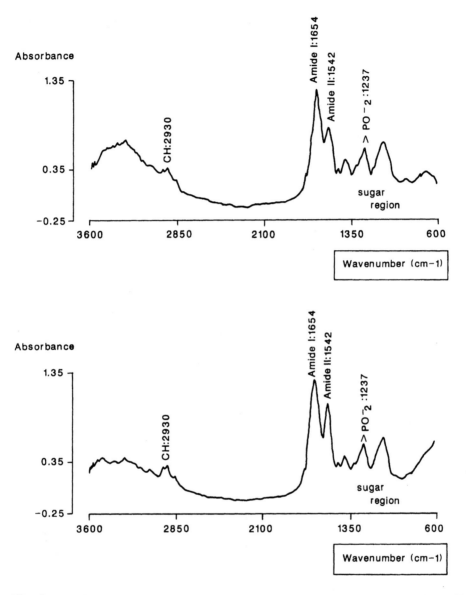

Fig. 9 FTIR spectra of a nonencapsulated (upper line) and an encapsulated (lower line) *Staphylococcus* species.

IV. AN (EXTENDED) DLVO THEORY

The classical DLVO theory describes the interaction energy between a microorganism and a substratum surface as a function of their separation distance and comprises two main contributions $\Delta G^{LW}(d)$ and $\Delta G^{EL}(d)$ (9). Consequently,

$$\Delta G^{DLVO}(d) = \Delta G^{LW}(d) + \Delta^{EL}(d) \tag{8}$$

The distance dependence of $\Delta G^{EL}(d)$ is given by Eq. (7), and the LW component decays

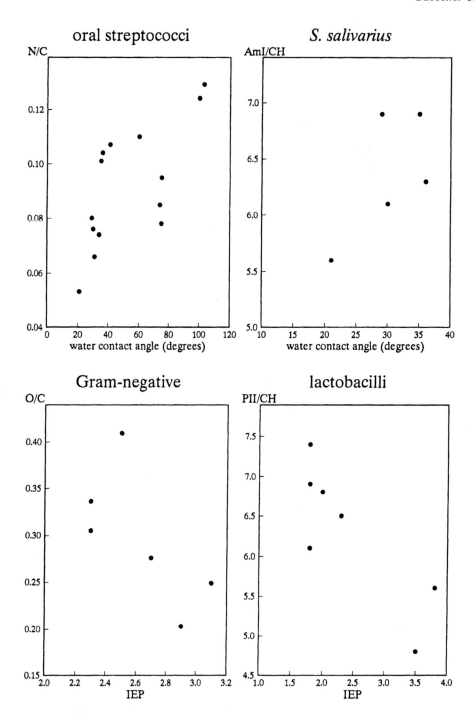

Fig. 10   Relationships between physical cell surface properties and compositional data for collections of microbial strains. Water contact angles versus N/C for oral streptococci (top left); water contact angles versus AmI/CH for *S. salivarius* strains (top right); isoelectric points versus O/C for gram-negative bacteria (bottom left); isoelectric points versus PII/CH for lactobacilli (bottom right).

with separation distance according to

$$\Delta G^{LW}(d) = \frac{A_{slm}}{6}\left\{\frac{a}{d} + \frac{a}{d+2a} + \ln\left(\frac{d}{d+2a}\right)\right\} \tag{9}$$

in which A_{slm} is the Hamaker constant. Assuming that the LW component of the interfacial free energy of adhesion ΔG^{LW} represents the LW interaction at closest approach ($d_0 = 1.57$ Å), the Hamaker constant A_{slm} can be calculated from contact angle data according to

$$A_{slm} = \frac{\Delta G^{LW}_{adh}}{12\pi d_0^2} \tag{10}$$

From Eq. (10), it can be seen that the apolar liquid (α-bromonaphthalene and methyleneiodide) contact angles are the main determinants of A_{slm}. Therewith the classical DLVO theory does not account for an influence of hydrophobicity (45), as probed by water and formamide contact angles and expressed by ΔG^{AB}_{adh} [Eq. (5)]. By associating ΔG^{AB}_{adh} with the AB interaction at closest approach, similar to what is being assumed for the LW interaction, a distance dependence for the sphere-plate configuration can be obtained (44,45)

$$\Delta G^{AB}(d) = 2\pi a\lambda\Delta G^{AB}_{adh}\exp\{(d_0 - d)/\lambda\} \tag{11}$$

in which λ is the correlation length of molecules in the liquid and varies with ionic strength and, as sometimes suggested, with the sign of ΔG^{AB}_{adh}. For $\Delta G^{AB}_{adh} < 0$ ("hydrophobic attraction"), λ may be as long as 13 nm, making AB interactions similar to long-range, hydrophobic attraction. "Hydrophilic repulsion," $\Delta G^{AB}_{adh} > 0$, is generally taken to be short range with $\lambda = 0.6$ nm.

Continuing this discussion of the decay length of AB interactions, an extended DLVO theory distinguishes three main components:

$$\begin{aligned}
G^{EXT}(d) &= G^{DLVO}(d) + \Delta G^{AB}(d) \\
&= \Delta G^{LW}(d) + \Delta G^{EL}(d) + \Delta G^{AB}(d)
\end{aligned} \tag{12}$$

A. Model Calculations

The Lifshitz–van der Waals, electrostatic, and acid-base components of the interaction energy in the extended DLVO theory are illustrated in Fig. 11 for the interaction of a moderately hydrophilic organism approaching a hydrophilic and hydrophobic substratum surface under high- and low-ionic-strength conditions. On glass, the Lifshitz–van der Waals forces are attractive and the acid-base interaction is repulsive. Combined with the repulsive electrostatic interaction, the sum total according to Eq. (12) yields a secondary interaction minimum that is deeper and positioned closer to the substratum surface in the high-ionic-strength solution than in the low-ionic-strength solution. On Teflon, the Lifshitz–van der Waals and the electrostatic interactions are both repulsive, but an acid-base attraction exists. Because of strong electrostatic repulsion in the low-ionic-strength solution, the total interaction energy according to Eq. (12) is repulsive, but in the high-ionic-strength solution elec-

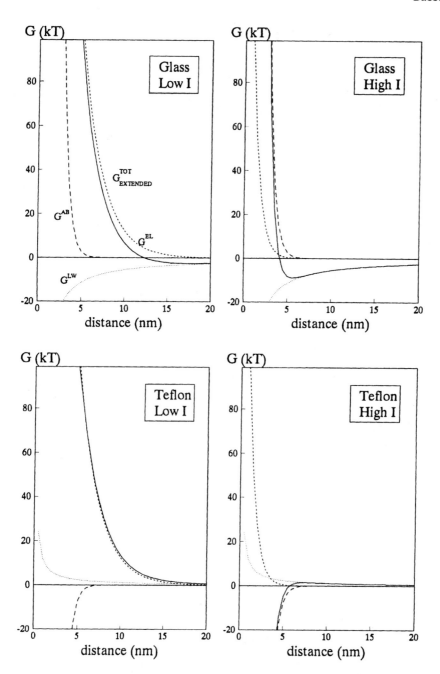

Fig. 11 Extended DLVO interaction energies versus distance for a relatively hydrophilic organism approaching a hydrophilic glass surface and a hydrophobic Teflon surface from solutions of low (Low I) and high (High I) ionic strength. DLVO input data: Hamaker constants, $0.73\ kT$ (glass) and $-0.15\ kT$ (Teflon); bacterial cell radius, 500 nm; acid-base interaction energy at closest approach 19.8 mJ m^{-2} (glass) and -52.2 mJ m^{-2} (Teflon); zeta potentials -20 mV (High I) and -30 mV (Low I) with corresponding κ^{-1} values of 1.0 nm and 2.3 nm, respectively, for a potassium phosphate solution. The decay length λ for acid-base interactions was assumed to be 0.6 nm.

trostatic repulsion is less and the attractive acid-base interactions cause an attractive total interaction at close approach.

Figure 12 shows results of model calculations according to the extended DLVO theory comparing the interaction energies of this organism and of possible hydrophobic tufts with substratum surfaces. The effects of the tufts as compared with the hydrophilic cell surface are twofold. In the low-ionic-strength solution, the tufts experience less electrostatic repulsion and the potential energy barrier is clearly less high for the tufts. In the high-ionic-strength solution the hydrophobic tufts assist the total attraction, especially to the hydrophilic substratum surface.

Clearly, model calculations will become increasingly difficult when appendages with different lengths and chemical compositions are present on a cell surface.

V. AN UNDERSTANDING OF MICROBIAL ADHESION THROUGH PHYSICOCHEMISTRY

A. Microbial Adhesion to Hydrocarbons and Organic Solvents

Hydrocarbon droplets in suspension are hydrophobic and negatively charged (59–61). Consequently, microbial adhesion to hydrocarbons (MATH) provides an interesting method for demonstrating physicochemical interactions in microbial adhesion (62,63), especially because the hexadecane interface against an aqueous microbial suspension is almost ideally smooth and homogeneous.

Hydrophilic microorganisms, i.e., organisms with a strong AB interaction with water, do not adhere to hexadecane in MATH. In fact, Van Loosdrecht et al. (64) described a "threshold hydrophobicity" for adhesion to hexadecane that amounted to a water contact angle of 20°. The study of Van Loosdrecht et al. (64) involved 20 different strains; a similar study by Reid et al. (65), involving only lactobacilli (23 isolates), indicated a threshold hydrophobicity by water contact angles of 60° (see Fig. 13).

The involvement of AB interactions in microbial adhesion can also be demonstrated by comparing adhesion to hexadecane with adhesion to chloroform (66), which can be studied in essentially the same way as MATH. Although the AB interactions between microorganisms and water are not affected by replacing hexadecane by chloroform, the microorganisms experience an additional AB interaction with the electron-accepting chloroform interface that is absent in the case of hexadecane. Because most microbial cell surfaces are predominantly electron donating, their adhesion to chloroform exceeds their adhesion to hexadecane.

Hydrophobic microorganisms, i.e., organisms with a water contact angle above the threshold value, do not necessarily adhere to hexadecane due to repulsive EL interactions (61,63). Consequently only hydrophobic strains with a low negative surface charge adhere to hexadecane (see Fig. 14).

B. Adhesion of Colloidal Gold to Tufted Streptoccal Strains

Physicochemical properties of bacteria may be asymmetrically distributed over the cell surface. Tufted organisms such as *S. sanguis* CR311 and PSH 1b with two lengths of fibrils in their tufts can be demonstrated to carry charge and hydrophobicity on the short and long fibrils, respectively (67). This asymmetry of surface properties was detected using colloidal gold, which is both hydrophobic and negatively charged.

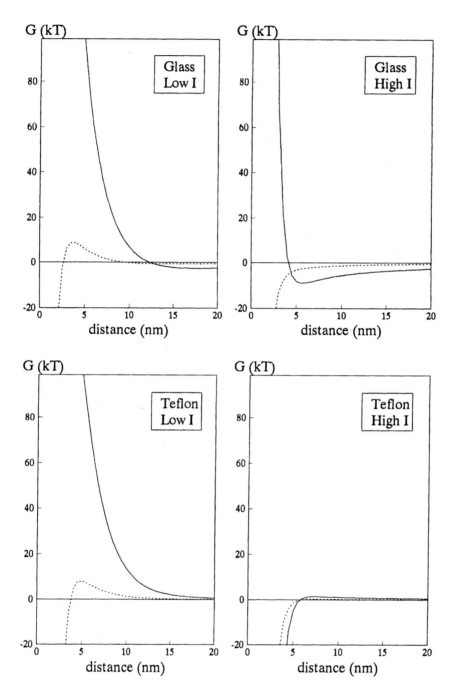

Fig. 12 Extended DLVO interaction energies versus distance for a relatively hydrophilic organism (—) and possible hydrophobic tufts (– – –) approaching a hydrophilic glass surface and a hydrophobic Teflon surface from solutions of low (Low I) and high (High I) ionic strength and for 200-nm-wide, hydrophobic tufts. DLVO input data for the organism, see Fig. 11. DLVO input data for the tufts: Hamaker constants, 1.02 kT (glass) and -0.20 kT (Teflon); tuft width, 200 nm; acid-base interaction energy at closest approach -10 mJ m^{-2} (glass) and -70 mJ m^{-2} (Teflon); zeta potentials -10 mV (High I) and -20 mV (Low I).

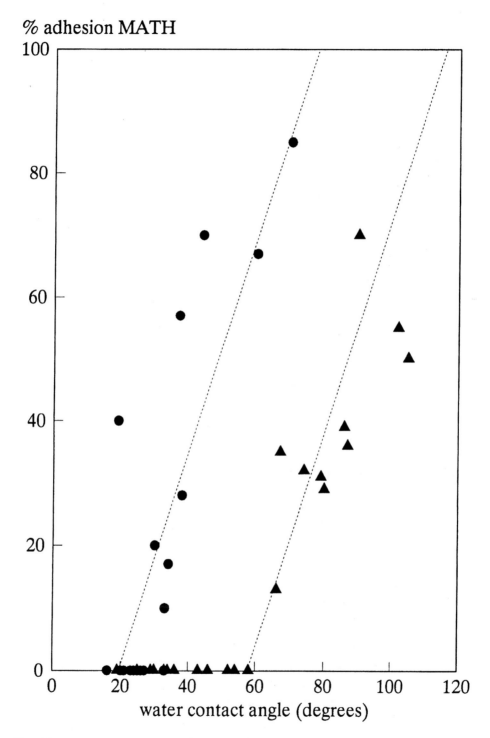

Fig. 13 Microbial adhesion to hexadecane as a function of the intrinsic cell surface hydrophobicity of the organisms by water contact angles. (●) Results of Van Loosdrecht et al. (64). (▲) Results of Reid et al. (65).

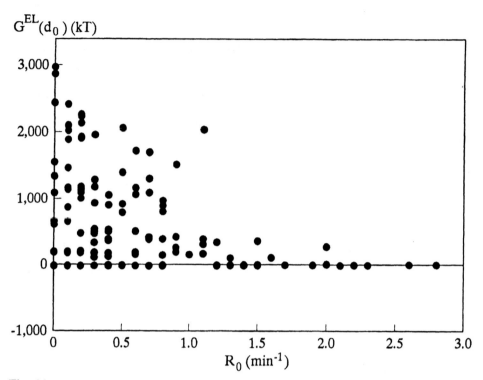

Fig. 14 Microbial adhesion to hexadecane in a 10 mM potassium phosphate solution as a function of the EL interaction energy at closest approach (1.57 Å) for a collection of 23 staphylococci.

Electron microscopy revealed that colloidal gold did not adhere to the bald part of the cell surface (Fig. 15a) of tufted *S. sanguis* strains over the pH range 3.7–9.0 because of the hydrophilic nature of this portion of the cell surface. Adhesion of colloidal gold to the long tuft fibrils occurred irrespective of pH (Fig. 15b), confirming conclusions from contact angle measurements that these fibrils convey hydrophobicity to the cell surface. Adhesion of colloidal gold to short fibrils occurred at pH 3.7 (Fig. 15c) but not at elevated pH because of increased EL repulsion between the gold and the negatively charged short fibrils with increasing pH.

These observations demonstrate that microbial adhesion to solid substrata cannot be considered from a physicochemical perspective without accounting for structural and chemical features of the cell surfaces as done by the DLVO-type analysis presented in Fig. 12. However, often the proper input data on the physicochemical nature of the structural and chemical cell surface features involved in adhesion are lacking and the analysis is based on unverifiable assumptions.

C. Adhesion of Coagulase-Negative Staphylococci to Silicone Rubber

Coagulase-negative staphylococcal (CNS) strains can be either hydrophobic or hydrophilic, more or less electron donating, and more or less negatively charged. Recently, it was established for a collection of 23 different CNS isolates (34) that their physicochemical cell surface properties were not related to the absence or presence

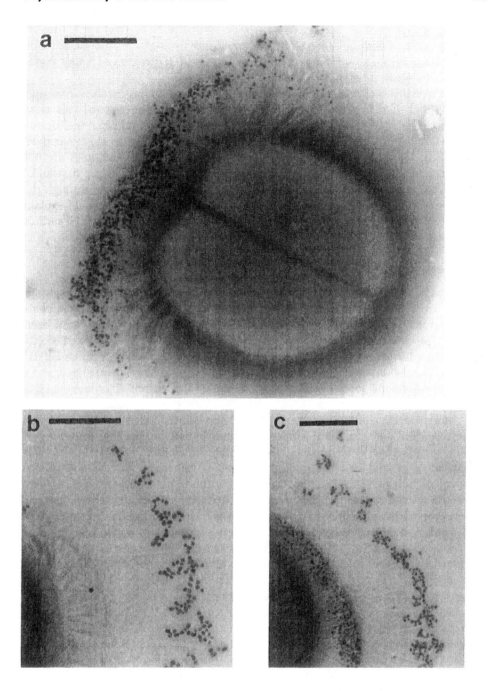

Fig. 15 (a) *S. sanguis* CR311 labelled with colloidal gold at pH 3.7. Gold particles are attached to the short tuft fibrils only. The long fibrils are not visible. (b) *S. sanguis* PSH 1b labeled with colloidal gold at pH 10.2. The gold is attached only to the ends of the long fibrils. (c) *S. sanguis* PSH 1b labeled with colloidal gold at pH 3.7. Gold is attached to the ends of both long and short fibrils. All cells were stained with 1% methylamine tungstate. Bar markers are all 200 nm.

of slime production or a capsule, demonstrable with India ink staining. Moreover, adhesion of the strains to silicone rubber could not be explained on the basis of the physicochemical cell surface properties, and in fact, irrespective of contact angles and zeta potentials, all staphylococci evaluated adhered.

Because this is contrary to observations on, e.g., oral streptococcal adhesion to inert substrata, it is hypothesized that all CNS isolates have a capsule. This must be envisaged as a polysaccharide gel around the cell body, as could also be inferred from the lack of an isoelectric point of all CNS strains involved in the cluster analysis presented in Fig. 6. Unfortunately, depending on their thickness, capsules are not always demonstrable by staining. Although in adhesion the capsule may first make contact with a substratum surface, the cell body presumably displaces through the gel to become the determinant factor in adhesion. Such a hypothesis would explain why adhesion of CNS cannot be explained on the basis of physicochemical cell surface properties.

D. Miscellaneous Examples

Applications of physicochemistry to explain and predict microbial adhesion to surfaces are numerous; recently, for instance, preadsorption of hydrophobic components to oral streptococcal cell surfaces and the effects on adhesion have been studied (68). Another example involves an analysis of the role of hydrophobicity in adhesion of dissimilatory Fe(III)-reducing *Shewanella alga* to amorphous Fe(III) oxide (69). However, it is considered beyond the scope of this chapter to give an extensive and completeness-approaching review of the current literature on this point.

More relevant examples in the context of this chapter concern studies in which physicochemical and structural aspects of microbial cell surfaces are dealt with in relation to adhesion. Unfortunately, these studies are scarce. Rijnaarts et al. (70) postulated in general that the isoelectric point of bacteria is an indicator of the presence of polymeric cell surface appendages inhibiting adhesion. Strains with isoelectric points below pH 2.8 were said to have polysaccharide-rich surface appendages, inhibiting adhesion to both hydrophilic and hydrophobic surfaces. Bacteria with an isoelectric point above pH 3.2 were suggested to be free of inhibiting surface appendages and to adhere to any substratum. Bacteria with intermediate isoelectric points were envisaged to have amphiphilic appendages facilitating adhesion to hydrophobic but not to hydrophilic surfaces. Although it is possibly valid for coryneform bacteria and pseudomonads, several studies on, e.g., *E. coli* (35) and *Streptococcus mitis* (71) strains contradict the general validity of such an attractive predictive model based on isoelectric points as an indicator of the adhesive ability of microorganisms.

VI. SUMMARY OF CONCLUSIONS

Physicochemical surface properties of microorganisms can vary widely and generalizations at the species or even strain level are virtually impossible. Application of physicochemical models to explain microbial adhesion to solid substrata has been successful for a limited number of strain collections, because of the overall nature of the input data (i.e., Hamaker constants, acid-base properties, and zeta potentials).

Although structural and chemical heterogeneities on microbial cell surfaces have an impact on overall cell surfaces, methods for obtaining detailed knowledge of cell surface heterogeneities are lacking. Considering the importance of structural and chemical heterogeneities in microbial adhesion, the development of a generalized model for microbial adhesion to surfaces is still far beyond reach. However, because even the most specific interactions are governed by physicochemical forces at the microscopic level, the analysis as discussed here will undoubtedly contribute to the development of a generalized model.

REFERENCES

1. JW Costerton, K-J Cheng, GG Geesey, TI Ladd, JC Nickel, M Dasgupta, TJ Marrie. Bacterial biofilms in nature and disease. Annu Rev Microbiol 41:435, 1987.
2. GK Richards, RF Gagnon, RJ Morcos. An assay to measure antibiotic efficacy against *Staphylococcus epidermidis* biofilms on implant surfaces. ASAIO J 40:M570, 1994.
3. KE Cooksey, B Wigglesworth-Cooksey. Adhesion of bacteria and diatoms surfaces in the sea: A review. Aquat Microb Ecol 9:89, 1995.
4. RMM Diks, SPP Ottengraf. Verification studies of a simplified model for the removal of dichloromethane from waste gases using a biological trickling filter. Bioprocess Eng 6:93, 1991.
5. AG Gristina. Biomaterial-centered infection. Microbial adhesion versus tissue integration. Science 237:1588, 1987.
6. DR Absolom, FV Lamberti, Z Policova, W Zingg, CJ van Oss, AW Neumann. Surface thermodynamics of bacterial adhesion. Appl Environ Microbiol 46:90, 1983.
7. HJ Busscher, AH Weerkamp, HC van der Mei, AWJ van Pelt, HP de Jong, J Arends. Measurements of the surface free energy of bacterial cell surfaces and its relevance for adhesion. Appl Environ Microbiol 48:980, 1984.
8. S Friberg. Colloidal phenomena encountered in the bacterial adhesion to the tooth surface. Swedish Dent J 1:207, 1977.
9. PR Rutter, B Vincent. Attachment mechanisms in the surface growth of microorganisms. In: RCW Berkeley, JM Lynch, J Melling, PR Rutter, B Vincent, eds. Microbial Adhesion to Surfaces. London: Ellis Horwood, 1980, p 79.
10. J Sjollema, HJ Busscher. Deposition of polystyrene latex particles towards polymethylmethacrylate in a parallel plate flow cell. J Colloid Interface Sci 132:382, 1989.
11. JE Seebergh, JC Berg. Evidence of a hairy layer at the surface of polystyrene latex particles. Colloids Surf A Physicochem Eng Aspects 100:139, 1995.
12. W Paranchych, LS Frost. The physiology and biochemistry of pili. Adv Microbial Physiol 29:53, 1971.
13. PS Handley, PL Carter, J Fielding. *Streptococcus salivarius* strains carry either fibrils and/or fimbriae on the cell surface. J Bacteriol 157:64, 1984.
14. F Yoshimura, K Takahashi, Y Nodasaka, T Susuki. Purification and characterisation of novel type of fimbriae from the oral anaerobe *Bacteroides gingivalis*. J Bacteriol 160:949, 1984.
15. PS Handley, LS Tipler. An electron microscope survey of the surface structures and hydrophobicity of oral and non-oral species of the bacterial genus *Bacteroides*. Arch Oral Biol 31:325, 1986.
16. PS Handley. Structure, composition and functions of surface structures on oral bacteria. Biofouling 2:239, 1990.
17. MW Jameson, HF Jenkinson, K Parnell, PS Handley. Polypeptides associated with tufts of cell surface fibrils in an oral *Streptococcus*. Microbiology 141:2729, 1995.

18. PS Handley, PL Carter, JE Wyatt, LM Hesketh. Surface structures (peritrichous fibrils and tufts of fibrils) found on *Streptococcus sanguis* strains may be related to their ability to co-aggregate with other oral genera. Infect Immun 47:217, 1985.

19. HJ Busscher, PS Handley, PG Rouxhet, LM Hesketh, HC van der Mei. The relationship between structural and physico-chemical surface properties of tufted *Streptococcus sanguis* strains. In: N Mozes, PS Handley, HJ Busscher, PG Rouxhet, eds. Microbial Surface Analysis: Structural and Physico-Chemical Methods New York: VCH, 1991, p 317.

20. C Mouton, HS Reynolds, EA Gasiecki RJ Genco. In vitro adhesion of tufted oral streptococci to *Bacterionema matruchotii*. Curr Microbiol 3:181, 1979.

21. PL Cuperus, HC van der Mei, G Reid, AW Bruce, AE Khoury, PG Rouxhet, HJ Busscher. The effect of serial passaging of lactobacilli in liquid medium on their physico-chemical and structural surface characteristics. Cells Mater 2:271, 1992.

22. JH Luft. Ruthenium red and violet. I. Chemistry, purification, methods used for electron microcopy and mechanism of action. Anat Rec 171:347, 1971.

23. PS Handley, J Hargreaves, DWS Harty. Ruthenium red staining reveals surface fibrils and a layer external to the cell wall in *Streptococcus salivarius* HB and adhesion deficient mutants. J Gen Microbiol 134:3165, 1988.

24. GI Geertsema-Doornbusch, J Noordmans, AW Bruce, G Reid, AE Khoury, HC van der Mei, HJ Busscher. Quantitation of microbial cell surface heterogeneity by microelectrophoresis and electron microscopy—Application to lactobacilli after serial passaging. J Microbiol Methods 19:269, 1994.

25. JL van Haecht, C Defosse, B van den Bogaert, PG Rouxhet. Surface properties of yeast cells: Chemical composition by XPS and isoelectric point. Colloids Surf 4:343, 1982.

26. DE Amory, MJ Genet, PG Rouxhet. Application of XPS to the surface analysis of yeast cells. Surf Interface Anal 11:478, 1988.

27. HC van der Mei, AJ Leonard, AH Weerkamp, PG Rouxhet, HJ Busscher. Properties of oral streptococci relevant for adherence: Zeta potential, surface free energy and elemental composition. Colloids Surf 32:297, 1988.

28. HC van der Mei, J Noordmans, HJ Busscher. Molecular surface characterization of oral streptococci by Fourier transform infrared spectroscopy. Biochim Biophys Acta 991:395, 1989.

29. HC van der Mei, J Noordmans, HJ Busscher. The influence of a salivary coating on the molecular surface composition of oral streptococci as determined by Fourier transform infrared spectroscopy. Infrared Phys 30:143, 1990.

30. CJ van Oss, CF Gillman. Phagocytosis as a surface phenomenon. I Contact angles and phagocytosis of non-opsonized bacteria. J Reticuloendothel Soc 12:283, 1972.

31. HC van der Mei, HJ Busscher. On the difference between water contact angles measured on partly dehydrated and on freeze-dried oral streptococci J Colloid Interface Sci 136:297, 1990.

32. HC van der Mei, AJ Leonard, AH Weerkamp, PG Rouxhet, HJ Busscher. Surface properties of *Streptococcus salivarius* HB and nonfibrillar mutants: Measurements of zeta potential and element composition with X-ray photoelectron spectroscopy. J Bacteriol 170:2462, 1988.

33. MM Cowan, HC van der Mei, PG Rouxhet, HJ Busscher. Physico chemical and structural properties of the surfaces of *Peptostreptococcus micros* and *Streptococcus mitis* as compared to those of mutans streptococci, *Streptococcus sanguis* and *Streptococcus salivarius*. J Gen Microbiol 138:2707, 1992.

34. HC van der Mei, B van de Belt-Gritter, G Reid, H Bialkowska-Hobrzanska, HJ Busscher. Adhesion of coagulase-negative staphylococci grouped according to physico-chemical surface properties. Microbiology 143:3861, 1997.

35. G Harkes, HC van der Mei, PG Rouxhet, J Dankert, HJ Busscher, J Feijen. Physicochemical characterization of *Escherichia coli*. A comparison with gram-positive bacteria. Cell Biophys 20:17, 1992.

36. G Reid HC van der Mei, C Tieszer, HJ Busscher. Uropathogenic *Escherichia coli* adhere to urinary catheters without using fimbriae. FEMS Immun Med 16:159, 1996.

37. KW Millsap, G Reid HC van der Mei, HJ Busscher. Cluster analysis of genotypically characterized *Lactobacillus* species based on physicochemical cell surface properties and their relationship with adhesion to hexadecane. Can J Microbiol 43:284, 1997.

38. HC van der Mei, MM Cowan MJ Genet, PG Rouxhet, HJ Busscher. Structural and physicochemical surface properties of *Serratia marcescens* strains. Can J Microbiol 38:1033, 1992.

39. AW Neumann, RJ Good, CJ Hope, M Sejpal. An equation-of-state approach to determine surface tensions of low-energy solids from contact angles. Colloid Interface Sci 49:291, 1974.

40. JK Spelt, D Li, AW Neumann. The equation of state approach to interfacial tensions. In: ME Schrader, G Loeb, eds. Modern Approaches to Wettability: Theory and Applications. New York: Plenum, 1992, p 101.

41. DK Owens, RC Wendt. Estimation of the surface free energy of polymers. J Appl Polym Sci 13:1741, 1969.

42. JR Dann. Forces involved in the adhesive process. I. Critical surface tensions of polymeric solids as determined with polar liquids. J Colloid Interface Sci 32:302, 1970.

43. CJ van Oss, MK Chaudhury, RJ Good. Interfacial Lifshitz–van der Waals and polar interactions in macroscopic systems. Chem Rev 88:927, 1988.

44. CJ van Oss, RJ Good, MK Chaudhury. Additive and nonadditive surface tension components and the interpretation of contact angles. Langmuir 4:884, 1988.

45. CJ van Oss. Hydrophobicity of biosurfaces—origin, quantitative determination and interactions energies. Colloids Surf B Biointerfaces 5:91, 1995.

46. AM James. Charge properties of microbial cell surfaces. In: N Mozes, PS Handley, HJ Busscher, PG Rouxhet, eds. Microbial Surface Analysis: Structural and Physico-Chemical Methods. New York: VCH, 1991, p 221.

47. A van der Wal, M Minor, W Norde, AJB Zehnder, J Lyklema. The electrokinetic potential of bacterial cells. Langmuir 13:71, 1997.

48. HC van der Mei, JM Meinders, HJ Busscher. The influence of ionic strength and pH on diffusion of micro-organisms with different structural surface features. Microbiology 140:3413, 1994.

49. BA Jucker, H Harms, AJB Zehnder. Adhesion of the positively charged bacterium *Stenotrophomonas* (*Xanthomonas*) *maltophilia* 70401 to glass and Teflon. J Bacteriol 178:5472, 1996.

50. PG Rouxhet, N Mozes, PB Dengis, YF Dufrene, PA Gerin, MJ Genet. Application of X-ray photoelectron spectroscopy to microorganisms. Colloids Surf B Biointerfaces 2:347, 1994.

51. KC Marshall, R Pembrey, RP Schneider. The relevance of X-ray photoelectron spectroscopy for analysis of microbial cell surfaces: A critical review. Colloids Surf B Biointerfaces 2:371, 1994.

52. PFG Herben, N Mozes, PG Rouxhet. Variation of the surface properties of *Bacillus licheniformis* according to age, temperature and aeration. Biochim Biophys Acta 1033:184, 1990.

53. PB Dengis, PA Gerin, PG Rouxhet. X-ray photoelectron spectroscopy of biosurfaces: Examination of performances with yeast cells and related model compounds. Colloids Surf B Biointerfaces 4:199, 1995.

54. YF Dufrene, A van der Wal, W Norde, PG Rouxhet. X-ray photoelectron spectroscopy of whole cells and isolated cell walls of gram-positive bacteria: Comparison with biochemical analysis. J Bacteriol 179:1023, 1997.

55. M van Raamsdonk, HC van der Mei, JJ de Soet, HJ Busscher, J de Graaff. Effect of polyclonal and monoclonal antibodies on surface properties of *Streptococcus sobrinus*. Infect Immun 63:1698, 1995.

56. HC van der Mei, MJ Genet, AH Weerkamp, PG Rouxhet, HJ Busscher. A comparison between the elemental surface compositions and electrokinetic properties of oral streptococci with and without adsorbed salivary constituents. Arch Oral Biol 34:889, 1989.

57. HC van der Mei, JJ de Soet, J de Graaff, PG Rouxhet, HJ Busscher. Comparison of the physicochemical surface properties of *Streptococcus rattus* with those of other mutans streptococcal species. Caries Res 25:415, 1991.

58. PL Cuperus, HC van der Mei, G Reid, AW Bruce, AH Khoury, PG Rouxhet, HJ Busscher. Physico-chemical surface characteristics of urogenital and poultry lactobacilli. J Colloid Interface 156:319, 1993.

59. KB Medrzycka. The effect of particle concentration in extremely dilute solutions. Colloid Polym Sci 269:85, 1991.

60. GI Geertsema-Doornbusch, HC van der Mei, HJ Busscher. Microbial cell surface hydrophobicity. The involvement of electrostatic interactions in microbial adhesion to hydrocarbon (MATH). J Microbial Methods 18:61, 1993.

61. HJ Busscher, B van de Belt-Gritter, HC van der Mei. Implications of microbial adhesion to hydrocarbons for evaluating cell surface hydrophobicity. 1. Zeta potentials of hydrocarbon droplets. Colloids Surf B Biointerfaces 5:111, 1995.

62. HC van der Mei, J de Vries, HJ Busscher. Hydrophobic and electrostatic cell surface properties of thermophilic dairy streptococci. Appl Environ Microbial 59:4305, 1993.

63. HC van der Mei, B van de Belt-Gritter, HJ Busscher. Implications of microbial adhesion to hydrocarbons for evaluating cell surface hydrophobicity. 2. Adhesion mechanisms. Colloids Surf B Biointerfaces 5:117, 1995.

64. MCM van Loosdrecht, J Lyklema, W Norde, G Schraa, AJB Zehnder. The role of bacterial cell wall hydrophobicity in adhesion. Appl Environ Microbiol 53:1893, 1987.

65. G Reid, PL Cuperus, AW Bruce, HC van der Mei, L Tomeczek, AH Khoury, HJ Busscher. Comparison of contact angles and adhesion to hexadecane of urogenital, dairy, and poultry lactobacilli: Effect of serial culture passages. Appl Environ Microbiol 58:1549, 1992.

66. MN Bellon-Fontaine, J Rault, CJ van Oss. Microbial adhesion to solvents: A novel method to determine the electron-donor/electron acceptor or Lewis acid-base properties of microbial cells. Colloids Surf B Biointerfaces 7:47, 1996.

67. PS Handley, LM Hesketh, RA Moumema. Charged and hydrophobic groups are localised in the short and long tuft fibrils on *Streptococcus sanguis* strains. Biofouling 4:105, 1991.

68. T Saito, T Takatsuke, T Kato, K Okuda. Adherence of oral streptococci to an immobilized antimicrobial agent. Arch Oral Biol 42:539, 1997.

69. F Caccavo Jr, PC Schramberger, K Keiding, PH Nielsen. Role of hydrophobicity in the adhesion of the dissimilatory Fe(III)-reducing bacterium *Shewanella alga* to amorphous Fe(III) oxide. Appl Environ Microbiol 63:3837, 1987.

70. HHM Rijnaarts, W Norde, J Lyklema, AJB Zehnder. The iso-electric point of bacteria as an indicator for the presence of cell surface polymers that inhibit adhesion. Colloids Surf B Biointerfaces 4:191, 1995.

71. IH Pratt-Terpstra, AH Weerkamp, HJ Busscher. Microbial factors in a thermodynamic approach of oral streptococcal adhesion to solid substrata. J Colloid Interface Sci 129:568, 1989.

14

Cell-Cell Interactions

A. Pierres, A. Benoliel, and P. Bongrand
Hôpital de Sainte-Marguerite, Marseille, France

I. INTRODUCTION

A. Importance of Cell-Cell Interactions

Most important biological and pathological phenomena are heavily dependent on interactions between cells and other cells or tissues. Indeed, proliferation and differentiation, arguably the most essential properties of living cells, are often regulated by adhesive interactions. Thus, most cells need to adhere to a surface in order to divide (Folkman and Moscona, 1978) and this requirement, known as "anchorage dependence," may be lost during tumoral transformation. Conversely, intercellular contact may inhibit proliferation (Caveda et al., 1996) and this "contact inhibition" may also be lost in tumoral cells. Further, gene expression was found to be modulated by adhesive interactions (Cunningham, 1995).

Cell migrations play an important role in development. Thus, brain properties are largely determined by the extensive connections formed between neurons. These connections are due to the formation of so-called growth cones that emerge from neural cells and travel over long distances to find their specific targets; these displacements are guided by adhesive ligand-receptor interactions as well as chemotactic stimuli (Tessier-Lavigne and Goodman, 1996). The function of leukocytes is to patrol throughout the organism in order to detect and subsequently destroy abnormal cells or potentially harmful invaders such as bacteria or parasites. The migration of blood leukocytes toward peripheral tissues is exquisitely orchestrated by intercellular adhesive interactions involving a variety of specialized molecules expressed by these leukocytes and endothelial cells (Springer, 1995).

Pathological processes are also heavily dependent on cell-cell interactions; indeed, the first step of infection is usually the adhesion of an invading pathogen to a specific tissue (Klotz, 1992). Metastasis formation in tumor-bearing patients results from the capacity of transformed cells to detach from the primitive tumor and bind to distant tissues (Pauli et al., 1990). Thrombus formation, a major component of cardiovascular diseases, is strongly dependent on platelet activation and adhesion to other platelets or endothelial cells.

Thus, it is clearly of paramount theoretical and practical importance to understand the mechanisms of cell-cell interactions. The aim of this review is to provide the information required to apply the physical chemical data described in other chapters of this book to actual biological systems. For this purpose, we shall first recall some experimental data on biological adhesion. Then, we shall give a quantitative description of the various steps of cell-cell adhesion and deadhesion with an emphasis on the processes that seem amenable to physical-chemical study.

B. Cell-Cell Interactions are Essentially Mediated by Dedicated Adhesion Molecules

In view of the remarkable success met by the DLVO theory in accounting for the properties of colloidal suspensions, it was tempting to apply this theoretical framework to cell-cell interactions (Curtis, 1967). Indeed, some experimental data suggested that nonspecific cell surface properties might determine adhesive behavior. The uptake of different bacterial species by blood phagocytes was reported to be highly correlated with the hydrophobicity of these microorganisms, as determined with contact angle measurements (van Oss et al., 1975). Also, the adhesion efficiency between aldehyde-treated erythrocytes and macrophages was dramatically increased when negative charges borne by red cells were removed or masked (Capo et al., 1981), which might give some support to the concept that the net negative charge found on the surface of essentially all cell species (Sherbet, 1978) might be a means of preventing undesirable adhesion.

However, as previously emphasized (Bongrand, 1998), during the past 20 years, cell biologists provided overwhelming evidence to support the view that in most cases, cell-cell adhesion is dominated by specific interactions formed between dedicated adhesion molecules expressed by interacting membranes. The most compelling evidence was obtained (a) by inhibiting adhesion with antibodies specific for particular molecules of the cell surface and (b) by provoking adhesion with transfection-induced expression of adhesion receptors. Following are representative examples.

Lo et al. (1989) studied the adhesion of stimulated polymorphonuclear leukocytes to monolayers of endothelial cells. Adhesion was decreased by 50% when leukocytes were pretreated with monoclonal antibodies directed against either LFA-1 or MAC-1, two well-known species of adhesion molecules (see later). Further, binding was abolished when both antibodies were added at the same time. The authors concluded that LFA-1 and MAC-1 accounted for all of the binding behavior of leukocytes. Conversely, Norment et al. (1988) transfected Chinese hamster ovary (CHO) cells to make them express high levels of CD8, a molecular species found on a lymphocyte subpopulation. They demonstrated that transfected cells, not controls, adhered to different cell lines expressing class I major histocompatibility complex molecules (MHC class I). Adhesion did not occur when MHC class I molecules were not expressed by the lines used to interact with transfected CHO cells. This set of experiments provided clear-cut support for the hypothesis that cell-cell adhesion might be mediated by specific interactions between CD8 and MHC class I molecules. Similarly, Patel et al. (1995) allowed CHO cells to bind polymorphonuclear leukocytes by transfecting the former population with the P-selectin gene (see later for a definition of this class of adhesion molecules). The main feature of *specific* cell-cell interactions is that they may involve only a limited fraction of the cell surface area. Thus, T lymphocytes may recognize abnormal cells displaying less than 100 foreign peptides presented by histocompatibility molecules, which may occupy less than 1/100,000 of the total membrane area (Harding and Unanue, 1990). This explains why cell-cell interactions may not in general be related to bulk membrane properties. On the contrary, it may be more rewarding to study the molecular properties of adhesion molecules and then predict the outcome of cell interactions from the pattern of expressed adhesion molecules. It may thus be useful to review some basic information on the main properties of biological adhesion molecules.

C.   Main Classes of Cell Adhesion Molecules

Several tens or even hundreds of adhesion receptors have been characterized with powerful biological tools such as monoclonal antibodies or genetic engineering techniques. Interestingly, it appeared that most of these molecules belong to a small number of protein families, and most of the adhesive phenomena observed with various cellular species seemed to be due to members of these families. Therefore, it seems warranted to give a brief description of these molecules for two reasons. First, this information may be useful for any physicist willing to accede to the huge biological litterature relevant to cell adhesion. Second, only an accurate knowledge of some properties of adhesion molecules may provide a feeling for the orders of magnitude of physical-chemical parameters likely to play a role in cell interactions. We shall consider five main classes of cell receptors.

1.   The Immunoglobulin Superfamily

The immunoglobulin superfamily has been described in a review (Williams and Barclay, 1988). When the structure of antibody molecules was elucidated, it appeared that these were made of polypeptide chains that might be divided into characteristic domains of about 110 amino acids with both sequence homologies and a characteristic spatial organization that was later called the "immunoglobulin fold," with two β-sheets and a characteristic intrachain disulfide bond. These domains behaved as compact, independent units with a characteristic size of about $2.5 \times 2.5 \times 4$ nm. It was later demonstrated that many molecular species, in addition to antibodies, were endowed with one or several immunoglobulin-like domains; this property made them members of the immunoglobulin superfamily. There are nearly 100 known members in this family, with a number of domains ranging between 1 and more than 10, yielding a molecular length between about 4 and more than 40 nm. Some rotation may be possible between adjacent domains. Thus, when ICAM-1 (intercellular adhesion molecule 1, CD54) a ubiquitous adhesion receptor made of five immunoglobulin domains, was studied with electron microscopy by Staunton et al. (1990), it appeared as a rod about 19 nm long that seemed straight or with a single bend.

Many members of the immunoglobulin superfamily are receptors for molecules of the same group. An important example is the recognition of antigen-presenting cells by T lymphocytes. Thus, a virus-infected cell often bears fragments of viral proteins on its histocompatibility molecules, which are members of the immunoglobulin superfamily. These complexes may be recognized and specifically bound by T lymphocyte receptors that are heterodimers made of two chains consisting of two immunoglobulin domains each. As a rule, this interaction is strengthened by so-called accessory molecules such as T lymphocyte CD8, which binds to histocompatibility molecules, or T lymphocyte CD2, which binds to the ubiquitous molecule LFA-3 (lymphocyte function-associated molecule 3, CD58). All these molecules, and many others, belong to the immunoglobulin superfamily. Interestingly, the length of ligand+receptor couples is about 16 nm, allowing all these adhesive bonds to gather into the same membrane areas. The importance of molecular length was rightly emphasized by Springer (1990).

In some cases, adhesion is homotypic; this means that the molecule is its own ligand. Important examples are NCAM (neural cell adhesion molecule) which plays an important role in neural development. There exist several isoforms of these molecules, with different amounts of polysialic acid (sialic acid, or *N*-acetylneuraminic acid, is a small negatively charged carbohydrate). A developmentally regulated increase in NCAM-mediated adhesion seems to be correlated with removal of polysialic acid (Rutishauser et al., 1988). Another example is provided by phagocytes. An important task of these cells is to bind and engulf antibody-coated particles such as bacteria. This interaction may involve immunoglobulin receptors that belong to the immunoglobulin superfamily. Recent evidence suggested that the efficiency of these receptors may be dramatically enhanced if the sialic acid content of leukosialin (CD43, a mucin-like molecule to be described later) is decreased following cell activation (Soler et al., 1997; Sabri et al., in preparation).

An important point is about *regulation*. In most cases, all membrane-bound adhesion molecules of the immunoglobulin superfamily seem endowed with complete functional activity, and adhesion may be controlled by up- or down-regulation of the density of these molecules. Thus, resting endothelial cells are not expected to express VCAM-1 (vascular cell adhesion molecule 1, CD106). When endothelium is activated by a suitable inflammatory stimulus, VCAM-1 appears within about 24 h, allowing some interaction with its integrin ligand, which is borne by some lymphocyte subpopulations (Harlan and Liu, 1992). However, recent experimental data suggest that a general means of regulating adhesion mediated by these molecules consists of removing bulky negatively charged antiadhesive structures (see Soler et al., 1997, for more information). Also, the binding affinity of a particular immunoglobulin receptor on monocytes, which is a member of the immunoglobulin superfamily, was reported to be regulated by a proteolytic event (van de Winkel et al., 1989).

Finally, it must be emphasized that membrane molecules belonging to the immunoglobulin superfamily are often responsible for the triggering of important signaling phenomena, as exemplified in recent reviews (Alberola-Ila et al., 1997; Daeron, 1997).

Many important counterreceptors of members of the immunoglobulin superfamily are integrins. These are described in the next section.

2. Integrins

Integrins are ubiquitous adhesion molecules of prominent importance (Hynes, 1992). These molecules are heterodimers made of two subunits that are called, as usual, α and β. There are more than 20 known integrins. For convenience, these are often grouped into subfamilies sharing a common β chain. Thus, β_1 integrins (also called VLAs, or very late antigens) are made of a common β_1 chain (also called CD49) and an α chain that may be α_1, α_2, ... (or CD29a, CD29b, ...). The ligands of β_1 integrins are often components of extracellular matrices such as fibronectin, laminin, or collagen. However, CD29dCD49 (VLA-4) can interact with both fibronectin and VCAM-1, an immunoglobulin superfamily member mentioned earlier. Whereas β_1 integrins are ubiquitous, β_2 integrins are restricted to leukocytes. The main members of this family are LFA-1 (also called $\alpha_L\beta_2$, CD11a/CD18, or lymphocyte function-associated molecule 1), a receptor for the ubiquitous immunoglobulin superfamily member ICAM-1, and MAC-1 ($\alpha_M\beta_2$, CD11b/CD18, macrophage-1), a receptor of ICAM-1 and C3bi, a degraded form of the C3 component of complement that may bind to bacteria or antibody-coated particles and enhance their phagocytosis.

Integrins are fairly large molecules, with a length of about 20 nm. Their chains are endowed with transmembrane and intracytoplasmic regions, and they can interact with cytoskeletal elements in a highly regulated way. *Regulation* is indeed an essential property of integrins. During the past 10 years, it appeared that integrin molecules expressed on the surface of resting cells were often unable to bind to their ligand. Functional activity may be obtained by combination of two classes of phenomena (Mould, 1996; Sánchez-Mateos et al., 1996):

(1) Integrin molecules were found to display conformational changes: indeed, an increase of binding activity may be correlated with the acquisition of new antigenic sites (Keizer et al., 1988, van Kooyk et al., 1991), and antibodies directed at suitable sites may activate integrins. Also, experiments performed in vitro with soluble recombinant ICAM-1 molecules and LFA-1–expressing T lymphocytes suggested that cell activation might induce the appearance of a subpopulation of LFA-1 molecules with 200-fold higher affinity (Lollo et al., 1993). The molecular mechanisms of integrin activation remain, however, incompletely understood, and some evidence suggests that activation may in some cases involve the release of a kinetic barrier rather than affinity increase (Cai and Wright, 1995). In any case, there is evidence suggesting that integrin molecules may exhibit more than two conformational states with varying affinity and/or binding efficiency, as shown by studying the interaction of soluble VCAM-1–immunoglobulin fusion protein with cells expressing VLA4 integrin (Jakubowski et al., 1995).

(2) However, the capacity of integrins to mediate intercellular adhesion may be increased without detectable affinity change (Danilov and Juliano, 1989; Jakubowski et al., 1995). This emphasizes the importance of so-called *postreceptor* events. Several mechanisms were suggested to account for these phenomena. Detmers et al. (1987) reported that the activation of the MAC-1 integrin on neutrophil granulocytes was related to clustering of these adhesion receptors in small aggregates of a few molecules. More recently, Erlandsen et al. (1993) observed that neutrophil activation resulted in clustering of the Mac-1 integrin with more frequent occurrence of this molecule on the tip of surface projections (called microvilli or ruffles) rather than on the cell

body as found in resting cells. Thus, the adhesion efficiency of integrins might be increased by suitable redistribution on the cell membrane. Another important postreceptor event may be integrin association with intracytoplasmic elements, particularly the actin cytoskeleton (Kupfer and Singer, 1989). This association may increase the mechanical strength of receptor association with the cell, which is important, because cell detachment may often be associated with uprooting of membrane molecules (Regen and Horwitz, 1992). Also, cell activation may allow integrins to trigger cell spreading, i.e., an active increase of cell-cell contact area, which is often required to achieve strong adhesion (Bragina et al., 1976; see Sec. IV). However, integrin binding to immobile structures may reduce lateral diffusion and hamper adhesion in some circumstances (Sanguedolce et al., 1992).

Finally, it must be emphasized that integrin receptors are endowed with prominent signaling capacity (Schwartz et al., 1995).

3. Selectins

Selectins (Bevilacqua and Nelson, 1993) are recently described adhesion molecules that seem restricted to the intravascular compartment because they are essentially expressed by endothelial cells, leukocytes, and platelets. There are three known selectins, denominated L-selectin (CD62L, LECCAM-1; this is expressed on most leukocyte subpopulatons), E-selectin (CD62E; this is the endothelial selectin, also called ELAM-1, for endothelium leukocyte adhesion molecule 1), and P-selectin (CD62 P; this may be expressed by platelets or endothelial cells and was called PADGEM or GMP 140). Each selectin is a type I membrane glycoprotein (i.e., with an extracellular NH_2 terminal residue). The outer region is a lectin domain (related to animal calcium-dependent lectins) bound to a domain with homology to epidermal growth factor and a variable number of repeats related to complement regulatory proteins. The number of repeats is respectively 2, 6, and 9 for L-, E-, and P-selectin, resulting in a total length between about 10 and 40 nm. Although there seems to be some overlap in selectivity among the three selectins, with a common affinity for sialylated-fucosylated sugars, there are some differences between the fine specificities of the molecular species (Varki, 1994). The main ligand of P-selectin is a mucin-like molecule about 40 nm long, found on many leukocyte species (mucins are described in the next section); this is called PSGL-1 (for P-selectin glycoprotein ligand 1). E selectin also binds to PSGL-1 as well as to another molecule called ESL-1 (E selectin ligand 1). Finally, L selectin binds to CD34 mucin that is expressed by endothelial cells in some particular sites of leukocyte emigration from blood to lymph nodes (see Springer, 1995, for more details).

The main role of selectins is to allow adhesion between blood cells and endothelium, when their relative velocity is of the order of several hundreds of micrometers per second. Thus, selectins were found to induce a very particular interaction between leukocytes and activated endothelium (VonAndrian et al., 1991; Lawrence and Springer, 1991). Leukocytes were found to *roll* along endothelial cells with 50- to 100-fold decreased velocity compared with unattached cells. This interaction was ascribed to the rapid formation and dissociation of bonds between selectins and their ligands. The remarkable capacity of selectins to mediate this rolling interaction was ascribed to a combination of specific molecular properties, including the following:

1. An especially high rate of bond formation and dissociation, whereas the affinity constant is not unusual. Indeed, the lifetime of selectin-ligand bonds is of the order of 1 s (Kaplanski et al., 1993; Alon et al., 1995).
2. High molecular length to rise above antiadhesive glycocalyx elements.
3. High tensile strength.
4. Lack of bond strengthening.

Note that the rolling phenomenon does not seem to require any active cell participation, because it was not abolished when neutrophils were fixed with aldehydes (Lawrence and Springer, 1993), and it was observed in a cell-free system made of sialyl Lewis X-coated microspheres moving along substrates coated with E-selectin–immunoglobulin G (IgG) chimeras (Brunk et al., 1996). An intriguing report by Finger et al., (1996) suggested that the interaction between leukocyte L-selectin and endothelial ligand required that the hydrodynamic flow be higher than a threshold value.

The *regulation* of selectin function is essentially related to membrane expression. Thus, resting endothelial cells are not expected to express any selectin on their membrane. However, P-selectin is stored in Weibel-Palade granules and may appear on cell membranes within a few minutes after stimulation with factors such as histamine. Other agents such as cytokines induce E-selectin synthesis with a membrane expression peaking within a few hours. Down-regulation may then be obtained by endocytosis of surface selectins: P-selectin may return to storage granules for future use, whereas E-selectin is degraded into lysosomes (Subramaniam et al., 1993). Note that soluble forms of selectins may also be shed into blood, where they may be used as markers of inflammatory reactions. In contrast, L-selectin is constitutively expressed on the membrane of many leukocyte subpopulations and may be shed within a few seconds by a protease. This might play a role in regulating the rolling phenomenon (Walcheck et al., 1996). Despite a previous report by Spertini et al. (1991), there is little support for the importance of functional selectin regulation in addition to the modulation of expression. However, it was reported by VonAndrian et al. (1995) that the ability of L-selectin to mediate leukocyte adhesion under flow was related to the presence of these molecules on microvilli (see later for more information on these protrusions of the cell membrane), and the authors suggested that control of receptor topography might provide a means for cells to regulate adhesive behavior.

A final point about selectins is their capacity to generate intracellular signals after interacting with their ligands. Cross-linking E- or P-selectins on the surface of activated endothelial cells resulted in cell rounding, as demonstrated by scanning electron microscopy and confocal fluorescence microscopy (Kaplanski et al., 1994), and cross-linking E-selectins resulted in a transient increase of the cytosolic calcium concentration. Stimulation of L-selectin was reported to activate leukocyte integrins (Crockett-Torabi et al., 1995) and generate an intracellular calcium rise (Waddell et al., 1995) as well as tyrosine phosphorylation events (Brenner et al., 1996).

As already mentioned, selectin ligands are often mucins. This class of molecules will now be briefly reviewed.

4. Mucins

Mucins may be found in extracellular regions or tightly associated with cell membranes. They are very heavily glycosylated proteins with O-linked carbohydrates bound to serine or threonine residues. These carbohydrate chains impose on the protein an extended configuration (with a length of about 0.25 nm per residue). In addition to their mass (often more than 50% of the total molecular weight), they often bear a high negative charge due to the presence of numerous sialic acid residues (Jentoft, 1990; Cyster et al., 1991; Lasky et al., 1992).

Mucins may have an antiadhesive function, as discussed later. A prominent example is CD43 (leukosialin), a major component of the lymphocyte surface. As already indicated, they may also act as selectin ligands. Indeed, PSGL-1 (P-selectin glycoprotein ligand 1) is a long mucin-like molecule of about 40 nm (Sako et al., 1993; Li et al., 1996). Thus, there is some ambiguity in the function of mucins, as they may be pro- or antiadhesive.

Few studies have been devoted to the *regulation* of mucin function. It was reported that neutrophil activation resulted in redistribution of PSGL-1 to an uropod (Lorant et al., 1995), supporting the generality of receptor regulation by lateral redistribution on the cell membrane. Another mechanism of control is alteration of carbohydrate chains. Thus, Soler et al. (1997) reported that cytokine activation of cells from the THP-1 monocytic line resulted in dyssialylation of CD43/leukosialin with a concomitant increase of adhesive capacity. Further, the shedding of mucin-like molecules by a proteolytic mechanism was described in activated neutrophils (Rieu et al., 1992).

Finally, some data support the view that mucin might generate *intracellular signals*. Thus, interaction of monocytes with P-selectin stimulated chemokine production, and this was inhibited by an antibody specific for PSGL-1, the mucin-like P-selectin ligand (Weyrich et al., 1995). Also, CD44, a protein with a mucin-like segment, was reported to generate intracellular signals (Taher et al., 1996). However, more data are required to achieve a clear understanding of mucin signaling capacity.

5. Cadherins

Cadherins are homotypic adhesion molecules that play an important role in the cohesion of many tissues (Takeichi, 1991). They are made of domains with a size comparable to that of the characteristic immunoglobulin domain. Further, they are associated with intracellular components (α- and β-catenins, plakoglobin) that react with their intracytoplasmic domain and provide a link with cytoskeletal elements. These components and interactions are required for adhesive function and they play a role in regulating cadherin-mediated adhesion. As compared with the previous classes of adhesion molecules, cadherins may be considered rather slow receptors because several tens of minutes are required to achieve strong adhesion and extensive contact zones (Angres et al., 1996; Adams et al., 1996).

Finally, there is little information available on the signaling capacity of cadherins (Suzuki, 1996). However, interactions between cadherins and molecules with demonstrated signaling capacity such as Shc have been demonstrated (Xu et al., 1997).

We shall now describe the sequential steps of cell adhesion. The data we have just described will help in understanding the peculiar features of the representative examples we use to illustrate general principles. Some typical couples of adhesion receptors are depicted in Fig. 1.

II. THE FIRST STEP OF CELL ADHESION: INITIAL APPROACH

If it is accepted that cell adhesion is essentially mediated by specific cell receptors, it is understandable that the critical step of attachment might be the formation of the first

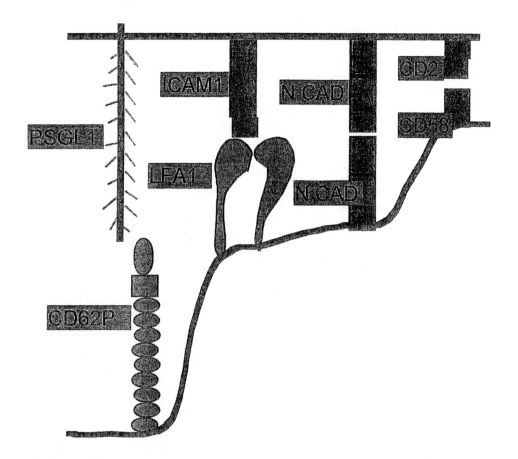

Fig. 1 Some ligand-receptor couples. Some ligand-receptor couples are displayed. The P-selectin (CD62P)–PSGL-1 couple has a total length of about 0.8 μm, allowing efficient tethering of flowing leukocytes to activated endothelial cells. The interaction between LFA-1 integrin and ICAM-1, a member of the immunoglobulin superfamily, or the homotypic interaction between N cadherins (NCad) requires that the distance between phospholipid bilayers be smaller than about 40 nm. Finally, the scanning of antigen-presenting cells by T lymphocytes is facilitated by the interaction between CD2 and CD58 members of the immunoglobulin superfamily: the receptor-ligand length is only about 16 nm.

bond between approaching surfaces. This requires that these surfaces remain at binding distance for a sufficient time. We shall show in the present section that the fulfillment of this requirement may be hampered by the presence on most cells of a repulsive structure surrounding the plasma membrane and called the *glycocalyx, cell coat,* or *pericellular matrix*. In order to allow a quantitative assessment of the importance of repulsion, we shall first give some information about the conditions of cell-cell encounter under physiological conditions. We shall then review some quantitative experimental data in order to demonstrate the reality and importance of repulsion. Finally, we shall discuss the relevance of currently available physical-chemical knowledge to these phenomena.

A. Conditions of Cell-Cell Encounter

The forces described in other chapters of this book may vary over several orders of magnitude. In order to assess their relevance to cell-cell interactions, it is important to have some information about the magnitude of the forces occurring during these interactions. This point was discussed in previous reviews (Bongrand and Bell, 1984; Bongrand, 1995). We shall give some additional information. We may consider two extreme cases of cell adhesion and detachment.

Prolonged contact is expected when adhesion is driven by fairly slow cell movements. Cells usually probe their environment by sending forward protrusions that may be flat lamellipodia or cylindrical filopodia. The forward velocity of these protrusions may vary in different cell types. As reviewed by Condeelis (1993), a convenient order of magnitude for the rate of protrusion growth in cells such as fibroblasts or macrophages is several micrometers per minute. Note, however, that most data were obtained by observations made with conventional optical microscopy, and there is little available information on the rate of deformation of the cell surface at the submicrometer scale.

The size of the protrusions may display wide variations. A convenient order of magnitude for the initial contact area is about $0.1 \times 0.1 \ \mu m^2$, because the thickness of a lamellipodium or the diameter of a filopodium is of order of 0.1 μm.

It is not easy to give a general estimate of the order of magnitude of the forces exerted by cell protrusions. We shall give a few examples. Using a double-chamber device, Inouyé and Takeuchi (1980) found that the *force* required to prevent the progression of *Dictyostelium discoideum* was proportional to the slug *volume* with a value of about 58 pN/μm^3. Usami et al. (1992) determined the pressure required to prevent the progression of a blood leucocyte into a micropipette containing a chemoattractant oligopeptide; the maximum force resisted by cells was about 30,000 pN. A similar value was reported by Vereycken et al. (1995). Also, Oliver et al. (1995) found that a force of about 45,000 pN was required to stop a moving keratocyte.

Other authors studied the traction exerted by cells adhering to a surface. By studying the wrinkles formed by fibroblasts bound to a silicone rubber surface, Harris et al. (1980) were able to estimate as 10,000 pN per μm of cell boundary the traction exerted by the cell advancing margin. More recently, Kolodney and Wysolmerski (1992) estimated as 4500 and 6100 pN/μm^2 the traction exerted on their substrate by adherent fibroblasts and endothelial cells. Note that these figures are not easy to compare, because there is no general agreement concerning the expression of experimental data. It is not clear whether exerted forces are proportional to the cell volume, section area of protrusions, or linear dimensions of these protrusions.

In any case, a cell can exert forces of the order of several tens of thousands of pN. Assuming that the force exerted by a protrusion is proportional to its area (which may be reasonable if this force is proportional to the number of actin filaments), it is concluded that a filopodium 0.1 μm thick may exert at most a force of the order of 100 pN. Interestingly, as discussed later, this is of the order of magnitude of the mechanical resistance of a typical bond between adhesion receptors as well as the force required to uproot a molecule embedded in the cell membrane.

Transient intercellular contacts may occur when adhesion is driven by external forces generating a stronger stress than described above. An important example that has attracted much attention is the adhesion of flowing leukocytes to blood vessel walls as a preliminary step in migration toward peripheral tissues during inflammatory reactions. The wall shear rate may be several hundreds of seconds^{-1} or higher (Atherton and Borne, 1972). The velocity of cells passing within binding distance of the vessel walls is of the order of several hundreds of μm/s, and the time of contact between endothelium and the tip of leukocyte microvilli may not be higher than 1 ms with an approach velocity of order of several mm/s (taking the wall shear rate as a convenient order of magnitude for the cell rotational velocity; see Goldman et al., 1968). The hydrodynamic force acting on a surface-bound cell is about

$$F = 32\mu a^2 G \tag{1}$$

where μ is the medium viscosity (about 0.001 Pa s in aqueous medium), a is the cell radius (about 4 μm for a blood neutrophil), and G is the wall shear rate. The force is thus about 50 pN for a wall shear rate of 100 s^{-1}. Note that the force experienced by bonds located in the contact area may be significantly higher due to a lever effect (Alon et al., 1995; Pierres et al., 1995).

Note also that other interactions may occur under intermediate conditions. Thus, lymphocytes scanning living organisms in order to detect foreign substances or antigens may interact with antigen-bearing cells during their passage through lymphoid organs such as lymph nodes (see Gretz et al., 1996, for a discussion of this process). Because the time required to traverse a lymph node 1 cm in diameter is of order of 18 h (Sprent, 1977), the mean lymphocyte velocity must be at least ≈ 1 cm/10 h, i.e., about 0.15 μm/s.

B. The Cell Membrane is Coated with a Carbohydrate-Rich Layer

Following the seminal paper of Singer and Nicolson (1972), the core of the plasma membrane is usually viewed as a fluid phospholipid bilayer about 4.5 nm thick binding to a variety of intrinsic (often membrane-crossing) or peripheral glycoproteins. However, it is now well known that this basic structure is coated with an electron-light layer of widely varying thickness; it is usually some tens of nanometers thick, but values as high as several micrometers were reported in some extreme cases (Lee et al., 1993a). This structure may be visualized with optical microscopy as a pericellular region excluding small particles or cells such as erythrocytes (Clarris and Fraser, 1968; Lee et al., 1993a). Frey and colleagues (1996) concluded that a glycocalyx 20 nm thick would prevent the interaction between membrane glycolipids and particles 1 μm in diameter. When regions of intercellular contact are studied with electron microscopy, the osmiophilic phospholipid bilayers may appear as parallel lines separated by a light zone of several tens of nanometers, depending on the binding mechanism (Easty and Mercer, 1962; Heaysman and Pegrum, 1973; Benoliel et al., 1980; Capo et al., 1982). It can also be stained with carbohydrate-specific procedures such as periodic acid–Shiff treatment (Bennett, 1963) or cationic substances such as lanthanum, terbium, alcian blue, or ruthenium red (e.g., see Martinez-Palomo, 1970; Foa et al., 1994, 1996). Note that the apparent thickness of this coat may depend on the dye used for staining (Foa et al., 1996). Further, treating cells with glycosidases such as hyaluronidase (Clarris and Fraser, 1968) or neuraminidase (Sabri et al., in preparation) reduced the apparent density of the cell coat.

Different molecular species may be found in the pericellular matrix. A detailed description of these components would not fall within the scope of the present chapter, and we shall give only a crude description. Standard membrane glycoproteins may be viewed as compact proteic cores of a few nanometers diameter with several carbo-hydrate chains typically made of about 10 hexasaccharides, often terminated with a negatively charged sialic acid. Note that the length of an *extended* polyhexasaccharide chain is about 0.57 nm per hexose (Kabat, 1968). Such short carbohydrate chains are also bound to membrane-embedded lipid moieties. The NCAM molecule is a member of the immunoglobulin superfamily bearing five Ig domains and various amounts of polysialic acid. Using dynamic light scattering, Yang et al. (1992) reported that the molecule radius was decreased from 17.3 to 12.2 nm after desialylation. Mucin-like molecules may be substantially larger: the antiadhesive CD43/leukosialin may be viewed as a rod of about 45 nm length (Cyster et al., 1991). The mucin-like domain of episialin may be between 200 and 500 nm long (Hilkens et al., 1992), assuming an average length of 0.25 nm per amino acid in a mucin-like segment (Jentoft, 1990). Glycosaminoglycans are negatively charged polysaccharides made of repeating disac-charide units that can be bound to the cell surface (Höök et al., 1984) or be part of extracellular matrices. Their length may display considerable variations; thus, hyaluronic acid is a linear polysaccharide that may contain several thousands of units of glucuronic acid and N-acetylglucosamine. Hyaluronic acid might be synthesized on the cell surface (Höök et al., 1984). Other glycosaminoglycans such as heparan sulfate or chondroitin sulfate are also made of disaccharide repeating units (made of a uronic acid and a hexosamine). They bear numerous negative charges (namely the carboxyl groups of uronic acids or sulfate groups) as well as positive amino groups. Proteoglycans are made of a proteic core bearing several glycosaminoglycan chains. As an example, Oldberg et al. (1979) reported the presence on rat hepatocyte surface of a heparan sulfate proteoglycan made of a 75,000-dalton protein core bearing four or five polysaccharide chains of about 14,000 daltons (i.e., of the order of 70 saccharide residues).

An essential point is to know how these molecules behave in solution. Must glycosaminoglycans be viewed as flexible coils, or does the presence of electric charges impart substantial rigidity, making these molecules resemble rigid rods? Much valu-able information was summarized by Phelps (1983). Studying the viscosity of polymer solutions, it is possible to find an experimental relationship between the intrinsic vis-cosity $[\eta]$ (i.e., the limit for low concentration of the ratio between the relative viscosity increase due to the polymer and the polymer concentration) and the molecular weight M. It is usually found that $[\eta]$ is proportional to some power α of M (this is the Mark-Houwink equation). The parameter α is expected to be lower than 0.8 for a flexible polymer, and higher values are found when molecules are rigid. Experimental data suggest that glycosaminoglycans display fairly intermediate behavior.

The application of physical knowledge to a quantitative evaluation of intercellular forces requires *quantitative* modeling of the cell surface. Such estimates were previously reported for erythrocytes (Levine et al., 1983) or nucleated cells (Bongrand et al., 1982; Bongrand and Bell, 1984; Bongrand, 1995). We suggest the following procedure:

Although the precise chemical composition of cell plasma membranes may dis-play wide variations, it is useful to look for an average order of magnitude.

First, the two main components of cell membranes are *proteins* and *lipids*. We shall now try to estimate the average mass of these molecular species per unit of membrane area. First, Law and Snyder (1972) estimated that isolated membranes are composed of about 40% lipids and 60% proteins. Following Gennis (1989), the percentage of proteins may vary from 20% (myelin) to 80% (in mitochondria). Thus, we may consider that *a typical cell plasma membrane contains comparable amounts of lipids and proteins.* We must determine the fraction of membrane area contributed by these molecules (see Fig. 2).

The *mass per nm^2 of lipid bilayer* may be calculated as follows: Bilayers are essentially made of cholesterol and lipids with a cholesterol/phospholipid ratio of the order of 0.72. Further, the area occupied per unit of (1 phospholipid + 0.72 cholesterol molecule) is 0.765 nm^2 (Levine, 1972). Because, the corresponding molecular weight is about 1000, the mass per nm^2 of phospholipid *bilayer* is 4.34 × 10^{-21} g.

The *mass per nm^2 of protein* may now be calculated as follows. As a very crude estimate, we consider membrane-associated proteins as composed of domains of molecular weight 20,00 and density 1.31 g/cm^3, yielding a radius of 1.8 nm (Bongrand et al., 1982). Further, we assume that about a quarter of these domains contribute the membrane area, as most domains are in the extracellular space. The protein mass per nm^2 is thus 12.8 × 10^{-21} g. This would imply that about 25% of the membrane area is occupied by proteins, with a total amount of proteins and lipids equal to 6.5 × 10^{-21} g/nm^2. This estimate is consistent with an interesting report of Ryan et al. (1988), who took advantage of a quantitative analysis of the surface redistribution of fluorescein-labeled receptors of cells subjected to an electric field to estimate as about 45% the effective fraction of membrane area occupied by proteins. Also, it compares well with a report from Golan et al. (1984), who estimated the fraction of total area occupied by transmembrane proteins on erythrocytes to be 0.17 by estimating the size and number of band 3 molecules, considered as the major protein species. Saxton (1990) obtained a slightly lower estimate of about 0.10 by considering the size and density of intramembrane particles in reported electron microscopic images of freeze-fracture samples. Note that it would not be warranted to look for a more precise estimate, as only part of membrane-associated proteins cross the phospholipid bilayer, and the radius of the transmembrane part may be different from the size of extra- and intracellular domains.

Second, *carbohydrates* represent up to 10% of the membrane dry weight (Harrison and Lunt, 1975; Cook, 1976; Levine et al., 1983; Gennis, 1989). These carbohydrates may compose the short oligosaccharide chains linked to integral membrane proteins and lipids. Indeed, the carbohydrate content of membrane proteins is of the order of 10% (Sharon, 1981), and these carbohydrates form fairly short chains (less than about 20 residues, or about 12 nm). Carbohydrates may also form the longer chains found in glycosaminoglycans and proteoglycans. These structures are of particular interest because they probably compose the outer region of the cell coat, thus accounting for the major part of repulsion between approaching membranes. It is thus of interest to estimate the amount of cell surface glycosaminoglycans. As already mentioned, these molecules are made of disaccharide units including one uronic acid each. Taniguchi et al. (1974) estimated at about 36 μg the uronic acid content of 100 mL of total blood leukocytes. Approximating at 6 × 10^8 cells/per 100 mL the blood leukocyte concentration, 400 μm^2 the average leukocyte membrane area, and 200 the molecular weight of uronic acids, we may estimate at 2 nm^2 the mean surface area per uronic acid molecule (as 100 mL of blood contains 2.4 × 10^{17} nm^2 leukocyte plasma membrane and 1.1 × 10^{17} uronic acid molecule). Thus, the surface amount of glycosaminoglycans may be about *one monosaccharide per nm^2*. This estimate may be compared with the following one: estimating at 45%, 45% and 10% the relative amount per weight of proteins, lipids, and carbohydrates in a typical plasma membrane, we may calculate that to an area of 1 nm^2 of plasma membrane (i.e. 6.5×10^{-21} g of lipid + protein) correspond about 0.72×10^{-21} g carbohydrate, i.e. about 2.2 monosaccharide units of molecular weight 200. This would suggest that glycosaminoglycans might contribute a significant part to cell surface carbohydrates. Thus, all aforementioned estimates are consistent with the view that there is of the *order* of two monosaccharide units per square nanometer of plasma membrane. The volume concentration of carbohydrates in the glycocalyx (with a thickness ranging between 10 and 50 nm) would thus be of the order of 0.04–0.2 monosaccharide unit per nm^3, i.e. between 60 and 300 mM.

We shall now consider the *surface charge* of mammalian cells. Despont et al. (1975) estimated at 1.9 nmole/per 10^8 cells the sialic acid content of thymocytes. About half this amount could be released by neuraminidase treatment. Estimating at 400 μ m^2 the average thymocyte membrane area, the surface amount of sialic acid residues is thus about 0.03/nm^2. Although most negative charges of human erythrocytes may be ascribed to sialic acid, other molecular species (including negative sulfate groups and uronic acids or positive amino groups) contribute the total charge of most cells. It would be tempting to derive the total cell surface charge from electrophoretic mobility measurements; however, this is not an entirely reliable approach. Indeed, the interpretation of experimental data would require precise assumptions concerning the glycocalyx structure. This must be modeled as a permeable polyelectrolyte layer, and the actual surface charge might be two- to threefold higher that estimated by considering all membrane-associated charges as concentrated on a smooth surface (Levine et al., 1983). However, mobility measurements (Eylar et al., 1962; Sherbet, 1978; Lackie, 1980) and biochemical studies (Eylar et al., 1962) are consistent with the view that the average density of bound charges on cell membranes is *of the order* of 0.1 negative electronic charge per nm^2. The proportion of total charge contributed by sialic acid may be nearly 100%, as reported for human erythrocytes, or less than 50% on cells bearing many sulfated and carboxylic groups contributed by glycosaminoglycans.

Another important quantitative property of the pericellular coat is its *viscosity*. Lee et al. (1993b) estimated the viscosity of the glycocalyx of keratocytes by studying the diffusion of fluorescein-derivatized phosphatidylethanolamine molecules bound by antifluorescein antibodies tagged with colloidal gold particles: the viscosity ranged between 0.05 and 0.09 Pas (as compared with a water viscosity of 0.001 Pas), and this dropped to 0.01 Pas when cells were treated with heparinase; this treatment was expected to remove heparan sulfate proteoglycan, an important proteoglycan family found on cell surfaces.

Now, we shall review selected experimental data suggesting that the pericellular coat may hamper cell adhesion and that this phenomenon is especially significant when cell-cell encounters occur under dynamic conditions. The data we discussed are summarized in Fig. 2.

C. Impairment of Cell Adhesion by Components of the Pericellular Coat

1. Inverse Relationship Between Cell Surface Charge and Adhesivity

It has long been found that cell agglutinability is inversely related to sialic acid content. Thus, platelet ADP-induced aggregation was decreased when sialic acid content was

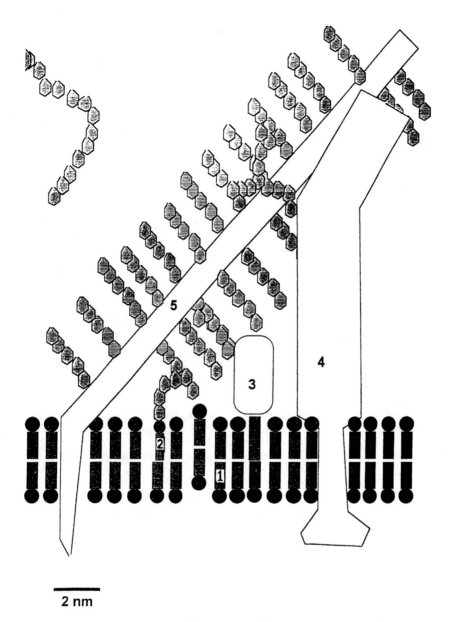

2 nm

Fig. 2 The cell surface. A general view of the cell surface is shown. The basic structure is made of a phospholipid bilayer (1) where proteins are embedded (4). The area contributed by proteins was very crudely estimated by modeling these molecules as spheres of 4 nm diameter, half of them being embedded in the bilayer and contributing the total membrane area and the other half in the pericellular zone. The thickness of the phospholipid bilayer is about 45 nm. Proteins may display a wide size variety: Thy-1 is made of one immunoglobulin-like domain and may resemble (3). Protein (4) may represent ICAM-1, which appears as a bent rod made of five immunoglobulin-like domains. CD43 (5) has an elongated peptide backbone bearing many tens of oligosaccharide chains; a hexagon may represent a hexose. Finally, huge glycosaminoglycans may be linked to the membrane through proteic cores that may be inserted in the bilayer or bound to intrinsic proteins.

increased by a factor of about 2 (Mester et al., 1972). On the contrary, removing surface sialic acid groups on erythrocytes by neuraminidase treatment enhanced dextran-mediated agglutination (Jan and Chien, 1973), and treating platelets with neuraminidase made them bind to macrophages by a process that was inhibited when high amounts of D-galactose were added in the medium (Kluge et al., 1992). Finally, Capo et al. (1981) reported that the uptake by macrophages of erythrocytes made hydrophobic by aldehyde treatment was increased sixfold when the cell surface charge was reduced by removing sialic acid or coating with positively charged polylysine molecules.

Many electron microscopic studies confirmed the antiadhesive potential of cell surface charges. Sugimoto (1981) used transverse sectioning to study the interaction between murine fibroblasts and a surface coated with negatively charged polymerized albumin; only 2% of the cell contour in the bottom side seemed to be involved in adhesion with a cell-to-substrate gap less than 100 nm thick. This proportion increased to 40% when the negative charge of the substrate was neutralized with polylysine. Mège and colleagues (1987) used computer-assisted analysis of digitized electron micrographs to study the interaction area between macrophages and glutaraldehyde-treated erythrocytes. The percentage of the contour length where erythrocyte and macrophage surfaces were separated by a gap less than 50 nm thick increased from 31 to 80% when the negative surface charge of erythrocytes was neutralized with polylysine. In another analysis performed on the same set of images (Foa et al., 1996), the mean width of the gap separating the erythrocyte and macrophage surfaces was 21.5 nm ± 0.5 (standard error), 18.1 ± 0.65 nm, and 11.7 ± 0.25 nm with control, neuraminidase-treated, or polylysine-coated erythrocytes, respectively. Finally, Rutishauser et al. (1988) reported that the fraction of the contour of aggregated neural cells appearing involved in adhesion increased from 3.5 to 7.6% when cells were treated with an endoglycosidase to remove polysialic moieties bound to NCAM adhesion molecules.

2. Inverse Relationship Between Cell Adhesivity and Amount of Surface-Bound Mucin-Like Molecules

Many authors have taken advantage of molecular biology techniques to modulate the amount of cell surface mucins. Ardman and colleagues (1992) studied the binding of HeLa cells to T lymphocytes. This interaction was mediated by the interaction between HeLa cell ICAM-1 and lymphocyte LFA-1, a ligand-receptor couple about 40 nm long. Adhesion was impaired when HeLa cells were made to express CD43/leukosialin by transfection. Similarly, transfecting mammary epithelial cells with episialin inhibited homologous aggregation (Ligtenberg et al., 1992). Conversely, the adhesiveness of a population of osteosarcoma cells was increased when the expression of a surface proteoglycan was repressed with an antisense, strategy (Yamagata and Kimata, 1994). When CD43 expression was inhibited by homologous recombination, homotypic adhesion was increased together with adhesion to molecules such as fibronectin, ICAM-1, and VCAM-1 (Manjunath et al., 1995).

3. The Antiadhesive Potential of the Cell Surface Glycocalyx Is Dependent on the Relative Length of Adhesive and Repulsive Molecules

The following examples are intended to emphasize the importance of the relative length of adhesive and antiadhesive molecules. First, when macrophages were made to bind glutaraldehye-treated or immunoglobulin-coated erythrocytes under static conditions (i.e., by coincubating cells in a tube or a petri dish), adhesion was increased in the former case when erythrocytes were treated with neuraminidase (Capo et al., 1981; Foa et al., 1996). Second, in an elegant study, Chan and Springer (1992) investigated the interaction between T lymphocytes (expressing CD2, a molecule about 8 nm long) and CHO cells transfected with wild-type LFA-3 (a ligand of CD2 about 8 nm long) or engineered LFA-3 molecules with a length increased by about 16 nm by incorporation of four immunoglobulin-like domains from ICAM-1. The length of the ligand-receptor couple was thus increased from about 16 to 31 nm. The adhesion efficiency of engineered LFA-3 was higher than that of wild-type molecules by a factor ranging between 4 and 20. Third, Patel et al. (1995) studied the adhesion of flowing neutrophils to transfected cells expressing P-selectin; the adhesion efficiency was drastically reduced when the lenth of P-selectin was decreased by genetic engineering, and binding was restored when cell surface glycosylation was impaired by a metabolic inhibitor such as tunicamycin. Fourth, when E-cadherin–mediated homotypic cell aggregation was studied, adhesion was more efficiently decreased when cells expressed wild-type rather than shortened episialin molecules (Wesseling et al., 1996).

4. The Glycocalyx Antiadhesive Potency Is More Important When Cell-Cell Contact Is Transient

It was reported that the efficiency of glycocalyx-mediated cell-cell repulsion was more apparent when cell-cell contacts occurred under dynamic conditions. Thus, Sabri et al. (1995) studied the interaction between cells from the human THP-1 monocytic line and spheres coated with monoclonal antibodies specific for various cell surface antigens; particles were efficiently bound when they were made to sediment for several minutes on THP-1 cell monolayers. However, when they were driven along similar monolayers in a laminar flow chamber with a wall shear rate of 10–20 s^{-1}, only minimal adhesion was observed, and adhesive efficiency displayed about a 10-fold increase when cells were treated with neuraminidase. Also, when Patel et al. (1995) studied the interaction between human neutrophils and transfected cell monolayers expressing wild-type or shortened P-selectin molecules, the latter molecules displayed highly reduced binding efficiency under dynamic, not static, conditions. Finally, when Foa et al. (1996) studied the interaction between macrophages and immunoglobulin-coated erythrocytes, they found that adhesion was quite efficient under static contitions. When cell-cell encounters occurred in a laminar flow chamber operated at a low wall shear rate of 10–20 s^{-1}, adhesion was quite inefficient, and it was dramatically increased when erythrocytes were treated with neuraminidase. In other experiments, McFarland et al. (1995) studied the killing of different target cells by cytolytic T lymphocytes, a process that is highly dependent on lymphocyte-to-target contact interaction and adhesion (Bongrand et al., 1983). Target cells expressing CD43 were less sensitive to lysis than others, and desialylation increased this sensitivity. However, these phenomena became less apparent when the duration of lymphocyte-to-target contract was increased.

All these results suggest that cell-cell adhesion may require a dynamic reorganization of membrane repulsive layers. A first hypothesis to account for these findings would be that bond formation might require some compression of large repeller molecules, and the repulsive force is expected to be dependent on the rate of approach (Dimitrov, 1983; Foa et al., 1996). Alternatively, repellers might need exit from contact regions through random diffusion, which is clearly a time-dependent process. Indeed, Soler et al. (1997) reported that the region of contact between macrophages and immunoglobulin-coated erythrocytes was partially deprived of

CD43 molecules. Yamagata and Kimata (1994) noticed that PG-M/versican proteoglycan was excluded from cell-substrate contact areas. Other experiments showed that cell adhesiveness could be increased by provoking a lateral redistribution of large repeller molecules. Thus, Kemperman et al. (1994) reported that KTA3/HA murine carcinoma cells did not adhere to laminin-coated substrates, although they expressed the laminin receptor integrin $\alpha_6\beta_4$. However, adhesion was obtained when they treated cells with antibodies specific for the mucin-like molecule epiglycanin expressed by carcinoma cells, which induced a gathering of this molecule on a localized area of the cell surface (a process called capping). Also, treating leukocytes with anti-CD43 antibodies induced a surface redistribution of this mucin and simultaneously increased the cell capacity to bind fibronectin, VCAM-1, or ICAM-1 molecules (SanhcezMateos et al., 1995). Similarly, treating THP-1 cells with anti-CD43 increased their capacity to bind immunoglobulin-coated erythrocytes (Soler et al., 1997).

5. Modulating Glycocalyx Organization Is a Physiological Means of Regulation Cell Adhesiveness

Several reports convincingly demonstrated that cells can regulate their adhesiveness through an active modification of the glycocalyx. This provides an alternative to the regulation of adhesion by modulating the activity of a precise receptor species.

As already mentioned, NCAM is a homotypic adhesion molecule found on neural cells. The amount of polysialic acid borne by NCAM molecules is developmentally regulated, and a decrease of these negatively charged chains is thought to increase binding efficiency (Rutishauser et al., 1988). More rapid regulatory mechanisms are displayed by leukocytes. Activated neutrophils were found to release 80% of surface CD43 through proteolytic cleavage (Rieu et al., 1992; Remold-O'Donnell and Parent 1994) and this release seemed to be associated with increased adhesion and spreading to foreign surfaces (Nathan et al., 1993). Alternatively, adhesion might be increased through decreased sialylation. The binding of B lymphocytes to T-lymphocyte clones was increased when the former cells were activated with lipopolysaccharide (LPS), an important component of bacterial cell walls. This increase was associated with sialic acid release and was prevented by sialidase inhibition (Guthridge et al., 1994). Also, activating monocytic THP-1 cells with cytokines, which are physiological mediators of phagocyte activation, resulted in dyssialylation of CD43 molecules together with an increased capacity to bind immunoglobulin-coated erythrocytes (Soler et al., 1997).

6. Summary

Much experimental evidence has shown that

1. The cell surface is coated with a carbohydrate-rich electron-light area called the glycocalyx, cell coat, or pericellular coat, with an apparent thickness of the order of 20–50 nm.
2. This cell coat may prevent cell-to-cell approach, because adhesion is increased when the cell coat is reduced by enzymatic or metabolic treatment or when the length of adhesion molecules is increased. Alternatively, adhesiveness is decreased when the density of the cell coat is increased or the length of adhesion receptors is decreased.

3. The cell coat must be viewed as a dynamic repulsive barrier. Indeed, during prolonged cell contact, substantial glycocalyx reorganization must result in a decrease of repulsion.
4. Finally, modulating the glycocalyx is a physiological way to regulate cell adhesiveness.

D. Relevance of Currently Available Models of Surface Interactions to Cell Adhesion

Clearly, it would be highly desirable to find a suitable framework for deriving the force experienced by approaching cells from structural properties. However, this seems a quite difficult task because problems of two different kinds must be solved: first, there is a need for a workable description of the cell surface. This is difficult to achieve, for it may not be sufficient to retain average properties. Second, there is a need for physical procedures allowing one to deal with these cell models. Probably neither of these two requirements is satisfied at the present time. However, an impressive amount of data has been gathered and we shall emphasize the most promising approaches.

Although cell-cell interactions essentially involve electrodynamic forces, it is not feasible to use basic physical equations to perform ab initio calculation of intercellular forces. Even if this were feasible, it would result in formulas that would be too complicated to be of practical use. It is therefore convenient to split overall interaction into simpler components susceptible to intuitive representation. It may thus be found enlightening to discriminate between electrostatic repulsion and van der Waals attraction. It must be understood that there is no unique procedure for classifying these phenomena, and the choice of "primary" interactions is essentially a matter of convenience. The problem, as emphasized by van Oss (1991), who found 17 varieties of elementary interactions in recent scientific papers (e.g., electrostatic forces, hydrogen bonding, Brownian forces), is that there is a risk of counting the same primary force twice. We shall now give arguments suggesting that the current theoretical framework of polymer science may provide a convenient way to understand long-distance cell interactions. First, we shall discuss the van der Waals and electrostatic forces that form the basis of the DLVO theory, which met remarkable success in colloid science.

1. Electrodynamic Forces

We refer the reader to a previous review (Bongrand et al., 1982) for more information on these forces and we give only a summary. Electrodynamic (or van der Waals) forces stem from correlations between electronic motions in interacting molecules. It was found convenient to discriminate between Debye forces (due to the mutual orientation of permanent dipoles), Keesom forces (corresponding to the interaction between a dipolar molecule and the dipole induced in a polarizable structure), and London or dispersion forces (involving two molecules without any need for permanent polarization). The interaction between two parallel slabs of thickness d_1 and d_2, separated by a gap of width h in vacuum, is always attractive with an energy

$$W_{12} = -[A_{12}/12\pi][1/h^2 - 1/(h + d_1)^2 - 1(h + d_2)^2 + 1/(h + d_1 + d_2)^2] \qquad (2)$$

where A_{12} is the Hamaker constant for materials 1 and 2 interacting in a vacuum. This formula was obtained by Hamaker (1937) by summing the interactions between individual molecules in media 1 and 2. A macroscopic theory of electrodynamic forces was developed by Lifshitz (1956), thus raising uncertainty about the additivity of the interactions between individual molecules. It was possible to achieve a direct experimental check of the theory when Israelachvili and Tabor (1972) developed the surface forces apparatus and measured the van der Waals forces between crossed cylindrical mica surfaces up to 130 nm separation. Note that Eq. (2) is no longer valid when distances are greater than some tens of nanometers, due to retardation effects, resulting in an h^{-3} decay of interaction energy (Casimir and Polder, 1948).

When interaction occurs in a material medium (named 0), the Hamaker constant must be replaced with the following effective constant:

$$A_{102} = A_{12} - A_{10} - A_{20} + A_{00} \tag{3}$$

The effective constant may be negative, depending on the relative values of the four terms on the right-hand side of Eq. (3).

Now, Hamaker constants may in principle be estimated from material properties of interacting molecules (namely polarizability and light absorption properties or refractive indices) and some practical schemes were used by different authors (Gingell and Parsegian, 1972; Nir and Andersen, 1977; Hough and White, 1980). The following formula was suggested by Israelachvili (1991) and used by later authors (e.g., Yu et al., 1998):

$$A_{102} = \frac{3}{4}kT\left(\frac{\varepsilon_1 - \varepsilon_0}{\varepsilon_1 + \varepsilon_0}\right)\left(\frac{\varepsilon_2 - \varepsilon_0}{\varepsilon_2 + \varepsilon_0}\right)$$
$$+ \frac{3hv_e}{8\sqrt{2}} \frac{(n_1^2 - n_0^2)(n_2^2 - n_0^2)}{(n_1^2 + n_0^2)^{0.5}(n_2^2 + n_0^2)^{0.5}\left\{(n_1^2 + n_0^2)^{0.5} + (n_2^2 + n_0^2)^{0.5}\right\}} \tag{4}$$

where ε_i is the static dielectric constant of medium i, n_i is the refractive index of medium i, k is Boltzmann's constant, T is the absolute temperature, h is Planck's constant, and v_e is an absorption frequency, which should not be very different from 3×10^{15} s^{-1}. Now, the first term represents the contribution of permanent dipoles and should be strongly screened in concentrated electrolyte solutions (Israelachvili, 1991). The second term is strongly decreased when the distance is greater than about 50 nm, due to the finiteness of the velocity of light (which reduces the correlation between electron oscillations when atoms are too distant—Casimir and Polder, 1948).

We made use of Eq. (4) to estimate the Hamaker constant for the interaction between components of the cell membrane. The cell coat was modeled as a region 20 nm thick containing 36.5 g/L carbohydrates and 82.5 g/L protein (assuming that 50% of membrane-associated proteins were in the glycocalyx). Because to a first approximation the contributions of these components to the refractive index may be considered as additive and linearly dependent on solute concentration when this is low (e.g., see van Holde, 1971), we added the contributions of a 36.5 g/L glucose solution (Weast et al., 1986) and a 82.5 g/L albumin (Mishell and Shiiqi, 1980) or urea (Fasman, 1975) solution, yielding an approximate value of 1.352 for the refractive index, as compared with 1.333 for water and 1.45 for a lipid bilayer (Yu et al., 1998). The calculated Hamaker constant was 3.65×10^{-21}, 0.6×10^{-21}, and 0.1×10^{-21} joule, respectively, for the bilayer-bilayer, bilayer–cell coat, and cell coat–cell coat

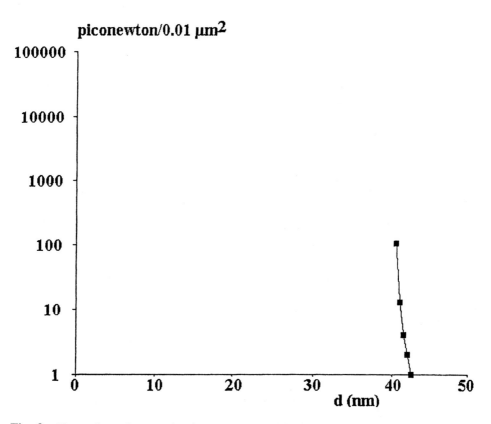

Fig. 3 Electrodynamic attraction between approaching membranes. Cell membranes were modeled as phospholipid bilayers associated with a glycocalyx 20 nm thick. Hamaker constants were calculated from Eq. (4) for glycocalyx-glycocalyx, glycocalyx-bilayer, and bilayer-bilayer interactions. Even when retardation effects were neglected, the interaction between glycocalyces was dominant when the distance between bilayer surfaces was decreased from 50 nm (i.e., 10 nm distance between glycocalyces) to about 40 nm. The corresponding interaction force is shown for a contact area of 0.01 μm^2.

interactions in water. The interactions between cell coats might thus be considered as dominant during cell-cell approach; the estimated interaction force is shown in Fig. 3 with a distance between cell coats ranging between 40 and 50 nm.

Note that there is definite experimental support for the validity of Eq. (4). Yu et al. (1998) studied the interaction between lipid layers containing various amounts of glycosphingolipids in a medium containing 20 mM $NaNO_3$. Using Eq. (4), they estimated at 7×10^{-21} and 0.04×10^{-21} J, respectively, the Hamaker constant for the interaction between lipid layers interacting across 6% sugar solution and two slabs of 6% sugar solution interacting across water. Experimental data were consistent with theoretical estimates except at short distance, when molecular contact was possible. In this case, contact interactions resulted in a twofold to fourfold increase of attraction. It must be emphasized that these short-distance effects should be more properly accounted for by parameters included in polymer theories, such as the Flory interaction parameter χ.

2. Electrostatic Repulsion

According to DLVO theory, electrostatic forces are mainly responsible for colloid stability. The interaction between two charged surfaces in a medium of low ionic strength may be safely estimated with standard Debye-Hückel theory (e.g., see Bockris and Reddy, 1970). However, there are two problems when cell interactions in physiological media are considered. First, Debye-Hückel theory is no longer strictly valid in a solution of high ionic strength such as 0.15 M NaCl. Second, cell surface charges are not distributed on a smooth surface but they are scattered on the surface of macromolecules spanning the entire glycocalyx.

The following method was suggested for obtaining a crude estimate of the interaction between fixed charges bound to approaching surfaces (Bongrand et al., 1982; Bongrand and Bell, 1984). We make use of the Debye-Hückel approximation. The interaction energy between two charges q_1 and q_2 at distance r in an ionic solution is written as

$$W = q_1 q_2 \exp(-\kappa r)/4\pi\varepsilon r \tag{5}$$

where ε is the dielectric permittivity (about 6.9×10^{-10} in water, using MKS units) and the Debye-Hückel length $1/\kappa$ is about 0.8 nm in physiological media. A straightforward integration yields the energy density per unit volume in a medium constaining a density ρ of fixed charges. We obtain

$$W = (1/2)\rho \int_0^\infty \rho \exp(-\kappa r) r \, dr/\varepsilon = \frac{\rho^2}{2\varepsilon\kappa^2} \tag{6}$$

This equation was used to estimate the energy increase resulting from the interaction between two surfaces coated with a layer of uniform charge density and thickness L. Because L is expected to be much greater than the Debye-Hückel length, the range of repulsion is about $2L$. Assuming that charged layers do not interpenetrate (this is not an essential point because we are interested only in orders of magnitude), it is easily shown that, when the distance d between surfaces is smaller than $2L$, the energy per unit area is

$$\frac{\rho^2 L(2L/d - 1)}{\varepsilon\kappa^2} \tag{7}$$

Numerical values are shown in Fig. 4, using estimates discussed earlier (i.e., 0.1 electronic charge per nm^2 and a cell coat thickness L of 20 nm). Clearly, electrostatic repulsion may be quite efficient in preventing adhesion if the total length of receptor-ligand couples is substantially lower than L.

It may be useful to discuss a potential problem in the use of the preceding formula. In his well-known theory of polyelectrolyte solutions, Manning (1969) showed that the effective charge of linear polyelectrolytes is decreased by tight binding of counterions when the linear density of fixed charge is higher than about 0.7 nm in water. This phenomenon, called "counterion condensation," might be neglected if we estimate at about one charge per disaccharide, i.e., about one charge per nm, the density of ionized groups on proteoglycans.

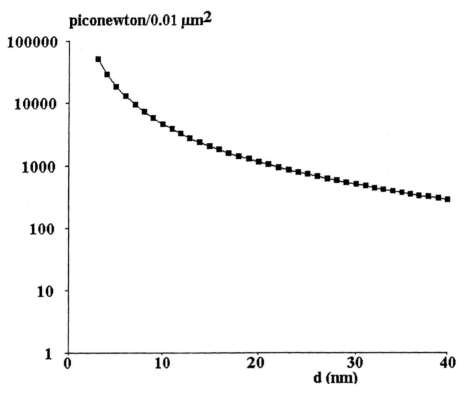

Fig. 4 Electrostatic repulsion between approaching membranes. Cell membranes were modeled as phospholipid bilayers coated with a glycocalyx 20 nm thick and with 0.1 electronic charge per nm^2. The repulsive force between parallel membrane patches of area 0.01 μm^2 is shown as a function of the distance between the bilayer surfaces.

3. Steric Stabilization

Colloid scientists have long recognized that, although colloidal suspensions of charged particles may be quite stable in pure water, increasing the ionic strength of the surrounding medium may result in flocculation. However, efficient stabilization can be achieved by coating particles with suitable polymeric molecules. This phenomenon is usually called steric stabilization (Napper, 1977). Numerous models were elaborated to account for steric stabilization in quantitative terms, as exemplified in the book by Napper (1983). However, despite the major clarification brought by de Gennes' work, as summarized in several reviews (de Gennes, 1979, 1987, 1988), we are still a long way from accounting quantitatively for the interaction between cell surface polymers. In the present section, we recall some fundamental concepts and discuss their relevance to cell-cell interaction

The basic problem is, of course, to derive a theory allowing us to calculate the partition function (or, which is equivalent, the free energy) of an assembly of polymer molecules. The partition function may be written as

$$Z = -F/kT = \sum_i \exp(-E_i/kT) \tag{8}$$

where the summation is extended over all possible conformations of energy E_j, and k is Boltzmann's constant, T is the absolute temperature, and F is the free energy. The basic problem thus consists of counting available states and calculating their energy. The summation may be replaced with an integral if a continuous representation is preferred.

The simplest system is an assembly of identical flexible chains (made of N freely jointed segments of length a) with negligible thickness (which means that all segments may rotate independently without any risk of overlap) and no interaction between themselves or with the solvent molecules (which means that all conformations have the same energy). In this case, the partition function can be calculated analytically. The polymer conformation may be compared with a random walk, and the root-mean-square end-to-end distance R (i.e., the square root of the average value of the squared distance between polymer ends) is proportional to the square root of the number of segments.

$$R = a(N)^{1/2} \tag{9}$$

Note that a related quantity is the radius of gyration R_g; this is the root-mean-square value of the distance between polymer segments and the center of gravity. The radius of gyration is equal to $R/6$ (Flory, 1953).

Extending this simple model to actual systems requires that the following problems be considered:

1. Rotation of intersegment bonds may be hindered by structural constraints. This is not a serious problem, and this may be accounted for by considering an equivalent chain with lower N and higher a (de Gennes, 1979).
2. Conformations involving an overlap between nonneighboring segments are forbidden; thus a fraction of the solution volume is not accessible to the center of any segment (excluded volume effect).
3. Intermolecular forces may be different between polymer segments, solvent molecules, or between a polymer segment and a solvent molecule. Thus, available conformations may have different energies. Flory, who pioneered polymer theory (Flory, 1953), defined an empirical interaction parameter χ to account for this possibility. If χ is higher than 0.5, solvent-polymer interactions are energetically unfavorable. The polymer is said to be in a bad solvent, and a phase separation between polymer and solvent may occur. If χ is 0.5, the polymer is said to be in a θ solvent. If χ is lower than 0.5, polymer-solvent interactions will be preferred, and the polymer will swell: R is proportional to $N^{0.6}$ instead of $N^{0.5}$. The latter result was first obtained by Flory. The statistics of chain conformation are somewhat simpler in a θ solvent, corresponding to a random walk.

In view of (2) and (3), the partition function cannot be determined analytically, and several approximations were devised. The standard way may be to neglect molecular correlations and use mean field theories: the polymer density around a given polymer segment is taken as the mean polymer density in the considered zone. This approximation may be improved by sequential iteration: the polymer conformation is deduced from a "reasonable" field expression, and an improved field function is derived from the conformation obtained. This procedure is repeated until "consistency" is obtained. This "self-consistent mean field approach" is similar in spirit to the Hartree-Fock self-consistent molecular field method that has long been used by quantum chemists. However, this approach is quite complicated and may yield only numerical results or fairly intricate formulas that are rather difficult to interpret.

The discovery of scaling laws considerably facilitated a general understanding of the behavior of polymer solutions (de Gennes, 1979). The general principle consists of using fairly simple (although not always easy) reasoning to derive a simple form for the dependence of some parameter (such as the radius of gyration or osmotic pressure) on quantities such as polymer concentration or segment number. The price to be paid is often lack of accurate determination of numerical coefficients. The simplest example may be the scaling law of the radius of gyration, which is proportional to $N^{1/2}$ in a θ solvent and $N^{3/5}$ in a good solvent.

The aforementioned principles may now be applied to cell-cell interactions. We are interested in the interaction between two plates coated with flexible homopolymers of uniform length. Two different situations must be considered.

1. Polymers are reversibly adsorbed on surfaces, and the amount of bound molecules may vary during surface approach. This situation can probably be ruled out during short-term interactions because the exchange of macromolecules between extracellular medium and two parallel plates may be exceedingly slow (up to 10^{10}-fold slower than in solution, as emphasized by Israelachvili, 1991).

2. Polymers do not interact with the surfaces excepted on one end that is bound irreversibly. We shall consider this situation. Two alternative cases may then be considered (Fig. 5):

2a. There is substantial interaction between polymers, because the mean distance D between neighboring anchoring points is less that twice the radius of gyration. The polymer layer was compared with a brush (de Gennes, 1987, 1988). Neglecting excluded volume effects, the scaling law for the thickness L in a good solvent is

$$L \sim N a^{5/3}/D^{2/3} \tag{10}$$

and the law for the interaction between surfaces at distance h is

$$F \sim kT/D^3[(2L/h)^{9/4} - (h/2L)^{3/4}] \tag{11}$$

However, this situation is not fully consistent with our quantitative model of the cell coat. Indeed, assuming a mean number of one monosaccharide unit per nm^2, considering the case of short chains with $N = 50$ segments, the mean chain spacing of about 7 nm would be higher than the Flory radius because $a \times N^{3/5}$ is about 6 nm if the chain length is 0.57 nm, as estimated for a monosaccharide unit. Numerical results are shown in Fig. 6.

2b. Alternatively, surface-bound polymers may be considered as independent units because they are separated by a distance larger than their diameter. This is the "mushroom" situation (de Gennes, 1979). An analytic treatment of this situation was provided by Dolan and Edwards (1974), who neglected excluded volume effects. A good approximation of these authors' results was worked out by Israelachvili (1991), yielding a simple expression for

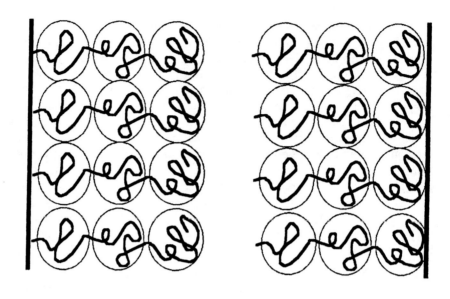

A

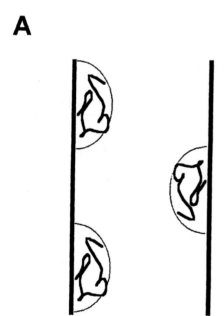

B

Fig. 5 Possible patterns for polymers bound to approaching surfaces. Two limiting cases may be considered. (A) Dense coating: chains are extended and may be considered as linear arrays of spherical "blobs." The polymer coating may be compared with a brush. (B) Sparse chains: interactions between chains may be neglected. Each chain is compared with a mushroom (de Gennes, 1987).

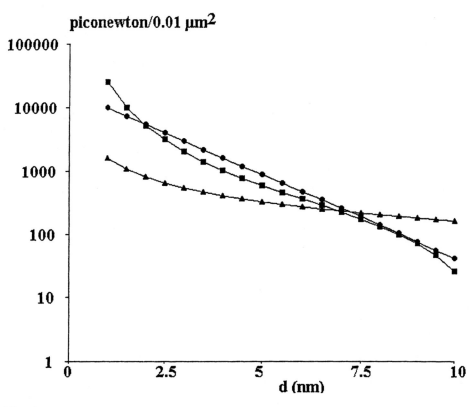

Fig. 6 Steric repulsion between approaching membranes. Cell surfaces were modeled as phospholipid bilayers coated with carbohydrate chains made of 50 segments of 0.57 nm each and anchored on the surface with a spacing of 7 nm. The repulsion was calculated with approximate formulas devised for interacting brushes in a good solvent (squares) and mushrooms in a theta solvent (circles). The case of rigid rods is also shown (triangles).

the repulsive free energy w generated by a chain of radius of gyration R_g squeezed between two surfaces at distance h:

$$w = 36kT \exp(-h/R_g) \tag{12}$$

Numerical results are shown in Fig. 6, estimating a at 0.57 nm and calculating the radius of gyration as $a(N/6)^{1/2}$, i.e., about 4 nm. Note that Dolan and Edwards (1975) later reported a numerical treatment of excluded volume effects. The range of estimated repulsion is lower than the experimental value of the cell coat thickness. This means that the preceding models are probably inadequate to provide a full description of the conformation of cell surface carbohydrates. Thus, it is possible that some long chains are anchored at some distance from the phospholipid bilayers, due to the presence of a branched scaffold on the cell surface. Indeed, several problems are raised by the application of these models to the cell surface:

May cell surface carbohydrates be modeled as linear polymers? The presence of branching points might substantially alter the properties of polymer molecules.

De Gennes (1988) pointed out that in this case the molecular size might vary as $N^{1/2}$ rather than $N^{3/5}$ in a good solvent. This prediction is supported by some experimental data. Dextrans are glucose polymers with 1–4% branching points (Kabat, 1968). Extensive information is provided by sigma on the hydrodynamic radius of a series of fluoresceinated dextran preparations with a molecular weight ranging between 3000 and 150,000: the Stokes radius (in nm) is close to $0.3 \times N^{1/2}$.

As reviewed earlier, the proteoglycan carbohydrate chains found on the cell surface are usually linear. However, these chains may be linked by a variety of interactions with proteins or other carbohydrates. Thus, glycosaminoglycans synthesized by transformed fibroblasts were reported to bind fibronectin, a protein bound to the surface of many cell populations (Latena et al., 1980). The core protein of several cartilage proteoglycans was shown to bind hyaluronic acid (Hascall and Heingard, 1974). Multiple interactions between carbohydrate chains were also described. Thus, Fransson (1976) and Fransson et al. (1979) described interactions between dermatan sulfate chains and galactosaminoglycans. These interactions should probably be described as local binding events rather than being viewed as a general attraction accounted for by the Flory interaction parameter, as these sugar chains are heterogeneous and interactions might involve localized areas. Thus, the capacity of galactomannans to enhance the formation of gels by carrageenan, a sulfated algal polysaccharide, was ascribed to interactions between smoother (unbranched) regions of these galactomannans and carrageenan (Dea et al., 1972).

May the cell surface be modeled as a homogeneous assembly of polymers? Our calculations were focused on the longer glycosaminoglycan chains. However, many cell surface carbohydrates occur as short oligosaccharide chains bound to lipids or proteins. Among these glycoproteins, the extended mucin-like molecules such as leukosialin (described in a previous section) may play the most important role in steric stabilization. Thus, the cell coat might probably be viewed as a mixture of rigid rods and flexible coils. The repulsion generated by an assembly of rigid rods of 15 nm length was estimated by Foa et al. (1996) by estimating as $k \ln(d/L)$ the entropy variation due to the confinement of an anchored rod of length L in a space of thickness d. Numerical results are shown in Fig. 6.

Interestingly, following the pioneering studies of Luckham and Klein (1985) on the interaction between approaching surfaces in the presence of polymers, some authors obtained direct information on the repulsive potential of biological macromolecules. Perez and Proust (1987) used the surface forces apparatus to measure the interaction between mica surfaces coated with bovine submaxillary mucin; this molecule was modeled as a proteic rod of ~ 800 nm length studded with multiple oligosaccharide chains. The radius of gyration was 140 nm, as determined by light scattering, and the molecular weight was about 4×10^6. When large amounts of this mucin were adsorbed on mica surfaces (about 675 molecules/μm^2), a repulsive interaction was detected with a range of 230 nm and an exponential dependence on distance with a decay length of 41 nm. The repulsive energy was about 360 kT units per molecule at 20 nm separation. The authors concluded that the standard DLVO theory could not account for the experimental data. In another study, Malmsten et al. (1992) used a similar approach to compare repulsive interactions generated by rat and pig gastric mucin. They also concluded that the decay length of interaction (30 nm with rat gastric mucin) could not be accounted for by DLVO theory.

4. Dynamical Aspects of Cell-Cell Interactions

As previously emphasized, repulsive intercellular forces are time dependent. Thus, equilibrium models of colloid stabilization cannot account for all cell interactions. Several phenomena might be considered:

(1) *Desorption of loosely bound repellers* in contact areas. As discussed earlier, this might be too slow to influence significantly the first step of adhesion.

(2) *Hydrodynamic repulsion.* The force is strongly dependent on the shape of approaching surfaces. Indeed, the force between two approaching disk of radius a at distance d and relative velocity v may be approximated as (Bongrand et al., 1988a, equations 7–9):

$$F = (3/2)\pi\mu a^4/d^3 v \tag{13}$$

where μ is the medium viscosity. Assuming that a cell approaches a flat surface, and modeling the interaction area as a flat disks of radius 50 nm (corresponding to the tip of microvilli), the repulsive force at 10 nm separation and 10 μm/s relative velocity (which is the order of magnitude of the velocity of a blood leucocyte rolling on activated endothelium) is 0.3 pN in aqueous medium. However, if the glycocalyx is modeled as a fluid of about 0.05 Pa s viscosity (Sec. II), the calculated repulsion is 15 pN, which may be quite significant.

However, if the cell region approaching the surface is modeled as a sphere of radius a rather than a flat disk, the repulsive force is decreased by a factor $(a/d)^2$, i.e., about 25 following our numerical example (Dimitrov, 1983; Bongrand et al., 1988a, equations 7–9).

Hydrodynamic effects might also influence the topography of cell contacts; this point was clearly explained in a review by Gallez and Coakley (1996), who contributed much to the understanding of this phenomenon. Using a far-from-equilibrium approach, they studied the stability of the liquid film separating two flexible membranes with different kinds of forces (van der Waals, electrostatic, steric). They demonstrated that in some cases periodic rupture of the film should occur, in accordance with experimental data. Indeed, electron microscopic studies were performed on erythrocytes agglutinated with dextran, polylysine, or lectins. In some cases, periodic contacts with a separation ranging between several tenths of a μm and more than 1 μm were reported (Darmani and Coakley, 1990; Darmani et al., 1990). More work is required to assess the relevance of these concepts to nucleated cells because (a) the latter have a much more irregular surface, (b) bending behavior is quite different, and (c) glycocalyx elements are probably different. These points are important in view of the influence of repulsive forces on intercellular contact pattern (Darmani and Coakley, 1990).

(3) *Polymer compression.* Cell surface approach may require extensive deformation of surface polymers. Note that this phenomenon is partly accounted for by previous considerations, because these molecules are responsible for the glycocalyx viscosity. However, the viscosity estimates we cited relied on the motion of small spheres embedded in the pericellular matrix. It is difficult to estimate the mechanical properties of cell surface macromolecules; they may be heavily dependent on branching and transient molecular interactions that might induce a gel-like structure.

(4) *Diffusion of surface molecules out of contact areas.* The diffusion of membrane molecules might be significantly reduced in contact areas because of crowding effects. However, minimal molecular freedom is probably retained in this area, and it was recently demonstrated by fluorescence microscopy that the antiadhesive mucin-like CD43 molecule was substantially depleted in the area of contact between monocytic THP-1 cells and immunoglobulin-coated red cells. Further, some evidence suggested that this depletion was induced by binding because the distribution of CD43 was fairly homogeneous on cells that had not bound any red cell (Soler et al., 1997).

III. THE FIRST BOND

A. Theoretical Discussion

The formation of the first molecular bond between approaching surfaces is a critical step of adhesion. Two questions are of particular importance:

1. When two cells managed to overcome repulsive interactions and bring their surfaces within a distance d, how long will it take for the first bond to be formed?
2. When the first bond is formed, will it maintain surfaces in contact for sufficient time to allow the formation of additional bonds? Intuitively, it may be expected that if the formation of a second bond is possible, adhesion will then undergo irreversible strengthening. However, whereas the formation of the first bond may be amendable to a physical-chemical approach (Pierres et al., 1994), bond strengthening is highly dependent on active cell functions. These problems will be discussed in the next section.

Numerous adhesion molecules have been characterized and prepared in soluble form, and many authors have studied their behavior with currently available methods of biochemical investigation. It is usually felt that an exhaustive description of molecular interactions can be achieved by determining the kinetic constants of bond formation and dissociation. Considering the following reaction:

$$A + B \underset{k_{off}}{\overset{k_{on}}{\rightleftharpoons}} AB \tag{14}$$

where k_{on} is the on-rate of association, expressed in $mol^{-1}\ s^{-1}$, and k_{off} is the rate of bond dissociation, expressed in s^{-1}, we may write

$$d[AB]/dt = k_{on}[A][B] - k_{off}[AB] \tag{15}$$

where the square brackets are used to represent the molar concentrations of different species. The use of surface plasmon resonance (see Schuck, 1997, for a review) proved a rapid and efficient way to study the dynamic aspects of interactions between soluble receptors and surface-bound ligands. An interesting example is the quantitative study of the short-term interactions between recombinant CD2 and CD48 molecules that are involved in lymphocyte function (van der Merwe et al., 1993). However, these conventional approaches could not yield any direct information on interactions between surface-bound molecules.

This emphasizes the interest in a theoretical framework described by Bell (1978) in a seminal paper that remains widely cited. We shall now discuss two important points considered in this paper.

1. Kinetics of Bond Formation and Dissociation Between Surface-Attached Molecules

The primary goal of the author was to estimate the kinetics of bond formation and dissociation between membrane-attached adhesion receptors. For this purpose, he made use of the concept of "encounter complex" by splitting the interaction between molecules A and B in two steps.

$$A + B \underset{d_-}{\overset{d_+}{\rightleftharpoons}} C \underset{r_-}{\overset{r_+}{\rightleftharpoons}} AB \tag{16}$$

The *first step* is the formation of a so-called encounter complex between A and B. This involves only translational diffusion of molecules A and B, and the complex is said to be formed when the centers of gravity of the molecules are separated by a distance smaller than some arbitrary value R_{AB} that is expected to be close to the sum of the radii of these molecules. The kinetic constants d_+ and d_- can be calculated by standard methods of statistical mechanics. When molecules are bound to surfaces, three-dimensional diffusion is replaced by two-dimensional motion. Now, the diffusion constant of a free protein such as an immunoglobulin in water is of order of 10^{-6} cm^2/s, and the diffusion constant of many membrane-bound molecules ranges between about 10^{-9} and 10^{-10} cm^2/s (see Bongrand, 1995, for a short review and Schlessinger et al., 1976, for an early description of membrane mobility determinations with the method of fluorescence recovery after photobleaching). Thus, at first sight it might seem that surface reactions should be 1000-fold slower than encounters between soluble molecules. However, this conclusion is not warranted. As pointed out by Bell, the steady-state encounter rate d_+ is of order of $D \times R_{AB}$ in solution and D on a surface. Indeed, the dimension of d_+ is the inverse of a concentration. Thus, 3D constants should be expressed in liter/mole/second, whereas 2D constants are often expressed in μm^2/molecule/second. It is therefore meaningless to compare the kinetic rates of 2D and 3D reactions, as their ratio is dependent on length units.

However, the dissociation rate d_- is expressed in second^{-1} and is independent of the complex concentration. It may thus be stated that the separation of membrane-attached molecules should be 1000-fold slower than that of soluble molecules.

The *second step* was postulated to proceed with similar kinetics under 3D and 2D conditions (Bell, 1978). This step includes two sequential phases:

1. Molecular rotation in order that binding sites might get into close contact
2. Bond formation

As previously discussed (Pierres et al., 1998), the kinetics of this step is highly dependent on the flexibility of adhesion molecules. Indeed, if these molecules are flexible, the rotation of the domains bearing binding sites may display similar kinetics in free and bond molecules. Further, because many reactions between biomolecules, including the antigen-antibody association, are diffusion limited (Bell, 1978), moderate differences concerning the kinetics of this second step are not expected to alter the overall reaction rate. On the contrary, if attached molecules are rigid, it may be difficult for complementary sites to encounter each other unless the distance and orientation of the membranes fall within a very restricted domain.

A major assumption of Bell's model is that interacting membranes are within binding distance. This assumption may be considered reasonable if membranes are bound by at least one ligand-receptor couple, which may be sufficient to maintain surfaces at binding distance (see later). Otherwise, the binding kinetics are expected to be heavily dependent on the width of the intercellular gap. Thus, Bell's theoretical framework may be better suited to the prediction of the rate of bond accumulation when the first binding event has occurred rather than the rate of formation of the first bond.

2. Mechanical Strength of Intercellular Bonds

As shown earlier it is important to know the mechanical strength of ligand-receptor bonds. As clearly shown by Bell (1978), the concept of bond strength has no accurate intrinsic significance. Only, the off-rate k_- is excepted to be influenced by the presence of a distractive force Bell suggested applying to individual bonds an empirical formula that was reported by Zurkhov (1956), who studied the time to rupture of macroscopic material samples subjected to a wide range of forces. Bell's formula reads

$$k_-(F) = k_-(0)\exp(\gamma F/kT) \tag{17}$$

where k is Boltzmann's constant, T is the absolute temperature, and γ is a constant with the dimension of a length that was suggested to be close to the range of the interaction (or, equivalently, the depth of the binding sites). Note that the force required to increase the off-rate by a factor of 2 is $kT\ln(2)/\gamma$. Thus, estimating as 0.1 nm the range of the interaction, the force required to halve the bond lifetime would be about 30 pN. Interestingly, Bell noticed that a similar force might be sufficient to uproot a molecule embedded in the plasma bilayer through a hydrophobic domain.

As previously shown (Pierres et al., 1998), Eq. (16) may be easily derived within the framework of Eyring's theory of the absolute rate of chemical reactions (Eyring, 1935). A thorough discussion of the physical significance of bond rupture was provided by Evans and Ritchie (1997). Note that the physical significance of the experimental bond lifetime is not always straightforward. First, this depends on the rate of thermal energy accumulation by the ligand-receptor complex, which may depend on the molecular details of the binding-complex. Also, if the rate of rebinding of a broken complex is not negligible, the rate of dissociation will depend on the dynamics of surface separation.

B. Experimental Results

During the past 10 years, several powerful experimental approaches yielded a wealth of information on the kinetics and force dependence of ligand-receptor dissociation. Much less information is available on association kinetics.

1. Kinetics and Force Dependence of Bond Rupture

Essentially three methodologies yielded quantitative information on bond rupture. These will now be described.

a. Use of Hydrodynamic Flow

The first experimental check of Bell's theory was provided by Tha et al. (1986), who took advantage of the traveling microtube technique devised by Goldsmith (see Goldsmith et al., 1994, for a brief description of the apparatus). They studied the agglutination of erythrocytes that had been made spherical by exposure to hypoosmotic solutions followed by aldehyde fixation. Particles were agglutinated with minimal amounts antibodies and driven with low pressure into a vertical capillary tube mounted on a moving stage. Cells were examined with a microscope that had been rotated by 90° to make the optical axis horizontal. The velocity of the moving stage was chosen such that selected erythrocytes might remain in the microscope field for a prolonged period of time. Theoretical analysis of the particle motion showed that these particles occasionally came into close contact, forming rotating doublets that were first subjected to a compressive and later to a disruptive force during rotation. The force could be deduced from the rotation velocity and local shear rate.

Assuming that particles were bound by a low number of bonds whose number followed a Poisson distribution, the authors derived the mechanical strength of single bonds from the distribution of forces at the time of breakup; their data were found to be consistent with a minimum binding strength of 24 pN for a polyclonal antibody. A similar value (20 pN) was later reported when erythrocytes were agglutinated with a monoclonal antibody (Tees et al., 1993). However, particle monitoring could not be carried out for a sufficient time to observe both doublet formation and rupture. Thus, the authors subsequently made use of a Couette viscometer (this apparatus is made of two coaxial rotating cylinders, allowing easy generation of a constant shear flow in the separating space). They performed continuous monitoring of flowing cells in a particular region; revolving doublets could thus be followed for a sufficient time to determine bond duration. The authors estimated the natural lifetime of the inter-action between immunoglobulin G and protein G, a natural IgG receptor of bacterial origin, to be 175 s. The force required to increase the off-rate by a factor of 2 was about 7 pN (Kwong et al., 1996). In another study, the lifetime of the association between a polysaccharide antigen and an antibody was 25 s, and the force required to double the off-rate was about 24 pN (Tees and Goldsmith 1996).

Increased monitoring period and temporal resolution could be obtained with a parallel-plate flow chamber (see Pierres et al., 1996a, for a review). The principle was to drive receptor-bearing cells or particles along ligand-coated glass coverslips forming the bottom of a flow chamber fixed on the stage of an inverted microscope. The hydrodynamic drag was lower than 1 pN. A single bond was thus sufficient to stop a cell or a particle (note that inertial forces were estimated to be less than 1 pN). By recording particle motion in a selected microscopic field, it was thus expected to monitor the formation and dissociation of individual bonds. Bond lifetime and force dependence could thus be calculated by determining the distribution of arrest duration for a varying shearing rate.

Kaplanski et al. (1993) first used this apparatus to determine the interaction of flowing neutrophils and monolayers of activated endothelial cells; they were thus able to estimate as 2.4 s the median duration of the association between endothelial cell E-selectin and its neutrophil ligand. In a later study, Alon et al. (1995) deter-mined the interaction between flowing neutrophils and surface-bound P-selectin molecules. The natural bond lifetime was of the order of 2 s, and the force required to double the rate-off was about 60 pN. The authors emphasized the importance of high mechanical strength to allow selectin-mediated bonds to resist the high shear forces exerted by flowing blood. A problem with these studies was that due to the irregularity of cell shape, it might be difficult to discriminate between actual cell arrest and a transient velocity decrease (see also Tissot et al., 1992) or a deformation of the cell image associated with rotation. Thus, it was found useful to replace cells with model particles. Pierres et al. (1995) studied the interaction between spheres coated with antirabbit immunoglobulin and surfaces derivatized with rabbit antibodies. Binding sites were diluted until the frequency of particle arrest was proportional to the surface density of these sites, so that most arrests might be ascribed to single binding events. Interestingly, quantitative analysis of arrest duration showed that antigen-antibody association could not be considered as a monophasic event. Four kinetic constants were needed to describe the initial interaction between antigen (Ag) and antibody (Ab) sites:

$$\mathrm{Ag + Ab} \underset{k_{\mathrm{d}}}{\overset{k_{\mathrm{a}}}{\rightleftharpoons}} \mathrm{C} \underset{k_{-}}{\overset{k_{+}}{\rightleftharpoons}} \mathrm{(AgAb)} \tag{18}$$

Thus, the initial interaction resulted in the formation of a transient complex with a dissociation rate k_{d} of about 0.8 s^{-1} and a "strengthening rate" k_{+} of the order of 0.5 s^{-1} resulting in the appearance of a long-lived complex with a lifetime of several hours or more, corresponding to the expected duration of a strong antigen-antibody association. The force required to double the dissociation rate k_{d} of the transient binding state was about 36 pN (unpublished results). This emphasizes the lack of correlation between the natural lifetime and mechanical strength of a given bound state. In a further development of this technique, Pierres et al. (1996b) adapted an image analysis procedure allowing real-time determination of the sphere position with about 0.05 μm spatial accuracy and 5 ms temporal resolution. This allowed easy detection of very transient interactions. The association between recombinant CD2 and CD48 molecules could thus be studied. The estimated off-rate was 7.8^{-1} and it was estimated that this constant was doubled by a disruptive force of 22 pN (Pierres et al., 1996c).

b. Use of Soft Vesicles as Transducers

The use of soft vesicles to study surface adhesion with very high sensitivity was pioneered by Evans (e.g., see Evans 1980). The starting point was an analysis of the mechanical properties of erythrocyte membranes by studying the deformation of single cells aspirated into micropipettes with calibrated pressure. The adhesive energy between erythrocytes and/or lipid vesicles was then studied by monitoring the deformation of two of these particles when they were brought into contact with two micropipettes and then progressively separated. This technique was applied (Evans et al., 1991) to the study of erythrocytes agglutinated with minimal densities of various adhesion molecules; because of the limited lateral mobility of the red blood cell membrane, it might be expected that adhesion was mediated by a few or even one molecular bond. When adhesion was mediated by different antibodies or lectins, rupture occurred between a few seconds and several tens of seconds after the onset of separation, and the force at the time of rupture was of order of 19–20 pN, whatever the ligand-receptor couple. However, as emphasized by the author, rupture is a stochastic event. The rupture force is not an intrinsic parameter and depends on the rate of force increase.

This technique was later improved by coupling rigid microbeads to the vesicles and mounting the pipette on a piezoelectric device. This apparatus was used to study the interaction between microbeads and plane glass surfaces derivatized with adhesion molecules. The advantage of this apparatus was that the bead-to-surface distance might in principle be determined with nanometer resolution, using interference reflection microscopy, and the displacements were performed with high accuracy. It was thus conceivable to measure interacting forces ranging between 0.01 pN and more than 1000 pN by adapting the vesicle stiffness through controlled variations of aspiration pressure (Evans et al., 1994).

c. Measurement of Bond Strength with Atomic Force Microscopy

The basis of atomic force microscopy (AFM) consists of scanning a surface with a very sharp tip mounted on a cantilever. The force exerted on the tip is measured with piconewton accuracy by determining the reflection of a laser beam on the cantilever. A piezoelectric device allows controlled displacement of the surface with better than nanometer resolution. The first reports on the strength of ligand-receptor bonds were authored by Florin et al. (1994) and Lee et al. (1994), who studied the interaction between streptavidin and biotin-coated surfaces. This ligand-receptor couple displays unusually high binding affinity of the order of 10^{-15} M. When the AFM tip was coated with small amounts of biotin molecules and then repeatedly pushed against a streptavidin surface and pulled with increasing strength, the cantilever displayed progressive bending events terminated by a sharp jump that was ascribed to bond rupture. Interestingly, the histogram of the distribution of rupture forces displayed a series of quantized peaks, tentatively ascribed to the simultaneous rupture of one, two, three, or more bonds (Florin et al. 1994). The rupture force ranged between 160 pN (Florin et al., 1994) and 300–400 pN (Lee et al., 1994). Interestingly, a rupture force as low as 50 pN was reported by Merkel et al. (1995), who studied the avidin-biotin interaction by a soft vesicle technique and used a very slow increase of distractive force, thus demonstrating the influence of the loading rate on the measured force. Other models were later studied; the force required to separate paired adenine and thymine molecules was estimated as about 54 pN (Boland and Ratner, 1995). A force of about 400 pN was required for rapid separation of two cell adhesion proteoglycans (Dammer et al., 1995); the authors estimated that about 10 weaker interactions of the order of 40 pN might be involved in this association. Finally, the rupture force of the bond formed between human albumin and specific antibodies was about 240 pN (Hinterdorfer et al., 1996).

d. Other Methods

Although the surface forces apparatus readily detected ligand-receptor interactions (Helm et al., 1991; Leckband et al., 1992, 1995), this yielded energy-distance relationships rather than lifetime values. Thus, this approach must be considered complementary to the aforementioned methods. Miyata et al. (1996) used optical tweezers to separate bound actinin and α-actinin molecules; a wide range of values was found, with an average lifetime of 0.5 s. The force required to double the off-rate ranged between about 10 and 60 pN.

2. Kinetics of Bond Formation Between Surface-Attached Molecules

It is not an easy task to determine the rate of bond formation as a function of the distance between surfaces; an ideal experiment would consist of approaching a ligand

and a receptor molecule at known distance d, waiting for some time t, and pulling to determine whether binding occurred. This procedure should be repeated a sufficient number of times to know the binding probability, which is in principle equal to $(1 - \exp(-k_{on}t))$. These series of experiments should then be repeated for a range of different values of d (and also of t). Although such an approach has not yet been carried out, some attempts at achieving a direct determination of k_{on} were reported.

Hinterdorfer et al. (1996) studied the interaction between the (antibody-bearing) tip of an atomic force microscope and an antigen-coated surface. Estimating the tip velocity as 200 nm/s and the dynamical reach r_{eff}, as about 6 nm, they concluded that the tip remained at binding distance of the surface for 60 ms (i.e., $2r_{eff}/200$ nm/s) s) and that the screened volume V_{eff} was about $2/3\ \pi r_{eff}^3$, i.e., about 450 nm³. Because the binding probability was about 0.5, the association constant was about $0.5/(60\ ms) \times V_{eff}/P_A$, assuming that there was a single antibody molecule on the AFM tip and estimating as P_A the probability that there was an antigen molecule in the screened volume. The mean separation distance between surface-bound antigen molecules was about 100 nm, yielding an estimated association constant of about $5 \times 10^4\ M^{-1}\ s^{-1}$. Note that there is some difficulty in applying macroscopic kinetic equations to volumes comparable to molecular dimensions. As shown in Fig. 7, an

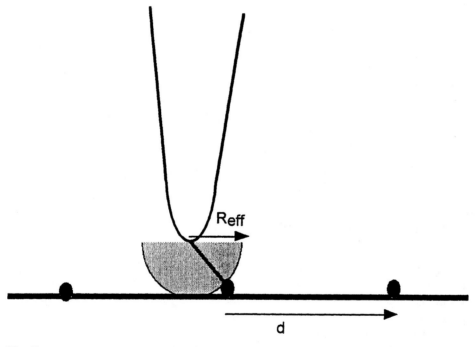

Fig. 7 Estimating the rate of bond formation with atomic force microscopy. The tip of the AFM bears a fairly mobile molecule that can bind to any ligand entering the half-sphere of radius R_{eff} centered on its anchoring point (shaded area). If the surface density of ligands is $1/d^2$, the probability P_A that the half-sphere contains a ligand molecule when the distance between the tip and surface is z is simply $\pi(R_{eff}^2 - z^2)/d^2$. The mean value of P_A when z is lower than R_{eff} is $\langle P_A \rangle = \int_0^{R_{eff}} [\pi(R_{eff}^2 - z^2)/R_{eff}f^2]dz$ and the association rate can be estimated by using as the ligand concentration the ratio $\langle P_A \rangle/(2\pi R_{eff}^3/3)$.

estimate of 3×10^{-1} M^{-1} s^{-1} for k_{on} may be derived from the authors' data. The essential approximation was indeed to represent $k_{on}(d)$ as a step function equal to k_{ass} for r lower than a threshold value (r_{eff}) and zero for other values.

Pierres et al. (1997) used a laminar flow chamber to study the frequency of bond formation between CD48-coated spheres and CD2-derivatized surfaces. They monitored hundreds of beads and hundreds of thousands of sequential positions in order to obtain an experimental plot of the binding frequency (per second) versus sphere velocity. Then they used theoretical data relating the particle velocity to the particle-to-surface separation (Goldman et al., 1967) in order to extract a plot of the adhesion frequency versus distance. Finally, using experimental determination of the binding site density on interacting surfaces, they obtained a theoretical plot of the binding frequency (in second^{-1}) per molecule versus distance. They found that the binding probability was inversely related to the cube of the distance. However, there were two problems with this approach. First, surface rugosity was neglected. Thus, binding events might occur on surface asperities, which would lead to an overestimate of the binding probability at large distances. Second, thermal motion perpendicular to the surface was neglected, which might also lead to an overestimate of the binding probability, because even fairly rapid particles might spend some time very close to the surface.

Thus, more experiments are needed to achieve accurate determination of the absolute binding rate. Note, however, that all experimental data stressed the importance of molecular length and flexibility. Hinterdorfer et al. (1996) used a spacer 8 nm long between the AFM tip and antibodies. Florin et al. (1994) emphasized the interest of using soft surfaces. Pierres et al. (1996c, 1997) used an immunoglobulin G molecule (longer than 20 nm) to link CD4-CD48 chimeras (about 16 nm long) to flowing spheres.

IV. EXTENSION OF THE CONTACT AREA

The initial steps of cell-cell interactions that have just been described might be satisfactorily modeled with standard physical-chemical models, provided the cell structure is described with sufficient accuracy. However, the completion of adhesion and subsequent deadhesion are heavily dependent on active cell functions. A complete understanding of the multiple intracellular events responsible for structural reorganization and adhesion strengthening is certainly not available at the present time. Thus, in the next sections, we shall focus on biophysical aspects of adhesion, but we shall try to delineate the limits of a physical approach.

We shall now describe some selected experimental data in order to facilitate a quantitative discussion of biological adhesion mechanisms.

A. Experimental Data

As summarized in Fig. 8, the strengthening of cell adhesion after the formation of the first contact involves three concomitant phenomena. As shown in Fig. 8A, a nonadherent cell may often be compared with a sphere studded with numerous microvilli. *First*, during the first minutes following contact, the sphere will progressively flatten with an apparent area increase at the micrometer level (Fig. 8B and C). *Second*, if the cell-cell or cell-substratum contact area is observed by electron

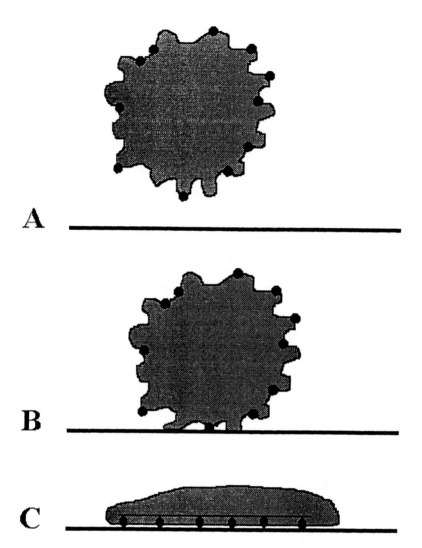

Fig. 8 Active adhesion strengthening. After the formation of the first bond, the following events are expected. First, cells spread at the μm level in order to create a nonzero contact area (B and C). Second, the molecular contact area is increased by smoothing of the cell surface at the submicrometer level (C). Third, membrane receptors (black disks) are concentrated in the cell-surface contact area. Fourth, cytoskeletal elements display marked organization in the cytoplasmic region underlying contact areas (C).

microscopy, it is frequently found (Fig. 8B) that only a limited fraction of contours are separated by a gap compatible with molecular interactions (say less than 50 nm separation, assuming that the fixation procedure did not drastically alter this distance, as supported by some evidence from Heath, 1982). This fraction increases with concomitant smoothing of the membrane at the submicrometer level (Fig. 8C). *Third*, the binding receptors that might be scattered on the cell surface may become concentrated in the contact area with concomitant organization of submembranar cytoskeletal elements (Fig. 8C).

1. Morphological Aspects of Cell Spreading

Follett and Goldman (1970) reported a very interesting electron microscopic study of the interaction between BHK fibroblasts and a glass surface. They noticed that microvilli (0.1 μm diameter, up to 5 μm length) were more abundant on spherical cells than on cells that were completely spread on a surface. When recently trypsinized cells were made to adhere to a flat substrate, the percentage of cells coated with microvilli dropped from about 90% to 20–30% 24 h later. Further, the apparent area of spread cells was about fourfold higher than that of rounded ones, suggesting that microvilli acted as reserve membrane area. Spreading might thus induce concomitant smoothing at the submicrometer level, in accordance with the concept that plasma membrane bilayers can be deformed only with essentially constant area. In another study, Grinnell et al. (1976) reported the smoothing of the electron microscopic contour of BHK cells during the first tens of minutes following deposition on a epoxy substrate. Similarly, in different models of cell-cell aggregation, several authors reported that first contacts involved microvilli with a gradual increase of interaction area, as found, e.g., on chicken embryo retina cells (Ben Shaul and Moscona, 1975) or *Dictyostelium discoideum* (Choi and Siu, 1987).

2. Cell Spreading Is an Active Cell Process

Because there is some similarity between the spreading of a cell and that of a mineral liquid droplet on a surface, it is tempting to ask whether cell spreading may be viewed as a passive consequence of the equilibrium between cell mechanical resistance and adhesion force. We shall give some experimental information on this point before a more quantitative discussion is presented.

First electron microscopic data suggest that spreading is initiated by thin cell protrusions (Witkowski and Brighton, 1971; Heaysman and Pegrum, 1973; see also Fig. 1b of Mège et al., 1986a).

Second, some experimental data suggested that spreading is a metabolically active phenomenon that could be dissociated from adhesion. Thus, spreading, not adhesion, of melanoma cells on laminin-coated substrates required the presence of galactosyl groups on laminin (Runyan et al., 1988). Further, chelating intracellular calcium inhibited spreading, not adhesion, of monocytes deposited on plastic surfaces (Lefkowitz et al., 1992). Finally, many authors were able to inhibit spreading with metabolic inhibitors such as cold (Grinnell and Hays, 1978), microfilament inhibitors such as cytochalasin B (Grinnell and Hays, 1978) or cytochalasin D (Leven and Nachmias, 1982), microtubule inhibitors such as colchicin (Grinnell and Hays, 1978; Leven an Nachmias, 1982), proton ionophores such as monensin (Pizzey et al., 1983), and dibutyryl cyclic adenosine monophosphate (cAMP) (Leven and Nachmias, 1982).

3. Strength of the Adhesive Stimulus and Contact Area

Because spreading seems to be an active process, this might be an all-or-none phenomenon that is spontaneously completed when cells perceive a suitable triggering signal. Indeed, some authors reported that the spreading of cells deposited on surfaces coated with adhesion molecules occurred when the density of binding sites was higher than some threshold value (Aplin and Hughes, 1981; Massia, 1991). However, definite experimental evidence suggests that the extent of cell-cell or cell-substratum apposition is positively correlated with the intensity of adhesive interactions. We describe some examples (some of them were already mentioned in Sec. II).

Capo et al. (1981) studied the interaction between rat peritoneal macrophages and erythrocytes that had been made hydrophobic by glutaraldehyde treatment. These red cells were efficiently bound and subsequently ingested, but electron microscopic observation of contact areas showed that the percentage of cell contour length that was involved in close interaction was fairly low. However, when the erythrocyte negative charge was reduced by removing sialic acid groups with neuraminidase or adding positively charged polylysine molecules, membranes became tightly apposed along extensive areas. A similar increase of fibroblast-to-substratum apposition was found by Sugimoto (1981), who studied the interaction between fibroblasts and negative or neutral surfaces. In another study, Capo et al. (1982) studied the aggregation of rat thymocytes that were coated with various concentrations of concanavalin A, a multivalent ligand for some cell surface carbohydrate group. When the concentration of concanavalin A was increased from 0.125 to 8 μg/mL, the mean area of close apposition between the surfaces of bound cells increased from about 0.3 to 6.5 μm^2. Also, the interaction between neural cells was tightened when negatively charged polysialic acid molecules were removed (Rutishauser et al., 1988; Yang et al., 1992).

4. Cell Adhesion Usually Involves a Concentration of Binding Molecules and Cytoskeletal Elements in Contact Areas

Many experimental reports support the general concept that cell adhesion induces a concentration of several molecular species, including adhesion receptors, in contact areas. Thus, when macrophages were deposited on surfaces coated with immunoglobulins, cell antibody receptor activity disappeared from the upper side, suggesting that these receptors had been concentrated into the contact area (Rabinovitch et al., 1975). This phenomenon allowed Michl et al. (1983) to estimate the diffusion coefficient of macrophage Fc receptors at about 2.5×10^{-9} cm^2/s, in accordance with the view that this migration might be driven by thermal motion. Also, when Singer (1976) studied with electron microscopy lymphocytes that had been aggregated with a lectin, he found that this lectin was concentrated in contact regions. In a later study, McCloskey and Poo (1986) studied the interaction between dinitrophenol (DNP)-derivatized lipid vesicles and rat basophilic leukemia (RBL) cells coated with fluorescent anti-DNP antibodies. Upon cell-to-vesicle contact, fluorescent antibodies began accumulating at contact sites and at the same time the strength of adhesion displayed a time-dependent increase, as assayed by a micropipette aspiration method. Interestingly, redistribution occurred in plasma membrane blebs that were devoid of cytoskeletal elements, and this redistribution was not inhibited by metabolic inhibitors, in accordance with the concept of passive diffusion. Finally, in very elegant work, Dustin et al. (1996) studied the interaction of cells from the Jurkat human T-lymphocyte line, expressing CD2 adhesion receptors, and planar phospholipid bilayers containing freely diffusing fluorescent LFA-3 molecules (LFA-3 is a counterreceptor for CD2). By performing quantitative fluorescence imaging, the authors demonstrated a progressive concentration of LFA-3 in cell surface contact areas. This redistribution occurred rapidly during the first 30 min after contact, and the authors were able to estimate the two-dimensional dissociation constant for CD2/LFA-3 association ($K_d = 21$ molecules/μm^2).

Note, however, that active cellular processes may play a role in the concentration of adhesion molecules in contact areas. An important example is the interaction between lymphocytes and antigen-bearing cells (Kupfer et al., 1987; Kupfer and Singer, 1989). In our laboratory, André et al. (1990) studied the interaction between cytotoxic T lymphocytes and tumor cells bearing the specific antigenic determinants recognized by T-cell receptors or deprived of these antigens. In all cases, some adhesion occurred and it was possible to observe so-called conjugates made of bound lymphocytes and target cells. A quantitative immunofluorescence method (reviewed by Sabri et al., 1997) was used to quantify the redistribution of cell molecules in the contact area, and electron microscopy was used to quantify the fraction of cell area involved in binding. It was found that (a) the cell-cell contact area was about fourfold higher in specific conjugates than in nonspecific ones and (b) the surface *density* of molecules involved in adhesion (including CD8, LFA-1, and possibly T-cell receptor) exhibited a similar increase (between 30 and 80%) in the contact areas of both specific and nonspecific conjugates. Thus, the *amount* of redistributed molecules was dependent on the extent of contact formation, which was considered a metabolic event. Note that actin microfilaments were redistributed in the cytoplasmic zone underlying contact areas. In a later study, André et al. (1991) studied the mechanisms involved in the redistribution of lymphocyte surface molecules that were cross-linked with suitable antibodies. They studied the redistribution of wild-type CD8 and genetically engineered molecules that were essentially deprived of cytoplasmic tail. Quantitative immunofluorescence microscopy was used to study the kinetics of redistribution of CD8 molecules and underlying actin microfilaments. The following conclusions were reached:

1. The cross-linking of both wild-type and tailless molecules induced concentration of these molecules on a pole of the cell (a well-known phenomenon called capping; see Bourguignon and Bourguignon, 1984, for a review).
2. The capping of both wild-type and tailless CD8 induced a concomitant concentration of actin microfilaments, suggesting that interactions involving extracellular and/or transmembrane parts of CD8 and other molecular species provided a link with intracellular cytoskeletal elements.
3. Cytochalasin D, a microfilament inhibitor, substantially decreased the redistribution of CD8. This strongly suggested that concentration of surface molecules involved both passive and active, cytoskeleton-dependent, cell participation.

Note that the accumulation of adhesion molecules and cytoskeletal elements in contact areas is not restricted to cells from the immune system and was reported, for example, in neural cells (Bloch, 1992).

B. Models for the Extension of Contact Area During Cell Adhesion

Although a molecular description of all steps involved in the extension of cell-to-cell contact is presently unavailable, it would be desirable to obtain some semiquantitative understanding of the parameters responsible for the structural reorganization involved in adhesion strengthening. Many theoretical models were elaborated to account for cell binding behavior and adhesion-induced structural changes. For clarity, we shall somewhat arbitrarily discriminate between thermodynamic models, aiming at determining the equilibrium state of adherent cells, and kinetic models, yielding information on the rapidity of cell adhesion and deformation.

1. Thermodynamic Modeling

As mentioned earlier, there seems to be a quantitative relationship between the adhesive stimulus and the extent of cell reorganization. It is therefore tempting to speculate that the final structure of bound cells is a consequence of a balance between several interactions, with minimization of the total free energy. An alternative view would be that the association of cell adhesion molecules with exogenous ligand might generate a signal resulting in active deformation. We shall consider these hypotheses sequentially.

a. Biophysical Modeling

The following parameters may be involved in determining the shape of adherent cells:

1. *Adhesive energy.* Formation of intercellular bonds results in a free energy decrease. This might be quantified by defining a "work of adhesion" (per unit area), using the formalism of the physical chemistry of surfaces (Adamson, 1976; Bongrand et al. 1988b; Bongrand, 1998). However, more information may be obtained by discriminating between adhesion molecules and "surface repellers," antiadhesive structures described in previous sections of this chapter.
2. *Cell mechanical resistance to deformation.* As mentioned previously, the formation of an extensive contact area requires both cell deformation at the micrometer level and smoothing of microvilli at the submicrometer level. A detailed discussion of cell mechanical properties would not fall within the scope of this review, but a brief recall of basic concepts seems essential. We refer the interested reader to a review by Richelme et al. (1996) for more information on cell shape control.

First, although much work was done on the red cell membrane, we shall focus on nucleated cells, whose mechanical properties are strikingly different. Indeed, the erythrocyte surface may deform more rapidly (by several orders of magnitude) than that of nucleated cells. Among these, leukocytes have been extensively studied, and two major approaches have been developed. Schmid-Schönbein et al. (1981) studied the short-term deformations of leukocytes sucked into micropipettes with calibrated pressure, with a time resolution of 30 ms. They accounted for experimental data with a standard viscoelastic model. However, this model is not well suited to the description of slow and large deformations found on adherent cells. Evans and Kukan (1984) also used the micropipette aspiration technique, and they reported that leukocytes might be viewed as viscous droplets (with a viscosity 100,000-fold higher than that of water) surrounded by a membrane under constant tension of 0.035 mN/m. When the suction force was stronger than this tension, cells could thus display extensive deformation and adopt a sausage shape, allowing complete penetration into the pipette if this was of sufficient radius. Note that the tension value may be compared with the energy provided by a density of 1000 bonds/μm^2, with a free energy of 10 kT each (yielding about 0.041 mJ/m^2). We shall now describe three theoretical approaches based on these data.

Bell et al. (1984) neglected cell mechanical properties and considered the interaction between surfaces coated with fixed repellers and mobile tensile adhesion molecules; the balance between attraction and repulsion determined the equilibrium distance between interacting membranes. Increasing the contact area simultaneously increased the entropy of the assembly of ligand-receptor complexes, but this resulted in a decrease of bond density with increase of the tension per bond and concomitant energy increase. An equilibrium contact area could thus be obtained, yielding reasonable agreement with selected experimental data. Torney et al. (1986) later extended this model to the case of free mobile repellers; interestingly, they concluded that coexisting equilibrium states could occur, with either unstretched bonds and repeller redistribution or stretched bonds and partial repeller redistribution. Note that experimental demonstration of repeller redistribution was reported by Soler et al. (1997).

A mechanical study of adhesion was reported by Evans (1985a,b), who considered the equilibrium between an elastic shell and a rigid surface linked by adhesive bonds. He concluded that the classical Young equation was consistent with his model. Also, when he included the discreteness of bonding (Evans, 1985b), he could account for the experimental difference between the minimum tension required to separate membranes and the maximum tension induced in membranes during contact formation (i.e., hysteresis effect). However, his model might be better suited to lipid vesicles or erythrocytes than nucleated cells in view of the importance of bending moments in this system.

In a later study, Mège et al. (1987) attempted to account for both micrometer and submicrometer deformations accompanying adhesion. They studied the interaction between rat macrophages and different erythrocyte samples. In some cases, the erythrocyte negative charge was reduced by treatment with polylysine, a polycation that was adsorbed on the red cell surface. Electron microscopy and digital image processing were combined to analyze the cell rugosity in contact areas as compared with free areas. Micropipette aspiration was used to estimate the relationship between applied tension and apparent area increase after about 30–45 s. The authors assumed that the resistance to membrane apparent extension (at the micrometer level) was accounted for by the concomitant smoothing of microvilli. They used a mechanical equilibrium relationship to derive a tentative "work of adhesion" from the experimental smoothing of membranes in contact areas, yielding respective values of 0.0084 and 0.018 mJ/m^2 for control and polylysine-coated erythrocyte samples.

b. Limits of Biophysical Models

Although the preceding models were not disproved by experimental evidence, they relied on too many poorly determined experimental parameters to be amenable to rigorous experimental check. Perhaps some progress will be achieved when experimental studies of individual ligand-receptor bonds have yielded more precise estimates of some of these unknown parameters. We shall make only a few remarks.

That adhesion and spreading may be impaired by metabolic inhibitors does not prove that the equilibrium binding state is strongly influenced by the cell metabolism. Thus, random cell deformation might provide only a way to accelerate the passage to a bound state. This is the "tar baby effect" mentioned by Bell et al. (1984). It would be useful to know whether spontaneous thermal fluctuations of nucleated cell membranes might provide a rapid enough mechanism for cell deformation, as suggested in a model of lamellipodium formation (Peskin et al., 1993).

If cell membrane tension sets a limit to spreading, it is expected that adherent cells should exert a tension on their substratum. Indeed, Harris et al. (1980) demonstrated that fibroblasts deposited on flexible surfaces exerted a tension, as evidenced by the formation of ripples around these cells. However, the universality of this finding is far from proved. Thus, when the shape of (flexible) erythrocytes bound by phagocytic cells was studied by electron microscopy, only in some cases (Tizard et al., 1974) was erythrocyte deformation reported. Also, when Foa et al. (1988) performed electron microscopic studies of the shape of cytotoxic T lymphocytes binding target tumor cells, they found that the rugosity of the killer cells changed in the contact area to match the target, while micropipette aspiration experiments suggested that cytotoxic cells were more resistant to deformation than their targets. It is indeed quite likely that in many cases cell contact extension is an active phenomenon.

2. Kinetic Models

It has long been recognized that cell adhesion does not behave as an equilibrium phenomenon (Evans, 1985b). Indeed, in his general model of receptor-mediated cell adhesion, Bell (1978) provided an estimate of the initial rate of bond formation. A few years later, Hammer and Lauffenburger (1987) elaborated a dynamical model for receptor-mediated adhesion in a shear field. The contact area was considered constant. Important parameters were the densities of receptor and ligand molecules, the kinetic constants of bond formation and dissociation, and the force dependence of the off-rate. Depending on the numerical values of these parameters, it was possible to predict the occurrence of a rate-controlled, high-affinity regime and an affinity-controlled, low-affinity regime. This model was intended to account for cell adhesion under flow conditions.

The same general framework was rapidly extended to the problem of cell detachment kinetics (Dembo et al., 1988). This required accounting for membrane deformation, because detachment was viewed as a peeling process. Also, the authors considered the possibility that stress might *decrease* the rate of bond dissociation, a thermodynamically conceivable possibility. The authors thus discriminated between "slip bonds," whose dissociation is increased by exogenous forces, and "catch bonds," which might display the reverse behavior. They emphasized that only the ratio between the rates of association and dissociation must be decreased by a disrupting force, as expected from basic thermodynamic principles. This model was extended to account for the possibility that adhesion receptors might form clusters rather than being uniformly distributed (Ward et al., 1994). Experimental data supporting the need to discriminate between cell attachment and detachment will be described in the following section.

A limitation of the foregoing models was that they used a continuous description of bond density. This may be a problem if only a few bonds are involved. This point was addressed by Cozens-Roberts et al. (1990), who presented a probabilistic model of receptor-ligand bond formation between a cell and a surface. By comparing their results with predictions based on the previous model from Hammer and Lauffenburger, they concluded that the continuous approach might underestimate the time needed for cell attachment and overestimate the time and force necessary for cell detachment.

A particular interest of the modeling approach might concern the *rolling phenomenon* that is an important step of inflammatory reactions. When endothelial cells are locally activated following infection or injury, they may express selectin molecules that mediate a very peculiar interaction with leukocytes: cells flowing with high velocity of several hundreds of μm/s (as expected because the wall shear rate is usually higher than 100 s^{-1}) display a drastic reduction of their velocity (by a factor of 50 or more). This phenomenon was demonstrated both in vivo (Von Andrian et al., 1991) and in vitro (Lawrence and Springer, 1991) in a reconstituted model of leukocyte flow on surfaces derivatized with P-selectins. This phenomenon does not require active cell participation because it was obtained with fixed leukocytes flowing on adsorbed selectins (Lawrence and Springer, 1993) and it was reproduced with a totally artificial system made of adsorbed selectin sites and flowing microspheres coated with sialyl Lewis$x$, which is a part of the binding site of leukocyte P-selectin ligand (Brunk et al., 1996). It was found that rolling may be facilitated when the wall shear flow is higher than a threshold value (Finger et al., 1996; Lawrence et al., 1997). Also, it could be found when selectin-bearing surfaces were replaced with suitable antibodies immobilized on the wall of a flow chamber (Chen et al., 1997). However, antibodies supported rolling only within a restricted range of site densities and wall shear stresses; if these requirements were not satisfied, firm adhesion or detachment occurred.

It is not surprising that rolling attracted the interest of biophysicists. Hammer and Apte (1992) performed extensive computer simulation based on previously described kinetic models of adhesion. They added the condition that interaction occurred only through the tip of leukocyte microvilli, in accordance with later experimental data, because the main leukocyte selectin ligand was reported to be concentrated on microvilli (Moore et al., 1995). The main conclusion is that the high capacity of selectins to mediate rolling might be due to a particularly low dependence of the rate of dissociation on applied force, a conclusion supported by later results from Alon et al. (1995). Another model was elaborated by Tözeren and Ley (1992), who concluded that the molecular properties responsible for rolling behavior might be a high rate of bond formation and high bond length and flexibility.

V. CELL DETACHMENT

Although detachment might be considered the last step of cell adhesion, this may be the first process to be studied in this field. Indeed, the only means of proving that a cell is attached to a surface is to exert a well-chosen disrupting force for a given amount of time and determine whether this procedure is sufficient to displace the cell. It is therefore not surprising that many experimental methods were elaborated to quantify the strength of cell-cell adhesion. In this section, we shall first describe selected experimental data on cell detachment by exogenous forces. Then we shall discuss the physiological mechanisms of deadhesion. It is hoped that this presentation will help in assessing the importance of purely mechanical and biological phenomena in the determination of adhesive strength.

A. Measuring the Binding Strength

Three complementary approaches will be described: centrifugation-based methods, micropipette experiments, and use of hydrodynamic forces.

1. Measuring Adhesive Strength with Centrifugal Forces

Nearly 40 years ago, Easty et al. (1960) made carcinoma cells adhere to rectangular coverslips that were then subjected to tangential centrifugation. Substantial detachment was obtained with a tangential disrupting force (i.e., centrifugal force minus Archimedes force) of order of 100 pN that was applied for about 2 min. The next step consisted of assessing the importance of the duration of centrifugation. Corry and Defendi (1981) showed that the percentage of glass-bound macrophages that were detached by a centrifugal force was proportional to both the force and duration of centrifugation. About 50% of these cells were detached when a force of 50,000 pN was applied for about 1000 s. Note that they used virally transformed murine peritoneal macrophages, and macrophages are known to be highly adherent cells. In a later study, Lepoivre and Lemaire (1986) found that glass-bound macrophages resisted a force of 1000 pN.

The importance of time and metabolic events in adhesion was then demonstrated by McClay et al. (1981). They prepared monolayers of chicken retina cells in polylysine-coated multiwell plastic plates. Then they deposited different samples of radioactively labeled cells on these monolayers and, after incubation at 4°C, they reverted and centrifuged the plates and determined the fraction of bound radioactivity. They concluded that a force of the order of 100 pN might be sufficient to disrupt cell-cell adhesion. Further, the fraction of cells that were detached by centrifugation increased when adhesion was made to proceed at 37°C instead of 4°C. Also, this strengthening was impaired by dinitrophenol, an inhibitor of oxidative phosphorylation. The authors concluded that this strengthening involved a passive and an active, metabolically dependent step. The occurrence of a two-step binding process was also found by Sommers et al. (1986), who studied the interaction between macrophages and tumor cells. Finally, Lotz et al. (1989) found that fibroblasts deposited on fibronectin layers at 4°C could resist a disrupting force of 100 pN, and binding strength was increased 10-fold after incubation at 37°C, following a mechanism requiring cytoskeletal integrity.

2. Micropipette Experiments

Early experiments reported by Coman (1944) yielded a very crude estimate of the force required to detach cells from surrounding tissues. This consisted of inserting a microneedle into a single cell and applying a mechanical force by progressive bending of the needle. Proper calibration allowed quantitative determination of the rupture force. A force of the order of 10 μN was required to disrupt normal lip tissue, and the cohesion of carcinoma cells was threefold lower. However, many years were needed before precise calibration of adhesion forces was obtained by Evans and Leung (1984), who monitored the separation of two erythrocytes aggregated by wheat germ agglutinin (WGA, a multivalent ligand of carbohydrates found on the erythrocyte membrane). Red cells were maintained on the tip of micropipettes and controlled suction allowed accurate determination of the membrane tension. The separation force was of the order of 5000 pN, and this increased when the WGA concentration was increased. Also, the force increased during erythrocyte separation, suggesting that the concentration of adhesive bonds increased in the contact zone when this was progressively reduced. This technique was later used by Sung et al. (1986), who estimated at about 4500 pN the strength of the interaction between cytolytic lymphocytes and target cells. Because the radius of the contact zone was of the order of 1 μm, the adhesive energy might be estimated as about 7.5×10^{-4} J/m^2. Further analysis of this experimental model (Tözeren, 1990) suggested that the adhesive bond density increased during detachment, which was consistent with the model of an assembly of freely diffusing bonds considered by Bell et al. (1984).

Finally, Tözeren et al. (1992) studied the interaction between cells from the CD2-positive lymphocyte line Jurkat and a planar bilayer containing LFA-3, a counterreceptor for CD2. Interestingly, the authors compared two isoforms of LFA-3. The first isoform was endowed with a transmembrane component that impaired lateral diffusion, presumably by interacting with the glass surface underlying the bilayer. The second isoform was bound to membranes through a phosphadidylinositol group and exhibited free diffusion. Interestingly, Jurkat cells were strongly bound to monolayers of diffusible LFA-3 molecules, with an adhesive strength comparable to the force measured on previously studied lymphocyte models. In contrast, adhesion on fixed molecules was much weaker, emphasizing the importance of molecule mobility in achieving substantial binding strength.

3. Measurement of Cell Adhesion with Hydrodynamic Flow

As described earlier, the use of hydrodynamic flow proved an efficient way of studying the formation and dissociation of *individual* bonds. However, this approach was first introduced to study *cell* adhesion. Interestingly, following the order of magnitude of applied forces, quite different information could be obtained.

a. Rapid Disruption of Cell Adhesion

A very simple method for measuring adhesive strength was described by Bongrand et al. (1979). This consisted of driving cell aggregates through a syringe needle with controlled pressure (this was achieved by depositing a small weight on the piston of the syringe, which was maintained vertical). The percentage of detached cells was determined by microscopic count after a few passages through the needle. Quantification was achieved by assuming that most cell multiplets displayed fairly random motion in the needle and had the opportunity to experience the maximal shearing forces that occurred near the wall (whereas the shear rate was zero on the needle axis). Now, known results from fluid mechanics indicate that the duration of application of shearing force was of the order of the inverse of shear rate, because doublets exhibited rapid rotation. Because shear rates of the order of $100,000 \text{ s}^{-1}$ had to be applied to detach cells, separation was thus quite rapid. It was found that a force of the order of 100,000 pN was required to release immunoglobulin-coated erythrocytes from macrophages. It was concluded that van der Waals forces were unlikely to account for this high adhesive strength.

This technique was later used to study the mechanical strength of rat thymocyte aggregates obtained with concanavalin A, a multivalent ligand of cell surface carbohydrates. When the concentration of concanavalin A was increased from 0.125 to 8 μg/mL, the force required to disrupt 50% of aggregates displayed a threefold increase. The surface density of bound concanavalin A concomitantly displayed more than a 10-fold increase, and the contact area exhibited more than a 20-fold increase (Capo et al., 1982). In a later study, Bongrand and Golstein (1983) studied the mechanisms of dissociation of conjugates made between cytotoxic T lymphocytes and specific targets. Using chromium-labeled target cells, they were able to demonstrate that disruption of conjugates often resulted in destruction of the target, suggesting that the cell membrane was the weakest link of the adhesive chain.

Note that simple methods of force calibration were reviewed by Benoliel et al. (1994), allowing the implementation of this approach with standard laboratory equipment. Amblard et al. (1992, 1994) described a very efficient method for rapid determination of adhesive forces. This consisted of using the flow generated by a modified flow cytometer, allowing rapid analysis of samples at flow rates extending over three orders of magnitude. They were thus able to study the weak antigen-independent interaction between T and B lymphocytes.

b.   *Progressive Disruptive of Cell-Substrate Attachment*

Flow methods could be used to study cell-surface adhesion as well as cell-cell adhesion. Pioneering studies were reported by Weiss (1961), who subjected adherent rat fibroblasts to a tangential shear force generated by a rotating disk. A force of the order of 100 pN could detach cells when applied for a period of 210 seconds, and the author emphasized the influence of force duration on detachment efficiency. In a later study, Weiss and Coombs (1963) demonstrated the presence on the substrate of surface antigens from mechanically separated cells.

Mège et al. (1986b) studied the adhesion of macrophage-like P388D1 cells to glass capillary tubes with a moderate flow rate. Cells were made to adhere at a low flow rate and high shearing forces were then applied for detachment. The following conclusions were reached: (a) The maximum flow rate compatible with attachment was about 1000-fold lower than that required to detach glass-bound cells. This emphasizes the importance of the bond strengthening process. (b) The minimal time required for the initiation of this strengthening process was estimated to be a few seconds, and strengthening seemed to proceed for a few tens of minutes. (c) A tangential force of about 50,000 pN per cell was required for rapid detachment (say less than a second). (d) There was a reciprocal relationship between the minimal intensity and duration of the force required to detach cells. Indeed, a force of about 500 pN cell could detach a substantial fraction of tested cells within a few tens of seconds. (e) Detachment forces were able to induce marked cell deformation, because cells remaining bound displayed a spindle shape immediately after being sheared. In a later study, Gallik et al. (1989) reported that a force of the order of 20 pN could detach about 50% of neutrophils bound to endothelial cells within a minute.

In an elegant study Chan et al. (1991) explored the influence of molecular mobility on adhesion strengthening by measuring the force required to detach human Jurkat CD2-positive lymphocytes from supported phospholipid bilayers containing a mobile or an immobile form of the CD2 ligand LFA-3. The following conclusions were drawn: (a) At high ligand density (1500 molecules per μm^2), a similar force of the order of 100 pN (corresponding to a wall shear stress of about 12 dyne/cm^2) was required to detach 50% of cells bound to mobile or immobile LFA-3 isoform. (b) When the surface density of LFA-3 was reduced by a factor of about 6, the binding strength was drastically decreased with the immobile ligand and only marginally diminished (by about one third) with the mobile LFA-3 isoform. (c) The binding strength exhibited a threefold increase when the adhesion step was prolonged from 5 to 20 min on a surface containing 50 mobile LFA-3 sites per μm^2. (d) Maximum binding strength was obtained within 5 min when the ligand density was 1000 molecules/μm^2. These conclusions very convincingly demonstrated the importance of receptor concentration in contact areas for efficient adhesion strengthening.

Another model was used by van Kooten et al. (1992), who monitored the detachment of individual surface-bound fibroblasts under flow. A wall shear stress of 35 N/m^2 (corresponding to a force of 14,000 pN per cell, assuming an area of 400 μm^2) could detach 50% of bound cells within 75–90 min. Further, detachment was preceded by cell rounding, emphasizing the influence of cell mechanical properties on adhesive strength. This point is supported by a report from Truskey and Proulx (1993), who studied the shear-induced detachment of fibroblasts adhering to glass or silane surfaces; when shear stress was higher than 4 N/m^2, significant membrane rupture occurred on detachment of glass-bound cells.

In conclusion, during the few minutes following initial attachment, adhesion strengthening results from an extension of the contact area and local concentration of adhesion receptors and submembrane cytoskeletal elements with possible exit of repeller molecules. Detachment may be obtained through membrane rupture and extraction of adhesion receptors, which may happen within a fraction of a second in response to a force of several tens of thousands of pN. Alternatively, deadhesion may proceed within a few minutes, in response to a separating force of the order of 100 pN. This requires progressive cell deformation and sequential rupture of individual bonds or shedding of membrane receptors that may remain bound to the substratum. Release of an active cell "grip" therefore strongly facilitates detachment (Rees et al., 1977).

B. Deadhesion as an Active Cellular Process

Cell function may require spontaneous rupture of adhesive interactions. Thus, migration on a surface involves sequential attachment of the cell leading edge and detachment of the posterior side (e.g., see Stossel, 1993, for a review). The destruction of specific target cells by cytotoxic lymphocytes occurs during a transient adhesion that is initiated and terminated by the killer cell (Rothstein et al., 1978). It is of obvious interest to know how adherent cells manage to terminate adhesion. Some alternative hypotheses might indeed be considered: (a) cells might negatively regulate the affinity of adhesion receptors such as integrins; (b) cells might concentrate repulsive elements in the region where deadhesion is desirable; (c) cells might actively contract, which might lead to mechanical disruption of adhesion; and (d) cells might increase local membrane flexibility, thus allowing easier rupture of individual bonds and progressive detachment.

We shall now review some experimental evidence in order to discuss the physiological relevance of these potential mechanisms.

1. Deadhesion May Be Triggered by Suitable Cell Stimulation

Convincing evidence that deadhesion may be triggered by stimulation of cell surface molecules was provided by Mazerolles et al. (1990). These authors studied antigen-independent adhesion between T lymphocytes (expressing CD4$^+$, CD2, and LFA-1 molecules on their surface) and B lymphocytes expressing the corresponding counterreceptors for these molecular species, i.e., human leukocyte antigen (HLA) class II molecules, LFA-3, and ICAM-1, respectively. Adhesion dropped off about 20 min after contact formation. However, when B-lymphocyte mutants that did not express HLA class II were used, adhesion lasted more than 60 min. Further, the spontaneous deadhesion of the conjugates formed with normal control cells was prevented when CD4–HLA class II interaction was prevented with a peptide inhibitor. The authors concluded that this interaction between CD4 and HLA class II molecules was responsible for triggering deadhesion. Further work suggested that deadhesion required the activation of a kinase ($p56^{lck}$) known to be associated with CD4 (Mazerolles et al., 1994). Evidence provided by Luescher et al. (1995) is consistent with the hypothesis that CD8, a ligand of class I histocompatibility molecules that is expressed by a T-lymphocyte subset, might trigger a similar deadhesion event.

2. Shedding of Cell Surface Receptors Is a Common Means of Achieving Deadhesion

Regen and Horwitz (1992) studied the motion of chick fibroblast migration on a laminin substrate. Cell adhesion was mediated by β_1 integrins. The authors showed that upon detachment at the cell rear, an important amount of β_1 chains remained on the surface, and this was not due to cleavage of this chain because the cytoplasmic domain was not separated from the remainder of the protein.

3. Cell Surface Separation May Be Induced by a Contractile Event

Crowley and Horwitz (1995) showed that adherent fibroblasts that had been permeabilized with digitonin and stimulated with ATP displayed both contraction and detachment. Further, integrins could be detected on the regions where detached cells had adhered. Also, this detachment might be impaired by a peptide inhibiting actin-myosin interaction. These results strongly suggest that detachment was induced by active cell contraction, resulting in uprooting of some integrin molecules. This view is supported by the following three reports. First, as mentioned before, mechanically detached fibroblasts left membrane components on their substrate (Weiss and Coombs, 1963). Second, it was reported that myosin II–defective mutants of *D. discoideum* displayed decreased migration velocity, and this defect was more apparent when substratum adhesiveness was increased (Jay et al., 1995). Third, uprooting of cell surface molecules is a fairly common event, as suggested by the observation that in the neighborhood of the contact area between attached lymphocytes and target cell, either membrane bore antigenic determinants specific for the other cell (Foa et al., 1985).

4. Deadhesion May Involve Cell Surface Repulsive Elements

If the mechanical hypothesis of deadhesion is retained, it is of interest to investigate the precise mechanism relating cytoskeletal contraction to detachment. Evidence suggested that antiadhesive cell surface structures such as CD43/leukosialin might be involved in this process. Indeed, Yonemura et al. (1993) described a concentration of CD43 in the cleavage furrow of dividing leukocytes and demonstrated that this was due to an interaction between the cytoplasmic domain of CD43 and actin filaments, possibly involving molecules of the ERM (ezrin-radixin-moesin) family. More recently, Seveau et al. (1997) reported that CD43 was redistributed in the uropod of polarized neutrophils, and they suggested that this accumulation might allow the uropod detachment during cell migration. Similarly, polysialylated NCAM might allow neural cell migration during development (Ono et al., 1994).

5. Cell Surface Separation May also Be Triggered by a Proteolytic Event

Alternatively, when apoptosis (i.e., programmed cell death) was induced in adherent carcinoma cells by treatment with anti-Fas antibodies, cells separated from their hyaluronate substratum through proteolysis of CD44 surface molecule, a well-known hyaluronan receptor (Gunthert et al., 1996). Also, Cai and Wright (1996) reported that antibodies specific for human leukocyte elastase inhibited neutrophil migration on surfaces coated with fibrinogen, a ligand of MAC-1/CR3 integrin. Because elastase purified from these leukocytes could cleave MAC-1, the authors concluded that migration required rearward rupture of adhesion through a proteolytic event. This conclusion was in line with their previous study (Cai and Wright, 1995) suggesting that deadhesion could not be ascribed to a modulation of the kinetics of ligand binding by MAC-1.

6. Which Intracellular Events Are Associated with Cell Detachment?

In a most informative series of experiments, Maxfield and colleagues studied leukocyte migration on diverse substrates (Marks et al., 1991; Hendey et al., 1992). They found that (a) transient increases of cytosolic calcium concentration seemed required for migration on polylysine-coated substrates, not on less adhesive albumin-bearing surfaces; (b) migration on vitronectin was impaired by inhibiting the intracellular phosphatase calcineurin, and motility was restored by blocking vitronectin-receptor interaction with a soluble peptide; and (c) motility was also prevented by okadaic acid, a phosphatase inhibitor. In line with these results, Nakamura et al. (1995) reported that the neutrophil-derived oxidant NH2Cl induced the detachment of cultured myocytes; this process was inhibited by sodium orthovanadate, a phosphatase inhibitor, and involved intracellular increase of the free calcium concentration. Detachment was also induced by combining calcium increase and inhibition of tyrosine phosphorylation.

Another report by Rees et al. (1977) suggests that cell detachment may be preceded by intracellular events leading to release of the cytoskeleton "grip," which might involve an increase of cell deformability and release of cytoskeleton bound molecules. Such phenomena might be triggered by aforementioned biochemical events.

VI. CONCLUSION

Cell adhesion and deadhesion are clearly a multistep process. A satisfactory understanding of these phenomena requires a thorough integration of physical-chemical and cell biological results. Obviously, some steps of cell adhesion (e.g., formation of the first bond) seem amenable to biophysical modeling, whereas others (such as bond strengthening) cannot be understood without a proper account of intracellular processes. It is hoped that this review will help scientists of different backgrounds to enter this fascinating field.

ACKNOWLEDGMENT

Part of the work described in this chapter was supported by the A.R.C (Association pour la Recherche sur le Cancer).

REFERENCES

Adams CL, Nelson WJ, Smith SJ. Quantitative analysis of cadherin-catening-actin reorganization during development of cell-cell adhesion. J Cell Biol 135:1899–1911, 1996.

Adamson AW. Physical Chemistry of Surfaces. New York: Wiley, 1976.

Alberola-Ila J, Takaki S, Kerner JD, Perlmutter R. Differential signaling by lymphocyte antigen receptors. Annu Rev Immunol 15:203–234, 1997.

Alon R, Hammer DA, Springer TA. Lifetime of P-selectin-carbohydrate bond and its response to tensile force in hydrodynamic flow. Nature 374:539–542, 1995.

Amblard F, Cantin C, Durand J, Fisher A, Sékaly R, Auffray C. New chamber for flow cytometric analysis over an extended range of stream velocity and its application to cell adhesion measurements. Cytometry 13:15–22, 1992.

Amblard F, Auffray C, Sékaly R, Fisher A. Molecular analysis of antigen-independent adhesion forces between T and B lymphocytes. Proc Natl Acad Sci USA 91:3628–3632, 1994.

André P, Benoliel AM, Capo C, Foa C, Buferne M, Boyer C, Schmitt-Verhulst AM, Bongrand P. Use of conjugates made between a cytolytic T cell clone and target cells to study the redistribution of membrane molecules in contact areas. J Cell Sci 97:335–347, 1990.

André P, Gabert J, Benoliel AM, Capo C, Boyer C, Schmitt-Verhulst AM, Malissen B, Bongrand P. Wild type and tailless CD8 display similar interaction with microfilaments during capping. J Cell Sci 100:329–337, 1991.

Angres B, Barth A, Nelson WJ. Mechanism for transition from initial to stable cell-cell adhesion: Kinetic analysis of E-cadherin–mediated adhesion using a quantitative adhesion assay. J Cell Biol 134:549–557, 1996.

Aplin ID, Hughes RC. Cell adhesion on model substrata: Threshold effects and receptor modulation. J Cell Sci 50:89–103, 1981.

Ardman B, Sikorski MA, Staunton DE. CD43 interferes with T-lymphocyte adhesion. Proc Natl Acad Sci USA 89:5001–5005, 1992.

Atherton A, Born GVR. Quantitative investigations on the adhesiveness of circulating polymorphonuclear leukocytes to blood vessel walls. J Physiol (Lond) 222:447–474, 1972.

Bell GI. Models for the specific adhesion of cells to cells. Science 200:618–627, 1978.

Bell GI, Dembo M, Bongrand P. Cell adhesion: Competition between nonspecific repulsion and specific bonding. Biophys J 45:1051–1064, 1984.

Bennett HS. Morphological aspects of extracellular polysaccharides. J Histochem Cytochem 11:14–23, 1963.

Benoliel AM, Capo C, Bongrand P, Ryter A, Depieds R. Nonspecific binding by macrophages: Existence of different adhesive mechanisms and modulation by metabolic inhibitors. Immunology 41:547–560, 1980.

Benoliel AM, Capo C, Mège JL, Bongrand P. Measurement of the strength of cell-cell and cell-substratum adhesion with simple methods. In: Bongrand P, Claesson P, Curtis A, eds. Studying Cell Adhesion. Heidelberg: Springer, 1994, pp 81–92.

Ben Shaul Y, Moscona AA. Scanning electron microscopy of aggregating embryonic neural retina cells. Exp Cell Res 95:191–204, 1975.

Bevilacqua MP, Nelson RM. Selectins. J Clin Invest 91:379–387, 1993.

Bloch RJ. Clusters of neural cell adhesion molecules at sites of cell-cell contact. J Cell Biol 116:449–463, 1992.

Bockris JOM, Reddy AKN. Modern Electrochemistry. New York: Plenum, 1970.

Boland T, Ratner BD. Direct measurement of hydrogen bonding in DNA nucleotide bases by atomic force microscopy. Proc Natl Acad Sci USA 92:5297–5301, 1995.

Bongrand P. Adhesion of cells. In: Lipowsky R, Sackmann E, eds. Structure and Dynamics of Membranes. Vol 1B. Amsterdam: Elsevier/North Holland, 1995, pp 755–803.

Bongrand P. Specific and nonspecific interactions in cell biology. J Dispersion Sci Technol, 19:963–978, 1998.

Bongrand P, Bell GI. Cell-cell adhesion: Parameters and possible mechanisms. In: Perelson AS, DeLisi C, Wiegel FW, eds. Cell Surface Dynamics: Concepts and Models. New York: Marcel Dekker, 1984, pp 459–493.

Bongrand P, Golstein P. Reproducible dissociation of cellular aggregates with a wide range of calibrated shear forces: Application to cytolytic lymphocyte-target cell conjugates. J Immunol Methods 58:209–224, 1983.

Bongrand P, Capo C, Benoliel AM, Depieds R. Evaluation of intercellular adhesion with a very simple technique. J Immunol Methods 28:133–141, 1979.

Bongrand P, Capo C, Depieds R. Physics of cell adhesion. Prog Surf Sci 12:217–286, 1982.

Bongrand P, Pierres M, Golstein P. T cell–mediated cytolysis: On the strength of effector-target cell interaction. Eur J Immunol 13:424–429, 1983.

Bongrand P, Capo C, Mège JL, Benoliel AM. Use of hydrodynamic flows to study cell adhesion. In: Bongrand P, ed. Physical Basis of Cell-Cell Adhesion. Boca Raton, FL: CRC Press, 1988a, pp 125–156.

Bongrand P, Capo C, Mège JL, Benoliel AM. Surface physics and cell adhesion. In: Bongrand P, ed. Physical Basis of Cell-Cell Adhesion. Boca Raton, FL: CRC Press, 1988b, pp 61–90.

Bourguignon LYW, Bourguignon G. Capping and the Cytoskeleton. Int Rev Cytol 87:195–224, 1984.

Bragina EE, Vasiliev M, Gelfand IM. Formation of bundles of microfilaments during spreading of fibroblasts on the substrate. Exp Cell Res 97:241–248, 1976.

Brenner B, Gulbins E, Schlottmann K, Koppenhoefer U, Busch GL, Walzog B, Steinhausen M, Coggeshall KM, Linderkamp O, Lang F. Proc Natl Acad Sci USA 93:15376–15381, 1996.

Brunk DK, Goetz DJ, Hammer DA. Sialyl Lewis$x$/E-selectin–mediated rolling in a cell free system. Biophys J 71:2902–2907, 1996.

Cai TO, Wright SD. Energetics of leukocyte integrin activation. J Biol Chem 270:14358–14365, 1995.

Cai T-Q, Wright SD. Human leukocyte elastase is an endogenous ligand for the integrin CR3 (CD11b/CD18, Mac-1, $\alpha_M\beta_2$) and modulates polymorphonuclear leukocyte adhesion. J Exp Med 184:1213–1223, 1996.

Capo C, Garrouste F, Benoliel AM, Bongrand P, Depieds R. Nonspecific binding by macrophages: Evaluation of the influence of medium-range electrostatic repulsion and short-range hydrophobic interaction. Immunol Commun 10:35–43, 1981a.

Capo C, Bongrand P, Benoliel AM, Ryter A, Depieds R. Particle-macrophage interaction: Role of surface charges. Ann Immunol (Inst Pasteur) 132D:165–173, 1981b.

Capo C, Garrouste F, Benoliel AM, Bongrand P, Ryter A, Bell GI. Concanavalin A–mediated thymocyte agglutination: A model for a quantitative study of cell adhesion. J Cell Sci 26:21–48, 1982.

Casimir HBG, Polder D. The influence of retardation on the London–van der Waals forces. Phys Rev 73:360–372, 1948.

Caveda L, Martinpadura L, Navarro P, Breviario F, Corada M, Gulino D, Lampugnani MG, Dejana E. Inhibition of cultured cell growth by vascular endothelial cadherin (cadherin-5 VE-cadherin). J Clin Invest 98:886–893, 1996.

Chan PY, Springer TA. Effect of lengthening lymphocyte function associated antigen 3 on adhesion to CD2. Mol Biol Cell 3:157–176, 1992.

Chan PY, Lawrence MB, Dustin ML, Ferguson LM, Golan DE, Springer TA. Influence of receptor lateral mobility on adhesion strengthening between membranes containing LFA-3 and CD2. J Cell Biol 115:245–255, 1991.

Chen S, Alon R, Fuhlbrigge RC, Springer TA. Rolling and transient tethering of leukocytes on antibodies reveal specializations of selectins. Proc Natl Acad Sci USA 94:3172–3177, 1997.

Choi AHC, Siu CH. Filopodia are enriched in a cell cohesion molecule of Mr 80,000 and participate in cell-cell contact formation in *Dictyostelium discoideum*. J Cell Biol 104:1375–1387, 1987.

Clarris BJ, Fraser JRE. On the pericelular zone of some mammalian cells in vitro. Exp Cell Res 49:181–193, 1968.

Coman DR. Decreased mutual adhesiveness, a property of cells from squamous cell carcinomas. Cancer Res 4:625–629, 1944.

Condeelis J. Life at the leading edge: The formation of cell protrusions. Annu Rev Cell Biol 9:411–444, 1993.

Cook GMW. Techniques for the analysis of membrane carbohydrates. In: Maddy AH, ed. Biochemical Analysis of Membranes. London: Chapman, 1976, pp 283–351.

Corry WD, Defendi V. Centrifugal assessment of cell adhesion. J Biochem Biophys Methods 4:29–38, 1981.

Cozens-Roberts C, Laufenburger DA, Quinn JA. Receptor-mediated cell attachment and detachment kinetics. I. Probabilistic model and analysis. Biophys J 58:841–856, 1990.

Crockett-Torabi E, Sulenbarger B, Wayne-Smith C, Fantone JC. Activation of human neutrophils through L-selectin and Mac-1 molecules. J Immunol 154:2291–2302, 1995.

Crowley E, Horwitz AF. Tyrosine phosphorylation and cytoskeletal tension regulate the release of fibroblast adhesions. J Cell Biol 131:525–537, 1995.

Cunningham BA. Cell adhesion molecules as morphoregulators. Curr Opin Cell Biol 7:628–633, 1995.

Curtis ASG. The Cell Surface: Its Molecular Role in Morphogenesis. London: Logos Press, Academic Press, 1967.

Cyster JC, Shotton DM, Williams AF. The dimensions of the T lymphocyte glycoprotein leukosialin and identification of linear protein epitopes that can be modified by glycosylation. EMBO J 10:893–902, 1991.

Daeron M. Fc receptor biology. Annu Rev Immunol 15:203–234, 1997.

Dammer U, Popescu O, Wagner P, Anselmetti D, Güntherodt H-J, Misevic GN. Binding strength between cell adhesion proteoglycans measured by atomic force microscopy. Science 267:1173–1175, 1995.

Danilov YN, Juliano RL. Phorbol ester modulation of Integrin-mediated cell adhesion: A postreceptor event. J Cell Biol 108:1925–1933, 1989.

Darmani H, Coakley WT. Membrane-membrane interactions: Parallel membrane or patterned discrete contacts. Biochim Biophys Acta 1021:182–190, 1990.

Darmani H, Coakley WT, Hamm AC, Brain A. Spreading of wheat germ agglutinin–induced erythrocyte contacts by formation of spatially discrete contacts. Cell Biophys 10:105–126, 1990.

De Gennes PG. Scaling Law in Polymer Physics. Ithaca, NY: Cornell University Press, 1979.

De Gennes PG. Polymers at an interface—A simplified view. Adv Colloid Interface Sci 27:189–209, 1987.

De Gennes PG. Model polymers at interfaces. In: Bongrand P, ed. Physical Basis of Cell-Cell Adhesion. Boca Raton, FL: CRC Press, 1988, pp 39–60.

Dea ICM, McKennon AA, Rees DA. Tertiary and quaternary structure in aqueous polysaccharide. Systems which model cell wall cohesion: Reversible changes in conformation and association of agarose. J Mol Biol 68:153–172, 1972.

Dembo M, Torney DC, Saxman K, Hammer D. The reaction-limited kinetics of membrane-to-surface adhesion and detachment. Proc R Soc Lond Ser B 234:55–83, 1988.

Despont JP, Abel CA, Grey HM. Sialic acid and sialyl transferases in murine lymphoid cells: Indicators of T cell maturation. Cell Immunol 17:487–494, 1975.

Detmers PA, Wright SD, Olsen E, Kimball B, Cohn ZA. Aggregation of complement receptors of human neutrophils in the absence of ligand. J Cell Biol 105:1137–1145, 1987.

Dimitrov DS. Dynamic interactions between approaching surfaces of biological interest. Prog Surf Sci 14:295–423, 1983.

Dolan AK, Edwards SF. Theory of the stabilization of colloids by adsorbed polymer. Proc R Soc 1 337:509–516, 1974.

Dolan AK, Edwards SF. The effect of excluded volume on polymer dispersant action. Proc R Soc Lond Ser A 343:427–442, 1975.

Dustin ML, Ferguson LM, Chan P-Y, Springer TA, Golan DE. Visualization of CD2 interaction with LFA-3 and determination of the two-dimensional dissociation constant for adhesion receptors in a contact area. J Cell Biol 132:465–474, 1996.

Easty GC, Mercer EH. An electron microscope study of model tissues formed by the agglutination of erythrocytes. Exp Cell Res 28:215–227, 1962.

Easty GC, Easty DM, Ambrose EJ. Studies of cellular adhesiveness. Exp Cell Res 19:539–548, 1960.

Erlandsen SL, Hasslen SR, Nelson RD. Detection and spatial distribution of the β_2-integrin (Mac-1) and L-selectin (LECAM-1) adherence receptors on human neutrophils by high-resolution field emission SEM. J Histochem Cytochem 41:327–333, 1993.

Evans EA. Minimum energy analysis of membrane deformation applied to pipet aspiration and surface adhesion of red blood cells. Biophys J 30:265–284, 1980.

Evans EA. Detailed mechanics of membrane-membrane adhesion and separation. I. Continuum of molecular cross-bridges. Biophys J 48:175–183, 1985a.

Evans EA. Detailed mechanics of membrane-membrane adhesion and separation. II. Discrete kinetically trapped molecular cross-bridges. Biophys J 48:185–192, 1985b.

Evans EA, Kukan B. Passive material behavior of granulocytes based on large deformations and recovery after deformation tests. Blood 64:1028–1035, 1984.

Evans E, Leung A. Adhesivity and rigidity of erythrocyte membrane in relation to wheat germ agglutining binding. J Cell Biol 98:1201–1208, 1984.

Evans E, Ritchie K. Dynamic strength of molecular adhesion bonds. Biophys J 72:1541–1555, 1997.

Evans E, Berk D, Leung A. Detachment of agglutinin-bonded red blood cells. I—Forces to rupture molecular-point attachments. Biophys J 59:838–848, 1991.

Evans E, Merkel R, Ritchie K, Tha S, Zilker A. Picoforce method to probe submicroscopic actions in biomembrane adhesion. In: Bongrand P, Claesson PM, Curtis ASG, eds. Studying Cell Adhesion. Heidelberg: Springer Verlag, 1994, pp 125–139.

Eylar EH, Madoff MA, Brody OV, Oncley JL. The contribution of sialic acid to the surface charge of the erythrocyte. J Biol Chem 237:1992–2000, 1962.

Eyring H. The activated complex in chemical reactions. J Chem Phys 3:107–115, 1935.

Fasman GD. Handbook of Biochemistry and Molecular Biology. Boca Raton, FL: CRC Press, 1975, p 387.

Finger EB, Puri KD, Alon R, Lawrence MB, VonAndrian UH, Springer TA. Adhesion through L-selectin requires a threshold hydrodynamic shear. Nature 379:266–269, 1996.

Florin EL, Moy VT, Gaub HE. Adhesion forces between individual ligand-receptor pairs. Science 264:415–417, 1994.

Flory PJ. Principles of Polymer Chemistry. Ithaca, NY: Cornell University Press, 1953.

Foa C, Bongrand P, Galindo JR, Golstein P. Unexpected cell surface labeling in conjugates between cytotoxic T lymphocytes and target cells. J Histochem Cytochem 33:647–657, 1985.

Foa C, Mège JL, Capo C, Benoliel AM, Galindo JR, Bongrand P. T-cell-mediated cytolysis: Analysis of killer and target cell deformability and deformation during conjugate formation. J Cell Sci 89:561–573, 1988.

Foa C, Soler M, Fraterno M, Passerel M, Lavergne JL, Martin JM, Bongrand P. Electron microscopical analysis of cell-cell and cell-substrate interactions: Use of image analysis, X-ray microanalysis and EFTEM. In: Bongrand P, Claesson P, Curtis A, eds. Studying Cell Adhesion. Heidelberg: Springer Verlag, 1994, 219–241.

Foa C, Soler M, Benoliel AM, Bongrand P. Steric stabilization and cell adhesion. J Mater Sci Mater Med 7:141–148, 1996.

Folkman J, Moscona A. Role of cell shape in growth control. Nature 273:345–349, 1978.

Follett EAC, Goldman RD. The occurrence of microvilli during spreading and growth of BHK 21/C13 fibroblasts. Exp Cell Res 59:124–136, 1970.

Fransson LA. Interaction between dermatan sulphate chains. I—Affinity chromatography of copolymeric galactosaminoglycans or dermatan sulphate–substituted agarose. Biochim Biophys Acta 437:106–115, 1976.

Fransson LA, Nieduszynski IA, Phelps CF, Sheehan JK. Interactions between dermatan sulphate chains. III—Light scattering and viscometry study of self association. Biochim Biophys Acta 586:179–188, 1979.

Frey A, Grannasca KT, Weltzin R, Grannasca PJ, Reggio H, Leucer WI, Neutra MR. Role of the glycocalyx in regulating access of microparticles to apical plasma membranes of intestinal epithelial cells: Implications for microbial attachment and oral vaccine targeting. J Exp Med 184:1045–1059, 1996.

Gallez D, Coakley WT. Far-from-equilibrium phenomena in bioadhesion processes. Heterogeneous Chem Rev 3:443–475, 1996.

Gallik S, Usami S, Jan K-M, Chien S. Shear stress–induced detachment of human polymorphonuclear leukocytes from endothelial monolayers. Biorheology 26:823–834, 1989.

Gennis RB. Biomembranes—Molecular structure and function, Heidelberg: Springer Verlag, 1989, p 20.

Gingell D, Parsegian VA. Computation of van der Waals interactions in aqueous systems using reflectivity data. J Theor Biol 36:41–52, 1972.

Golan DE, Alecio RM, Veatch WR, Rando RR. Lateral mobility of phospholipid and cholesterol in the human erythrocyte membrane: Effects of protein-lipid interaction. Biochemistry 23:332–339, 1984.

Goldman AJ, Cox RG, Brener H. Slow viscous motion of a sphere parallel to a plane wall. II—Couette flow. Chem Eng Sci 22:653–660, 1967.

Goldsmith HL, Takamura K, Tha S, Tees D. Study of cell-cell interactions with the travelling microtube. In: Bongrand P, Claesson P, Curtis ASG, eds. Studying Cell Adhesion. Heidelberg: Springer Verlag, 1994, pp 141–156.

Gretz JE, Kaldjian EP, Anderson AO, Shaw S. Sophisticated strategies for information encounter in the lymph node. J Immunol 157:495–499, 1996.

Grinnell F, Hays DG. Induction of cell spreading by substratum-adsorbed ligands directed against the cell surface. Exp Cell Res 116:275–284, 1978.

Grinnell F, Tobleman MQ, Hackenbrock I. J Cell Biol 70:707–713, 1976.

Gunthert AR, Strater J, von Reyher U, Henne C, Joos S, Koretz K, Moldenhauer G, Krammer PH, Moller P. Early detachment of colon carcinoma cell during CD95 (APO-1/Fas)-mediated apoptosis. 1—De-adhesion from hyaluronate by shedding of CD44. J Cell Biol 134:1089–1096, 1996.

Guthridge JM, Kaplan AM, Cohen DA. Regulation of B cell: T-cell interaction: Potential involvement of an endogenous B cell sialidase. Immunol Invest 23:393–411, 1994.

Hamaker HC. The London–van der Waals attraction between spherical particles. Physica IV:1058–1072, 1937.

Hammer DA, Apte SM. Simulation of cell rolling and adhesion on surfaces in shear flow: General results and analysis of selectin-mediated neutrophil adhesion. Biophys J 63:35–57, 1992.

Hammer DA, Lauffenburger DA. A dynamical model for receptor-mediated cell adhesion to surfaces. Biophys J 52:475–487, 1987.

Harding CV, Unanue ER. Quantitation of antigen-presenting cell MHC class II/peptide complexes necessary for T-cell stimulation. Nature 346:574–576, 1990.

Harlan JM, Liu DY. Adhesion—Its role in Inflammatory Diseases. New York: WH Freeman, 1992.

Harris AK, Wild P, Stopak D. Silicone rubber substrata: A new wrinkle in the study of cell locomotion. Science 208:177–179, 1980.

Harrison R, Lunt GG. Biological Membranes—Their Structure and Function. Glasgow: Blackie, 1975, p 117.

Hascall VC, Heingard D. Aggregation of cartilage proteoglycans. I.—The role of hyaluronic acid. J Biol Chem 249:4232–4241, 1974.

Heath JP. Adhesions to substratum and locomotory behaviour of fibroblastic and epithelial cells in culture. In: Bellairs R, Curtis A, Dunn G, eds. Cell Behaviour. Cambridge: Cambridge University Press, 1982, pp 77–108.

Heaysman JEM, Pergrum SM. Early contacts between fibroblasts. Exp Cell Res 78:71–78, 1973.

Helm CA, Knoll W, Israelachvili JN. Measurement of ligand-receptor interactions. Proc Natl Acad Sci USA 88:8169–8173, 1991.

Hendey B, Klee CB, Maxfield FR. Inhibition of neutrophil chemokinesis on vitronectin by inhibitors of calcineurin. Science 258:296–299, 1992.

Hilkens J, Ligtenberg MJL, Vos HL, Livitnov SV. Cell membrane–associated mucins and their adhesion-modulating property. Trends Biochem Sci 17:359–363, 1992.

Hinterdorfer P, Baumgartner W, Gruber HJ, Schilcher K, Schindler H. Detection and localization of individual antibody-antigen recognition events by atomic force microscopy. Proc Natl Acad Sci USA 93:3477–3481, 1996.

Höök M, Kjellén L, Johansson S. Cell-surface glycosaminoglycans. Annu Rev Biochem 53:847–869, 1984.

Hough DB, White LR. The calculation of Hamaker constants from Lifshitz theory with applications to wetting phenomena. Adv Colloid Interface Sci 14:3–41, 1980.

Hynes RO. Integrins: Versatility, modulation and signaling in cell adhesion. Cell 69:11–25, 1992.

Inouyé K, Takeuchi I. Motive force of the migrating pseudoplasmodium of the cellular slime mould *Dictyostelium discoideum*. J Cell Sci 41:53–64, 1980.

Israelachvili JN. Intermolecular and Surface Forces. 2nd ed. New York: Academic Press, 1991, pp 288–311.

Israelachvili JN, Tabor D. The measurement of van der Waals dispersion forces in the range 1.5 to 130 nm. Proc R Soc A 331:19–38, 1972.

Jakubowski A, Rosa MD, Bixler S, Lobb R, Burkly LC. Vascular cell adhesion molecule (VCAM)–Ig fusion protein defines distinct affinity states of the very late antigen-4 (VLA-4) receptor. Cell Adhesion Commun 3:131–142, 1995.

Jan KM, Chien S. Role of surface electric charge in red blood cell interactions. J Gen Physiol 61:638–654, 1973.

Jay PY, Pham PA, Wong SA, Elson EL. A mechanical function of myosin II in cell motility. J Cell Sci 108:387–393, 1995.

Jentoft N. Why are proteins O-glycosylated? Trends Biochem Sci 15:291–294, 1990.

Kabat EA. Structural Concepts in Immunology and Immunochemistry. New York: Holt, Rinehart & Winston, 1968, pp 86–87.

Kansas GS. Selectins and their ligands: Current concepts and controversies. Blood 88:3259–3287, 1996.

Kaplanski G, Farnarier C, Tissot O, Pierres A, Benoliel A-M, Alessi M-C, Kaplanski S, Bongrand P. Granulocyte-endothelium initial adhesion. Analysis of transient binding events mediated by E-selectin in a laminar shear flow. Biophys J 64:1922–1933, 1993.

Kaplanski G, Farnarier C, Benoliel AM, Foa C, Kaplanski S, Bongrand P. A novel role for E- and P-selectins: shape control of endothelial cell monolayers. J Cell Sci 107:2449–2457, 1994.

Keizer GD, Visser W, Vliem M, Figdor CG. A monoclonal antibody (NKI-L16) directed against a unique epitope on the α-chain of human leukocyte function-associated antigen 1 induces homotypic cell-cell interactions. J Immunol 140:1393–1400, 1988.

Kemperman H, Wijnands Y, Wesseling J, Niessen CM, Sonnenberg A, Roos E. The mucin epiglycanin on TA3/HA carcinoma cells prevents $\alpha_6\beta_4$-mediated adhesion to laminin and kalinin and E-cadherin–mediated cell-cell interaction. J Cell Biol 127:2071–2080, 1994.

Klotz SA. Fungal adherence to the vascular compartment: A critical step in the pathogenesis of disseminated candidiasis. Clin Infect Dis 14:340–347, 1992.

Kluge A, Reuter G, Lee HY, Rucheeger B, Schauer R. Interaction of rat peritoneal macrophages with homologous sialidase-treated thrombocytes in vitro. Biochemical and morphological studies. Eur J Cell Biol 59:12–20, 1992.

Kolodney MS, Wysolmerski RB. Isometric contraction by fibroblasts and endothelial cells in tissue culture: A quantitative study. J Cell Biol 117:73–82, 1992.

Kupfer A, Singer SJ. Cell biology of cytotoxic and helper T cell functions: Immunofluorescence microscopic studies of single cells and cell couples. Annu Rev Immunol 7:309–337, 1989.

Kupfer A, Singer SJ, Janeway CA, Swain SL. Coclustering of a CD4 (L3TE) molecule with the T-cell receptor is induced by specific direct interaction of helper T cells and antigen-presenting cells. Proc Natl Acad Sci USA 84:5888–5892, 1987.

Kwong D, Tees DFJ, Goldsmith HL. Kinetics and locus of failure of receptor-ligand-mediated adhesion between latex spheres. II. Protein-protein bonds. Biophys J 71:1115–1122, 1996.

Lackie JM. The structure and organization of the cell surface. In: Bittar EE, ed. Membrane Structure and Function. Vol 1. New York: Wiley, 1980, pp 73–102.

Lasky LA, Singer MS, Dowbenko D, Imai Y, Henzel WJ, Grimley C, Fennie C, Gillett N, Watson SR, Rosen SD. Cell 69:927–938, 1992.

Latena J, Ansbacher R, Culp LA. Glycosaminoglycans that bind cold-insoluble globulin in cell-substratum adhesion sites of murine fibroblasts. Proc Natl Acad Sci USA 77:6662–6666, 1980.

Law JH, Snyder WR. Membrane lipids. In: Fox CF, Keith A, eds. Membrane Molecular Biology. Stamford: Sinauer, 1972, pp 3–26.

Lawrence MB, Springer TA. Leukocytes roll on a selectin at physiologic flow rates: Distinction from and prerequisite for adhesion through integrins. Cell 65:859–873, 1991.

Lawrence MB, Springer TA. Neutrophils roll on E-selectin. J Immunol 151:6338–6346, 1993.

Lawrence MB, Kansas GS, Kunkel EJ, Ley K. Threshold levels of fluid shear promote leukocyte adhesion through selectins (CD62L, P, E). J Cell Biol 136:717–727, 1997.

Leckband DE, Israelachvili JN, Schmitt FJ, Knoll W. Long range attraction and molecular rearrangements in receptor-ligand interactions. Science 255:1419–1421, 1992.

Leckband DE, Kuhl T, Wang HK, Herron J, Müller W, Ringsdorf H. 4-4-20 anti-fluoresceyl IgG recognition of membrane bound hapten: Direct evidence for the role of protein and interfacial structure. Biochemistry 34:11467–11478, 1995.

Lee GM, Johnstone B, Jacobson K, Caterson B. The dynamic structure of the pericellular matrix on living cells. J Cell Biol 123:1899–1907, 1993a.

Lee GM, Zhang F, Ishihara A, McNeil CL, Jacobson KA. Unconfined lateral diffusion and estimate of pericellular matrix viscosity revealed by measuring the mobility of gold-tagged lipids. J Cell Biol 120:25–35, 1993b.

Lee GU, Kidwell DA, Colton RJ. Sensing discrete streptavidin-biotin interactions with atomic force microscopy. Langmuir 10:354–357, 1994.

Lefkowitz JB, Lennartz MR, Rogers M, Morrison AR, Brown EJ. Phospholipase activation during monocyte adherence and spreading. J Immunol 149:1729–1735, 1992.

Lepoivre M, Lemaire G. Quantification of intercellular binding strength by disruptive centrifugation: Application to the analysis of adhesive interactions between P815 tumour cells and activated macrophages. Ann Immunol Inst Pasteur 137C:329–344, 1986.

Leven RM, Nachmias VT. Cultured megakaryocytes: Changes in the cytoskeleton after ADP-induced spreading. J Cell Biol 92:313–323, 1982.

Levine S, Levine M, Sharp KA, Brooks DE. Theory of the electrokinetic behavior of human erythrocytes. Biophys J 42:127–135, 1983.

Levine YK. Physical studies of membrane structure. Prog Biophys Mol Biol 24:1–74, 1972.

Li F, Erickson HP, James JA, Moore KL, Cummings RD, McEver RP. Visualization of P-selectin glycoprotein ligand-1 as a highly extended molecule and mapping of protein epitopes for monoclonal antibodies. J Biol Chem 271:6342–6348, 1996.

Lifshitz EM. The theory of molecular attractive forces between solids. Sov Phys JETP 2:73–83, 1956.

Ligtenberg MJ, Buijs F, Vos HL, Hilkens J. Suppression of cellular aggregation by high levels of episialin. Cancer Res 52:2318–2324, 1992.

Lo SK, Van Seventer GA, Levin SM, Wright SD. Two leukocyte receptors (CD11a/CD18 and CD11b/CD18) mediate transient adhesion to endothelium by binding to different ligands. J Immunol 143:3325–3329, 1989.

Lollo BA, Chan KWH, Hanson EM, Moy VT, Brian AA. Direct evidence for two affinity states for lymphocyte-function-associated antigen 1 on activated T cells. J Biol Chem 15:21693–21700, 1993.

Lorant DE, McEver RP, McIntyre TM, Moore KL, Prescott SM, Zimmerman GA. Activation of polymorphonuclear leukocytes reduces their adhesion to P-selectin and causes redistribution of ligands of P-selectin on their surfaces. J Clin Invest 96:171–182, 1995.

Lotz MM, Burdsal CA, Erickson HP, McClay DR. Cell adhesion to fibronectin and tenascin: Quantitative measurements of initial binding and subsequent strengthening response. J Cell Biol 109:1795–1805, 1989.

Luckham P, Klein J. Interactions between smooth solid surfaces in solutions of adsorbing and nonadsorbing polymers in good solvent conditions. Macromolecules 18:721–728, 1985.

Luescher IF, Vivier E, Layer A, Mahiou J, Godeau F, Malissen B, Romero P. CD8 modulation of T-cell antigen receptor-ligand interactions on living cytotoxic T lymphocytes. Nature 373:353–356, 1995.

Malsmsten M, Blomberg E, Claesson P, Carlstedt I, Lusegren I. Mucin layers on hydrophobic surfaces studied with ellipsometry and surface force measurements. J Colloid Interface Sci 151:579–590, 1992.

Manjunath N, Correa M, Ardman M, Ardman B. Negative regulation of T-cell adhesion and activation by CD43. Nature 317:535–538, 1995.

Manning GS. Limiting laws and counterion condensation in polyelectrolyte solutions. I. Colligative properties. J Chem Phys 51:924–933, 1969.

Marks PW, Hendey B, Maxfield FR. Attachment to fibronectin or vitronectin makes human neutrophil migration sensitive to alterations in cytosolic free calcium concentration. J Cell Biol 112:149–158, 1991.

Martinez-Palomo A. The surface coats of animal cells. Int Rev Cytol 29:29–75, 1970.

Massia SP. An RGD spacing of 440 nm is sufficient for integrin $\alpha_v\beta_3$-mediated fibroblast spreading and 140 nm for focal contact and stress fiber formation. J Cell Biol 114:1089–1100, 1991.

Mazerolles F, Amblard F, Lumbroso C, Lecomte O, van de Moortele P-F, Barbat C, Piatier-Tonneau D, Auffray C, Fischer A. Regulation of T helper–B lymphocyte adhesion through CD4-HLA class II interaction. Eur J Immunol 20:637–644, 1990.

Mazerolles F, Barbat C, Meloche S, Gratton S, Soula M, Fagard R, Fischer S, Hivroz C, Bernier J, Sekaly RP, Fischer A. LFA-1 mediated antigen-independent T cell adhesion regulated by CD4 and p56[lck] tyrosine kinase. J Immunol 152:5670–5679, 1994.

McClay DR, Wessel GM, Marchase RB. Intercellular recognition: Quantitation of initial binding events. Proc Natl Acad Sci USA 78:4975–4979, 1981.

McCloskey MA, Poo M-M. Contact-induced redistribution of specific membrane components: Local accumulation and development of adhesion. J Cell Biol 102:2185–2196, 1986.

McFarland TA, Ardman B, Manjunath N, Fabry JA, Lieberman J. CD43 diminishes susceptibility to T lymphocyte–mediated cytolysis. J Immunol 154:1097–1104, 1995.

Mège JL, Capo C, Benoliel AM, Foa C, Galindo R, Bongrand P. Quantification of cell surface roughness; a method for studying cell mechanical and adhesive properties. J Theor Biol 119:147–160, 1986a.

Mège JL, Capo C, Benoliel Am, Bongrand P. Determination of binding strength and kinetics of binding initiation. A model study made on the adhesive properties of P388D1 macrophage-like cells. Cell Biophys 8:141–160, 1986b.

Mège JL, Capo C, Benoliel AM, Bongrand P. Use of cell contour analysis to evaluate the affinity between macrophages and glutaraldehyde-treated erythrocytes. Biophys J 52:177–186, 1987.

Merkel R, Ritchie K, Evans E. Slow loading of biotin-streptavidin bonds yields unexpectedly low detachment forces. Biophys J 68:A404, 1995.

Mester L, Szabados L, Born GVR, Michal F. Changes in the aggregation of platelets enriched in sialic acid. Nature New Biol 236:213–214, 1972.

Michl J, Pieczonka MM, Unkeless JC, Bell GI, Silverstein SC. Fc receptor modulation in mononuclear phagocytes maintained on immobilized immune complexes occurs by diffusion of the receptor molecule. J Exp Med 157:2121–2139, 1983.

Mishell BM, Shiigi SM. Selected Methods in Cellular Immunology. San Francisco: WH Freeman, 1980, pp 190–191.

Miyata H, Yasuda R, Kinosita K Jr. Strength and lifetime of the bond between actin and skeletal muscle α-actinin studied with an optical trapping technique. Biochim et Biophys Acta 1290:83–88, 1996.

Moore KL, Patel KD, Bruehl RE, Fugang L, Johnson DA, Lichenstein HS, Cummings RD, Bainton DF, McEver RP. P-selectin glycoprotein ligand-1 mediates rolling of huyman neutrophils on P-selectin. J Cell Biol 128:661–671, 1995.

Mould AP. Getting integrins into shape: Recent insights into how integrin activity is regulated by conformational changes. J Cell Sci 109:2613–1618, 1996.

Nakamura TY, Yamamoto I, Nishitani H, Matozaki T, Suzuki T, Wakabayashi S, Shigekawa M, Goshima K. Detachment of cultured cells from the substratum induced by the neutrophil-derived oxidant NH2Cl: Synergistic role of phosphotyrosine and intracellular Ca^{2+} concentration. J Cell Biol 131:509–524, 1995.

Napper DH. Steric stabilization. J Colloid Interface Sci 58:390–407, 1977.

Napper DH. Polymeric Stabilization of Colloidal Dispersions. London: Academic Press, 1983.

Nathan C, Xie Q, Halbwachs-Mecarelli L, Jin W. Albumin inhibits neutrophil spreading and hydrogen peroxide release by blocking the shedding of CD43 (sialophorin, leukosialin). J Cell Biol 122:243–256, 1993.

Nir S, Andersen M. van der Waals interactions between cell surfaces. J Membr Biol 31:1–18, 1977.

Norment AM, Salter RD, Parham P, Engelhard V, Littman DR. Cell-cell adhesion mediated by CD8 and MHC class I molecules. Nature 336:79–81, 1988.

Oldberg A, Kjellén L, Höök M. J Biol Chem 254:8505–8510, 1979.

Oliver T, Dembo M, Jacobson K. Traction forces in locomoting cells. Cell Motil Cytoskeleton 31:225–240, 1995.

Ono K, Tomasiewicz H, Magnuson T, Rutishauser U. N-CAM mutation inhibits tangential neuronal migration and is phenocopied by enzymatic removal of polysialic acid. Neuron 13:595–609, 1994.

Patel KD, Nollert MU, McEver RP. P-selectin must extend a sufficient length from the plasma membrane to mediate rolling of neutrophils. J Cell Biol 131:1893–1902, 1995.

Pauli BU, Augustin-Voss HG, El-Sabban ME, Johnson RC, Hammer DA. Organ-preference of metastasis. The role of endothelial adhesion molecules. Cancer Metastasis Rev 9:175–189, 1990.

Perez E, Proust JE. Forces between mica surfaces coveret with adsorbed mucin across aqueous solution. J Colloid Interface Sci 118:182–191, 1987.

Peskin CS, Modell GM, Oster GF. Cellular motions and thermal fluctuations: The Brownian ratchet. Biophys J 65:316–324, 1993.

Phelps CF. The dilute solution properties of glycosaminoglycans and proteoglycans. In: Arnott S, Rees A, Morris R, eds. Molecular Biophysics of the Extracellular Matrix. Clifton, New Jersey: Humana Press, 1983, pp 21–39.

Pierres A, Tissot O, Malissen B, Bongrand P. Dynamic adhesion of CD8-positive cells to antibody-coated surfaces: The initial step is independent of microfilaments and intracellular domains of cell-binding molecules. J Cell Biol 125:945–953, 1994.

Pierres A, Benoliel AM, Bongrand P. Measuring the lifetime of bonds made between surface-linked molecules. J Biol Chem 270:26586–26592, 1995.

Pierres A, Benoliel AM, Bongrand P. Measuring bonds between surface-associated molecules. J Immunol Methods 196:105–120, 1996a.

Pierres A, Benoliel AM, Bongrand P. Experimental study of the rate of bond formation between individual receptor-coated spheres and ligand-bearing surfaces. J Phys III 6:807–824, 1996b.

Pierres A, Benoliel AM, Bongrand P, van der Merwe PA. Determination of the lifetime and force dependence of interactions of single bonds between surface-attached CD2 and CD48 adhesion molecules. Proc Natl Acad Sci USA 93:15114–15118, 1996c.

Pierres A, Benoliel AM, Bongrand P. The dependence of the association-rate of surface-attached adhesion molecules CD2 and CD48 on separation distance. FEBS Lett 403:239–244, 1997.

Pierres A, Benoliel AM, Bongrand P. Studying receptor-mediated cell adhesion at the single molecule level. Cell Adhesion Commun, 5:375–395, 1998.

Pizzey JA, Bennett FA, Jones GF. Monensin inhibits initial spreading of cultured human fibroblasts. Nature 305:315–317, 1983.

Rabinovitch M, Manjjas RE, Nussenzweig V. Selective phagocytosis paralysis induced by immune complexes. J Exp Med 142:827–838, 1975.

Rees DA, Lloyd CW, Thom D. Control of grip and stick in cell adhesion through lateral relationships of membrane glycoproteins. Nature 267:124–128, 1977.

Regen CM, Horwitz AF. Dynamics of β_1 integrin–mediated adhesive contacts in motile fibroblasts. J Cell Biol 119:1437–1359, 1992.

Remold-O'Donnell E, Parent D. Two proteolytic pathways for down-regulation of the barrier molecule CD43 of human neutrophils. J Immunol 152:3595–3605, 1994.

Richelme F, Benoliel AM, Bongrand P. The leucocyte actin cytoskeleton. Bull Inst Pasteur 94:257–284, 1996.

Rieu P, Porteu F, Bessou G, Lesavre P, Halbwachs-Mecarelli L. Human neutrophils release their major membrane sialoprotein, leukosialin (CD43), during cell activation. Eur J Immunol 22:3021–3026, 1992.

Rothstein TL, Mage M, Jones G, McHugh LL. Cytotoxic T lymphocyte sequential killing of immobilized allogeneic tumor target cells measured by time-lapse microcinematography. J Immunol 121:1652–1656, 1978.

Runyan RB, Versakovic J, Shur BD. Functionally distinct laminin receptors mediate cell adhesion and spreading: The requirement for surface galactosyltransferase in cell spreading. J Cell Biol 107:1863–1871, 1988.

Rutishauser U, Acheson A, Hall AK, Mann DM, Sunshine J. The neural cell adhesion molecule (NCAM) as a regulator of cell-cell interactions. Science 240:53–57, 1988.

Ryan TA, Myeres J, Holowka D, Baird B, Webb WW. Molecular crowding on the cell surface. Science 239:61–64, 1988.

Sabri S, Pierres A, Benoliel AM, Bongrand P. Influence of surface charges on cell adhesion: Difference between static and dynamic conditions. Biochem Cell Biol 73:411–420, 1995.

Sabri S, Richelme F, Pierres A, Benoliel AM, Bongrand P. Interest of image processing in cell biology and immunology. J Immunol Methods 208:1–27, 1997.

Sako D, Chang X-J, Barone KM, Vachino G, White HM, Shaw G, Veldman GM, Bean KM, Ahern TJ, Furie B, Cumming DA, Larsen GR. Expression cloning of a functional glycoprotein ligand for P-selectin. Cell 75:1179–1186, 1993.

Sánchez-Mateos P, Cabañas C, Sáchez-Madrid F. Regulation of integrin function. Semin Cancer Biol 7:99–109, 1996.

Sánchez-Mateos P, Campanero MR, Delpozo MA, Sanchez-Madrid F. Regulatory role of CD43/leukosialin on integrin-mediated T cell adhesion to endothelial and extracellular matrix ligands and its polar redistribution to a cellular uropod. Blood 86:2228–2239, 1995.

Sanguedolce MV, Capo C, Bongrand P, Mège JL. Zymosan-stimulated tumor necrosis factor-α production by human monocytes. Down-modulation by phorbol ester. J Immunol 148:2229–2236, 1992.

Saxton MJ. Lateral diffusion in a mixture of mobile and immobile particles. A Monte Carlo Study. Biophys J 58:1303–1306, 1990.

Schlessinger J, Koppel DE, Axelrod D, Jacobson K, Webb WW, Elson EL. Lateral transport on cell membranes: mobility of Concanavalin A receptors on fibroblasts. Proc Natl Acad Sci (USA) 73:2409–2413, 1976.

Schmid-Schönbein GW, Sung KLP, Tozeren H, Skalak R, Chien S. Passive mechanical properties of human leukocytes. Biophys J 36:243–256, 1981.

Schuck P. Use of surface plasmon resonance to probe the equilibrium and dynamic aspects of interactions between biological macromolecules. Annu Rev Biophys Biomol Struct 26:541–566, 1997.

Schwartz MA, Schaller MD, Ginsberg MH. Integrins: Emerging paradigms of signal transduction. Annu Rev Cell Dev Biol 11:549–599, 1995.

Seveau S, Lopez S, Lesavre P, Guichard J, Cramer EM, Halbwachs-Mecarelli L. Leukosialin (CD43, sialophorin) redistribution in uropods of polarized neutrophils is induced by CD43 cross-linking by antibodies, by colchicine or by chemotactic peptides. J Cell Sci 110:1465–1475, 1997.

Sharon N. Glycoproteins in membranes. In: Balian R, Chabre M, Devanx PF, eds. Membranes and Intracellular Communication. Amsterdam: North Holland 1981. pp 117–182.

Sherbet GV. The Biophysical Characterisation of the Cell Surface, London: Academic Press, 1978.

Singer SJ. The fluid mosaic model of membrane structure: Some applications to ligand-receptor and cell-cell interactions. In: Bradshaw RA, Frazier WA, Merrell RC, Gottlieb DL, Hogue-Angeletti RA, eds. Surface Membrane Receptors. New York: Plenum, 1976, pp 1–23.

Singer SJ, Nicolson GL. The fluid mosaic model of the structure of cell membranes: Cell membranes are viewed as two-dimensional solutions of oriented proteins and lipids. Science 175:720–731, 1972.

Soler M, Merant C, Servant C, Fraterno M, Allasia C, Lissitzky JC, Bongrand P, Foa C. Leukosialin (CD43) behavior during adhesion of human monocytic THP-1 cells to red blood cells. J Leukoc Biol 61:609–618, 1997.

Sommers SD, Whisnant CC, Adams DO. Quantification of the strength of cell-cell adhesion: The capture of tumor cells by activated murine macrophages proceeds through two distinct stages. J Immunol 136:1490–1496, 1986.

Spertini O, Kansas GS, Munro JM, Griffin JD, Tedder TF. Regulation of leukocyte migration by activation of the leukocyte adhesion molecule-1 (LAM-1) selectin. Nature 349:691–694, 1991.

Sprent J. Migration and lifespan of lymphocytes. In: Loor F, Roelands GE, eds. B and T Cells in Immune Recognition. New York: Wiley, 1977, pp 59–82.

Springer TA. Adhesion receptors of the immune system. Nature 346:425–434, 1990.

Springer TA. Traffic signals on endothelium for lymphocyte recirculation and leukocyte emigration. Annu Rev Physiol 57:827–872, 1995.

Staunton DE, Dustin ML, Erickson HP, Springer TA. The arrangement of the immunoglobulin-like domains of ICAM-1 and the binding sites for LFA-1 and rhinovirus. Cell 61:243–254, 1990.

Stossel TP. On the crawling of animal cells. Science 260:1086–1094, 1993.

Subramaniam M, Koedam JA, Wagner DD. Divergent fates of P- and E-selectins after their expression on the plasma membrane. Mol Biol Cell 4:791–801, 1993.

Sugimoto Y. Effect on the adhesion and locomotion of mouse fibroblasts by their interacting with differently charged substrates. A quantitative study by ultrastructural method. Exp Cell Res 135:39–45, 1981.

Sung KLP, Sung LA, Crimmins M, Burakoff SJ, Chien S. Determination of junction avidity of cytolytic T cell and target cell. Science 234:1405–1408, 1986.

Suzuki ST. Structural and functional diversity of cadherin superfamily: Are new members of cadherin superfamily involved in signal transduction pathway? J Cell Biochem 61:531–542, 1996.

Taher TEI, Smit L, Griffioen AW, Schilderto EJM, Borst J, Pals ST. Signaling through CD44 is mediated by tyrosine kinases—Association with p56(lck) in T lymphocytes. J Biol Chem 271:2863–2867, 1996.

Takeichi M. Cadherin cell adhesion receptors as a morphogenetic regulator. Science 251:1451–1455, 1991.

Taniguchi N, Nanba I, Kozumi S. Characterization of glycosaminoglycans in human leukocytes: Enzymatic subunit assay of isomeric chondroitin sulfate. Biochem Med 11:217–226, 1974.

Tees DFJ, Coenen O, Goldsmith HL. Interaction forces between red cells agglutinated by antibody. IV. Time and force dependence of breakup. Biophys J 65:1318–1334, 1993.

Tees DFJ, Goldsmith HL. Kinetics and locus of failure of receptor-ligand-mediated adhesion between latex spheres. I—Protein-carbohydrate bond. Biophys J 71:1102–1114, 1996.

Tessier-Lavigne M, Goodman CS. The molecular biology of axon guidance, Science 274:1123–1133, 1996.

Tha SP, Shuster J, Goldsmith HL. Interaction forces between red cells agglutinated by antibody. IV Time and force dependence of break-up. Biophys J 50:1117–1126, 1986.

Tissot O, Pierres A, Foa C, Delaage M, Bongrand P. Motion of cells sedimenting on a solid surface in a laminar shear flow. Biophys J. 61:204–215, 1992.

Tizard IR, Holmes WL, Parappally NP. Phagocytosis of sheep erythrocytes by macrophages: A study of the attachment phase by scanning electron microscopy. J Reticuloendothel Soc 15:225–231, 1974.

Torney DC, Dembo M, Bell GI. Thermodynamics of cell adhesion. II. Freely mobile repellers. Biophys J 49:501–507, 1986.

Tözeren A. Cell-cell conjugation. Transient analysis and experimental implications. Biophys J 58:641–652, 1990.

Tözeren A, Ley K. How do selectins mediate leukocyte rolling in venules? Biophys J 63:700–709, 1992.

Tözeren A, Sung KLP, Sung LA, Dustin ML, Chan P-Y, Springer TA, Chien S. Micromanipulation of adhesion of a Jurkat cell to a planar bilayer membrane containing lymphocyte function–associated antigen 3 molecules. J Cell Biol 116:997–1006, 1992.

Truskey GA, Proulx TL. Relationship between 3T3 cell spreading and the strength of adhesion on glass and silane surfaces. Biomaterials 14:243–254, 1993.

Usami S, Wung SL, Skierczynski BA, Skalak R, Chien S. Locomotion forces generated by a polymorphonuclear leucocyte. Biophys J 63:1663–1666, 1992.

van de Winkel JG, van Omnen R, Huizinga TWJ, de Raad MAHVM, Tuijman WB, Groenen PJTA, Capel PJA, Koene RAP, Tax WJM. Proteolysis induces increased binding affinity of the monocyte type II FcR for human IgG. J Immunol 143:571–578, 1989.

van der Merwe PA, Brown MH, Davis SJ, Barclay AN. Affinity and kinetic analysis of the interaction of the cell adhesion molecules rat CD2 and CD48. EMBO J 12:4945–4954, 1993.

van Holde KE. Physical Biochemistry. Englewood Cliffs, NJ; Prentice-Hall, 1971, p 189.

van Kooten TG, Schakenraad JM, van der Mei HC, Busscher HJ. Development and use of a parallel-plate flow chamber for studying cellular adhesion to solid surfaces. J Biomed Mater Res 26:725–738, 1992.

van Kooyk Y, Weder P, Hogervorst F, Verhoeven AJ, van Seventer G, te Velde AA, Borst J, Keizer GD, Figdor CF. Activation of LFA-1 through a Ca^{2+}-dependent epitope stimulates lymphocyte adhesion. J Cell Biol 112:1345–1354, 1991.

van Oss CJF. Interaction forces between biological and other polar entities in water: How many different primary forces are there? J Dispersion Sci Technol 12:201–219, 1991.

van Oss CJ, Gillman CF, Neumann AW. Phagocytic Engulfment and Cell adhesiveness as Cellular Surface Phenomena. New York: Marcel Dekker, 1975.

Varki A. Selectin ligands. Proc Natl Acad Sci USA 91:7390–7397, 1994.

Vereycken V, Gruler H, Bucherer C, Lacombe C, Lelièvre JC. The linear motor in the human neutrophil migration. J Phys III Fr 5:1469–1480, 1995.

Von Andrian UH, Chambers JD, McEvoy LM, Bargatze RF, Arfors KE, Butcher EC. Two-step model of leukocyte–endothelial cell interaction in inflammation: Distinct roles for LECAM-1 and the leukocyte β_2 integrins in vivo. Proc Natl Acad Sci USA 88:7538–7542, 1991.

Von Andrian UH, Hasslen SR, Nelson RD, Erlandsen SL, Butcher EC. A central role for microvillus receptor presentation in leukocyte adhesion under flow. Cell 82:989–999, 1995.

Waddell TK, Fialkow L, Chan CK, Kishimoto TK, Bowney GP. Signaling functions of L-selectin. Enhancement of tyrosine phosphorylation and activation of MAP kinase. J Biol Chem 270:15403–15411, 1995.

Walcheck B, Kahn J, Fisher JM, Wang BB, Fisk RS, Payan DG, Feehan C, Betageri R, Darlak K, Spatola AF, Kishimoto TK. Neutrophil rolling altered by inhibition of L-selectin shedding in vitro. Nature 380:720–717, 1996.

Ward MD, Dembo M, Hammer DA. Kinetics of cell detachment: Peeling of discrete receptor clusters. Biophys J 67:2522–2534, 1994.

Weast RC, Astle MJ, Beyer WH. Handbook of Chemistry and Physics. Boca Raton, FL: CRC Press, 1986, p D231.

Weiss L. The measurement of cell adhesion. Exp Cell Res 8(Suppl):141–153, 1961.

Weiss L, Coombs RRA. The demonstration of rupture of cell surfaces by an immunological technique. Exp Cell Res 30:331–338, 1963.

Wesseling J, van der Valk SW, Hilkens J. A mechanism for inhibition of E-cadherin-mediated cell-cell adhesion by the membrane associated mucin episialin MUC-1. Mol Biol Cell 7:565–577, 1996.

Weyrich AS, Mcintyre TM, McEver RP, Prescott SM, Zimmerman GA. Monocyte tethering by P-selectin regulates monocyte chemotactic protein-1 and tumor necrosis factor-alpha secretion—Signal integration and NF-kappa B translocation. J Clin Invest 95:2297–2303, 1995.

Williams AF, Barclay AN. The immunoglobulin superfamily—Domains for cell surface recognition. Annu Rev Immunol 6:181–405, 1988.

Witkowski JA, Brighton WD. Stages of spreading of human diploid cells on glass surfaces. Exp Cell Res 68:372–380, 1971.

Xu Y, Guo DF, Davidson M, Inagami T, Carpenter G. Interaction of the adaptor protein Shc and the adhesion molecule cadherin. J Biol Chem 272:13463–13466, 1997.

Yamagata M, Kimata K. Repression of a malignant cell-substratum adhesion phenotype by inhibiting the production of the anti-adhesive proteoglycan PG-M/versican. J Cell Sci 107:2581–2590, 1994.

Yang P, Yin X, Rutishauser U. Intercellular space is affected by the polysialic acid content of NCAM. J Cell Biol 116:1487–1496, 1992.

Yonemura S, Nagafuchi A, Sato N, Tsukita S. Concentration of an integral membrane protein, CD43 (leukosialin, sialophorin), in the cleavage furrow through the interaction of its cytoplasmic domain with actin-based cytoskeletons. J Cell Biol 120:437–449, 1993.

Yu ZW, Calvert TL, Leckband D. Molecular forces between membranes displaying neutral glycosphingolipids: Evidence for carbohydrate attraction. Biochemistry, 37:1540–1550, 1998.

Zurkhov SN. Kinetic concept of the strength of solids. Int J Fract Mech 1:311–323, 1956.

15

Axisymmetric Drop Shape Analysis*

P. Chen, O. I. del Río, and A. W. Neumann
University of Toronto, Toronto, Ontario, Canada

I. INTRODUCTION

In physical chemistry of biological surfaces, one of the most sought after properties is surface (interfacial) tension (see preceding chapters). Surface tension is not only an intrinsic thermodynamic quantity but also closely related to surface adsorption (1–4). Different techniques have been employed to measure the surface tension of biomolecules at air-water and oil-water interfaces, such as the Wilhelmy plate (5,6), the du Noüy ring tensiometer (7), and those based on the volume (5,8), weight, or shape of a pendant drop (9,10). In the ring method, the force required to pull a ring from the surface of a liquid is determined. This method has the disadvantage of enlarging the surface area during the measurement process, which leads to alteration of the adsorption state of biomolecules. Viscoelastic effects in addition to surface tension effects may also come into play. The Wilhelmy plate technique requires the establishment of a zero contact angle, which is difficult to guarantee with systems involving biomolecular solutions because of adsorption onto the plate. Moreover, this is even more difficult in liquid-liquid systems, which are relevant to many biological processes. The ring method also suffers further complications in liquid-liquid systems. The calculation of interfacial tension with the du Noüy tensiometer requires a correction factor for the weight of the column of liquid while the ring is removed. In liquid-liquid systems, consideration of the density difference across the interface is required for an accurate correction. The drop volume technique relies on the volume of a liquid drop detaching from a capillary tube to determine the interfacial tension.

---

* Dedicated to the memory of R. M. Prokop.

Although it is applicable to liquid-liquid systems, it requires extremely careful manipulations for the determination of the volume of the detaching drop. Also, to perform time-dependent studies, the detachment of the drop at the desired time must be elicited by the rate at which the drop is grown. This in itself inflicts an added disturbance to the system.

An alternative approach to obtaining the liquid-vapor or liquid-liquid interfacial tension is based on the shape of a pendant or sessile drop. In essence, the shape of a drop is determined by a combination of surface tension and gravity effects. Surface forces tend to make drops spherical, whereas gravity tends to elongate a pendant drop or flatten a sessile drop. When gravitational and surface tension effects are comparable, then, in principle, one can determine the surface tension from an analysis of the shape of the drop. Figure 1 shows two pendant drop images of a 0.02 mg/mL bovine serum albumin aqueous solution at 37°C: image (A) was acquired at time zero, with a corresponding surface tension of 70.24 mJ/m^2, and image (B) was acquired at time 400 s, with a corresponding surface tension of 54.22 mJ/m^2.

The advantages of pendant and sessile drop methods are numerous. In comparison with the aforementioned methods such as the Wilhelmy plate technique, only small amounts of the liquid are required. Drop shape methods easily facilitate the study of both liquid-vapor and liquid-liquid interfacial tensions. Also, the methods have been applied to materials ranging from organic liquids to molten metals and from pure solvents to concentrated solutions. There is no limitation to the magnitude of surface or interfacial tension that can be measured: the methodology works as well at 10^3 mJ/m^2 as at 10^{-3} mJ/m^2. Measurements have been satisfactorily made over a range of temperatures and pressures (3). In addition, because the profile of the drop may be recorded by photographs or digital image representation, it is possible to study interfacial tensions in dynamic systems, where the properties are time dependent. Furthermore, drop shape methods measure not only surface tension but also contact angle in the case of a sessile drop.

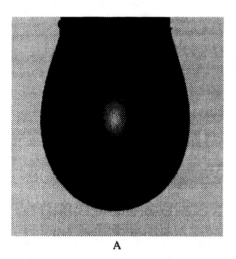 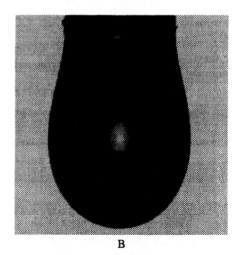

A   B

Fig. 1 Pendant drop images of a 0.02 mg/mL bovine serum albumin aqueous solution at 37°C; image (A) was acquired at time zero, with a corresponding surface tension of 70.24 mJ/m^2, and image (B) was acquired at time 400 s, with a corresponding surface tension of 54.22 mJ/m^2.

Mathematically, the balance between surface tension and external forces, such as gravity, is reflected in the so-called Laplace equation of capillarity. The Laplace equation is the mechanical equilibrium condition for two homogeneous fluids separated by an interface. It relates the pressure difference across a curved interface to the surface tension and the curvature of the interface:

$$\gamma\left(\frac{1}{R_1} + \frac{1}{R_2}\right) = \Delta P$$

where γ is the interfacial tension, R_1 and R_2 are the two principal radii of curvature, and ΔP is the pressure difference across the interface. In the absence of any external forces other than gravity, ΔP may be expressed as a linear formation of the elevation:

$$\Delta P = \Delta P_0 + (\Delta\rho)gz$$

where ΔP_0 is the pressure difference at a reference plane, $\Delta\rho$ is the density difference between the two bulk phases, g is the gravitational acceleration, and z is the vertical height of the drop measured from the reference plane. For a given γ, the shape of a drop may be determined readily. The inverse, i.e., determination of the interfacial tension γ from the shape, is also possible in principle, although this is a much more difficult task.

Mathematically, the integration of the Laplace equation is straightforward only for cylindrical menisci, i.e., menisci for which one of the principal curvatures, $1/R$, is zero. For a general irregular meniscus, mathematical analysis would be very difficult. For the special case of axisymmetric drops, numerical procedures have been devised. Fortunately, axial symmetry is not a very significant restriction for most sessile drop and pendant drop systems.

The earliest efforts in the analysis of axisymmetric drops were those of Bashforth and Adams (11). Hartland and Hartley (12) collected numerous solutions for determining the interfacial tensions of axisymmetric fluid-liquid interfaces of different shapes. A computer program was used to integrate the appropriate form of the Laplace equation and the results were presented in tables. The major shortcoming of these methods is in data acquisition. The description of the surface of the drop is accomplished by the measurement of a few preselected points. These points are critical because they correspond to special features, such as inflection points on the interface, and must be measured with a high degree of accuracy. However, these measurements are not easily obtained. In addition, the use of these tables is limited to drops of a certain size and shape range.

Maze and Burnet (13,14) developed a more satisfactory scheme for the determination of interfacial tension from the shape of sessile drops. They utilized a numerical nonlinear regression procedure in which a calculated drop shape is made to fit a number of arbitrarily selected and measured points on the drop profile. In order to start the calculation, reasonable estimates of the drop shape and size are required, otherwise the calculated curve will not converge to the measured one. The initial estimates are obtained, indirectly, using values from the tables of Bashforth and Adams. Despite the progress in strategy, there are deficiencies in this algorithm. The error function is computed by summing the squares of the horizontal distances between the measured points and the calculated curve. This measure may not be adequate, particularly for sessile drops whose shapes are strongly influenced by gravity. In addition, the identification of the apex of the drop is of paramount importance because it acts as the origin of the calculated curves (3).

Rotenberg et al. (15) developed a technique, called axisymmetric drop shape analysis-profile (ADSA-P), that is superior to the preceding methods and does not suffer from their deficiencies. ADSA-P fits the measured profile of a drop to a Laplacian curve. An objective function is formed that describes the deviation of the experimental profile from the theoretical profile as the sum of the squares of the normal distances between the experimental points and the calculated curve. This function is minimized by a nonlinear regression procedure, yielding the interfacial tension and the contact angle in the case of a sessile drop. The location of the apex of the drop is assumed to be unknown and the coordinates of the origin are regarded as independent variables of the objective function. Thus, the drop shape can be measured from any convenient reference frame and any measured point on the surface is equally important. A specific value is not required for the surface tension, the radius of curvature at the apex, or the coordinates of the origin. The program requires as input several coordinate points along the drop profile, the value of the density difference across the interface, and the magnitude of the local gravitational constant (and the distance between the base of the drop and the horizontal coordinate axis). Initial guesses of the location of the apex and the radius of curvature at the apex are not required. The solution of the ADSA-P program yields not only the interfacial tension and contact angle but also the volume, surface area, radius of curvature, and contact radius of the drop. Essentially, ADSA-P employs a numerical procedure that unifies the sessile and pendant drop methods. There is no need for any table, nor is there any drop size restriction on the applicability of the method.

Cheng et al. (16) automated the methodology by means of digital image acquisition and image analysis. Pictures of sessile or pendant drops are acquired using a video camera attached to a computer, where image analysis software automatically extracts several hundred coordinates of the drop profile, which in turn are analyzed by ADSA-P to compute surface tension.

Recently, ADSA-P has been rewritten (17), implementing more efficient, accurate, and stable numerical methods in order to overcome convergence problems of the original program for very low interfacial tensions with well-deformed drop shapes (17). Also, two additional optimization parameters were introduced: the angle of vertical misalignment of the camera and the aspect ratio of the video image. With these revisions, the accuracy and the range of applicability of ADSA-P have been further improved.

For contact angle determinations, with most techniques it becomes increasingly difficult to make measurements for flat sessile drops with very low contact angles, say below 20°. The accuracy of ADSA-P also decreases under these circumstances because it becomes more difficult to acquire accurate coordinate points along the edge of the drop profile. For these situations, it is more useful to view a drop from above and determine the contact angle from the contact diameter of the drop (18). A modified version of ADSA, called axisymmetric drop shape analysis–contact diameter (ADSA-CD), developed by Rotenberg and later implemented by Skinner, does not ignore the effects of gravity (19). ADSA-CD requires the contact diameter, the volume, and the liquid surface tension of the drop, the density difference across the liquid-fluid interface, and the gravitational constant as input to calculate the contact angle by means of a numerical integration of the Laplace equation of capillarity.

It has been found that drop shape analysis utilizing a top view is quite useful for the somewhat irregular drops that often occur on rough and heterogeneous surfaces. In these cases, an average contact diameter leads to an average contact angle. The usefulness of ADSA-CD for averaging over irregularities in the three-phase contact line proved to be such an asset that is became desirable to use it instead of ADSA-P for large contact angles as well. Unfortunately, for contact angles above 90° the three-phase line is not visible from above. For such cases, yet another version of ADSA has been developed by Moy et al. (20). called axisymmetric drop shape analysis–maximum diameter (ADSA-MD). ADSA-MD is similar to ADSA-CD; however, it relies on the maximum equatorial diameter of a drop to calculate the contact angle. ADSA-CD and ADSA-MD have been unified into a single program called axisymmetric drop shape analysis-diameter (ADSA-D).

The following sections provide an account of these ADSA methodologies. The first section contains a description of the numerical algorithms and their implementation. Later sections discuss the applicability of ADSA and illustrations of the investigation of surface tension measurements with pendant and sessile drops and contact angle experiments with sessile drops using ADSA-D.

II. AXISYMMETRIC LIQUID-FLUID INTERFACE (ALFI)

The classical Laplace equation of capillarity describes the mechanical equilibrium conditions for two homogeneous fluids separated by an interface. For axisymmetric interfaces it can be written as the following system of ordinary differential equations as a function of the arc length s, as shown in Fig. 2 (15):

$$\frac{dx}{ds} = \cos\theta \tag{1a}$$

$$\frac{dz}{ds} = \sin\theta \tag{1b}$$

$$\frac{d\theta}{ds} = 2b + cz - \frac{\sin\theta}{x} \tag{1c}$$

$$\frac{dV}{ds} = \pi x^2 \sin\theta \tag{1d}$$

$$\frac{dA}{ds} = 2\pi x \tag{1e}$$

$$x(0) = z(0) = \theta(0) = V(0) = A(0) = 0 \tag{1f}$$

where b is the curvature at the origin of coordinates and $c = (\Delta\rho)g/\gamma$ is the capillary constant of the system. θ is the tangential angle, which, for sessile drops, becomes the contact angle at the three-phase contact line. Although the surface area A and the volume V are not required to define the Laplacian profile, they are included here because of their importance and the fact that they can be integrated simultaneously without a significant increase of computational time.

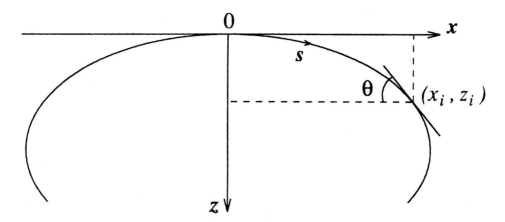

Fig. 2 Coordinate system used in the numerical solution of the Laplace equation for axisymmetric liquid-fluid interfaces (ALFI).

For given values of b and c, a unique shape of a Laplacian axisymmetric fluid-liquid interface can be obtained by simultaneous integration of the preceding initial value problem. However, there is no known analytical solution for this problem except for very limited cases, and a numerical integration scheme must be used. There exist several numerical methods to solve systems of ordinary differential equations for initial value problems and considerable research is still devoted to this subject (21). One of the most efficient and flexible methods is the fifth- and sixth-order Runge-Kutta-Verner pair, DVERK, written by Hull et al. (22,23).

A computer called ALFI was written (17), implementing the DVERK numerical integration scheme to generate Laplacian profiles of pendant and sessile drops of any size (controlled by the apex curvature b) and surface tension (specified by the capillary constant c) by integrating problem (1). Some of the features of ALFI are:

1. The volume V and surface area A are computed simultaneously with the drop profile.
2. The integration can be stopped when any given values of s, θ x, z, V, or A are reached, allowing the computation of drop profiles of any specified contact angle, volume, surface area, or size. The integration also terminates if θ reaches 180° (sessile drops) or becomes negative (pendant drops).
3. The inflection point of pendant drops is accurately computed, which is useful for testing and evaluating drop profile methods.
4. The origin of the coordinate system can be translated and rotated arbitrarily, and the coordinates can be scaled in the horizontal and vertical directions. This feature permits comparison between theoretical and experimental drop profiles by matching the origins of the coordinate systems, the magnification, and the vertical alignment.
5. The profile coordinates can be randomly perturbed in the normal direction, allowing the simulation of experimental error, which can be used to evaluate ADSA methods.

As mentioned before, ALFI generates complete Laplacian profiles from values of b and c by integrating problem (1). The inverse process of determining b and c (from which γ and contact angle θ can easily be computed) based on drop profile characteristics is a more difficult task and forms the basis of the ADSA methods described in the following sections.

A. Axisymmetric Drop Shape Analysis-Profile (ADSA-P)

The ADSA-P methodology for determining interfacial properties by means of a numerical fit of several arbitrary drop profile coordinates to the Laplace equation was originally developed by Rotenberg et al. (15). The current version of ADSA-P (17) uses the same strategy as the original, i.e., a nonlinear least-squares optimization, but with a slightly different definition of the objective function (see below) and implementation of more advanced numerical methods, as described in the following. The method is applicable to sessile and pendant drops.

The strategy utilized is to construct and minimize an objective function E, defined as the sum of the weighted squared normal distances between any N profile coordinates and the Laplacian profile (IVP 1), as seen in Fig. 3:

$$E = \sum_{i=1}^{N} w_i e_i \tag{2a}$$

$$e_i = \tfrac{1}{2} d_i^2 = \tfrac{1}{2}\left[(x_i - X_i)^2 + (z_i - Z_i)^2\right] \tag{2b}$$

where w_i is a weighting factor, (X_i, Z_i) are the measured drop coordinates, and (x_i, z_i) are the Laplacian coordinates closest to (X_i, Z_i). Currently w_i is set equal to 1.0 until more studies are available on the effect of weighting factors. By introducing the generally unknown origin (x_0, z_0) and angle of rotation of the system of coordinates α and scaling factors on both coordinates (X_s, Z_s), the individual error can be written

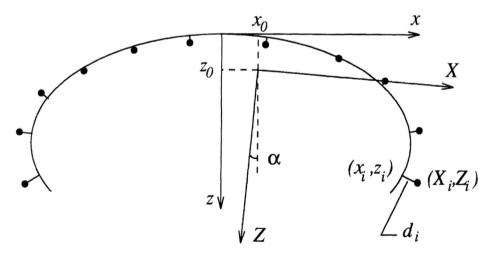

Fig. 3 Definition of error function parameters for the ADSA-P optimization problem.

(dropping the subscript *i*) as

$$e = \tfrac{1}{2}\left(e_x^2 + e_z^2\right) \tag{2c}$$

$$e_x = x - x_0 - X_s X \cos\alpha + Z_s Z \sin\alpha \tag{2d}$$

$$e_z = z - z_0 - X_s X \sin\alpha - Z_S Z \cos\alpha \tag{2e}$$

The objective is therefore to compute the set of M optimization parameters a that minimizes (2), where $a = [b\ c\ x_0\ z_0\ \alpha\ X_s\ Z_s]^T$ or any subset of it. It should be noted, though, that only one of the scaling factors, X_s or Z_s, can be optimized simultaneously with b and c for the solution to be unique. Generally, one of the scaling factors is known from the experimental setup and can be held constant while optimizing the other to correct for the aspect ratio to calibrate the optical system. The rotational angle α can also be optimized to correct for the rotational misalignment of the camera for calibration purposes.

The optimization problem can be written as

$$\min_{a}\ E(a) = \sum_{i=1}^{N} w_i e_i(a)$$

which is a multidimensional nonlinear least-squares problem that requires an iterative optimization procedure. When the minimum has been found, the optimization parameters determine the Laplacian profile that best fits the given profile, from which γ and other properties can be readily computed.

Evaluating E for a trial set of a, i.e., for each optimization iteration, involves determining the minimum (normal) distance from the Laplacian curve to each experimental point. This is done using a one-dimensional Newton-Raphson iteration to solve, for each ith point,

$$f(s) = \frac{de}{ds} = e_x \cos\theta + e_z \sin\theta = 0$$

There exist several numerical methods for solving optimization problems. Among them, Newton's method is well known for its second-order convergence if the initial values are very close to the solution, but it is unpredictable otherwise, particularly for multidimensional problems. To overcome this problem, several Newton-like algorithms have been developed with more advanced convergence strategies. The original ADSA-P used Newton's method with incremental loading to approach the solution, but this approach is computationally expensive and its convergence is not guaranteed. A more efficient and globally convergent method for nonlinear least-squares optimization is the Levenberg-Marquardt method, as implemented in the MINPACK library by Moré and Wright (24). The current version of ADSA-P employs a combination of Newton and Levenberg-Marquardt methods. Very often, as in the case of time-dependent studies, the results from a previous run can be used as initial values and Newton's method can be used to take advantage of its fast convergence, but it is aborted as soon as divergence is detected. It good initial values are not available or if Newton's method fails, the Levenberg-Marquardt is used.

As with any nonlinear numerical method, the optimization parameters must be initialized with approximate values of the solution. Good initial values for the curvature at the apex b and the origin of the system of coordinates x_0 and z_0 can be found by a least-squares elliptical fit of several points near the drop apex, and the rotational angle α and the scaling factors X_s and Z_s are generally known from the experimental setup. The capillary parameter c is initialized using an estimated surface tension value, but the method will converge even with a bad initial guess.

In practice, the drop profile coordinates (X_i, Z_i) are extracted from digital images of pendant or sessile drops using edge detection techniques as implemented by Cheng et al. (16): By applying the well-known Sobel operator on the digital image, the pixel coordinates of the drop edge can be obtained as the pixels with a maximum Sobel value, following the contour of the drop from one end of the drop to the other. This procedure yields profile coordinates with pixel resolution, which are limited by the resolution of the digital image (usually 640 by 480 pixels). A more accurate subpixel resolution can be obtained by means of a cubic spline fit to the pixel values across the interface to find the position of the interface as implemented by Cheng or by a quadratic polynomial fit of the Sobel values across the interface to find the position with maximum Sobel value that represents the drop edge.

B. Axisymmetric Drop Shape Analysis-Diameter (ADSA-D)

The ADSA-D methodology for computing contact angles θ from the contact or maximum diameter D (usually measured from a picture of the drop looking from above) and volume V of sessile drops with known surface tension γ was originally developed by Skinner et al. (19) and Moy et al. (20). The current implementation by del Río and Neumann (17) uses the numerical solution of the Laplace equation as a boundary value problem, as described in the following.

There are two cases to consider, depending on the contact angle: (a) contact angles greater than or equal to 90° and (b) contact angles less than 90°, which represent two separate problems. In the first case the maximum diameter corresponds to the equatorial diameter of the drop (at $\theta = 90°$) and in the second case the maximum diameter corresponds to the three-phase contact line (see Fig. 4).

1. Contact Angle Greater Than or Equal to 90°

Rewriting Eqs. (1a)–(1c) as functions of x, considering the curvature b as a new variable and with the boundary conditions as seen in Fig. 4a, the Laplace equation can be written as the following boundary value problem for contact angles greater than or equal to 90°:

$$\frac{d\theta}{dx} = \frac{1}{\cos\theta}\left(2b + cz - \frac{\sin\theta}{x}\right) \tag{3a}$$

$$\frac{dz}{dx} = \tan\theta \tag{3b}$$

$$\frac{db}{dx} = 0 \tag{3c}$$

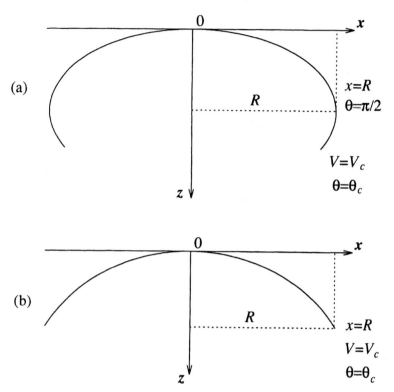

Fig. 4 Boundary conditions for ADSA-D boundary value problems. (a) Contact angle greater than or equal to 90°; (b) contact angle less than 90°.

$$z(0) = \theta(0) = 0 \qquad \theta(R) = \frac{\pi}{2} \tag{3d}$$

where $R = D/2$ is the maximum (equatorial) radius. Problem (3) completely defines the Laplacian shape; its solution gives directly the profile shape for $0 \leq x \leq R$ and the constant value of the apex curvature b. The contact angle can then be computed by integrating the initial value problem (IVP) (1), for the known values of b and γ, past the maximum diameter, stopping when the computed volume reaches the drop volume, V_c.

2. Contact Angle Less Than 90°

Similarly, as seen in Fig. 4b, the Laplace equation for contact angles less than 90° can be written as the following:

$$\frac{d\theta}{dx} = \frac{1}{\cos\theta}\left(2b + cz - \frac{\sin\theta}{x}\right) \tag{4a}$$

$$\frac{dz}{dx} = \tan\theta \tag{4b}$$

$$\frac{dV}{dx} = \pi x^2 \tan\theta \tag{4c}$$

$$\frac{db}{dx} = 0 \tag{4d}$$

$$z(0) = \theta(0) = V(0) = 0 \qquad V(R) = V_c \tag{4e}$$

where V_c is the total volume of the drop. Problem (4) completely defines the Laplacian shape. There is no need for an additional numerical integration because the contact angle can be obtained simply from the value of θ at $x = R$.

To initialize ADSA-D it is necessary to determine, for given values of V_c and R, whether the contact angle is greater than or equal to 90° or less than 90°. On occasion, the user may be able to give this information as input, but in many cases, especially for contact angles near 90°, the distinction may not be known a priori. Therefore, the following approach is implemented in the program: (a) If the user knows whether the drop is wetting or nonwetting, solve the respective problem and exit; otherwise (b) assume that the contact angle is greater than or equal to 90°, solve problem (3) for the given R, and compute volume V_{90} at $\theta = 90°$ by integrating problem (1). If $V_{90} \le V_c$ the initial assumption was correct, compute the contact angle and exit; otherwise, (c) solve problem (4) for a contact angle less than 90°.

The nonlinear problems (3) and (4) must be solved numerically. The current version of ADSA-D uses a finite-difference method with collocation formulas, as implemented in the COLSYS library by Ascher et al. (25). The program must be initialized with approximate values of the solution. For the case of contact angles greater than or equal to 90°, the profile and the curvature are initialized with an elliptical approximation, and for contact angles less than 90° the solution is initialized with zero values and the curvature with $1/R$. It was found that these rough approximations are sufficient for the method to converge in most cases. However, a continuation algorithm (25) was implemented to guarantee convergence to a solution in case of an initial failure of COLSYS, using the capillary constant c as the continuation parameter. Care is taken in the numerical implementation to avoid the discontinuity of problem (3) at $\theta = 90°$, and the algorithm succeeds for any contact angle, including $\theta = 90°$.

In practice, the drop diameter can be obtained from digital images of the drop, acquired with the camera positioned vertically, looking at the sessile drop from above. The drop volume can be measured with a micrometer syringe.

III. APPLICATIONS

In the following, we demonstrate selected applications of ADSA to studies of biological interfaces: (A) concentration dependence of interfacial tension of human serum albumin at the water-decane interface; (B) dynamic surface tension response to changes in surface area of solutions containing a protein and lipids, as a means of studying molecular interactions and their specificity and dose effects; (C) ADSA–captive bubble method for measuring low surface tension of lung surfactant solutions; and (D) contact angle measurements on rough and heterogeneous biological solid surfaces by ADSA-D. More applications of ADSA can be found in Refs. 3 and 26–29.

A schematic of the ADSA-P experimental setup is given in Fig. 5; the basic components are as follows: With the use of a microsyringe (Hamilton Gastight syringe, Chromatographic Specialties Inc., Brockville, ON, Canada), a pendant drop of the

ADSA-P Pendant Drop

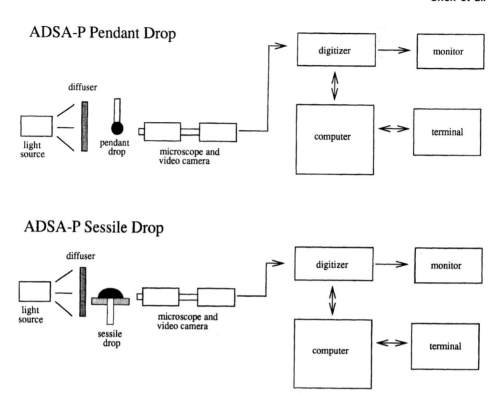

ADSA-P Sessile Drop

Fig. 5 Schematic of an experimental setup for pendant drop and sessile drop experiments.

sample solution was formed at the tip of a vertical Teflon capillary of circular cross section (inner diameter, 1.5 mm), thus producing an axisymmetric boundary for the drop. The drop was enclosed in a sealed quartz cuvette (model 330984, $10 \times 10 \times 30$ mm^3, Hellma Canada Ltd., Concord, ON, Canada) that contained decane or air. The cuvette was mounted in an environmental chamber (model 100-07, Ramé Hart, Inc., Mountain Lakes, NJ). The chamber was linked to a thermostatted water bath (Lauda K-2/R, Brinkmann Instruments) maintaining the temperature of the setup to the accuracy of $\pm 0.1°C$. In the case of a sessile drop, the drop was generated from the bottom on a flat solid surface, and the experiment was conducted at room temperature. The drops were then illuminated with a white light source (model V-WLP 1000, Newport Corp., Irvine, CA) shining through a heavily frosted diffuser. Images of the drop were obtained with a microscope (Leitz Apozoom, Leica, Willowdale, ON, Canada) linked to a monochrome charge-coupled device video camera (Cohu 4810, Infranscan, Inc., Richmond, BC, Canada). The video signal of the drop was transmitted to a digital video processor (Xvideo board, Parallax Graphics Inc., Santa Clara, CA), which performed the frame grabbing and digitization of the image to 640×480 pixels with 256 gray levels.

In measuring equilibrium surface tension, the experiment was continued until an approximately constant surface (interfacial) tension was obtained. In each run, images were captured at 1-s intervals initially and progressively less rapidly (up to 150-s intervals) near the end of the run. In dynamic surface tension studies, the experiment was continued until repeated cycles were observed in the surface tension response to surface area perturbation. For each run, images were captured at a reasonably fast pace (up to 0.5-s, intervals) so that the features of dynamic surface tension could be obtained. To produce a controlled surface area perturbation, the microsyringe was connected to a stepper motor (model 18515, Oriel Corp., Stratford, CT), which was computer controlled. The motion of the syringe plunger changed the volume of the drop and hence changed the surface area (3,26).

A workstation (Sun SPARCstation 10, Sun Microsystems, Mountain View, CA) was used to acquire the images from the digitization board. Image analysis schemes were used to determine the drop profile coordinates with subpixel resolution and to correct for optical distortion (4). The entire setup, except for the water bath and the workstation, was placed on a vibration-free table (Technical Manufacturing Corp., Peabody, MA) to isolate the system from external disturbances. Each single image of a drop was analyzed 10 times with 20 different randomly chosen profile coordinate points each time. The average resulting 95% confidence limit for each measurement is better than ± 0.2 mJ/m^2 in this work [although greater accuracy, approximately ± 0.04 mJ/m^2, has been obtained routinely using this procedure for nonprotein solutions (3,26)].

A. Concentration Dependence of the Interfacial Tension of Human Serum Albumin at the Water-Decane Interface

Axisymmetric drop shape analysis (ADSA-P) was employed to obtain highly accurate measurements of the concentration dependence of the interfacial pressure of human serum albumin (HSA) at the water-decane interface. The effect of concentration in the surface activity of proteins has been well documented (4,6,30–35). In general, with an increase in the bulk protein concentration, a decrease in the interfacial tension, i.e., a positive interfacial pressure, has been reported. This reflects increased diffusion of protein to the interface, followed by unfolding and molecular rearrangements of adsorbed molecules (9). Here, we report the measurement of negative interfacial pressures at very low concentrations (1×10^{-4} and 1×10^{-3} mg/ml). An interpretation of this finding in terms of the effects of electrical charges and pH is attempted.

1. Materials

The human serum albumin (Sigma Chemical Corp., St. Louis, MO) contained 15.4% nitrogen, was free from fatty acids, and had an average molecular weight of 65,000. The decane (Caledon Laboratories Ltd., Georgetown, ON, Canada) was distilled in glass and certified for gas chromatography (Code 3301-2). Fifteen aqueous solutions of the albumin were prepared with distilled water. The protein concentrations ranged from 1×10^{-4} to 5 mg/mL. For a limited number of experiments at a protein concentration of 1×10^{-4} mg/mL, 20 mM Trizma Base buffers (Sigma catalog No. T-1503, Sigma, Mississauga, ON, Canada), pH adjusted with HCl, were used to produce aqueous solutions with a pH of 3.5, 4.8 (the isoelectric point of albumin), and 5.6.

2. Results

Figure 6 illustrates the time dependence of the interfacial tension of aqueous HSA solution of 15 concentrations at a decane interface. In all cases, a reduction of the interfacial tension to a relatively constant value is observed with the passage of time,

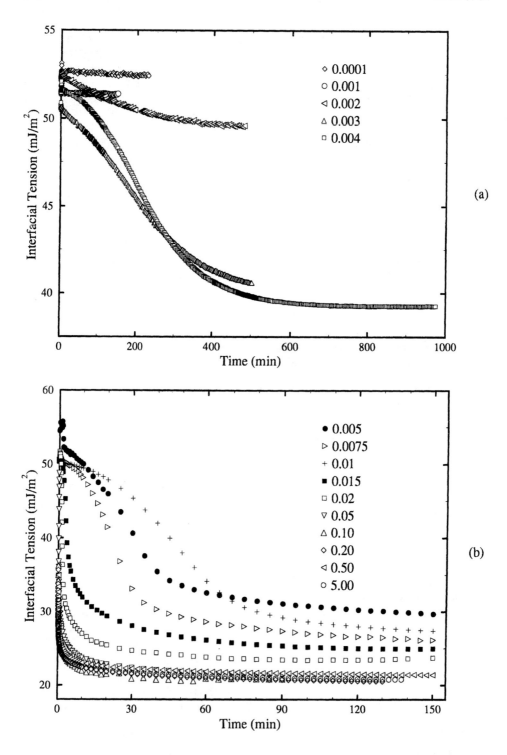

Fig. 6 The interfacial tension between the aqueous HSA solution and decane versus time for various bulk concentrations of the HSA: (a) 0.0001 to 0.004 mg/mL; (b) 0.005 to 5.0 mg/mL.

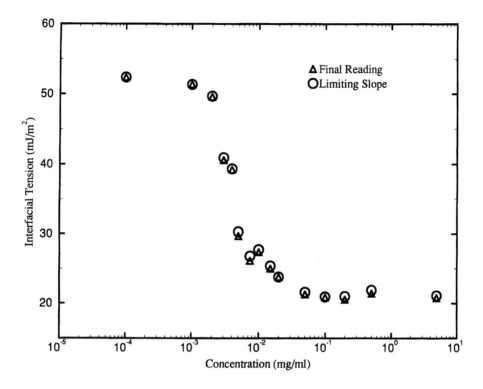

Fig. 7 The "equilibrium" interfacial tension obtained by the minimum slope criterion and the last experimental value for various bulk concentrations of the HSA.

t. In general, the higher the bulk protein concentration, the lower the observed equilibrium value. Also, the rate of decrease in γ at early times increases with increasing bulk protein concentration. As a consequence of this last observation, the duration of each experiment varied with its protein concentration. That is, for low concentrations (from 2×10^{-3} to 4×10^{-3} mg/mL) the time required for an experiment is more than 8 h (Fig. 6a), and for high concentrations (from 5×10^{-3} to 5.00 mg/mL) the time for each experiment is about 2 h (Fig. 6b). An exception is found for the two lowest concentrations (Fig. 6a), at which the interfacial tensions are rather constant with the passage of time and stay above the value of the pure solvent interfacial tension [about 50.4 mJ/m^2 (27)].

In order to establish an "equilibrium" interfacial tension, γ_∞, a method of "limiting slope" (27) was adopted. First, for each concentration, the derivative of interfacial tension with respect to time was calculated by the forward difference method for the experimental curve. This results in a calculated slope for each experimental point on the curve. The final, minimum slope was then found for each concentration. Next, the largest value among these minimum slopes was identified and used as the cutoff value for all concentrations and the corresponding interfacial tension values were regarded as the equilibrium interfacial tension of the protein solution.

Figure 7 illustrates the "equilibrium" interfacial tension as a function of concentration. The largest minimum slope among 15 concentrations was found to be 5×10^{-4} mJ/m^2 s; this value was used as the cutoff value. The times corresponding to this cutoff point were all above 1 h, with higher concentrations requiring less time to reach this cutoff point. The final interfacial tension value obtained in the time-dependent measurement is also given in Fig. 7. The two types of interfacial tension values agree quite well. The "final readings" are only slightly lower than the values at $|d\gamma/dt| = 5 \times 10^{-4}$ mJ/m^2 s. The duration of the kinetic interfacial tension measurement at each concentration had been chosen on the basis of preliminary measurements and was nothing more than a rough optimization of the experimentation. It is therefore significant that the two types of data agree so well. The agreement suggests that we have obtained a reasonable approximation to the true equilibrium interfacial tension.

The interfacial tension γ remains almost constant at approximately 52 mJ/m^2 over more than 3 h for the two lowest concentrations, 1×10^{-4} and 1×10^{-3} mg/mL (Fig. 6a). With further increases in bulk concentration, the interfacial tension declines sharply to approximately 20 mJ/m^2. However, at concentrations above 0.05 mg/mL, γ remains essentially unchanged at that level with an increase in concentration (Fig. 7).

The results may also be interpreted in terms of interfacial pressure, π, which is defined as the difference in interfacial tension between water-decane and HSA solution–decane interfaces. The interfacial tension of the pure water-decane interface, γ_0, was measured as 50.4 mJ/m^2 at 25°C. The π values calculated by employing the equilibrium interfacial tension obtained by both the final reading and the minimum slope criterion are given in Fig. 8. Interestingly, at the two lowest concentrations (1×10^{-4} and 1×10^{-3} mg/mL}), a negative interfacial pressure of up to approximately 2 mJ/m^2 was obtained.

In order to investigate whether the observed negative interfacial pressures are electrostatic in nature, the interfacial tension of the aqueous HSA solution–decane interface was measured at pH values of 3.5, 4.8, and 5.6; a pH of 4.8 represents the isoelectric point of HSA. This was accomplished by preparing the aqueous HSA solution in Trizma buffer. The measurements were performed at an protein concentration of 1×10^{-4} mg/mL. The interfacial tension of the buffer (without HSA) and decane was also measured. The results are given in Fig. 9. At a pH value of 3.5 (Fig. 9a), HSA decreased the interfacial tension approximately from 53.8 to 53.0 mJ/m^2. Therefore, a positive interfacial pressure of 0.8 mJ/m^2 was obtained. A positive interfacial pressure (1.6 mJ/m^2) was also obtained at the pH value of 4.8 (Fig. 9b); the interfacial tension was decreased approximately from 52.4 to 50.8 mJ/m^2. However, at a pH value of 5.6 (Fig. 9c), a negative interfacial pressure (about 0.4 mJ/m^2) was measured. The interfacial tension increased from 53.6 to 54.0 mJ/m^2. The pH value of the protein solutions without buffer was found to be 5.5 (and a negative interfacial pressure was found).

3. Discussion

In view of the concentration dependence of the interfacial tension, three domains may be identified in Fig. 7: (a) the slow change in γ at low bulk concentrations ($C < 10^{-3}$ mg/mL); (b) the sharp decline in γ within the region of intermediate concentrations ($10^{-3} < C < 10^{-2}$ mg/mL); (c) the time-independent region of γ ($C > 10^{-2}$ mg/mL). The slow change in the interfacial tension at low concentrations may be indicative of relatively weak interaction between the adsorbed protein molecules in the adsorbed surface layer. Upon reaching the region of intermediate concentrations, the strong interactions between protein molecules in the adsorbed

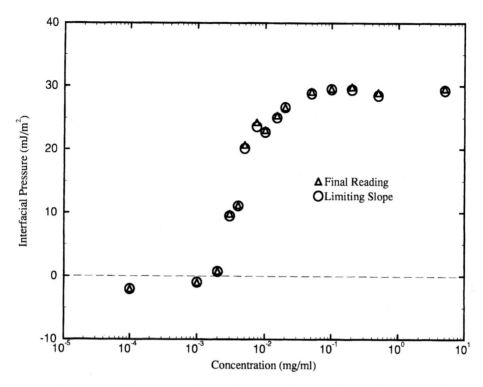

Fig. 8 The interfacial pressure obtained by the minimum slope criterion and the last experimental value for various bulk concentrations of the HSA.

surface layer induce a sharp decline in the interfacial tension. As the bulk concentration increases further, the surface layer will be saturated with protein molecules forming a close-packed monolayer, resulting in a constant value for the interfacial tension. It can be postulated that the close-packed monolayer has a comparatively stable structure and the interfacial tension does not decrease noticeably with increasing concentration at these high protein concentrations.

Our experiments show a negative interfacial pressure for the aqueous human serum albumin solution–decane interface at low bulk albumin concentrations (i.e., 1×10^{-4} and 1×10^{-3} mg/mL). Negative surface pressures have been reported in the past for organic and inorganic solutes in water. For example, the amino acid glycine increases the surface tension of water: For weight percentages (w/v) of 3.62, 6.98, 10.12, and 13.10, surface tensions of 72.54, 73.11, 73.74, and 74.18 mJ/m^2 have been reported, respectively (36). We have performed measurements at weight percentages of 6.98 and 13.10 of glycine in double-distilled water by ADSA-P and have found close agreement with the literature values. These increases in interfacial tension are thought to have electrostatic origins (37). When charged particles approach an interface from solution, particles with the same sign of charge will repel one another. This repulsion hinders ions from adsorbing to the interface. Thus, a depletion layer is formed that results in an increase in interfacial tension and hence in a negative interfacial pressure. In case of the small amino acid glycine, repulsive interactions may occur between the dipolar amino acid molecules (38,39).

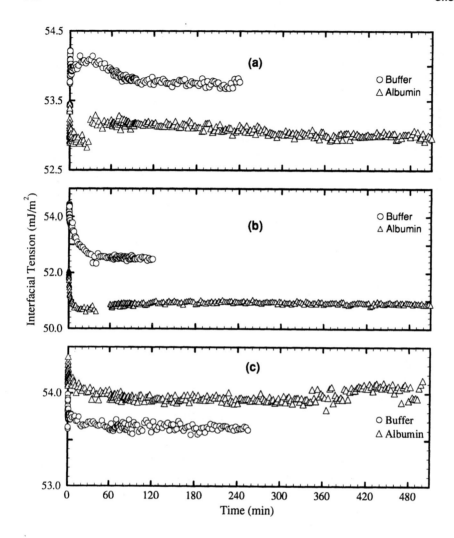

Fig. 9 The interfacial tension versus time at a bulk concentration of 0.0001 mg/mL. Measurements were performed at three pH values: (a) 3.5; (b) 4.8; (c) 5.6.

In the pH experiments, at the isoelectric point (pH 4.8) and below this point (pH 3.4), a positive interfacial pressure was obtained. However, a negative interfacial pressure was measured for a pH above the isoelectric point. Similarly, in the measurements without the Trizma buffer at a concentration of 1×10^{-4} mg/mL, a surface pressure of -1.7 mJ/m^2 (Fig. 8) and a pH of 5.5 were recorded. It is known that albumin molecules are negatively charged when the pH exceeds the isoelectric point because the side chains, which have slightly more carboxyl groups than amino groups, are hydrolyzed and become negatively charged. A charged albumin molecule at the interface induces a repulsive image potential. The resulting electrostatic repulsion will result in a depletion layer at the interface (37) and an increase in the interfacial tension.

At lower pH values, the albumin molecule exists in a fast-migrating and expanded from where most tyrosines and other hydrophobic residues are exposed to the solvent (40). The hydrophobicity of the exposed residues provides the driving force for the albumin molecules to adsorb at the interface. Therefore, a surface depletion layer does not form and a rise in interfacial tension does not occur. At pH values above the isoelectric point, the protein has a different expanded form. Increased accessibility of the hydrogen atoms for exchange, increased mobility of the thiol group, and a slight loss of the helical structure are observed (40). This variant expanded form exposes less of the hydrophobic residues, and as a result the negative charges of the side chains play a more dominant role in dictating the behavior of the molecule at the interface. Hence, negative interfacial pressures are observed at a pH value above the isoelectric point.

The preceding explanation of charge effects may also be supported or supplemented by observations of the charge properties of hydrocarbon surfaces in aqueous solutions (41–43). The measurements of the zeta potential at relatively high pH indicate that some hydrocarbons are negatively charged, just as albumin. Hence, there is an electrostatic repulsion between the hydrocarbon and the protein. This is in line with the supposition of a repulsive image potential induced by the protein adsorbed at the interface. It is then reasonable to expect that the observed negative interfacial pressure is electrostatic in nature. However, one has to be cautious about the role that the hydrocarbon plays in the negative interfacial pressure. The presence of hydrocarbon is not essential in obtaining negative interfacial potential. Experiments (44) have shown that, at a water-air interface, human serum albumin of low concentrations also has negative interfacial pressures. For example, in an HSA aqueous solution at a concentration of 1×10^{-5} mg/mL, a surface tension of 73.5 mJ/m^2 is observed at 20°C, and this value is steady for more than 4 h after an initial equilibration period of about 14 h. Nevertheless, further experiments must be performed for the system presented here, so that a direct correlation between zeta potentials and negative interfacial pressures may be obtained.

To conclude, negative interfacial pressures were observed only at the two lowest bulk concentrations used in our experiments. There is a concentration region where a transition occurs from negative interfacial pressure to positive interfacial pressure, i.e., zero interfacial pressure. It can be postulated that, with an increase in the bulk albumin concentration, the conformation of the protein at the interface is altered. With increasing concentration, close packing of the molecular segments at the interface occurs. Therefore, repulsive electrostatic forces are overcome by the close packing of the protein molecules. As a result, rather than formation of a depletion layer leading to an elevation of the interfacial tension, a reduction in the interfacial tension and a positive surface pressure ensue.

B. Dynamic Surface Tension of Mixed Solutions of a Protein and Lipids

Biologically, the study of the molecular interactions between lipids and protein has important practical implications. The binding of lipid to a membrane protein would significantly alter the surface properties of the membrane, such as hydrophobicity, and hence affect membrane-assisted enzymatic reactions involving, e.g., phospholipases (45,46), possibly ion channel behavior, and adhesion of cells. Indeed, 12S-HETE, which is used in this demonstration, has been shown to affect phospholipases (45,46) and potently causes tumor cells to adhere to the vascular endothelium (47).

The objectives here are (a) to measure the dynamic surface tension response to a sawtooth change in the surface area of solution drops using ADSA; (b) to probe the molecular interactions or binding between bovine serum albumin (BSA), as a model protein, and three lipids; and (c) to investigate the concentration dependence and specificity of the molecular binding.

1. Materials

The sample of bovine serum albumin (Sigma Chemical Co., St. Louis, MO) was essentially fatty acid and globulin free, with an average molecular weight of 67,000. It was used without further purification. Deionized and glass distilled water was used.

Three similar lipids were chosen for this study: 12S-hydroxy-5Z,8Z,10E,14Z-eicosatetraenoic acid (12S-HETE–free acid), methyl 12S-hydroxy-5Z,8Z10E,14Z-eicosatetraenoate (12S-HETE–methyl ester) and 5Z,8Z,11Z,14Z-eicosatetraenoic acid (arachidonic acid–free acid). In molecular structure, 12S-HETE–free acid differs from arachidonic acid by having an additional hydroxyl group and a *cis-trans* conjugated diene system, and the difference between 12S-HETE–free acid and 12S-HETE–methyl ester is at the carboxyl end. It was interesting to investigate whether these subtle structural differences would distinguish one molecule from another in terms of their interactions with bovine serum albumin at the surface. 12S-HETE–free acid and arachidonic acid were purchased from Cayman Chemicals, Ann Arbor, MI. The methyl ester derivative was prepared with an ether solution of diazomethane (48). The molecular weights of these compounds are 320, 304, and 334, respectively. Because these lipids are not soluble in water, they have been initially dissolved in 1 μL dimethyl sulfoxide (DMSO) before addition to 1 mL of BSA aqueous solutions.

Three types of mixed solutions were prepared: (a) 0.02 mg/mL BSA aqueous solution containing 12S-HETE–free acid at a concentration ranging from 0.001 to 1.0 μg/mL (note that, within this range, a concentration of 0.1 μg/mL corresponded to a molecular ratio between 12-HETE–free acid and BSA of approximately 1:1); (b) 0.02 mg/mL BSA aqueous solution containing 12S-HETE–methyl ester at a concentration ranging from 0.01 to 1.0 μg/mL; (c) 0.02 mg/mL BSA aqueous solution containig arachidonic acid at a concentration ranging from 0.001 to 1.0 μg/mL. Two control experiments were performed using the following two solutions: (a) a pure BSA aqueous solution at a concentration of 0.02 mg/mL and 1 μL DMSO; (b) a BSA-free solution of 1 μg 12S-HETE–free acid in a mixture of 1 μL DMSO in 1 mL (instead of a BSA solution). All the experiments were conducted at 37°C.

2. Results and Discussion

    a. *Molecular Interactions and Dose Effects*

Figure 10 shows the dynamic surface tension response to the sawtooth change in surface area in the time range from 240 to 300 s from the beginning of the experiment with the aqueous BSA solution at a concentration of 0.02 mg/mL and 1μL DMSO. It has been shown (49) that at early times (less than 60 s) the tension response reflects the initial adsorption process of BSA to the surface and does not repeat itself from cycle to cycle. After 120 s, the tension response starts showing constant cycles, as shown in Fig. 10. Each cycle shows a characteristic skewed shape, with two kinks occurring in the two branches corresponding to surface expansion and compression.

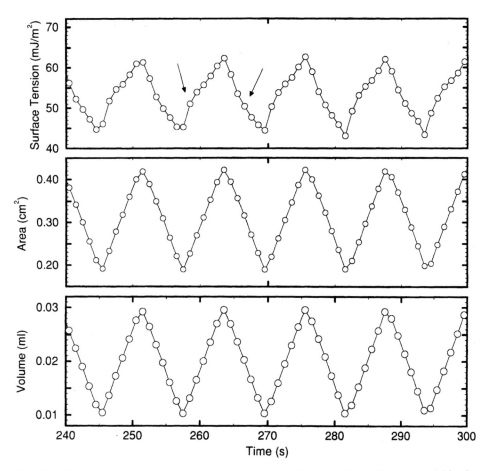

Fig. 10 Dynamic surface tension response to a sawtooth change in surface area, within the time range of 240 to 300 s from the beginning of the experiment with the BSA aqueous solution at a concentration of 0.02 mg/mL and 1 µL DMSO. Each cycle shows a characteristic skewed shape in the tension response, with two kinds in the two branches corresponding to surface expansion and compression, respectively.

Figure 11 shows the dynamic surface tension response to the same sawtooth variation in surface area as in Fig. 10, for the BSA solution to which 12S-HETE–free acid had been added. A series of 12S-HETE–free acid concentrations were used: 0.001, 0.005, 0.01, 0.02, 0.05, 0.1, and 1.0 µg/mL BSA aqueous solution. It can be seen that at low 12S-HETE–free acid concentrations the tension response to the area variation is similar to that observed in Fig. 10 for the pure BSA solution: the characteristic skewed shape indicates that the surface is covered mainly with BSA molecules. However, at the concentration of 0.1 µg/mL, a distinct pattern change is observed in the dynamic surface tension response: the skewed pattern of the BSA solution is replaced by a rather symmetric one. This indicates that the surface properties are not determined solely by BSA; i.e., the added 12S-HETE–free acid plays a role. It is noted that the 0.1 µg/mL concentration corresponds to a molecular ratio between 12S-HETE–free acid and BSA of approximately 1:1. As the concentration of

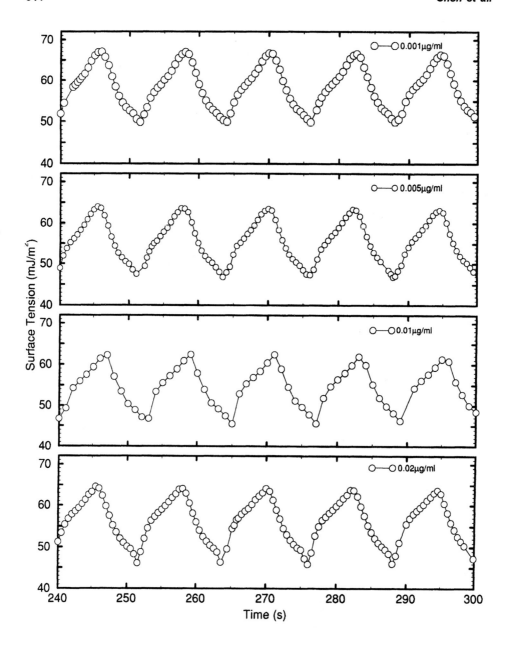

Fig. 11 Dynamic surface tension response to the same sawtooth variation in surface area as in Fig. 10, for the BSA solution to which 12S-HETE–free acid had been added. A series of 12S-HETE–free acid concentrations were used: 0.001, 0.005, 0.01, 0.02, 0.05, 0.1, and 1.0 μg/mL BSA aqueous solution. It is seen that at low 12S-HETE–free acid concentrations the tension response to the area variation is similar to that observed in Fig. 10. However, at the concentration of 0.1 μg/mL, a distinct pattern change is observed in the dynamic surface tension response: the skewed pattern in the tension response of the BSA solution is replaced by a rather symmetric one. As the concentration of 12S-HETE–free acid increases to 1.0 μg/mL, yet another different, symmetric pattern is observed in the tension response, indicating a dose effect on the surface tension behavior and hence on the surface physicochemical properties.

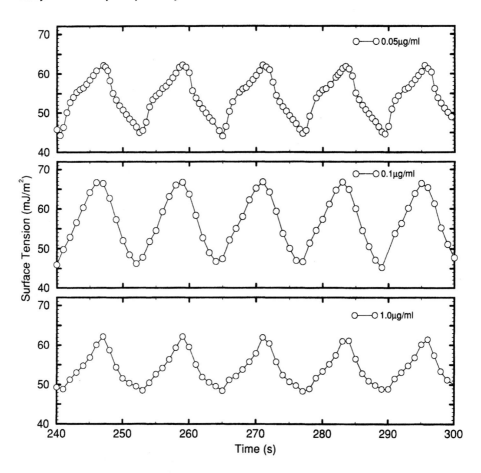

Fig. 11 Continued.

12S-HETE–free acid increases to 1.0 μg/mL, yet another asymmetric pattern emerges in the tension response. Clearly, there is a dose effect on the surface tension behavior and hence on the physicochemical properties. At concentrations above 0.1 μg/mL, the tension response to the area change provides a useful tool for probing possible molecular interactions between 12S-HETE–free acid and BSA.

To investigate further the effects of 12S-HETE–free acid, a control experiment was performed in which 1 μg 12S-HETE–free acid dissolved in 1 μL DMSO was added to 1.0 mL water, i.e., in the absence of BSA. The results are shown in Fig. 12, where minimal changes in the dynamic surface tension pattern are observed in response to the same sawtooth variation in surface area as in Figs. 10 and 11. If there were no interaction between 12S-HETE–free acid and BSA in the mixed solution, the resulting surface tension would have to be a superposition of the surface tensions of the individual lipid and protein solution. However, from Figs. 10–12, the tension response of the mixture at a concentration of 12S-HETE–free acid of 0.1 mu;g/mL does not reflect the pattern of the pure BSA solution at all. Therefore, molecular interactions must exist between 12S-HETE–free acid and BSA, probably formation of lipid-protein complexes. These complexes, being species different from albumin alone, no longer

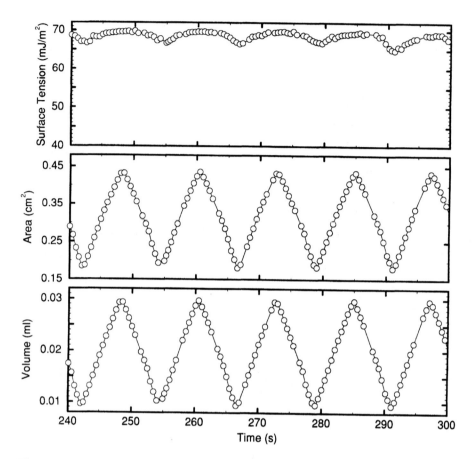

Fig. 12 Dynamic surface tension response to the same area variation as in Fig. 10, for a control experiment in which 1 μg 12S-HETE–free acid dissolved in 1 μL DMSO was added to 1.0 mL water in the absence of BSA. Rather symmetric cycles with small amplitudes are observed for the tension response.

show the skewed shape of BSA, presumably caused by conformational changes. Because the distinct change in the tension response pattern of BSA occurs at the concentration of 0.1 μg/mL of 12S-HETE–free acid, corresponding to a molecular ratio of 1:1, such interactions between 12S-HETE–free acid and BSA are presumably connected with a single binding site (see also later).

Another observation in Fig. 11 is that the tension value reached at the peaks for the concentration 0.1 μg/mL is significantly higher than that for the other concentrations including the pure BSA solution. This indicates that the mixture, at the 1:1 molecular ratio, is more hydrophilic than BSA itself; therefore, the molecular interactions between 12S-HETE–free acid and BSA may be hydrophobic in nature, such that the hydrophobic end of the lipid attaches to a similar part of the protein, leaving the hydrophilic end of the lipid exposed to the surrounding water environment (50).

At the 12S-HETE–free acid concentration of 1.0 μg/mL (Fig. 11), the molecular ratio between lipid and protein is roughly 10:1. One might think that the resulting dynamic surface tension should be due predominantly to the presence of

12S-HETE–free acid at the surface. However, the tension of the lipid alone (Fig. 12) shows a very different response, with a much smaller amplitude. If we were to assume that BSA has only one binding site for 12S-HETE–free acid, then 9 out of 10 lipid molecules would exist in water freely, and the resulting surface properties of the mixture would have to be dominated by the surface properties of the free lipid. However, from Fig. 12, the free lipid solution has high surface tension values, above 65 mJ/m^2; from Fig. 11, the maximum tension value is also above 65 mJ/m^2 for the mixture of the lipid and the protein at the concentration of 0.1 μg/mL. If there were no interaction between lipid and protein-lipid complex, the tension response of the combination would be expected to be at the same high level. This, however, is not the case; the maxima for 1.0 μg/mL in Fig. 11 are clearly below 65 mJ/m^2. Therefore, one needs to conclude that not all additional lipid molecules remain free, but rather bind to BSA, at least to such a degree that the surface tension is significantly lowered. Because the maximum tension value at the 1.0 μg/mL concentration is smaller than that at the 0.1 μg/mL concentration, the new complex, as a result of the new binding of the lipid to the protein, is more hydrophobic, compared with the protein-lipid complex of 1:1 ratio, i.e., 0.1 μg/mL of 12S-HETE. Hence, the additional binding of the lipid to the protein is presumably hydrophilic in nature (50).

In the preceding analysis leading to the conclusion of 12S-HETE–free acid binding to BSA and the dose dependence of such molecular binding, we tacitly assumed that the DMSO, used as dissolving agent for the lipid, does not interact with BSA, nor does the DMSO play a role in the surface tension response after the initial few cycles. It has been established that, for a mixture of DMSO and BSA, the surface molecular population is dominated by small DMSO molecules only at early stages of the cycling experiment before 60 s, due to DMSO's much higher diffusion coefficient (49). With the passage of time, BSA gradually adsorbs at the surface, and BSA molecules stay at the surface once they adsorb. This leads to a squeeze-out of the DMSO molecules from the surface. After repeated cycles, the surface properties are essentially determined by the BSA molecules adsorbed at the surface, and no DMSO contribution to the surface tension response can be detected. This indicates that DMSO is merely a vehicle for carrying lipids, and it does not contribute to the tension response at late stages of the cycling experiment, as shown in Figs. 10–12.

b. Specificity of Molecular Interactions

To study the possible specificity of the molecular binding of 12S-HETE–free acid to BSA, two similar lipids, 12S-HETE–methyl ester and arachidonic acid–free acid, were used to perform the same tension response experiment as for 12S-HETE–free acid. The results for 12S-HETE–methyl ester are shown in Fig. 13 for three concentrations, 0.01, 0.1, and 1.0 μg/mL. Again, 0.1 μg/mL corresponds to a molecular ratio of lipid to protein of approximately 1:1. All three concentrations show a skewed pattern, similar to that of the pure BSA solution (Fig. 10), and hence there is no indication of molecular interactions, i.e., binding. Figure 13 suggests the dominance of protein adsorption at the surface, while 12S-HETE–methyl ester molecules play little role and are possibly squeezed out of the surface by the cycling of surface area (49).

Figure 14 shows the dynamic surface tension response to the same sawtooth area variation that was used before, with arachidonic acid–free acid added to the BSA solution. A series of arachidonic acid concentrations were used: 0.001, 0.005, 0.01,

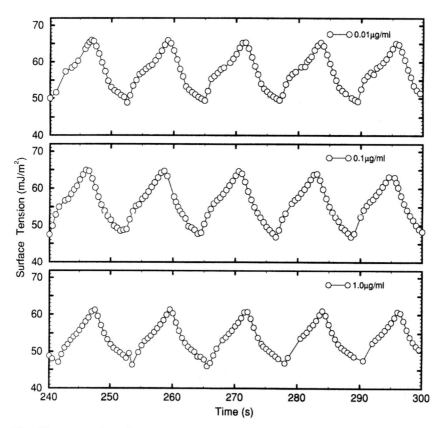

Fig. 13 Dynamic surface tension response to the same area variation as in Fig. 10, for the BSA solution to which 12S-HETE–methyl ester had been added at three concentrations of 0.01, 0.1, and 1.0 μg/mL. At all three concentrations a skewed pattern is seen, similar to that of the pure BSA solution.

0.05, 0.1, and 1.0 μg/mL. Throughout these concentrations, a skewed shape is always observed in the cycling tension response. In contrast to the observation in Fig. 11 for 12S-HETE–free acid, there is no distinct change in the tension response as the arachidonic acid concentration passes 0.1 μg/mL, corresponding to a molecular ratio between lipid and protein of approximately 1:1. As previously shown with 12S-HETE–methyl ester, Fig. 14 indicates that there is no molecular interaction between arachidonic acid and BSA as reflected in the surface tension behavior. This implies that the detected molecular binding of 12S-HETE–free acid to BSA is rather specific and that a subtle variation in the lipid molecular structure leads to significantly different interaction properties between the lipid and the protein.

C. ADSA-CB: Captive Bubble Method in Lung Surfactant Studies

As illustrated in the previous section, in the presence of a surface film, surface tension will change as the interfacial area is decreased and increased. Without proper design, at low surface tension, this type of system could suffer from film leakage (51–55). Film

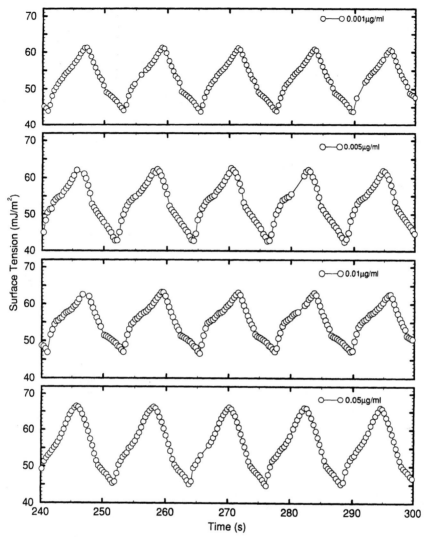

Fig. 14 Dynamic surface tension response to the same area variation as in Fig. 10, for the BSA solution to which arachidonic acid–free acid had been added at concentrations of 0.001, 0.005, 0.01, 0.05, 0.1, and 1.0 μg/mL. At all these concentrations, a skewed shape is always observed in the cycling tension response.

leakage is due to a fundamental surface thermodynamic principle: at low surface tension the surface-active molecules can spread from the liquid-air interface onto the surrounding solid, thereby decreasing the free energy of the system. Film leakage has been demonstrated to occur in the Langmuir-Wilhelmy film balance (56) and has been observed in the pulsating bubble surfactometer (57). Film leakage can lead to surface behavior that has been erroneously thought to be intrinsic to the system under study.

The only way to eliminate film leakage is by removing the potential pathway through which the surface-active molecules can leave the air-liquid interface. The

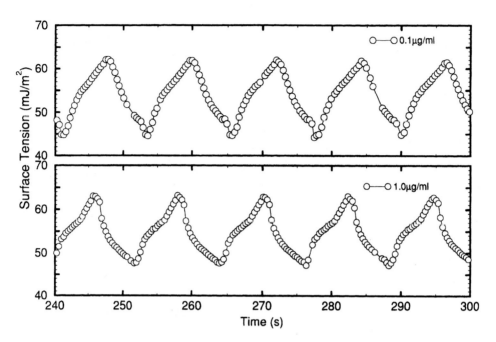

Fig. 14   Continued.

captive bubble geometry accomplishes this by holding a bubble of air captive at the top of a chamber filled with sample liquid. In this system, there is no need to pierce the bubble with any capillaries. A hydrophilic ceiling ensures an aqueous layer between the solid ceiling and the bubble, leaving the air-liquid interface completely intact (55).

1.   The ADSA Captive Bubble Chamber

The ADSA captive bubble chamber (55) comprises two quartz viewing windows that are secured on both sides of a metal plate. A section of this plate has been removed, forming the side walls of the chamber (the end walls are the viewing windows). The windows are placed in between the metal plate and two metal end plates. Seals are ensured by O-rings on both sides of the windows. Four sets of bolts are used to fasten the whole assembly together. Two lateral holes were drilled through the end plates to allow water circulation for temperature control of the chamber. As shown in Fig. 15, the section hole of the middle metal plate has straight edges on the sides and bottom. The top was designed such that a glass piece with a concave surface could be held in place, thereby providing a glass "ceiling" for the chamber. Glass was chosen to ensure an aqueous layer between the captive bubble and the ceiling, leaving the air-liquid interface completely intact. The glass piece was obtained by cutting an optical lens.

Three ports were made to provide access to the chamber (Fig. 15). One port was designed for the temperature probe, which remains in place during the experiment. Fittings are used to connect a Teflon capillary to the second port of the chamber. The capillary, in turn, is attached to a motorized syringe. The chamber internal pressure is changed by pumping liquid in or out. The last port is used to form an air bubble in the sample chamber using a microsyringe.

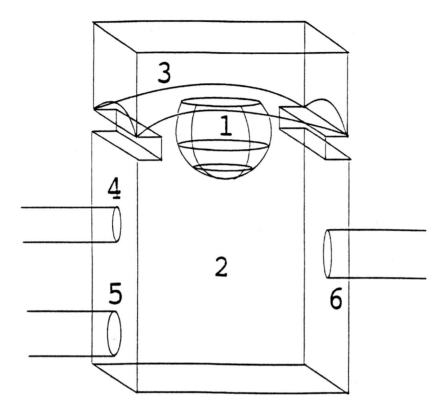

Fig. 15 Schematic of the section hole of the metal plate forming the side walls of the ADSA-CB test chamber. (1) Captive bubble, (2) pulmonary surfactant solution, (3) glass ceiling, (4) microsyringe port, (5) syringe port, (6) temperature probe port (55).

2. Surface Tension Measurements of Lung Surfactant

The surfactant used for this demonstration was bovine lipid extract surfactant (BLES, Biochemicals Inc., London, ON, Canada). BLES was supplied as a suspension, containing the phospholipids of natural surfactant (27 mg/mL) and surfactant-associated proteins: SP-B and SP-C. The suspension was gently stirred and 0.15 mL was diluted in a 10-mL flask with 0.9% NaCl solution, resulting in a phospholipid concentration of 400 μg/mL in the diluted surfactant solution.

The cycling of the interface between the air bubble and the lung surfactant solution was performed at a rate of 25 s per cycle. The amount of compression was approximately 80% by volume, corresponding to about 65% change in surface area. The experiment was conducted at a temperature of 37.0±0.1°C.

Figure 16 shows three dynamic cycles of a typical experiment in which the interface was compressed sufficiently to achieve collapse of the surface film. Although not entirely obvious in Fig. 16, close examination of a region of minimum surface tension (Fig. 17) showed an intricate pattern. During compression, at 523.5 s, the surface tension γ was approximately 0.5 mJ/m^2, and half a second later it increased to approximately 1.5 mJ/m^2. This indicates that expulsion of dipalmitoylphosphatidylcholine (DPPC) molecules from the film (i.e., collapse) occurred, resulting in a surface tension

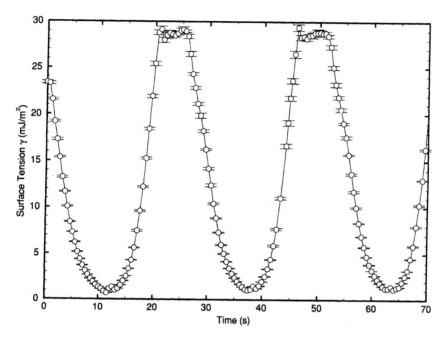

Fig. 16 Surface tension as a function of time for the first three cycles of a captive bubble experiment. Note that these compressions were sufficient to achieve collapse. The error limits shown are the 95% confidence levels (55).

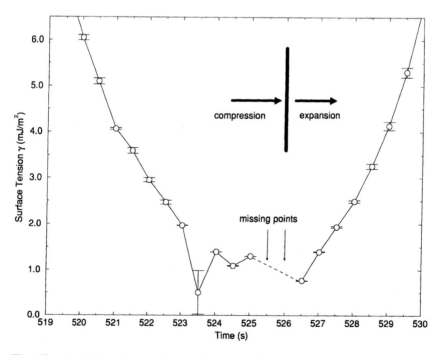

Fig. 17 Detailed surface tension as a function of time of a collapsing film. The missing points are probably due to the non-Laplacian shape of the bubble (see text). The error limits shown are the 95% confidence levels (55).

increase of about 1 mJ/m^2. At 524 s, the interface was still being compressed, and between this time and 1 s later the surface tension decreased to, and remained at, approximately 1.0 mJ/m^2. Then between 525 and 526.5 s, when the bubble was between compression and expansion, ADSA failed to provide data, despite the fact that images were acquired. This might be due to the bubble being non-Laplacian in shape (55). At these very low surface tensions, sudden oscillations of the drop were observed, presumably due to collapse of the film. The rather large error limits at $\gamma = 0.5$ mJ/m^2 in Fig. 17 may indicate that the bubble was in the midst of one of these sudden movements and not in an equilibrium shape. In addition, in the interlude between the compression and expansion, the images could not be processed by ADSA, again indicating deviation of the bubble shape from mechanical equilibrium (55).

D. Contact Angle Measurements on Rough and Heterogeneous Solid Surfaces by ADSA-D

Figure 18 shows a schematic of the experimental setup for ADSA-CD and MD, which have been unified into a single program, ADSA-D. ADSA-D can be used for contact angle measurements not only on smooth but also on rough and heterogeneous solid surfaces. Most biological surfaces present not only small contact angles but also morphological and energetic imperfections, leading to irregularities of the three-phase contact line, as seen in Fig. 19, where sessile drops of water on a layer of *T. ferrooxidans*. It will be difficult and dubious to measure contact angles on such drops by finding a tangent of the drop profile at a three-phase contact point. However, ADSA-D avoids this problem and provides averaging by analyzing drop contact area.

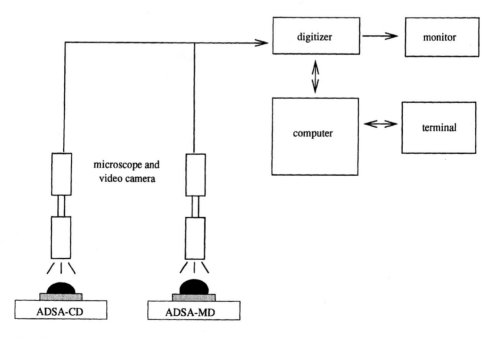

Fig. 18 Schematic of the experimental setup for ADSA-CD and MD.

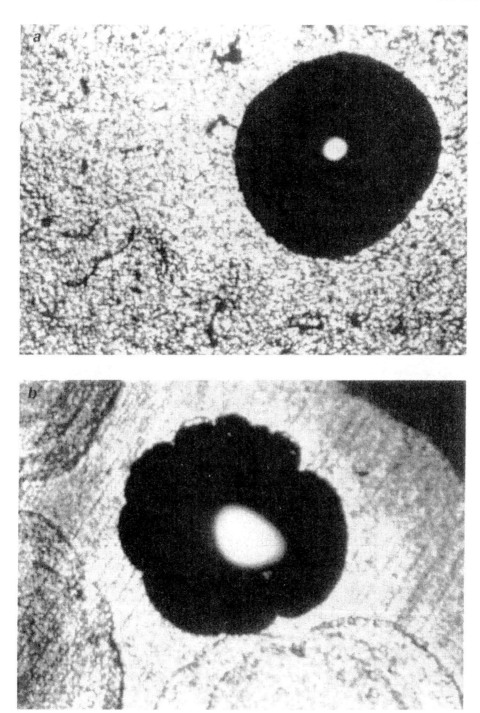

Fig. 19 Sessile drops of water on a layer of *T. ferrooxidans*. The contact angles calculated for two different drops using ADSA-CD were (a) 12.7° and (b) 11.3°.

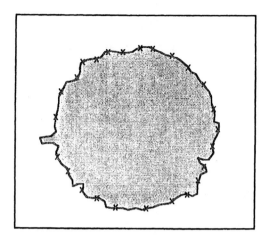

Fig. 20 Schematic of the determination of the perimeter of a fictitious sessile drop on a video screen using a cursor controlled by a mouse.

As an illustration, contact angle measurements on layers of several bacterial strains were measured using ADSA-D. A video image of the drop contact area can be digitized semiautomatically, using a cursor controlled by a "mouse" on the video screen. The pattern of selection of points on the three-phase contact line is characterized schematically in Fig. 20. An average drop diameter was calculated from these values. ADSA-D then calculates the contact angle based on such a drop diameter and the input values for the drop volume and the density and surface tension of the liquid. It was found that contact angles of water on piliated *Escherichia coli* RDEC-1 and nonpiliated RDEC-1 strains are 20.8±0.9° and 10.6±0.5°, respectively (58).

Another illustration of the use of ADSA-D to measure contact angles on imperfect solid surfaces is shown as follows: It was proposed to investigate the hydrophobicity of the intestinal tract at different sections for rabbits of different ages (59). Because the surfaces of the intestine are, of course, imperfect and rough, a conventional goniometer technique cannot be used. ADSA-D was employed to measure water contact angles on these surfaces. It was found that the intestinal hydrophobicity is altered by maturational changes, regional differences, and mucosal inflammation. For example, a water contact angle of 53.2±8.4° was obtained on the proximal colon of suckling rabbits and of 93.2±6.7° on proximal colon of adult rabbits (59).

ACKNOWLEDGMENTS

Financial support for this project was provided by the Medical Research Council of Canada (grant MT-5462).

REFERENCES

1. JL Brash, TA Horbett. Protein at interfaces: An overview. In: TA Horbett, JL Brash, eds. Proteins at Interface II. Fundamentals and Applications. Washington, DC: American Chemical Society, 1995, pp 1–25.

2. HB Callen. Thermodynamics and an Introduction to Thermostatics. 2nd. ed. New York: Wiley, 1985.

3. P Chen, DY Kwok, RM Prokop, OI del Río, SS Susnar, AW Neumann. Axisymmetric liquid-fluid interfaces (ALFI). In: D Möbius, R Miller, eds. Drops and Bubbles in Interface Research. Amsterdam: Elsevier, 1997, Chap 2.

4. DE Graham, MC Phillips. Proteins at liquid interfaces: I. Kinetics of adsorption and surface denaturation. J Colloid Interface Sci 70:415, 1979.

5. M Paulsson, P Dejmek. Surface film pressure of β-lactoglobulin, α-lactalbumin and bovine serum albumin at the air/water interface studied by Wihelmy plate and drop volume. J Colloid Interface Sci 150:394, 1992.

6. DE Graham, MC Phillips. Proteins at liquid interfaces: II. Adsorption isotherms. J Colloid Interface Sci 70:403, 1979.

7. P Suttiprasit, V Krisdhasima, J McGuire. The surface activity of α-lactalbumin, β-lactoglobulin, and bovine serum albumin. J Colloid Interface Sci 154:316, 1992.

8. E Tornberg. The application of the drop volume technique to measurements of the adsorption of proteins at interfaces. J Colloid Interface Sci 64:391, 1978.

9. AJI Ward, LH Regan. Pendant drop studies of adsorbed films of bovine serum albumin. J Colloid Interface Sci 78:389, 1980.

10. A Voigt, O Thiel, D Williams, Z Policova, W Zingg, AW Neumann. Axisymmetric drop shape analysis (ADSA) applied to protein solutions. Colloids Surf 58:315, 1991.

11. F Bashforth, JC Adams. An Attempt to Test the Theory of Capillary Action. Cambridge: Cambridge University Press and Deighton Bell, 1883.

12. S Hartland, RW Hartley. Axisymmetric Fluid-Liquid Interfaces. Amsterdam: Elsevier, 1976.

13. C Maze, G Burnet. A non-linear regression method for calculating surface tension and contact angle from the shape of a sessile drop. Surf Sci 13:451, 1969.

14. C Maze, G Burnet. Modification of a non-linear regression technique used to calculate surface tension from sessile drops. Surf Sci 24:335, 1971.

15. Y Rotenberg, L Boruvka, AW Neumann. Determination of surface tension and contact angle from the shapes of axisymmetric fluid interfaces. J Colloid Interface Sci 93:169, 1983.

16. P Cheng, D Li, L Boruvka, Y Rotenberg, AW Neumann. Automation of axisymmetric drop shape analysis for measurements of interfacial tensions and contact angles. Colloids Surf 43:151, 1990.

17. OI del Río, AW Neumann. Axisymmetric drop shape analysis: Computational methods for the measurement of interfacial properties from the shape and dimensions of pendant and sessile drops. J Colloid Interface Sci, 196:136, 1997.

18. JJ Bikerman. Ind Eng Chem Anal Ed 13:443, 1941.

19. FK Skinner, Y Rotenberg, AW Neumann. Contact angle measurements from the contact diameter of sessile drops by means of a modified axisymmetric drop shape analysis. J Colloid Interface Sci 130:25, 1989.

20. E Moy, P Cheng, Z Policova, S Treppo, D Kwok, DR Mack, PM Sherman, AW Neumann. Measurement of contact angles from the maximum diameter of nonwetting drops by means of a modified axisymmetric drop shape analysis. Colloids Surf 58:215, 1991.

21. TE Hull, WH Enright, KR Jackson. Runge-Kutta research at Toronto. Appl Numer Math 22:225, 1996.

22. TE Hull, WH Enright, KR Jackson. User's Guide for DVERK—A Subroutine for Solving Non-Stiff ODEs. Technical Report 100, Department of Computer Science, University of Toronto, 1976.

23. WH Enright, DJ Higham, B Owren, PW Sharp. A Survey of the Explicit Runge-Kutta Method. Technical Report 291, Department of Computer Science, University of Toronto, 1995.

24. JJ Moré, SJ Wright. Optimization Software Guide. Philadelphia: SIAM, 1993.

25. UM Ascher, RM Mattheij, RD Russell. Numerical Solution of Boundary Value Problems for Ordinary Differential Equations. Englewood Cliffs, NJ: Prentice Hall, 1988.
26. S Lahooti, OI del Río, P Cheng, AW Neumann. Axisymmetric drop shape analysis (ADSA). In: AW Neumann, JK Spelt, eds. Applied Surface Thermodynamics. New York: Marcel Dekker, 1996, Chap 10, pp 441–507.
27. MA Cabrerizo-Vílchez, Z Policova, DY Kwok, P Chen, AW Neumann. The temperature dependence of the interfacial tension of aqueous human albumin solution/decane. Colloids Surf B Biointerfaces 5:1, 1995.
28. P Chen, S Lahooti, Z Policova, MA Cabrerizo-Vílchez, AW Neumann. Concentration dependence of the film pressure human serum albumin at a water/decane interface. Colloids Surf B Biointerfaces 6:279, 1996.
29. R Miller, S Treppo, A Voigt, W Zingg, AW Neumann. Contact angle kinetics of human albumin solutions at solid surfaces. Colloids Surf 69:203, 1993.
30. G Gonzalez, F MacRitchie. Equilibrium adsorption of proteins. J Colloid Interface Sci 32:55, 1970.
31. R Miller, G Kretzschmar. Adsorption kinetics of surfactants at fluid interfaces. Adv Colloid Interface Sci 37:97, 1991.
32. R Miller, K Lunkenheimer. On the determination of equilibrium surface tension values of surfactant solutions. Colloid Polym Sci 261:585, 1983.
33. R Miller, Z Policova, R Sedev, AW Neumann. Relaxation behaviour of human albumin adsorbed at the solution/air interface. Colloids Surf A Physiochem Eng Aspects 76:179, 1993.
34. JA de Feijter, J Benjamin. Adsorption kinetics of proteins at the air-water interface. In: E Dickinson, ed. Food Emulsions and Foams. London: Royal Society of Chemistry, 1987, p 72.
35. F MacRitchie. In: JL Brash, TA Horbett, eds. Proteins at Interfaces: Physicochemical and Biochemical Studies. Washington, DC: American Chemical Society, 1987, Chap 11.
36. RC Weast, ed. Handbook of Chemistry and Physics. Cleveland: Chemical Rubber Company, 1971–72, pp 28–29.
37. C Ramachandran, RA Pyter, P Mukerjee. Microenvironmental effects on energies of visible bands of nitroxides in electrolyte solutions and when solubilized in micelles of different charge types. Significance of effective polarity estimates. Implications for spectroscopic probe studies in lipid assemblies. J Phys Chem 86:3198, 1982.
38. L Onsager, NNT Samaras. The surface tension of Debye-Hückel electrolytes. J Chem Phys 2:528, 1934.
39. FP Buff, NS Goel. Image potentials for multipoles embedded in a cavity and their applications to the surface tension of aqueous solutions of amino acids. J Chem Phys 56:2405, 1972.
40. T Peters Jr, RG Reed. Serum albumin: Conformation and active sites. In: T Peters, I Sjoholm, eds. Albumin Structure, Biosynthesis, Function. Proceedings of the 11th FEBS Meeting, Copenhagen, 1977, Vol 50, Colloquium B9. Oxford: Pergamon Press, 1978, p 11.
41. KB Medrzycka. The effect of particle concentration on zeta potential in extremely dilute solutions. Colloid Polym Sci 269:85, 1991.
42. HC van der Mei, J de Vries, HJ Busscher. Hydrophobic and electrostatic cell surface properties of thermophilic dairy streptococci. Appl Environ Microbiol 59:4305, 1993.
43. GI Geertsema-Doornbusch, HC van der Mei, HJ Busscher. Microbial cell surface hydrophobicity—The imvolvement of electrostatic interactions in microbial adhesion to hydrocarbons (MATH). J Microbiol Methods 18:61, 1993.
44. P Chen, Z Policova, AW Neumann. In preparation.
45. A Zakaroff, N Meskini, C Joulain, G Nemoz, M Lagarde, AF Prigent. 12(*S*)-HETE primes a phospholipase D pathway in activated human blood mononuclear cells. In: JY Vanderhoek, ed. Frontiers in Bioactive Lipids. New York: Plenum, 1996, pp 291–297.

46. N Meskini, A Zakaroff, C Joulain, G Nemoz, M Lagarde, AF Pregent. Triggering of a phospholipase D pathway upon mitogenic stimulation of human peripheral blood mononuclear cells enriched with 12(*S*)-hydroxyicosatetraenoic acid. Eur J Biochem 233:907, 1995.
47. KV Honn, DG Tang, X Gao, IA Butovich, B Liu, J Timar, H Hagmann. Cancer Metastasis Rev 13:365, 1994.
48. CR Pace-Asciak, E Granström, B Samuelsson. Arachidonic acid epoxides. Isolation and structure of two hydroxy epoxide intermediates in the formation of 8,11,12- and 10,11,12-trihydroxyeicosatrienoic. J Biol Chem 258:6835, 1983.
49. P Chen, Z Policova, SS Susnar, CR Pace-Asciak, PM Demin, AW Neumann. Dynamic surface tension responses to surface area change of mixed solutions of a protein and small or medium-sized organic molecules. Colloids Surf A Physicochem Eng Aspects 114:99, 1996.
50. P Chen, Z Policova, CR Pace-Asciak, AW Neumann. A study of binding of 12*S*-hydroxy-5*Z*,8*Z*,10*E*,14*Z*-eicosatetraenoic acid to bovine serum albumin using dynamic surface tension measurements. J Pharm Sci, submitted.
51. S Schürch, H Bachofen, J Goerke, F Possmayer. A captive bubble method reproduces the in situ behavior of lung surfactant monolayers. J Appl Physiol 67:2389, 1989.
52. S Schürch, F Possmayer, S Cheng, AM Cockshutt. Pulmonary SP-A enhances adsorption and appears to induce surface sorting of lipid extract surfactant. Am J Physiol 263:L210, 1992.
53. S Schürch, H Bachofen, J Goerke, F Green. Surface properties of rat pulmonary surfactant studies with the captive bubble method: Adsorption, hysteresis, stability. Biochim Biophys Acta 1103:127, 1992.
54. S Schürch, D Schürch, T Curstedt, B Robertson. Surface activity of lipid extract surfactant in relation to film area compression and collapse. J Appl Physiol 77:974, 1994.
55. RM Prokop, A Jyoti, M Eslamian, A Garg, M Mihaila, OI del Río, SS Susnar, Z Policova, AW Neumann. A study of captive bubbles with axisymmetric drop shape analysis. Colloids Surf A Physicochem Eng Aspects, 131:231, 1998.
56. J Goerke, J Gonzales. Temperature dependence of dipalmitoyl phosphatidylcholine monolayer stability. J Appl Physiol 51:1108, 1981.
57. G Putz, J Goerke, HW Taeusch, JA Clements. Comparison of captive and pulsating bubble surfactometers with use of lung surfactants. J Appl Physiol 76:1425, 1994.
58. B Drumm, AW Neumann, Z Policova, PM Sherman. Bacterial cell surface hydrophobicity properties in the mediation of in vitro adhesion by the rabbit enteric pathogen *Escherichia coli* strain RDEC-1. J Clin Invest 84:1588, 1989.
59. DR Mack, AW Neumann, Z Policova, PM Sherman. Surface hydrophobicity of the intestinal tract. Am J Physiol 262:171, 1992.

16

Brewster Angle Microscopy

C. Lheveder and J. Meunier
*Laboratoire de Physique Statistique de l'ENS, UMR 8550 du CNRS,
Universités Paris VI et Paris VII, Paris, France*

S. Hénon
*Laboratoire de Biorhéologie et d'Hydrodynamique Physicochimique, CNRS,
Universités Paris VI et Paris VII, Paris, France*

Brewster angle microscopy is a recent technique for the study of thin films on a flat surface. It is based on the properties of reflectivity of light at interfaces. It is able to make images of films as thin as monomolecular films, for example, monolayers of amphiphilic molecules at the free surface of water (Langmuir films), a domain where it is of current use.

I. PRINCIPLE OF THE TECHNIQUE

Let us consider a flat interface between two transparent and nonadsorbing media. An incident light beam of intensity I_0 is partially reflected and partially transmitted at the interface. The reflected intensity, I_R, is a function of the incidence angle θ_i, of the polarization of the light and of the details of the interface. The reflectivity is the ratio of the reflected intensity over the incident one: $R = I_R/I_0$.

For a Fresnel interface, i.e., a flat interface between two isotropic media, in which the refractive index changes abruptly at the level $z = 0$ from n_1, the refractive index of the incidence medium, to n_2, the refractive index of the second medium, the reflectivity is given by the Fresnel formulas:

For a polarization p of the incident beam (electric field in the plane of incidence):

$$R_p^F = \left(\frac{\tan(\theta_i - \theta_r)}{\tan(\theta_i + \theta_r)} \right)^2 \tag{1a}$$

For a polarization s of the incident beam (electric field perpendicular to the plane of incidence):

$$R_s^F = \left(\frac{\sin(\theta_i - \theta_r)}{\sin(\theta_i + \theta_r)}\right)^2 \tag{1b}$$

in which θ_r is the angle of refraction: $n_1 \sin \theta_i = n_2 \sin \theta_r$

Figure 1 gives the reflectivities R_s^F and R_p^F as a function of the incidence angle θ_i, for $n_1 = 1$ (the refractive index of air) and $n_2 = 1.33$ (the refractive index of water). R_s^F increases from $[(n_2 - n_1)/(n_2 + n_1)]^2$ to 1 with the incidence angle, and R_p^F first decreases to 0, then increases to 1. R_p^F vanishes at an angle of incidence θ_B called the Brewster angle. Equation (1a) shows that this happens when $\tan(\theta_i + \theta_r)$ becomes infinite, i.e., when $\theta_i + \theta_r = \pi/2$, or when the reflected and the refracted beams are perpendicular to each other. At this angle, the dipolar moments induced by the electric field of the incident beam in the second medium point out in the direction of reflection and consequently do not radiate in this direction. The two equations $n_1 \sin \theta_i = n_2 \sin \theta_r$ and $\theta_i + \theta_r = \pi/2$ at the Brewster angle allow us to write $\tan \theta_B = n_2/n_1$.

For real interfaces, R_p decreases to a minimum value at the Brewster angle but does not vanish. The origin of this discrepancy is the structure of the interfacial zone, which can be thick (the refractive index varies smoothly from n_1 to n_2 on a distance l), rough (the refractive index in the interface depends on the coordinates x and y), or optically anisotropic (the molecules in the interface have a preferential orientation). In the simplest cases, when the interfacial zone is much thinner than the wavelength of the light, $l \ll \lambda$, and has no optical anisotropy or is optically uniaxial

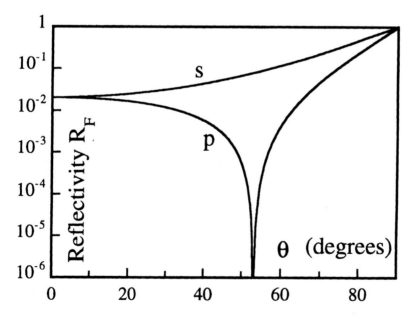

Fig. 1 The reflectivities R_s^F and R_p^F as a function of the incidence angle θ_i for a Fresnel interface between two media of refractive indices $n_1 = 1$ and $n_2 = 1.33$.

with a symmetry axis normal to the interface, the optical properties of real interfaces can be characterized by one number, the ellipticity (1) $\bar{\rho}$. For $\bar{\rho} \ll 1$ and in the vicinity of the Brewster angle the reflectivity differs from the Fresnel reflectivity by

$$R_p = R_p^F + R_s^F \bar{\rho}^2 \tag{2}$$

$\bar{\rho}$ can be measured by ellipsometry. For instance, the measured value at the Brewster angle for the air-water interface is -40×10^{-5}. This means that the p-reflectivity of the air-water interface at the Brewster angle is very small ($R_p \sim 10^{-8}$) but does not vanish.

In the following we examine the different contributions to R_p (or $\bar{\rho}$) at the Brewster angle for real interfaces.

A. Thick Interfaces

Drude (2) was the first to give an explanation for the discrepancy between the Fresnel formula and the measurements on real interfaces. He calculated the ellipticity $\bar{\rho}_l$ for a smooth but thin interface in which the refractive index $n(z)$ varies along the normal z of the interface on a distance $l \ll \lambda$:

$$\bar{\rho}_l = \frac{\pi}{\lambda} \frac{\sqrt{n_1^2 + n_2^2}}{n_1^2 - n_2^2} \int_{-\infty}^{+\infty} \frac{[n_1^2 - n(z)^2][n_2^2 - n(z)^2]}{n(z)^2} \, dz \tag{3a}$$

In a model in which the interfacial film is a homogenous medium of thickness l and refractive index n, Eq. (3a) reduces to

$$\bar{\rho}_l = \pi \frac{l}{\lambda} \frac{\sqrt{n_1^2 + n_2^2}}{n_1^2 - n_2^2} \frac{[n_1^2 - n^2][n_2^2 - n^2]}{n^2} \tag{3b}$$

A surfactant monolayer at the free surface of water with a typical refractive index $n \sim 1.4$ and a typical thickness $l \sim 20$ Å increases significantly the reflectivity R_p at the Brewster angle of the surface of water: $R_p(\theta_B) \sim 5 \times 10^{-5}$. R_p as a function of the incidence angle θ_i for this interface is shown in Fig. 2. It is significantly different from R_p^F only in the vicinity of the Brewster angle.

B. Interfacial Anisotropy with a Symmetry Axis Normal to the Interface

Nonspherical molecules in an interfacial layer have a preferential orientation that induces an optical anisotropy. In the simplest cases the interfacial medium is uniaxial and its optical axis is perpendicular to the interface. The ellipticity becomes (3):

$$\bar{\rho}_a = \pi \frac{l}{\lambda} \frac{\sqrt{n_1^2 + n_2^2}}{n_1^2 - n_2^2} \left[\frac{(n_1^2 - n_e^2)(n_2^2 - n_e^2)}{n_e^2} + (n_o^2 - n_e^2) \right] \tag{4}$$

where n_o is the ordinary refractive index and n_e the extraordinary refractive index. For instance, the orientation of the molecules of water at its free surface can in part explain that its reflectivity does not vanish at the Brewster angle. However, another important origin of this discrepancy is the interfacial roughness.

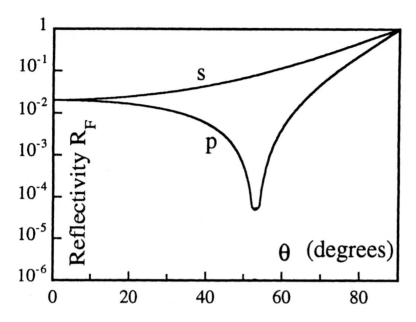

Fig. 2 The reflectivity R_p as a function of the incidence angle θ_i for a flat layer of refractive index 1.5 and thickness $l = 20$ Å at an interface between two media of refractive indices $n_1 = 1$ and $n_2 = 1.33$.

C. Interfacial Roughness

A third origin that explains a nonvanishing reflectivity at the Brewster angle is the interfacial roughness. The position ζ of the interface along the normal Oz to the interface is a function of the coordinates $\mathbf{r}(x,y)$. In the following we suppose that the position of the interface is $z = 0$, which means that the mean value of ζ is zero: $\langle\zeta\rangle = 0$. We consider only a flat interface in which the roughness is very small: $|\zeta| \ll \lambda$. By Fourier transform, the roughness is written as a sum of modes of wave vectors $\mathbf{q}$:

$$\zeta(\mathbf{r}) = \int_{-\infty}^{+\infty} \zeta_\mathbf{q} \exp(i\mathbf{q}\mathbf{r}) \frac{d\mathbf{q}}{(2\pi)^2} \tag{5}$$

The surface roughness of a liquid such as water is due to the thermal fluctuations and is very small ($|\zeta| \sim 3$ Å for pure water). The variation of the surface slope due to this roughness is also very small: $|d\zeta/d\mathbf{r}| \ll 1$. The amplitude of a mode $\mathbf{q}$ depends on a small number of constants that can be determined independently:

$$\langle\zeta_\mathbf{q}^2\rangle = \frac{k_\mathrm{B}T}{\rho g + \gamma q^2 + \kappa q^4} \tag{6}$$

where k_B is the Boltzmann constant, T the temperature, ρ the density of the liquid, g the gravitational acceleration, λ the surface tension, and κ the elastic constant of mean curvature of the interface (4).

Modes of long wavelengths ($q\lambda < 1$) change slightly the value of the Brewster angle but do not change the value of the minimum of reflectivity of the interface (5).

Modes of short wavelength ($q\lambda > 1$) modify the reflectivity of the interface at the Brewster angle, and the modification can be deduced from the ellipticity calculated for a rough interface without an interfacial layer (6–8):

$$\bar{\rho}_R = -\frac{3\pi}{2\lambda}\frac{n_1^2 - n_2^2}{\sqrt{n_1^2 + n_2^2}}\int_{1/\lambda}^{q_e}\langle\zeta^2 q\rangle\frac{d\mathbf{q}}{(2\pi)^2} \tag{7}$$

where q_e is a cutoff. Taking into account Eqs. (6) and (7), one obtains for a liquid surface (9,10):

$$\bar{\rho}_R = -\frac{3\pi}{2\lambda}\frac{n_1^2 - n_2^2}{\sqrt{n_1^2 + n_2^2}}q_e$$

q_e is the higher of the two values $(\pi/2)\sqrt{\gamma/\kappa}$ and $q_{mol} = \pi/a$, where a is the intermolecular distance.

For rough and thick interfaces (a rough interfacial layer), the ellipticity is to a first approximation the sum of the ellipticity due to the roughness and the ellipticity due to the thickness: $\bar{\rho} = \bar{\rho}_l + \bar{\rho}_R$.

On solid surfaces, the shape and amplitude of the roughness depend on the means of preparation of the surface and are generally unknown. The reflectivity due to the roughness cannot be calculated. Moreover, the slope of the roughness can be large (scratches on polished surfaces) and Eq. (7) does not apply (11).

D. General Interfacial Anisotropy

The interfacial structures examined in Secs. I.A, B, and C have a symmetry axis O_Z perpendicular to the interface. The reflected light on the interface has the same polarization as the incident beam. This is not the case when an interfacial medium is optically anisotropic and its symmetry axes are not in the plane of incidence. When the interface is illuminated with a beam of polarization p, the reflected beam has a component of polarization s and a component of polarization p. The 4×4 matrix method (12–14), or a direct electromagnetism calculation, modeling the interfacial film by a film of radiating dipoles at the interface (15,16), allows one calculate the reflection coefficients, r_{pp} and r_{sp}, which give the electric components p and s of the reflected field. The reflectivity of the interface, $R = |r_{ss} + r_{sp}|^2$, depends on the orientation of the optical axes of the interfacial medium with respect to the plane of incidence. This effect can be enhanced by adding an analyzer in the path of the reflected light. The measured reflectivity becomes a function of the angle α that the axis of the analyzer makes with the plane of incidence (the component of the electric field parallel to this axis is the only one to be transmitted through the analyzer): $R(\alpha) = |r_{pp}\cos\alpha + r_{sp}\sin\alpha;|^2$.

For example, in the simple case of a uniaxial interfacial medium of uniform thickness l at the air-water interface ($n_1 = 1$, $n_2 = 1.33$), with ordinary refractive index n_o and extraordinary refractive index n_e, the optical symmetry axis making an angle t with the normal to the interface and an angle ϕ with the plane of incidence (see Fig.

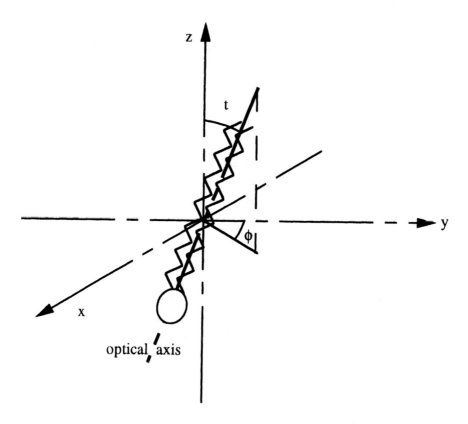

Fig. 3 Anisotropic uniaxial medium at the interface between two isotropic transparent media. Along the optical axis, the refractive index is n_e, in the orthogonal directions it is n_o. The optical axis is tilted by an angle t from the normal (Oz) to the interface and makes an angle ϕ with the plane of incidence (yOz). For a monolayer of amphiphile molecules, the optical axis corresponds to the direction of the molecules.

3), the reflectivity is given by (15):

$$R = X^2 \left\{ A_p^2 + A_s^2 \right\}$$

with

$$X = \pi \frac{l}{\lambda} \sqrt{1 + n_2^2}$$

$$A_p = [(n_o^2 - 1) + \frac{n_o^2(n_e^2 - n_o^2)\sin^2 t}{n_o^2 + (n_e^2 - n_o^2)\cos^2 t} \cos^2 \phi] \cos^2 \theta_B$$

$$- \frac{(n_o^2 - 1) + (n_e^2 - n_o^2)\cos^2 t}{n_o^2 + (n_e^2 - n_o^2)\cos^2 t} \sin^2 \theta_B$$

$$A_s = \frac{2}{1 + n_2^2} \frac{n_e^2 - n_o^2}{n_o^2 + (n_e^2 - n_o^2)\cos^2 t} \sin t [\cos t \sin \theta_B + n_o^2 \sin t \cos \theta_B \cos \phi] \sin \phi$$

When an analyzer is added in the path of the reflected light, with its axis making an angle α with the plane of incidence, the measured reflectivity becomes

$$R(\alpha) = X^2 (A_p \cos \alpha + A_s \sin \alpha)^2$$

Figure 4 shows the reflectivity of a thin anisotropic uniaxial layer as a function of the tilt azimuthal angle ϕ when it is observed without and with an analyzer and for different orientations α of the axis of this analyzer. The analyzer makes it possible to enhance the contrast between the different tilt azimuthal orientations. The contrast is maximum when the camera receives only the polarization s ($\alpha = 90°$), but the maximum intensity is low.

II. PRINCIPLE AND REALIZATION OF A BREWSTER ANGLE MICROSCOPE

Brewster angle microscopy is based on the reflectivity properties of surfaces developed in the previous paragraphs. At the Brewster angle, and with light polarized in the plane of incidence, an interface has a very low reflectivity that is highly sensitive to the interfacial structures.

This low reflectivity $R_p(\theta_B)$ can be observed directly, but this requires sufficient accuracy of the incidence angle and of the incident polarization so that the increase of the Fresnel reflectivity due to an error in the characteristics of the incident light [Eqs. (1)] does not exceed $R_p(\theta_B)$. For instance, for the free surface of pure water $R_p(\theta_B) \approx 10^{-8}$. An error $\delta\theta_i$ from θ_B in the incidence angle induces an error $\delta\theta_r = (n_1/n_2)^2 \delta\theta_i$ in the refraction angle, and the Fresnel reflectivity becomes

$$R_p^F \approx [\tan(\theta_i - \theta_r) \times (\delta\theta_i + \delta\theta_r)]^2 = \left[\frac{n_2^2 - n_1^2}{2 n_1 n_2} \left\{ 1 + \left(\frac{n_1}{n_2} \right)^2 \right\} \right]^2 (\delta\theta_i)^2$$

For R_p^F to be smaller than 10^{-8}, $\delta\theta_i$ has to be smaller than $10^{-4} \, 2 n_1 n_2^3 / |n_2^4 - n_1^4|$, which gives for water ($n_1 = 1$, $n_2 = 1.33$), $\delta\theta_i < 0.01°$, an accuracy easy to obtain with a laser beam.

An error $\delta\beta \ll 1$ in the angle if the polarizer introduces a small s component in the incident beam. Its intensity is $\delta\beta^2$ times the intensity of the p component. The reflectivity of the interface for this s component is R_s^F. $R_s^F(\delta\beta)^2$ is smaller than 10^{-8} when $\delta\beta < 0.02°$. A good polarizer is needed to reach this accuracy.

In a Brewster angle microscope, an image of the surface is formed by an objective with the reflected light and part of the light scattered around the direction of reflection by the surface structures. The reflected light determines the brightness of large interfacial structures, and the scattered light gives the details of the image. The resolution in the plane of the surface is as good as the numerical aperture of the objective is high.

Very high sensitivity to the interfacial thickness (a few angstroms) is obtained with high intensity, good polarization, and good parallelism of the incident beam. The incident beam is a laser beam, a Glan-Taylor prism increases its polarization ratio, and the incidence angle is carefully adjusted at the Brewster angle (minimum of reflectivity of a water surface) either by an optical fiber or by plane mirrors.

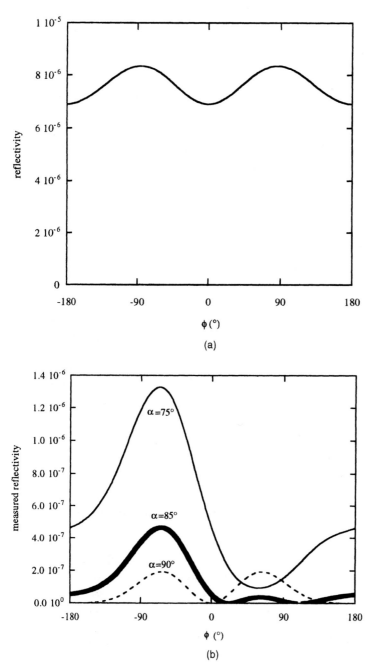

Fig. 4 Reflectivity of a flat anisotropic uniaxial layer of thickness $l = 20$ Å at the interface between the media of refractive indices $n_1 = 1$ and $n_2 = 1.33$ as a function of the azimuthal angle ϕ of the tilted optical axis. The ordinary index of the layer is $n_o = 1.50$, the extraordinary index is $n_e = 1.56$. The tilt angle of the optical axis is $t = 20°$. (a) The interface is observed without an analyzer. (b) The interface is observed with an analyzer in the path of the reflected light, the axis of this analyzer making an angle α with the plane of incidence; three curves are drawn for three different values of α, $\alpha = 75°$, $\alpha = 85°$, and $\alpha = 90°$ (analyzer crossed with the polarizer).

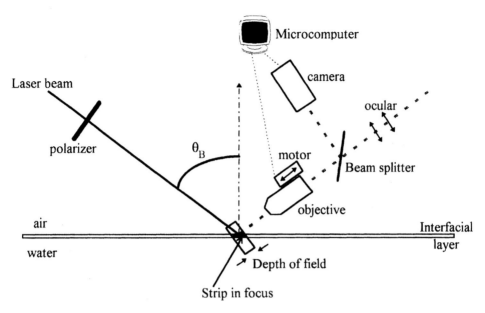

Fig. 5 Schematic drawing of a scanning Brewster angle microscope (17).

The intensity of the light reflected at the Brewster angle on low-density Langmuir films is very low, so high laser intensity and a sensitive video camera are required, and parasitic light must be avoided. However, this reflected intensity increases as the square of the interfacial film thickness, l^2, or the square of the molecular density of the film, and images of high-density films are relatively easy to obtain.

In the first Brewster angle microscope (17) (Fig. 5), the objective was a long frontal microscope objective with a large numerical aperture (0.4). The objective receives the reflected and scattered light to form an image on the sensitive tube of a video camera. However, as the objective axis is tilted to receive the reflected light, the image is in focus only along a narrow strip (see Fig. 5). A motor is used to move the objective in order to scan the interface. The images formed on the video camera are transferred to a microcomputer, and the strips successively in focus are placed side by side to reconstitute the full image. About 25 strips are needed to obtain a image of size 730×650 μm, with a resolution close to 1 μm. It takes about 2.5 s to make such an image.

A more simple setup was realized a few months later (18) in which a lens with a small aperture takes the place of the microscope objective. Because of this small aperture the resolution is less than with the microscope objective but at the same time the depth in focus is increased. Consequently, the width of the strip in focus is increased, and no scanning is needed; the image is obtained at the video rate, but it is out of focus on its borders. The resolution in the center of the image is probably of the order of 5 to 10 μm. Many setups of this type and similar simple ones have been largely developed because they are low in price and easy to build.

A third setup that avoids the drawbacks of the previous ones (good resolution but long time to reconstitute an image, or video rate but bad quality of the images) has been built (19). A homemade objective with a very large numerical aperture (0.97) is placed,

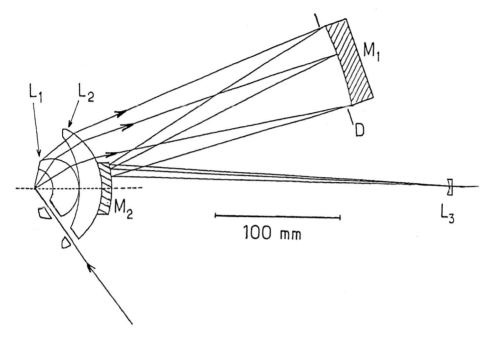

Fig. 6 Schematic drawing of the objective of the new Brewster angle microscope allowing one to obtain an image in focus in the whole plane without scanning (18).

with its axis vertical, above the surface of the water of a Langmuir trough. It gives an image in focus in the whole surface without scanning (Fig. 6). A long frontal distance (3 mm) is obtained by the use of a large lens at the objective entrance. A hole in the objective lenses allows illumination of the water surface at the Brewster angle. The scattered light is collected around the direction of reflection in a cone with a large aperture (about 47°). The measured in-plane resolution is 0.8 μm in one direction and 1 μm in the other.

III. WHAT CAN WE OBSERVE WITH BREWSTER ANGLE MICROSCOPY?

Brewster angle microscopy has been used mainly to get images of monolayers or multilayers at the free surface of water. Brewster angle microscopy has several advantages over fluorescence microscopy (20). First, no fluorescent probe, i.e., no impurity, is added to the film: the film does not have to solubilize a fluorescent probe, any film having a refractive index different from that of water can be observed, and the monolayer is not perturbed by the absorption of light by the fluorescent probes, except when the studied molecules themselves adsorb the light. Second, Brewster angle microscopy allows a more direct and easy study of the optical anisotropy of monolayers than polarized fluorescence microscopy (21). The power of this technique in observing phase transitions was demonstrated by the discovery of a new phase that had not been previously found by other techniques, the "Overbeck phase" (22). However, Brewster angle microscopy requires plane surfaces. This means that layers in a meniscus cannot be observed with this technique.

In the following, we first study the sensitivity of the technique to interfacial thickness or roughness (A) and then give examples of images that can be obtained on monolayers of a phospholipid and showing inhomogeneities of the film density (B) and the surface film anisotropy (C).

A. Sensitivity of Brewster Angle Microscopy

Figure 7 is an image of a monolayer of a polymer, polydimethylsiloxane (PDMS), on water at low surface density showing the coexistence of dark and bright domains. This coexistence indicates a first-order phase transition in the monolayer between a dense phase (liquid in this case) and a phase of lower density (gaseous phase) (23). An increase in the amount of PDMS at the interface increases the area covered by the dark domains. This means that in this case the surface liquid phase appears dark while the gaseous one appears bright: the reflected intensity is not proportional to the square of the film density as expected. To understand that, we must notice that the reflectivity of this film is very low, explaining the bad quality of the images. Consequently, the reflectivity of the pure water has to be taken into account. The ellipticity of the surface of pure water is $\bar{\rho}_{W} = -40 \times 10^{-5}$. The PDMS has a refractive index higher than that of water, $n \approx 1.4$. The ellipticity $\bar{\rho}_{PDMS}$ due to a thin dense film of PDMS is given by Eq. (3b); it is positive. The total ellipticity in presence of a dense

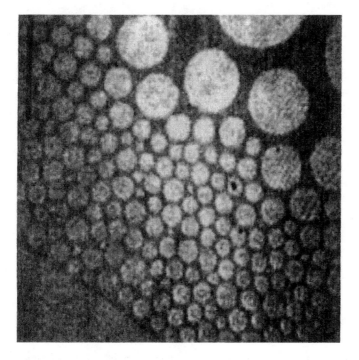

Fig. 7 Image of a polymer monolayer (PDMS) on the surface of water at low surface density. Bright domains are very dilute in PDMS, and dark regions are dense in PDMS. The bar corresponds to 100 μm.

film of PDMS is $\bar{\rho} = \bar{\rho}_W + \bar{\rho}_{PDMS}$. As $\bar{\rho}_W$ and $\bar{\rho}_{PDMS}$ have opposite signs but approximately the same amplitude, a dense monolayer of PDMS at the free surface of water decreases the reflectivity, which is proportional to $\bar{\rho}^2$. The reflectivity at the Brewster angle of the pure water surface, which can be observed here, is due partly to the surface anisotropy and mainly to the surface roughness (10). The amplitude of this roughness is about 3 Å: Brewster angle microscopy is sensitive to structures as thin as a few angstroms.

B. Surface Density

The amplitude of thermal fluctuations of a Langmuir film is generally small because the surface tension is high. Images, such as that obtained with PDMS, in which the roughness of the surface has a large influence on the image contrast, are exceptional. This happens only for very thin monolayers or monolayers having a refractive index close to that of water, such as some fluorinated molecules. For most of the films the roughness can be ignored compared with the effect of the film thickness. The reflected intensity is, to a first approximation, proportional to the square of the film thickness [see Eq. (3b)] or to the square of the surface density. This means that a contrast in density of the monolayer, even if it is small, gives a high contrast in Brewster angle microscopy images. Such a contrast is observed for first-order phase transitions in monolayers. During a first-order phase transition,

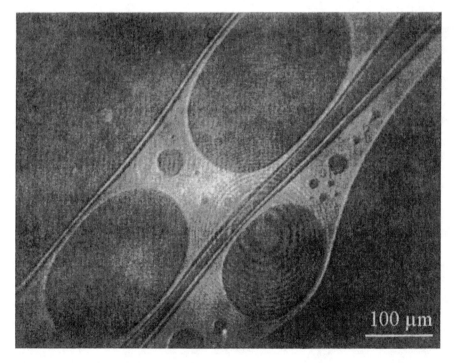

Fig. 8   Coexistence of dark and bright domains in a monolayer of a phospholipid (1,2-dimyristoyl-*sn*-glicero-3-phosphoethanolamine at 20°C). The monolayer appears as bright as the phospholipid monolayer is dense; the LE (liquid expanded), or L_1, phase appears bright and the gaseous phase (G) appears dark.

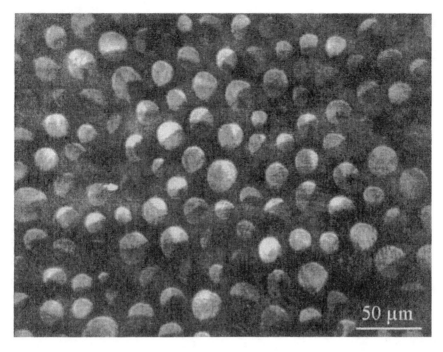

a

Fig. 9 Sequence of images obtained with an analyzer in the path of the reflected light during the compression of a monolayer of 1,2-dimyristoyl-*sn*-glicero-3-phosphoethanolamine at 20°C and high density. Image (a) shows the coexistence of the anisotropic tilted phase (L_2) in the L_1 phase.

b

Fig. 9 (b) shows the coexistence of the anisotropic tilted phase (L_2) in the L_1 phase.

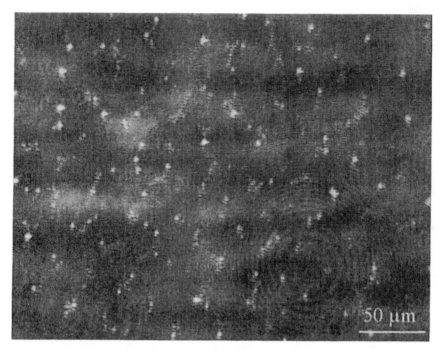

c

Fig. 9 Image (c) was obtained at higher pressure, when the state of the film was a dense nontilted phase; small three-dimensional domains have nucleated, and they are very bright because they are much thicker than the monolayer.

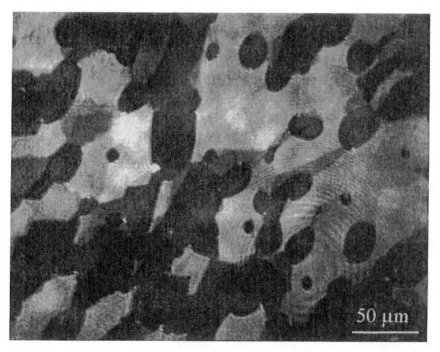

d

Fig. 9 Image (d) was obtained during a decompression and shows bubbles of L_1 in the L_2 phase.

two phases with different densities coexist, which can be observed by Brewster angle microscopy if the domains of these phases are larger than the resolution of the microscope (17,18). As an example, Fig. 8 shows the coexistence of a gaseous phase and a liquid expanded phase (LE or L_1) in a Langmuir monolayer of 1,2-dimyristoyl-*sn*-glicero3-phosphoethanolamine at 20°C and low surface pressure.

A contrast in the thickness of a layer also gives a contrast in Brewster angle microscopy, and the formation of multilayers, or collapse, can be observed. An example of collapse is shown in Fig. 9c.

C. Optical Anisotropy

Optical anisotropy is frequently observed in Langmuir films. It originates from preferential orientation at the interface of the amphiphile molecules, which are in general elongated because they consist of a polar head and carboxylic chains. The refractive index has a different value in the direction of the chains (n_e) and in the orthogonal direction (n_o). In Langmuir films, there are many mesophases or crystalline phases in which the chains are tilted from the vertical, and such a Langmuir film is a uniaxial medium with its axis tilted from the normal to the interface (see Fig. 3).

In-plane anisotropy is also observed in the mesophases and crystalline phases that have a nonhexagonal lattice. When the molecules are untilted, such a monolayer is biaxial, with one refractive index n_e along the vertical direction (direction of the chains) and two different refractive indices, n_o and n_o', in the plane of the surface.

The anisotropy of tilted mesophases or crystalline phases in which the molecular tilt azimuthal angle is preserved at macroscopic distances can be observed in Langmuir films. The polarization of the reflected light is a function of the orientation of the film birefringence, i.e., of the tilt azimuthal angle, and consequently the way this angle varies in the plane of the film, commonly called the texture of the phase, can be observed. It is revealed by an analyzer (21,22) but can also be observed without an analyzer when it is strong. Figure 9a and b show domains of the L_2 (or LC) phase in the L_1 phase for a monolayer of 1,2-dimyristoyl-*sn*-glicero-3-phosphoethanolamine on water at 20°C. The tilt angle t being constant in a phase, the different gray levels in the picture correspond to different values of the tilt azimuthal angle ϕ, assuming a uniform laser beam intensity in the image. In principle, one can deduce the molecular azimuthal direction from images obtained with an analyzer. For instance, for a crossed polarizer and analyzer ($\alpha = 90°$) it can be seen from Fig. 4b that a region that appears dark corresponds to molecules in the plane of incidence ($\phi = 0$ or $180°$).

In nontilted crystalline phases and mesophases, a low residual anisotropy can be observed that is due to the lattice anisotropy of these phases (14,21). This cannot be observed for 1,2-dimyristoyl-*sn*-glicero-3-phosphoethanolamine at 20°C, either because this anisotropy is too small or, more probably, because small collapsed pieces increase the parasitic light too much (Fig. 9c).

The sequence in Fig. 9 shows the evolution of a dense film of 1,2-dimyristoyl-*sn*-glicero-3-phosphoethanolamine (at 20°C) when the pressure is increased (from a to c) and then decreased (d). In Fig. 9a one observes the coexistence of small domains of the anisotropic L_2 phase in the L_1 phase that collide when the area per molecule decreases (Fig. 9b). A first-order phase transition to a nontilted phase is then observed (Fig. 9c). In this phase bright points appear, which are very small three-dimensional crystals (collapse). This collapse is reversible. When the pressure is decreased, they disappear as soon as the transition to the tilted phase is observed. At a lower pressure, in Fig. 9d, bubbles of L_1 (dark) nucleate in the L_2 phase during the pressure decrease. The range of area per molecule where the pure L_2 phase is observed at 20°C is very small.

IV. CONCLUSION

Brewster angle microscopy is a powerful technique for studying Langmuir films or any thin film formed at the free surface of a plane transparent medium. It gives direct information on the density of the interfacial layer and on its optical anisotropy, contrary to fluorescence microscopy, which gives the density of the added fluorescent impurity in the layer or the orientation of the absorption dipoles of the fluorescent molecules. Some information on the lattice orientation of the nontilted phase can also be extracted from images obtained by Brewster angle microscopy. The technique can be used on plane surfaces when the surface layer has a refractive index different from that of the substrate, and the simplest setups are easy to build and to handle.

REFERENCES

1. RM Azzam, NM Bashara. Ellipsometry and Polarized Light. Amsterdam: North Holland, 1977.
2. P Drude. The Theory of Optics. New York: Dover, 1959.
3. F. Abèlès. Recherches sur la propagation des ondes sinusoïdales dans les milieux stratifiés. Application aux couches minces. Ann Phys (Paris) 5:596, 1950.
4. W Helfrich. Blocked lipid exchange in bilayers and its possible influence on the shape of vesicles. Z Naturforsch 28c:693, 1973.
5. J Meunier. Light reflectivity and ellipsometry. In: D Langevin, ed. Light Scattering by Liquid Surfaces and Complementary Techniques. Surfactant Science Series Vol 41. New York: Marcel Dekker, 1992.
6. P Croce. Sur l'effet des couches très minces sur la réflexion, la transmission et la diffusion de la lumière sur un dioptre. J Opt (Paris) 8:127, 1977.
7. D Beaglehole. Ellipsometric study of the surface of simple liquids. Physica 100B:163, 1980.
8. BJA Zielinska, D Bedeaux, J Vliegler. Electric and magnetic susceptibilities for a fluid-fluid interface; the ellipsometric coefficient. Physica 107A:91, 1981.
9. J Meunier. Measurement of the rigidity coefficient of a surfactant layer and the structure of the oil or water microemulsion interface. J Phys Lett (Fr) 46:1005, 1985.
10. J Meunier. Liquid interfaces: Role of the fluctuations and analysis of ellipsometric and reflectivity measurements. J Phys (Fr) 48:1819, 1987.
11. J Meunier. Optical properties of plane and rough interfaces in the electrostatic approximation. Physica 230A:27, 1996.
12. GA Overbeck, D Hönig, L Wolthaus, M Gnade, D Möbius. Observation of bond orientational order in floating and transferred monolayers with Brewster angle microscopy. Thin Solid Films 242:26, 1994.
13. GA Overbeck, D Hönig, D Möbius. Stars, stripes and shells in monolayers: Simulation of the molecular arrangement in Schlieren structures. Thin Solid Films 242:213, 1994.
14. MW Tsao, TM Fischer, C Knobler. Quantitative analysis of Brewster angle microscope images of tilt order in Langmuir monolayer domains. Langmuir 11:3184, 1995.
15. Hénon. Microscopie à l'angle de Brewster: Transitions de phase et défauts d'orientation dans des films monomoléculaires. Thesis, Université Paris VI, 1993 (unpublished).

16. Y Tabe, H Yokoyama. Fresnel formula for optically anisotropic Langmuir monolayers: An application to Brewster angle microscopy. Langmuir 11:699, 1995.
17. S Hénon, J Meunier. Microscope at the Brewster angle: Direct observation of first-order phase transitions in monolayers. Rev Sci Instrum 62:936, 1991.
18. D Hönig, D Möbius. Direct visualization of monolayers at the air/water interface by Brewster angle microscopy. J Phys Chem 95:4590, 1991.
19. C Lheveder, S Hénon, R Mercier, G Tissot, P Fournet, J Meunier. A new Brewster angle microscope. Rev Sci Instrum 69:1446, 1998.
20. M Lösche, E Sackmann, H Möhwald. A fluorescent study concerning the phase diagram of phospholipids. Ber Busenges Phys Chem 87:848, 1983.
21. S Rivière, S Hénon, J Meunier, DK Schwartz, M-W Tsao, CM Knobler. Textures and phase transitions in Langmuir monolayers of fatty acids. A comparative Brewster angle microscope and fluorescence microscope study. J Chem Phys 101:10045, 1994.
22. GA Overbeck, D Höning, D Möbius. Visualization of first- and second-order phase transitions in eicosanol monolayers using Brewster angle microscopy. Langmuir 9:555, 1993; GA Overbeck, D Möbius. A new phase in the generalized phase diagram of monolayer films of long-chain fatty acids. J Phys Chem 97:7999, 1993.
23. EK Mann, S Hénon, D Langevin, J Meunier. Molecular layers of a polymer at the free water surface: Microscopy at the Brewster angle. J Phys II (Fr) 2:1683, 1992.

17

Ellipsometry

H. Arwin
Linköping University, Linköping, Sweden

I. ELLIPSOMETRY IN INTERFACIAL BIOLOGY

Ellipsometry is based on analyzing polarization changes caused by reflection of light and allows detailed optical and microstructural characterization of surfaces, thin films, and multilayers. The thickness resolution is 0.01 nm or better, and studies of layers of molecular dimensions on solid substrates is possible. As ellipsometry is based on photons in and photons out, a vacuum ambient is not needed. The technique can therefore be used directly at a solid-liquid interface, where many reactions in biology and physical chemistry take place. Besides being useful for nondestructive analysis of thin layers, ellipsometry also allows dynamic studies of film growth with a time resolution relevant for many biological processes.

A large number and variety of applications of ellipsometry are found in the area of physical chemistry as well as organic layers in general (1–5). Early pioneering work was done by Rothen (6) and Vroman and Lukosevicius (7). However, very few studies deal with adsorption on real biological surfaces. In some cases macromolecular adsorption on inorganic model surfaces prepared to mimic a natural interface is measured, but in most cases the scope is more narrow in the sense that focus is on investigations of the dependence of adsorption on only one surface parameter such as wettability, surface charge, or a specific chemical surface modification.

Recent developments in ellipsometry, such as computer control of the angle of incidence, spectroscopy, imaging, time-resolved spectroscopy and powerful software for analysis, have added new possibilities. By employing spectroscopic ellipsometry, the optical functions of thin layers become accessible and details of the interaction mechanisms between a substrate and an adsorbed layer can be studied. However, of even greater importance is that the increased amount of information obtainable from spectroscopy allows microstructural modeling of thin layers. The number of parameters in the analysis can be extended beyond layer thickness or surface concentration to include phenomena such as layer reorganization (surface dynamics), layer density effects, molecular orientation, clustering, multilayering, and combinations of these phenomena.

The purpose of this chapter is to introduce the reader to ellipsometry, to point on some of its applications in the area of physical chemistry and discuss some future possibilities. The technique is widely used, but this is not intended to be a complete review and most of the examples and applications are therefore chosen from the author's own experience. Further, general input to the area can be found in the proceedings of the two last ellipsometry conferences (8,9). The outline and limitations are as follows. First, the principles of ellipsometry and the most basic instrumental designs are briefly described. The theoretical framework is then given in a section devoted to optical models of thin-film samples followed by a description of methods for extracting relevant physical information. The importance of familiarizing oneself with these two sections cannot be overemphasized, as ellipsometry is an indirect technique and therefore the experimentalist must have access to a toolbox and know-how for data analysis. For the potential user we have included a short section on methodology. Selected applications are described with focus on protein layers at the solid-liquid interface. Finally, some thoughts about the future are communicated.

II. PRINCIPLES OF ELLIPSOMETRY

Ellipsometry is based on the measurement and analysis of the polarization changes occurring upon reflection at oblique incidence of a polarized monochromatic plane wave. The basic quantity measured is the complex reflectance ratio defined as

$$\rho = \frac{R_p}{R_s} \tag{1}$$

where R_p and R_s are the reflection coefficients for light polarized parallel (p) and perpendicular (s) to the plane of incidence, respectively. Usually ρ is written in complex polar form as

$$\rho = \tan \psi \exp(i\Delta) \tag{2}$$

where $i = \sqrt{-1}$. The two parameters ψ and Δ are called the ellipsometric angles and are the experimentally determined quantities obtained from an ellipsometer. A closer examination of Eqs. (1) and (2) shows that tan ψ contains the *ratio* between the electric field amplitudes of two perpendicular components (in the p and s directions) of the same light beam. The fact that ellipsometry is based on *amplitude ratios* and not on absolute amplitudes means that there is an inherent reference in the principle compared with related optical techniques based on reflectance and transmittance, in which a second beam is normally used for reference purposes. Ellipsometry is therefore not sensitive to drift and absolute intensity of the light source. Furthermore, Δ is the reflection-induced *difference* in phase change between the two field components, and the phase information in the light beam can thus be taken advantage of with a very high precision. In practice, Δ is obtained from an ellipsometer as an azimuthal reading (a mechanical angle) of an optical component or as a phase of an electrical signal. Such quantities are easy to measure very accurately (better than 0.01°), explaining the high precision of ellipsometry. The obtainable resolutions in Δ and ψ lead to very high surface sensitivity, especially with respect to the thickness of a thin surface layer. Simple model calculations show that, under typical measurement conditions,* a change of 0.1° in Δ corresponds to growth of a 0.03-nm thin organic layer on a silicon substrate. On metal surfaces the sensitivity is slightly lower, and 0.1° in Δ then corresponds to 0.05 nm on gold as an example. One also finds that on silicon, mainly Δ changes with film thickness, which is an example of the ellipsometric rule that a transparent layer on a transparent substrate mainly affects Δ but not ψ.

With spectroscopic ellipsometry, it is possible to determine the complex reflectance ratio as a function of photon energy in the near-infrared–visible–near ultraviolet optical range, opening up possibilities for analysis in the area of material optics. A detailed theoretical background on polarized light, ellipsometry, and technical details of basic ellipsometer systems can be found in the ellipsometer reference book by Azzam and Bashara (10).

III. ELLIPSOMETER SYSTEMS

A. Classification and General Measurement Principles

An ellipsometer is a measurement system capable of determining the complex reflectance ratio ρ. In principle, an ellipsometer consists of a light source, a polarization state generator, an arrangement to hold and adjust the position of the sample, a polarization state detector, a photosensitive detector, apertures, wavelength filters, necessary control electronics, and a computer. The components are arranged in an optical path so that the light emerging from the polarization generator is reflected at the sample surface and then enters into the polarization state detector. Ellipsometer systems may be classified in *null ellipsometers* and *photometric ellipsometers*. Sometimes the more general terms compensating and noncompensating ellipsometers are used instead.

In a null ellipsometer system, the strategy is that the polarization changes due to reflection at the sample should be compensated by the optical components in the instrument to achieve extinction at the detector. The compensation is done by adjusting their azimuths, phase retardation, or some other parameter. The compensation can be done before or after reflection. The readings of azimuths, retardation, etc. at "null" constitute the experimental data, which in most configurations have simple relations to ψ and Δ.

In a photometric ellipsometer system, the state of polarization before reflection is normally predetermined and after reflection it is measured by analyzing the intensity variation recorded by the detector as a function of azimuth, retardation, etc. of one or several of the optical components in the optical path. In many photometric ellipsometers, the measured quantities are tan ψ and cos Δ instead of ψ and Δ. This will have some implications for precision, which will then depend on the actual values of Δ and ψ. In both null and photometric ellipsometers different modulation techniques are normally used to enhance sensitivity.

---

* Wavelength 633 nm, angle of incidence 70°, refractive index of layer 1.5, and ambient being air.

A way to increase the amount of data is to vary the wavelength, that is, to do spectroscopy. However, null ellipsometry is seldom used in a spectroscopic mode because one of its optical components, the compensator, is wavelength dependent. On the other hand, many types of photometric ellipsometers are well adopted to spectroscopy. Besides spectroscopy, one can increase the amount of data by doing variable angle of incidence ellipsometry. In in situ applications this can be complicated because of cell limitations, however.

In the techniques just discussed, the beam diameter is of the order of a few millimeters. If analysis is required in smaller spots, one can include a focusing lens in the light beam. At the expense of thickness resolution, a lateral resolution down to $50\,\mu$m can then be obtained and two-dimensional mechanical scanning to map a larger surface area can be performed to determine, e.g., a lateral thickness distribution. A more time-efficient way to map a surface ellipsometrically is to employ *imaging ellipsometry*. Most such systems are based on using a charge-coupled device (CCD) camera as a light detector.

In this section, one type of null ellipsometer will be discussed in some detail, the so-called *polarizer-compensator-sample-analyzer* ellipsometer or the PCSA system. Also, for photometric ellipsometers restriction is made to one type, the *rotating-analyzer ellipsometer* or the RAE system. An imaging ellipsometer setup suitable for imaging of patterns of adsorbed proteins is also briefly described, and finally some consideration of in situ measurements are made. High-precision ellipsometer systems of several types and for different applications are commercially available and further information is readily found by a search on www.

B. The PCSA Ellipsometer

The PCSA ellipsometer is the classical ellipsometer system used in numerous applications. Figure 1 shows the principle of this system. The light source is often an HeNe laser with a wavelength of 633 nm or a mercury lamp with a filter transmitting the intense emission line at 546 nm. In the latter case a collimating lens system is included in the optical path. The collimated light beam is made linearly polarized with a polarizer. The polarizer azimuth is denoted *P*, which is the angle between the transmission axis of the polarizer and the *p* axis of a coordinate system as shown in Fig. 1. Next the beam passes through a compensator with azimuth *C* as defined by its fast axis. The compensator is a quarter-wave plate and induces a phase shift between the p and s components in the beam. After reflection from the sample, which causes an additional phase shift Δ and a relative amplitude change tan ψ, the light passes through another polarizer with azimuth *A*. This second polarizer is normally called the analyzer. Finally, a detector measures the intensity transmission through the whole system.

The strategy of operation is to adjust the optical components so that the light is extinguished at the detector. This can be achieved only if the light is linearly polarized before the analyzer. As the polarizer-compensator combination in general generates elliptically polarized light, this means that the reflection from the sample must compensate for the phase difference between the p and s components in the light beam to generate linearly polarized light after reflection. In practice, it works in the following way. The compensator is normally fixed at $C = -45°$ ($C = +45°$ can also be used but then other relations similar to the following are used). The advantage of using this specific compensator position is that the p and s components before reflection will have

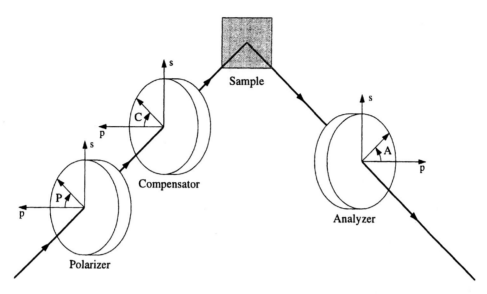

Fig. 1 Schematic diagram of a polarizer-compensator-sample-analyzer (PCSA) ellipsometer system. The two reference directions (*p* and *s*) are indicated and the direction of the transmission axes of the polarizer and analyzer are given by *P* and *A*, respectively. *C* shows the direction of the fast axis of the compensator. (From Ref. 13.)

the same magnitude independent of *P* but the phase difference between the p and s components will be 90°−2*P*. See Ref. 10 for a detailed derivation. The polarizer can now be rotated so that the reflected light will be linearly polarized. This will occur when the total phase shift between the p and s components is an integer multiple of 180°:

$$90° - 2P + \Delta = n180° \tag{3}$$

Here $n = 0$ means that the phase shifts from the compensator and the sample have opposite signs and thus cancel, and $n = 1$ corresponds to the case when they add to 180°. That the correct polarizer position has been found is checked by adjusting the analyzer position because extinction at the detector can be achieved only when the light is linearly polarized before the analyzer, that is, when Eq. (3) is fulfilled. At extinction, the azimuth of the analyzer equals ψ (not shown here). From Eq. (3) we can solve for Δ and at null conditions ($n = 1$) we thus have

$$\psi = A \tag{4a}$$

$$\Delta = 2P + 90° \tag{4b}$$

For increased precision, a measurement can be repeated for $n = 0$ resulting in the readings A' and P'. The following relations are then in effect (10):

$$\psi = 180° - A' \tag{4c}$$

$$\Delta = 2P' - 90° \tag{4d}$$

On averaging the two readings (called zone averaging), several types of instrumental imperfections cancel. Further details about null ellipsometers can be found elsewhere (10).

The ellipsometric parameters ψ and Δ are thus obtained from the azimuthal positions A and P, which can be determined with a precision of 0.01° in standard routine instruments and at least one order of magnitude better in research instruments. In modern designs the polarizer and analyzer azimuths are computer controlled with stepping motors and an automatic search algorithm is used to find the intensity minimum. As the compensator is a monochromatic device, a null ellipsometer is not suited for spectroscopy and is therefore used mainly in single-wavelength applications. Its strength lies in its robustness, simplicity, and high precision.

C. The RAE System

The rotating-analyzer-ellipsometer or RAE is the basic spectroscopic ellipsometer system with a principle as outlined in Fig. 2. In contrast to the null ellipsometer, it requires a computer for its basic operation and came into use first in the 1970s (11). The light source in an RAE system is a white light source such as a xenon lamp and the wavelength is selected with a monochromator. A typical wavelength range is 190–1000 nm. The monochromatic light is polarized with a polarizer and reflected at the sample surface. After reflection, the light is in general elliptically polarized and cannot be extinguished with a polarizer. Instead, the state of polarization of the reflected beam is measured by using a rotating polarizer (called analyzer) located between the sample and the detector. Only the component of the polarization ellipse of the reflected wave parallel to the transmission axis of the analyzer is passed, but, as the analyzer is rotating, the whole polarization ellipse is traced. If the analyzer prism rotates at a constant speed, the output signal is ideally sinusoidal with the minimum and maximum corresponding to the minor and major axes of the polarization

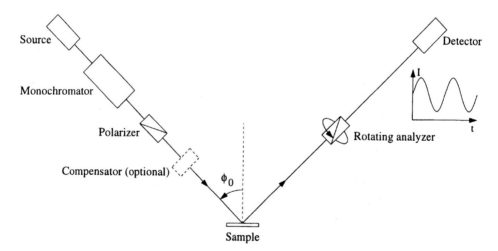

Fig. 2   Schematic diagram of the optical system of a rotating analyzer ellipsometer (RAE). The insert shows the time variation of the detector signal according to Eq. (5).

ellipse, respectively. The signal I versus time t can formally be expressed as

$$I = I_0\left[1 + \hat{\alpha}\cos 2A(t) + \hat{\beta}\sin 2A(t)\right] \tag{5}$$

where I_0 is the average intensity, $\hat{\alpha}$ and $\hat{\beta}$ are the normalized Fourier coefficients of the signal, and $A(t)$ is the angular position of the rotating analyzer. From $\hat{\alpha}$ and $\hat{\beta}$ the ellipsometric angles Δ and ψ, and thereby the complex reflectance ratio ρ, can be determined provided that the polarization of the incident beam is known. The relations can be written as (11)

$$\tan\psi = \tan P\sqrt{\frac{1 + \hat{\alpha}}{1 - \hat{\alpha}}} \tag{6a}$$

$$\cos\Delta = \frac{\hat{\beta}}{\sqrt{1 - \hat{\alpha}^2}} \tag{6b}$$

The polarization of the incident beam is controlled by a polarizer with azimuth P followed by an optional compensator (quarter-wave plate). The latter is, however, not normally used because a compensator is a chromatic device and is thereby not suited for spectroscopy. Despite this, it is in some cases, i.e., for measurements on dielectrics, necessary to use a compensator to improve the sensitivity of the instrument. The polarizer can be fixed at $P = 45°$ to ensure equal intensity in the p and s directions of the sample but can also be under dynamic control to increase precision. A completely automatic RAE system also requires an electronic detection system capable of monitoring the output signal and computer overhead to control the wavelength of the light and the azimuths of the two polarizers.

D. Variable Angle Ellipsometry and Time-Resolved Spectroscopy

Since the pioneering instrumental development on RAE systems by Aspnes and Studna (11), driven by the need for improved analytic tools in the semiconductor area, several spectroscopic ellipsometer systems have been developed. Variants with a rotating polarizer (RPE) instead of a rotating analyzer as well as with polarization modulation using electro-optical modulators are commercially available. With these systems it is often possible to vary the angle of incidence under computer control to allow measurements at multiple angles of incidence or to avoid low reflecting conditions near the Brewster or pseudo-Brewster angle of the sample. In combination with spectroscopy, one then uses the term VASE, variable-angle-spectroscopic-ellipsometry. State of the art in instrumentation is time-resolved spectroscopy (12). With such instruments it is possible to record ellipsometric spectra with hundreds of wavelengths in the range 300 to 1000 nm with a rate of several spectra per second. Fourier transform–based ellipsometers are also being developed for use in the infrared.

E. Imaging Ellipsometry

A null ellipsometer can be operated in an off-null mode, meaning that the intensity is monitored and taken as a measure of the thickness of a layer on a surface (13). This can be taken advantage of in a simplified but fast imaging ellipsometer system.

The major disadvantage of the off-null mode is that the independence of lamp intensity is lost. However, the technical development is fast and in the near future we will probably see imaging systems having both speed and precision. Here we will limit the description to an off-null system useful for the determination of the lateral thickness distributions of protein layers on silicon substrates. In this particular case, the recorded intensity I is proportional to the square of the layer thickness d according to (13)

$$I = \text{const} \cdot d^2 \tag{7}$$

The experimental set up is shown in Fig. 3. It is based on a PCSA ellipsometer with a collimated expanded beam and a CCD camera as the detector. Special considerations must be taken of the stability of lamp intensity, detector linearity, parallelism of the light beam, and uniform intensity over the beam. A suitable light source is a 75-W xenon arc lamp. A collimated beam can be obtained using a pinhole configuration. With a 0.6-mm pinhole a parallelism of 0.3° and a nonuniformity of 2% can be achieved over a beam diameter of 25 mm (14). Monochromatic light is obtained with an interference filter (633 nm, bandwidth 10 nm). The detector is a Sony XC-73CE CCD B/W connected to a frame grabber in the computer. In normal operation, the compensator is fixed at 45° to the plane of incidence. The analyzer and polarizer are adjusted so that the null conditions are fulfilled [Eqs. (4)] on a bare part of the surface under investigation. The intensity distribution measured in other areas can then be used to calculate the layer thickness pixel by pixel from Eq. (7). With a simple system as described above it is possible to obtain a thickness resolution of 0.5 nm with a lateral resolution of 5μm over a sample area of 15×30 mm^2. Some applications are presented in the following and further technical details can be found in Ref. 14.

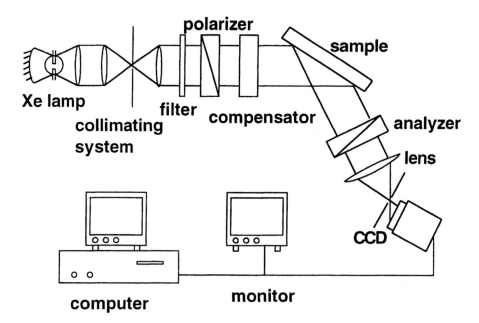

Fig. 3 Imaging ellipsometry based on a PCSA system with a CCD camera and an expanded beam. Further details are given in the text. (From Ref. 14.)

F. In Situ Considerations

It is strongly recommended that in situ experiments be done whenever possible. In dynamic studies, e.g., when protein adsorption is followed, in situ measurement is of course necessary, but in the case in which static parameters are measured, an in situ approach is also superior. The most important aspect is that an in situ measurement guarantees that the substrate properties are the same before and after layer deposition. Other advantages are that sample alignment is invariant and that the sample surface does not have to be transferred through an air-liquid interface. If measurements are done at the solid-liquid interface, a special cell to hold the liquid is necessary. The windows of this cell should be free from birefringence and the incident and reflected beams must be perpendicular to the windows to avoid polarization effects. Figure 4a shows a basic type of cell suitable for ellipsometric studies at the solid-liquid interface. A flow cell arrangement is also possible as shown in Fig. 4b.

When additional components, such as cell windows, are introduced in the light path, errors arise because of imperfections in these components. In ellipsometry a small intensity loss is of minor importance but birefringence effects can be troublesome, especially if they are stress induced. It is therefore of crucial importance to select windows that are free of such effects, e.g., quartz windows. For vacuum applications there are special window designs. However, it is recommended, if possible, to determine the properties of the windows and to include them as a correction in the data analysis. This is readily done, although the mathematics becomes a little more complicated (10,15). Correction for component imperfections are very important when absolute measurements are performed. However, in the context discussed here, we are mostly concerned with relative measurements in the sense that changes in the ellipsometric parameters Δ and ψ are recorded before and after an event such as protein adsorption. Cell window imperfections are then important to second order and can in most cases be neglected.

IV. OPTICAL MODELS

To interpret ellipsometric data, a reasonable optical model has to be assumed. The three basic optical models are shown in Fig. 5. The index of refraction is denoted N_j and the angle of incidence ϕ_j in medium (phase) j, starting with $j = 0$ in the ambient. The thickness of a layer is denoted d_j. Notice that for a nontransparent medium, N is complex valued and written as $N = n + ik$, where n is the real part of the refractive index, k is the extinction coefficient, and $i = \sqrt{-1}$ is the complex entity. The extinction coefficient is related to the absorption coefficient α as $\alpha = 4\pi k/\lambda$, where λ is the wavelength of the light. As an alternative, the optical properties can be expressed in terms of the complex dielectric function $\varepsilon = N^2$. The ambient-substrate model or the two-phase model is appropriate if bulk materials are studied. The ambient-film-substrate model or the three-phase model is the basic model for studies of thin layers. For multilayered structures the n-phase model is the choice. In all models, the layers and the substrate are assumed to have plane-parallel and abrupt boundaries and, in the most simple cases, to be homogeneous and optically isotropic.

The purpose of this section is to provide the workbench for analysis of ellipsometry data. Often it is advantageous to use the most simple model in order

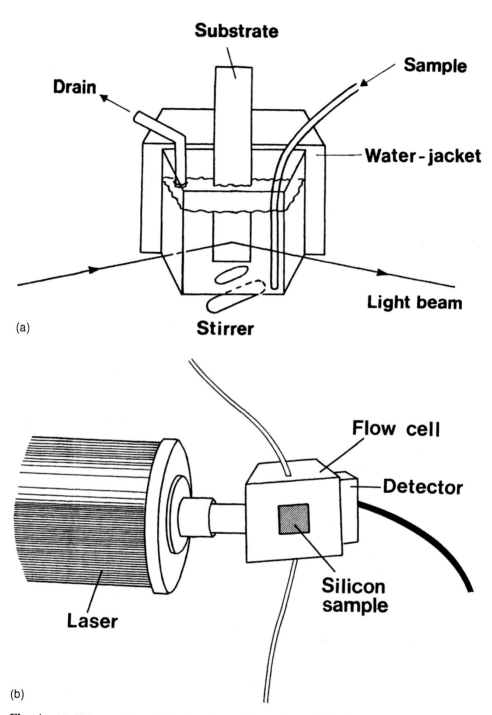

Fig. 4 (a) Ellipsometric cell for in situ studies at the solid-liquid interface. The sample (substrate) is mounted on the sample holder in the ellipsometer and inserted into the liquid. (b) Ellipsometric flow cell for in situ studies at the solid-liquid interface. The sample is mounted over a hole in the cell and thus constitutes one of the cell walls. Liquids are flowing through the cell in the tubings.

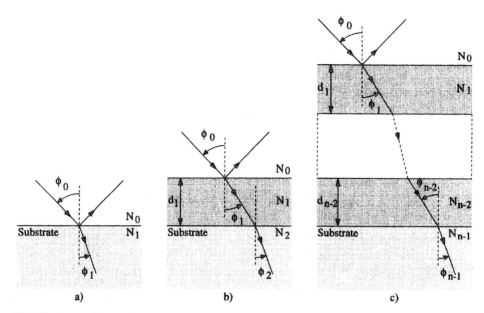

Fig. 5 Isotropic stratified planar structures. (a) The two-phase model. (b) The three-phase model. (c) The n-phase model.

to work with a small number of parameters. However, the most appropriate optical model is determined by the sample under investigation, and much can be gained by using a well-controlled sample preparation. For example, if the surface topography is of minor interest for the biological conclusions, one should work with a surface that is as smooth as possible. Also included is a short description of a method useful for modeling composite materials including porous layers and surface roughness.

A. The Two-Phase Model

In the two-phase model, Fig. 5a, the substrate and the ambient constitute the two phases. In this model, the overall complex reflection coefficients R_p and R_s in Eq. (1) equal the Fresnel reflection coefficients r_{01p} and r_{01s}, respectively, where index 01 indicates the ambient-substrate interface. The explicit expressions for R_p and R_s can be found elsewhere (10; this volume, chapter by Lheveder et al.). We thus have

$$\rho = \frac{r_{01p}}{r_{01s}} \tag{8}$$

This expression can be solved for the optical properties of the substrate (10) either in terms of N_1

$$N_1 = N_0 \sin \phi_0 \sqrt{1 + \left(\frac{1-\rho}{1+\rho}\right)^2 \tan^2 \phi_0} \tag{9a}$$

or in terms of the complex dielectric function $\varepsilon_1 = N_1^2$

$$\varepsilon_1 = \varepsilon_0 \sin^2 \phi_0 \left(1 + \left(\frac{1-\rho}{1+\rho} \right)^2 \tan^2 \phi_0 \right) \tag{9b}$$

where $\varepsilon_0 = N_0^2$ is the dielectric function of the ambient. As ρ is generally a complex-valued number, it is clear from Eqs. (9) that both the real and imaginary parts of N_1 or ε_1 can be determined directly from ρ provided that N_0 and ϕ_0 are known.

For real samples the assumptions of ideal boundaries or total absence of overlayers cannot always be made. A very useful parameter for such nonideal samples is the pseudodielectric function $\langle \varepsilon_1 \rangle$ (and corresponding pseudo-refractive index $\langle N_1 \rangle = \sqrt{\langle \varepsilon_1 \rangle}$, which is defined as

$$\langle \varepsilon_1 \rangle = \varepsilon_0 \sin^2 \phi_0 \left(1 + \left(\frac{1-\rho}{1+\rho} \right)^2 \tan^2 \phi_0 \right) \tag{10}$$

where the measured ellipsometric data ρ in addition to the information about bulk optical properties are also influenced by these nonideal effects. In practice, one often uses Eq. (10) and determines $\langle \varepsilon_1 \rangle$ before deposition of a layer. The assumed second-order effects of, e.g., a thin oxide layer are neglected. However, whenever possible one should strive for exact modeling.

B. The Three-Phase Model

In the three-phase model, shown in Fig. 5b, there is a layer of thickness d having its own optical identity, N_1, between the ambient and the substrate. In this case, the complex reflection coefficients are given by (10)

$$R_p = \frac{r_{01p} + r_{12p}e^{(-i2\beta)}}{1 + r_{01p}r_{12p}e^{(-i2\beta)}} \tag{11a}$$

$$R_s = \frac{r_{01s} + r_{12s}e^{(-i2\beta)}}{1 + r_{01s}r_{12s}e^{(-i2\beta)}} \tag{11b}$$

where r_{01p}, r_{01s}, r_{12p}, and r_{12s} are the Fresnel reflection coefficients at the ambient-film (0–1) and film-substrate (1–2) interfaces for p- and s-polarized light, respectively, and β is the film phase thickness given by

$$\beta = \frac{2\pi d}{\lambda} N_1 \cos \phi_1 \tag{12}$$

By substituting the values of R_p and R_s into Eq. (1), the complex reflectance ratio is obtained:

$$\rho = \frac{r_{01p} + r_{12p}e^{(-i2\beta)}}{1 + r_{01p}r_{12p}e^{(-i2\beta)}} \cdot \frac{1 + r_{01s}r_{12s}e^{(-i2\beta)}}{r_{01s} + r_{12s}e^{(-i2\beta)}} \tag{13}$$

Equation (13) depends, in general, on nine real-valued arguments: the real and imaginary part of the three complex refractive indices N_0, N_1, N_2; the film thickness d; the angle of incidence ϕ_0; and the wavelength λ. In the framework of this chapter, the number of parameters reduces to seven or eight because the ambient is normally either air or a water-based solution (N_0 real valued) and the layer is often transparent, implying that N_1 is real valued. Nevertheless, Eq. (13) is quite complicated and cannot be analytically inverted for any of the unknown parameters, except for some special cases.

C. The *n*-Phase Model

In the *n*-phase model (Fig. 5c) with $m = n - 2$ layers, Eqs. (11) can be used recursively to obtain ρ but the expressions soon become long and complicated to handle. A more efficient way is to employ a 2×2 matrix formalism in which each layer and each interface is represented with a layer matrix $\mathbf{L}_j$ and interface matrix $\mathbf{I}_{ij}$, respectively. The total system can then be described by a scattering matrix $\mathbf{S}$

$$\mathbf{S} = \begin{pmatrix} S_{11} & S_{12} \\ S_{21} & S_{22} \end{pmatrix} = \mathbf{I}_{01}\mathbf{L}_1\mathbf{I}_{12}\mathbf{L}_2 \ldots \mathbf{L}_m\mathbf{I}_{m,m+1} \tag{14}$$

The interface and layer matrices contain the different interface (Fresnel) reflection coefficients, the phase thicknesses, and the layer thicknesses. After multiplication of the matrices in Eq. (14), the overall reflection for a multilayer system is obtained from the ratio of two of the elements of the scattering matrix:

$$R = \frac{S_{21}}{S_{11}} \tag{15}$$

The same formalism is applicable to both p and s polarizations. The scattering matrix formalism is an extremely powerful tool in analysis of ellipsometric data and is simple to implement in computer programs. A detailed description can be found elsewhere (10).

D. Models Containing Anisotropic Layers

Some molecules form ordered layers by self-assembly. Examples are layers of liquid crystals, thiol layers on gold, and Langmuir-Blodgett layers. These layers can be anisotropic; that is, they may have different optical properties in different directions. For such layers the preceding formalism must be extended to include anisotropy. Recent theoretical work has led to development of a general matrix formalism suitable for this (16).

E. Models Containing Composite Layers

Layers can be nonideal in different ways. A few examples are discontinuous layers, inhomogeneous and porous layers, layers with density gradients, layers with an inhomogeneous thickness, and layers with interface roughness. The success of analyzing ellipsometric data relies on whether such effects can be neglected or can be included in the analysis. Here we address modeling of one type, namely inhomogeneous layers in the sense that they contain several constituents. Such layers can in general be described by effective medium approximations (EMAs). These theories are based on solving the electric field equations in the microstructure of the material. This involves several assumptions and generalizations, and different final expressions are found and used depending on applications. One of the more general and commonly used approximation is the Bruggeman EMA. In this theory the dielectric function ε of a two-composite material is obtained from

$$f_A \frac{\varepsilon_A - \varepsilon}{\varepsilon_A + 2\varepsilon} + (1 - f_A) \frac{\varepsilon_B - \varepsilon}{\varepsilon_B + 2\varepsilon} = 0 \tag{16}$$

where ε_A and ε_B are the dielectric functions of materials A and B, respectively, and f_A is the volume fraction of material A. The volume fraction of material B is then $1 - f_A$. Equation (16) can be extended to more than two materials and can also be used for porous materials using $\varepsilon_A = 1$. The porosity is then given by f_A. If the porous material is immersed in a liquid medium filling the pores, ε_A must be assigned the value corresponding to the liquid.

Surface roughness can to first approximation be modeled using Eq. (16) to determine the optical properties of the rough layer assuming $f_A = 0.5$, where medium A corresponds to the ambient. The thickness of the layer is then fitted to the experimental data. Also, a layer with a density or compositional gradient can be modeled. The layer is then divided in sublayers and the formalism in Sec. IV.C is employed with sublayer optical functions determined from Eq. (16).

V. EVALUATION OF ELLIPSOMETRY DATA

A. Parameter Estimation Using Inversion and Fitting

Once the optical model is specified, the ellipsometry data can be inverted or more generally fitted to the model. Depending on the type of measurements done, the level of complexity varies. However, it is important to note that, for a given optical system, a measurement of the complex reflectance ratio ρ at a specific angle of incidence and wavelength provides information to determine two real-valued optical parameters of the system, assuming that all the remaining parameters are known. For example, the complex refractive index N_1 can be determined if the film thickness d and the refractive indices of the ambient, N_0, and the substrate, N_2, are known or can be determined independently. In single-wavelength in situ studies it may therefore often be possible only to determine one or at most two parameters of the sample. In the framework of this chapter, the single parameter in most cases is a layer thickness. However, if data from VASE measurements are used, it may be possible to determine optical properties of one or more layers at the same time that wavelength-independent parameters such as layer thicknesses are determined.

The fitting problem is basically nonlinear but considerable knowledge and powerful mathematical tools are available but unfortunately not always taken advantage of. In some simple cases it can be sufficient to generate tables or graphs by forward calculations based on the theories in the preceding section. Furthermore, for thin layers on near-dielectric substrates, e.g., protein layers adsorbed on silicon, one often finds that most changes occur in Δ and that the relation between layer thickness and Δ is linear. In routine measurements it may then be adequate to use a simplified data analysis by determining a sensitivity factor $\partial d / \partial \Delta$. From a measured change $\delta\Delta$ the thickness is obtained as $d = \delta\Delta \, \partial d / \partial \Delta$.

In a more complete and also more correct approach, the first step is to define a figure of merit. A recommended form is the biased square error

$$\chi^2 = \frac{1}{2P - M + 1} \sum \left[\left(\frac{\psi_{exp} - \psi_{calc}(\mathbf{z})}{\delta\psi} \right)^2 + \left(\frac{\Delta_{exp} - \Delta_{calc}(\mathbf{z})}{\delta\Delta} \right)^2 \right] \qquad (17)$$

where $2P$ is the total number of data points, M the number of parameters in the optical model, $\delta\psi$ and $\delta\Delta$ are experimental errors, and the sum is over all experimental data ψ_{exp} and Δ_{exp}. The $\psi_{calc}(\mathbf{z})$ and $\Delta_{calc}(\mathbf{z})$ are data calculated with the optical model employed. The parameters to be determined are contained in the M-component vector $\mathbf{z}$. The parameters may be layer thicknesses, constituent fractions, optical parameters, etc. The inclusion of the experimental errors in Eq. (17) has the advantage that data points with larger errors get lower weight and thus have less influence on the results.

The next step is to use a fitting algorithm to minimize χ^2. Usually the Levenberg-Marquardt method is used (17,18). With this method it is also possible to obtain the parameter correlation matrix and 90% confidence limits, which, besides the value of χ^2, are very useful for estimating the quality of the data analysis. Further details about data analysis for ellipsometry are given by Jellison (19).

B. From Physics to Biology

We have described data analysis in terms of extracting optical and microstructural parameters from ellipsometric data using optical models. The parameters are essentially refractive indices, layer thicknesses, and volume fractions. This approach is motivated from a methodological point of view because these parameters bear a simple relationship to the microscopic polarizability of dipoles and basically the photons probe the number of dipoles and their strengths in the probe depth of the sample. However, when biological processes are discussed, these terms are not always appropriate and conversion to other forms or further modeling is therefore requested.

Let us first focus on a few simple cases in which it is of interest to determine the surface concentration Γ, i.e., the mass per unit area. A common situation, especially when using single-wavelength ellipsometry, is that only the thickness d of a layer can be determined because of correlation between (a real-valued) refractive index n of the layer and d. A value on n must be taken from the literature or assumed. After extracting d, the surface concentration is then obtained from

$$\Gamma = d\rho_{layer} \qquad (18)$$

where ρ_{layer} is the density of the layer material. An independent determination of ρ_{layer} must be done if it is not available in the literature. A comparison with other techniques such as radioactive labeling or fluorescence labeling will strengthen such an analysis.

If the ellipsometric data also allow the refractive index n of the layer to be determined, one can use de Feijter's formula (20)

$$\Gamma = \frac{d(n - n_0)}{dn/dc} \qquad (19)$$

where n_0 is the refractive index of the ambient, and dn/dc is the refractive index increment for the molecules in the layer. Using this approach requires that it is possible to determine *dn/dc* independently (19) and also that the linearity between concentration *c* and *n* holds in the dense limit. One can go one step further by starting from the polarizability of the atoms of the molecules under study as described by Cuypers et al. (21).

These very simplified optical models may be questioned. First of all, it is not easy to convince a layman about their correctness because often the layers are formed by adsorption of macromolecules in submonolayers and can therefore be considered discontinuous, and issues such as clustering and anisotropy are seldom addressed. Second, the evaluation is often based on numerical inversion of a limited amount of data and no figure of merit is obtained as in parameter-fitting routines based on regression. Experimental and model errors may therefore propagate unnoticed through the analysis. However, comparative calibration measurements have been done by means of radio-labeling (22) and in most cases good correlations are found, and ellipsometric results are therefore in general used with rather high confidence in protein adsorption studies.

However, in many applications it is of interest to do further modeling to arrive at parameters of more biological relevance. In more advanced studies such parameters could be the tilt angle of molecules in a monolayer, density of a protein layer, mass distribution (density gradient) over a protein layer, lateral mass distribution over a surface (clustering), etc. Such studies can be performed in a few cases and represent the state of the art of ellipsometric analysis on thin layers. Spectroscopic ellipsometry is then necessary, and when it is combined with advanced optical modeling, more details about the microstructure and optical properties of monolayers of organic films are obtainable (3,5,9). In this chapter a few examples are briefly described to indicate the type of information it is possible to determine with ellipsometry.

VI. METHODOLOGY

It is of vital importance to be aware that in many of the ellipsometric studies discussed here, one essentially determines the amount of material adsorbing on a surface and in some cases the layer microstructure, but fingerprinting is normally not possible. Unwanted adsorption may occur and one can therefore not distinguish between adsorption by contaminants and the molecules under study. A good understanding of surface cleaning is therefore extremely important. A protocol for cleaning relevant for the substrate material under study must be developed and tested thoroughly. In this testing phase, ellipsometry can be a very useful tool due to its capability of detecting very small amounts of surface contaminations. Special care must also be taken when surfaces are transferred through liquid-air interfaces and during rinsing and drying procedures. Whenever possible, statistics should be included to identify sample-to-sample variations. One of the main advantage of ellipsometry is the in situ possibility. However, such measurements are time consuming and for practical reasons the number of experiments is normally considerably reduced in in situ studies. It therefore becomes even more important to establish the correct starting conditions in terms of a clean surface and absence of a baseline drift due to unwanted adsorption from contaminated solutions. The actual procedures differ from system to system and depend on the substrate, type of molecules, and type of problem being addressed. It is always advisable to consult the literature for surface cleaning recipies.

This brings us to the choice and role of the substrate. Silicon has been and is still widely used for several reasons. First, silicon dioxide, which is always present, is a good model for a glass surface, which is commonly in contact with solutions containing biomolecules. Second, the surface energy, polarity, and surface chemistry of silicon dioxide can easily be controlled by silanization. Another reason for using silicon is that optically the silicon–silicon dioxide system is very close to an ideal substrate for ellipsometric studies. It is almost atomically flat and the refractive index of silicon is high, and thus a good optical contrast to organic materials is obtained, which is a guarantee for high resolution. Also, metal substrates are used, especially in biomaterials research. Gold substrates modified by thiol chemistry and titanium substrates are frequently used (23,24).

When the number of unknown parameters exceeds two, the number of independent ellipsometric measurements has to be increased from one ψ-Δ pair in order to determine all unknowns. This can be done in several ways. If, for instance, the main interest is in the optical properties of the film, measurement can be performed with different ambient or substrate media, provided that the optical properties of the film are not disturbed (25). Alternatively, measurement can be done at different film thicknesses, assuming that the refractive index of the film is thickness independent. However, these methods rely on parameter invariance for the changes made and are not always applicable. To overcome this problem, measurements with multiple angles of incidence can be used in order to acquire more independent experimental data from one and the same sample. An even more powerful approach is provided by photon energy variation or spectroscopy.

VII. APPLICATIONS

Ellipsometry has found numerous applications in several areas of research. The applications may be classified as *single-wavelength* studies and *spectroscopic* studies. Single-wavelength applications may be further divided into ex situ routine measurements of layer thicknesses in air and in situ *kinetic* (dynamic) studies of film growth, macromolecular adsorption, or other surface or thin-film phenomena. In the second class, spectroscopy, we find applications in solid-state material optics and microstructural characterization. State of the art is *time-resolved spectroscopic* ellipsometry, in which dynamic and spectroscopic measurements are done simultaneously. Here we will illustrate the use of ellipsometry in physical chemistry by showing selected applications mainly in protein adsorption.

A. Protein Adsorption on Solid Substrates

A common application of ellipsometry is the measurement of thicknesses of thin layers in the semiconductor processing industry. The thicknesses of silicon dioxide on silicon can be determined very precisely down to subnanometer resolution with standard push-button instruments that are fully automated, including software for on-line evaluation. The operator, or a robot in a process line, inserts the sample, and within a few seconds a value of the thickness is obtained. Complete scanning systems for two-dimensional mapping of film thicknesses over a silicon wafer are also commercially available.

Analogous methodology can be applied to protein layers on solid substrates. Silicon and metals such as gold, chromium, and titanium are suitable substrates and often used as model surfaces. Therefore, ex situ (in air) determination of the amount of adsorbed protein and in situ (in liquid) monitoring of protein adsorption have been traditional applications of ellipsometry. Representative examples including modeling of kinetics and conformation can be found in Ref. 2. Examples of investigations are protein adsorption from blood plasma on titanium surfaces (23,24). More general issues can be found in Refs. 26 and 27. To give the reader a taste of the possibilities, two examples are shown in Fig. 6. The first example is ferritin adsorption on gold at two different concentrations (28). Here the primary data, Δ versus time at $\lambda = 620$ nm, are shown to demonstrate the noise level and time resolution. The concentration dependence of the rate of adsorption is evident. The protein molecules must diffuse over an unstirred layer, explaining the rate difference. By extrapolation to large times one also finds a difference in the final thickness. For the $1000\,\mu g/mL$ case the thickness was 9.5 nm ($\Gamma = 0.55\,\mu g/cm^2$), corresponding to a monolayer of ferritin. The second example is adsorption of lactoperoxidase (LPO) on oxidized silicon with different wettabilities (29). These measurements were performed at $\lambda = 546$ nm at different protein concentrations (mg/mL) as indicated in Fig. 6b, which shows $\delta\Delta$, the decrease in Δ versus time after addition of LPO. The maximum thickness corresponds to a monolayer of LPO. Notice that $\delta\Delta$ is larger on a hydrophilic surface than on a hydrophobic surface, which was interpreted in terms of different orientations of the ellipsoidal protein molecules in the surface layer. The corresponding thicknesses were 3.7 and 2.5 nm, respectively. Also observe that in this particular case the difference in thickness is small between adsorption from an initial high concentration (1.1 mg/mL) and adsorption at 0.74 mg/mL followed by an increase to 1.1 mg/mL.

It is also of interest to compare protein adsorption and layer dynamics under static conditions with those under flow conditions: in vivo (e.g., in blood vessels) and in vitro (as in tubing for blood handling). Phenomena such as fouling and surface friction then play important roles. Special cell design considerations are needed to make sure that the flow can be controlled in a proper manner (30). Work is in progress in this challenging area (31).

B. Protein Monolayer Spectroscopy

Spectroscopy on monolayers of biological molecules is essential for determining their optical properties in thin-film form. Early studies ex situ on protein monolayers showed that this is fully possible using RAE on several types of substrates including both metals and semiconductors (32). Figure 7 shows two examples of ellipsometrically determined optical properties of thin protein layers. The general problem of a large correlation between the refractive index and layer thickness was found not to be a main problem. The analysis is eased by the fact that most protein layers can be assumed transparent ($\varepsilon_2 = 0$) in most of the spectral region studied. An appropriate strategy for extracting the optical data is to determine the thickness in the wavelength region of transparency (below 3.5 eV) and then determine the dielectric function $\varepsilon = \varepsilon_1 + i\varepsilon_2 (= N^2)$ on a wavelength-by-wavelength basis in the whole photon energy range. Notice that the absorption of the polypeptide backbone is seen in the ultraviolet (UV) range above 4 eV. The absorption peak at 280 nm (≈ 4.4 eV) used for determination of protein concentration in solution is also resolved. A main motivation for these studies is that reference values for the refractive indices are obtained. These values are very useful in single-wavelength studies with a focus on thickness or surface

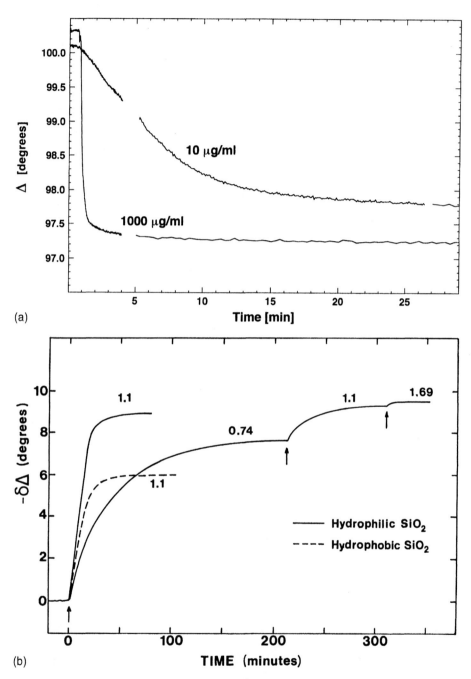

Fig. 6   (a) Δ versus time for ferritin adsorption on a gold substrate measured at $\lambda = 620$ nm at two concentrations in solution (0.02 M phosphate buffer, pH 7.3, 0.15 M NaCl). (From Ref. 28.) (b) Change in Δ versus time after addition of lactoperoxidase to a solution in which hydrophilic (solid lines) and hydrophobic (dashed line) silicon dioxide surfaces were inserted. The arrows show the points in time when lactoperoxidase was added. The bulk concentrations of lactoperoxidase in μg/mL are given by the numbers at the plateau of each curve. (From Ref. 29.)

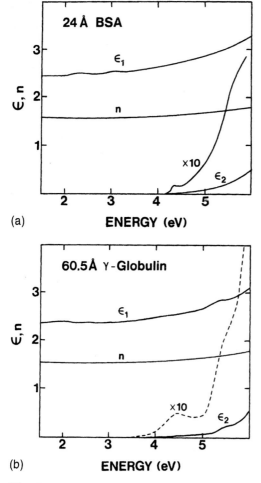

(a)

(b)

Fig. 7 (a) The complex dielectric function ($\varepsilon = \varepsilon_1 + \varepsilon_2$) and real part of the refractive index of a 24-Å layer of bovine serum albumin (BSA) adsorbed on a HgCdTe substrate. ε_2 is also shown magnified by a factor of 10. (b) Same as in (a) but for γ-globulin adsorbed on a gold substrate. (From Ref. 32.)

concentration measurements, as described in the previous section. For bovine serum albumin and γ-globulin n is 1.572 and 1.542, respectively, at $\lambda = 633$ nm.

The preceding studies were extended to the solid-liquid interface (28,29,33) as shown by the two examples in Fig. 8. The procedure is basically first to measure an ellipsometric spectrum of the substrate in solution in a cell like that in Fig. 4a. From this spectrum, $\langle \varepsilon \rangle$ is calculated using Eq. (10). Then proteins are added and after adsorption a new ellipsometric spectrum is recorded. From the two spectra, the thickness and optical properties of the protein layer are calculated as described earlier. The main advantage is that these results more closely represent the native state of the protein layer because drying the sample is not necessary. A further advantage is that the data on the protein layer are recorded at the same spot as the reference data on the substrate. Observe that the absorption due to the heme group in lactoperoxidase

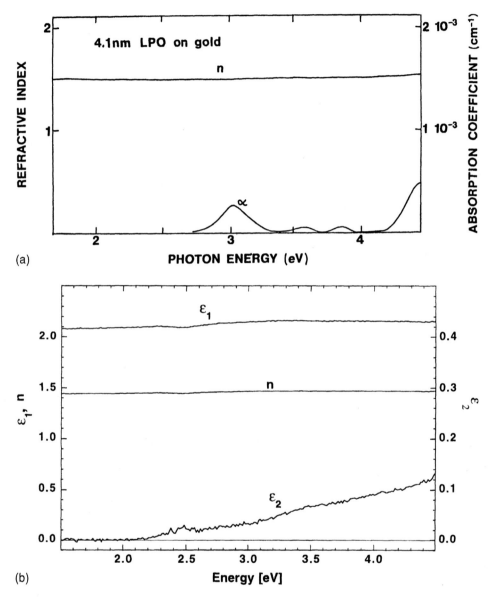

Fig. 8 Ellipsometrically determined optical spectra of protein monolayers on gold substrates measured at a solid-liquid interface. (a) The refractive index n and the absorption coefficient α for a 4.1-nm layer of lactoperoxidase (LPO). (From Ref. 29.) (b) The dielectric function ε and refractive index n for a 9.5-nm layer of ferritin.

is resolved. Also observe that the ferritin layer due to the iron content only is transparent in the left part of the photon energy region.

Other biologically and also technically important layers are Langmuir-Blodgett layers and self-assembly monolayers (SAMs), which can be used to modify and functionalize surfaces, e.g., on biomaterials. Ellipsometry has been applied to study such layers but mainly for thickness determinations (34–38). The analysis is complicated because these layers often show optical anisotropy. Work on thin polymer films shows, however, that such studies are feasible (39) because of new developments in the analysis of anisotropic media (16), and spectroscopic ellipsometry studies of SAMs will probably be seen in the literature in the near future.

C. Macromolecular Adsorption on Complex Substrates

Important new developments include studies of nonideal surfaces and layers. Of special interest are surfaces with microroughness, which may be random or micro-fabricated. Such surfaces are believed to improve the interface between biomaterials and tissue because of similarities in dimensions between macromolecules and surface features that will have effects on recognition, surface mobility of molecules, and mechanical properties (40). Work is in progress on titanium and silicon, and one of the more fascinating challenges is to develop micropatterned surfaces that promote nerve cells to connect to an artificial substrate, thus enabling electronic interfacing on a microscopic level.

Also, porous layers are of interest especially for biosensing purposes and for biomaterials research. In biosensor applications porous surfaces provide a very large area for immobilization of reagents and porous silicon has been suggested as a support in an enzyme-based glucose sensor (41). In biomaterials a porous surface provides a gradual interface between the tissue and the solid substrate, and the interactions with biomolecules are expected to be influenced by the surface topography. Studies of protein adsorption on porous silicon using spectroscopic ellipsometry show very interesting results (42). Figure 9 shows examples of ellipsometric spectra

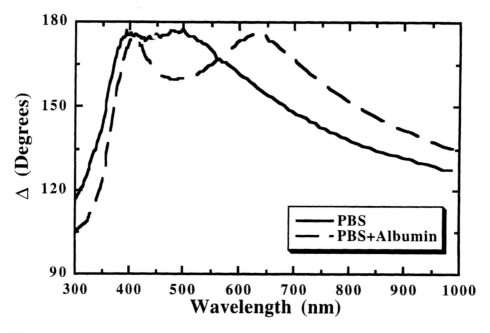

Fig. 9   Spectral changes in Δ due to adsorption of human serum albumin in a 255-nm-thick oxidized porous silicon layer. The measurements were done in phosphate-buffered saline (PBS) at a protein concentration of 1 mg/mL.

(Δ only) taken before and after adsorption of human serum albumin (concentration 1 mg/mL) on or in a 255-nm-thick oxidized porous silicon layer. These data were analyzed in a three-layer (five-phase) model with each layer having different porosity. By comparing adsorption of albumin and fibrinogen it was found that the adsorbed amount depends strongly on molecular size. It was also possible to resolve the depth in the porous layer where the protein molecules adsorbed. Another interesting observation was that fibrinogen adsorbed in the near-surface region of the porous layer whereas albumin penetrated further in. The ability to determine the penetration of protein molecules dynamically in thin layers of controllable microstructure opens up new possibilities.

D. Advanced Analysis of Thin Organic Layers

As already discussed, protein adsorption on solid surfaces has been studied by single-wavelength ellipsometry for a long time and has generated valuable information about the kinetics of mass uptake. Here we focus on developments including advanced studies dealing with layer density as well as with the interplay between adsorption and layer dynamics. The demand is to characterize layer microstructure in more detail than just to determine surface mass or layer thickness. Spectroscopic ellipsometry is necessary in this case.

The first example demonstrates how information about layer structure can be obtained by comparing optical data from layers adsorbed on different substrates (32). This is illustrated in Fig. 10 by modeling the dielectric function of a layer of bovine serum albumin (BSA) on platinum using the Bruggeman EMA and the optical properties of BSA in Fig. 7a. The idea is to use the data in Fig. 7a as reference data. The conclusion is that the BSA layer on platinum is less dense, with a porosity of 30% relative to the BSA layer on the HgCdTe substrate.

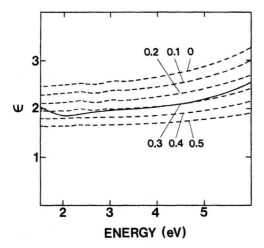

Fig. 10 ε_1 (solid curve) of a 24-Å-thick bovine serum albumin layer on platinum. The dashed curves show calculated ε_1 obtained with the Bruggeman EMA assuming void fractions as given by the numbers at each curve. The top curve (void fraction = 0) is identical to that in Fig. 7a. (From Ref. 32.)

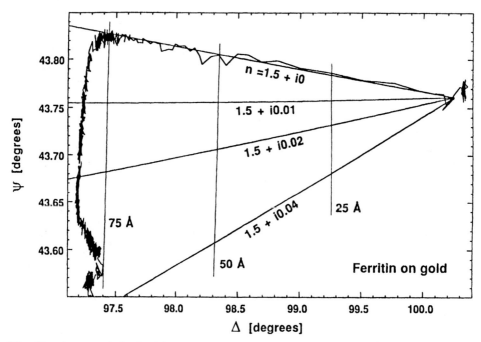

Fig. 11 ψ versus Δ for ferritin adsorption (curve) from a 1 mg/mL solution together with simulations (lines) in the three-phase model, for a layer with $n = 1.5$ and four different values on the imaginary part of the refractive index. The vertical lines are lines of constant thickness. (From Ref. 28.)

Valuable information about adsorption mechanisms can be obtained by detailed analysis of ellipsometric data. A model system for such studies is ferritin on gold. Ferritin is a rigid globular protein with a diameter of ≈ 10 nm. Figure 11 shows ψ versus Δ for adsorption of ferritin on gold measured at a photon energy of 2 eV ($\lambda = 620$ nm). The adsorption trajectory starts to the right in Fig. 11 and follows the "noise" curve to the lower left corner. The total time of the experiment was around 20 h and the knee in the upper left corner occurs after approximately 1 min. Modeling in a three-phase model (ambient-layer-substrate) is not sufficient to explain the experimental trajectory in the ψ-Δ plot. The growth of a layer with a constant refractive index would give almost straight lines in the ψ-Δ plot, as also shown in Fig. 11. It is necessary to incorporate a second layer with thickness d_{EMA} in between the protein layer and the gold substrate in the model. This layer is given optical properties using the Bruggeman effective medium model (33). Furthermore, the interface layer is not static but developing in time, and the interpretation is that the protein layer induces electronic shifts in the near-surface region of the gold substrate, thus altering the optical properties near the surface. By means of spectroscopic ellipsometry, the thickness of the ferritin layer and the interface layer as well as the optical properties (Fig. 8b) of the ferritin layer could be determined. Table 1 summarizes the parameters obtained. It is seen that over a 46-h time period thickness decreases 33% and the refractive index increases. This shows that the layer densifies. Part of the changes can be explained by desorption because Γ as determined from Eq. (19) decreases around 14%.

Table 1 Parameters for a Ferritin Layer Evaluated in a Four-Phase Model

Time (h)	d_{EMA} (nm)	d (nm)	n (at 1.96 eV)	Γ ($\mu g/cm^2$)
0.5	0.15	13	1.428	0.64
3.4	0.15	12	1.432	0.61
22	0.24	11	1.435	0.58
46	0.31	9.5	1.446	0.55

The ellipsometric results led to a suggestion for a two-state model for protein adsorption as illustrated in Fig. 12. The precursor state corresponds to loosely bound protein molecules with lateral mobility. From this state molecules can transfer to the second energetically lower state in which the molecules interact more strongly with the gold substrate, induce a decrease in electron density of the gold substrate, and thus lower the refractive index in a near-surface region. A more extensive discussion of the analysis and the two-state model can be found elsewhere (33).

Initial adsorption processes are another example of an important aspect of protein adsorption. The question at issue is what happens during the first few seconds when a surface comes into contact with a protein solution. To avoid mass transport limitations, a special flow cuvette with a volume of 50 μL was developed to enable fast exchange of solution close to a substrate surface. Off-null ellipsometry was used to allow measurements with a time resolution of 0.1 s (13). In Fig. 13, an example of initial adsorption of ferritin is shown. A delay in the initial kinetics is clearly seen. This delay cannot be explained by diffusion over an unstirred layer. Assuming that the initial lag phase corresponds to a nucleation process, a kinetic model including cluster formation was developed (43).

E. Imaging Ellipsometry—A New Tool for Surface Investigation

Imaging ellipsometry provides an additional tool for studying biological interfaces. The possibility of laterally resolving adsorption patterns down to dimensions of the order of micrometers complements scanning probe techniques. Even simple sys-

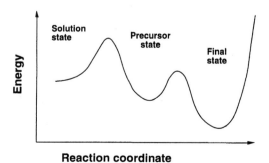

Fig. 12 Schematic illustration of the energetics of a two-state model for protein adsorption. (From Ref. 28.)

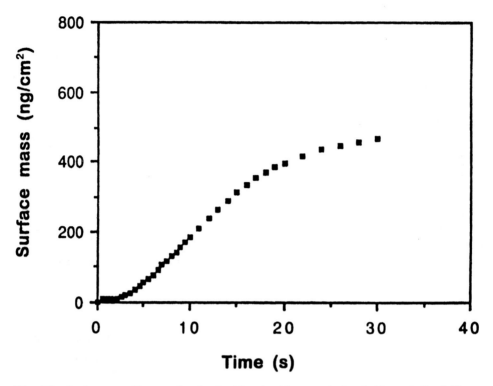

Fig. 13 Surface mass Γ versus time for ferritin adsorbing onto hydrophobic methylized silicon dioxide measured with off-null ellipsometry.

tems based on off-null techniques such as that described in Sec. III. E can give valuable contributions to a detailed understanding of macromolecular behavior on surfaces. Examples of issues that can be addressed are adsorption on small surfaces, particles, etc.; adsorption on microscopically patterned surfaces; dynamics of cluster formation in protein adsorption; and border effects. Also, imaging ellipsometry has potential for developing biosensor systems. The high surface sensitivity implies that it can be used as a readout principle in surface and thin film–based sensor systems. The specificity is provided, for example, by utilizing biological recognition combined with multisensor approaches. Here an illustrative example from pilot surface structure analysis and an idea for a biosensor concept based on imaging ellipsometry are described.

The first example is an ellipsometric image of the lateral thickness distribution of fibrinogen adsorbed in a dot on a hydrophobic silicon surface as shown in Fig. 14. The layer is prepared by putting a small drop of buffer containing fibrinogen directly on a silicon surface. After an incubation time of 30 min, the protein solution is removed and the surface is rinsed in distilled water. The resulting protein layer is very nonuniform in thickness, with a thickness variation between 6 and 25 nm. It is crater-like with dramatically increased thicknesses along the border of the original drop. One explanation is that it is an artifact such as a drying phenomenon due to the sample preparation. However, it is not observed if similar experiments are performed with human serum albumin or γ-globulin. Another possibility is that proteins during the incubation denature on the surface of the drop and then move down

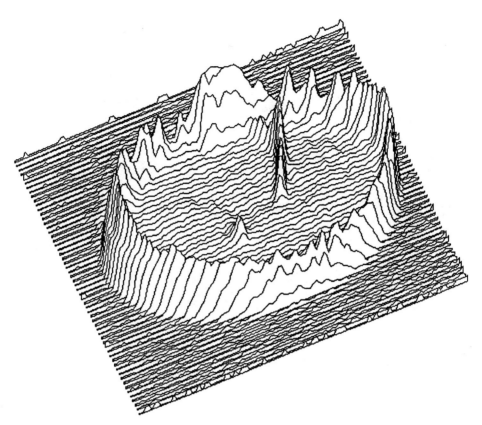

Fig. 14 Thickness distribution of a fibrinogen layer adsorbed in a 3-mm spot on a hydrophobic oxidized silicon surface.

along the liquid-air interface to accumulate at the silicon interface. A third very interesting explanation is that the three-medium interface (air-solid-liquid) at the border along the periphery of the drop catalyzes coagulation and that the phenomenon seen in Fig. 14 is conversion from fibrinogen to fibrin. Further work is necessary to understand the details.

For ellipsometrically based biosensor systems, the second example, different sensing principles can be used. The simplest is to utilize specific binding to a surface-bound reagent. The ellipsometer will then measure the increase in surface mass or thickness of a second layer on the reagent. Another possibility is to use a sensing layer, eg., a polymer layer, that can absorb the substance to be detected. This absorption can lead to swelling of the layer and thereby to a change in the layer optical properties and thickness (44). These changes can be determined ellipsometrically and used as sensor outputs. In porous materials, pore filling by adsorption on the inner walls of pores and capillary condensation are useful mechanisms (45). Also, electronic changes chemically induced by gas absorption in thin organic layer can be utilized (46).

Here the principles of an affinity biosensor system based on imaging ellipsometry are described briefly. The image in Fig. 15 demonstrates that it is possible to visualize the thickness (or surface mass) distribution of protein layers selectively adsorbed in specific areas on a surface, in this case on a 15 × 25 mm hydrophobic

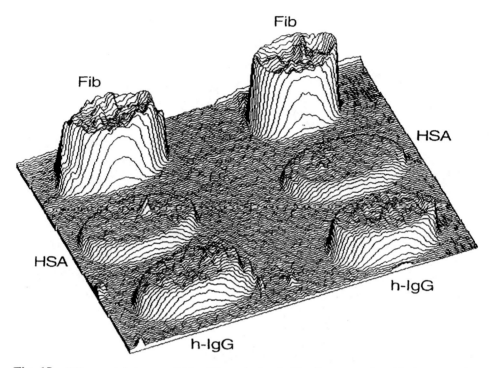

Fig. 15 Ellipsometric image of 15 × 25 mm hydrophobic silicon sample with three proteins adsorbed in 4-mm spots in duplicate.

silicon sample. Three proteins were used: fibrinogen (Fib), human serum albumin (HSA), and human immunoglobulin G (h-IgG). The different layer thicknesses in the different areas are due to the different sizes of the protein molecules. The HSA layer is thinnest with a thickness of around 2 nm. The present thickness resolution is around 0.3 nm but can be improved by further developments of the optics and the detection system. A sample such as that in Fig. 15 can then be used as a "biochip" by incubating it in a solution containing molecules that have affinity for one of the preadsorbed layer. In the corresponding area the thickness will then increase and a new ellipsometric image is recorded and compared with the first one. The differences in ellipsometric signals from the different areas are the biosensor output. Table 2 shows results of an experiment in which antigen-antibody specificity was used as the recognition principle (47). It is seen that the thickness increase in

Table 2 Average Layer Thickness (in Arbitrary Units) in IgG Areas and Fib Areas and for the Background Before and After Incubation in Anti-IgG serum

	Before incubation	After incubation
IgG area	55±8	93±6
Fib area	121±22	123±24
Background	34±4	40±3

areas with specificity is considerable larger than in other areas. These results are intended for demonstration purposes only. In a clinical application more relevant molecules must be selected. Applications of an affinity sensor can be, eg., for screening tests in immunology or for specific analysis of a blood sample to investigate whether a patient has a certain disease.

VIII. SUMMARY AND FUTURE PERSPECTIVES

Ellipsometry has sufficient sensitivity for studies of layers of molecular dimensions on solid substrates. It can be applied in a nonvacuum ambient such as air or liquids, and appropriate models and powerful algorithms for thin-film analysis have been developed. Thus, there are many research opportunities for ellipsometry for interfacial examination in physical chemistry. However, extended cooperation with biochemists and physicians would be very fruitful. This would add expertise to the key scientific issues and make possible a focus on long-term clinical and community needs.

The traditional use of ellipsometry in physical chemistry is in monitoring surface mass, but increased used of ellipsometry, especially in its spectroscopic form, can be foreseen in the following areas:

Surface dynamics in terms of the evolution of the microstructure of thin films
Biosensor systems utilizing sensor arrays and imaging ellipsometry
Imaging of molecular superstructures with predetermined biological functions
Questions related to interfacing biology with electronics

An important aspect to consider is that answers to many basic questions are found not only in the static but also in the dynamic properties of layers, that is, in the microstructure as well as in its time evolution. An important development is therefore real-time spectroscopic ellipsometry. Finally, it is important to follow up the methodological development with new instrumental designs. The complex systems currently being developed for in situ monitoring and growth control of inorganic thin films are not well suited for the laymen in the area of physical chemistry. The potential users are non physicists or instrumentally untrained personnel who need advanced but easy-to-operate instruments.

ACKNOWLEDGEMENTS

I want to acknowledge Robert Bjorklund, Shuwen Guo, Roger Jansson, Gang Jin, Ingemar Lundström, Jan Mårtensson, Håkan Nygren, Leif Pettersson, Pentti Tengvall, Fredrik Winquist, and Shahin Zangooie for collaboration. Financial support has been obtained from the Swedish Research Council for Engineering Sciences and from the Swedish National Board for Industrial and Technical Development.

REFERENCES

1. MK Debe. Optical probes of organic films: Photons-in, photons out. In: SG Davison, ed. Progress in Surface Science, Vol. 24. New York: Pergamon Press, 1987, p 54.
2. I Lundström, B Ivarsson, U Jönsson, H Elwing. Protein adsorption and interaction at solid surfaces. In: WJ Feast, HS Munro, eds. Polymer Surfaces and Interfaces. New York: Wiley, 1987, p 201.

3. H Arwin. Application of spectroscopic ellipsometry on thin organic films, Trends Appl Spectrosc, 1:79, 1993.

4. PA Cuypers, WT Hermens, HC Hemker. Ellisometry as a tool to study protein films at liquid-solid interfaces. Anal. Biochem. 84:56, 1978.

5. H Arwin. Spectroscopic ellipsometry and biology: Recent developments and challenges. Thin Solid Films, 764:313–314, 1998.

6. A Rothen. The ellipsometer, an apparatus to measure thickness of thin surface films. Rev Sci Instrum 16:26, 1945.

7. L Vroman, A Lukosevicius. Ellipsometer recordings of changes in optical thickness of adsorbed films associated with surface activation of blood clotting, Nature 204:701, 1964.

8. AC Boccara, C Pickering, J Rivory, eds. Proceedings of the 1st International Conference on Spectroscopic Ellipsometry, Paris, France, 1993. Thin Solid Films 233/234, 1993.

9. Proceedings of the 2nd International Conference on Spectroscopic Ellipsometry, Charleston, 1998. Thin Solid Films 313/314, 1998.

10. RMA Azzam, NM Bashara. Ellipsometry and Polarized Light. Amsterdam: North-Holland, 1977.

11. DE Aspnes, AA Studna. High precision scanning ellipsometer. Appl Opt 14:220, 1975.

12. RW Collins, I An, HV Nguyen, Y Lu. Real time spectroscopic ellipsometry for characterization of nucleation, growth, and optical function of thin films. Thin Solid Films 233:244, 1993.

13. H Arwin, S Welin-Klintström, R Jansson. Off-null ellipsometry revisited: Basic considerations for measuring surface concentrations at solid/liquid interfaces. J Colloid Interface Sci 156:377, 1993.

14. G Jin, R Jansson, H Arwin. Imaging ellipsometry revisited: Developments for visualization of thin transparent layers on silicon substrates. Rev Sci Instrum 67:2930, 1986.

15. RW Collins. Automatic rotating element ellipsometers: Calibration, operation and real-time applications. Rev Sci Instrum 61:2029, 1990.

16. M Schubert. Polarization-dependent optical parameters of arbitrarily anisotropic homogeneous layered systems. Phys Rev B 53:4265, 1996.

17. PR Bevington. Data Reduction and Error Analysis for the Physical Sciences. New York: McGraw-Hill, 1969.

18. WH Press, BP Flannery, SA Teukolsky, WT Vetterling. Numerical Recipes. Cambridge: University Press, 1986.

19. GE Jellison. Data analysis for spectroscopic ellipsometry. Thin Solid Films 234:416, 1993.

20. JA De Feijter, J Benjamins, FA Veer. Ellipsometry as a tool to study the adsorption behavior of synthetic and biopolymers at the air-water interface. Biopolymers 17:1759, 1978.

21. PA Cuypers, JW Corsel, MP Janssen, JMM Kop, WT Hermens, HC Hemker. The adsorption of prothrombin to phosphatidylserine multilayers quantitated by ellipsometry. J Biol Chem. 258:2426, 1983.

22. U Jönsson, M Malmqvist, I Rönnberg. Adsorption of immunoglobulin G, protein A, and fibronectin in the submonolayer region evaluated by a combined study of ellipsometry and radiotracer techniques. J Colloid Interface Sci 103:360, 1985.

23. B Wälivaara, I Lundström, P Tengvall. An in-vitro study of H_2O_2-treated titanium surfaces in contact with blood plasma and a simulated body fluid. Clin Mat 12:141, 1993.

24. B Wälivaara, A Askendal, I Lundström, P Tengvall. Blood protein interactions with titanium surfaces. J Biomater Sci Polymer Ed 8:41, 1996.

25. F Tiberg, M Landgren. Characterization of thin nonionic surfactant films at the silica/water interface by means of ellipsometry. Langmuir 9:927, 1993.

26. B Ivarsson, I Lundström. Physical characterization of protein adsorption on metal and metaloxide surfaces. CRC Crit Rev Biocompat 2:1, 1986.

27. P Tengvall, I Lundström, B Liedberg. Protein adsorption studies on model organic surfaces. Biomaterials, 19:407, 1998.

28. J Mårtensson, H Arwin, H Nygren, I Lundström. Adsorption and optical properties of ferritin layers on gold studied with spectroscopic ellipsometry. J Coll Int Sci 174:79 1995.

29. J Mårtensson, H Arwin, I Lundström, T Ericson. Adsorption of lactoperoxidase on hydrophilic and hydrophobic silicon dioxide surfaces: an ellipsometric study. J Colloid Interrace Sci 155:30 1993.

30. S Rekveld. Ellipsometric Studies of Protein Adsorption onto Hard Surfaces in a Flow Cell. Febodruk Enschede, Netherlands, 1997.

31. JL Ortega-Vinuesa, P Tengvall, B Wälivaara, I Lundström. Stagnant versus dynamic conditions: A comparative adsorption study of blood proteins. Biomaterials 19:251, 1998.

32. H Arwin. Optical properties of thin layers of bovine serum albumin, γ-globulin, and hemoglobin. Appl Spectrosc 40:313, 1986.

33. J Mårtensson, H Arwin. Interpretation of spectroscopic ellipsometry data on protein layers on gold including substrate-layer interactions. Langmuir 11:963, 1995.

34. MJ Dignam, M Moskovits, RW Stobie, Specular reflectance and ellipsometric spectroscopy of oriented molecular layers. Trans Faraday Soc 67:3306, 1971.

35. RC Thomas, L Sun, RM Crooks, AJ Ricco. Langmuir 7:620, 1991.

36. A Tronin, T Dubrovsky, C Nicolini. Deposition, molecular organization and functional activity of IgG Langmuir films. Thin Solid Films 284/285:894, 1996.

37. AY Tronin, AF Konstantinova. Ellipsometric study of the optical anisotropy of lead arachidate Langmuir films. Thin Solid Films 177:305, 1989.

38. MD Porter, TB Bright, DL Allara, CED Chidsey. Spontaneously organized molecular assemblies. 4. Structural characterization of *n*-alkyl thiol monolayers on gold by optical ellipsometry, infrared spectroscopy, and electrochemistry. J Am Chem Soc 109:3559, 1987.

39. LAA Pettersson, F Carlsson, O Inganäs, H Arwin. Spectroscopic ellipsometry studies of the optical properties of doped poly (3,4-ethylenedioxythiophene): An anisotropic metal. Thin Solid Films 313/314:357, 1998.

40. B Kasemo, J Lausmaa. Material-tissue interface: The role of surface properties and processes. Environ Health Perspect 102(5):41, 1994.

41. T Laurell, J Drott, L Rosengren. Silicon wafer integrated enzyme reactors. Biosensors Bioelectron 10:289, 1995.

42. S Zangooie, R Bjorklund, H Arwin. Protein adsorption in thermally oxidized porous silicon layers. Thin Solid Films 313/314:827, 1998.

43. H Nygren, H Arwin, S Welin-Klintström. Nucleation as the rate-limiting step in the initial adsorption of ferritin at a hydrophobic surface. Colloids Surf A 76:87, 1993.

44. S Guo, R Rochotzki, I Lundström, H Arwin. Ellipsometric sensitivity to halothane vapors of hexamethyldisiloxane plasma polymer films. Sensor Actuators B Chem 44:243, 1997.

45. S Zangooie, R Bjorklund, H Arwin. Vapor sensitivity of thin porous silicon layers. Sensors Actuators B43:168, 1997.

46. J Mårtensson, H Arwin, I Lundström. Thin films of phthalocyanines studied with spectroscopic ellipsometry: An optical gas sensor? Sensor Actuators B1:134, 1990.

47. G Jin, P Tengvall, I Lundström, H Arwin. A biosensor concept based on imaging ellipsometry for visualization of biomolecular interactions. Anal Biochem 232:69, 1995.

18

The Application of Neutron and X-Ray Specular Reflection to Proteins at Interfaces

J. R. Lu
University of Surrey, Guildford, United Kingdom

R. K. Thomas
University of Oxford, Oxford, United Kingdom

I. INTRODUCTION

Neutron reflection and X-ray reflection have been widely used for studies of adsorption at flat surfaces, both air-liquid and solid-liquid, and have proved to be valuable techniques for providing detailed information about the structure and composition of adsorbed layers (1,2). Because many biological processes occur at interfaces, the possibility of using reflection to study structural and kinetic aspects of model biological systems should be of considerable interest. However, the number of such experiments so far performed is small. The main reasons for this may be that the range of types of surface suitable for exploration is perceived to be limited, especially in relation to interfacial systems of biological interest, and also that the most effective use of neutron reflection is assumed to involve extensive deuterium substitution. Not only is a high level of deuteration not usually an available option in biological molecules but also there is always concern that deuteration may affect the behavior of the system. Although these are reasons for taking care in the choice of system for study, they do not preclude the effective use of either X-ray or neutron reflection and it would be a pity not to take advantage of their excellent sensitivity to structure and composition at interfaces. In this introduction to the two specular reflection techniques we focus particularly on situations in which neutron reflection may be sensitive to nondeuterated biological materials, how to enhance this sensitivity, and some of the possibilities for mimicking interfaces of biological interest. We illustrate these ideas with experimental work on adsorbed proteins. Other experimental techniques for studying biological interfaces are described elsewhere in this volume and, for reasons of space, we generally avoid direct comparisons with other techniques.

II. THE BASIS OF NEUTRON AND X-RAY REFLECTION

When radiation is reflected off a flat surface covered with a uniform thin film it may undergo interference. The phase of the radiation reflected off the upper part of the film is shifted with respect to that from the lower part of the film (Fig. 1). For monochromatic radiation the phase shift varies systematically with the angle of reflection, leading to alternate constructive and destructive interference as the two partially reflected beams move in and out of phase. Thus, provided there is sufficient reflection from either surface, a measurement of the reflected monochromatic radiation as a function of angle of reflection gives a series of interference fringes with constructive peaks at angular spacings determined by

$$\sin \theta = n \frac{\lambda}{2d} \tag{1}$$

where θ is the glancing angle of reflection as defined in Fig. 1, λ is the wavelength of the radiation, n is an integer, and d is the thickness of the thin film. The physics of the experiment is the same for all radiation and is the basis of the use of X-ray and neutron reflection for the study of interfacial layers at flat surfaces. Thus, the thickness of the film can easily be deduced from the known wavelength of the radiation and the separation of the fringes. The magnitude of the intensity fluctuations in the fringes depends on the refractive index of the film relative to that of the bulk phases on either

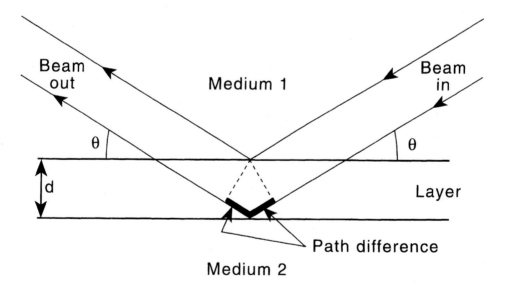

Fig. 1 Reflection of radiation from a thin layer deposited on a substrate. Some radiation is reflected at each interface but that reflected from the lower interface has a longer path length of $2 \times d \sin \theta$. Constructive interference occurs when the path difference is a whole number of wavelengths.

side. For X-rays and neutrons the refractive index is directly related to composition and, because composition of the two bulk phases is usually known, it is then possible to determine the composition of the thin film from the amplitude of the fringes. The separation of the fringes and their amplitude are independent and therefore the thickness and composition of the film are determined independently, although this is seldom quite as ideally achieved in practice.

Several technical factors degrade the ideal experiment just described. The main problem is that the refractive indices for X-rays and neutrons differ from unity by only parts per million, and this has the consequence that the reflectivity is measurable only at low angles of grazing incidence. The refractive index is related to composition in two steps. First, it is related to a quantity known as the scattering length density by

$$\eta^2 = 1 - \frac{\lambda^2}{\pi} \rho \tag{2}$$

where η is the refractive index, λ the wavelength of the radiation, and ρ the scattering length density. The scattering length density of the material, ρ, is then related to the composition, for neutrons by

$$\rho = \sum_i b_i n_i \tag{3}$$

where b_i is the scattering length of nuclear species i and n_i is its number density, and for X-rays by

$$\rho = \sum_i n_i \frac{f_i e^2}{mc^2} \tag{4}$$

where f_i is the atomic form factor, c is the speed of light, and e and m are the charge and mass of the electron. The sums in Eqs. (3) and (4) are over all the nuclei or atoms in the material.

For X-rays the atomic form factor, and hence the scattering length, is proportional to the number of electrons in the atom and therefore to the atomic number. For neutrons the scattering length varies erratically with atomic number and may change sign both between elements and, more importantly, between different nuclei of the same element. The most important isotopic difference from a biological point of view is between H and D, for which the scattering lengths are -3.74×10^{-6} nm and 6.67×10^{-6} nm, respectively. After allowance has been made for the contribution from the oxygen atom, the scattering length densities of H_2O and D_2O are of opposite sign and the same would be true for alkyl chains and many other hydrogen-containing fragments. The possibility of using isotopic substitution to change the reflectivity is of crucial importance for the neutron experiment, as will be seen later, but is not, of course, possible in X-ray reflection, although the X-ray refractive index can be varied by working near an X-ray absorption edge.

There are two further important differences between X-rays and neutrons. The first is that X-rays are strongly absorbed by matter, and this makes it more difficult to penetrate through to buried interfaces. Thus, if the layer shown in Fig. 1 is between solid and liquid bulk phases, most of the X-ray intensity will be lost by absorption, whereas in a suitably designed neutron experiment the loss of beam may be less than

50%. A large loss of X-ray beam does not mean that the experiment is impossible because the high fluxes available in synchrotron sources can more than compensate for this loss. The advantage of X-ray reflection over neutrons comes in part from the high beam intensities now available and in part from the high absorption. To obtain useful information about a thin layer it is necessary to measure the reflectivity down to very low values, from about 10^{-5} to 10^{-10}, and the reflectivity is therefore being measured in circumstances in which most of the incident beam passes into the second bulk medium, where it will be scattered and/or absorbed. Although this scattering may not have any particular directional characteristics, some of it will be along the direction of the specularly reflected beam. This creates a background that interferes with the measurement. For X-rays the stronger absorption is now an advantage because it means that there is less scattered background. Thus, when all other factors are equal the extra intensity of the beam and the lower background make X-ray reflection much more sensitive than neutron reflection.

To demonstrate the simplicity of the reflection experiment, we consider two examples of reflectivity profiles simulated for situations that are easily realized experimentally using neutrons. The reflectivity profile in Fig. 2a is simulated with parameters approximately appropriate for a spread phospholipid monolayer on water, where the isotopic composition of the water has been adjusted so that its refractive index is the same as that of air. In this case there would be no reflected signal at all in the absence of the monolayer; there would be just the background from scattering of the neutrons from the interior of the aqueous subphase. If the phospholipid layer is deuterated, the scattering length density profile then has the simple form shown in Fig. 2b and this gives rise to a large signal whose general level is approximately proportional to the square of the surface density of the phospholipid and whose shape is dependent on the thickness of the layer. To illustrate this, the reflectivity profiles

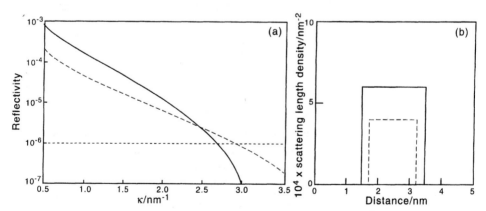

Fig. 2 (a) Neutron reflectivity profiles calculated for parameters approximately appropriate to a fully deuteriated phospholipid layer spread on null reflecting water at two surface coverages. The scattering length density and thickness of the layer are $6 \times 10^{-4}\,\mathrm{nm}^{-2}$ and $2.0\,\mathrm{nm}$ (continuous line) and $4 \times 10^{-4}\,\mathrm{nm}^{-2}$ and $1.5\,\mathrm{nm}$ (large dashes). The line with small dashes in (a) shows the typical level of background to be expected from scattering by the bulk solution. Thus, it rapidly becomes more difficult to measure neutron reflectivities below about 5×10^{-7}. (b) The scattering length density profiles along the surface normal direction for the two profiles calculated in (a).

in Fig. 2 have been calculated for two typical conditions of the layer with different coverage and thickness. Note that, once the flat background, for which the typical level is shown as a dotted line in Fig. 2, has been subtracted, the signal arises entirely from the phospholipid layer; i.e., the technique is surface specific. The second example, shown in Fig. 3, shows the reflectivity profile simulated for a globular protein layer adsorbed at a hydrophilic solid-water interface, where the H/D ratio of the aqueous solution has been adjusted so that the refractive indices of solution and solid are such that reflection is only from the protein layer; i.e., the scattering length densities of solution and solid are identical. Exactly how this is achieved is discussed in the following, as is the question of what the isotopic composition of the protein is under these circumstances. The reflectivity is determined only by the average refractive index of the layer. Thus, when the coverage of protein decreases, the reflected signal decreases. If the protein is not denatured by the surface, two results are possible; first, the thickness of the layer will correspond approximately to one of the bulk dimensions of the aqueous protein, and second, although dilution of the protein may reduce its surface coverage, the thickness of the layer will always be consistent with the bulk dimensions of the protein. Figure 3 compares the two different effects of decrease of layer thickness (dashed line) and decrease in the coverage at constant thickness (dotted line). Although the profiles are simulated, the dimensions and coverages chosen are those typical of many protein layers.

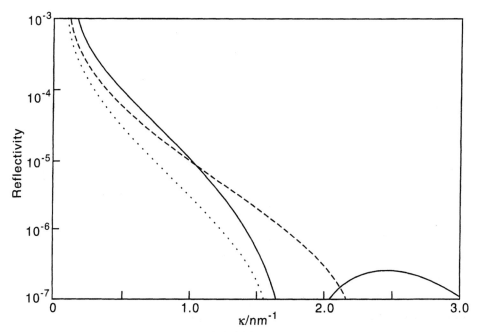

Fig. 3 Neutron reflectivity profiles calculated for a typical layer of a globular protein adsorbed from aqueous solution onto a solid substrate with the isotopic composition of the water adjusted so that the scattering length densities of water and solid are matched. The protein layer is taken to have a thickness and volume fraction of 3.5 nm and 0.6 (continuous line), 2.5 nm and 0.6 (dashed line), and 3.5 nm and 0.3 (dotted line). The calculation approximately matches the conditions for protein adsorption on quartz.

When an interference fringe, or part of one, is observed, it becomes particularly easy to derive the thickness and coverage of the layer independently of one another. Thus, the positions of the minima in Figs. 2 and 3 are related to the thickness of the interfacial layer by $d = \lambda/(2 \sin \theta)$. The composition of the adsorbed layer mainly determines the amplitude of the fringe. It is not always the case that interference is observed, but the dimensions of many layers of biological interest are such that the shape of the reflectivity profile is often sufficiently sensitive to the thickness that thickness and composition can be distinguished, although if there is poor contrast between the layer and its surroundings (i.e., not much difference in scattering length density) this breaks down. The snag in the determination of the composition of the protein layer under discussion is that the measurement has been made in a mixture of D_2O and H_2O and the scattering length density of the layer depends on the isotopic composition of the protein, which in turn depends on the extent of exchange of deuterium with the solvent. We discuss this question in detail later, but here we note that this exchange does not interfere at all with the determination of the thickness of the layer. To calculate the surface density of protein molecules it is, however, necessary to have an approximate estimate of the H-D exchange. For a typical protein the scattering length density of the monolayer will be about $2.0 \times 10^{-4}\,nm^{-2}$ if all the labile hydrogens are fully protonated and $3.5 \times 10^{-4}\,nm^{-2}$ if they are fully deuterated. The exchange of about half the labile protons is expected to be sufficiently rapid on the time scale of the experiment (typically a few hours) that these will certainly have equilibrated with the solvent (3,4). For the isotopic composition under consideration, this would change the two extreme values for the scattering length density of the protein layer by only about 5%. In calculating the coverage at the air-water interface from neutron reflectivity data the uncertainty in the extent of exchange for the isotopic composition therefore introduces a maximum uncertainty of 10% into the surface coverage determination. This is the maximum uncertainty in a situation in which one has limited knowledge of the extent of exchange. In practice, sufficiently accurate information is available that the uncertainty in the derived surface coverage will be less than 5%. The error here has been calculated for the particular case of protein adsorbed at the air-water interface; it will be different in other situations, and we discuss these as they arise in the following.

The simulated profiles in Figs. 2 and 3 are for situations in which there is no difference in refractive index between the two bulk phases; i.e., there is no contrast between the two bulk phases. The system is null reflecting unless material giving rise to a layer of different refractive index is adsorbed at the interface, and then the reflected signal is only from the adsorbed layer. The scattering length densities of most materials fall between the two limits of D_2O and H_2O and therefore it is nearly always possible to generate this null reflecting condition and maintain the surface specificity of the technique. For example, the scattering length densities of H_2O and D_2O are -0.56×10^{-4} and $6.35 \times 10^{-4}\,nm^{-2}$, respectively, so that water of any scattering length density intermediate between these two values is easily prepared. Thus, the air-water interface can be made null reflecting (the scattering length density of air is 0), as can the interfaces between water and many common solid materials such as silicon, amorphous silica, and quartz with scattering length densities of 2.1×10^{-4}, 3.4×10^{-4}, and $4.2 \times 10^{-4}\,nm^{-2}$, respectively. The difference between the solid-liquid and air-liquid interfaces is, however, that the value of the scattering length density at the null reflecting condition (often referred to as contrast matched) is quite different from those of the hydrogenated and the deuterated forms of most surface-active materials. It is then not so important to use isotopic labeling to make the layer visible.

The null reflecting condition cannot be realized in X-ray reflection but direct structural information can still be obtained about the layer. The scattering length density profile for X-rays at an air-water interface consists to a first approximation of an abrupt step and this generates a quite different reflectivity profile from that of the water alone, as shown in Fig. 4. The comparable reflectivity profile for neutrons is at the air-D_2O interface because D_2O has a large positive scattering length density, and this is shown in Fig. 5. The adsorption of material at the interface will modify this sharp step differently for the two techniques. For phospholipid molecules partially embedded in water the layer can be divided into two regions. The outer layer contains only hydrocarbon chains, and these are likely to have a density somewhat lower than that of liquid hydrocarbon (depending on the spreading pressure). The inner layer consists of the headgroup of the phospholipid and water, if any water is present in this layer. An approximation to the X-ray scattering length density profile is that it consists of a step of lower scattering length density corresponding to the outer hydrocarbon layer and that the inner layer is indistinguishable from water because of the similarity between the scattering length densities of headgroup and water. For a typical surface density the reflectivity shown as a continuous line in Fig. 4 is obtained; the X-ray reflectivity profile is evidently very sensitive to the composition and thickness of

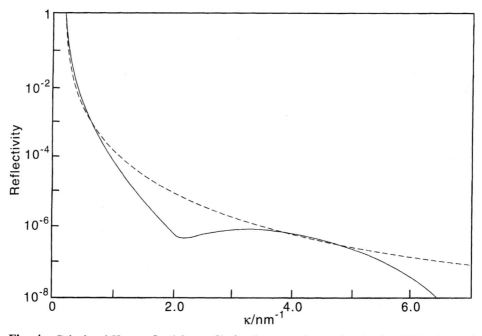

Fig. 4 Calculated X-ray reflectivity profile for the approximate situation in which a layer of phospholipid is spread on water at an intermediate surface coverage. The scattering length density of the hydrocarbon layer protruding out of the water is taken to be 6×10^{-4} nm^{-2} (the X-ray scattering length density for water is 9.5×10^{-4} nm^{-2} and its thickness 1.5 nm). The headgroup region of the layer is taken to have the same scattering length density as the underlying aqueous subphase. The X-ray reflectivity profile of water alone is shown as a dashed line.

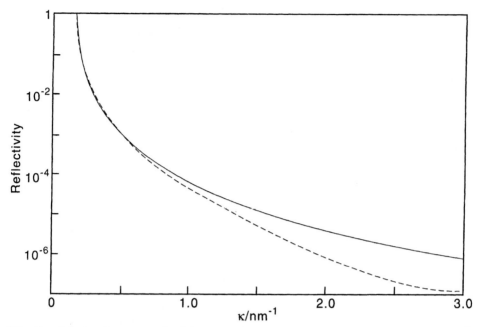

Fig. 5 Calculated neutron reflectivity profile for the same approximate situation as in Fig. 4, where a layer of phospholipid is spread on water at an intermediate surface coverage. The neutron scattering length density of the hydrocarbon layer (protonated) protruding out of the water is 0 and the headgroup region of the layer is taken to have a thickness of 1.2 nm and a scattering length density about one third that of the underlying aqueous subphase, which is D_2O. The neutron reflectivity profile of D_2O alone is shown as a continuous line.

the layer protruding from the water. For neutrons the situation is quite different. If the phospholipid is protonated, the protruding hydrocarbon layer has approximately the same scattering length density as air and so is invisible. In contrast, the headgroup, which is also approximately null scattering, displaces D_2O from the second layer, and it is the depressed scattering length density of this layer that dominates the reflectivity profile (dashed line in Fig. 5). Thus, the two reflection techniques are sensitive to different features of the layer. In practice, further detail may be revealed about the layer structure because the assumptions made to describe the profiles here are too simple. For example, the X-ray reflectivity will probably have some sensitivity to the hydrated layer, depending on the exact composition of the headgroup. X-ray reflectivity is intrinsically more sensitive to structural detail within a layer because of its higher resolution, which results from the ability to make measurements down to lower reflectivities.

The X-ray and neutron profiles of Figs. 4 and 5 give separately the thickness of the region of the phospholipid lying outside the water and the thickness of the hydrated region. This gives the impression that both techniques are needed to determine the structure of such a layer. This need not be the case. As already indicated, the difference between the scattering length densities of the phospholipid headgroup and water may be sufficient for X-ray reflection to be sensitive to the hydrated region, as well as to the top layer. In neutron reflection the use of a deuterated phospholipid will make the experiment sensitive to both layers. Thus, the combination of the measurement of the deuterated phospholipid in null reflecting water (Fig. 2), which gives the thickness of the whole phospholipid layer, with the measurement of the protonated phospholipid in D_2O (Fig. 5), which gives the thickness of that part of the layer that is immersed in water, characterizes the basic structure of the layer. Such a set of measurements may, however, be prevented from being useful by isotope effects, which are all too common in biologically interesting systems. In general, when an adsorbed layer can be described in terms of two layers of different scattering characteristics, some care has to be taken in the design and analysis of the experiment in order to elucidate these details.

III. QUANTITATIVE ANALYSIS OF REFLECTION DATA

There are two methods commonly used for analyzing reflection data. The first follows the standard method for calculating the optical reflectivity of thin films. Provided the appropriate changes are made in the refractive indices, the X-ray reflectivity is exactly the same as that of light and the neutron reflectivity is the same as that of light polarized at right angles to the reflection plane. At the low angles at which the reflected signal is significant, the two states of polarization of X-rays have identical reflectivities and only the one state need be considered. The reflectivity from a single thin film is obtained by summing over all the contributions shown in Fig. 6 for which the exact expression is (5,6)

$$R = |\mathbf{R}|^2 = \frac{r_{01}^2 + r_{12}^2 + 2r_{01}r_{12}\cos(2\beta)}{1 + r_{01}^2 r_{12}^2 + 2r_{01}r_{12}\cos(2\beta)} \tag{5}$$

where r_{ij} are the Fresnel reflection coefficients and β is the phase shift on traversing the layer once and equals $q_1\tau\sin\theta_1$, where q_1 is the wave vector of the radiation normal to the interface in the monolayer, τ is the thickness of the layer, and θ_1 is the glancing angle of incidence at the interface between layers 1 and 2. It is customary in neutron reflection to express the reflectivity in terms of the momentum transfer κ rather than the angle of incidence, where κ is defined in terms of the grazing angle of incidence by

$$\kappa = \frac{4\pi\sin\theta_0}{\lambda} = 2q_0 \tag{6}$$

Defining the momentum transfer at which total reflection occurs between layers i and j as κ_{ci}, which is given by

$$k_{ci}^2 = 4q_{ci}^2 = 16\pi(\rho_{i+1} - \rho_i) \tag{7}$$

the Fresnel coefficients are

$$r_{ij} = \frac{k_c^2}{[\kappa + \sqrt{(\kappa^2 - \kappa_c^2)}]^2} \tag{8}$$

It is important to realize that Eq. (5) is exact; it includes multiple scattering and refractive index effects. Furthermore, it has been extended to multilayer films to give an exact method for calculating the reflectivity of any film (5,6). The only assumption that has to be made is that each component layer is laterally homogeneous. What is meant by this is that the inhomogeneities are less than a few 100 nm in extent, although the exact dimension depends on the particular experimental configuration. The procedure for doing the calculation is then to guess a structure, divide the adsorbed layer up into as many component layers as are thought necessary, and to calculate the characteristic matrix for each layer, which is given by

$$[M_j] = \begin{bmatrix} \cos \beta_j & -\left(\frac{i}{q_j}\right) \sin \beta_j \\ -iq_j \sin \beta_j & \cos \beta_j \end{bmatrix} \tag{9}$$

where q_j and β_j are as defined before in connection with Eq. (5). The characteristic matrices for the various component layers are multiplied together and the reflected amplitude is given by

$$\mathbf{R} = \frac{(M_{11} + M_{12}q_{n+1})q_0 - (M_{21} + M_{22})q_{n+1}}{(M_{11} + M_{12}q_{n+1})q_0 + (M_{21} + M_{22})q_{n+1}} \tag{10}$$

where M_{ij} are the elements of the final 2×2 matrix and q_0 and q_{n+1} refer to the initial and bulk phases. Recurrence relations between the Fresnel coefficients are used to speed up the calculation (5). The software is widely available for such calculations, and they can be done conveniently on a PC.

The optical matrix method is easy to use and exact but it takes much experience before a clear understanding is obtained of how different structural features affect a reflectivity profile. It is useful to consider an alternative, approximate expression for the reflectivity, which gives a more direct relation between structure and reflectivity. This is the kinematic approximation, which neglects both the refractive index effects and the multiple scattering of Fig. 6, to give the approximate reflectivity profile from

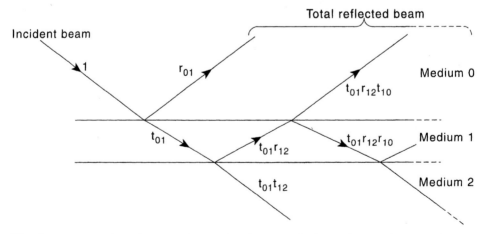

Fig. 6 The multiple reflection of radiation from a thin layer deposited on a substrate. Radiation is reflected and transmitted at each interface so that the resultant reflected beam contains many multiply reflected components. The optical matrix calculation takes this multiple reflection fully into account.

a flat surface of (7,8)

$$R = \frac{16\pi^2}{\kappa^2} |\rho(\kappa)|^2 \tag{11}$$

where κ is the momentum transfer defined as $(4\pi \sin \theta)/\lambda$, and $\rho(\kappa)$ is the one-dimensional Fourier transform of the mean scattering length density profile normal to the interface, $\rho(z)$,

$$\rho(\kappa) = \int_{-\infty}^{\infty} \rho(z) \exp(-i\kappa z)\, dz \tag{12}$$

This is similar to the relationship between structure and observation that is found in all scattering or diffraction experiments; i.e., the measured quantity is proportional to the squared modulus of the Fourier transform of the desired structural quantity. It is relatively straightforward to show that for a uniform monolayer on a substrate the expression for the reflectivity becomes

$$R = \frac{16\pi^2}{\kappa^4} [(\rho_1 - \rho_0)^2 + (\rho_2 - \rho_1)^2 + 2(\rho_1 - \rho_0)(\rho_2 - \rho_1) \cos \kappa d] \tag{13}$$

where d is the thickness of the layer. This is the approximate form of Eq. (5) and describes the reflectivities in Figs. 4 and 5 with adequate accuracy once the reflectivity has dropped below about 10^{-2}. The reflectivity profiles in Figs. 2 and 3 are for the situation in which H/D substitution was used to match the scattering length densities of the two bulk media on either side of the layer and then, since $\rho_0 = \rho_2(= \rho)$, the reflectivity from the uniform layer becomes

$$R = \frac{16\pi^2}{\kappa^4} \left[4(\rho_1 - \rho)^2 \sin^2\left(\frac{\kappa d}{2}\right) \right] \tag{14}$$

which is the equation for a set of regular interference fringes of amplitude $(\rho_1 - \rho)^2$ superimposed on a signal that decays rapidly as κ increases, just as shown in Figs. 2 and 3. Because the reflectivity in this situation is always low, the kinematic equation (14) is accurate. As discussed in the context of Figs. 2 and 3, the thickness of the layer is obtained directly from the position of the fringes without any reference to $(\rho_1 - \rho)$; i.e., the thickness is obtained independently of composition. As also discussed in the previous section, it is possible to derive the surface coverage independently of the thickness of the layer by a suitable extrapolation of the reflectivity to $\kappa = 0$. This can be demonstrated as follows. We rearrange Eq. (14)

$$\kappa^2 R = 16\pi^2 d^2 (\rho_1 - \rho)^2 \left[\sin^2\left(\frac{\kappa d}{2}\right) \Big/ \left(\frac{\kappa d}{2}\right)^2 \right] \tag{15}$$

which tends to a limiting value of

$$\lim_{\kappa \to 0} \kappa^2 R = 16\pi^2 d^2 (\rho_1 - \rho)^2 \tag{16}$$

At the air-water interface when null reflecting water (NRW) is used $\rho = 0$ and

$$\lim_{\kappa \to 0} k^2 R = 16\pi^2 d^2 b^2 n^2 = 16\pi^2 b^2 \Gamma^2 \qquad (17)$$

where b and n are the scattering length and number density of the whole molecule (or fragment) and Γ is the surface coverage. This has been shown to be a general result for the case in which the two bulk phases on either side of the monolayer have identical scattering length densities (7).

A. Neutron Reflection from Lysozyme at the Air-Water Interface

There are two main situations in reflection in which a large signal is obtained from an interfacial layer. One is the obvious one when the two bulk phases are matched and the signal is only from the interfacial layer [Eq. (14)]. In this case the signal is maximized by making the scattering length density of the layer as different as possible from those of the surrounding bulk phases. The second, which can be equally effective, is to have a large difference between the scattering length densities of the two bulk phases and to adjust the scattering length density of the layer so that it is close to the average of that of the two bulk phases. Examination of Eq. (13) shows that the maximum amplitude of the interference fringes is obtained when the scattering length density, ρ_1, is exactly $(\rho_2 + \rho)/2$. We start with the first situation, which can be realized only in neutron reflection.

At the air-liquid interface the bulk phases are matched by adjusting the H/D composition of the water (about 10% D_2O) so that it has the same zero scattering length density as air. Apart from the incoherent background, which may easily be subtracted, the reflected signal is then entirely from the adsorbed layer and the technique is truly surface specific. The optimum signal in this situation is obtained by using deuteriated interfacial material. This is easily done for phospholipids, but other biological materials are difficult or impossible to deuteriate and an adsorbed protonated layer gives rather a small signal. However, the dimensions of biological materials are often large enough that, even with the relatively low scattering length density of the protonated material, the extra thickness of the layer can give an adequate signal. This is the situation for many proteins, and Dickinson et al. (9) have used neutron reflection to study the adsorption of β-casein at the air-water interface and Lu et al. (10) have studied lysozyme adsorbed at the same interface. Figure 7 shows that the signal from lysozyme is indeed adequate and Lu et al. were able to show from the thickness of the adsorbed lysozyme layer that it retains its globular structure, i.e., it is not denatured at the surface, and that it orients with the most hydrophobic domains projecting toward the air. They were also able to follow changes in the orientation of the molecule with surface coverage. Thus, at the lowest concentration studied, the thickness of the lysozyme layer is found to be 3.0 ± 0.3 nm, to be compared with the crystalline dimensions of lysozyme of $3.0 \times 3.0 \times 4.5$ nm^3 (11), suggesting that the lysozyme molecules adopt a sideways-on orientation at this concentration. An increase in the bulk concentration from 9×10^{-4} to 3×10^{-2} g dm^{-3} results in an increase in the adsorbed amount, but the protein molecules still retain their sideways-on orientation. The measurements therefore suggest that deformation of the globular structure is negligible at these surface concentrations, as might be expected from the known robustness of the globular structure of lysozyme. A further

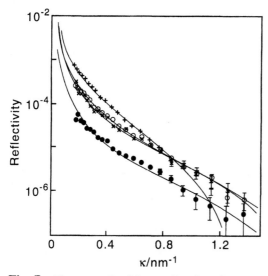

Fig. 7 Neutron reflectivity profiles from lysozyme adsorbed on the surface of null reflecting water at (●) 9×10^{-4}, (×) 0.03, (○) 0.1, and (+) $1.0\,\mathrm{g\,dm^{-3}}$. The solution pH was 7 and the buffer ionic strength 0.02 M. The continuous lines were calculated using the optical matrix formalism and the fitted surface excesses and thicknesses are given in Table 1.

increase of bulk concentration to $0.1\,\mathrm{g\,dm^{-3}}$ leads to an increase of the thickness of the protein layer to 3.4 nm with an accompanying decrease in the area per protein molecule to $14.0 \pm 1.0\,\mathrm{nm^2}$, which is close to the minimum area of $13.5\,\mathrm{nm^2}$ needed to accommodate sideways-on lysozyme molecules. A final increase of bulk lysozyme concentration to $1\,\mathrm{g\,dm^{-3}}$ produces a layer 4.7 nm thick, which is close to the axial length of 4.5 nm in the bulk form, suggesting that the molecules have aligned with the long axis normal to the interface at this high concentration and that the earlier intermediate value of the thickness was caused by a slight tilt resulting from geometric constraints or electrostatic repulsion within the layer, or a combination of these factors. Apart from the intrinsic dimensions of the lysozyme, there will always be a small contribution to the thickness from thermal motion (capillary waves) at the air-water interface. The area per molecule at this highest concentration was found to be $9.5 \pm 0.5\,\mathrm{nm^2}$, which is close to the theoretical limit for accommodating the longways-on packing. These structural parameters are summarized in Table 1.

Such measurements give no information about the position of the protein with respect to water. The second method of obtaining a good signal from an adsorbed

Table 1 Structural Parameters for the Lysozyme Layer Adsorbed on the Surface of Water Obtained from the Measurements in Null Reflecting Water

Concentration (g dm^{-3})	A (nm^2)	$\Gamma \pm 0.2$ (mg m^{-2})	$\tau \pm 0.3$ (nm)	$10^4\,\rho$ (nm^{-2})
9×10^{-4}	30.0 ± 5.0	0.8	3.0	0.40
3×10^{-2}	16.6 ± 2.0	1.4	3.0	0.72
0.1	14.1 ± 1.0	1.6	3.3	0.77
1.0	9.5 ± 1.0	2.4	4.7	0.80

layer is to have a layer with a scattering length density intermediate between two bulk phases of very different scattering length densities. The optimum situation is almost what is realized when a protein is adsorbed at the air-D_2O interface. Thus, lysozyme has a scattering length density of $2 \times 10^{-4}\,nm^{-2}$ in its protonated form. Depending on the extent of exchange of its labile protons, this increases up to a possible maximum value of $3.66 \times 10^{-4}\,nm^{-2}$ in D_2O. Thus, if the lysozyme formed a layer just above the D_2O, i.e., none of it was immersed, a broad interference fringe would be observed because the optimum condition of Eq. (13) is close to being satisfied. The calculated profile for this position of the lysozyme is shown as a dashed line in Fig. 8, where the lysozyme orientation has been chosen to satisfy the observed thickness of Fig. 7. The scattering length density of this layer changes if the lysozyme is completely immersed in the D_2O because it becomes the average of that of the lysozyme and that of the D_2O and the calculated profile is shown for this case as a dash-dotted line in Fig. 8b [it is so similar to the dashed line in (a) that it is not shown]. To calculate this profile, some assumptions have to be made about the molar volume of the lysozyme in order to calculate the amount of D_2O in the layer. The true position of the lysozyme in the layer will be intermediate between the two extremes giving two layers of different scattering length densities and the experimental results are compared with the calculated profiles (continuous lines) in Fig. 8. Using this two-layer model, Lu et al. were able to show that over the whole concentration range both the fraction of protein and the total amount immersed in water increase steadily with surface concentration, most of the thickness of the layer being accounted for by the immersed layer of protein. Thus, at a bulk concentration of $9 \times 10^{-4}\,g\,dm^{-3}$ the protein is about 50% immersed, and this increases to about 75% at a bulk concentration of $1\,g\,dm^{-3}$. Table 2 summarizes the results of this analysis.

Figure 9 shows two views of lysozyme with the charged residues at pH 7 labeled. The region of the protein with the lowest charge density is sited toward the N-terminus of the protein and is opposite to the cleft at the active site. This should be the most hydrophobic region of the protein and therefore the part that protrudes out of the water. Taking into account the observed variation of thickness and immersion of the layer with concentration, the two orientations shown in Fig. 9 probably represent the mean orientation of the protein at high and low coverages. The region of least charge density is indicated approximately by shading, but it can be seen that any significant protrusion of the protein into the air will ensure that a small number of

Table 2 Structural Parameters for the Lysozyme Layer Adsorbed on the Surface of Water Obtained from the Combined Fittings to the Measurements in D_2O and Null Reflecting Water

c (g dm^{-3})	Water	A (nm^2)	$\tau_1 \pm 0.3$ (nm)	$10^4\,\rho_1$ (nm^{-2})	$\tau_2 \pm 0.2$ (nm)	$10^4\,\rho_2$ (nm^{-2})	f[a]
9×10^{-4}	D_2O	30.0	1.5	0.68	1.5	5.9	0.5 ± 0.2
3×10^{-2}	D_2O	16.6	0.9	1.21	2.1	5.5	0.7 ± 0.1
0.1	D_2O	14.1	0.9	1.30	2.4	5.6	0.76 ± 0.1
1.0	D_2O	9.5	1.1	1.36	3.6	5.6	0.76 ± 0.1

[a] f denotes the fraction of lysozyme immersed in the water.

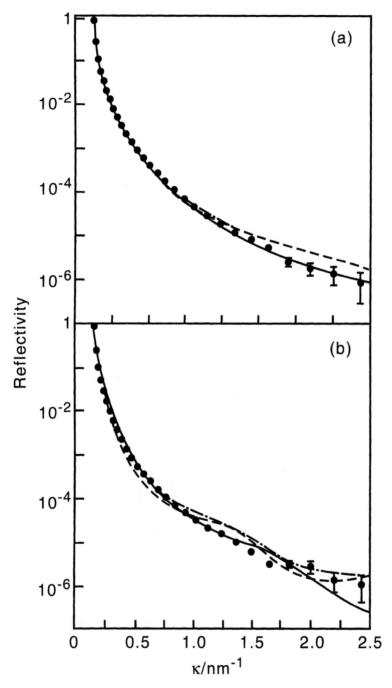

Fig. 8 Reflectivity profiles measured (points) from lysozyme layers adsorbed on the surface of D_2O at 9×10^{-4} g dm^{-3} (a) and 1 g dm^{-3} (b). The continuous line in (a) was calculated for a layer protruding out of the water of 1.5 nm thickness and that immersed in water also 1.5 nm. The continuous line in (b) assumes the protruding layer to be 1.1 nm with an immersed layer of 3.6 nm. The dashed lines assume the whole lysozyme layer to be out of water [3.0 nm in (a) and 4.7 nm in (b)], and the dash-dotted line in (b) assumes the lysozyme layer to be just fully immersed.

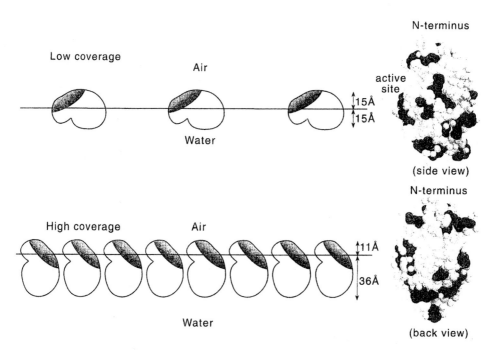

Fig. 9 A space-filling view of the lysozyme molecule showing the charge distribution (dark shading indicates the whole of a charged residue) and surface arrangements of adsorbed lysozyme consistent with maintaining the observed structural parameters while minimizing the number of charged groups exposed to air. In the schematic surface layer, the region of the protein estimated to have the lowest charge density is shaded.

charged groups will be out of the water, especially at the lower coverage. This is somewhat surprising, even though these charged groups will be in the form of ion pairs. However, in surface-active diblock copolymers where one block is a polyelectrolyte, it has been found that a much higher proportion of the charged groups lie out of the water (12–14). Whether or not these charged groups are hydrated is less easy to detect, and we cannot rule out the possibility that they are hydrated to a limited extent. The gradual shift from sideways-on to longways-on adsorption of the protein is driven by the increasing surface concentration, which must be accompanied by an increase in the lateral electrostatic repulsion. Normally, as the surface concentration of a surface-active species increases, the extent of immersion in the water subphase decreases. That the lysozyme layer changes in the opposite direction must reflect the change in the balance of hydrophobic and charged group–water interactions as the orientation of the molecule changes.

The use of variation of isotopic composition raises two important issues. The first is the extent to which isotopic substitution may change the structure of a biological system. If a structure changes significantly, it would clearly be invalid to make use of a comparisons of structural features obtained at different isotopic compositions. In the particular case of lysozyme the robust globular structure would be expected to be altered only in very minor ways by the change to D_2O and, bearing in mind that reflectivity gives only relatively low-resolution structural information, any such changes probably fall within the errors of the experiment. This might not be the case with proteins that are denatured at a surface, and then care has to be taken to ensure that adequate allowance has been made for isotope-dependent structural changes. Denaturation is more common at the solid-liquid interface and fortunately it is possible to apply more checks at this interface, as will be discussed in the following. The most obvious way to circumvent the problem of isotope effects, which has been used on a combined phospholipid-protein system at the air-water interface, is to combine neutron and X-ray reflectivity measurements on a single isotopic species (15). The combination that would be most effective in such a combined measurement is to have a deuterated outer layer (16). No one has yet used X-ray reflection to examine proteins adsorbed on their own at the air-water interface.

The second issue is the sensitivity of the derived structure to assumptions about the extent of isotope exchange. In the brief discussion given in an earlier section, we showed that the uncertainty in the determination of the surface coverage using null reflecting water is no greater than that arising from other experimental errors, i.e., about 5%, and the exchange has no effect on the derived thickness of the protein layer. Where isotope exchange is more significant is in the determination of the extent of immersion of the protein in D_2O. Thus, for lysozyme at a bulk concentration of $1\,g\,dm^{-3}$ the air-water interface consists of a lysozyme layer protruding out of the water with thickness 1.1 nm and scattering length density $1.36 \times 10^{-4}\,nm^{-2}$ and an immersed layer of thickness 3.6 nm and scattering length density $5.6 \times 10^{-4}\,nm^{-2}$. The amount of protein and the total thickness are fixed by the measurement in null reflecting water. In fitting the D_2O data there are then two adjustable parameters, the extent of exchange and the division between the two layers. The preceding values were calculated assuming complete exchange with D_2O. Taking the more pessimistic view that exchange might be only 50% complete, the scattering length densities of the two layers would be 1.03×10^{-4} and $5.35 \times 10^{-4}\,nm^{-2}$. With these scattering lengths the best fit to the data requires the dividing surface to be moved so that the two layer thicknesses are 0.9 and 3.8 nm and the degree of immersion of the layer would change from 75 to 80%. This is one of the situations in which the results are most sensitive to isotopic substitution and yet the error is only slightly larger than those arising from other errors in the experiment.

B. Neutron Reflection from Proteins at the Water–Hydrophilic Solid Interface

At the solid-liquid interface the situation regarding choice of contrast is changed because the scattering length densities of solids are generally nonzero. The two solids most suitable for neutron reflection experiments are silicon ($\rho = 2.1 \times 10^{-4}\,nm^{-2}$) and quartz ($\rho = 4.2 \times 10^{-4}\,nm^{-2}$ for crystalline quartz). In the situation where the signal is predominantly from the protein layer only, i.e., the H/D ratio of the water has been adjusted to match the solid (comparable to the situation in Fig. 3), the signal from an adsorbed protein layer will be weak because the layer already contains a significant fraction of water, which gives no signal. Nevertheless, there is a large difference in the sensitivity to the layer in the two contrast matching situations for silicon and for quartz. When the contrast of water is adjusted to be between 2×10^{-4} and $4 \times 10^{-4}\,nm^{-2}$, the scattering length density for most proteins is around $2.5 \times 10^{-4}\,nm^{-2}$. Because the protein layer contains a significant fraction of water, the scattering length density of the hydrated protein layer will be the appropriate average between water and protein. In the case of the silicon substrate the resultant scattering length density of the layer is between 2.1×10^{-4} and $2.5 \times 10^{-4}\,nm^{-2}$ and this is too low a contrast to be easily detected. The sensitivity is further reduced because the oxide layer on the silicon cannot be neglected and, indeed, the dimensions of the two layers contributing to the signal are such that in practice the oxide layer dominates the reflectivity. The effects of the adsorbed protein layer using silicon and quartz are compared in Fig. 10a, where a realistic oxide layer has been included for the silicon case. The situation in quartz can be seen to be much better (Fig. 10b), partly because the situation is closer to a perfect contrast match in the absence of protein and partly because of the added contrast between the layer and the two bulk phases. The structural parameters used for the calculations in Fig. 10 were chosen from known parameters for lysozyme layers (17), although no original measurements at this contrast are yet available for quartz.

The alternative method of obtaining a good signal from the protein layer, i.e., generating a high contrast between the two bulk phases with the protein layer at an intermediate level of scattering length density, has been found to be the more effective method for studying proteins at the silica-water interface. Figure 11 shows the neutron reflectivity of a lysozyme layer adsorbed on silica-silicon from D_2O, and the same layer when the deuterium content of the water has been adjusted to give water scattering length densities of 4.0×10^{-4} and $2.9 \times 10^{-4}\,nm^{-2}$ (17). In each case the adsorbed layer has a pronounced effect on the reflectivity. This makes it possible to follow with some accuracy the effects of parameters such as concentration, pH, and ionic strength on both surface coverage and layer thickness.

In terms of surface coverage, proteins are often found to adsorb irreversibly from aqueous solution at a hydrophilic surface; i.e., there is hysteresis in the adsorption isotherm (18). Hysteresis in the amount adsorbed is probably related to the number of contacts between protein and surface. Although the fraction of segments in contact with the surface may be small and typically less than a few percent, the adsorption energy can easily be in excess of $100\,kJ\,mol^{-1}$ because of the large total number of contacts. This is not easily overcome by the weak driving force of a concentration gradient. Thus, irreversible adsorption is not necessarily associated with denaturation of the protein. Change of pH may have a large enough effect on the interaction of protein with the surface to provide a driving force for desorption and should therefore be a more thorough test of irreversible adsorption, although it has seldom been used in this manner. This is particularly the case when the protein in bulk solution is stable with respect to pH change.

Lysozyme has a well-defined equilibrium structure in bulk solution within a significant pH range (19). The effects of pH on lysozyme adsorption at a fixed protein concentration of $0.03\,g\,dm^{-3}$ are shown in Fig. 12 (20). The measurements were made by varying the solution pH in two cycles. In the first cycle the pH started at 4, increased to 7, then 8, and back to 4, and Fig. 12 shows the reflectivity profiles in D_2O measured at these four values of the pH in this cycle. The close resemblance of the two reflectivity curves at pH 4 shows that adsorption is reversible with respect to this particular pattern of pH variation. Figure 12 also shows quite clearly that the adsorbed amount increases at pH 7 and reaches its highest value at pH 8. The effects of the reversed cycle, i.e., starting at pH 7, moving down to 4, and returning to 7, are similar, again showing that adsorption is completely reversible and that the adsorbed amount at pH 4 is less than that at pH 7. Neutron reflectivity therefore shows that the adsorption depends only on the actual pH and not on the route to a given pH. The reversibility with respect

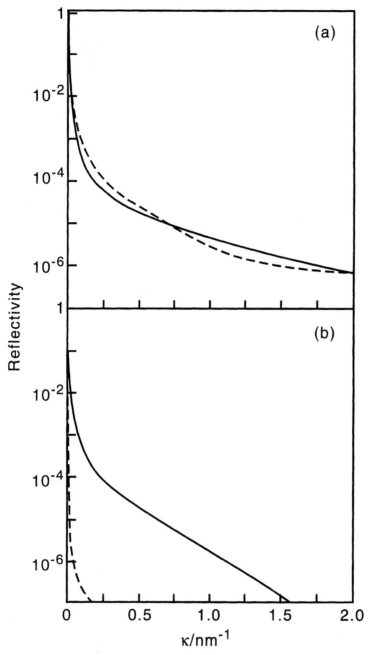

Fig. 10 Comparison of the reflectivity signals for the adsorption of a 3.0-nm layer of protein (volume fraction = 0.4) onto (a) a silica layer on silicon in water of the same scattering length density as silicon ($\rho = 2.07 \times 10^{-4}\,\text{nm}^{-2}$) and (b) a crystal quartz surface in water contrast matched to quartz ($\rho = 4.2 \times 10^{-4}\,\text{nm}^{-2}$). The dashed line in (a) was calculated for the bare silica-water interface assuming that typical parameters for the oxide layer, i.e., a thickness of 1.2 nm. The dashed line in (b) was calculated assuming an interfacial roughness of 1.2 nm, although the roughness has little effect on the level of the reflectivity in the κ range of the calculation.

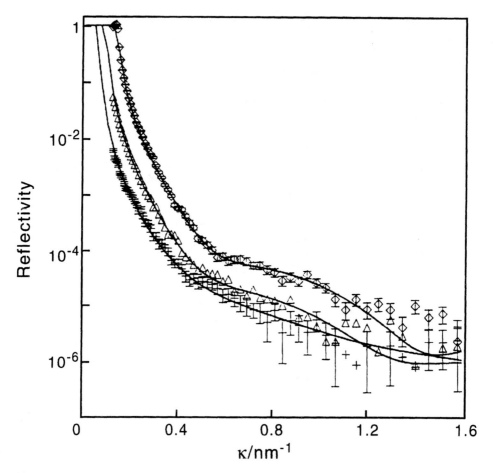

Fig. 11 The fits of a two-layer model to reflectivity profiles at pH 7 in the presence of 1.0 g dm^{-3} lysozyme: the scattering length densities of the water are ($\diamondsuit$) 6.35 × 10^{-4} nm^2 (D$_2$O), ($\triangle$) 4.00×10^{-4} nm^2, (+) 2.90×10^{-4} nm^2. The continuous lines were calculated using $\Gamma = 3.6$ mg m^{-2}, $\tau_1 = 3.0$ nm, and $\tau_2 = 3.0$ nm.

to pH suggests that no denaturation of lysozyme occurs at this interface. It is interesting to note that the surface excess at pH 8 is 1.9 mg m^{-2}, which is almost twice the value at pH 4, showing that pH has a strong effect on lysozyme adsorption at the silica surface. However, the decrease of surface excess with pH is opposite to what one might expect just from electrostatic considerations. The negative charge density on the silica surface is approximately constant between pH 4 and 8 (21). The isoelectric point for lysozyme is at a pH of about 11 (22) and the net positive charge on lysozyme increases from 8 at pH 8 to 10 at pH 4 (23,24), hence the electrostatic attraction between lysozyme and surface should increase with decreasing pH. That the coverage does not increase as the pH is lowered then suggests that the electrostatic attraction between surface and lysozyme is less important than the lateral electrostatic repulsion between protein molecules within the layer; i.e., the pH dependence of lysozyme adsorption is governed more by protein-protein interaction than by protein-surface interaction.

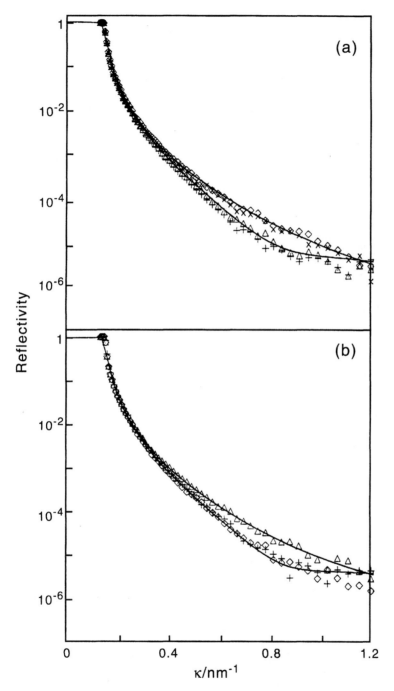

Fig. 12 The variation of neutron reflectivity from a lysozyme layer adsorbed on silica from D_2O at a bulk concentration of $0.03\,g\,dm^{-3}$ with pH: (a) the solution pH was initially 4 ($\diamond$), then raised to 7 ($\triangle$), followed by 8 ($+$), before returning to 4 ($\times$), and (b) the pH was initially 7 ($\diamond$), lowered to 4 ($\triangle$), and finally returned to 7 ($+$). The continuous lines are single uniform layer fits to the measured reflectivities at pH 4 and pH 7. The identical reflectivity profiles obtained at a given pH regardless of the route show that adsorption is reversible.

   That adsorption is reversible with respect to pH does not necessarily mean
that the structure of the protein is not affected by adsorption. Direct contacts
between protein and solid surface may lead to partial breakdown of fragments
of α-helix or β-sheet, which may generate further contacts with the surface. However,
the native state of lysozyme in aqueous solution is highly ordered with most of the
polypeptide backbone having little or no rotational freedom and, although structural rearrangement may occur upon adsorption, the internal coherence of the
lysozyme should prevent it from unfolding into loose random structures on the surface (19,23). Just as at the air-liquid interface, the thicknesses of adsorbed layers
of reversibly adsorbed globular proteins should therefore be comparable to their
dimensions in aqueous solution and measurements of the layer thickness can then
be used to assess the orientation of the protein molecules on the surface and
whether they are distorted to any extent. Claesson et al. (24) have, for example,
made this assumption in their determination of the conformation of lysozyme
at the solid-water interface using a combination of surface force apparatus and
ellipsometry. The advantage of neutron reflection is that it is more sensitive than
other techniques to the structural distribution normal to the interface. Figure 13

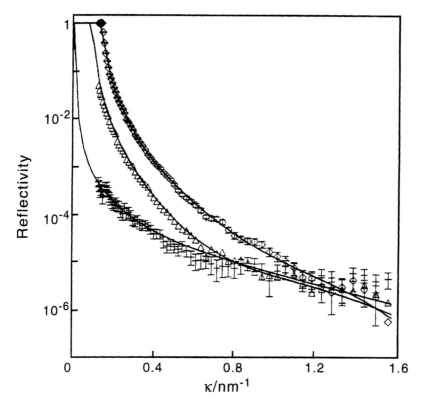

Fig. 13   Single uniform layer fits to the neutron reflectivity profiles from lysozyme adsorbed
on silica at pH 4 at a bulk concentration of $0.03\,g\,dm^{-3}$ lysozyme. The profiles are measured
in water of scattering length density (◇) $6.35 \times 10^{-4}\,nm^2$ (D_2O), (△) $4.00 \times 10^{-4}\,nm^2$, (+)
$2.07 \times 10^{-4}\,nm^2$ (matched to silicon). The continuous lines were calculated using a surface
excess of $0.85\,mg\,m^{-2}$ and a thickness of $3.0\,nm$.

shows reflectivity measurements from a $0.03 \, g \, dm^{-3}$ lysozyme solution at pH 7 and at different water contrasts. The fits of the set of reflectivity profiles to a single uniform layer model for the protein layer, after allowance has been made for the exchange of labile hydrogen atoms, are shown as continuous lines. The thickness of the layer at pH 7 was found to be $3.0 \pm 0.3 \, nm$, suggesting that the protein molecules are adsorbed sideways on (long axis parallel to the surface) at this concentration. Similar measurements at different water contrasts at pH 4 also show that the thickness of the layer remains $3.0 \, nm$ at this pH, although the amount of protein in the layer is less, again suggesting sideways-on adsorption (20).

Claesson et al. (24) have suggested that adsorption should change from sideways on at very low lysozyme concentration to end on (long axis normal to the surface) at higher concentrations, arguing that the changeover enables the surface to accommodate more protein molecules. As already shown, this does indeed happen at the air-water interface. However, an end-on monolayer would give a monolayer thickness of $4.5 \, nm$, and no such change was observed in the neutron reflectivity from lysozyme at the silica-water interface at the higher coverage at pH 7. As can be deduced from Figs. 2 and 3, the sensitivity of neutron reflection to the difference between 3.0 and $4.5 \, nm$ is large and, even when the surface excess is allowed to vary in order to obtain the best possible match between calculated and observed reflectivities, the fit for a thickness of $4.5 \, nm$ is quite unacceptable.

The variation of the structure of the adsorbed lysozyme layer with pH was also investigated by neutron reflection at the higher concentration of $1 \, g \, dm^{-3}$. At pH 7 a single uniform layer model does not fit the reflectivity profiles at all satisfactorily in the high-κ range. A two-layer model with a volume fraction of protein of 0.65 in the inner $3.0 \, nm$ layer and a fraction of 0.35 in the outer $3.0 \, nm$ gives the best fit to the data, which suggests that the layer consists of two protein layers each with side-on molecules, not the end-on adsorption proposed from the ellipsometric measurements (24). The high-coverage data at pH 4 give a thickness of $3.0 \, nm$ for the inner layer and $1.0 \, nm$ for the outer layer, suggesting a combination of end-on and side-on orientations, with the fraction of end-on molecules being 0.37 ± 0.05. These observations indicate that lateral electrostatic repulsion within the adsorbed protein layer determines not only the level of adsorption but also the structural orientation of the molecules inside the layer. The exact location of the charge in the protein is probably an important factor in determining any tilting of the protein molecules in the layer.

The effects of pH on lysozyme adsorption on the hydrophilic surface are summarized schematically in Fig. 14. The surface coverage decreases with pH as a result of increased repulsion between the molecules inside the monolayer, and this increased level of repulsion is also reflected in the increased percentage of end-on orientations. At higher bulk concentrations, adsorption produces a side-on bilayer at pH 7 and the measurements suggest that there is less lysozyme in the outer layer. Only monolayer adsorption occurs at pH 4 because of the strong electrostatic repulsion within the protein layer, but the higher coverage induced by higher bulk lysozyme concentration leads to a higher fraction of molecules adopting the end-on orientation.

The adsorption of bovine serum albumin (BSA) at the hydrophilic silica-water interface has similarly been investigated using specular neutron reflection and was also found to be reversible with respect to pH, indicating that there is no denaturation

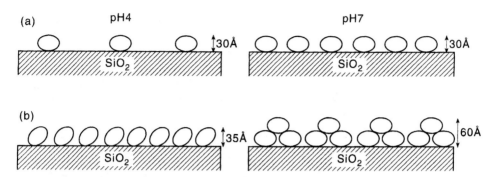

Fig. 14 Schematic diagram of the surface coverage and orientation of lysozyme molecules adsorbed at the silica-water interface at bulk solution concentrations of (a) $0.03 \, g \, dm^{-3}$ and (b) $1 \, g \, dm^{-3}$.

(25). This was confirmed by being able to model the reflectivity using a single uniform layer model. Denaturation would lead to a more fragmented peptide distribution and hence layers of different density rather than a single uniform layer. Comparison of the layer thickness with the dimensions of the ellipsoidal structure of the globular solution structure again indicates that the molecules adsorb sideways on. Nevertheless, the layer thickness is always less than the relevant globular dimension of 4.0 nm (26), suggesting that adsorption onto the hydrophilic surface results in some structural deformation. An increase of layer thickness with bulk concentration also suggests that the extent of the distortion is reduced as the lateral repulsion between protein molecules increases.

For both the lysozyme and BSA systems the question of isotopic exchange is again an important consideration. Lysozyme has a total of about 260 labile hydrogens, of which about half are on the amino acid side chains and are expected to exchange completely with D_2O on the time scale of preparation of samples for the neutron experiment. The remainder are on the backbone and their exchange rates vary widely because of the possible encapsulation in the hydrophobic region of the protein. From nuclear magnetic resonance (NMR) data (27) it can be estimated that within about 20 min lysozyme will have exchanged all but about 50 of the 260 labile hydrogens. The remainder exchange over a period of days. In the calculation of scattering length density it has been assumed that exchange is complete, giving the value of scattering length density of $3.7 \times 10^{-4} \, nm^{-2}$ in D_2O as compared with $2.0 \times 10^{-4} \, nm^{-2}$ in H_2O. The uncertainty about the final 50 hydrogens leads to a maximum error of 9% and 3% in the scattering length densities of the protein. However, the reflectivity profiles are determined by the average scattering length densities of each layer, and because the layers contain appreciable fractions of water the error in the scattering length density drops still further to below 5%, often down to 2–3%, depending on the isotopic composition. Differences at this level are within the other uncertainties of the experiment, for example, in the estimated volume of the protein used in calculating the scattering length density. We have also made a direct test of the time scale of labile hydrogen exchange by measuring neutron reflectivity of lysozyme adsorbed at the hydrophilic silica interface from D_2O, the isotopic composition at which any effect is a maximum. One sample was left in D_2O for 4 days and the reflectivity of a second was determined within 20 min of preparation of the solution.

The two reflectivity profiles were found to be identical, indicating either that exchange was already complete within 20 min or that the remaining labile hydrogens take longer to exchange than 4 days or that the reflectivity is not sufficiently sensitive to isotopic change at this level. The important result is that the issue of uncertainties in the extent of isotopic exchange is not important in the interpretation of neutron reflection measurements.

The extent of exchange of the labile hydrogens in BSA with D_2O has also been extensively investigated (3) and the conclusions are similar. The only additional difference is that the distortion of the BSA structure on adsorption may facilitate complete exchange. It is known that folding and unfolding of protein molecules under different solution conditions enhance the rate of exchange of labile hydrogens. This is thought to be the mechanism behind the increase in exchange rate with changes in pH and temperature.

C. Neutron Reflection from Proteins at the Water–Hydrophobic Solid Interface

One of the interests in adsorbing proteins onto surfaces is to examine the way specific interactions with the surface modify the adsorbed layer. The easiest way to achieve this is to graft an organic layer onto either quartz or silicon and to confer on the outer part of this layer an appropriate chemical functionality. Although this may produce the right type of surface for studying the protein adsorption, it may be difficult to obtain an adequate signal from the protein. The presence of the additional layer means that the situation in which the signal is dominated by the protein, i.e., contrast match of the two bulk phases, cannot easily be realized. Thus Fig. 15a shows the calculated reflectivity from a typical protein layer adsorbed on an organic layer 18 carbon atoms thick grafted onto silicon with the scattering length density of the water matched to silicon, and Fig. 15b shows the reflectivity from the same layer in the maximum contrast condition, i.e., D_2O as substrate. In both these situations the effect of the protein layer is small and this impairs the resolution of any structural determination. Such an experiment has been done by Liebmann-Vinson et al. (28,29) on human serum albumin, and although their signal from the adsorbed protein was slightly larger than shown in Fig. 15b, it was far from what one might wish.

In the experiment of Liebmann-Vinson et al. protonated material was grafted onto the silica to generate the surface for protein adsorption. The contrast situation changes dramatically if deuteriated material is used to coat the layer instead of protonated material. Figure 16a shows that with the grafted layer deuteriated, although the protein remains more or less invisible in the null situation in which the water is matched to the substrate, there is a very large contribution to the reflectivity from the protein in D_2O (Fig. 16b). Thus specific deuteriation of the grafted layer may be used to optimize the signal. This method has been used by Fragneto et al. (30) in a preliminary study of the structure of β-casein adsorbed at the hydrophobic solid–water interface. They obtained sufficient sensitivity to be able to show that the protein could be divided into two layers, one with a relatively high volume fraction next to the surface and a more diffuse outer layer, a typical result for a denatured protein layer.

The protein layer used to illustrate the contrast situations in Figs. 15 and 16 was taken to be a simple monolayer, but in many real protein layers the structure will be more complex, especially if there is any denaturation, and the question arises

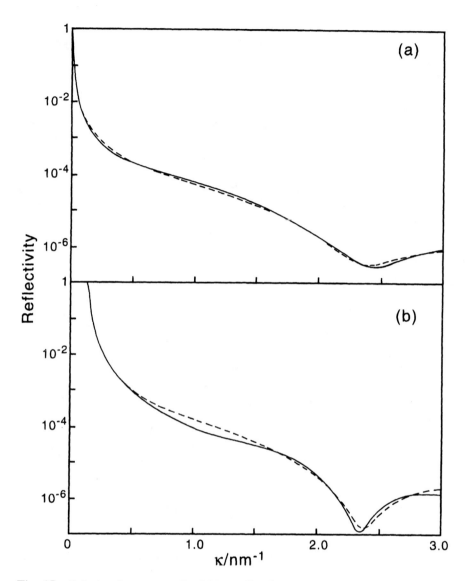

Fig. 15 Calculated neutron reflectivity profiles for a model protein layer adsorbed on a protonated self-assembled monolayer on silicon. In (a) the scattering length density of the water is matched to the silicon and in (b) it is D_2O. The protein layer is taken to be 3 nm thick with a volume fraction of 0.5 and complete exchange of the labile hydrogens with water is assumed. The self-assembled monolayer corresponds to an octadecyl trichlorosilane monolayer with scattering length density $-0.5 \times 10^{-4} \, \text{nm}^{-2}$ and thickness 2.7 nm. The silicon is assumed to have an oxide layer 1.2 nm thick. The dashed lines are the reflectivity profiles when no protein is adsorbed.

of what resolution could be expected in the reflection experiment and how it can be improved. The resolution of the neutron experiment is inherently low (of the order of 1 nm), but if contrast variation techniques are used this can be improved significantly (16). We now illustrate a strategy for improving the resolution in experiments

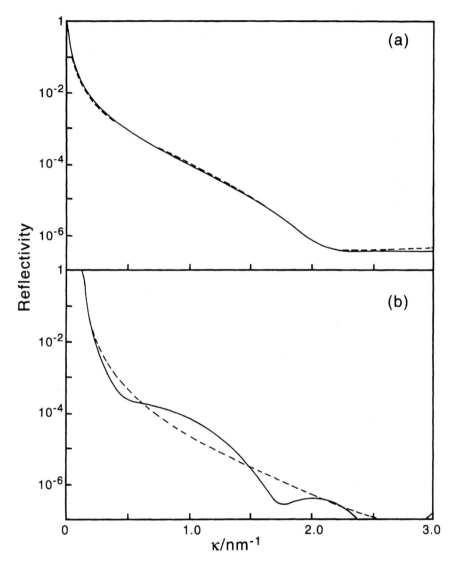

Fig. 16 Calculated neutron reflectively profiles for a model protein layer adsorbed on a deuterated self-assembled monolayer on silicon. In (a) the scattering length density of the water is matched to the silicon and in (b) it is D_2O. The protein layer is taken to be 3 nm thick with a volume fraction of 0.5 and complete exchange of the labile hydrogens with water is assumed. The self-assembled monolayer corresponds to an octadecyl trichlorosilane monolayer with scattering length density $6.4 \times 10^{-4} \, nm^{-2}$ and thickness 2.7 nm. The silicon is assumed to have an oxide layer 1.2 nm thick. The dashed lines are the reflectivity profiles when no protein is adsorbed.

on protein layers. It is clear from Fig. 16b that the presence of a fully deuterated grafted layer of hydrocarbon between the substrate and protein layer greatly enhances the signal from the protein. A further option is to deuterate only part of the grafted layer. The effect of deuteriating a C_{18} layer in progressively increasing fragments of C_6, C_{12},

C_{18} with the hydrogenated part always on the protein side of the interface is shown in Fig. 17a for the same single-layer protein. In each case large differences are observed in the signal, but if the protein forms a single layer the extra profiles would add no new information to that which could have been obtained from Fig. 16b. However,

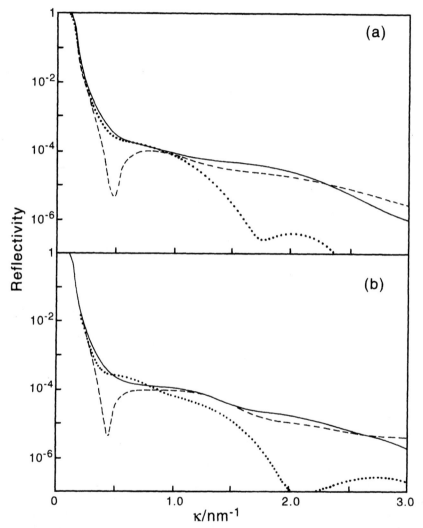

Fig. 17 Calculated neutron reflectivity profiles from a model protein layer adsorbed on silicon coated with partially deuterated self-assembled monolayers in D_2O. The layers consist of octadecyl trichlorosilane with the first 6 methylene groups next to the solid surface deuterated (continuous line), the first 12 methylene groups deuterated (dashed line), and all 18 methylene groups deuterated (dotted line). The hydrophobic layer is taken to be 2.7 nm thick. In (a) the protein layer is taken to be 3.0 nm thick with a protein volume fraction of 0.5, and in (b) the protein is divided into two layers, the first next to the hydrophobic surface 2.0 nm thick with volume fraction 0.8 and the second, outer, layer 4.0 nm thick of volume fraction 0.2. The protein is assumed to have exchanged all its labile hydrogens with the D_2O. The differing interference effects resulting from each layer are clearly observed in the profiles in (b).

in the case of a two-layer system, such as typically adopted by β-casein, the effect of changing the width of the deuterated label is to probe different Fourier components of the scattering length density profile of the protein; i.e., the extra experiments now give new information about the two layers. The model calculations are shown in Fig. 17b. Thus, by using more subtle, but still chemically accessible, labeling it becomes possible to enhance the ability of the experiment to resolve the different protein layers considerably. Given the range of grafting chemistry that is already well established [e.g., (31,32)], the number of possibilities for designing reflection experiments to answer specific problems about protein layers is greatly increased. We have given one option here but, of course, the labeling could be reversed to have the deuterated fragment on the outside of the grafted layer. Another contrast that we have found useful is to use water of scattering length density $4 \times 10^{-4}\,\mathrm{nm}^{-2}$ with a grafted deuterated layer, which, by lowering the critical angle for total reflection, gives access to reflectivities at lower values of κ, which contain information about any longer dimensions in a protein layer.

We illustrate some of the features of the preceding method in a study of lysozyme adsorption on a hydrophobic surface prepared by self-assembly of a monolayer of octadecyltrichlorosilane (OTS) (33). For this experiment the hydrophobic layer was prepared from partially deuterated OTS, $\mathrm{hC_6dC_{12}SiCl_3}$. The effect of pH on lysozyme adsorption at the hydrophobic-water interface was investigated in a manner similar to that described in the previous section for the hydrophilic-water interface. Thus, the lysozyme concentration was fixed at $0.03\,\mathrm{g\,dm}^{-3}$ with a constant ionic strength of $0.02\,\mathrm{M}$ and the layer structure studied as a function of pH. The first part of the experiment established that adsorption of lysozyme on this surface is irreversible with respect to concentration and pH, in contrast to its behavior at the bare silica surface. Thus, when the pH was varied in two opposite directions, from 7 to 4 and back to 7, and from 4 to 7, different reflectivity profiles were obtained, corresponding to different final states depending on the path taken.

Figure 18a shows the reflectivity profiles in $\mathrm{D_2O}$ measured for the first three values of the pH with the profile for the partially deuterated OTS-$\mathrm{D_2O}$ surface included for comparison. The shift in the minimum toward lower κ in the reflectivity at the initial pH of 7 from that for OTS on its own indicates that there is adsorption of lysozyme onto the hydrophobic surface and a further shift in reflectivity upon the first change in pH from 7 to 4 suggests an increase in the adsorbed amount of lysozyme. The change in pH from 4 back to 7 has little further effect on the adsorbed amount, suggesting that the adsorbed protein layer is no longer able to respond to a change in pH. Thus the protein has been irreversibly adsorbed to a saturation limit. Two of the reflectivity profiles for the reverse cycle of pH change are shown in Fig. 18b. Although there is some response of the adsorbed layer upon pH change from 4 to 7, the level of the change is less than for the first pH change in the opposite direction shown in Fig. 18a. The small difference in this case is caused by either adsorption, desorption, or structural rearrangement, and the exact structural change upon pH shift can be obtained by modeling the surface layer distribution. The reflectivity profiles at the initial pH 7 and at the initial pH 4 are similar, which suggests that the initial adsorption is largely driven by the hydrophobic surface and is indifferent to solution pH. Although in some cases the uniform layer model does not fit the intensity at the minimum in the reflectivity profile exactly, it does reproduce the position of the minimum accurately. In general, the uniform layer model usually offers an accurate

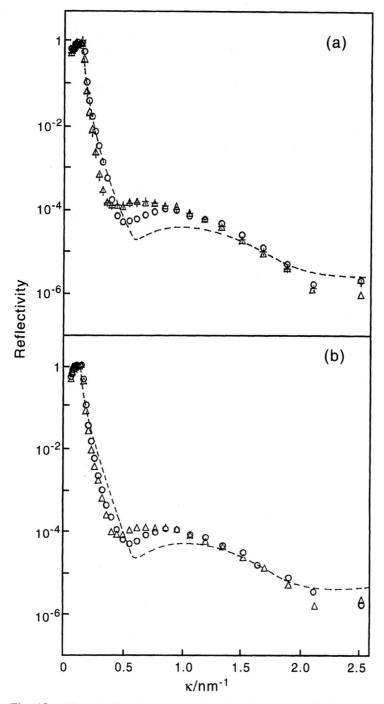

Fig. 18   Effect of pH on the neutron reflectivity from a partially deuterated OTS-D$_2$O interface in the presence of lysozyme at a bulk concentration of 0.03 g dm^{-3}. In (a) the solution pH was initially 7 (○), then lowered to 4 (△), and finally back to 7 (+); in (b) the pH was initially 4 (○), then increased to 7 (△). The reflectivity from the OTS-D$_2$O interface in the absence of protein is included as a dashed line.

estimate of the surface excess, even when it does not account too well for the detailed shape of the reflectivity. The conclusion from the monolayer fit is that the thickness of the layer is between 1.1 and 1.5 ± 0.3 nm for all the pH conditions studied.

The strong hydrophobic interaction between fragment and surface may lead to breakdown of the globular assembly. If the denatured protein completely unfolds, there will be regions of loose random structure on the surface. Depending on the extent of denaturation, hydrophobic fragments will be segregated at the OTS surface, allowing hydrophilic fragments to extend into the aqueous solution. Information regarding the state of the adsorbed lysozyme can be obtained by examining the structural distribution of the protein layer on the OTS surface. The work on the adsorption of lysozyme at the hydrophilic surface showed that in all circumstances the thickness of the layer could be related to the known dimensions of the protein in solution, and this was taken to be evidence for the tertiary structure of the protein remaining essentially intact on adsorption. However, for the hydrophobic surface no model of the interfacial layer where the tertiary structure of the lysozyme was retained could be made to fit the observed profiles. This, even without more detailed fits, further confirms that lysozyme is denatured at the hydrophobic OTS surface.

Once the constraint of using the dimension of the unperturbed globular protein was discarded, it was found to be relatively straightforward to fit the reflection data to a single uniform layer model but with a thickness that best fits the data. The calculated reflectivities for the three water contrasts at the initial pH of 7, shown as continuous lines in Fig. 19, are for a layer thickness of 1.0 ± 0.3 nm and a protein volume fraction of 0.85 ± 0.1. Note that an added advantage of the labeled hydrophobic layer is that the composite thickness of protein and hydrophobic layer gives interference features in an accessible range of momentum transfer. A layer only 1 nm thick would normally be too thin to measure with any accuracy. The thickness of 1.0 nm is equivalent to the length of the side chains of two average amino acids. This suggests that the adsorbed protein layer is in the form of peptide chains with the hydrophobic side chains adsorbed on the surface of OTS and the hydrophilic side chains extending into aqueous solution. That this structural model fits three separate water contrasts is a good indication of its correctness. Once again, in calculating the scattering length density of the protein we have assumed that the labile hydrogens on all the amide groups exchange with the surrounding water. This is likely to be correct because denaturation has led to the breakdown of the globular structure and all the labile hydrogens are in direct contact with water.

Evidence that part of the lysozyme can extend into the water comes from a more careful examination of the data at low momentum transfer and the correlation of this with pH. Because the scattering length density distribution and reflectivity profile are related by a Fourier transform [Eqs. (11) and (12)] any distribution of the protein that extends into the solution will produce its effect at low momentum transfer and it is this region that is not fitted very well by the thin uniform layer so far assumed. This is particularly the case at pH 4 and 7 following initial deposition of the protein. Figure 20 compares the fits of uniform and two-layer models for the data at different water contrasts. Bearing in mind that the reflectivity is on a logarithmic scale, there are significant differences in the fit to the steep portion of the reflectivity at values of κ below the destructive part of the interference fringe at about 0.5 nm^{-1}. The importance of these small, but measurable, differences between the two models illustrates the importance of maximizing the contrast between the protein layer and the two bulk

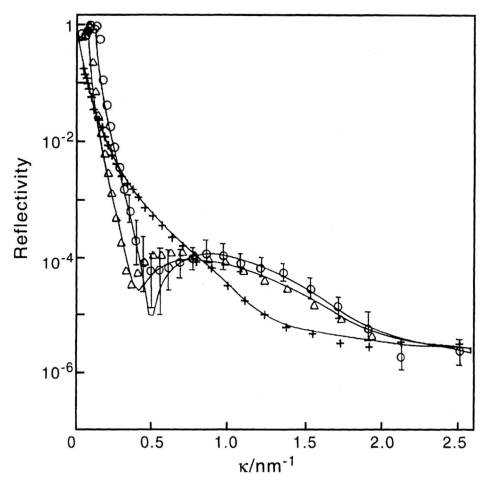

Fig. 19 The fits of a single uniform layer model to neutron reflectivity profiles from a hydrophobic surface at initial pH 7 in the presence of $0.03 \, \text{g dm}^{-3}$ lysozyme. The profiles are measured in water of scattering length density ($\Diamond$) $6.35 \times 10^{-4} \, \text{nm}^2$ (D_2O), ($\triangle$) $4.00 \times 10^{-4} \, \text{nm}_2$, and (+) $-0.58 \times 10^{-4} \, \text{nm}^2$ (H_2O). The continuous lines were calculated using $\Gamma = 1.4 \, \text{mg m}^{-2}$ and $\tau = 1.0 \pm 0.3 \, \text{nm}$.

phases. The contrast of Fig. 15 could not possibly reveal such differences. A summary of the best two-layer model fits to all the data is given in Table 3. For this table it can be seen that, although the coverages of lysozyme at pH 4 and 7 are the same, the outer layer is thicker at pH 4. This variation of the diffuse layer with solution pH suggests that the tail region contains charged groups. As described earlier, the net charge on a lysozyme molecule depends on solution pH and the number of charges decreases from 10 at pH 4 to 8 at pH 7 (23, 34). Thus, at pH 4, the charge density within the protein layer is higher and, once the protein is denatured, the greater freedom of movement will allow the stronger repulsion to cause the layer to be more diffuse. At pH 7, thus repulsion is reduced and the protein layer becomes more dense. The structural changes within this lysozyme layer are quite different from those on the hydrophilic surface and are further strong evidence that the protein is denatured.

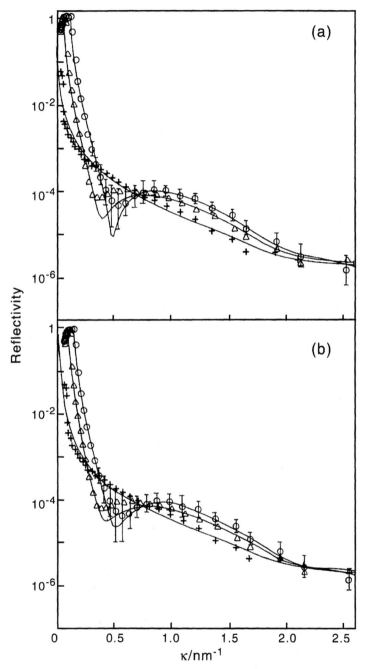

Fig. 20 The fits of (a) a single uniform layer model and (b) a two-layer model to neutron reflectivity profiles from a hydrophobic surface at an initial pH of 4 in the presence of 0.03 g dm^{-3} lysozyme. The profiles were measured in water of scattering length density ($\Diamond$) 6.35 × 10^{-4} nm$_2$ (D$_2$O), ($\triangle$) 4.00 × 10^{-4} nm^2, (+) 2.07 × 10^{-4} nm^2 (matched to silicon). The continuous lines were calculated using (a) $\Gamma = 1.6$ mg m^{-2} and $\tau = 1.0 \pm 0.3$ nm and (b) $\Gamma_1 = 1.6$ mg m^{-2}, $\tau_1 = 1.1 \pm 0.3$ nm, $\Gamma_2 = 0.4$ mg m^{-2} and $\tau_2 = 7.0 \pm 2.0$ nm. The final set of structural information is given in Table 3.

Table 3 The Effect of pH on the Structure of the Lysozyme Layer Adsorbed at the Hydrophobic OTS Surface Obtained from a Two-Layer Model

	7	4 (from 7 to 4)	4	7 (from 4 to 7)
Lysozyme layer 1 (± 0.2 nm)	10	15	10	13
Lysozyme layer 2 (± 2.0 nm)	40	70	60	40
$\phi_1 \pm 0.1$	0.93	0.97	0.93	0.95
$\phi_2 \pm 0.03$	0.04	0.08	0.04	0.08
Γ_1 (± 0.3 mg m^{-2})	1.4	2.2	1.6	1.9
Γ_2 (± 0.2 mg m^{-2})	0.25	0.8	0.45	0.5
Γ_{total} (0.2 mg m^{-2})	1.65	3.0	2.0	2.3

Although the two-layer model fits all the reflectivity profiles from lysozyme on the hydrophobic surface at different pH, it is a coarse model and one would expect the actual segment distribution profiles to be smoother, although the intrinsic resolution of the neutron reflection experiment is not great enough to distinguish the two-layer model of Table 3 from a continuous distribution. In previous work on the adsorption of polyethylene oxide (PEO) at the air-water interface, Lu et al. (35) have shown that an equivalent effect can be obtained by further subdivision of the two layers into a smooth distribution. Figure 21 shows volume fraction profiles that also fit the observed reflectivities and where the subdivision into six or seven layers is further smoothed by joining the midpoints of each block in the histogram. Not surprisingly, these protein distributions fit all the sets of data at least as well as the simpler structure. The diffuse region of such distributions is composed either of hydrophobic fragments that are forced into the diffuse layer because of steric constraints or of peptide chains that contain more polar or charged groups and that dislike the OTS surface. The response of the diffuse layer distribution to the change in solution pH suggests the latter. If polar and charged groups dominate the composition of the diffuse layer, the scattering length for the protein fragments in this region would be different from the average value (it is the average value that has been used in the fitting described before). In the case of D_2O the scattering length of the protein fragments in the diffuse layer will tend to be higher due to exchange of a greater number of labile hydrogens and the current treatment might then underestimate the amount of protein in the diffuse region. It is difficult to estimate such errors because they depend on precisely which component amino acids form the diffuse layer, which is not known. However, because the fraction of protein in the diffuse layer is in all cases less than 0.3 of the total surface excess, the uncertainty will not lead to serious errors in the values of the total surface composition.

There have been no directly comparable X-ray reflection studies of these systems, but an illustration of the potentially greater resolution of the X-ray experiment is given by the work of Petrash et al. (36) on the adsorption of human serum albumin (HSA) on a self-assembled monolayer of hexadecyl trichlorosilane on silica (silicon wafer). Figure 22 shows the clear fringes obtained from such a system and the changes in those fringes on adsorption of the protein. Clearly, the extra signal, extending out to a much higher momentum transfer (0.7 nm^{-1}) and much lower reflectivity (10^{-8} than any of the neutron reflectivity profiles so far shown, should enable a better determination of the layer structure to be made. The difficulty is that at present such experiments

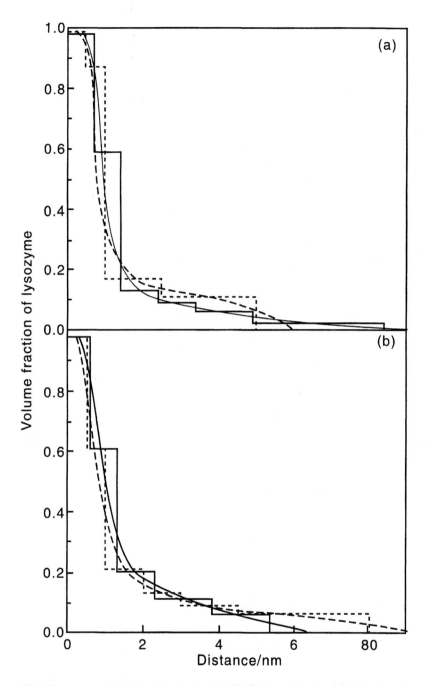

Fig. 21 Volume fraction distributions of lysozyme at a hydrophobic interface obtained by subdividing the two-layer model of Fig. 20b into a smoother distribution of six or seven layers. The conditions correspond to the solution pH being changed from (a) 7 (dotted line) to 4 (continuous line) and (b) from 4 (dotted line) to 7 (continuous line). The histograms were smoothed by drawing a curve through the middle position of each individual slab at a given pH. Each structural profile fits three to four reflectivity profiles from different isotopic contrasts at the same pH.

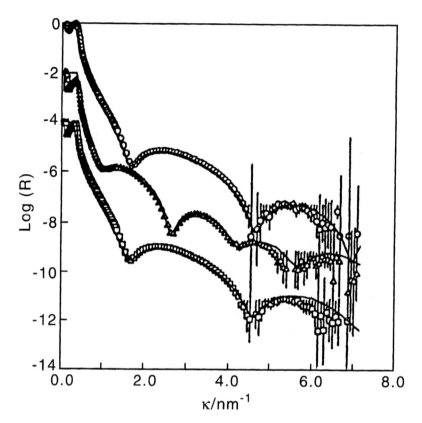

Fig. 22 X-ray reflectivity profiles from human serum albumin (HSA) on a self-assembled monolayer of hexadecyltrichlorosilane on silicon (○) before adsorption, (△) after adsorption, and (□) after desorption. The last two are displaced down the log(R) axis by 2 and 4 units, respectively. The reflectivity profiles were taken after removal of the silicon sample from aqueous solution and they show both the greater sensitivity of X-ray reflection in this situation and that adsorption of the HSA is reversible (29).

can only be done ex situ and the subtle changes characteristic of proteins in an aqueous environment are lost.

D. Protein-Surfactant Binding at the Hydrophilic Solid–Water Interface

Protein-surfactant interactions are widely used as a means of isolating pure proteins and their interactions in bulk solution have been well studied (see, for example, Refs. 37 and 38), but there is relatively little information about how they interact at a solid surface, although this has a number of implications for the deposition of blood proteins during cardiovascular implants and in many biomedical separations. Neutron reflection offers excellent sensitivity for the study of mixtures at an interface, especially when one of the components can be deuterated. We illustrate this with some measurements on BSA–sodium dodecyl sulfate (SDS) at the silica-water interface (39).

The key to the experiment is that the scattering length density of the mixed BSA-SDS layer can be varied by doing the experiment with protonated SDS (denoted hSDS) or deuterated SDS (dSDS) or any intermediate composition, and it is possible to devise a combination of the isotopic composition of water and SDS such that the SDS remains invisible. At this contrast, the SDS may change the structure of the BSA at the surface by interacting with it but only this change is observed; the SDS itself remains invisible. In such an experiment any changes in surface coverage or thickness of the BSA layer can be monitored directly with no assumptions. Alternatively, an isotopic composition of water can be chosen that is matched to the scattering length density of the protein and then the isotopic composition of the SDS changed so that only the SDS in the layer gives rise to the reflected signal. Hence the distribution of SDS in the protein layer can be determined, again with almost no assumptions. Figure 23a shows the reflectivity of a BSA layer with and without hSDS in D_2O. Both surfactant and protein contribute to the signal and there is a large change in the reflectivity, corresponding to an expansion of the layer. The protein becomes more or less invisible if the scattering length density of the water is adjusted to about the same value as for the silicon supporting the interface. In these circumstances adsorbed dSDS gives a significant signal, as can be seen in Fig. 23b, from which the coverage of SDS in the layer is easily deduced. Finally, Fig. 23c shows the reflectivity profiles of BSA in D_2O with and without dSDS. The dSDS is now invisible and the coverage of BSA in the layer can be determined independently. With these isotopic compositions and others, Lu et al. were able to show that the preadsorbed layer of BSA expands from 3.5 to 5.0 ± 0.3 nm in contact with 0.1 mM SDS with no change in coverage. This is a quite different structure from that observed in bulk solution, where the aggregate is an ellipsoid with a long-axis radius of 10.0 nm and a short-axis radius of 1.8 nm. On the other hand, the surface composition of 0.43 SDS:BSA by weight is almost exactly the same as in the bulk complex. The difference is probably the close packing in the preadsorbed BSA layer, which prevents in-plane expansion of the protein.

E. Neutron and X-Ray Reflection from Proteins Adsorbed at Phospholipid Monolayers

Phospholipid monolayers spread at the air-water interface can be considered as mimicking one half of a membrane and adsorption of proteins in and at this surface from the underlying aqueous phase may create situations that are of potential biological interest. We do not consider here the structure of the phospholipid layers themselves, which, because they commonly have long-range order in many of their phases, are better studied by grazing incidence X-Ray diffraction, which gives very precise information about the in-plane and out-of plane structure (see, for example, Refs. 40 and 41).

The interaction of species other than proteins with phospholipid layers is also of some interest and deuteration can be used in conjunction with neutron reflection to determine the compositions of the separate components and their contribution to the structure of the mixed layer, just as for BSA-SDS in the previous section. This has been done by Majewski et al. (42) for phospholipid monolayers containing poly(ethylene glycol) lipids at the air-water interface, although their use of isotopic substitution was not as extensive as that used earlier to study the incorporation of nonionic surfactants into phospholipid monolayers by Naumann et al. (43). The more interesting systems from a biological point of view are those where the binding of proteins to the phospholipid monolayer has been studied. Johnson et al. (44) used a range of isotopic substitution in conjunction with just neutron reflection to study the binding of spectrin or polylysine to phospholipid layers of different charge type. However, as we have not yet considered any application of reflection where use has been made of the combination of X-rays and neutrons, we have chosen to focus

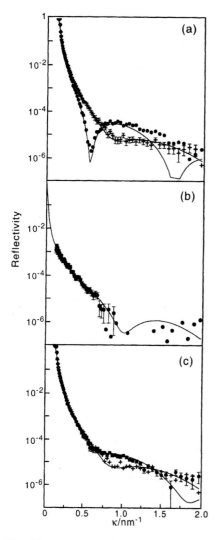

Fig. 23 Neutron reflectivity profiles from SDS bound to a BSA layer preadsorbed at the hydrophilic silicon oxide–water interface. The BSA layer was preadsorbed onto the solid surface from $0.15\,\mathrm{g\,dm^{-3}}$ buffered D_2O solution. (a) Binding of h-SDS to BSA from buffered D_2O (●) compared with the preadsorbed BSA layer on its own (+). (b) Binding of d-SDS to BSA from buffered water whose scattering length density is matched to silicon, and (c) binding of d-SDS to BSA (●) from buffered D_2O compared with the preadsorbed BSA layer on its own (+). The continuous lines were calculated using thicknesses of 5.0 nm for the mixed layer and 3.5 nm for the preadsorbed BSA layer. The volume fractions in the mixed adsorbed layer are 0.38, 0.17, and 0.45 for BSA, SDS, and water, respectively.

on an example in which these two techniques have been combined to study the binding of the protein streptavidin to biotin chemically bound to phospholipid monolayer (45).

There are two advantages in using the combination of X-rays and neutrons. First, a single isotopic composition may be used, thereby avoiding any problems of isotopic effects, which may be significant for phospholipid monolayers, and this choice can be used to obtain the maximum benefit from the neutron measurement, which would normally be to use the chain or fully deuterated phospholipid. Second, once the chain (or whole phospholipid) thickness is known from the neutron measurement, the constraint of using this measurement in the fitting allows the high resolution of X-ray reflection to be used to understand the part of the composition layer that is immersed in the aqueous subphase. Vaknin and Losche et al. (15,45–47) have used the combination of neutron and X-ray reflection to study the binding of streptavidin to phospholipids that have biotin groups chemically bound to their headgroups. Streptavidin interacts very strongly with the biotin group and, on injection of streptavidin into the subphase under the spread phospholipid monolayer, the lipid layer changes its thickness quite drastically, from 3.0 to 1.3 nm. Initially, it was thought that the streptavidin, which is organized in the layer in macroscopic domains, is differently organized in H_2O and D_2O, i.e., there is a significant isotope effect, but this was later found not to be the case. Apart from determining the effect of the streptavidin binding on the structure of the phospholipid layer, the authors were able to determine the fractional composition of water in the protein layer and to show that the thickness of the layer at 4.2 nm is close to one of the dimensions of the streptavidin in solution ($5.4 \times 5.4 \times 4.8$ nm^3). In this respect the results are of a similar type to those obtained for the more generally accessible solid-liquid interface, which were described earlier. However, the choice of functionalizing the air-liquid rather than the solid-liquid interface does open up some extra opportunities for studying protein binding to functionalized surfaces.

No attempt has been made to use a phospholipid *bi*layer as a model substrate using the same type of techniques, although such a bilayer is easily generated at the hydrophilic solid-liquid interface and has, indeed, been studied by neutron reflection (48).

IV. CONCLUSIONS

Although the study of adsorption at macroscopically flat surfaces would not seem to be of obvious general interest in biology, the inaccessibility of interfacial systems to techniques capable of resolving structure at the level of molecular fragments and the obvious power of the two reflection techniques to do just this makes reflection measurements potentially valuable for obtaining indirect information about biologically interesting systems from studies of model systems at flat surfaces. In cases in which the flat surface is intrinsically important, e.g., the degradation of protein systems at surfaces during blood transfusions, there is no more powerful techniques for examining the underlying molecular processes than neutron reflection. In the case of neutron reflection, which relies heavily on isotopic substitution to reveal the interfacial composition and structure, we have shown that such difficulties as isotope exchange and isotope effects on structure and composition are relatively easily overcome. In comparison with the uncertainties in many other surface techniques, e.g., ellipsometry, isotope effects should cause only minor difficulties. We have given greater emphasis to neutron reflection because it has so far been more widely applied to systems of biological interest. This is because these systems generally involve an interface between two condensed phases and the strong absorption of X-rays by matter makes access to this interface very difficult. However, one of the developments to be expected is the wider application of X-rays to the solid-liquid interface because the loss of beam intensity from the intense sources now available is less of a problem. As we have tried to make clear, the lower contrast between the different interfacial components in an X-ray reflection experiment may be offset by the greater sensitivity.

REFERENCES

1. J Penfold, RM Richardson, A Zarbakhsh, JWP Webster, DG Bucknall, AR Rennie, RAL Jones, T Cosgrove, RK Thomas, JS Higgins, PDI Fletcher, EJ Dickinson, SJ Roser, IA McLure, AR Hillman, RW Richards, EJ Staples, AN Burgess, EA Simister, JW White. Recent advances in the study of chemical interfaces by specular neutron reflection. J Chem Soc Faraday Trans 93:3899, 1997.
2. JR Lu, RK Thomas. Neutron reflection from wet interfaces. J Chem Soc Faraday Trans 94:995, 1998.
3. A Hvidt, SO Nielsen. Hydrogen exchange in proteins. Adv Protein Chem 21:287, 1966.
4. SE Radford, M Buck, KD Topping, CM Dobson, PA Evans. Hydrogen exchange in native and denatured states of hen egg-white lysozyme. Proteins Struct Funct Genet 14:237, 1992.
5. OS Heavens. Optical Properties of Thin Solid Films. New York: Dover, 1965.
6. M Born, E Wolf. Principles of Optics. Oxford: Pergamon, 1970.
7. TL Crowley. A uniform kinematic approximation for specular reflection. Physics A195:354, 1993.
8. J Als-Nielsen. The Liquid-vapour interface. Z Phys B61:411, 1985.
9. E Dickinson, DS Horne, JS Phipps, RM Richardson. A neutron reflection study of the adsorption of β-casein at fluid interfaces. Langmuir 9:242, 1993.
10. JR Lu, TJ Su, RK Thomas, J Penfold, J Webster. Structural conformation of lysozyme layers at the air-water interface studied by neutron reflection. J Chem Soc Faraday Trans 94:3279, 1998.
11. D Voet, JG Voet. Biochemistry New York: Wiley, 1990.
12. Z Li, W Zhao, J Quinn, MH Rafailovich, J Sokolov, RB Lennox, A Eisenberg, XZ Wu, MW Kim, Sinha, M Talon. X-ray reflectivity of diblock copolymer monolayer at the air-water interface. Langmuir. 11:4785, 1995.
13. SW An, TJ Su, RK Thomas, FL Baines, NC Billingham, SP Armes, J Penfold. Neutron reflectivity of an adsorbed water block copolymer: A surface transition to micelle-like aggregates at the air-water interface. J Phys Chem 102:387, 1998.
14. SW An, RK Thomas, FL Baines, NC Billingham, SP Armes, J Penfold. Neutron reflectivity of an adsorbed water soluble block copolymer at the air/water interface: the effects of pH and ionic strength. J Phys Chem B: 102:5120, 1998.
15. D Vaknin, J Als-Nielsen, M Piepenstock, M Losche. Recognition processes at a functionalized lipid surface observed with molecular resolution. Biophys J 60:1545, 1991.
16. JR Lu, EM Lee, RK Thomas. The analysis and interpretation of neutron and x-ray specular reflection. Acta Crystallogr. A52:11, 1996.
17. TJ Su, JR Lu, RK Thomas, ZF Cui, J Penfold. The effect of solution pH on the structure of lysozyme layers adsorbed at the silica-water interface studied by neutron reflection. Langmuir 14:438, 1998.
18. TA Horbett, J Brash. Proteins at Interfaces II. ACS Symp Ser 602, 1995.
19. CA Haynes, W Norde. Globular proteins at solid/liquid interfaces. Colloids Surf B Biointerfaces 2:517, 1994.

20. TJ Su, JR Lu, RK Thomas, ZF Cui, J Penfold. The adsorption of lysozyme at the silica-water interface: A neutron reflection study. J Colloid Interface Sci 203:419, 1998.

21. RK Iler. The Chemistry of Silica. New York: Wiley, 1979.

22. C Tanford, R Roxby. Interpretation of protein titration curves: Application to lysozyme. Biochemistry 11:2192, 1972.

23. CA Haynes, E Sliwinsky, W Norde. Structural and electrostatic properties of globular proteins at a polystyrene-water interface. J Colloid Interface Sci 164:394, 1994.

24. PM Claesson, E Blomberg, JC Froberg, T Nylander, T Arnebrant. Protein interactions at solid surfaces. Adv Colloid Interface Sci 57:161, 1995.

25. TJ Su, JR Lu, RK Thomas, ZF Cui, J Penfold. The conformational structure of bovine serum albumin layers adsorbed at the silica-water interface. J Phys Chem B: 102:8100, 1998.

26. T Peters. Serum albumin. Adv Protein Chem 37:161, 1985.

27. AD Miranker, CM Dobson. Collapse and cooperativity in protein folding. Curr Opin Struct Biol 6:31, 1996.

28. A Liebmann-Vinson, LM Lander, MD Foster, WJ Brittain, EA Vogler, CF Majkrzak, SK Satija. A neutron reflectometry study of human serum albumin adsorption. Langmuir 12:2256, 1996.

29. S Petrash, A Liebmann-Vinson, MD Foster, LM Lander, WJ Brittain, EA Vogler, CF Majkrzak. Neutron and x-ray reflectivity studies of human serum albumin adsorption onto functionalized surfaces of self-assembled monolayers. Biotechnol Prog 13:635, 1997.

30. G Fragneto, RK Thomas, AR Rennie, J Penfold. Neutron reflection study of bovine β-casein adsorbed on OTS self-assembled monolayers. Science 267:657, 1995.

31. R Maoz, J Sagiv, D Degenhardt, H Mohwald, P Quint. Hydrogen bonded multilayers of self assembled silanes: Structure elucidation by combined Fourier transform infra-red spectroscopy and x-ray scattering techniques. Supramol Sci 2:9, 1995.

32. R Maoz, J Sagiv. In A Ulman, ed. Organic Thin Films and Surfaces: Directions for the Nineties. San Diego: Academic Press, 1995.

33. JR, Lu, TJ Su, RK Thomas, AR Rennie, R Cubitt. Denaturation of lysozyme at the hydrophobic solid-water interface studied by neutron reflection. J Colloid Interface Sci 206:212, 1998.

34. C Tanford, ML Wagner. Hydrogen ion equilibria of lysozyme. J Phys Chem 76:3331, 1954.

35. JR Lu, TJ Su, RK Thomas, J Penfold, RW Richards. The determination of segment density profiles of polyethylene oxide layers adsorbed at the air-water interface. Polymer 37:109, 1996.

36. S Petrash, NB Sheller, W Dando, MD Foster. Variation in tenacity of proteins adsorbed on self assembled monolayers with monolayer order as observed by x-ray reflectivity. Langmuir 13:1881, 1997.

37. C Tanford. Hydrophobic free energy, micelle formation and the association of proteins with amphiphiles. J Mol Biol 67:59, 1972.

38. JA Reynolds, JP Gallagher, J Steinhardt. Effect of pH on the binding of N-alkyl sulphates to bovine serum albumin. Biochemistry 9:1232, 1970.

39. JR Lu, TJ Su, RK Thomas, J Penfold. Binding of sodium dodecyl sulphate to bovine serum albumin layers adsorbed at the silica-water interface. Langmuir, 14:6261, 1998.

40. J Als-Nielsen, D Jaquemain, K Kjaer, F Leveiller, M Lahav, L Leiserowitz. Principles and applications of grazing incidence x-ray and neutron scattering from ordered molecular monolayers at the air-water interface. Phys Rep 246:251, 1994.

41. H Mohwald. Phospholipid and phospholipid-protein monolayers at the air-water interface. Annu Rev Phys Chem 41:441, 1990.

42. J Majewski, TL Kuhl, MC Gerstenberg, JN Isrealichvili, GS Smith. Structure of phospholipid monolayers containing polyethylene glycol lipids at the air-water interface. J Phys Chem 101:3122, 1997.

43. C Naumann, C Dietrich, JR Lu, RK Thomas, AR Rennie, J Penfold, TM Bayerl. Structure of mixed monolayers of dipalmitoyl glycerophosphocholine and polyethylene glycol monododecyl ethers at the air-water interface determined by neutron reflection and film balance techniques. Langmuir 10:1919, 1994.

44. SJ Johnson, TM Bayerl, W Weihan, H Noack, J Penfold, RK Thomas, D Kanelleas, AR Rennie, E Sackmann. Coupling of spectrin and lysine to phospholipid monolayers studied by specular reflection of neutrons. Biophys J 60:1017, 1991.

45. M Losche, C Erdelen, E Rump, H Ringsdorf, K Kjaer, D Vaknin. On the lipid head group hydration of floating surface monolayers bound to self assembled molecular protein layers. Thin Solid Films 242:112, 1994.

46. D Vaknin, K Kjaer, H Ringsdorf, R Blankenburg, M Piepenstock, A Diederich, M Losche. X-ray and neutron reflectivity studies of a protein monolayer adsorbed to a functionalized aqueous surface. Langmuir 9:1171, 1993.

47. M Losche, M Piepenstock, A Diederich, T Grunewald, K Kjaer, D Vaknin. Influence of surface chemistry on the structural organization of monomolecular protein layers adsorbed to functionalized aqueous surfaces. Biophys J 65:2160, 1993.

48. SJ Johnson, TM Bayerl, DC McDermott, GW Adam, AR Rennie, RK Thomas, E Sackmann. The structure of an adsorbed dimyristoyl phosphatidylcholine bilayer measured with specular reflection of neutrons. Biophys J 59:289, 1991.

19

Time-Resolved Fluorescence Techniques Applied to Biological Interfaces

A. van Hoek and A. J. W. G. Visser

Wageningen University and Research Centre, Wageningen, The Netherlands

I. INTRODUCTION

All interaction and assembly mechanisms in living matter start at a molecular scale. This fact makes it quite relevant to study and understand the underlying principles of molecular properties. Because of the small (nanometer) scale, the direct observation of molecular mechanisms is never a straightforward matter. Rather complex setups are required in these studies, in which spectroscopy is used as an analytical tool. The molecules under study are subjected to continuous and/or alternating magnetic fields and/or light. The response of the molecular system to that excitation is then detected and analyzed. From the results a molecular model is proposed that is in agreement with the spectroscopic observations. Complementary conclusions can be drawn and the model can be refined by variation of external parameters such as temperature, pressure, molecular concentration, and type of environment or solvent. Depending on the applied stimuli (magnetic fields, photons, etc.), spectroscopic techniques can be divided into magnetic resonance techniques (1) and optical spectroscopy (2).

The spectroscopic techniques discussed in this chapter make use of the absorption of light by molecules. Light absorption brings the molecules instantaneously into an excited state, which is followed by the emission of fluorescence photons in arbitrary directions. Molecular interactions, (radiationless) energy transfer, environment (solvent) dynamics, and structural fluctuations have a direct influence on the fluorescence properties of molecules. These processes can then be described by careful observation and analysis of the fluorescence emission. The study of fluorescence dynamics, in particular, will provide an extra tool and dimension for the interpretation of fluorescence data. Examples of different excited-state processes that influence the fluorescence dynamics have been schematized in Fig. 1. In Refs. 3 and 4 an extensive overview of fluorescence techniques and a wide range of applications are presented. Numerous other books on fluorescence applications have appeared but are not listed here.

An advantage of the application of fluorescence spectroscopy for studies on biological interfaces is the high sensitivity in combination with the very specific information content. Even the tiniest amount of fluorescence light, the single photon, can be detected. The fluorescence photon carries information on (a) the photon energy, (b) the polarization direction, (c) the timing moment, which can be used to elucidate details of molecular interactions. In some advanced optical spectroscopic techniques, such as coherent anti-Stokes Raman scattering, even the direction of propagation of the emitted photon contains information. In addition to this information content, a large variety of natural and artificial fluorescence probes is available that can be used to investigate a broad range of molecular phenomena (5). These specific fluorescent probes can be incorporated into self-aggregating molecular systems such as mono-, bi-, and multilayers, artificial interfaces, or micelles. The whole system can then be effectively studied using time-resolved fluorescence spectroscopy.

Another important reason to use fluorescence techniques is that molecular processes take place on the same time scale as the fluorescence phenomenon itself. This aspect can be illustrated using the Einstein relation:

$$\langle \Delta x^2 \rangle = 2Dt \tag{1}$$

stating that the mean square root molecular displacement Δx is related to the translational diffusion coefficient D and the traveling time t. If we substitute a time of 150 ns (order of magnitude of the time constant of the fluorescence decay of pyrene) and a diffusion constant $D = 3 \times 10^{-12} \, \text{m}^2/\text{s}$ (lateral diffusion over a bilayer membrane), the displacement ($\langle \Delta x^2 \rangle^{1/2}$) is about 1 nm, which is a significant microscopic distance over the surface of such a membrane. This relatively simple example is taken to illustrate that fluorescence spectroscopy is a sensitive method for registering dynamic events in biological systems.

The information retrieved from molecules can be even more specific when the influence of their environment can be modulated, minimized, or well described. Alternatively, the heterogeneity of the spectroscopic data (and thus the complexity) can be reduced by minimizing the number of molecules under study. Examples are (a) the loading of single molecules in a host matrix with silent spectroscopic properties; (b) using an ultrasmall experimental volume by applying evanescent-field, near-field, or multiphoton excitation; or (c) decreasing the concentration of molecules in solution or other matrices. The application of site-selective excitation and the freezing of thermal movements are other methods. One should keep in mind, however, that it is not always realistic to cool biological samples down to cryogenic temperatures, because their properties might be drastically altered.

Fluorescence investigations of interfaces and microemulsions require extra attention to the experimental conditions. In a fluorescence experiment with molecules in transparent and clear solution in a standard cuvette, the position of the effective measurement volume is chosen such that it is well separated from the cuvette walls.

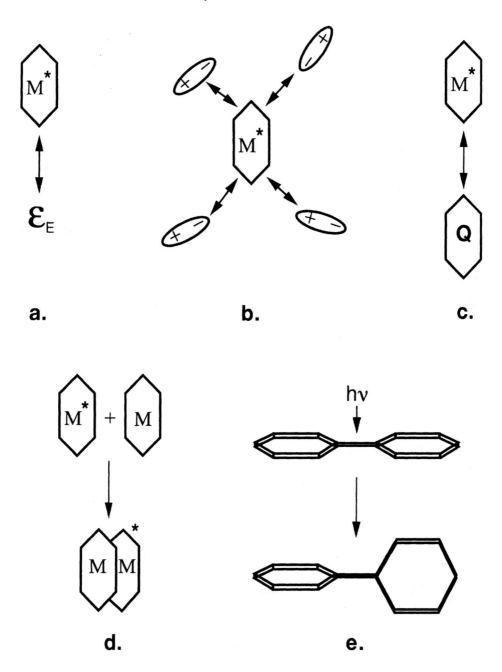

Fig. 1 The decay time of fluorescence of a chromophore can be influenced by different factors, for instance, the dielectric properties ε_E of the environment of the chromophore (a), polarity interactions of the excited molecule with the environment (b), interaction of the chromophore with a quencher molecule (c), excimer formation (d), or conformational changes of the excited molecule upon excitation (e), and that again may influence the fluorescence dynamics. All these different molecular effects can be studied by careful registration and analysis of fluorescence dynamics.

The reason is that one should avoid detection of fluorescence of probe molecules that may interact with the cuvette walls or fluorescence components from other objects or scatter. Therefore the experimental conditions must be adapted when fluorescence measurements are carried out at interfaces or in emulsions. In the case of emulsions, front surface excitation and detection should be applied to avoid artifacts from scattered light in both excitation and detection pathways. Scattering has a direct influence on the polarization properties of that light. In the case of interface studies, not only should the interaction of the fluorescent molecules with the interface be investigated and included in the applied physical model but also the surfaces should be scrupulously cleaned to avoid impurity fluorescence.

II. TIME-RESOLVED SPECTROSCOPIC TECHNIQUES

In this section the time-resolved fluorescence techniques for the study of biological interfaces are briefly explained. The design of different fluorescence probes is very important in the application of fluorescence spectroscopy. In principle, the design should be such that the system under investigation is minimally perturbed after introduction of the fluorescent marker. The spectroscopic methods discussed here are time-resolved fluorescence quenching, time-resolved fluorescence anisotropy, fluorescence decay measurements, fluorescence resonance energy transfer, multiphoton excitation, and single-molecule fluorescence spectroscopy. A review of these techniques as applied to self-aggregating molecular systems can be found in Ref. 6. In this chapter some recent applications are subsequently discussed. The experimental systems range from adsorbed molecular layers on artificial interfaces; micelles from anionic, cationic or nonionic surfactants; bilayer membranes; lipid monolayers; and reversed micelles to single molecules adsorbed on solid-air interfaces. A summary of time-resolved fluorescence techniques applied to these biological interfaces is provided. Some of our own results will be used as illustrations in this survey.

A. Time-Resolved Fluorescence Quenching

The method of time-resolved fluorescence quenching (TRFQ) is a rapid and popular method for obtaining the size of an aggregate, and it can be considered a classical method. The principle is as follows. Fluorophores and quencher molecules are found to be distributed in the micellar compartments in a Poissonian way. In most applications with aqueous micellar solutions, pyrene has been selected as fluorescent probe because of its long fluorescence lifetime, and alkylpyridinium cations frequently serve as a quencher. In reverse micelles a ruthenium-bipyridyl quencher pair, solubilized in the aqueous core, is often chosen. Methylviologen and potassium hexacyanoferrate(III) can act as quenchers. The quenching occurs only in an intramicellar reaction, because when fluorophore and quencher are in separate micelles, the distance for effective fluorescence quenching is too large. Therefore, by choosing sufficiently low concentrations, both fluorophores and quenchers are supplied from different micelles and collide with each other via intermicellar exchange (see Fig. 2). Depending on the concentration ratios of fluorescent probe and quenching molecules, there is a certain distribution of quenched (short time part of the fluorescence decay) and unquenched fluorophores (long time part of that decay). The flu-

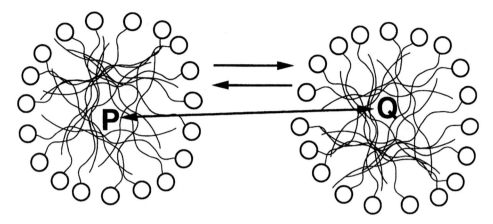

Fig. 2 When chromophores as well as quencher molecules are added to a micellar solution, where both chromophore and quencher prefer the inner environment of the micelle, the probability for quenching is, among other parameters, dependent on the concentration of both components. By using a concentration series and comparing the contributions of the fast part (quenching) and the slow part (no quenching) of the fluorescence decay, the size of the aggregates can be determined (see text for details).

orescence decay $F(t)$ can be modelled by the Infelta-Tachiya equation (7,8):

$$F(t) = A_1 \exp\{-A_2 t + A_3[\exp(-A_4 t) - 1]\} \qquad (2)$$

where

$A_1 = F(0)$, the fluorescence intensity at $t = 0$
$A_2 = 1/\tau_0 + k_q k_- \langle n \rangle / A_4$
$A_3 = k_q^2 \langle n \rangle / A_4^2$
$A_4 = k_q + k_-$

k_q and k_- are the pseudo-first-order quenching rate constant and quencher exit rate constant, respectively, τ_0 is the fluorescence lifetime in the absence of quenching, and $\langle n \rangle$ is equal to the quencher and micellar molar ratio and is the average number of quencher molecules per micelle or the occupation number. An important assumption is that the probe remains stationary in the same compartment during the observation time. In most studied cases it is valid that $k_q \gg k_-$, leading to $A_2 = 1/\tau_0 + k_e$, $A_3 = \langle n \rangle$, and $A_4 = k_q$. Such simplification implies that probe and quencher distributions are frozen during the fluorescence observation time. By fitting Eq. (2) to the experimental decay curves, one can finally obtain all rate constants and $\langle n \rangle$, from which the micellar aggregation number N_M can be determined from

$$N_M = \langle n \rangle / X = \langle n \rangle([S] - CMC)/[\text{quencher}] \qquad (3)$$

where X is the mole fraction of quenchers in the micelles, $[S]$ the surfactant concentration, and CMC the critical micelle concentration.

Two-dimensional diffusion of amphiphiles in phospholipid [dimyristoyl phosphatidylcholine (DMPC), dipalmitoyl phosphatidylcholine (DPPC)] monolayers at the air-water interface has been determined by studying the fluorescence quenching of pyrene-labeled phospholipid [pyrene–dipalmitoyl phosphatidylethanolamine (DPPE)] (9). Without the presence of quencher, the fluorescence decay of pyrene-DPPE was found to be exponential, which indicates a homogeneous distribution of probe lipid over the two-dimensional lipid surface. Addition of quencher to the monolayer produced a nonexponential decay. The experimental decay could be adequately described by diffusion-controlled quenching in a two-dimensional environment.

A few remarks concerning TRFQ should be made. In most cases studied, the technique is not the only applied one but is complemented by other methods such as viscosimetry, conductivity, potentiometry, self-diffusion measurements, light scattering, calorimetry, or electron microscopy (cryo–transmission electron microscopy, TEM) to provide a full characterization of the system and to check hypotheses concerning the size and shape of the aggregates. In order to apply TRFQ in a strict sense, the micelles should have a spherical or spheroidal shape. Cryo-TEM has shown to be a powerful tool to provide the morphology of surfactant systems and to decide whether TRFQ can be safely applied (10). A more advanced TRFQ model accounting for rodlike micelles has been described (11). The same research group also made a critical assessment of TRFQ in a nonideal mixed micellar system in which the apparent aggregation number significantly deviates from the true one (12).

B. Time-Resolved Fluorescence Anisotropy

Time-resolved fluorescence anisotropy (TRFA) is a technique capable of detecting motional behavior of probe molecules in macromolecules (see Fig. 3) and bilayer membranes (see Fig. 4). In the case of bilayer membranes, probe molecules often have a cylindrical shape. From TRFA, information can be retrieved about the order and reorientational dynamics of these symmetric probes embedded in membranes. In addition, changes in membrane structure or the occurrence of interactions with proteins can be followed when these parameters are changing. The fluorescence anisotropy is defined as

$$r(t) = \frac{I_{\parallel}(t) - I_{\perp}(t)}{I_{\parallel}(t) + 2I_{\perp}(t)} = \frac{I_{\parallel}(t) - I_{\perp}(t)}{F(t)} \tag{4}$$

where $I_{\parallel}(t)$ and $I_{\perp}(t)$ are the parallel and perpendicular polarized fluorescence intensity decays and $F(t)$ is the total fluorescence decay, from which the fluorescence lifetime(s) is (are) obtained. Equation (4) can be written in a form so that it is immediately apparent that one is dealing with the motion of the emission transition moment μ_e:

$$r(t) = \langle P_2[\mu_a(0)\mu_e(t)]\rangle \tag{5}$$

$P_2(x)$ has the form of a second-order Legendre polynomial, μ_a is the direction of the absorption transition moment at $t = 0$, and the brackets $\langle\cdot\cdot\rangle$ denote an ensemble average. When there is no motion at all, $r(t)$ becomes

$$r(t) = r_0 = \tfrac{2}{5} P_2(\cos \delta) \tag{6}$$

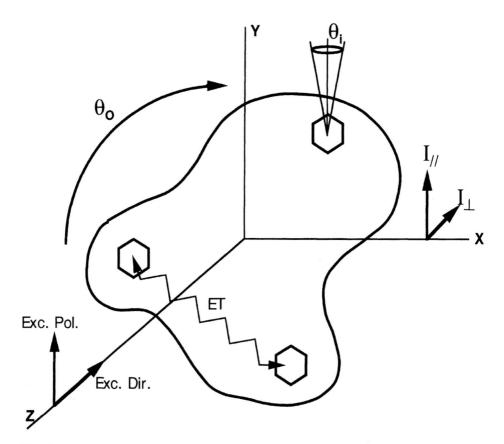

Fig. 3 Fluorescence anisotropy is measured by determining the (normalized) difference between fluorescence that is polarized parallel ($I_{\parallel}$) and that is polarized perpendicular ($I_{\perp}$) with respect to the polarization direction of the excitation light (see text). Fluorescence anisotropy can be initiated by local flexibility (θ_i) of the part of the molecule where the excited chromophore is situated, as well as by movement (θ_o) of the complete molecule. In addition, fluorescence anisotropy can be initiated by internal energy transfer (ET). The contributions of the different origins of fluorescence anisotropy can be separated by changing experimental parameters such as sample temperature or the wavelength of excitation and/or detection.

where r_0 is the fundamental anisotropy and δ the angle between absorption and emission transition dipole moments. If there are two rotational diffusion processes occurring on different time scales, Eq. (5) can be factorized in three terms:

$$r(t) = r_0 C_x(t) C_y(t) \tag{7}$$

where $C_x(t)$ and $C_y(t)$ are the correlation functions of two independent rotational diffusion processes.

For studies of phospholipid membranes the most popular probe has been (and still is) the rod-shaped 1,6-diphenyl-1,3,5-hexatriene (DPH), its charged derivative TMA-DPH (trimethylammonium-DPH), and the phosphatidylcholine (PC) analogue DPH-PC. From TRFA studies one can obtain information on the orientational order and restricted rotational dynamics of the DPH cylinder in phospholipid bilayer

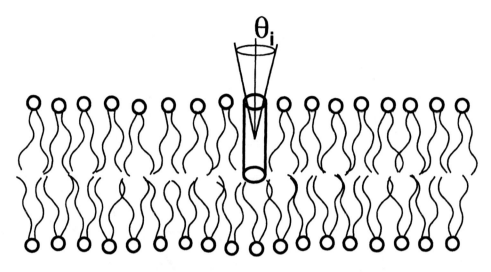

Fig. 4. When a chromophore is incorporated in a lipid bilayer, the fluorescence anisotropy of the chromophore originates from reorientation movements (θ_i) of the chromophore within the bilayer. The movement of the complete bilayer is beyond the nanosecond time scale of fluorescence and is found as a quasi-stead-state value in the experimental data of the anisotropy decay.

membranes such as vesicles. The rotational behavior [or one of the correlation functions in Eq. (7)] of the DPH cylinder in a membrane system can be adequately modeled using the concept of Brownian diffusion of the probe in an orienting potential (13,14). In most comparative TRFA investigations on membrane systems, relatively simple concepts such as the orientational order parameter S are used:

$$S = \sqrt{\frac{r_\infty}{r_0}} \tag{8}$$

where r_∞ is the residual anisotropy at the end of the observation time window. The average fluorescence lifetime of DPH is rather sensitive to the polarity of its immediate environment. Therefore comparative studies deal with the analysis of both $F(t)$ and $r(t)$. The effects studied are, among others, hydration as a function of degree of lipid unsaturation and sterol addition to lipid bilayers.

Another interpretation of TRFA was presented for DPH localized in two distinct populations in lipid vesicles (15). The first population was found to arise from DPH molecules residing in the core of the membranes with an orientation parallel to the membrane normal, while the other population was located at the interface with an average orientation perpendicular to the interface.

The TRFA of DPH-PC and DPH-DG (DPH-diacylglycerol) in dioleoyl phosphatidylcholine (DOPC) vesicles has been compared (16) (see Fig. 5). The TRFA of DPH-DG decays to a lower r_∞ value than that of DPH-PC, indicating a lower degree of ordering of the neighboring lipid chains experienced by DPH-DG. The lack of the phosphocholine headgroup in DG apparently induces a packing failure of close fit to the neighboring PC lipids. DPH-DG is a specific substrate for protein kinase C (PKC). Under optimum conditions for association of PKC to DOPC vesicles,

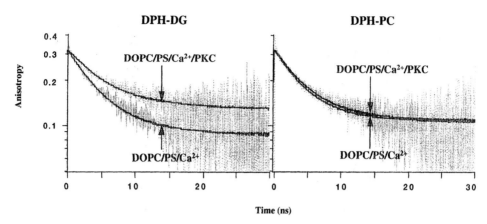

Fig. 5. Experimental and fitted fluorescence anisotropy decays of DPH-DG (left) and DPH-PC (right) in DOPC vesicles (containing brain PS in a DOPC/PS molar ratio of 4:1; total lipid concentration 15 μM) and either DPH-DG or DPH-PC (molar ratio of labeled lipid to unlabeled lipid = 1:150) and in the presence of 0.5 mM calcium. The temperature was 293 K. The buffer used was 20 mM Tris-HCl, 120 mM NaCl, pH 7.4. The fitted decays correspond to the optimal fit with a general diffusion model assuming a Gaussian orientation distribution. In the absence of protein kinase C (PKC) there is a distinct difference in the orientation distribution of both DPH lipids. DPH-PC is oriented with smaller angle (0.60 rad) to the membrane normal than DPH-DG (0.65 rad). When 0.4 μM PKC is present, DPH-DG is selectively bound to PKC, leading to less motional freedom and a smaller angle (0.54 rad). The control lipid DPH-PC is not bound to PKC. (Adapted from Ref. 16.)

DPH-DG is bound to PKC, which leads to a raise in r_{∞} value to a similar level as for DPH-PC. The TRFA then becomes indistinguishable from that of the control lipid DPH-PC, which does not interact with PKC.

The interaction of the channel-forming peptide gramicidin A with a wide range of bilayer vesicles has been studied with TRFA of TMA-DPH and DPH-PC (17) and with DPH (18). The nonchannel gramicidin conformation has an ordering effect on DPH in the bilayer, which lacks the channel conformation. An example of the application of other probes (in combination with DPH) is the use of the polarity probe dansyl-phosphatidylethanolamide (dansyl-PE), in conjunction with orientational order determinations using DPH-PC, to monitor hydration near the phospholipid headgroup (19). The membranes were composed of PC, but with acyl chains having varying degree of unsaturation either on the *sn*-2 chain or on both *sn*-1,2 chains.

Other investigations describe the combination of pyrene-PC with DPH probes (20,21). TMA-DPH is used as a surface probe to monitor the changes in fluorescence lifetime and rotational dynamics when cholesterol and β-sitosterol are incorporated in dipalmitoyl phosphatidylcholine (DPPC) bilayer vesicles or when measurements are performed below and above the phase transition temperature (20). Pyrene in Pyr-PC is located in the hydrophobic interior and registers lifetime and rotational changes upon sterol addition, which are significantly different from those observed via TMA-DPH (20). By measuring lateral diffusion of Pyr-PC in binary mixtures of PE (which tends to form a hexagonal ordered phase H_{II}) and PC in membranes and rotational diffusion of DPH-PC, the onset of packing defects at about 60%

PE in the mixture has been demonstrated (21). The rotational dynamics of DPH-PC in the PE-PC mixed membrane have been modeled using a rod-shaped probe in a curved matrix. TRFA can then be described by Eq. (7), in which $C_x(t)$ is the wobbling diffusion part and $C_y(t)$ the part describing the curvature-associated rotation.

Other reports mention the use of *trans*-parinaric acid as a natural fatty acid membrane probe for the investigation of cholesterol effects on artificial membranes or membranes isolated from blood platelets (22,23). *trans*-Parinaric acid has, like DPH, a more or less rodlike structure. The interaction between a labeled lyso-phospholipid [nitrobenzdiazole phosphatidylethanolamide (*N*-NBD-PE)] and the bile salt taurodeoxychelate (TDC) has been demonstrated with TRFA at concentrations below the CMC of both components (24). Evidence has been given that the lysophospholipid seeds the aggregation of TDC micelles. Porphyrins have been used as probes to characterize anionic (sodium dodecyl sulfate, SDS), cationic cetyltrimethyl ammonium bromide (CTAB), and nonionic micelles (Triton X-100) in water (25). The TRFA could be described by Eq. (7), where $C_x(t)$ can now be conceived as both rotational and translational diffusion inside the micelles and $C_y(t)$ as overall micelle tumbling. Fluorescence anisotropy decay measurements turn out to be an excellent method for directly observing a protein adsorbed to a hydrophobic latex (Teflon) (26). Because the Teflon particles are much larger than the protein (subtilisin), the TRFA of the tryptophan in subtilisin indicates complete immobilization of adsorbed protein on the observation time scale (15 ns). The TRFA of the protein in solution decays to zero.

A warning should be given in the case of fluorescence anisotropy decay investigations of pyrene-containing phospholipids in bilayer membranes. They should be only qualitatively interpreted. Some experiments using two different pyrene-labeled phopholipids DOPC vesicles and erythrocyte ghosts are described in Ref. 27. It can be clearly shown that the properties of pyrene are not so suitable here, because the fluorescence lifetime of pyrene is an order of magnitude longer than the reorientational dynamics of the probe in both membranes and the initial anisotropy (r_0) is rather low ($r_0 \approx 0.2$). In addition, the theoretical treatment of the rotational motion is extremely complex (see Ref. 13). In DOPC vesicles the fluorescence anisotropy of both pyrene lipids declines to zero at longer times (20 ns). This indicates that the rotation of the pyrene chains is not restricted on the time scale of the relatively long fluorescence lifetime (150 ns). However, in the protein-rich ghost cells, the fluorescence anisotropy of one of the lipids has a finite value at longer times (20 ns), indicating that its motion is restricted. In this particular case the acyl chains are apparently interacting with the protein surface. The control pyrene lipid is not restricted in its motions in the ghost membranes.

An interesting study of the distribution of hydrophobic probe molecules in lipid bilayers of palmitoyloleoyl phosphatidylcholine (POPC) vesicles was presented, where TRFA measurements were combined with Monte Carlo dynamics computer simulations (MCDCS) (28,29). It is found that with perylene (TRFA and MCDCS) and DPH (MCDCS) in the lipid bilayer two distinct orientational populations exist, in agreement with the two distinct environments of the middle of the bilayer and in regions near the bilayer interface. Ambiguities were found in comparing the results of TRFA and MCDCS. It was concluded that TRFA on macroscopically unoriented samples will not provide sufficient information for a full characterization of the orientational distribution of probe molecules in bilayers.

The long-lived fluorescence of coronene has been used as a probe for investigating submicrosecond lipid dynamics in DPPC bilayer membranes (small unilamellar vesicles, SUVs), below and through the gel-to-liquid crystalline phase transition (30). TRFA at increasing temperatures showed three rotational decay components: long correlation times of coronene located in a gel-like lipid phase, short correlation times representative of coronene in fluidlike regions, and intermediate correlation times characteristic of a coronene environment arising from fluidization of gel lipid. A Landau adapted expression for describing gated packing fluctuations has been developed, that could adequately describe both thermodynamics and (rotational and diffusion) kinetics of the anisotropy data.

C. Fluorescence Decay Measurements

The determination of fluorescence lifetimes of lipid probes in membranes systems has provided detailed insight into microenvironmental effects resulting from changes in membrane structure or phase. The most frequently used probes are DPH, pyrene-containing lipids, and NBD-lipids. In the following, the properties of these lipid probes in different membranes will be discussed. The remainder of this section is then devoted to some properties of a natural antibiotic molecule that possesses intrinsic fluorescence similar to that found in *trans*-parinaric acid and of fluorescent molecules that are characterized by a twisted intramolecular charge transfer (TICT) state.

As already stated in the previous section, the fluorescence lifetime (and quantum yield) of DPH is strongly dependent on the polarity of its microenvironment. This has been investigated for DPH and TMA-DPH in egg phosphatidylcholine vesicles in normal and deuterated water, the latter showing a dielectric constant 12% lower than that of normal water (31). The fluorescence decay was analyzed using a Lorentzian lifetime distribution. A decrease of the dielectric constant along the membrane normal shifts the centers of the distribution to longer lifetimes and sharpens the widths of the distribution. The addition of cholesterol results in a further shift and narrowing. The profile of the dielectric constant along the membrane normal turns out to be nonlinear.

Pyrene-containing lipids dispersed in membranes have been widely used to probe two-dimensional lateral diffusion in membranes via the propensity of pyrene for excimer formation. The application of a diffusion-controlled bimolecular reaction theory in two dimensions to fit the time-resolved monomer and excimer fluorescence intensity data results in very consistent values for the diffusion coefficient and monomer and excimer fluorescence lifetimes of pyr_{10}-PC(1-palmitoyl-2-(1-pyrenodecanoyl)-*sn*-glycero-3-phosphatidylcholine) in POPC multilamellar vesicles in the fluid phase (32). The diffusion coefficient (e.g., $D = 3.1 \times 10^{-12}\,m^2\,s^{-1}$ at 25°C) and diffusion activation energy ($E_a = 35\,kJ\,mol^{-1}$) obtained are in good agreement with the results of other techniques.

A Langmuir-type film balance was equipped with a fluorescence microscope for time-resolved fluorescence monitoring of membrane dynamics using pyrene excimer formation (33). The monolayers consisted of DMPC and the pyrene probe was attached to a lecithin. The fluorescence decay of the monomer-forming excimers was modeled in terms of the Smoluchowski theory for two-dimensional fluids. It is shown that the time course of excimer formation is strongly determined by a time-dependent association rate constant. The variation of lateral diffusion with lipid packing density has been measured and compared with theoretical predictions.

The important advantage of dipyrenyl lipids over monopyrenyl lipids is that excimer emission is formed at an extremely low probe-to-lipid molar ratio. A systematic fluorescence study of dipyr$_n$-PC (*n* from 4 to 14) in DOPC (L$_a$ phase) or DOPE (L$_a$ or H$_{II}$ phase depending on temperature) has been performed (34). The fluorescence dynamics of monomer and excimer emission were found to exhibit a chain length dependence in normal and curved bilayers. The internal dynamics of the pyrenyl chains could be modeled with a lattice model.

The photophysical properties of NBD attached to the *sn*-2 acyl chain of various phospholipids [PC, PE, phosphatidylserine (PS), phosphatidic acid (PA)] or to the PE headgroup and inserted into vesicles of PC, phosphatidylglycerol (PG), PA, or PS have been investigated (35). For various couples of probe and host lipids, the authors observed that the fluorescence decay was dominated by an increase in the nonradiative decay constant k_{nr}. NBD did exhibit a strong solvatochromism, and by comparing the fluorescence properties of *N*-propylamino-NBD in a set of organic solvents it was shown that NBD in the lipid environment experiences a dielectric constant of around 27–41, corresponding to a medium of rather high polarity found in the headgroup region. The fact that the fluorescence decay of NBD is very dependent on changes in k_{nr} renders the fluorescence very sensitive to thermal changes. This property of NBD-conjugated lipids (PC, PE) has prompted researchers to apply these lipids as optical thermometers in the membranes of living cells (36). When the NBD probe-to-phospholipid molar ratio is chosen relatively high (1–50 mol %), NBD molecules aggregate in the membrane, which leads to self-quenching of the fluorescence. An empirical model has been developed that accounts for changes in fluorescence lifetime and quantum yield of NBD with the fractional concentration of incorporated NBD-PE in the membrane matrix (egg phosphatidylcholine, soya bean L-α-lecithin, DPPC, DPPA) (37).

The self-association of the natural fluorescent antibiotic nystatin in gel-phase DPPC vesicles was demonstrated with time-resolved fluorescence techniques (38). In contrast to NBD-PE, nystatin exhibits no self-quenching of the fluorescence when it aggregates. The contribution of a long decay component of 33 ns increases, which was explained by a more rigid environment of the nystatin aggregates in the bilayer. Nystatin (which is supposed to form pores in a membrane bilayer) is more preferably partitioned in the gel phase than in the liquid crystalline phase.

A new method for the analysis of physicochemical properties of liquid water near surfaces [such as water entrapped in AOT reverse micelles (aerosol-OT or dioctylsulfosuccinate)] has been described (39). Fluorescent probe molecules were dissolved in the water phase. The ability of the probe to undergo nonradiative transitions after excitation with picosecond laser pulses depended on the reorientational relaxation time of the water: It can be orders of magnitude slower when the water is bound to the interface. It was possible to observe direct competition between probe diffusion and the nonradiative event and to determine both rates as a function of the distance of the probe from the interface.

The polarity probes Prodan and Patman have been incorporated into phospholipid vesicles with the aim of monitoring solvent relaxation (from time-resolved emission spectral shifts) at different depths in the bilayer (40). It was shown that Patman is localized deeper in the membrane and senses a less polar and more restricted probe environment.

TICT state fluorescent probes have been introduced in membranes and their time-resolved fluorescence properties studied. The potential-sensitive fluorescent styryl dye RH421 (*N*-(4-sulfobutyl)-4-(4-(*p*-dipentylaminophenyl)-butadienyl)-pyridinium inner salt), dissolved in organic solvents, cationic cetyltrimethyl ammonium chloride (CTAC), or anionic STS (sodium tetradecyl sulfate) micelles, is an example that exhibits can excited-state reaction because an initial rise in the fluorescence is observed in the red part of the emission spectrum (41). The results are discussed in terms of the formation of a fluorescent TICT state occurring simultaneously with the relaxation of the surrounding solvent cage. The effect of an intramembrane electric field on the photophysical properties of the same dye has been investigated by the binding of the hydrophobic ion tetraphenyl borate (TPB) to DMPC vesicles (42). TPB was found to increase significantly the average fluorescence lifetime, which is consistent with reorientation of the dye farther into the membrane interior.

The mesostructure of evaporated porphyrin films was studied, combining fluorescence decay measurements with atomic force microscopy, confocal fluorescence microscopy, and near-field scanning optical microscopy (NSOM) (43). On evaporation of the solvent, a thin film with rings of porphyrins (wheels) was formed on different substrates. It was found that the structure, size, and perfection of these rings depend mainly on the solvent ($CHCl_3$, CCl_4, MeOH, and mixtures of them) and the substrate (glass and graphite). The rings were composed of porphyrin molecules in a locally aggregated configuration, which was formed prior to complete evaporation of the solvent.

One technique of growing interest for probing molecular interactions at interfaces is time-resolved evanescent wave induced fluorescence spectroscopy (TREWIFS) (44). Using TREWIFS, the adsorption behavior of fluorescently labeled bovine serum albumin was studied as a function of penetration depth and concentration of adsorbed protein. The fluorescence kinetics strongly depend on the surface concentration of adsorbed protein. Hydrophilic and hydrophobic surfaces induced different effects, which were ascribed to either higher surface coverage or a closer packed protein structure.

D. Fluorescence Resonance Energy Transfer

Förster-type fluorescence resonance energy transfer (FRET) between a suitable chromophoric donor-acceptor pair has been extensively used as a "spectroscopic ruler" in self-assembly systems such as micelles, vesicles, and monolayers. Distance measurements using FRET are possible because the efficiency of energy transfer from donor to acceptor depends on the inverse sixth power of the donor-acceptor distance. The concepts of FRET in macromolecular assemblies were developed at the end of the 1970s (45–47). Depending on the nature of the applied donor-acceptor pair, the applicable range is between 1.0 and 5.0 nm (long range!), which spans more than the diameter of a bilayer membrane.

The theory of radiationless energy transfer was formulated by Förster (48). Important to note is that donor and acceptor molecular dipoles are very weakly interacting, meaning that they cannot be nearest neighbors in a self-aggregated system.

Förster has derived the following expression for the rate constant of transfer k_T:

$$k_T = 8.71 \times 10^{23} R^{-6} J \kappa^2 n^{-4} k_f \ (\text{s}^{-1}) \tag{9}$$

where R is the donor-acceptor distance, n the refractive index, k_f the radiative fluorescence rate constant, and J the overlap integral ($\text{M}^{-1}\,\text{cm}^3$), which is the degree of spectral overlap between donor fluorescence and acceptor absorption given by

$$J = \int_0^\infty F_D(\lambda) \varepsilon_A(\lambda) \lambda^4 \, d\lambda \tag{10}$$

where F_D is the normalized fluorescence spectrum of the donor (in such a way that $\int_0^\infty F_D(\lambda)\,d\lambda = 1$) and $\varepsilon_A(\lambda)$ is the extinction coefficient at wavelength λ. The so-called orientation factor κ^2 is

$$\kappa^2 = [\cos\theta_T - 3\cos\theta_D \cos\theta_A]^2 \tag{11}$$

where θ_D and θ_A are the angles between the emission transition moment of the donor and the absorption transition moment of the acceptor, respectively, and the donor-acceptor separation vector; θ_T is the angle between donor-acceptor transition moments.

The orientation factor in energy transfer is, for "unknown" systems, the indeterminate parameter in Eq. (9); all other parameters can be measured or evaluated. Furthermore, many proteins contain more than one donor or acceptor molecule. Dale and coworkers (49) have critically evaluated the possible cases encountered in biopolymers. For details we refer to the original paper. All factors in Eq. (9), except the distance, can be lumped together in the so-called critical transfer (or Förster) distance R_0:

$$R_0 = 9.79 \times 10^3 (\kappa^2 n^{-4} Q_D^0 J)^{1/6} \ (\text{Å}) \tag{12}$$

where Q_D^0 is the quantum yield of donor fluorescence in the absence of acceptor molecules and the superscript 0 denotes the absence of acceptor (note that $Q_D^0 = \tau_D^0/\tau_{D_r}^0$ where $\tau_{D_r}^0$ is the radiative lifetime and τ_D^0 the actual lifetime). The critical transfer distance is defined as the distance at which the transfer rate is equal to the donor fluorescence decay rate. Now Eq. (9) can be simplified to

$$k_T = \frac{1}{\tau_D^0}\left(\frac{R_0}{R}\right)^6 \tag{13}$$

The actual distance R between a donor and an acceptor can be determined by measuring the transfer efficiency E:

$$E = 1 - \frac{\tau_D}{\tau_D^0} = 1 - \frac{Q_D}{Q_D^0} \tag{14}$$

Thus, relative fluorescence lifetime or quantum yield measurements of the donor, in the absence and presence of the acceptor, will yield the efficiency of energy transfer. This experimental efficiency can be directly related to the distance R via

$$E = \frac{k_T}{(\tau_D^0)^{-1} + k_T} = \frac{R_0^6}{R_0^6 + R^6} \tag{15}$$

One practical drawback is that the donor fluorescence parameters in the absence of acceptor are always needed. This means that in proteins with natural donor-acceptor pairs the acceptor needs to be removed to have the fluorescence decay or intensity parameters in the absence of the acceptor.

In artificial membranes the donor-acceptor couples are usually lipophilic and are assumed to be uniformly distributed over both leaflets of the bilayer. In biological membranes, a significant fraction of membrane area is occupied by proteins and the assumption of a uniform lipid probe distribution is no longer valid in these crowded membranes. Zimet et al. (50) considered the situation of a fluorescent donor linked to a membrane protein and lipid-like acceptor molecules partitioned over both leaflets but excluded from the protein area. Monte Carlo simulations were used to generate donor and acceptor distributions that were linked numerically to FRET equations that determine the spectroscopic observables: the average donor quantum yield and fluorescence lifetime distribution as function of protein area fraction (up to 50%).

The problem of electronic energy transfer among chromophores randomly distributed within monodisperse micelles has been addressed (51). Because the energy transfer dynamics are well described by a superposition of pairwise interactions, the pairwise distribution function has been modeled under given geometric constraints and related to macroscopic FRET observables. Different types of energy transfer, application to concentrated micelles, and dynamic and static orientational averaging limits have been discussed in the developed model.

Both experimental and theoretical approaches have been followed to obtain information about the geometry of bilayers formed in nonionic surfactant aqueous dispersions (mixed membranes of POPC/$C_{12}EO_n$ with $n = 1$–8 ethylene oxide units) (52). Depending on the n value, nonlamellar structures ($n < 5$) or mixed micelles ($n > 5$ corresponding to a tendency to solubilize the membrane) are formed. In these mixed membranes the donor is NBD-PE and the acceptor is rhodamine-PE and both donor and acceptor molecules are assumed to be uniformly distributed. From FRET experiments and analyses, the surface area of the lamellae is determined and the area per detergent is found to increase with larger number of EO units. The area is expressed in terms of a critical packing parameter yielding a measure of the asymmetry of the membrane shape.

A purely experimental FRET approach has been developed to analyze the distance between several lipopolysaccharides (LPSs) of different molecular masses equipped with fluorescein isothiocyanate (FITC) as donor and two different lipophilic acceptor molecules (yielding two different critical transfer distances R_0 as internal standard) in sulfobetaine palmitate micelles (53). The short-chain FITC-LPS is located at the micellar surface, and the long-chain FITC-LPS is extended outside the micelle at a distance of $1.5R_0$ or more.

Quantitative studies of the binding of PKC to lipid cofactors in a membrane-mimetic system by monitoring resonance energy transfer have been described (54). For that purpose different headgroup lipids were labeled with a pyrenyl

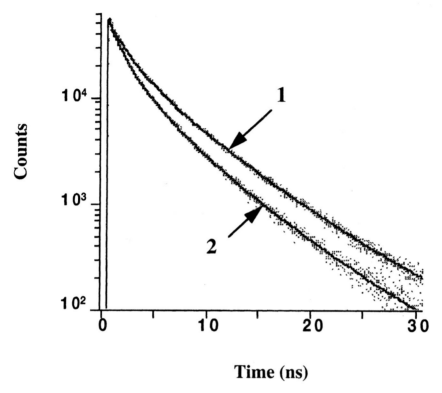

Time (ns)

Fig. 6.   Experimental and fitted decays of tryptophan fluorescence of 80 nM protein kinase C (PKC) in buffer (curve 1) and in the presence of 100 μM thesit [poly(oxyethylene)-9-lauryl ether] containing 5 mol % PS and 4 mol % pyrPIP$_2$ (*sn*-2-pyrenyldecanoyl-phosphatidylinositol-4,5-bisphosphate) (curve 2). The temperature was 293 K. The buffer used was 20 mM Tris-HCl, 50 mM KCl, and 1 mM calcium, pH 7.5. The fitted curves represent the fit to a lifetime distribution recovered with the maximum entropy method of analysis. Because PKC binds selectively to pyrPIP$_2$ in the thesit micelle, energy transfer from the tryptophan residues in PKC to pyrPIP$_2$ occurs, resulting in faster fluorescence decay than that in the absence of micelles. (From Ref. 54.)

decanoyl moiety at the *sn*-2 position of the lipid glycerol. These labeled lipids proved to be excellent energy acceptors of light-excited tryptophan residues in PKC. The quenching efficiency of the tryptophan fluorescence was determined as a function of lipid probe concentration in mixed micelles consisting of poly(oxyethylene)-9-lauryl ether, PS, and various mole fractions of probe lipid. The analysis of the progressively faster decaying tryptophan fluorescence in PKC allowed estimation of the binding constants of pyrene lipids to PKC (Figs. 6 and 7). Only the lipids that were also found to activate PKC in the standard enzymatic assay exhibit high affinity for PKC. In the latter case PKC is bound to the mixed micelle and energy transfer from the tryptophan residues to the pyrene acceptor can take place.

The antibiotic nystatin (see also Ref. 38) is also a good energy acceptor of tryptophan fluorescence, and its interaction with yeast plasma membranes has been investigated with FRET in order to assess whether membrane proteins interact with nystatin in the presence of ergosterol (55).

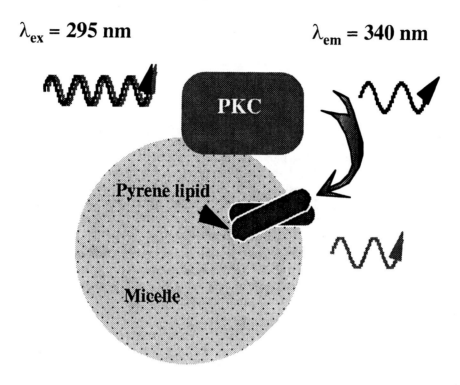

$\lambda_{ex} = 295$ nm $\qquad\qquad \lambda_{em} = 340$ nm

Fig. 7. Cartoon presentation of the binding experimnent in Fig. 6. When PKC selectively binds to the headgroup of a particular pyrenyl-phospholipid, the tryptophan residues in PKC will be close to or within the critical transfer distance (2.7 nm) of the tryptophan-pyrene pair and radiationless energy transfer from tryptophan residues to pyrene can occur, leading to faster tryptophan fluorescence decay. The diameter of PKC is about 6 nm and the diameter of the micelle about 7 nm. A pyrenyl-phospholipid that does not interact with PKC would be distributed over the micelle and energy transfer would be much less probable.

E. Multiphoton Excitation

Multiphoton excitation or MPE (56) can be used to excite molecules more selectively, both in spectral sense and spatially (57). It is a spectroscopic technique in which the energy of two (or more) photons is required for absorption by the chromophore. The flux of excitation light toward the molecule must be so high that, on the time scale of absorption (femtoseconds), multiple photons are available to be absorbed. The energy densities necessary for this can be obtained only from ultrashort (femtosecond) high peak power (sub-MW) laser pulses and only when the beam is focused to a micrometer spot. When two-photon excitation (2PE) is applied, the fluorescence yield has a quadratic dependence on the excitation power. On the other hand, the fluorescence yield, compared with the yield of background fluorescence, also has a quadratic dependence on the cross sections of the respective optical densities at half the excitation wavelength. Another important property of multiphoton excitation is the dependence of the fundamental anisotropy (r_0) on the applied transition moment. For reduced nicotinamide adenine dinucleotide (NADH) a 46% increase of the initial polarization was reported (58) in the case of 2PE compared with one-photon excitation (1PE). This will, of course, enhance the dynamic range and thus the accuracy of fluorescence anisotropy studies. For three-photon excitation (3PE), however, the initial anisotropy for *N*-acetyl tryptophanamide (NATA) was found to be 0.06 and negative, whereas it was 0.13 and positive for the case of 1PE (59). Therefore, different excitation conditions should be considered depending on the requirements of the experiment.

When a laser beam is focused into a sample, the beam travels through a cone, reaches the focal volume, and spreads out in another cone. When 1PE is applied, combined with confocal detection (60), (part of) the fluorescence of the focal point as well as the two cones will be excited and detected (see also Ref. 61). However, when 2PE is applied, the excitation energy density can be tuned in such a way that two-photon absorption can occur only in the very focus of the beam. At other places the energy density is then too low for two-photon absorption. Therefore no absorption and no fluorescence from places other than the focal volume (even smaller than in the case of 1PE) will be observed. Furthermore, Rayleigh scattering and Raman scattering usually have a quite different spectral position compared with a 1PE experiment and can be easily eliminated. In addition, with infrared light deeper penetration depths in biological materials can be obtained. A drawback for MPE can be the risk of continuum generation in the solvent when these high peak energies are applied (62).

The evanescent wave at 770 nm from a femtosecond Ti:sapphire laser was used to excite the calcium probe Indo 1 at a quartz-water interface (63). The fluorescence intensity of Indo 1 depended quadratically on the incident power slightly above and below the critical angle. Scattered light did not contribute to the emission signal. 2PE gave rise to significantly higher anisotropy.

The effect of near-infrared femtosecond lasers on the vitality and reproduction of Chinese hamster ovary cells (730–800 nm, 80 MHz, 80 μs pixel dwell time) has been investigated (64). Cells were unaffected by powers less than 1 mW. The cells did not divide (die or form giant cells) at laser powers > 6 mW and were destroyed at mean powers > 10 mW. Extremely high fields may induce destructive intracellular plasma formation.

MPE at 730 and 960 nm was used to image human skin autofluorescence near the surface (50 μm) or at greater depths (100–150 μm) (65). Using both excitation wavelengths, the fluorescence of reduced pyridine nucleotides (730 nm) and flavoproteins (960 nm) could be identified.

As already indicated, a number of MPE applications have been carried out in the past few years. An increasing number of applications can be foreseen because of the challenging possibilities in many fields of the biological sciences. The commercial availability of femtosecond pulsed lasers in the near-infrared wavelength region as well as complete microscope setups with 2PE will certainly catalyze MPE applications in biological interfaces.

F. Time-Resolved Single-Molecule Fluorescence

In the past few years, a number of different optical spectroscopic techniques have been developed that can be summarized as single-molecule spectroscopy. One goal of that idea is to isolate the molecule in such a way that the spectroscopic properties are not perturbed by the environment or are disturbed only in a well-described way. Furthermore, there has always been a quest for ultimate sensitivity in order to study the smallest amounts of material. There are in principle two directions to go in for exciting only a single molecule. These are to decrease the volume of excitation (MPE or near- or far-field excitation) and to decrease the concentration of chromophores. Only a few single-molecule techniques have been applied for time-resolved fluorescence studies of interactions of molecules with vesicles, membranes, or artificial interfaces but it is expected that many applications will be developed in the near future.

In scanning near-field optical microscopy (NSOM) the measured fluorescence lifetime of a single dye molecule can be shortened or lengthened depending on the distance between the molecule and an aluminum-coated fiber tip (66). The lifetime variations can be simulated and the computations provide insight into the spectroscopic properties in front of a metal-dielectric interface of arbitrary geometry.

Macklin et al. (67) applied far-field microscopy to probe the optical properties of single dye molecules at a polymer-air interface. Because of the isolated nature of the molecule in these experiments, more specific information can be retrieved. Shifts in the fluorescence spectrum and the orientation of the transition dipole moment were correlated with variation of the fluorescence lifetime caused by the frequency dependence of the spontaneous emission rate (in case of spectral shifts) and by the effects of electromagnetic boundary conditions on the fluorescent radiation at the polymer-air interface (in case of dipole orientation effects).

III. CONCLUSIONS

As is evident from this chapter, there are many interesting problems in the field of biological interfaces that can be effectively studied with time-resolved fluorescence techniques. Many specific questions concerning structure and dynamics of interfaces can be answered by the application of the different variations of time-resolved fluorescence techniques. Judging from other chapters in this book, different techniques have their own benefits, leading to information at other structural levels and in other time frames. A complete overview of all possibilities will lead to the most complete description of these interfaces.

ACKNOWLEDGEMENT

Part of this work was supported by the Netherlands Organization for Scientific Research (NWO). Dr. N. V. Visser is kindly acknowledged for preparing the figures.

REFERENCES

1. A Carrington, AD McLaughlan. Introduction to Magnetic Resonance. London: Harper, 1967.
2. JI Steinfeld. Molecules and Radiation. An Introduction to Modern Molecular Spectroscopy, New York: Harper & Row, 1974.
3. JR Lakowicz. Principles of Fluorescence Spectroscopy. New York: Plenum, 1983.
4. JR Lakowicz, ed.: Topics in Fluorescence Spectroscopy. Parts 1–4. New York: Plenum, 1991–1994.
5. RP Haugland. Handbook of Fluorescent Probes and Research Chemicals. 6th ed. Eugene, OR: Molecular Probes, 1996.
6. AJWG Visser. Time-resolved fluorescence on self-assembly membranes. Curr Opin Colloid Interface Sci 2:27, 1997.

7. PP Infelta, M. Grätzel, JK Thomas. Luminescence decay of hydrophobic molecules solubilized in aqueous micellar systems. A kinetic model. J Phys Chem 78:190, 1974.
8. M Tachiya. Application of a generating function to reaction kinetics in micelles. Kinetics of quenching of luminescent probes in micelles. Chem Phys Lett 33:289, 1975.
9. F Caruso, F Grieser, PJ Thistlewaite, M Almgren. Two-dimensional diffusion of amphiphiles in phospholipid monolayers at the air-water interface. Biophys J 65:2493, 1993.
10. D Danino, Y Talmon, R Zana. Alkanediyl- α,ω-bis(dimethylalkylammonium bromide) surfactants (dimeric surfactants). 5. Aggregation and microstructure in aqueous solutions. Langmuir 11:1448, 1995.
11. M Swanson-Vethamuthu, M Almgren, G Karlsson, P Bahadur. Effect of sodium chloride and varied alkyl chain length on aqueous cationic surfactant–bile salt systems. Cryo-TEM and fluorescence quenching studies. Langmuir 12:2173, 1996.
12. M Almgren, P Hansson, K Wang. Distribution of surfactants in a nonideal mixed micellar system. Effect of a surfactant quencher on the fluorescence decay of solubilized pyrene. Langmuir 12:3855, 1996.
13. C Zannoni, A Arcioni, P Cavatorta. Fluorescence depolarization in liquid crystals and membrane bilayers. Chem Phys Lipids 32:179, 1983.
14. A Szabo. Theory of fluorescence depolarization in macromolecules and membranes. J Chem Phys 81:150, 1984.
15. UA van der Heide, G van Ginkel, YK Levine. DPH is localised in two distinct populations in lipid vesicles. Chem Phys Lett 253:118, 1996.
16. EHW Pap, M Ketelaars, JW Borst, A van Hoek, AJWG Visser. Reorientational properties of fluorescent analogues of the protein kinase C cofactors diacylglycerol and phorbol ester. Biophys Chem 58:255, 1996.
17. JM Muller, G van Ginkel, E van Faassen. Effect of gramicidine A on structure and dynamics of lipid vesicle bilayers. A time-resolved fluorescence depolarization study. Biochemistry 34:3092, 1995.
18. JM Muller, G van Ginkel, E van Faassen. Effect of lipid molecular structure and gramicidine A on the core of lipid vesicle bilayers. A time-resolved fluorescence depolarization study. Biochemistry 35:488, 1996.
19. C Ho, SJ Slater, CD Stubbs. Hydration and order in lipid bilayers Biochemistry 34:6188, 1995.
20. C Bernsdorff, R Winter, TL Hazlett, E Gratton. Influence of cholesterol and β-sitosterol on the dynamic behavior of DPPC as detected by TMA-DPH and PyrPC fluorescence. A fluorescence lifetime distribution and time-resolved anisotropy study. Ber Bunsenges 99:1479, 1995.
21. S-Y Cheng, KH Cheng. Detection of membrane packing defects by time-resolved fluorescence depolarization. Biophys J 71:878, 1996.
22. CR Mateo, AU Acuna, JC Brochon. Liquid-crystalline phases of cholesterol/lipid bilayers as revealed by the fluorescence of *trans*-parinaric acid. Biophys J 68:978, 1995.
23. M Velez, MP Lillo, AU Acuna, J Gonzales-Rodriguez. Cholesterol effect on the physical state of lipid multibilayers from the platelet plasma membrane by time-resolved fluorescence. Biochim Biophys Acta 1235:343, 1995.
24. LJ DeLong, JW Nichols. Time-resolved fluorescence anisotropy of fluorescent-labeled lysophospholipid and taurodeoxycholate aggregates. Biophys J 70:1466, 1996.
25. NC Maiti, S Mazumdar, N Periasamy. Dynamics of porphyrin molecules in micelles. Picosecond time-resolved fluorescence anisotropy studies. J Phys Chem 99:10708, 1995.
26. MCL Maste, EHW Pap, A van Hoek, W Norde, AJWG Visser. Spectroscopic investigation of the structure of a protein adsorbed on a hydrophobic latex. J Colloid Interface Sci 180:632, 1996.

27. EHW Pap, A Hanicak, A van Hoek, KWA Wirtz, AJWG Visser. Quantitative analysis of lipid-lipid and lipid-protein interactions in membranes by use of pyrene-labeled phophoinositides. Biochemistry 34:9118, 1995.

28. MAMJ van Zandvoort, HC Gerritsen, YK Levine. Distribution of hydrophobic molecules in lipid bilayers. 1. Monte Carlo dynamics simulations. J Phys Chem B 101:4142, 1997.

29. MAMJ van Zandvoort, HC Gerritsen, G van Ginkel, YK Levine, R Tarroni, C Zannoni. Distribution of hydrophobic molecules in lipid bilayers. 2. Time-resolved fluorescence anisotropy study of perylene in vesicles. J Phys Chem B 101:4149, 1997.

30. L Davenport, P Targowski. Submicrosecond phopholipid dynamics using a long-lived fluorescence emission anisotropy probe. Biophys J 71:1837, 1996.

31. I Konopasek, P Kvasnicka, E Amler, A Kotyk, G Curatola. The transmembrane gradient of the dielectric constant influences the DPH lifetime distribution. FEBS Lett 374:338, 1995.

32. J Martins, WLC Vaz, E Melo. Long-range diffusion coefficients in two-dimensional fluid media measured by the pyrene excimer reaction. J Phys Chem 100:1889, 1996.

33. R Merkel, E Sackmann. Nonstationary dynamics of excimer formation in two-dimensional fluids. J Phys Chem 98:4428, 1994.

34. KH Cheng, P Somerharju. Effects of unsaturation and curvature on the transverse distribution of intramolecular dynamics of dipyrenyl lipids. Biophys J 70:2287, 1996.

35. S Mazères, V Schram, J-F Tocanne, A Lopez. 7-Nitrobenz-2-oxa-1,3-diazole-4-yl–labeled phospholipids in lipid membranes: Differences in fluorescence behavior. Biophys J 71:327, 1996.

36. CF Chapman, Y Liu, GJ Sonek, BJ Tromberg. The use of exogenous fluorescent probes for temperature measurements in single living cells. Photochem Photobiol 62:416, 1995.

37. JDA Shrive, JD Brennan, RS Brown, UK Krull. Optimization of the self-quenching response of nitrobenzoxadiazole dipalmitoylphosphatidylethanolamine in phospholipid membranes for biosensor development. Appl Spectrosc 49:304, 1995.

38. A Coutinho, M Prieto. Self-association of the polyene antibiotic nystatin in dipalmitoylphosphatidylcholine vesicles: A time-resolved fluorescence study. Biophys J 69:2541, 1995.

39. CH Cho, M Chung, J Lee, T Nguyen, S Singh, M Vedamuthu, S Yao, J-B Zhu, GW Robinson. Time- and space-resolved studies of the physics and chemistry of liquid water near a biologically relevant interface. J Phys Chem 99:7806, 1995.

40. R Hutterer, FW Schneider, H Sprinz, M Hof. Binding and relaxation behaviour of Prodan and Patman in phospholipid vesicles: A fluorescence and ^{1}H NMR study. Biophys Chem 61:515, 1996.

41. NV Visser, A van Hoek, AJWG Visser, RJ Clarke, JF Holzwarth. Time-resolved polarized fluorescence of the potential-sensitive dye RH421 in organic solvents and micelles. Chem Phys Lett 231:551, 1994.

42. NV Visser, A van Hoek, AJWG Visser, J Frank, H-J Apell, RJ Clarke. Time-resolved fluorescence investigations of the interaction of the voltage-sensitive probe RH421 with lipid membranes and proteins. Biochemistry 34:11777, 1995.

43. J Hofkens, L Latterini, P Vanoppen, H Faes, K Jeuris, S De Feyter, J Kerimo, PF Barbara, FC De Schryver, AE Rowan and RJM Nolte. Mesostructure of evaporated porphyrin thin films: Porphyrin wheel formation. J Phys Chem B 101:10588, 1997.

44. B Crystal, G Rumbles, TA Smith, D Phillips. Time-resolved evanescent wave induced fluorescence measurements of surface adsorbed bovine serum albumin. J Colloid Interface Sci. 155:247, 1993.

45. L Stryer. Fluorescence energy transfer as a spectroscopic ruler. Annu Rev Biochem 47:819, 1978.

46. RH Fairclough, CR Cantor. The use of singlet-singlet energy transfer to study macromolecular assemblies. Methods Enzymol 48:347, 1978.

47. BK-K Fung, L Stryer. Surface density determinations in membranes by fluorescence energy transfer. Biochemistry 17:5241, 1978.
48. T Förster. Zwischenmolekulare Energiewanderung und Fluoreszens. Ann Phys 2:55, 1948.
49. RE Dale, J Eisinger and WE Blumberg. The orientational freedom of molecular probes. The orientation factor in intramolecular energy transfer. Biophys J 26:161 1979.
50. DB Zimet, BJ-M Thevenin, AS Verkman, SB Shohet, JR Abney. Calculation of resonance energy transfer in crowded biological membranes. Biophys J 68:1592, 1995.
51. AV Barzykin, M Tachiya. Electronic energy transfer in concentrated micellar solutions. J Chem Phys 102:3146, 1995.
52. G Lantzsch, H Binder, H Heerklotz, M Wendling, G Klose. Surface areas and packing constraints in POPC/C12EOn membranes. A time-resolved fluorescence study. Biophys Chem 58:289, 1996.
53. CA Wiström, GM Jones, PS Tobias, LA Sklar. Fluorescence resonance energy transfer analysis of lipopolysaccharide in detergent micelles. Biophys J 70:988, 1996.
54. EHW Pap, PIH Bastiaens, JW Borst, PAW van den Berg, A van Hoek, GT Snoek, KWA Wirtz, AJWG Visser. Quantitation of the interaction of protein kinase C with diacyl glycerol and phosphoinositides by time-resolved detection of resonance energy transfer. Biochemistry 32:13310, 1993.
55. M Opekarova, P Urbanova, I Konopasek, P Kvasnicka, K Strzalka, K Sigler, E Amler. Possible nystatin-protein interaction in yeast membrane vesicles in the presence of ergosterol. A Förster energy transfer study. FEBS Lett 386:181, 1996.
56. WL Peticolas, JP Goldsborough, KE Rieckhoff. Double photon excitation in organic crystals. Phys Rev Lett 10:43 1963.
57. W Denk, JH Strickler, WW Webb. Two-photon laser scanning fluorescence microscopy. Science 248:73, 1990.
58. B Kierdaszuk, H Malak, I Gryczynski, P Callis, JR Lakowicz. Fluorescence of reduced nicotinamides using one and two photon excitation. Biophys Chem 62:1, 1996.
59. I Gryczynski, H Malak, JR Lakowicz, HC Cheung, J Robinson, PK Umeda. Fluorescence spectra properties of troponin C mutant F22w with one, two and three photon excitation. Biophys J 71:3448, 1996.
60. GJ Brakenhoff, P Blom, P Barends. Confocal scanning light microscopy with high aperture immersion lenses. J Microsc 117:219, 1979.
61. EHK Stelzer, S Lindek. Fundamental reduction of the observation volume by detection orthogonal to the illumination axis: Confocal theta microscopy. Opt Commun 111:536, 1994.
62. L Brand, C Eggeling, C Zander, KH Drexhage, CAM Seidel. Single molecule identification of Coumarine 120 by time-resolved fluorescence detection: Comparison of one and two photon excitation in solution. J Phys Chem A 101:4313, 1997.
63. I Gryczynski, Z Gryczynski, JR Lakowicz. Two-photon excitation by the evanescent wave from total internal reflection. Anal Biochem 247:69, 1997.
64. K König, PTC So, WW Mantulin, E Gratton. Cellular response to near infrared femtosecond laser pulses in two-photon microscopes. Opt Lett 22:135, 1997.
65. BR Masters, PTC So. Multiphoton excitation fluorescence microscopy and spectroscopy of in vivo human skin. Biophys J 72:2405, 1997.
66. RX Bian, RC Dunn, XS Xie. Single molecule emission characteristics in near field microscopy. Phys Rev Lett 75:4772 1995.
67. JJ Macklin, JK Trautman, TD Harris, LE Brus. Imaging and time-resolved spectroscopy of single molecules at an interface. Science 272:255, 1996.

20

Circular Dichroism of Proteins in Solution and at Interfaces

C. P. M. van Mierlo, H. H. J. de Jongh, and A. J. W. G. Visser
Wageningen University and Research Centre, Wageningen, The Netherlands

I. OUTLINE

Compared with circular dichroism (CD) investigations of proteins in solution, rather few CD studies have been made of proteins localized at interfaces. The majority of the systems studied consist of proteins (or peptides) either adsorbed to or incorporated into artificial biological membranes. Such systems can be considered as liquid-liquid interfaces with which proteins are in one way or another associated. The even more scarcely studied category comprises proteins adsorbed at microscopic solid interfaces that are dispersed in aqueous solution. With respect to both situations, the question arises of whether the protein retains or changes its conformation upon interaction with the interfaces. Answers can be obtained by using CD spectroscopy as illustrated in this chapter.

In order to understand the use of CD spectroscopy applied to the conformational analysis of proteins, some theory has to be given first. The basic aspects of the technique as treated in numerous physicochemical textbooks (e.g., 1–3) will be summarized first. Books that provide an updated treatise on a variety of applications of circular dichroism (4,5) are also consulted. They give an in-depth survey of various aspects of circular dichroism as applied to biological macromolecules that are not (or are hardly) covered in this chapter. After the basic introduction, a section is devoted to secondary structure analysis from far-ultraviolet (UV) protein CD spectra and to the fingerprint information that can be obtained from UV-visible (near-UV) CD spectra. Subsequently, the use of CD spectroscopy and of stopped-flow CD as an important tool to elucidate folding and unfolding of proteins is described. An updated survey of CD applied to proteins at interfaces is given in a separate section. Finally, the chapter closes with a summary of instrumental aspects of circular dichroism.

II. BASIC THEORY

A. Introduction to Optical Spectroscopy

Light is a rapidly oscillating electromagnetic field. The interaction of electromagnetic waves with matter can be used to extract information about biological macro-molecules. This is the basis of spectroscopy. Optical spectroscopy is concerned with wavelengths between about 150 nm (vacuum UV) and 6000 nm [far infrared (IR)]. Optical spectroscopy involves irradiation of the sample with some form of electro-magnetic radiation, followed by measurement of a spectral property such as scattering, absorption, or emission and, finally, by interpretation of the measured spectral par-ameters to obtain specific molecular information. Electromagnetic radiation has two waves at right angles to each other; one is electric ($\mathbf{E}$) and the other is magnetic ($\mathbf{H}$) (see Fig. 1). They are produced by oscillating electric or magnetic dipoles and are propagated through vacuum with the velocity of light $c(c = 3 \times 10^8 \, \mathrm{m\,s^{-1}})$. In Fig. 1 plane or linearly polarized light is shown, because the $\mathbf{E}$ vector oscillates in one plane. Unpolarized light contains oscillations of the $\mathbf{E}$ wave in all directions perpen-dicular to the direction of propagation (P). The frequency (v) and wavelength (λ) of a wave are related by the equation

$$v = c/\lambda \tag{1}$$

Frequency can be directly converted into energy units using the relationship

$$E = hv \tag{2}$$

where h is Planck's constant ($h = 6.63 \times 10^{-34} \, \mathrm{J\,s}$). Units of energy ($\mathrm{J\,mol^{-1}}$), fre-quency (Hz), and wavelength (m) are all used in optical spectroscopy. Also used is the wavenumber (v'), defined as the inverse of the wavelength in centimeters:

$$v' = 1/\lambda = v/c \tag{3}$$

The wave number is thus the number of waves per centimeter.

B. The Phenomenon of Optical Activity: Circularly Polarized Light

A molecule is optically active if it interacts differently with left- and right-handed circularly polarized light. Such interaction can be detected in two ways: either as a differential change in velocity of the two beams through the sample (optical rotatory dispersion) or as differential absorption of each beam (circular dichroism). These phenomena are usually measured in the wavelength range 180–700 nm connected with electronic transitions. Optical activity measurements applied to biological macro-molecules provide information on molecular conformation, conformational changes upon ligand binding, and secondary structure. Circular dichroism is more frequently

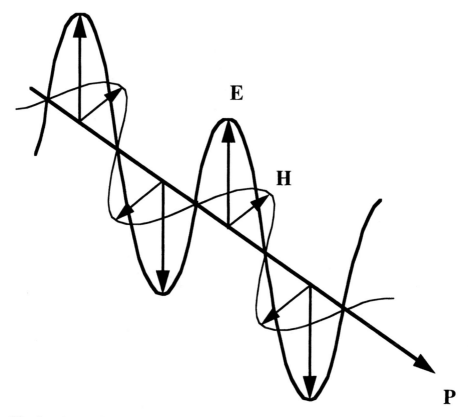

Fig. 1 Plane-polarized light.

used than optical rotatory dispersion (ORD). ORD will therefore not be treated in this chapter. To observe CD, a molecule must satisfy two prerequisites: it must have chirality (as in L-amino acids) and be chromophoric (the molecule must absorb light of certain wavelength). In biological macromolecules chirality is often induced by asymmetrically placed neighboring groups.

Circularly polarized light is shown in Fig. 2a. If one follows the point of the E vector, the movement is like a helical path around the axis of propagation (*P*). This motion can be left- or right-handed depending on the sense of rotation with respect to the direction of propagation. Plane-polarized light can be composed of a sum of left- and right-handed circularly polarized waves of equal amplitude (see Fig. 2b). When an optically active compound preferentially absorbs one of the circular components, the outgoing beam is elliptically polarized (Fig. 2c). This ellipticity is determined by the axial ratio of the ellipse. CD can be expressed in molar ellipticity [θ] (see later).

It is equally easy to represent a circularly polarized wave as the sum of two plane-polarized waves, 90° out of phase and with their electric vectors perpendicular (see Fig. 2d).

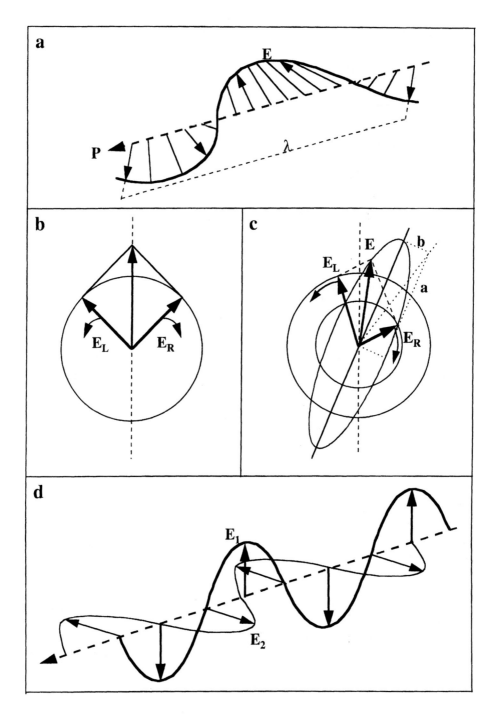

Fig. 2 (a) Circularly polarized light. Only the electric field is shown. (b) Plane-polarized light as the sum of two circularly polarized waves of equal amplitude. (c) Elliptically polarized light upon leaving an optically active, light-absorbing compound. (d) Circularly polarized wave represented as sum of two plane-polarized waves, shifted 90° out of phase and with their electric vectors perpendicular.

C. Circular Dichroism: The Molecular Basis of Optical Activity

CD at a given wavelength is defined as

$$\Delta\varepsilon = \varepsilon_L - \varepsilon_R \qquad (4)$$

$\Delta\varepsilon$, expressed in $(\text{mol dm}^{-3})^{-1}\,\text{cm}^{-1}$, is much smaller (approximately three to five orders of magnitude) than the molar extinction coefficient ε itself. The relation between molar ellipticity $[\theta]$, expressed in degree $\text{cm}^2\,\text{dmol}^{-1}$, and $\Delta\varepsilon$ is (the derivation can be found in Ref. 1 or 2):

$$[\theta] = 3300\Delta\varepsilon \qquad (5)$$

How can one obtain a molecular picture of CD? We know that when a molecule is excited with light, a displacement of charge takes place. This linear displacement is associated with an electric transition dipole moment, denoted by μ. The transition can also have a circular component, which in turn can generate a magnetic dipole moment **m** perpendicular to the circular plane (Fig. 3). Optical activity requires that circularly polarized light induces both dipole moments in more or less the same direction in the molecule. As the dipole strength is a measure of the intensity of an optical absorption transition, the rotational strength is the determining factor in CD. The rotational strength (R) is equal to the integrated dichroism over the absorption band:

$$R = \frac{3hc}{8\pi^3 N_0} \int \frac{\Delta\varepsilon}{\lambda}\, d\lambda \qquad (6)$$

(N_0 is Avogadro's number). The rational strength can be computed from quantum

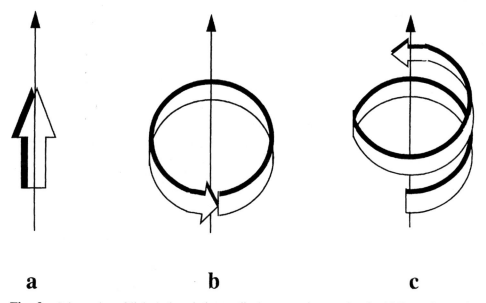

a   **b**   **c**

Fig. 3 Schematics of light-induced charge displacement in a molecule. (a) Pure electronic absorption: $\langle\Psi_0|\mu|\Psi_a\rangle$. (b) Pure magnetic absorption: $\langle\Psi_a|\mathbf{m}|\Psi_0\rangle$. (c) Optical activity: $\langle\Psi_0|\mu|\Psi_a\rangle \otimes \langle\Psi_a|\mathbf{m}|\Psi_0\rangle$.

mechanical principles and from knowledge of the wave functions describing ground (Ψ_0) and excited states (Ψ_a). The result is

$$R = \mathrm{Im}\left(\langle\Psi_o|\mu|\Psi_a\rangle\langle\Psi_a|\mathbf{m}|\Psi_o\rangle\right) \qquad (7)$$

The vector μ and $\mathbf{m}$ are now the electric dipole operator and magnetic dipole operator, respectively, because they operate on the wave functions. "Im" means the imaginary part of the expression that follows in Eq. (7). For each electron, $\mathbf{m}$ can be written as the following cross-product:

$$\mathbf{m} = \frac{e}{2mc}(\mathbf{r} \times \mathbf{p}) \qquad (8)$$

where e and m are the charge and mass of the electron, $\mathbf{p}$ is the momentum operator, and $\mathbf{r}$ is the position operator of the electron. The cross-product $\mathbf{r} \times \mathbf{p}$ is the orbital angular momentum of an electron. Therefore the magnetic transition dipole [right-hand part of right-hand side of Eq. (7)] corresponds to a circulation of charge or current loop (see Fig. 3). The dipole strength is an expression similar to Eq. (7) but with the electric dipole operator μ instead of the magnetic one $\mathbf{m}$. The dipole strength is then the square of the light-induced electric dipole (see Fig. 3).

Optical activity involves both electric and magnetic interactions. Equation (7) implies that R is a dot product and thus a scalar (a so-called physical observable, just a number). We can express this as follows:

$$R = |\mu||m|\cos\delta \qquad (9)$$

where $|\mu|$ and $|m|$ are the magnitudes of the vectors and δ is the angle between the two dipole moments. For a molecule to be optically active the magnetic dipole operator $\mathbf{m}$ must have a component parallel to the electric dipole operator μ. To accomplish this the molecule must be asymmetric. Many helical macromolecules have intense optical activity, because the helical structure facilitates a helical flow of charge.

Figure 4a shows a circularly polarized wave that interacts with a right-handed helical fragment. The circularly polarized light wave is represented by two plane-polarized components $90°$ out of phase. The electric field component $\mathbf{E}_1$ will induce an oscillating electric dipole moment μ with a component parallel to the helix axis. Similarly, an oscillating magnetic field component $\mathbf{H}_2$ will induce a current parallel to the helix axis. According to Maxwell's laws, this depends on $d\mathbf{H}_2/dt$ (first derivative with time). $d\mathbf{H}_2/dt$ is in phase with $\mathbf{E}_1$ and both contribute synchronously to electron displacement on the helical path or, in other words, create an oscillating electrical dipole moment μ and magnetic dipole moment $\mathbf{m}$ parallel to the helix axis. According to Eq. (9) this corresponds to maximal rotational strength because $\delta = 0°$. The components $\mathbf{E}_2$ and $\mathbf{H}_1$ do not contribute because they do not result in electron displacement.

One can now intuitively understand why a molecule such as coronene is optically inactive and why a molecule such as hexahelicene has intense optical activity (Fig. 4c). Coronene is symmetric (D_{6h} symmetry). Excitation with circularly polarized light of the proper energy induces a charge displacement of π^* electrons on a circular path (μ). From Maxwell's laws of electromagnetism we know that a magnetic moment

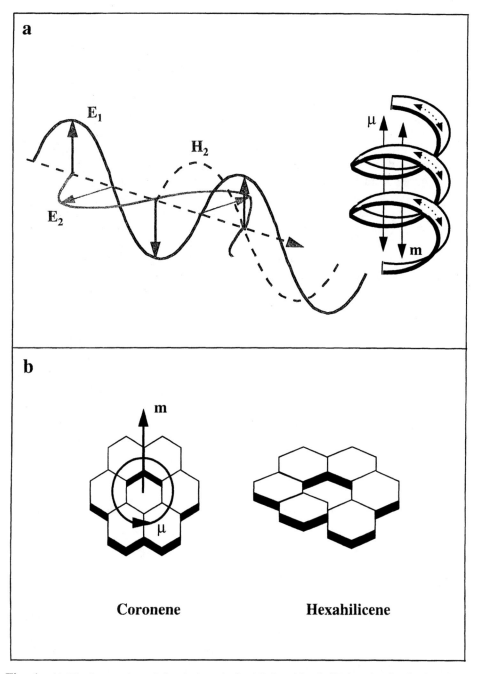

Fig. 4 (a) The interaction of circularly polarized light with a helical molecule. (b) Coronene and hexahelicene.

m will be induced that is perpendicular to the plane in which the electrons flow. The angle δ is 90° [see Eq. (9)] and $R = 0$. Hexahelicene is strongly optically active because the molecule adopts a helical configuration.

D. Exciton Splitting Illustrated with a Dimeric Molecule

When one makes, e.g., a 10 μM solution of reduced nicotinamide adenine dinucleotide (NADH) and subsequently measures its absorbance at 340 nm to determine its exact concentration, we rely on the assumption that the molecules do not interact with each other and on the extinction coefficient being the same for all molecules. We can safely assume this because, when the average distance between the molecules is calculated, it turns out that they are completely isolated and do not interact. In biological macromolecules, however, we are dealing with molecular assemblies. Examples are proteins in which peptide chromophores are linked together; nucleic acids, which are polymers of deoxyribonucleotide ribonucleotide units forming single or double-helical strands; and light-harvesting proteins in which many chlorophyll molecules are dispersed. In all these cases the chromophores are feeling each other's presence; they are definitely interacting. All biological aromatic molecules in the ground state have some permanent dipole moment (a few Debye). When these molecules are brought to the excited state by absorption of light of the proper energy, it can be generalized that these dipole moments become much larger (in some cases even 5 to 10 times). A molecule in the excited state has a different charge distribution and is much more polar than in the ground state. Consequently, the dipolar interactions are more pronounced in the excited state and strongly influence the absorption spectrum.

These effects are illustrated using two identical chromophores that are covalently linked together (see Fig. 5a). The two relative orientations of the dipoles in the two molecules correspond to a situation in which the dipoles oscillate in phase (plus state, left) or out of phase (minus state, right). Because of dipolar interaction an absorption spectrum is split into two bands corresponding with two transitions ν_+ and ν_- (Fig. 5b). This is called *exciton splitting* to emphasize the fact that the excited state is affected rather than the ground state. The transition probabilities (or dipole strength; see earlier) for the two states are given by

$$D_\pm = D_0(1 \pm \cos\alpha) \tag{10}$$

in which α is the angle between the transition dipoles μ_1 and μ_2 (see Fig. 5c). Note further that $D_+ + D_- = 2D_0$ for each angle α. The frequencies of the new absorption bands are

$$\nu_\pm = \nu_0 \pm \nu_{12} \tag{11}$$

where ν_{12} is a measure of the dipole-dipole interaction between μ_1 and μ_2.

$$\nu_{12} = D_0 = \frac{(\cos\alpha - 3\cos\gamma\cos\beta)}{r^3} \tag{12}$$

The dominant contribution to the rotational strengths R_+ and R_- of the two states arising from dipolar interaction is given by (see Ref. 2 for derivation)

$$R_\pm = \pm\left(\frac{\pi}{2\lambda}\right)r_{12}\cos\delta(\mu_1 \times \mu_2) \tag{13}$$

in which $\mu_1 \times \mu_2$ is a cross-product and δ is the angle between r_{12} (Fig. 5c) and the normal to the plane formed by μ_1 and μ_2. The product is a measure of helical twist in going from μ_1 to μ_2.

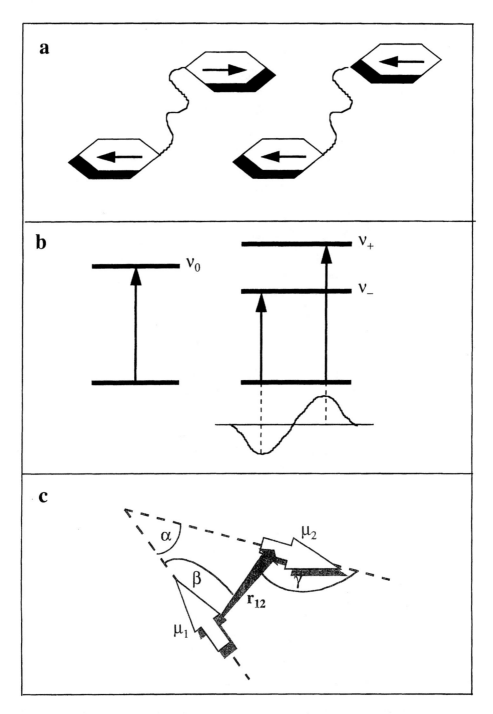

Fig. 5 (a) Orientation of transition dipole moments in a dimer. (b) Exciton splitting of an absorption spectrum into two bands by interacting transition dipoles. The CD spectrum shows a positive Cotton effect for the ν_+ transition and a negative one for the ν_- transition. (c) The interaction between two dipoles μ_1 and μ_2 can be described by the angles α, β, and γ and the position vector $\mathbf{r}_{12}$.

III. INFORMATION FROM CIRCULAR DICHROISM SPECTRA OF PROTEINS

A. Circular Dichroism of Proteins

All amino acids except glycine are asymmetric and hence optically active. Apart from aromatic amino acids, the chromophoric group in proteins is the peptide bond. The arrangement of these bonds in organized structures such as α-helix, β-sheets (parallel and antiparallel), and β-turns is the reason that such distinct and different CD spectra are observed between different proteins. Figure 6a shows an example of the CD spectrum of a synthetic polypeptide in α-helix conformation (adapted from Ref. 6). For comparison the absorption spectrum is also given.

Three electronic transitions can be observed:

1. At about 210–220 nm a weak absorption band can be seen in the absorption spectrum that has considerable optical activity in the case of an α-helix. This is the n,π^* transition, in which the nonbonding electrons of the carbonyl oxygen are promoted to the antibonding π orbital. The n,π^* transition has a low extinction coefficient (is forbidden) but high rotational strength. The latter can be visualized as a charge rotation about the $C-O$ bond accompanied by the generation of a magnetic moment (Fig. 6b).

2. In the absorption spectrum two shoulders are seen at 185 and 205 nm. These correspond to a single allowed π,π^* transition, which has a high extinction coefficient. The π-electrons are delocalized over the whole peptide fragment. In an α-helix the peptide chromophoric groups feel each other's presence and due to this interaction the π,π^* transition is split into two bands (positive and negative) with the sign change occurring at about 200 nm. This is a clear example of exciton splitting.

3. The spectrum below 180 nm is less well understood. There is probably another π,π^* transition at 160 nm. This spectral region (and lower) is called the vacuum UV and the optical energies are such that the localized σ-electrons can be excited into the antibonding σ-orbital (σ,σ^* transition).

B. Prediction of Protein Secondary Structure

Figure 6a shows the CD spectrum of a polypeptide in the α-helix conformation. Changing the orientation of the peptide bond into other organized or nonorganized structures has a drastic influence on the CD spectrum as illustrated for poly-L-lysine in Fig. 7a (adapted from Ref. 7).

In random coil conformation the n,π^* transition does not exhibit dichroism, and the exciton splitting of the π,π^* transition is completely absent. The β-sheet conformation shows a CD spectrum intermediate between the other two spectra. It has been pointed out by Johnson (8) that one should record CD spectra to about 180 nm in order to deduce spectral features that are characteristic for various conformations. In this way the information content of CD is increased and computer predictions can be made with higher confidence. For instance, it is possible to make distinctions between antiparallel or parallel β-sheets and a β-turn.

Nonetheless, in the 1970s the spectra shown in Fig. 7a served as a basis set for protein secondary structure prediction. As an approximation the protein is built up of a linear combination of α-helical, β-sheet, and random coil structures. A

a

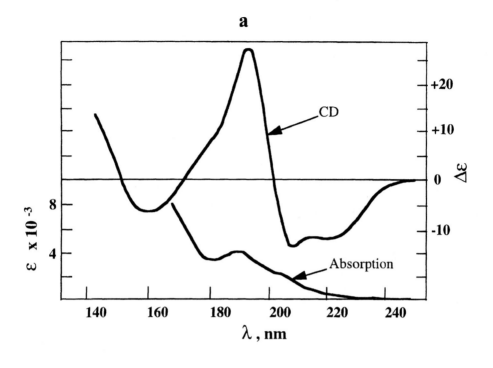

b

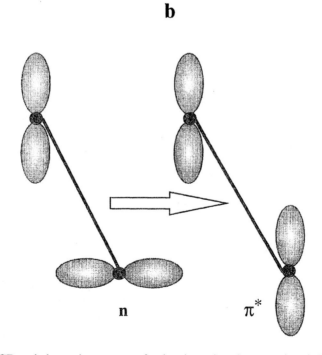

Fig. 6 (a) CD and absorption spectra of poly-γ-benzyl-L-glutamate in α-helix conformation adapted from Ref. 6. (b) n- and π*-orbitals of the carbonyl peptide bond. The magnetic moment is due to charge rotation about the C–O bond.

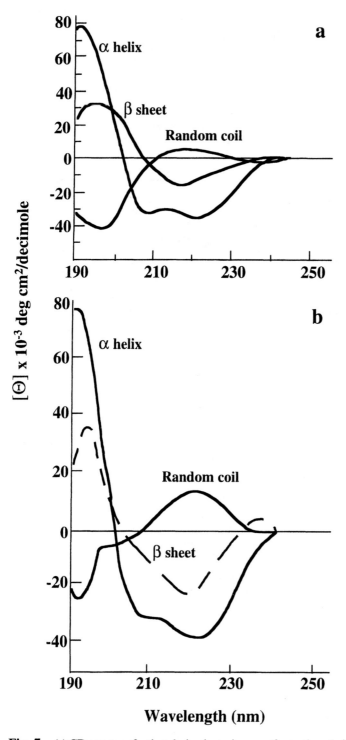

Fig. 7 (a) CD spectra of poly-L-lysine in various conformations (adapted from Ref. 7). (b) CD spectra for α-helix, β-sheet, and random coil conformations recovered from CD spectra of proteins of known 3D structure (adapted from Ref. 9).

multicomponent analysis is then made to match observed and calculated CD spectra according to the following linear equations for different λ:

$$\chi_{obs}(\lambda) = \alpha\chi_{\alpha H}(\lambda) + \beta\chi_{\beta S}(\lambda) + \gamma\chi_{RC}(\lambda) \tag{14}$$

where χ_{obs} is the observed CD spectrum at wavelength λ of the protein of interest. The χ values at the right side of the equation are the spectra of the basis set, and α, β, and γ are the fractions to be determined by solving simultaneously the set of linear equations given by Eq. (14) (of course, constraints may be applied such as $\alpha + \beta + \gamma = 1$ and $\alpha, \beta, \gamma > 0$). One should, however, be cautious in using this approach. Proteins do not contain large extended helices and sheets like those of the poly–amino acid model systems. In addition, helices in proteins can be densely packed and helix-helix interactions affect the CD spectra (the same applies for sheet-sheet interactions). Finally, as indicated earlier (8), several types of sheet conformations and turns are found in proteins. These aspects were noted in the beginning of the 1970s. Consequently, rather than using polypeptides, a set of "known" proteins is now chosen to compute the basis spectra. Known proteins are proteins whose three-dimensional (3D) structure has been determined using X-ray diffraction and whose fractions α, β, and γ can thus be evaluated. By solving a set of equations corresponding to Eq. (14), a CD spectral basis set can be recovered. A typical example of such a basis set is presented in Fig. 7b (adapted from Ref. 9).

A comprehensive review appeared (10) that compares the various methods of obtaining structural information from CD data. The computational methods, developed during the past two decades, include singular value decomposition, ridge regression, variable selection, self-consistent method, and neural networks. Because we have used the ridge regression method (CONTIN) in analyzing CD spectra, this method will be briefly outlined (details can be found in Ref. 11). The CD spectrum of an unknown protein is directly fitted to a linear combination of spectra from a large database of proteins with known conformations. CONTIN is a variation of the method of least-squares minimization. In this method, the contribution of each reference spectrum is kept small unless it contributes significantly in obtaining the optimal fit. Use is made of a so-called regularizor, which eliminates any biased contribution of a particular protein in the reference set. In this way the least-squares solution is stabilized, bearing in mind that many other solutions could be equally valid in multiparameter fitting.

An example of CD analysis to retrieve the secondary structure content of the flavoprotein lipoamide dehydrogenase (from *Azotobacter vinelandii*) using CONTIN is given in Fig. 8 (12). The experimental and calculated CD spectra and the residuals between both spectra are shown. The α-helix content determined (30%) is in excellent agreement with that from the X-ray structure. The β-sheet and β-turn contributions are somewhat higher than inferred from the 3D structure. The latter is due to the fact that the C-terminal domain of this protein is not well defined in the 3D structure. C-terminal deletion mutants of lipoamide dehydrogenase have the same secondary structure content as the X-ray structure (12).

C. Induced Circular Dichroism in Proteins: Fingerprint Function

CD of an optically active residue in a protein is very sensitive to its microenvironment. Therefore the technique can be exploited to detect conformational changes in proteins

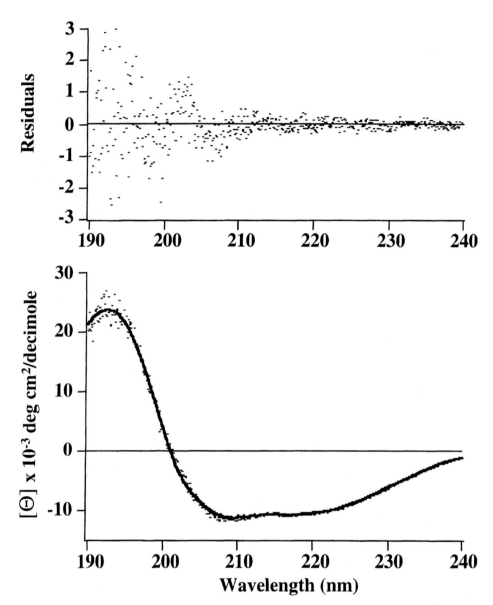

Fig. 8 Experimental and fitted far-UV CD spectra of *Azotobacter vinelandii* lipoamide dehydrogenase. The residuals between experimental and fitted curve (top) are randomly distributed around the zero line and indicate optimal fitting. (Redrawn from Ref. 12.)

resulting from ligand (substrate, inhibitor, effector) binding, variation in temperature, change in pH, addition of denaturant, etc. The aromatic amino acids tryptophan, tyrosine, and phenylalanine are examples of chiral, light-absorbing molecules that yield very distinct CD spectra in proteins. Other examples are the flavin prosthetic group in flavoproteins, heme in heme proteins, and pyridoxal phosphate and NADH in particular enzymes. All these molecules have in common that the chromophoric group is placed in an asymmetric protein environment. Chirality is then induced by the protein. In such a complex protein environment it is very difficult to derive the molecular origin of the optical activity. However, one can use CD as a convenient tool to monitor conformational changes and to establish thermodynamic information associated with these changes. A good review of the use of CD in conjunction with light absorption and fluorescence to investigate protein unfolding is given in Ref. 13.

In principle, disulfides in proteins also have some chirality. The optical activity arises from the electronic transition (near 250 nm), being dependent on the -S—S-dihedral angle.

As an example of the fingerprint function of CD, the CD spectra of flavin adenine dinucleotide (FAD) bound to wild-type lipoamide dehydrogenase and to a deletion mutant are shown in Fig. 9 (12). Free FAD exhibits very weak CD (not shown), and bound FAD shows high optical activity in the near-UV band. Deletion of 14 C-terminal residues of the protein changes the microenvironment of the flavin, which is reflected by a drastically changed CD spectrum. The CD spectrum of the deletion mutant resembles that of a reductase such as ferredoxin-NADPH oxidoreductase. Each flavoprotein with particular function has its own characteristic

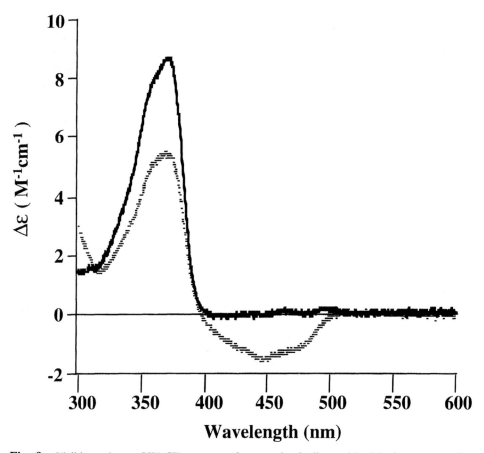

Fig. 9 Visible and near-UV CD spectra of *A. vinelandii* lipoamide dehydrogenase and a deletion mutant. Solid line, wild type; dashed line, deletion mutant. (Redrawn from Ref. 12.)

CD spectrum, whereas the light absorption spectra are much less distinctive. CD thus clearly has a fingerprint function.

D. CD Spectroscopy and the Study of Protein (Un)Folding

As described in the previous sections, CD spectroscopy provides information about the secondary (far-UV CD) and tertiary (near-UV CD) structure of a protein. The technique is therefore routinely applied to characterize the native state of a protein molecule. Estimates of the amounts of the various secondary structure elements (i.e., α-helices, β-strands, and coil regions) can be obtained using this technique. In the native state of a protein these secondary structure elements are ordered in a highly specific manner. The arrangement of these elements and the specific conformations of the side chains of the amino acid residues determine the three-dimensional structure of a protein. Formation of the complicated three-dimensional protein structures is essential for proteins to be biologically active. The functional properties of proteins depend on these specific structures (14).

Since the first three-dimensional structures of proteins were elucidated, the problem of how proteins fold to such complicated tertiary and quarternary structures has been debated. It was Anfinsen with his coworkers (15,16) who demonstrated that proteins can fold unassistedly and reversibly to their native three-dimensional state. Consequently, all the information required to define the tertiary fold is encoded in the amino acid sequence (17). It was readily realized that protein folding does not take place via a random sampling of conformational space. Even for small proteins, such as flavodoxin from *A. vinelandii* (179 residues), simple calculations show that this process would take longer than the lifetime of the universe (18). Protein folding, however, usually takes place on the millisecond to minute time scale. To explain the latter, the concept of protein folding pathways arose. A protein folds along a protein folding pathway from the fully unfolded to the fully folded native state. For most proteins under physiological conditions, the native three-dimensional structure is the thermodynamically most favorable state, which thus predominates.

The understanding of how proteins fold should answer a few related questions: what is the physical basis of the stability of the folded protein conformation; what processes determine that a protein adopts its native conformation; what are the rules that link the amino acid sequence with the three-dimensional structure of a protein; and finally, can the three-dimensional structure of a protein be predicted from its amino acid sequence (19–21)? Besides the fact that the answers to these questions are of academic interest, knowledge about protein folding is now being exploited in many practical applications in biotechnology and thus is also of industrial importance (22). General rules governing protein folding are now beginning to emerge (23,24). A "new view" on protein folding replaces the idea of distinct folding intermediates on a specific folding pathway with the idea of an ensemble of protein conformations that fold via parallel multipathway diffusion-like processes (25).

CD spectroscopy plays an important role in the study of protein folding as it allows the characterization of secondary and tertiary structures of proteins in native, unfolded, and partially folded states. As described earlier, the CD spectrum of a native protein reflects its three-dimensional structure. As a typical example, the far-UV CD spectrum of native *A. vinelandii* apoflavodoxin is shown in Fig. 10. Under physiological conditions apoflavodoxin adopts the α/β doubly wound topology, which consists of a

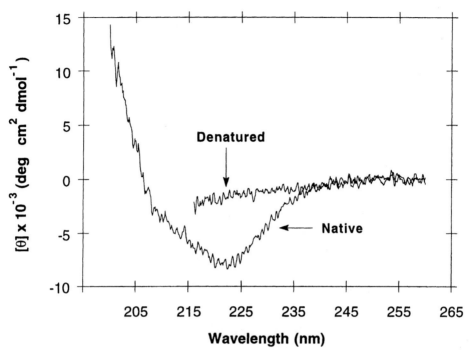

Fig. 10 Far-UV CD spectrum of native C69A apoflavodoxin from *A. vinelandii* and of apoflavodoxin in 4 M guanidinium hydrochloride, respectively. The protein concentration was 4 μM and the spectra were recorded at 25°C in 100 mM potassium pyrophosphate, pH 6.0. (Adapted from Ref. 113.)

five-stranded parallel β-sheet surrounded by α-helices at either side of the sheet (25a), as is also observed for holoflavodoxin (26). The presence of a sufficient amount of denaturant usually causes a protein to unfold. The CD spectrum of apoflavodoxin in 4 M guanidinium hydrochloride (Fig. 10) is typical for unfolded proteins exhibiting random coil behavior.

A combination of circular dichroism and fluorescence spectroscopy is often used to demonstrate whether equilibrium (un)folding of proteins takes place via a two-state mechanism or whether it involves protein folding intermediates (13). In case of apoflavodoxin, the combination of both techniques clearly demonstrates the presence of a relatively stable folding intermediate in the denaturant-induced equilibrium (un)folding of the protein (Fig. 11). The intermediate is characterized by loss of native tertiary interactions as its fluorescence emission intensity is decreased compared with that of the native state. The intermediate, however, has an appreciable amount of secondary structure as inferred from CD spectroscopy. Folding intermediates with such characteristics have been observed for a few globular proteins and are called molten globules (27). The occurrence of a molten globule–like folding intermediate is also observed during the thermal unfolding of apoflavodoxin: CD and fluorescence spectroscopy give drastically different thermal midpoints of unfolding (Fig. 12). Instead of using fluorescence spectroscopy, near-UV CD spectroscopy can be used as well to detect the immobilization of aromatic residues in the tertiary structure of a protein during folding. The molten globule state of α-lactalbumin, which is induced by

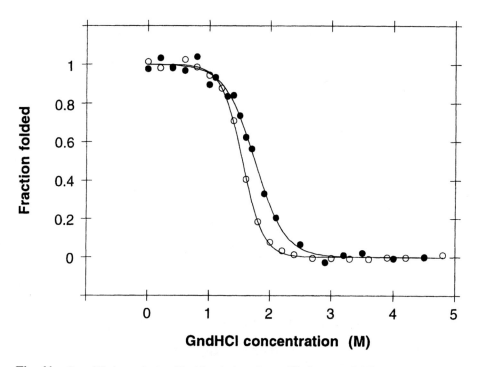

Fig. 11 Guanidinium hydrochloride induced equilibrium unfolding of $4\,\mu M$ C69A apoflavodoxin from *A. vinelandii* as monitored by fluorescence emission at 333 nm ($\bigcirc$) and by CD at 222 nm ($\bullet$). The fluorescence excitation wavelength was 280 nm. The spectra were recorded at 25°C in 100 mM potassium pyrophosphate, pH 6.0. The results for the simulated fits of the changes in fluorescence and ellipticity calculated for a three-state transition are also shown. (Adapted from Ref. 113.)

lowering the pH of the solution, is unfolded as inferred from near-UV CD but is native-like on basis of far-UV CD (28).

Stopped-flow CD studies have become increasingly popular in the kinetic analysis of protein unfolding and refolding. Modern CD instruments are useful for rapid reaction measurements as the lower limit of their time resolution, estimated to be as short as 0.2 ms, is short enough compared with the mixing dead time of the stopped-flow method (29). This mixing time is usually a few to ten milliseconds. The stopped-flow CD technique is widely used in studies of the kinetics of structural transitions of proteins and other biomolecules. Application of the stopped-flow CD technique to studies of protein folding is, however, rather new (29). The technique is effective in detecting and quantitating the secondary structure formed in the transient intermediates during protein folding (29).

If the denaturant-induced unfolding of a protein is reversible, as is the case for, e.g., the apoflavodoxin of *A. vinelandii* mentioned before, the refolding of an unfolded protein can be studied by diluting the denaturant. Stopped-flow CD spectroscopy then provides an indication of the overall extent of structural organization in the refolding molecules, in terms of both secondary structure (from far-UV CD) and tertiary structure (from near-UV CD) (17). To initiate protein refolding a concentration jump of the denaturant is created and special care is required to perform the stopped-flow

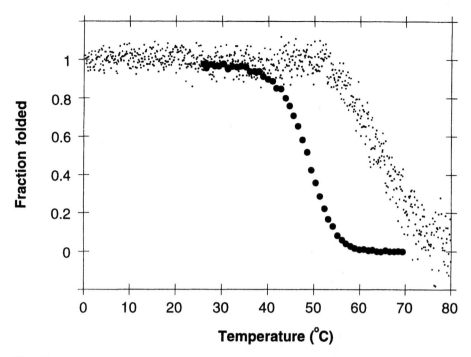

Fig. 12 Temperature-induced equilibrium unfolding of $2\,\mu M$ C69A apoflavodoxin from *A. vinelandii* as monitored by fluorescence emission at 350 nm (●) and by CD at 222 nm (·). The fluorescence excitation wavelength was 280 nm. The spectra were recorded in 100 mM potassium pyrophosphate, pH 6.0. (Adapted from Ref. 113.)

experiments. The leap to a native condition from the unfolded state requires a wide concentration jump, and a mixing apparatus with a high dilution ratio (10:1 or even more) is thus necessary (29). Mixing apparatus with two or more driving syringes have been designed in which plungers are driven by either stepping motors or pneumatics and control the solution delivery. In typical protein folding-unfolding experiments, a heavy solution of urea or guanidinium hydrochloride is mixed with pure aqueous buffer. The efficient mixing of solutions of different densities is a formidable challenge for a stopped-flow instrument. Various manufacturers have tackled this differently. Currently, several commercial stopped-flow instruments combined with CD spectrometers are available. Manufacturers are, among others, Biologic (France), Jasco (Japan), Aviv (USA) and Applied Photophysics (UK).

The information obtained so far by stopped-flow CD spectroscopy has been limited by the time and structural resolution of the technique. The time resolution is currently on the millisecond scale, which does not permit the study of early folding events, although further developments are expected (30). It is generally observed by CD spectroscopy that a large complement of secondary structure is formed within the first few milliseconds of protein folding (Fig. 13). This process takes place within the dead time of the measurement using the current stopped-flow techniques. As an example, Fig. 13 shows kinetic refolding curves of β-lactoglobulin measured by the ellipticities at 219 (far UV) and 293 nm (near UV) and a kinetic unfolding curve of the same protein measured by the ellipticity at 220 nm (31). Similar curves were

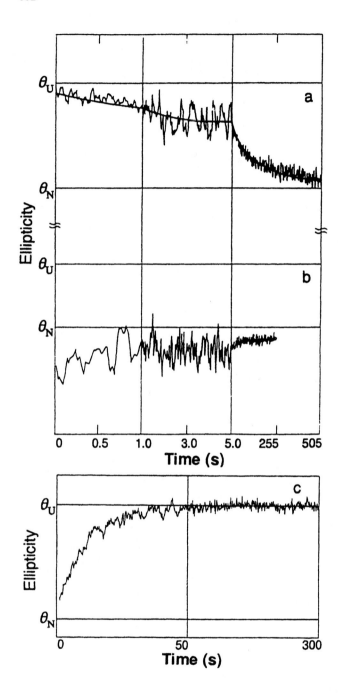

Fig. 13   Kinetic refolding (a,b) and unfolding (c) curves of bovine β-lactoglobulin A measured by the stopped flow CD at pH 3.2 and 4.5°C. The refolding was induced by a concentration jump on guanidinium hydrochloride from 4 to 0.4 M and measured at 293 nm (a) and 219 nm (b). The unfolding was induced by a concentration jump of guanidinium hydrochloride from 0 to 4 M and measured at 220 nm (c). θ_N and θ_U denote equilibrium CD values in the native (N) and unfolded (U) states, respectively. (Redrawn from Ref. 29.)

obtained by other groups who used the stopped-flow technique to study the refolding of small globular proteins from their unfolded state in either urea or guanidinium hydrochloride (e.g., 30,32–41). In the aromatic region, however, the changes in the CD spectra from the unfolded to the folded state (Fig. 13) can often be observed kinetically during refolding of proteins (29). The transient accumulation of an intermediate state I that contains secondary structure elements but is unfolded in terms of the aromatic CD spectra is observed for some proteins (27).

Characterization of the I state by CD spectroscopy is made possible by measurements of the kinetic progress curves of refolding at various wavelengths. Because the formation of the I state in the burst phase occurs much faster than the subsequent folding events, the wavelength dependence of the CD value obtained by extrapolation to the zero time of the observed refolding curve corresponds to the I state (29). It was demonstrated in this manner that the CD spectra of the I state for lysozyme and α-lactalbumin are similar to each other and to the equilibrium CD spectra of the molten globule state of α-lactalbumin. This suggests that the folding of both proteins takes place via the initial collapse of the polypeptide chain to a molten globule state. Subsequent relatively slow rearrangements of the protein structure lead to the fully folded native state of the protein.

CD spectroscopy provides average information about the secondary and tertiary structure of protein folding intermediates. However, detailed knowledge of the partially structured states of such folding intermediates is essential for understanding protein folding. CD spectroscopy cannot give any specific information about where the secondary structure such as an α-helix or a β-sheet is formed within a molecule (29). It has become possible, due to drastic improvements made in multidimensional nuclear magnetic resonance (NMR) spectroscopy, to characterize such intermediates using NMR (37,42–45). Use is made of differences in exchange rate of individual amide protons versus deuterium oxide in the folding intermediates. It is not necessary to study the folding intermediates themselves, as they exist only transiently on the time scale of milliseconds to seconds due to the high cooperativity of protein folding. Amide proton exchange pulse labeling against deuterium oxide (using rapid quench techniques) of the amide protons in the folding intermediates makes it possible to characterize the conformations of the intermediates using the two-dimensional NMR spectra of the native state. Pioneering initial structural characterizations of folding intermediates using this method have been obtained for a few proteins, including cytochrome c, ribonuclease A, guinea pig α-lactalbumine, and hen lysozyme (37,42–44). The most detailed structural characterization, however, has until now been possible only for stable analogues of the folding intermediates of bovine pancreatic trypsin inhibitor (BPTI) (46–51).

Stopped-flow CD data on protein folding are now used to complement results obtained via stopped-flow fluorescence and mass spectrometry experiments. The stopped-flow CD technique is of particular value in conjunction with the pulselabeling experiments. In the case of lysozyme, experiments in the far UV indicate that a large complement of secondary structure is formed within the first few milliseconds of folding as observed for several other proteins. The rate of secondary structure formation, however, is actually much higher than the rate at which exchange protection develops, suggesting that amide proton protection probably arises from stabilization of already formed helical structure in lysozyme (17).

In conclusion, both CD and stopped-flow CD are very valuable techniques for studying protein folding. They have significantly contributed in our understanding of how proteins fold. In combination with results from complementary techniques, both CD and stopped-flow CD are expected to play an important role during the coming years in the deciphering of the protein folding problem.

IV. CIRCULAR DICHROISM OF PROTEINS AT INTERFACES

In this section we describe CD studies of proteins located at membranes, solid carrier–air and solid carrier–water interfaces. The preparation of the samples and the influence of the system on the CD spectral characteristics is discussed. Some examples of reported CD studies will be presented to illustrate the strength of this technique to gain information on the structural properties of proteins at interfaces.

A. Proteins in Membranes or Membrane-Mimicking Systems

When protein-lipid complexes are the subject of a CD study, it is inevitable that light scattering caused by the relatively large particles ($\gg$ wavelength) can interfere with the interpretation of the data (52). The consequence of light scattering is primarily a strong reduction of the light intensity as sensed by the detection system, resulting in a lower signal-to-noise ratio. However, if the two rotations of light are scattered differently, a CD artifact can be created that could hinder the interpretation of protein CD. The presence of CD artifacts as a result of particle scattering can generally be identified by the appearance of CD outside the normal protein absorption bands.

The extent of differential light scattering can be minimized by sonication of the membrane particles, resulting in small unilamellar vesicles (SUVs) (53). It should be noted, however, that the stability of these small vesicles is low and spontaneous fusion to larger structures occurs in time. A more reproducible method for studying proteins in membrane systems is obtained by extrusion techniques, yielding large unilamellar vesicles (LUVs). The pore size of the polycarbonate filters used determines the size and homogeneity of the membrane structures (54). Due to the relatively high turbidity, the applicability of multilamellar vesicles, obtained by resuspension of lipid films in aqueous buffer followed by several heating-cooling cycles, for CD studies is less favorable. Final lipid concentrations of 5–10 mM SUVs or LUVs are feasible (using 0.01–0.02 cm cuvette path lengths) for reliable CD measurements of proteins in these systems (55,56), allowing systems with lipid-to-protein ratios higher than 100–150 to be studied by CD.

Micellar systems are an alternative widely used for CD studies. They mimic membrane environments and usually eliminate the problem of differential light scattering. In the literature a large variety of types of detergents have been reported. In some cases lysophospholipids have been used (57,58), but sodium dodecyl sulfate (SDS), a negatively charged surfactant, is the one most frequently used. Because the protein conformation might be strongly dependent on the membrane surface charges, neutral [for example, dodecylmaltoside, dimethyl dodecylamine-*N*-oxide (LDAO), octyl-glucoside, or Triton] or zwitterionic (e.g., dodecylphosphocholine) detergents have been used extensively as well. Both can be purchased or synthesized in an easy manner (59,60).

Besides light scattering due to particles, differential absorption flattening can also occur in parts of the far-UV CD spectra of membrane-embedded proteins (61–63). This spectral artifact is related to the deviation from an ideal homogeneous distribution of chromophores in the solution that is necessary to fulfill the criteria for Beer's law. When the chromophores are densely packed the total cross-sectional area is smaller than when the particles are uniformly dispersed. Consequently, the extent of flattening depends on both the size of the particles and the number of chromophores present in each particle. Especially in spectral regions where protein absorbance is strong, absorption flattening can contribute seriously. Hence, the extent of absorption flattening is wavelength dependent. Some studies have tried to determine the "flattening" factor for various wavelengths (53), but because these factors are system dependent a straightforward more general procedure to correct CD spectra for flattening effects cannot be given. A way to identify the presence of absorption flattening is by sonication of the membrane fragments as demonstrated by Mao and Wallace (53). They showed that disruption of purple membrane fragments into smaller particles reduced their size and the number of chromophores per particle. For this reason, detergent-solubilized membranes can be considered free of absorption flattening. All studies in which absorption flattening was apparent dealt with integral membrane proteins. In such systems the proteins are two-dimensionally concentrated in a membrane and often have a strong interaction with each other, forming, for example, defined hexagonal arrangements in the membrane, a case encountered in bacteriorhodopsin. However, it should be noted that absorption flattening effects can also occur in protein films or even in solutions containing densely packed proteins.

Because all protein absorption bands observed with far-UV CD reflect π,π^* and n,π^* transitions, it is obvious that the intensity and frequency position of these bands depend on the dielectric constant of the local environment. Comparable to the CONTIN method described earlier in this chapter, an effort has been made to establish reference spectra for integral membrane proteins based on the CD spectra of 30 different detergent-solubilized integral membrane proteins with known secondary structure content (64). From this analysis it was shown that two reference spectra are needed to describe the helical content of an integral membrane protein: one assigned to α-helices in soluble domains and the other accounting for transmembrane helices. In this latter spectrum the positive CD band below 200 nm (π,π^* transition) was red shifted approximately 5 nm, whereas the negative band at 222 nm (n,π^* transition) was slightly more negative than that at 208 nm (π,π^* transition). In addition, the rotational strength was greater for the transmembrane helices. The latter could be attributed to (a) the fact that helices spanning a membrane are generally two times larger than helices in soluble domains and (b) the lower dielectric constant.

1. Integral Membrane Proteins

In the literature only CD spectra of integral membrane proteins in detergent-reconstituted systems can be found. A few exceptions exist, with the CD spectrum of bacteriorhodopsin (BR) in purple membranes of *Halobacterium halobium* as a typical example (53). However, because of absorption flattening the secondary structure of BR determined by CD deviated strongly from the reported X-ray structure. The spectral distortion was reduced only slightly by sonication of the membranes. However, when changes in CD are monitored, as, for example, reported for the dependence of the kinetics of folding of BR in vitro on pH or membrane composition (65), the spectral distortions do not affect the results. Reconstitution of BR in dimyristoylphosphatidyl-choline (DMPC) membranes and subsequent extensive sonication of the sample resulted in CD spectra that revealed after analysis a secondary structure content comparable to the X-ray structure. Whereas BR is known to consist of seven transmembrane helices, almost no detailed information is available on β-structured integral membrane proteins. Only one study reported CD spectra of the voltage-gated channel VDAC (or mitochondrial porin) isolated from *Neurospora crassa* reconstituted in model membranes (66). Indeed, a large content of β-structure was found, in agreement with structure predictions and neutron diffraction data.

More quantitative CD analysis of integral membrane proteins has been carried out in micellar systems (for an extensive overview, see Ref. 64). An example is diacylglycerol kinase, a relatively small integral membrane protein (13 kDa) that spans the membrane three times. This protein is still active when reconstituted in particular mixed micellar systems. By variation of the molar ratio of β-maltoside and SDS this protein could be unfolded reversibly as monitored by CD, and the stabilities of the cytoplasmic domain and the membrane-embedded part could be established (67).

The treatment of membranes with detergents in order to break up the particles or to extract the integral membrane proteins should always be performed with great care. On treating purple membranes of *H. halobium* with Triton X-100, it was found that particular membrane lipids were removed. As a result, the kinetics of the photocycles of this proton pump were significantly altered (68). Both near- and far-UV CD analysis showed that this treatment affected both the secondary structure of BR and its ability to trimerize. This could be reversed by reintroduction of an extract of membrane lipids to the protein. Alternatively, the relatively small dimensions of micelles could prohibit such a solubilized integral membrane protein attaining the proper structure, as reported in some cases (53). In contrast, CD spectra of CHIP28, a water channel found in erythrocytes, solubilized either in octylglucoside micelles or in reconstituted proteoliposomes, were found to be very similar (69). To overcome the dimension problem of micelles with respect to the size of integral membrane proteins, one often synthesizes parts of these proteins that are predicted to be membrane spanning. The synthesis of 28-residue peptides resembling parts of phosphatidylglycero-phosphate synthase was successful in that respect. Their secondary structures in SDS micelles could be well determined by CD studies and confirmed by high-resolution NMR studies (70). Other examples are the synthetic peptides resembling isolated transmembrane helices of BR (71).

2. Membrane-Associated Proteins

When proteins associate with lipid-water interfaces the change in local environment can either destabilize them or introduce structure. This is caused by the difference in helical propensities of amino acids in membrane environments (72) compared with their conformational preferences in water (73). The propensity of proline for β-sheet formation is, however, not affected (74). Both situations, the folding and unfolding of proteins upon interaction with membranes, will be discussed in the next section.

The presence of a preestablished bilayer causes some proteins to adsorb to its interface but others to insert in the membrane. In general, it can be stated that poly-peptides that are destabilized on a tertiary and secondary level in aqueous solution exhibit a higher affinity to associate with membranes than the more globular ones do (75). This is best illustrated by apocytochrome *c* (55), melittin (76), or peptides resembling the protein kinase C binding domain of neuromodulin (77) or indolicidin (78). They were all shown to be non-structured in an aqueous environment, but adopted helical (apocytochrome *c* and melittin) or extended (the protein kinase C binding domain of neuromodulin and indolicin) structures upon binding to lipid bilayers. However, the binding of more globular proteins to vesicles can also result in an increased helix content, as demonstrated by CD, for example, for glucagon (79) or for rhodanese and its presequence (80).

Numerous studies have been reported on the conformation of water-soluble proteins in the presence of micelles. In some cases detergent concentrations higher than the critical micelle concentration were required to influence the conformational properties of, for example, (apo)cytochrome *c* (55) or yeast chaperonin protein cpn10 (81). Effects of the acyl chain length of the lipids used on the protein conformation have also been reported, as shown, for example, for myelin basic protein (82). In many cases the effect of the micelle on the structural properties of the associated protein has been shown to be strongly modulated by the nature of the polar lipid headgroups (55,83). The influence of water-micelle interfaces on the structural properties of proteins is best demonstrated in the case of apo- and holocytochrome *c*, shown in Fig. 14. Whereas in the absence of lipids apocytochrome *c* is a highly unstructured protein, as indicated by the single negative extreme below 200 nm, the holo protein, which differs only in the presence of a heme group covalently bound to the protein, possesses a CD spectrum having all features of a highly secondary folded protein (Fig. 14). However, after addition of detergent micelles to both proteins an identical CD spectrum is obtained, demonstrating that the water-lipid interface induces a second-

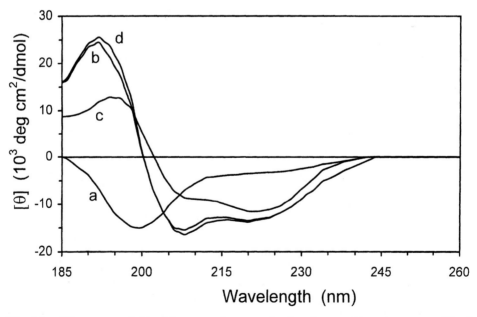

Fig. 14 CD spectra of 0.1 mM apocytochrome *c* in the absence (a) and presence (b) of dodecylphosphatidylcholine/dodecylphosphatidylglycol (9:1) micelles and of 0.1 mM cytochrome *c* in the absence (c) and presence (d) of these micelles in a lipid-to-protein ratio of 120 in a 10 mM phosphate buffer (pH 7.0.) (Redrawn from Ref. 60.)

ary fold in the apoprotein that is identical to that of the micelle-associated holo protein. From accompanying NMR data it is clear that a common folding intermediate is formed that is a secondary folded but tertiary destabilized protein (60).

B. Proteins at Solid-Air Interfaces

Protein samples can be transferred to a solid carrier in several manners. The most commonly used technique is the Langmuir-Blodgett technique (84). In a Langmuir trough a monolayer is prepared either by self-assembly of proteins at the air-water interface or by protein association with or insertion into a preestablished monomolecular layer of lipids. Subsequent collection of the material on quartz plates is generally achieved by moving the plate through the interface, thereby collecting materials on both sides of the plate. Maneuvering the plate through the monolayer repeatedly can yield a stack of layers on both sides of the plate. The displacement of the barrier of the Langmuir trough allows control of the amount of materials transferred to the plate (85). A more simple and efficient technique is solvent evaporation. The protein solution in the absence or presence of lipids is dried to the air on a quartz plate from an aqueous or organic solvent (86). In the case of pure protein samples, different types and concentrations of alcohol can be added to the solution to control the conformational properties of the adsorbed proteins upon drying of the film (87). In the presence of membranes, spontaneous multilayer arrangements are formed during the evaporation process due to capillary forces that flatten the membranes (86). Dichroism studies using infrared spectroscopy demonstrated that the membrane normal was perpendicular to the plate, yielding a well-oriented lipid-protein system (88). A third approach to transferring proteinaceous materials onto quartz plates is by isopotential spin-dry centrifugation. The solvent is then removed during centrifugation at high gravitational forces (89).

Proteins adsorbed on quartz plates can be placed in a CD spectropolarimeter, provided that the samples are oriented perpendicularly with respect to the incident light beam. Such a setup implies that polypeptides can adopt preferential orientations with respect to the incident light beam, in contrast to randomly dispersed complexes as described earlier. The CD spectrum of a polypeptide in one particular secondary structure type is a superposition of a defined number of (positive and negative) Gaussian absorption bands (90). In helical structures the so-called helix band is also present, which is positive in the long-wavelength region and negative at smaller wavelengths (91). It has to be noted that this latter band has been subject of much controversy (92). The origin of all these bands has been identified [for a review see Woody (93)] and correlated with the direction of absorption in the molecular axis system (56,94–96). Summarizing, two spectra are required to describe the helical contributions of an oriented sample, one corresponding to absorptions of light propagating parallel to the helix axis and the other corresponding to the perpendicular contributions. These two "reference" spectra are depicted in Fig. 15, together with the summation of these two spectra according to $\theta_{iso} = (\theta_{\parallel} + 2\theta_{\perp})/3$, where θ_{iso} represents the ellipticity in an isotropically distributed sample, and $\theta_{\parallel}$ and $\theta_{\perp}$ represent the parallel and perpendicular contributions, respectively. Figure 15 illustrates that any preferred orientation of helices with respect to the incident light beam does give rise to spectral "distortions." Because all absorptions (n,π^* and π,π^*) in a β-stranded peptide lie within the same plane parallel to the molecular axis of the β-strand (97,98),

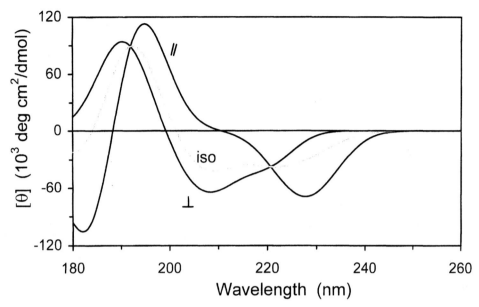

Fig. 15 Reconstructed CD spectra of the components representing helices oriented parallel and perpendicular with respect to the incident light beam and the spectrum reflecting the isotropic distribution of orientations according to $\theta_{iso} = (\theta_{\parallel} + 2\theta_{\perp})/3$. The spectra are obtained using the parameters describing the helix spectrum using four Gaussian bands and the helix band as reported by de Jongh et al. (56).

only its CD intensity is affected by a preferred orientation. Analysis of CD spectra of oriented protein systems can be performed in a way similar to that described previously with the CONTIN program but also via nonlinear regression curve-fitting procedures (56), by replacing the reference helix spectrum with the two helical contributions depicted in Fig. 15.

A possible spectral artifact that could interfere with a proper analysis of the protein CD on solid carriers is linear dichroism caused by inhomogeneities of the protein film or the quartz plates used. A general procedure to eliminate this artifact is via rotation of the quartz plate in a plane perpendicular to the incident light beam in cell holders especially designed for this purpose. In such a setup spectra recorded at every 11.25 or 22.5° of rotation are averaged (56,99).

An example of a protein studied at a solid-air interface is the pulmonary surfactant protein (SP-C). It is a small hydrophobic peptide that is palmitoylated at two adjacent cysteine residues. By site-directed mutagenesis these cysteines can be replaced by serines, prohibiting the in vivo acylation. In this way the influence of the acylation on the orientation of the (predominantly helical) peptide at the air--water interface has been studied. A monolayer of this protein was transferred onto quartz plates using the Langmuir-Blodgett technique and studied by CD (100). The palmitoylated SP-C was shown to be oriented preferably parallel to the interface, whereas the nonacylated peptide had a more isotropic orientation. A clear rise of helicity with increasing initial surface pressure of the monolayer was observed, but the preferred orientation was not affected. In another study, monolayers of poly-L-alanine and poly-γ-methyl-L-glutamate spread from different solvents in a

Langmuir trough were transferred to quartz plates and characterized by CD (101). Whereas the first polypeptide was found in each case to be α-helical with a strong preferred orientation parallel to the interface, the conformation of the second one could be modulated in either an α-helical or a β-stranded conformation by choice of the (organic) spreading solvent.

Several CD studies have been reported in which the orientation of polypeptides has been determined with respect to oriented membrane systems obtained by evaporation of the aqueous solvent. The most impressive work was presented by Bazzi and Woody (89). They reported on the orientation of the helices of cytochrome *c* oxidase, which is an integral membrane protein involved in the respiratory chain of mitochondria. An average angle of 39° of the helix axis with respect to the membrane normal was found for this protein. Even more extensive analyses of CD spectra for the preferred orientation of the small peptide alamethicin using oriented membranes have been reported by Vogel (99) and Wu and coworkers (95). Especially in the latter work a comprehensive description of the theory and experimental background of these studies is given. Both studies demonstrated that the orientation of the helix axis of alamethicin is parallel to the membrane normal. The peptide, however, tended to adopt a more perpendicular orientation when the degree of hydration of the membranes was limited. Besides the influence of the degree of hydration, the type of phospholipids differing in net charge or headgroup size was also demonstrated to be of great importance for the orientation of peptides at membrane interfaces (56,102). A method for analysis of oriented CD spectra is described enabling an accurate determination of the average angle that helical structures make with respect to the incident light beam. No knowledge of the actual peptide conformation is required beforehand, and therefore this method is suitable for more general applications to other protein systems (52,102). Using this approach, the domains responsible for the initial insertion of the apocytochrome *c* into membranes have been identified. The association-insertion behavior of the protein modulated by the composition of the membranes has been described and related to the spontaneous translocation of this protein over the outer mitochondrial membrane. CD on oriented membrane systems has been used more often to determine whether a polypeptide has the expected transmembrane orientation (103) or to study possible conformational changes upon association with monolayers of different composition. For β-lactoglobulin, for example, it has been shown that it adopts a preferred orientation parallel to the lipid interface (104).

C. Proteins at Solid-Water Interfaces

A few reports can be found in the literature in which CD has been used to establish the structural properties of proteins at solid carrier–water interfaces. Because hydrophobicity of the carrier is of great influence on the protein structure, both hydrophilic and hydrophobic carriers have been used. In the case of small particles stably dispersed in the aqueous phase, the particles are generally nonporous discrete spheres. The particles can be either of silica (hydrophilic) with a diameter ranging from 10 to 20 nm, thereby much smaller than the far-UV CD wavelength (105,106), or of Teflon (hydrophobic). The latter is a copolymer of tetrafluoroethylene and perfluorovinyl ether, has a refractive index close to that of water, and lacks UV-absorbing double bonds (106,107). Both materials are suitable as a sorbent for using CD to study adsorbed proteins, as they exhibit negligible light absorption and have very low light scattering contributions in far-UV CD. An alternative for these dispersed samples is the use of plane silica quartz plates. They can be made hydrophilic by extensive cleaning in chromic acid and rinsing with distilled water. Alternatively, they can be modified with, for example, C_{18} alkyl chains to make them hydrophobic (108). Several quartz plates can be stacked with small spaces in between the plates and incubated with a protein sample in a watertight sample holder. Subsequently, nonadsorbed proteins are removed by replacing the sample buffer by a protein-free buffer (109).

Analysis of CD spectra of adsorbed bovine serum albumin, egg lysozyme, α-chymotrypsin, and cutinase to hydrophilic silica particles revealed in all cases a loss of secondary structure as compared with the structure of these proteins in aqueous solution (105,106). This is illustrated for cutinase in Fig. 16 by the blue shift of the extreme from 208 to 205 nm and of the zero crossing from 203 to 194 nm upon adsorption to the silica particles. On the other hand, the adsorption of α-chymotrypsin and subtilisin to hydrophobic particles increased their secondary structure content (106,107). This is illustrated for cutinase by the red shift of the CD zero crossing and 207 extreme in Fig. 16. The specific enzymatic activities of α-chymotrypsin and cutinase were shown to be correlated with the degree of nativity of the proteins. A similar result was obtained for α-amylase adsorbed onto silica particles (110). Also, carbonic anhydrase and derived fragments did show a loss of structure upon binding to small silica particles (111). From both far- and near-UV CD data this protein was shown to maintain a secondary fold but to lose its tertiary interactions when adsorbed to this carrier, reflecting a molten globule–like conformation. On the other hand, however, the bee venom peptide melittin,

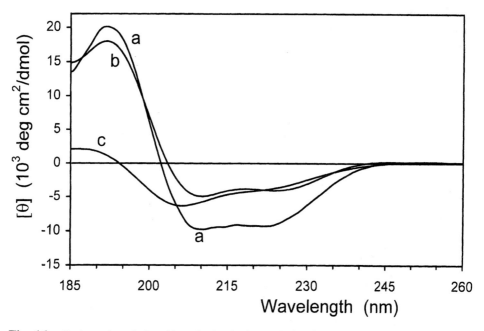

Fig. 16 Cutinase in solution (a) and adsorbed onto hydrophobic Teflon particles (b) or onto hydrophilic silica HS-40 particles (c). (Redrawn from Ref. 106.)

known for its ability to adopt under certain conditions amphipatic helical conformations in aqueous solution (112), lost part of its conformation upon adsorption on both the hydrophilic and hydrophobic quartz plates (109). These CD spectra, however, were difficult to analyze because of the spectral distortions due to orientation effects of the adsorbed materials as described earlier.

V. INSTRUMENTAL ASPECTS

The far UV (180–240 nm) is a spectral area in which everything starts to absorb light (also oxygen from air!). In addition, the light sources (usually high-pressure Xe arc lamps) do not have much intensity in this spectral region as compared with the longer wavelengths. One should therefore pay attention to the choice of buffers, salts, protein concentration, and optical path lengths of the cuvettes (1 mm and smaller). Care should also be taken in the preparation of the samples. This is discussed in detail in Refs. 8 and 10. For the near-UV and visible region, normal cuvettes with 1 cm path length are used.

How is circularly polarized light made? Certain electro-optic devices fabricated from, for instance, quartz show changes in birefringence upon experiencing stress caused by a voltage between two opposing faces of a block. Such a device is called a photoelastic modulator. The effect of a so-called quarter-wave plate causing birefringence is clarified in Fig. 17. Such an optical device is made from a crystal with the direction of the optic

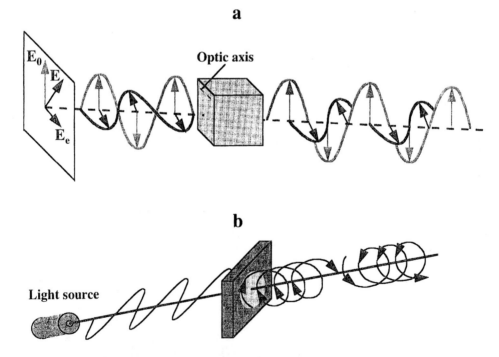

Fig. 17   Generation of circularly polarized light from plane-polarized light by a quarter-wave plate (a). The generation of left and right circularly polarized light by a photoelastic modulator (b).

axis as indicated. Plane-polarized light is represented in the figure by two plane-polarized waves perpendicular to each other. The component that has the electrical field component parallel to the optic axis propagates faster through the crystal than the perpendicular component does. The latter is retarded because it senses a higher refractive index. The outgoing beam consists of two perpendicular waves that are 90° out of phase (the thickness of the crystal is chosen such that one wave is one-quarter wavelength ahead of the other wave), which is, bearing Fig. 2d in mind, equivalent to circularly polarized light of a given handedness. Upon applying stress to such electro-optic material, the optic axis changes by 90° (called birefringence). In this way the other circularly polarized light wave is created. By modulating the voltage (which induces fluctuating birefringence) with a fixed frequency (for instance 20 kHz), alternately left and right circularly polarized light is created (Fig. 17). Figure 18 shows a simplified block diagram of a single-beam CD spectrometer.

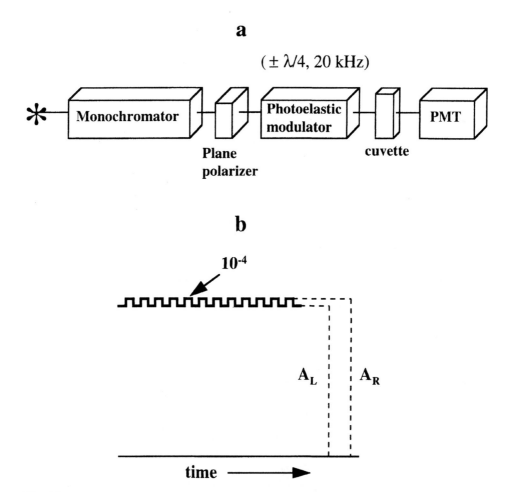

Fig. 18 (a) Block diagram of a CD spectrometer. (b) Illustration of the tiny dichroism effect to be measured. A_L and A_R are the absorbances for left and right circularly polarized light. The factor 10^{-4} illustrates that the signal related to the differential absorbance is 10^{-4} times smaller than the normal absorbance signal.

The extinction differences ($\Delta\varepsilon$) observed in a CD spectrometer are extremely small (Fig. 18). They are measurable only using phase-sensitive detection techniques. The signal coming from the detector has two components. One component is a DC signal (V_{DC}), which is related to the overall light throughput. The other AC component (V_{AC}) contains the differential absorption information $\Delta A(\Delta A = \Delta\varepsilon/(C.d))$, with C the concentration in $\text{mol}\,\text{dm}^{-3}$ and d the optical path length in cm):

$$\Delta A = k\frac{V_{AC}}{V_{DC}} \tag{15}$$

in which k is an instrumental constant. The synchronous detection of the signal relative to the modulation frequency of the photoelastic modulator determines V_{AC}. Both V_{AC} and V_{DC} are amplified (or attenuated) by an identical factor using a servo system in such a way that V_{DC} remains constant. The circular dichroism effect expressed in ΔA is

$$\Delta A = k'V_{AC} \tag{16}$$

with k' a simple calibration factor that can be set by the electronics.

ACKNOWLEDGMENT

We thank Nina Visser for her help with the preparation of figures.

REFERENCES

1. CR Cantor, PR Schimmel. Biophysical Chemistry. Part II. Chap 8. San Francisco: Freeman, 1980.
2. ID Campbell, RA Dwek. Biological Spectroscopy. Chap 10. Menlo Park: Benjamin-Cummings, 1984.
3. KE van Holde. Physical Biochemistry. 2nd ed. Chap 10. Englewood Cliffs, NJ: Prentice-Hall, 1985.
4. K Nakanishi, N Berova, RW Woody, eds. Circular Dichroism. Principles and Applications. New York: VCH, 1994.
5. GD Fasman, ed. Circular Dichroism and the Conformational Analysis of Biomolecules. New York: Plenum, 1996.
6. WC Johnson, I Tinoco. Circular dichroism of polypeptide solutions in the vacuum ultraviolet. J Am Chem Soc 94:4389–4890, 1972.
7. N Greenfield, GD Fasman. Computed circular dichroism spectra for the evaluation of protein conformation. Biochemistry 8:4108–4116, 1969.
8. WC Johnson. Protein secondary structure and circular dichroism: A practical guide. Proteins Struct Funct Gen 7:205–214, 1990.
9. VP Saxena, DB Wetlaufer. A new basis for interpreting the circular dichroic spectra of proteins. Proc Natl Acad Sci USA 66:969–972, 1971.
10. NJ Greenfield. Methods to estimate the conformation of proteins and polypeptides from circular dichroism data. Anal Biochem 235:1–10, 1996.
11. SW Provencher, J Glöckner. Estimation of globular protein secondary structure from circular dichroism. Biochemistry 20:33–37, 1981.
12. AJWG Visser, WJH van Berkel, A de Kok. Changes in secondary structure and flavin microenvironment between *Azotobacter vinelandii* lipoamide dehydrogenase and several deletion mutants from circular dichroism. Biochim Biophys Acta 1229:381–385, 1995.

13. GD Ramsay, MR Eftink. Analysis of multidimensional spectroscopic data to monitor unfolding of proteins. Methods Enzymol 240:615–645, 1994.

14. C Branden, J Tooze. Introduction to Protein Structure. New York: Garland, 1991.

15. CB Anfinsen. Principles that govern the folding of protein chains. Science 181:223–230, 1973.

16. CB Anfinsen, E Haber, M Sela, FH White Jr. The kinetics of formation of native ribonuclease during oxidation of the reduced polypeptide chain. Proc Natl Acad Sci 47:1309–1314, 1961.

17. CM Dobson, PA Evans, SE Radford. Understanding how proteins fold—The lysozyme story so far. Trends Biochem Sci 19:31–37, 1994.

18. C Levinthal. Are there pathways for protein folding? J Chim Phys 65:44–45, 1968.

19. TE Creighton. Protein Folding. New York: Freeman, 1992.

20. TE Creighton. The protein folding problem. In: RH Pain, ed. Mechanisms of Protein Folding. Oxford: IRL Press, 1994, pp 1–25.

21. E Steensma. Structure, stability and equilibrium (un)folding of flavodoxin. Thesis, Wageningen Agricultural University, Wageningen, The Netherlands, 1998.

22. DR Thatcher, A Hitchcock. Protein folding in biotechnology. In RH Pain, ed. Mechanisms of Protein Folding. Oxford: IRL Press, 1994, pp 229–261.

23. AR Fersht. Nucleation mechanisms in protein folding. Curr Opin Struct Biol 7:3–9, 1997.

24. H Roder, W Colón. Kinetic role of early intermediates in protein folding. Curr Opin Struct Biol 7:15–28, 1997.

25. KA Dill, HS Chan. From Levinthal to pathways to funnels. Nature Struct Biol 4:10–19, 1997.

25a. E Steensma, CPM van Mierlo. Structural characterisation of apoflavodoxin shows that the location of the most stable nucleus differs among proteins with a flavodoxin-like topology. J Mol Biol 282:653–666, 1998.

26. E Steensma, MJM Nijman, YJM Bollen, PA de Jager, WAM van den Berg, WMAM van Dongen, CPM van Mierlo. Apparent local stability of the secondary structure of *Azotobacter vinelandii* holoflavodoxin II as probed by hydrogen exchange: Implications for redox potential regulation and flavodoxin folding. Protein Sci, 7:306–317, 1998.

27. OB Ptitsyn. The molten globule state. In: TE Creighton, ed. Protein Folding. New York: Freeman, 1992, pp 243–300.

28. K Kuwajima. The molten globule state as a clue for understanding the folding and cooperativity of globular-protein structure. Proteins Struct Funct Gen 6:87–103, 1989.

29. K Kuwajima. Stopped-flow circular dichroism. In: GD Fasman, ed. Circular Dichroism and the Conformational Analysis of Biomolecules. New York: Plenum, 1996, pp 159–182.

30. E López-Hernández, P Cronet, L Serrano, V Muñoz. Folding kinetics of Che Y mutants with enhanced native α-helix propensities. J Mol Biol 266:610–620, 1997.

31. K Kuwajima, H Yamaya, S Miwa, S Sugai, T Nagamura. Rapid formation of secondary structure framework in protein folding studied by stopped-flow circular dichroism. FEBS Lett 221:115–118, 1987.

32. RI Gilmanshin, OB Ptitsyn. An early intermediate of refolding α-lactalbumin forms within 20 ms. FEBS Lett 223:327–329, 1987.

33. K Kuwajima, EP Garvey, BE Finn, CR Matthews, S Sugai. Transient intermediates in the folding of dihydrofolate reductase as detected by far-ultraviolet circular dichroism spectroscopy. Biochemistry 30:7693–7703, 1991.

34. AF Chaffotte, Y Guillou, ME Goldberg. Kinetic resolution of peptide bond and side chain far-UV circular dichroism during the folding of hen egg white lysozyme. Biochemistry 31:9694–9702, 1992.

35. J Mo, ME Holtzer, A Holtzer. Kinetics of folding and unfolding $\beta\beta$-tropomyosin. Biopolymers 32:1581–1587, 1992.

36. T Kiefhaber, FX Schmidt, K Willaert, Y Engelborghs, K Chafotte. Structure of a rapidly formed intermediate in ribonuclease T1 folding. Protein Sci 1:1162–1172, 1992.

37. SE Radford, CM Dobson, PA Evans. The folding of hen lysozyme involves partially structured intermediates and multiple pathways. Nature 385:302–307, 1992.

38. PA Jennings, PE Wright. Formation of a molten globule intermediate early in the kinetic folding pathway of apomyoglobin. Science 262:892–896, 1993.

39. CJ Mann, CR Matthews. Structure and stability of an early folding intermediate of *Escherichia coli* trp aporepressor measured by far-UV stopped-flow circular dichroism and 8-anilino-1-naphthalene sulfonate binding. Biochemistry 32:5282–5290, 1993.

40. K Chiba, A Ikai, Y Kawamurakonishi, H Kihara. Kinetic study on myoglobin refolding monitored by five optical probe stopped-flow methods. Proteins Struct Funct Gen 19:110–119, 1994.

41. SD Hooke, SE Radford, CM Dobson. The refolding of human lysozyme: A comparison with the structurally homologous hen lysozyme. Biochemistry 33:5867–5876, 1994.

42. H Roder, GA Elöve, SW Englander. Structural characterization of folding intermediates in cytocrome *c* by H-exchange labelling and proton NMR. Nature 335:700–704, 1988.

43. JB Udgaonkar, RL Baldwin. NMR evidence for an early framework intermediate on the folding pathway of ribonuclease A. Nature 335:694–699, 1988.

44. J Baum, CM Dobson, PA Evans, C Hanley. Characterization of a partly folded protein by NMR methods: Studies on the molten globule state of guinea pig α-lactalbumin. Biochemistry 28:7–13, 1989.

45. J Balbach, V Forge, WS Lau, NAJ van Nuland, K Brew, CM Dobson. Protein folding monitored at individual residues during a two-dimensional NMR experiment. Science 274:1161–1163, 1996.

46. CPM van Mierlo, NJ Darby, D Neuhaus, TE Creighton. Two-dimensional ^{1}H nuclear magnetic resonance study of the (5–55) single-disulphide folding intermediate of bovine pancreatic trypsin inhibitor. J Mol Biol 222:373–390, 1991.

47. CPM van Mierlo, NJ Darby, D Neuhaus, TE Creighton. (14–38, 30–51) double-disulphide intermediate in folding of bovine pancreatic trypsin inhibitor: A two-dimensional ^{1}H nuclear magnetic resonance study. J Mol Biol 222:353–371, 1991.

48. NJ Darby, CPM van Mierlo, GHE Scott, D Neuhaus, TE Creighton. Kinetic roles and conformational properties of the non-native two-disulphide intermediates in the refolding of bovine pancreatic trypsin inhibitor. J Mol Biol 224:905–911, 1992.

49. CPM van Mierlo, NJ Darby, TE Creighton. The partially folded conformation of the Cys30–Cys51 intermediate in the disulphide folding pathway of bovine pancreatic trypsin inhibitor. Proc Natl Acad Sci USA 89:6775–6779, 1992.

50. CPM van Mierlo, NJ Darby, J Keeler, D Neuhaus, TE Creighton. Partially folded conformation of the (30–51) intermediate in the disulphide folding pathway of bovine pancreatic trypsin inhibitor: ^{1}H and ^{15}N resonance assignments and determination of backbone dynamics from ^{15}N relaxation measurements. J Mol Biol 229:1125–1146, 1992.

51. CPM van Mierlo, J Kemmink, D Neuhaus, NJ Darby, TE Creighton. ^{1}H NMR analysis of the partly-folded non-native two-disulphide intermediates (30–51, 5–14) and (30–51, 5–38) in the folding pathway of bovine pancreatic trypsin inhibitor. J Mol Biol 235:1044–1061, 1994.

52. C Bustamante, I Tinoco, MF Maestre. Circular differential scattering can be an important part of the circular dichroism of macromolecules. Proc Natl Acad Sci USA 80:3568–3572, 1983.

53. D Mao, BA Wallace. Differential light scattering and absorption flattening optical effects are minimal in the circular dichroism spectra of small unilamellar vesicles. Biochemistry 23:2667–2673, 1984.

54. MJ Hope, MB Bally, G Webb, PR Cullis. Production of large unilamellar vesicles by a rapid extrusion procedure. Characterization of size distribution, trapped volume and ability to maintain a membrane potential. Biochim Biophys Acta 812:55–65, 1985.

55. HHJ de Jongh, B de Kruijff. The conformational changes of apocytochrome *c* upon binding to phospholipid vesicles and micelles of phospholipid based detergents; a circular dichroism study. Biochim Biophys Acta 1029:105–112, 1990.

56. HHJ de Jongh, E Goormaghtigh, JA Killian. Analysis of curcular dichroism spectra of oriented protein-lipid complexes: Toward a general application. Biochemistry 33:14521–14528, 1994.

57. A Gow, W Auton, R Smith. Interactions between bovine myelin basic protein and zwitterionic lysophospholipids. Biochemistry 29:1142–1147, 1990.

58. H Jung, R Windhaber, D Palm, KD Schnackerz. Conformation of a beta-adrenoreceptor–derived signal transducing peptide as inferred by circular dichroism and ¹H-NMR spectroscopy. Biochemistry 35:6399–6405, 1996.

59. N Dekker, AR Peters, AJ Slotboom, R Boelens, R Kaptein, R Dijkman, G de Haas. Two-dimensional ¹H-NMR studies of phospholipase A_2-inhibitor complexes bound to a micellar lipid-water interface. Eur J Biochem 199:601–607, 1991.

60. HHJ de Jongh, JA Killian, B de Kruijff. A water-lipid interface induces a highly dynamic folded state in apocytochrome *c* and cytochrome *c*, which may represent a common folding intermediate. Biochemistry 31:1636–1643, 1992.

61. DJ Gordon, G Holzworth. Artefacts in the measured optical activity of membrane suspensions. Arch Biochem Biophys 142:481–488, 1971.

62. RM Glaeser, BK Jap. Absorption flattening in the circular dichroism spectra of small membrane fragments. Biochemistry 24:6398–6401, 1993.

63. HA Swords, BA Wallace. Circular dichroism analyses of membrane proteins: examination of environmental effects on bacteriorhodopsin spectra. Biochem J 289:215–219, 1993.

64. GD Fasman. Differentiation between transmembrane helices and peripheral helices by the deconvolution of circular dichroism spectra of membrane proteins. In: GD Fasman, ed. Circular Dichroism and the Conformational Analysis of Biomolecules. New York: Plenum, 1996, pp 381–412.

65. ML Riley, BA Wallace, SL Flitsch, PJ Booth. Slow alpha-helix formation during folding of a membrane protein. Biochemistry 36:192–196, 1997.

66. L Shao, KW Kinnally, CA Mannella. Circular dichroism studies of the mitochondrial channel, VDAC, from *Neurosporra crassa*. Biophys J 71:778–786, 1996.

67. FW Lau, JU Bowie. A method for assessing the stability of a membrane protein. Biochemistry 36:5884–5892, 1997.

68. AK Mukhopadhyay, S Dracheva, S Bose, RW Hendler. Control of the integral membrane proton pump, bacteriorhodopsin, by purple membrane lipids of *Halobacterium halobium*. Biochemistry 35:9245–9252, 1996.

69. AN van Hoek, M Wiener, S Bicknese, L Miercke, J Biwersi, AS Verkman. Secondary structure analysis of purified functional CHIP28 water channels by CD and FTIR spectroscopy Biochemistry 32:11847–11856, 1993.

70. S Morein, TP Trouard, JB Hauksson, L Rilfors, G Arvidson, G Lindblom. Two-dimensional 1H-NMR of transmembrane peptides from *Escherichia coli* phosphatidylglycerophosphate synthase in micelles. Eur J Biochem 241:489–497, 1996.

71. KV Pervushin, VY Orekhov, AI Popov, LY Musina, AS Arseniev. Three dimensional structure of (1–71) bacterioopsin solubilized in methanol/chloroform and SDS micelles determined by ¹⁵N-¹H heteronuclear NMR spectroscopy. Eur J Biochem 219:571–583, 1994.

72. SC Li, CM Deber. A measure of helical propensity for amino acids in membrane environments. Nature Struct Biol 1:368–373, 1994.

73. PY Chou, GD Fasman. Prediction of protein conformation. Biochemistry 13:222–245, 1974.

74. SC Li, NK Goto, KA Williams, CM Deber. Alpha-helical, but not beta-sheet, propensity of proline is determined by peptide environment. Proc Natl Acad Sci USA 93:6676–6681, 1996.

75. RA Demel, W Jordi, H Lambrechts, H van Damme, R Hovius, B de Kruijff. Differential interactions of apocytochrome *c* and holocytochrome *c* with acidic membrane lipids in model systems and the implications for their import into mitochondria. *J Biol Chem 264:3988–3997, 1989.*

76. H Vogel, F Jähnig, V Hoffmann, J Stümpel. The orientation of melittin in lipid membranes; a polarized infrared spectroscopy study. Biochim Biophys Acta 733:201–209, 1983.

77. SL Wertz, Y Savino, DS Cafiso. Solution and membrane bound structure of a peptide derived from the protein kinase *c* substrate domain of neuromodulin. Biochemistry 35:11104–11112, 1996.

78. AS Ladokhin, ME Selsted, SH White. Bilayer interactions of indolicidin, a small antimicrobial peptide rich in tryptophan, proline and basic amino acids. Biophys J 72:794–805, 1997.

79. S Kimura, D Erne, R Schwyzer. Interaction of glucagon with artificial lipid bilayer membranes. Int J Pept Protein Res 39:431–442, 1992.

80. G Zardeneta, PM Horowitz. Analysis of the perturbation of phospholipid model membranes by rhodanese and its presequence. J Biol Chem 267:24193–24198, 1992.

81. HHJ de Jongh, S Rospert, CM Dobson. Comparison of the conformational state and in vitro refolding of yeast cpn10 with bacterial GroES, Biochem Biophys Res Commun 244:884–888, 1998.

82. GL Mendz, IM Jamie, JW White. Effects of acyl chain length on the conformation of myelin basic protein bound to lysolipid micelles. Biophys Chem 45:61–77, 1992.

83. TG Fletcher, DA Keire. The interaction of beta-amyloid protein fragment (12–28) with lipid environments. Protein Sci 6:666–675, 1997.

84. KB Blodgett. Films built by depositing successive monomolecular layers on a solid surface. J Am Chem Soc 57:1007–1022, 1935.

85. KB Blodgett, I Langmuir. Built up films of barium stearate and their optical properties. Phys Rev 51:964–976, 1937.

86. UP Fringeli, HH Güthard. Infrared membrane spectroscopy. In: E Grell, ed. Membrane Spectroscopy. Berlin: Springer-Verlag, 1981, pp 270–332.

87. DC Clark, LJ Smith. Influence of alcohol-containing spreading solvents on the secondary structure of proteins: A Circular dichroism investigation. J Agric Food Chem 37:627–633, 1989.

88. PW Yang, LC Stewart, HH Mantsch. Polarized attenuated total reflectance spectra of oriented purple membranes. Biochem Biophys Res Commun 145:298–302, 1987.

89. MD Bazzi, RW Woody. Oriented secondary structure in integral membrane proteins. I. Circular dichroism and infrared spectroscopy of cytochrome *c* oxidase in multilamellar films. Biophys J 48:957–966, 1985.

90. DW Urry. Circular dichroism of proteins. Annu Rev Phys Chem 19:477–530, 1968.

91. I Tinoco Jr. Circular dichroism and rotatory dispersion curves for helices. J Am Chem Soc 86:297–298, 1964.

92. RW Woody. Optical rotatory properties of biopolymers. J Polym Sci Macromol Rev 12:181–321, 1977.

93. RW Woody. Theory of circular dichroism of proteins. In: GD Fasman, ed. Circular Dichroism and the Conformational Analysis of Biomolecules. New York: Plenum, 1996, pp 25–67.

94. RW Woody, I Tinoco. Optical rotation of oriented helices. III. Calculation of the rotatory dispersion and circular dichroism of the alpha and 3_{10}-helix. J Chem Phys 46:4927–4945, 1967.

95. Y Wu, HW Huang, G Olah. Method of oriented circular dichroism. Biophys J 57:797–806, 1990.

96. RW Woody. The circular dichroism of oriented β-sheets: Theoretical predictions. Tetrahyd Asymm 4:529–544, 1993.

97. K Rosenheck, B Sommer. Theory of the far-ultraviolet spectrum of polypeptides in the β conformation. J Chem Phys 46:532–536, 1967.

98. MM Kelly, ES Pysh, GM Bonora, CJ Toniolo. Vacuum ultraviolet circular dichroism of protected homooligomers derived from L-leucine. J Am Chem Soc 99:3264–3266, 1977.

99. H Vogel. Comparison of the conformation and orientation of alamethicin and melittin in lipid membranes. Biochemistry 26:4562–4572, 1987.

100. LAJM Creuwels, RA Demel, LMG van Golde, BJ Benson, HP Haagsman. Effect of acylation on structure and function of surfactant protein c at the air-liquid interface. J Biol Chem 268:26752–26758, 1993.

101. DG Cornell. Circular dichroism of polypeptide monolayer. J Colloid Interface Sci 70:167–180, 1979.

102. HHJ de Jongh, R Brasseur, JA Killian. Orientation of the α-helices of apocytochrome c and derived fragments at membrane interfaces, as studied by circular dichroism. Biochemistry 33:14529–14535, 1994.

103. JA Killian, I Salemink, MRR De Planque, G Lindblom, RE Koeppe, DV Greathouse. Induction of non-bilayer structures in diacylphosphatidylcholine model membranes by transmembrane alpha-helical peptides: Importance of hydrophobic mismatch and proposed role of tryptophans. Biochemistry 35:1037–1045, 1996.

104. DG Cornell, DL Patterson. Interaction of phospholipids in monolayers with β-lactoglobulin adsorbed from solution. J Agric Food Chem 37:1455–1459, 1989.

105. W Norde, J-P Favier. Structure of adsorbed and desorbed proteins. Colloids Surf 64:87–93, 1992.

106. T Zoungrana, GH Findenegg, W Norde. Structure, stability, and activity of adsorbed enzymes. J Colloid Interface Sci 190:437–448, 1997.

107. MCL Maste, EHW Pap, A van Hoek, W Norde, AJWG Visser. Spectroscopic investigation of the structure of protein adsorbed on a hydrophobic latex. J Colloid Interface Sci 180:632–633, 1996.

108. CJ Brock, M Enser. A model system for studying protein binding to hydrophobic surfaces in emulsions. J Sci Food Agric 40:263–273, 1987.

109. LJ Smith, DC Clark. Measurement of the secondary structure of adsorbed protein by circular dichroism. 1. Measurement of the helix content of adsorbed melittin. Biochim Biophys Acta 1121:111–118, 1992.

110. A Kondo, T Urabe. Temperature dependence of activity and conformational changes in α-amylase with different thermostability upon adsorption on ultrafine silica particles. J Colloid Interface Sci 174:191–198, 1995.

111. P Billsten, P-O Freskgård, U Carlsson, B-H Jonsson, H Elwing. Adsorption to silica nanoparticles of human carbonic anhydrase II and truncated forms induce a molten-globule-like structure. FEBS Lett 402:67–72, 1997.

112. R Strom, F Podo, C Crifo, C Bertelet, M Zulauf, G Zaccai. Structural aspects of the interaction of bee venom peptide melittin with phospholipids. Biopolymers 22:391–397, 1983.

113. CPM van Mierlo, WMAM van Dongen, F Vergeldt, WJH van Berkel, E Steensma. The equilibrium unfolding of *Azotobacter vinelandii* apoflavodoxin II occurs via a relatively stable folding intermediate. Protein Science 7:2331–2344, 1998.

21

Infrared Spectroscopy of Biophysical Monomolecular Films at Interfaces: Theory and Applications

R. A. Dluhy
University of Georgia, Athens, Georgia

I. INTRODUCTION

Biophysical monomolecular films have received a great deal of attention in materials science, in part due to an increased interest in "soft matter systems" and in part due to a rapid development of novel surface-sensitive characterization techniques, which enable investigations of structural properties of such systems on the molecular level. The life sciences have not utilized these surface-sensitive spectroscopic techniques to the same extent as have the physical sciences, possibly because of the recent emphasis on physiological and molecular biological research in the modern life sciences. In addition, the skills and training needed for modern biological research are often at odds with those needed in the fields of surface science and physical chemistry, the disciplines that have driven the development of the surface-sensitive spectroscopies. Nevertheless, it is becoming progressively more obvious that many important biological processes occur at interfaces (1); as a consequence, there has been a resurgence of interest in the use of monolayer techniques for the investigation of biophysical systems. It is thus very important for the development of the field that both the physiological and physicochemical aspects of thin film analysis are merged in biophysical research.

For the study of biophysical monolayers directly at the air-water (A-W) interface (the so-called Langmuir monolayers), Figs. 1 and 2 present a schematic diagram of how the thermodynamic phases of Langmuir monolayers can be manipulated using

Manipulation of Amphipathic Monomolecular Films:
The Langmuir-Adam Film Balance

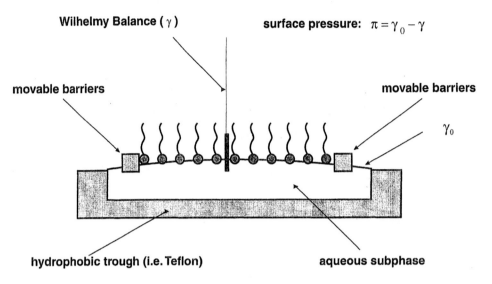

Fig. 1 Illustration of the methods used to manipulate monomolecular films at the air-water interface using the Langmuir-Adam film balance. An aqueous subphase is contained within a hydrophobic trough. Moveable barriers are used to compress and expand the surface area available to the surface-adsorbed film, thus changing its surface pressure, which is measured using a Wilhelmy plate method. In this case, surface pressure is defined as the difference in surface tension between the film-covered surface (γ) and the bare water substrate (γ_0) (i.e., $\pi = \gamma_0 - \gamma$).

the Langmuir-Adam film balance. The spectroscopic methods most widely used for the study of these aqueous monolayers include epifluorescence optical microscopy (1–3) and synchrotron X-ray diffraction (4). Unfortunately, even these powerful methods are subject to constraints. The epifluorescence microscopy studies incorporate a synthetic fluorescent lipid probe at low concentrations ($\sim$1–2 mol %) into the monolayer. The fluorescent probe molecules partition into one of the coexisting phases and can be used to visualize domains. One unresolved issue, however, in the use of this method is the role of the fluorescent amphiphile probe, which is essential to this method, in creating these images. Synchrotron X-ray diffraction has also been utilized to study the structures of lipid monolayers at the A-W interface (4,6–10). This is a powerful technique that has the ability to determine long-range positional and orientational order within the monolayer. Unfortunately, X-ray scattering from monolayers is very weak, so that even with high-intensity synchrotron sources, reliable measurements can be obtained only on high-density monolayer phases. For biomembrane studies, good data exist only for monolayers at high ($> \sim 40\,\mathrm{mN\,m^{-1}}$) surface pressures.

Many studies of biophysical interfaces have appeared in which the monolayer sample under study is transferred to a solid substrate for subsequent analysis. These

solid

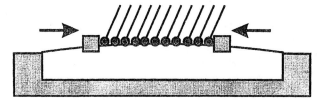

solid-condensed

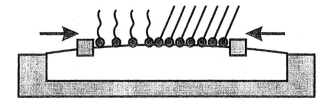

liquid-condensed

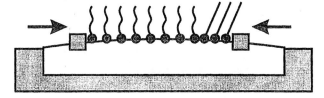

liquid-expanded

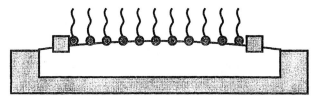

Fig. 2 Illustration of the various thermodynamic film states available to a Langmuir film. At low surface pressures (i.e., high surface tensions), the monolayer exists as a relatively disordered, expanded film. Upon compression of the trough barriers, the monolayer converts into a more ordered, condensed-phase film, until it reaches a two-dimensional solid state.

ordered monolayers are usually formed using some variation of the Langmuir-Blodgett (L-B) transfer method (11,12), which is presented schematically in Fig. 3. The advantage of L-B films is that this method is capable of assembling individual molecules into a highly organized molecular architecture, which can mimic biological systems. Diffraction studies of phospholipid monolayers transferred to solid substrates show

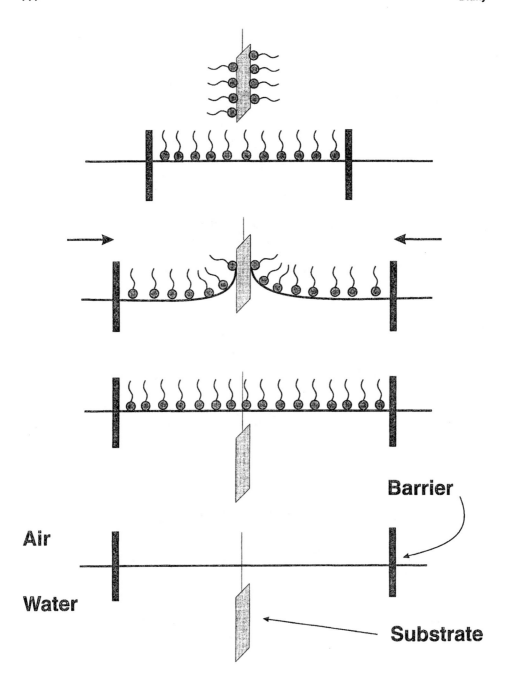

Fig. 3 Schematic diagram of the Langmuir-Blodgett deposition process for a monomolecular film onto a solid (hydrophilic) substrate. Starting with the substrate immersed in the subphase, the monolayer film is applied to the interface and the barriers compressed to the desired surface pressure. The substrate is then raised through the interface while the surface pressure is maintained, resulting in the transfer of a single monomolecular layer onto the substrate surface.

evidence of long-range ordering over long distances in the film (5,13). For optical wavelengths, fluorescent or other chromophoric dyes can be incorporated in the monolayer in order to study the incorporated molecules spectroscopically (11); alternatively, the naturally occurring pigment may be used as the chromophore (14). A wide variety of mono- and multilayer biomembrane systems have been studied using ultraviolet (UV)-visible wavelengths.

The development of surface-sensitive analytical methods has also been influential in the field of vibrational spectroscopy. Advances in instrumentation and methodology have enabled infrared and Raman spectroscopies to become true surface-sensitive analytical methods for thin films and monolayers on reflective or guiding substrates (e.g., 15,16). These new, surface-sensitive vibrational techniques are being increasingly applied to the study of biophysical monolayers.

Vibrational spectroscopy offers several advantages for the study of membrane molecular structure. Infrared absorptions are sensitive to displacements of the permanent bond dipole moment, which in turn depend on changes in conformation and configuration of the bonds making up the normal mode of vibration. Unlike other spectroscopic methods, which are restricted to monitoring certain molecular regions, such as ^{31}P nuclear magnetic resonance (^{31}P-NMR), or rely on the synthetic incorporation of an isotropic probe (e.g., ^{2}H-NMR or ^{19}F-NMR) or introduction of a possibly perturbing probe molecule (e.g., electron spin resonance and fluorescence), vibrational spectroscopy has the advantage of noninvasively monitoring absorptions by all regions of the lipid molecule. In addition, the time scale of the infrared experiment ($\sim 10^{-12}$ s) ensures that the interpretation of experimental results is not complicated by time scale averaging of anisotropic motions.

The aim of this chapter is to describe briefly the methodology and representative examples of the use of surface-sensitive vibrational spectroscopic methods for the study of biophysical interfaces. Two of these vibrational methods, in particular, will be surveyed: attenuated total reflectance infrared spectroscopy (ATR-IR) and external reflectance infrared spectroscopy.

II. INFRARED ATTENUATED TOTAL REFLECTANCE (ATR-IR) STUDIES OF BIOPHYSICAL MONOLAYERS

A. Attenuated Total Reflectance Spectroscopy: Theory and Methods

The potential of internal reflection (or, as it has come to be known, attenuated total reflection) spectroscopy to obtain spectra of interfaces was realized from its inception (17,18). Currently, ATR is one of the techniques most frequently used for characterizing the surfaces of materials with infrared spectroscopy (see, e.g., Ref. 19 and references contained within) because of its versatility, as well as the fact that information on the chemical composition, structure, orientation, and conformation of the thin film or interface may be obtained.

The theoretical background of ATR has been thoroughly described (20), so only a brief background discussion will be presented here. The reflection properties of light at an interface depend on the refractive index of the crystal (n_1), the refractive index of the sample (n_2), and the angles of reflection (θ_1) and refraction (θ_2) of the incoming radiation (note especially the requirement that n_1 must be greater than n_2 for total

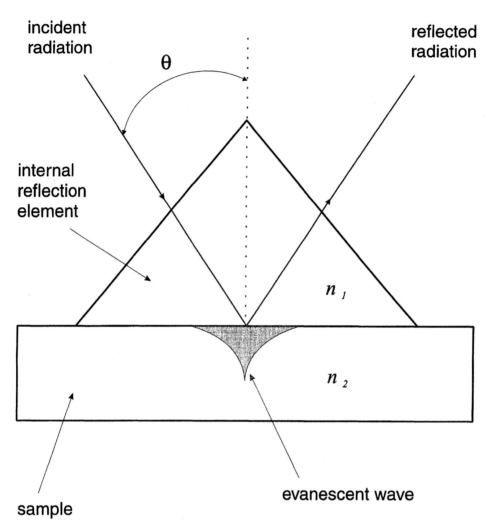

incident
radiation

θ

reflected
radiation

internal
reflection
element

n_1

n_2

evanescent wave

sample

Fig. 4 Schematic diagram of the internal reflection experiment. Radiation travels through an internal reflection element (of refractive index n_1) and strikes the element-sample interface at an incident angle θ. An exponentially decaying evanescent wave is created in the sample (of refractive index n_2). The internally reflected radiation (which is attenuated by any sample absorptions) is reflected from the crystal-sample interface and is detected.

internal reflection). Figure 4 illustrates the conventions and terms used in discussing an ATR spectroscopy experiment, and Fig. 5 illustrates the reflectance conditions under which total internal reflectance occurs.

As seen in Fig. 5, the refractive indices of the ATR crystal (n_1) and sample (n_2), as well as the incident angle of the incoming radiation (θ_1), control the properties of internal reflection in the following manner:

1. If $\sin \theta_1 < n_2/n_1$, some radiation will be reflected, but most of the radiation is refracted into phase 2 (i.e., the sample), because the angle of refraction (θ_2) will be smaller than the angle of reflection (θ_1) (see Fig. 5A).

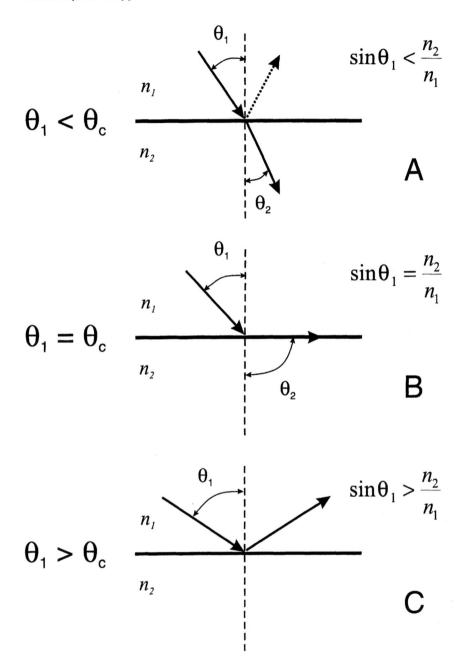

Fig. 5 Diagram illustrating how the angle of incidence (θ_1) and the refractive index of crystal (n_1) and sample (n_2) control the critical angle of total internal reflection (θ_c). (A) $\theta_1 < \theta_c$. Under these conditions, $\sin \theta_1 < n_2/n_1$ and most of the incident radiation penetrates into the sample. (B) $\theta_1 = \theta_c$. Under these conditions, $\sin \theta_1 = n_2/n_1$ and the incident radiation is reflected at an angle of 90° along the crystal-sample interface. (C) $\theta_1 > \theta_c$. Under these conditions, $\sin \theta_1 > n_2/n_1$ and the incident radiation is totally internally reflected from the crystal-sample interface.

2. If $\sin \theta_1 = n_2/n_1$, the angle of refraction (θ_2) equals $90°$ exactly, and the radiation travels along the interface. In this case, the angle of reflection (θ_1) is called the *critical angle* (θ_c) (see Fig. 5B).
3. If $\sin \theta_1 > n_2/n_1$, the angle of refraction (θ_2) is imaginary, and total internal reflection occurs (see Fig. 5C).

Therefore, the phenomena of total internal reflection at the crystal-sample interface will occur when the angle of incidence of radiation coming from the higher refractive index crystal (θ_1) exceeds the so-called critical angle θ_c, which is defined as

$$\theta c = \sin^{-1}\left(\frac{n_2}{n_1}\right) \tag{1}$$

The relationship between interface reflectivity, incident angle, and the critical angle is shown in Fig. 6 for a commonly used midinfrared ATR crystal (Ge) with a refractive index $n_1 = 4.0$ (e.g., 21).

Under the conditions in which $\theta_1 > \theta_c$, a standing electromagnetic wave exists at the crystal-sample interface, which is denoted as the "evanescent" (i.e., exponentially

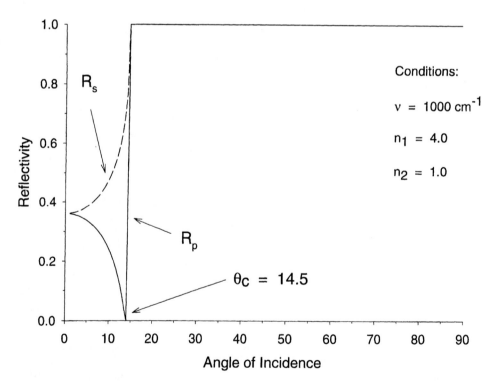

Reflectance versus Incident Angle for Ge ATR Element

Conditions:

$v = 1000 \text{ cm}^{-1}$

$n_1 = 4.0$

$n_2 = 1.0$

$\theta_c = 14.5$

Fig. 6 Calculated theoretical reflectance for a Ge ATR element as a function of the angle of incidence of the incoming radiation. Calculations were performed for both parallel (R_p) and perpendicular (R_s) polarized radiation at $\lambda = 10 \mu m$ (1000 cm^{-1}). The critical angle for Ge ($\theta_c = 14.5°$) is noted.

vanishing) wave. The amplitude of this evanescent wave at any point in the ambient medium (E) decays exponentially away from the crystal interface as described by the relationship

$$E = E_0 \exp\left(-\frac{z}{d_p}\right) \tag{2}$$

where E_0 is the maximum electric field intensity, z is the distance from the interface, and d_p is the maximum depth of penetration of the radiation into the second phase. If a sample is placed into contact with the higher refractive index crystal, attenuation of the incoming radiation occurs at the wavelengths of the sample's infrared absorption bands, with a concomitant decrease in the intensity of the reflected radiation, hence the name *attenuated total reflectance* (*ATR*).

The depth of penetration (d_p) in ATR spectroscopy is defined as the distance from the sample-crystal interface where the intensity of the evanescent wave falls to $1/e$ of its original value (e.g., $\sim 37\%$). The quantity d_p can be defined as

$$d_p = \frac{\lambda}{2\pi n_1 \sqrt{\sin^2 \theta - \left(\frac{n_2}{n_1}\right)^2}} \tag{3}$$

where λ represents the wavelength of the incoming radiation and θ is the angle of incidence of the incoming radiation. In the midinfrared spectral region, d_p corresponds to a distance of 1–$6\,\mu m$ from the ATR crystal surface. Figure 7 illustrates the values of d_p as a function of incident angle for a Ge ATR crystal calculated at a wavelength of $10\,\mu m$ ($1000\,cm^{-1}$). This value of d_p allows the experimenter to study very thick or very thin absorbing samples that cannot be studied by conventional transmission IR methods.

In the presence of an absorbing sample at the crystal interface, the evanescent wave interacts with this sample, leading to an attenuation of the totally reflected beam by absorption of the IR radiation from the sample. The result is an ATR spectrum of the sample that is similar to the standard IR absorbance spectrum obtained from transmission measurements, but with several important differences. Several of the differences between ATR and transmission spectra have been described (20). For example, the band intensities obtained in an ATR spectrum of a sample that is thicker than d_p are independent of the thickness of the sample, because the penetration depth of the radiation (which is on the order of several micrometers in the IR) will be less than the thickness of the sample. However, as Eq. (3) describes, d_p is also wavelength dependent, and the wavelength dependence of the band intensities must be taken into account whenever a quantitative analysis is needed. Figure 8 illustrates the wavelength dependence of d_p for a Ge ATR crystal with a $45°$ angle of incidence between 4000 and $400\,cm^{-1}$ in the midinfrared spectral region.

The absorption, A, of a thin film in contact with a crystal results in an ATR spectrum that can be theoretically described by a modification of Beer's law taking into account the exponential nature of the evanescent field (22):

$$A = \frac{n_2}{n_1} \frac{\alpha t}{\ln_{10} \cos \theta} \langle E^2 \rangle \tag{4}$$

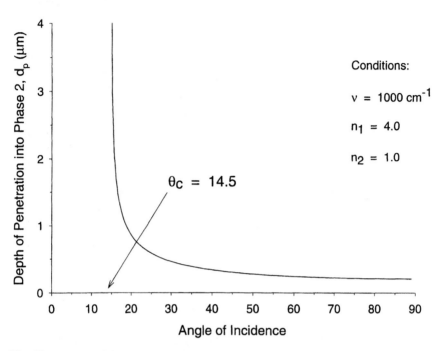

Fig. 7 Calculated depth of penetration (d_p) for a Ge ATR element as a function of the angle of incidence of the incoming radiation. Calculations were performed for radiation of $\lambda = 10\,\mu m$ (1000 cm^{-1}). The critical angle for Ge ($\theta_c = 14.5°$) is noted.

where t is the thickness of the surface layer and α is the absorption coefficient of the thin film. The value $\langle E^2 \rangle$ is the integrated average value of the electric filed intensity at the crystal surface. The electric field intensities are related to the polarization of the incoming radiation and the geometry of the ATR crystal. Figure 9 illustrates the laboratory-referenced geometry of the distribution of orthogonal electric field intensities for a common type of parallelogram ATR crystal. In this case, radiation polarized perpendicular to the plane of incidence (s polarization) contains only a component in the y direction along the surface of the ATR crystal. Radiation polarized parallel to the plane of incidence (p polarization), however, contains contributions from molecular dipole moments oriented in the x direction along the surface as well as in the z direction normal to the crystal surface. A quantitative description of the electric field distribution along the Ge crystal surface is given in Eq. (5) (22–24).

$$\langle E^2_{\perp 1}\rangle = (1+R_\perp) + 2R_\perp^{1/2}\cos\left(\delta^r_\perp - 4\pi\left(\frac{z}{\lambda}\right)\zeta_1\right)$$

$$\langle E^2_{\parallel 1x}\rangle = \cos^2\theta\left[(1+R_\parallel) - 2R_\parallel^{1/2}\cos\left(\delta^r_\parallel - 4\pi\left(\frac{z}{\lambda}\right)\zeta_1\right)\right] \tag{5}$$

$$\langle E^2_{\parallel 1z}\rangle = \sin^2\theta\left[(1+R_\parallel) + 2R_\parallel^{1/2}\cos\left(\delta^r_\parallel - 4\pi\left(\frac{z}{\lambda}\right)\zeta_1\right)\right]$$

In these equations, z is the distance from the first phase boundary (defined as negative

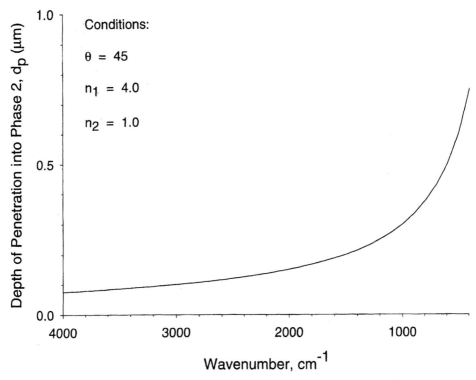

Fig. 8 Calculated depth of penetration (d_p) for a Ge ATR element as a function of the wave number of the incoming radiation. Calculations were performed for radiation at an incoming incident angle of 45° to the crystal-sample interface.

in phase 1), ξ is an angle-dependent refractive index term (defined in Sec. III), and δ is the phase change upon reflection at the Ge surface. Using a similar approach, analogous expressions may be written for the electric field intensities in any of the stratified layers of the reflectance system (23).

The relative intensity of each of these electric field components for a Ge ATR crystal is shown in Fig. 10; these data clearly indicate that above the critical angle, most of the electric field intensity present is oriented along the Ge surface, with very little present normal to the surface. This distribution of electric field intensities must be taken into account when orientation measurements using ATR polarization measurements are made on thin film samples (see later).

The influence of the relative intensities of the orthogonal electric field components on the theoretical absorbances for an isotropic thin film adsorbed onto an ATR crystal is presented in Fig. 11. These theoretical thin-film absorbances were calculated for each x, y, and z absorbance component using the expressions presented in Eq. (6), which are expansions of the generalized thin-film reflection-absorbance Beer's law analogy of Eq. (4) to the three orthogonal absorbance directions analogous to the E_x, E_y, and E_z electric field directors shown in Fig. 9 (24). As is obvious from an inspection of Fig. 11 and Eq. (6), the theoretical ATR absorbances track the same profile as the electric fields present at the crystal surface. It is also apparent from Fig. 11 that the observed experimental IR absorbances will be predominantly from mol-

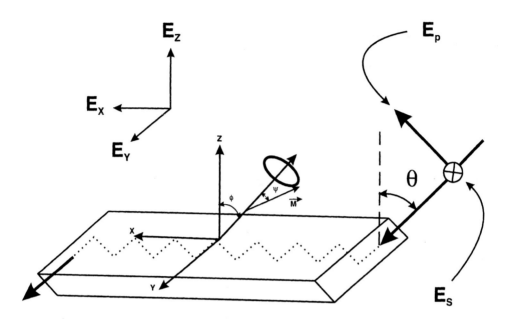

Fig. 9 Illustration of the optical path of radiation through a parallelogram ATR crystal. Radiation that is polarized either parallel (*p*-polarized, E_p) or perpendicular (*s*-polarized, E_s) to the plane of incidence enters the ATR crystal at an angle θ relative to the surface normal. The *s*-polarized radiation interacts with vibrational dipole moments oriented in the *y* coordinate geometry (E_y direction), while the *p*-polarized radiation interacts with vibrational dipole moments oriented in both the *x* and *z* coordinate geometry (E_x and E_z directions). Also illustrated in this figure is the uniaxial orientation of a molecular dipole moment (ϕ) superimposed on the ATR crystal geometry, which depicts the relationship between molecular orientation and the ATR dichroic ratio: $R = A_{\parallel}/A_{\perp}$.

ecular dipole moments oriented along the crystal surface (i.e., in the *x* and *y* directions), although a small but important fraction ($\sim 18\%$) of the p-polarized absorbance at a 45° incident angle will come from dipoles normal to the surface.

$$\Delta A_{\perp} = \frac{n_2 \alpha_2 t}{\ln 10 n_1 \cos \theta} \langle E_{\perp}^2 \rangle$$

$$\Delta A_{\parallel x} = \frac{n_2 \alpha_2 t}{\ln 10 n_1 \cos \theta} \langle E_{\parallel x}^2 \rangle \tag{6}$$

$$\Delta A_{\parallel z} = \frac{n_2 \alpha_2 t}{\ln 10 n_1 \cos \theta} \frac{n_1^4}{\left(n_2^2 + k_2^2\right)^2} \langle E_{\parallel z}^2 \rangle$$

Many different possible materials and geometries are used for ATR crystals (see, e.g., Ref. 20). Figure 9 portrays one of the types of ATR crystals more commonly used in monomolecular and thin-film studies, namely a parallelogram-shaped crystal with edges beveled at a particular angle of incidence greater than the critical angle. This type of crystal has the particular advantage in thin-film studies of being easy to use as a Langmuir-Blodgett deposition substrate. The radiation incident on the beveled edge of this type of ATR crystal is guided through the length of the crystal before exiting at the opposite end. A standard commercially available parallelogram ATR

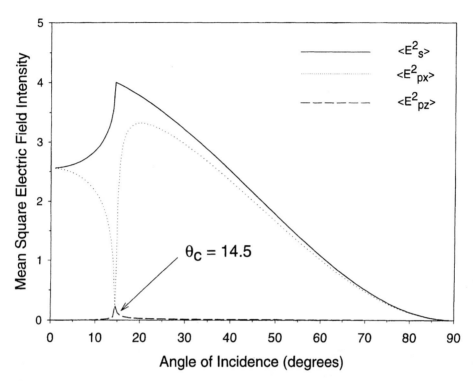

Electric Field Distribution For A 2-Phase System :
Ge (n_1=4.0) : air (n_2=1.0) : ν = 1000 cm^{-1}

Fig. 10 Calculated electric field distribution for a Ge ATR element as a function of the angle of incidence of the incoming radiation. Calculations were performed for electric field components oriented both perpendicular ($\langle E_s^2 \rangle$) and parallel ($\langle E_{px}^2 \rangle$ and $\langle E_{pz}^2 \rangle$) to the plane of incidence at $\lambda = 10\,\mu m$ (1000 cm^{-1}). The critical angle for Ge ($\theta_c = 14.5°$) is noted.

crystal is typically 50 mm long × 5 mm or 10 mm wide × 2 mm thick beveled at a 30 or 45° incident angle. From a spectroscopic viewpoint, this type of crystal also has the added advantage of multiplying the number of reflectances along the crystal surface, thus increasing the effective path length of the sample, which is an important consideration for monomolecular film studies. The number of reflections that the incident radiation makes during its travels down the crystal is given by

$$N = \frac{l}{t}\cot\theta \tag{7}$$

where l is the length of the crystal and t its thickness. Although the parallelogram-type crystal has undoubtedly been the most commonly used in monomolecular and Langmuir-Blodgett research, other ATR crystal geometries have also been successfully employed in biophysical thin-film studies (25).

One of the additional reasons for the popularity of ATR spectroscopy in the analysis of thin films has been its ability to determine the average molecular orientation of defined

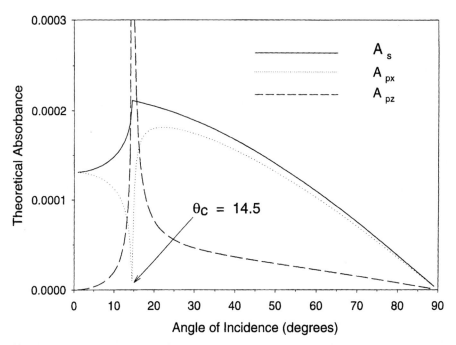

Fig. 11 Calculated theoretical absorbance for a hypothetical monomolecular film adsorbed on a Ge ATR element as function of the angle of incidence of the incoming radiation. Calculations were performed for theoretical absorbances oriented both perpendicular (A_s) and parallel (A_{px} and A_{pz} to the plane of incidence at $\lambda = 10\,\mu m$ ($1000\,cm^{-1}$). The critical angle for Ge ($\theta_c = 14.5°$) is noted. The thin film was assumed to have optical constants of $n = 1.5$, $k = 0.1$, and a thickness of $25\,\text{Å}$.

structural units within the thin film (26,27). There are several derivations of the orientation distribution function in the literature, but, in general, all rely on determining molecular orientation by the calculation of the order parameter or orientation function $S = \langle P_2(\cos\phi)\rangle$, which is the second-order Legendre polynomial obtained by series expansion of a uniaxial orientation distribution illustrated in Fig. 9:

$$S = \langle P_2(\cos\phi)\rangle = \tfrac{1}{2}\left(3\langle\cos^2\phi\rangle - 1\right) \tag{8}$$

The value obtained for the $\langle P_2\rangle$ order parameter is then used to describe the orientation of the individual molecular subgroups within the thin film. Experimentally, the value of $\langle P_2(\cos\phi)\rangle$ is related to molecular orientation through R, the dichroic ratio, obtained from polarized ATR measurements. The dichroic ratio is defined as the ratio of p-polarized and s-polarized absorption intensities as follows:

$$R = \frac{A_{\parallel}}{A_{\perp}} = \frac{A_x + A_y}{A_z} = \frac{\langle E_x^2\rangle\langle\sin^2\phi\rangle + 2\langle E_z^2\rangle\langle\cos^2\phi\rangle}{E_y^2\langle\sin^2\phi\rangle} \tag{9}$$

where $A_{\parallel}$ is the measured band absorbance when the incoming radiation is polarized parallel to the plane of incidence. The value $A_{\perp}$ is similarly defined as the measured absorbance when the incoming radiation is polarized perpendicular to the plane of incidence. The value of the order parameter (and hence the measure of molecular orientation) may be related to the values of the electric fields and the experimentally determined dichroic ratio as shown in Eq. (10) (27).

$$S = \frac{\langle E_x^2 \rangle - R\langle E_y^2 \rangle + \langle E_z^2 \rangle}{\langle E_x^2 \rangle - R\langle E_y^2 \rangle - 2\langle E_z^2 \rangle} \tag{10}$$

B. Attenuated Total Reflectance Spectroscopy: Applications

Attenuated total reflectance IR spectroscopy has been extensively used in the field of biophysical spectroscopy and has been especially widely used in the study of biomembranes and their interfacial interactions. Although it is beyond the scope of this chapter to survey the entire field, representative examples of the use of ATR-IR in phospholipid and phospholipid-peptide biomembrane interfacial studies will be briefly discussed. More thorough reviews of the use of the ATR-IR technique for other types of phospholipid and membrane protein studies are given in Refs. 28–31. In addition, further reading on the basic methodology of biophysical infrared spectroscopy (including transmission and reflectance methods, sample handling, data processing, as well as the assignment of vibrational frequencies to specific functional groups and molecular conformations) is presented in Refs. 32–40.

1. Phospholipid Model Membranes

The structure of dry oriented films of a homologous series of phospholipids was originally studied more than two decades ago by ATR (41). These lipids include 1,2-dipalmitoyl-*sn*-glycero-3-phosphoethanolamine (DPPE), naturally isolated sheep brain phosphoethanolamine, 1,2-dipalmitoyl-*sn-glycero-3-phospho-N*-methylethanolamine, 1,2-dipalmitoyl-*sn*-glycero-3-phospho-*N-N*-dimethylethanolamine, DPPC, and naturally isolated egg phosphocholine (egg-PC). Similar results were obtained for this series of compounds: (a) a 20–30° orientational deviation of the hydrocarbon chain from the normal to the plane of the bilayer was seen for these lipids; (b) the fatty acid ester groups in the β and γ positions of glycerol had different conformations; (c) the polar headgroups are oriented paralleled to the plane of the bilayer; and (d) the phosphate group of all phospholipids except DPPE were shown to exist in the ionized PO_2^- state, whereas the phosphate group of DPPE was concluded to exist in the O=P—OH state.

Polarized ATR spectroscopy has also been used to study the structure of hydrated phospholipid bilayers in order to better understand the function of more realistic biomembrane models. For example, the changes in structure and orientation that accompany the hydration of multibilayers of 1,2-dimyristoyl-*sn*-glycero-3-phosphocholine (DMPC) as compared with dry films has been examined (42). Bands at 3400 and 1650 cm^{-1} were assigned to the OH stretching and bending vibrations, not the result of bulk water but instead the result of water that penetrated into the hydrophilic part of the multibilayer. The antisymmetric and symmetric CH$_2$ stretching vibrations were shifted to higher frequency in the hydrated film. The weak CH$_2$

wagging band that was observed in the dry film was not seen in the hydrated film. The antisymmetric and symmetric PO_2^- stretching vibrations as well as the antisymmetric and symmetric C—O—C stretching vibrations and the ester C=O stretching mode are all shifted and broadened after hydration. It was concluded that the hydration caused changes in the spatial packing of the polar groups of DMPC. Hydration also resulted in a slight increase in the dichroic ratio of the CH_2 stretching and scissoring bands. The study also noted that the polar headgroups of DMPC are not completely disordered and retain a slight orientation in fully hydrated multibilayers.

Partially and fully hydrated multibilayers of 1,2-dipalmitoyl-*sn*-glycero-3-phosphocholine (DPPC) were examined using temperature-dependent ATR (43). A decrease in the order parameter of the hydrocarbon chains, as determined from the dichroic ratios of the antisymmetric and symmetric CH_2 stretching bands, was observed at the gel to liquid-crystalline transition temperature. This decrease was found to be independent of water content and implies that, because of the chain melting associated with an increase in the number of gauche conformers, the hydrocarbon chain is in a disordered state. At the pretransition temperature, there was an increase in the dichroic ratios of the hydrated DPPC assigned to the symmetric PO_2^- stretching and asymmetric $N^+(CH_3)_3$ stretching modes. There was a similar increase in the dichroic ratios of the OH stretching and H—O—H bending bands of water as compared with those of the polar bands. Therefore, it is believed that the pretransition temperature may be ascribed to the reorientation of the polar groups of DPPC and bound water. However, the orientational disorder of the hydrocarbon chains is the result of the main transition. The reorientation of the polar groups and water, in partially hydrated multibilayers, occurred in the temperature region near the gel to liquid-crystalline phase transition temperature.

Oriented multilayers of DMPC and DPPC that have had a ^{13}C atom synthetically incorporated in the *sn*-2 carbonyl carbon were investigated by polarized ATR (44). The synthetic incorporation of a ^{13}C label into one of the acyl chains shifted the vibrational frequency of this group enough that experiments were able to distinguish spectroscopically between the *sn*-1 and *sn*-2 carbonyls in these molecules. Analysis of the dichroic ratios for these two carbonyl groups showed that the ester double bond is aligned in the membrane plane with a tilt angle relative to the bilayer normal of greater than 60°. No differences were seen in the orientation of the *sn*-1 and *sn*-2 C=O bonds in dry films or in solid and liquid phases of hydrated films.

One of the more useful aspects of ATR-IR spectroscopy is its ability to analyze multicomponent films by incorporating synthetic, isotopically labeled lipids in the thin film in combination with the normal, protiated lipids. This method generally uses binary mixtures of perdeuterated DPPC with other defined protiated lipids as specific biomembrane models. Figure 12 is an example of an ATR-IR spectrum of an 88:12 (mol:mol) binary mixture of perdeuterated DPPC (DPPC-d_{62}) and DOPG, illustrating both the C—H region due to the DOPG component and the C—D region due to the DPPC-d_{62} component. The ability of the ATR-IR method to acquire good signal-to-noise (S/N) spectra of membrane components at even submonolayer coverage is well illustrated in this figure.

A method to determine the exact fractional composition of binary mixtures of phospholipids at the air-water interface by ATR-IR in combination with ^{31}P NMR spectroscopy has been reported (45,46). The C—H and C—D vibrational intensities obtained from the ATR spectra as well as data obtained from ^{31}P NMR experiments

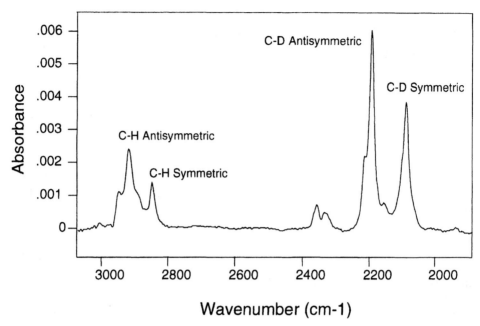

Fig. 12 ATR-IR spectrum of a two-component monomolecular phospholipid film transferred to a Ge ATR crystal. The sample in this spectrum was composed of a 88:12 (mol:mol) mixture of acyl chain perdeuterated 1,2-dipalmitoyl-*sn*-glycero-3-phosphocholine (DPPC-d_{62}) with 1,2-dioleoyl-*sn*-glycero-3-phosphoglycerol (DOPG) transferred to the Ge crystal at a surface pressure of $\sim 70\,\mathrm{mN\,m^{-1}}$.

were used to acquire quantitative information on the fractional composition of the binary mixtures of the phospholipids after monolayer compression and manipulation with subsequent transfer to ATR crystals. This system was used to evaluate monomolecular films of DPPC-d_{62} and DPPG transferred at high pressure (up to $70\,\mathrm{mN^{-1}}$) to germanium ATR crystals.

An oriented phospholipid bilayer consisting of 1,2-dipalmitoyl-*sn*-glycero-3-phosphoglycerol (DPPG) has also been studied using a combination of IR spectroscopy and scanning tunneling microscopy (47). In this case both L-B and self-assembly methods were used to construct the bilayer model. IR analysis showed the hydrocarbon portion of the bilayer to be conformationally ordered, packed in a hexagonal subcell, and oriented by 10–15° relative to the surface normal. Tunneling microscopy observed for the first time the presence of highly crystalline, nanometer-scale hexagonal domains within the DPPG monolayer, thus corroborating the IR results.

2. Lipid-Peptide Complexes

An additional area of study in which biophysical ATR-IR spectroscopy has been widely utilized has been in the analysis of the secondary structure of membrane peptides and proteins incorporated in phospholipid bilayer thin-film models. The usefulness of vibrational spectroscopy in this field comes from the fact that membrane-bound proteins and peptides are very difficult to crystallize or obtain in large enough quantities for solution analysis; therefore, there is a real need for surface-sensitive spectroscopic techniques to identify secondary structure in model systems that mimic a natural environment but that have proved difficult to study using modern methods of X-ray or NMR analysis.

The field of ATR-IR analysis of lipid-peptide or lipid-protein complexes is very large. Rather than attempt a comprehensive review, this section concentrates on a brief description of how ATR-IR methods have been used to study three different types of representative lipid-peptide classes, namely melittin, signal sequences, and pulmonary surfactant. The studies performed with these three types of lipid-peptide systems illustrate the power and utility of ATR-IR for biomembrane interfacial analysis. For a more comprehensive literature review of the field of IR spectroscopy as applied to lipid-protein systems in general, the reader is referred to the review articles mentioned earlier.

One of the most actively studied membrane-interacting peptides in the biophysical literature has been melittin, a hemolytic peptide isolated from bee venom that forms a highly amphipathic α-helix (48,49). Several thin-film IR studies have looked at the interaction of melittin with phospholipid bilayers (50–55). In one such study, the interaction of melittin (MLT), melittin hydrophobic fragment (residues 1–15, MLT_{1-15}), and melittin hydrophilic fragment (residues 16–26, MLT_{16-26}) with layers of DPPC and POPC has been studied (53). When incorporated in bilayers of DPPC, melittin had mostly α-helical secondary structure, although some β and/or aggregated structures were also observed. These α-helical portions were oriented perpendicular to the bilayer plane. Only a small amount of α-helical structure was found when MLT_{16-26} interacted with DPPC. It was also noted that there was only moderate penetration of MLT_{16-26} into the bilayer. A β-antiparallel pleated sheet structure with only a small amount of α-helical structure was found when MLT_{1-15} interacted with DPPC. This fragment also resulted in a disordering of the lipid acyl chains.

A study was performed whereby the structure of synthetic melittin and melittin analogues bound to monomyristoyl-*sn*-glycero-3-phosphocholine (MMPC) micelles, DMPC vesicles, and 1,2-diacyl-*sn*-glycero-3-phosphocholine films was investigated (55). ATR-IR was used to provide information about the secondary structure as well as a measure of the static orientational distribution of α-helical conformation to the bilayer plane and information about the phospholipid acyl chain configuration. It was concluded that there was an increase in α-helical conformation when all of the MLT peptides were bound to MMPC or DMPC lipid substrates. Of the residues studied that had α-helical structure, all of the MLT peptides except $([^{13}C\delta_1]\text{L-Trp9},[_{13}C\alpha]\text{Gly12,Leu19})$melittin), MLT-W9, showed a large degree of orientational order perpendicular to the plane of the lipid bilayer.

Although many of the IR studies performed during the 1980s using dry transferred films indicated that melittin was oriented perpendicular to the membrane surface, additional studies have used the ATR technique to study the orientation of melittin bound to hydrated single supported planar bilayers in addition to dry phospholipid multilayers (54). Melittin was found to have an α-helical conformation regardless of the phospholipid model system that was used. These models included multibilayers of DPPC, POPC, a 4:1 mixture of POPC and 1-palmitoyl,2-oleoyl-*sn*-glycero-3-phosphoglycerol (POPG), and single supported planar bilayers of 4:1 POPC:POPG. Unlike the conformation, however, the order parameter of the α-helix in the bilayers was found to be dependent on the type of membrane preparation.

Whereas a positive order parameter was observed for dry membrane preparations, a negative helical order parameter was observed for hydrated single supported planar bilayers. This observation resulted in the researchers concluding that the α-helix long axis of melittin is oriented parallel to the plane of the bilayer. It was found that the orientation of melittin in bilayers is dependent not on the technique used for the determination, but instead on the amount of hydration of the model membranes. The effect of membrane surface pressure on the conformation and orientation of a signal peptide from *Escherichia coli* in lipid monolayers has been determined using polarized ATR (56,57). The CH$_2$ stretching bands are used to determine the conformation and orientation of the phospholipid acyl chains. The researchers concluded that the hydrocarbon chain, in this case, is highly disordered. The study resulted in values ranging from 58 to 61° for the orientation of the peptide α-helix with respect to the membrane normal for low-pressure monolayers. Using high pressures, the amide I vibrations led to the conclusion that the peptide's conformation is largely β-structure. The amide II vibration is polarized perpendicularly, which suggests that the orientation of the β-structure is along the surface of the crystal and in the plane of the lipid monolayer. Figure 13 illustrates the polarized ATR-IR amide vibrations of the wild-type signal sequence peptide from *E. coli*. This figure is a good example of the type of data obtainable from the combination of surface biophysics and vibrational spectroscopy of peptides in membranes.

Additional studies of functional and nonfunctional permease signal sequences from *E. coli* have been published (58). These synthetic peptides were incorporated in single planar phospholipid bilayers and studied using polarized ATR-IR spectroscopy; order parameters were then derived from the dichroic ratios. It was found that the functional sequences were oriented along the plane of the membrane, while the nonfunctional peptides were disordered in the lipid bilayers.

The structure-function relationships between isolated and model systems that mimic pulmonary surfactant monolayers have been increasingly studied using a combination of surface chemistry and ATR spectroscopy. Several of these studies have focused on the peptide component of the surfactant. To date, three surfactant proteins have been found to be associated with pulmonary surfactant. These proteins, SP-A, SP-B, and SP-C, interact with pulmonary surfactant lipids and result in an enhancement of the surface properties of the lipids. SP-C, a hydrophobic protein, is a 35-residue polypeptide chain with a hydrophobic C-terminal region and a less hydrophobic N-terminal region. ATR has been used to examine the secondary structure and orientation of SP-C as well as its ability to alter the surface properties of phospholipids. Using SP-C isolated from bovine lung lavage and reconstituted into binary lipid mixtures of DPPC and DPPG, it was concluded that the orientation of helical segments is along a direction that is parallel to the phospholipid acyl chains (59). SP-C isolated from porcine lung lavage has also been the subject of a paper (60) in which the secondary structure of SP-C and its depalmitoylated form was studied in the absence and presence of lipids. It was found that the secondary structure of both forms of SP-C is primarily of an α-helical configuration. Interaction of SP-C with lipid bilayers resulted in an increase in the α-helical content and a decrease in the β-sheet portion. In addition, it was concluded that the α-helix axis is oriented parallel to the lipid chains and the β-sheet is parallel to the interface for the depalmitoylated SP-C.

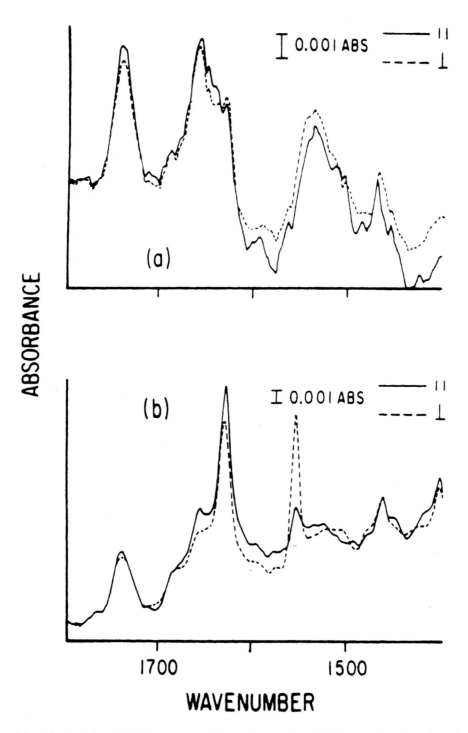

Fig. 13 Polarized ATR-IR spectra of the amide I and amide II spectral regions in a signal sequence peptide-phospholipid complex. The spectra illustrate the structural changes that occur when the lipid-peptide film was transferred at (a) low surface pressure and (b) high surface pressure. (From Ref. 57, © 1989 American Chemical Society, used with permission.)

A novel dimeric form of bovine SP-C (61) has been reported. The structure of this dimeric form of SP-C, [SP-C]$_2$, and the structure of the monomeric forms of SP-C and SP-B were examined in DPPC films. The results indicated a high α-helical ($\sim$47%) content for SP-C, which is consistent with previously published values. It was found that [SP-C]$_2$ was primarily (82%) β-sheet in structure with a small amount of β-turn and even less α-helical structure. SP-B contains approximately 59% β-sheet and β-turn as well as 27% α-helical structure and a small amount of random coil structure.

The secondary structure and orientation of porcine SP-B, with or without a lipid bilayer of DPPC or DPPG, was also studied (62). When SP-B was associated with a DPPC/PG (7:3 w/w) or a DPPG matrix, the secondary structure of the protein was conserved. The protein has no affect on the organization and orientation of lipid molecules in the bilayer. Results of this study indicated that 45% of SP-B is α-helical and 22% is β-sheet in structure. It was also concluded that a membrane-spanning segment is not needed to maintain this protein at the surface of a model membrane.

III. IN SITU EXTERNAL REFLECTION IR SPECTROSCOPY OF BIOMOLECULAR MONOLAYERS

A. Infrared External Reflection Spectroscopy: Introduction

Infrared external reflection spectroscopy, also sometimes referred to as infrared reflection-absorption spectroscopy (IRRAS), has for many years been successfully applied to the study of monomolecular thin films on metal surfaces (for a review, see Ref. 15). As shown diagrammatically in Fig. 14, an IR external reflection spectrum of a monomolecular film adsorbed on a reflective substrate is obtained by reflecting the incoming radiation from the three-phase ambient–thin film–substrate system and measuring the reflected intensity as a function of wavelength. Because this is a reflection experiment, the reflection-absorbance spectrum is created by ratioing the sample reflectance (R) against the reflectance of the film-free substrate (R_0) as

$$A = -\log\left(\frac{R}{R_0}\right) \qquad (11)$$

The reflection spectrum obtained by this process is a function of the wavelength, the state of polarization, the thin-film thickness, the angle of incidence of the reflected light, and the optical constants of the three phases involved (63,64).

Until just a few years ago, the published literature on external reflection IR spectroscopy dealt exclusively with thin films on reflective metals because of the low reflectivity of nonmetallic substrates, which leads to a low optical throughput and a degraded S/N ratio in the resulting spectrum. However, the use of modern Fourier transform infrared instrumentation and new experimental designs specifically adapted for thin films have changed this situation. It is now possible to acquire external reflectance spectra on a variety of dielectric substrates, including glassy carbon and water (65–67). As a consequence, researchers are able to study the structure of organic thin films on the most relevant substrates needed for their particular application. This has had a great impact on the use of reflection IR spectroscopy in biophysical research, because a large number of biologically relevant membrane surfaces may now be

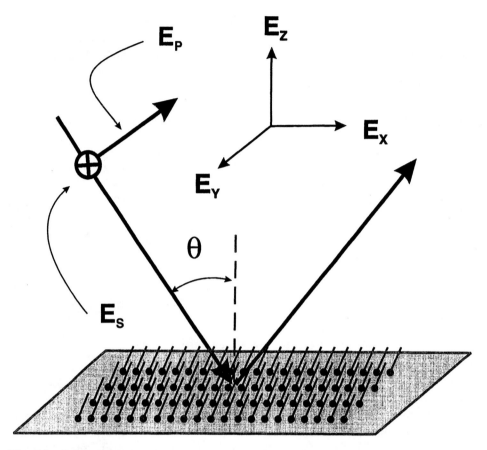

Fig. 14 Schematic diagram of the external reflection of polarized electromagnetic radiation from a thin film–covered surface. Parallel (E_p) and perpendicular (E_s) polarized components are shown for the incoming propagating wave. The coordinate geometry is indicated in the upper right of the diagram.

studied directly at an aqueous interface. This section describes the theory and methodology concerning the use of in situ external reflection IR spectroscopy at the air–water interface to study monomolecular films, with a brief discussion of some of the relevant published results in the area of biomembrane monolayers.

B. Infrared External Reflectance Spectroscopy: Theory

The substitution of a dielectric surface for a reflective metal surface in the external reflection IR experiment has profound consequences for the resulting spectra. The main differences can be understood by considering the underlying optical theory.

The theoretical description of external reflection spectroscopy is explicitly given by the Maxwell and Fresnel equations (68) and is based on the classical electromagnetic theory for an N-phase system of parallel, optically isotropic layers (23,24,63,64,69), which has been partially described for the ATR case (see earlier).

In an N-phase stratified layer system, the optical properties of the jth phase are characterized by the complex refractive index, defined as

$$\hat{n}_j = n_j + ik_j \tag{12}$$

where n_j is the real refractive index and k_j is the absorption constant of the jth phase.

For an N-phase system there are $N-1$ surface discontinuities at $z = z_j$ ($j = 1$ to $N-1$). For each space and final region there is a characteristic matrix M_j that can be used to relate the tangential fields at the first phase boundary to the final boundary as follows:

$$\begin{bmatrix} U_1 \\ V_1 \end{bmatrix} = M_2 M_3 \cdots M_{N-1} \begin{bmatrix} U_{N-1} \\ V_{N-1} \end{bmatrix} = M \begin{bmatrix} U_{N-1} \\ V_{N-1} \end{bmatrix} \tag{13}$$

where U_j and V_j are the tangential components of the field amplitude at the boundary. A matrix M that is characteristic of the stratified medium as a whole can be written as

$$M = M_2 M_3 \cdots M_{N-1} \tag{14}$$

The characteristic matrix M_j that defines the reflectance properties for each of the j layers is defined as follows for both polarizations:

$$\begin{aligned} M_j &= \begin{bmatrix} \cos \beta_j & \dfrac{-i}{q_j} \sin \beta_j \\ -i q_j \sin \beta_j & \cos \beta_j \end{bmatrix} \quad \text{for } p \text{ polarization} \\[2em] M_j &= \begin{bmatrix} \cos \beta_j & \dfrac{-i}{p_j} \sin \beta_j \\ -i p_j \sin \beta_j & \cos \beta_j \end{bmatrix} \quad \text{for } s \text{ polarization} \end{aligned} \tag{15}$$

In Eq. (15) the quantities p_j, q_j, and β_j are incident angle (θ)–dependent optical parameters relating to the d_j thickness of the jth phase and defined as

$$\begin{aligned} p_j &= \xi_j = \hat{n}_j \cos \theta_j = (\hat{n}_j^2 - n_1^2 \sin^2 \theta_1)^{1/2} \\ q_j &= \frac{\xi_j}{\hat{n}_j^2} \\ \beta_j &= \frac{2\pi d_j \xi_j}{\lambda} \end{aligned} \tag{16}$$

The Fresnel reflection coefficients for the stratified medium may be derived by using elements m_{ij} of overall matrix M [defined in Eq. (14)] in the following fashion:

$$\begin{aligned} r_\perp &= \frac{(m_{11} + m_{12} q_N) p_1 - (m_{21} + m_{22} p_N)}{(m_{11} + m_{12} q_N) p_1 + (m_{21} + m_{22} p_N)} \\ r_\parallel &= \frac{(m_{11} + m_{12} q_N) q_1 - (m_{21} + m_{22} q_N)}{(m_{11} + m_{12} q_N) p_1 + (m_{21} + m_{22} q_N)} \end{aligned} \tag{17}$$

The overall reflectance of the stratified medium for both polarizations is derived from the Fresnel reflectance coefficients by

$$R_\perp = |r_\perp|^2 \quad \text{and} \quad R_\| = |r_\||^2 \tag{18}$$

From the optical constants for each phase and the Fresnel reflection coefficients, the mean square electric field intensities at any point in the N-phase system can be calculated. Assuming that the incident amplitudes of the mean square electric fields for both polarizations equal one, the exact values of the mean square electric fields intensities at any point z in phase j are given by

$$
\begin{aligned}
E_{\perp jy}(z) &= U_j(z)e^{i(j \cdot r - \omega t)} \\
E_{\| jx}(z) &= V_j(z)e^{i(j \cdot r - \omega t)} \\
E_{\| jz}(z) &= W_j(z)e^{i(j \cdot r - \omega t)}
\end{aligned}
\tag{19}
$$

where

$$W_j(z) = \frac{n_1 \sin \theta_1 \, U_j(z)}{\hat{n}_j^2}$$

The use of Eqs. (12) through (19) allows one to calculate the optical properties of any thin-film system, assuming that the optical constants of the phases are known. Application of this theory to the case of monomolecular films in situ at the A-W interface has shown distinct differences between the spectra of these monolayers and those supported at an air-metal interface (70). For example, Fig. 15 shows the reflectance of

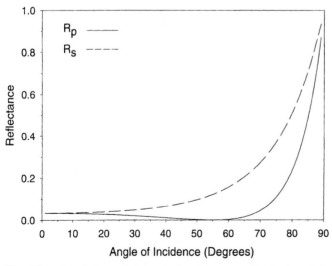

Reflectance of Light at the Air-Water Interface
3 Phase System: Air-Thin Film-H$_2$O

Fig. 15 Calculation of theoretical reflectance of radiation from the air-water interface as a function of the angle of incidence of the incoming radiation. Calculations were performed for both parallel (R_p) and perpendicular (R_s) polarized radiation at $\lambda = 10\,\mu$m (1000 cm^{-1}).

water in the infrared spectral region (i.e., $1000\,\text{cm}^{-1}$) as a function of the angle of incidence of the incoming radiation. Although the reflectivity of metals is always very high (>90%), Fig. 15 shows that the reflectance of IR radiation at the air-water interface is considerably weaker and approaches zero at the pseudo-Brewster angle (70). This difference has considerable implications for the appearance of the resulting reflectance spectra.

An additional difference between the two types of external reflection experiments is the polarization dependence of the spectra. When recording external reflectance spectra of thin films on metal substrates, only parallel polarized (*p*-polarized) light can be used, because there is a node at the surface for the *s*-polarized electric field. This is the origin of the so-called surface selection rule, which states that only vibrational dipole moments oriented perpendicular to the surface will be observed, as this is the only orientation that *p*-polarized radiation will excite (71).

This is not the case when using water as the reflective substrate, as finite values of the mean square electric fields for both *p*- and *s*-polarized radiation are present at the air-water interface (70). Figure 16 presents the results of calculations using Eqs. (12) through (19) to determine quantitatively the electrical field distribution at metal

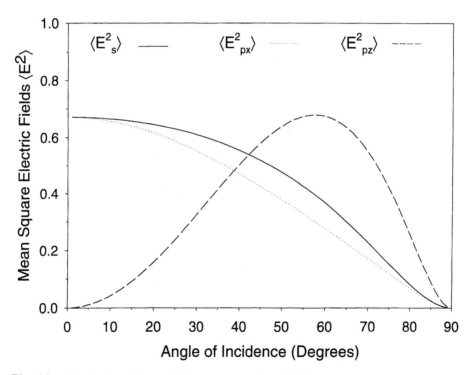

Mean Square Electric Field Intensities at the Air-Water Interface

Fig. 16 Calculation of theoretical mean square electric field intensities present at the air-water interface as a function of the angle of incidence of the incoming radiation. Calculations were performed for electric field components oriented both perpendicular ($\langle E_s^2 \rangle$) and parallel ($\langle E_{px}^2 \rangle$ and $\langle E_{pz}^2 \rangle$) to the plane of incidence at $\lambda = 10\,\mu\text{m}$ ($1000\,\text{cm}^{-1}$).

and water interfaces. The presence of electrical fields for both p and s polarization implies that the polarized external reflectance spectra contain information about all three orthogonal geometric orientations of the monolayer film, which is unlike the case of monolayer films on metal substrates.

Perhaps the most striking difference between IRRAS spectra of films on metals and on water is in the nature of the observed spectra. Because of the complex refractive index properties of water, the observed IR intensities of the monolayer vibrational bands will appear negative when plotted in absorbance mode, although positive absorbance bands are also possible, depending on the angle of incidence (70). Figure 17 is an example of the negative peaks observed in the reflection-absorbance spectrum of a monolayer film on water.

Using the equations concerning the optical physics of the reflection experiment, it is possible to describe quantitatively the theoretical absorbances expected for each of the geometric surface directions. For the case of external reflection, individual theoretical absorbances may be defined as follows (72):

$$\Delta A_\perp = -\frac{4}{\ln 10}\left[\frac{\cos\theta}{n_3^2 - 1}\right]n_2\alpha_{2s}d$$

$$\Delta A_{\parallel x} = -\frac{4}{\ln 10}\left[\frac{\cos\theta}{\dfrac{\xi_3^2}{n_3^4} - \cos^2\theta}\right]\left[-\frac{\xi_3^2}{n_3^4}\right]n_2\alpha_{2px}d \tag{20}$$

$$\Delta A_{\parallel z} = -\frac{4}{\ln 10}\left[\frac{\cos\theta}{\dfrac{\xi_3^2}{n_3^4} - \cos^2\theta}\right]\left[-\frac{\sin^2\theta}{n_2^2 + k_{2z}^2}\right]n_2\alpha_{2pz}d$$

Figure 18 presents the results of calculations using Eq. (20) that show how each geometric orientation contributes to the observed reflection-absorbance intensity for both s and p polarization of a monolayer film at the air-water interface as a function of the angle of incidence of the incoming radiation. The results of the calculations presented in Fig. 18 on the air-water interface and in previously published studies of other dielectric substrates (73,74) indicate that the theoretical absorbances for monomolecular films on nonmetallic substrates show significant changes in sign and magnitude as a function of incidence angle and polarization state of the incoming radiation. Using analyses similar to that presented here, several groups have used this polarization and angle dependence of the monolayer absorbance as the basis for attempts to determine thin-film orientation in Langmuir monolayers (75–78).

C. Infrared External Reflectance Spectroscopy: Instrumentation

As the use of IR external reflection spectroscopy applied to the study of Langmuir monolayers is a unique application with a recent history, some of the instrumental approaches to acquiring the IR spectra of these monomolecular films will be briefly described.

All of the published studies in this field utilize focusing lenses and/or mirrors, whether on axis or off axis, to deliver the source radiation to the A-W interface

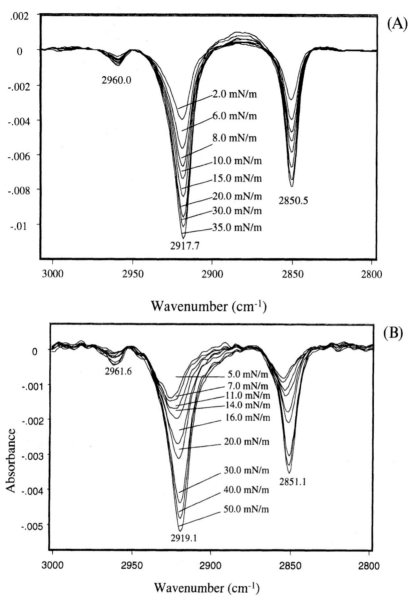

Fig. 17 IR external reflection-absorbance spectra in the C-H stretching region (3000–2800 cm^{-1}) of monomolecular films at the air-water interface recorded as a function of the surface pressure of the monolayer. The main spectral features present in these spectra are the symmetric CH$_2$ stretching vibration ($\sim$2850 cm^{-1}) and the antisymmetric CH$_2$ stretching vibration ($\sim$2920 cm^{-1}). The spectra are plotted in reflection-absorbance units [i.e., $R = -\log(R/R_0)$, where R = the reflectance of the film-covered substrate and R_0 = the reflectance of the pure water subphase]. The spectra presented here were recorded at 4 cm^{-1} resolution by coaddition of 1024 scans. (A) External reflection-absorbance IR spectra for a monolayer film of stearic acid (C$_{17}$COOH) at the air-water interface acquired between surface pressures of 2.0 and 35.0 mN m^{-1}. (B) External reflection-absorbance IR spectra for a monolayer film of 1,2-dipalmitoyl-*sn*-glycero-3-phosphocholine (DPPC) at the air-water interface acquired between surface pressures of 5.0 and 50.0 mN m^{-1}.

Theoretical IR Reflection-Absorbance of a Monomolecular Film
at the Air-Water Interface at 1000 cm^{-1}

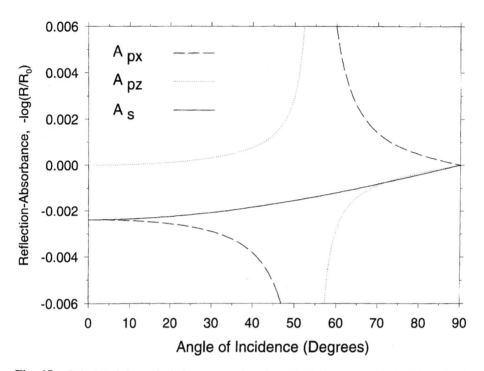

Fig. 18 Calculated theoretical absorbances for a hypothetical monomolecular film adsorbed at the air-water interface as function of the angle of incidence of the incoming radiation. Calculations were performed for theoretical absorbances oriented both perpendicular (A_s) and parallel (A_{px} and A_{pz}) to the plane of incidence at $\lambda = 10\,\mu m$ (1000 cm^{-1}). The monomolecular film was assumed to have optical constants of $n = 1.5$, $k = 0.1$, and a thickness of 2.5 nm.

and redirect the reflected radiation to the detector. The detector of choice in all these experiments has been a high-D^* HgCdTe photoconductive detector because of the very weak reflected signals that must be detected in these experiments. The mirrors used may deliver the radiation to the surface over a range of angles of incidence, depending on the experimental geometry and interest. Typical designs (78,79) use a Langmuir film constructed specifically to contain the monolayer film at the focal point of the IR beam, as well as to generate high surface pressures within the film itself. Figure 19 illustrates one approach to the optical interfacing of an IR spectrometer with a Langmuir film balance using a series of off-axis parabolic mirrors to focus the beam at the air-water interface.

Although the instrumental designs reported to date have proved to be successful in studying Langmuir monolayers, there are nonetheless several inherent fundamental physical and spectroscopic limitations that can still prove troublesome. These limitations include (a) very weak absolute reflectances, which restrict the possible S/N ratio in the final spectra; (b) interferences from H$_2$O vapor; and (c) baseline dispersion in regions of liquid H$_2$O absorbance bands because of the anomalous disper-

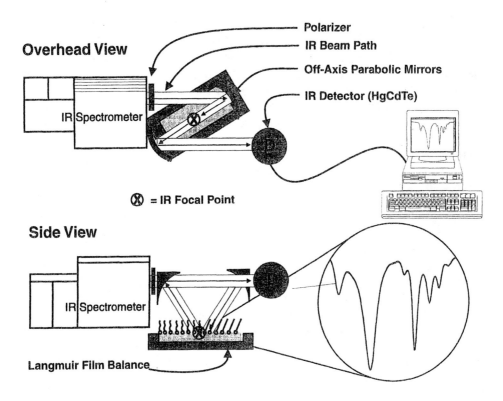

Fig. 19 Schematic diagram from the overhead and side perspectives of the optical interfacing used to redirect and focus the IR radiation for the external reflection experiment at the air-water interface.

sion of the H_2O reflectance. The last two points especially affect the spectral region from 1400 to 1900 cm^{-1}, which encompasses the conformationally sensitive amide I region of the IR spectrum. This is one of the major reasons why the IR external reflection method has not been routinely used to study peptide monolayer films on water.

In order to overcome several of these inherent limitations, polarization-modulation (PM) spectroscopy has been applied to monomolecular films at the air-water interface (80, 81). Polarization-modulation IR reflectance spectroscopy involves a fast modulation of the polarization of the incidence electric field between parallel and perpendicular linear polarization states (82). After electronic filtering and demodulation of the signal, the differential reflectivity spectrum is then computed. The PM reflectance spectrum is insensitive to isotropic absorptions (from, e.g., H_2O vapor) and reflects the *difference* between the parallel and perpendicular polarizations. Using PM reflectance spectroscopy of Langmuir films, the polar headgroup vibration of deuterated arachidic acid and arachidate monolayers interacting with the liquid H_2O substrate has been observed. This modulation technique has also been successfully employed in the study of peptide monolayers at the air-water interface (83,84).

D. Infrared External Reflectance Spectroscopy; Applications

Although the use of external reflectance IR spectroscopy is a relatively new approach to the study of Langmuir monolayers, there is already a growing body of literature reporting the results of the spectroscopic study of a variety of these monomolecular films. The majority of the work in this field has been in the conformational analysis and phase transitions of model monolayers (78,79,85–95), but there have also been several studies of naturally isolated pulmonary surfactant (96–98), cation interactions with monolayer headgroups (86,95), and peptide monolayer films (83,84,99,100). Figure 17 illustrates some representative external IR reflectance spectra of amphiphilic lipid monolayers obtainable at the air-water interface as a function of the surface pressure of the film. This figure is also a good illustration of the power of this technique: although a very weak reflector was used (water is only $\sim6\%$ reflective in the mid-IR region) and a monomolecular film approximately 2.5 nm thick was studied, there is enough detail evident in these spectra to assign conformational and thermodynamic significance.

A large number of studies have appeared in which external reflection IR spectroscopy was used to study phospholipid monolayers as models of biomembrane interfaces. Some of the earliest published IR spectra of Langmuir monolayers were of phospholipids at various surface pressures (70, 79, 85, 86, 101). It was shown that this IR reflectance method could identify vibrations due to the hydrocarbon acyl chains, carbonyl ester, and phosphate groups for these monolayer films at the air-water interface and that the external reflection method could differentiate between the physical conformation of phospholipid monolayer films at the air-water interface. In addition, the conformation-sensitive C—H stretching bands from the lipid's hydrocarbon chains could be used to monitor the expanded to condensed thermodynamic transition of the monolayer using this reflectance method.

Although the majority of studies published so far have focused on the hydrocarbon vibrations as indicators of monolayer phase states, it is possible to observe not only the hydrocarbon vibrations but also the vibrational modes due to the polar headgroups. One study (86) examined the spectral region between 1300 and $1000\,cm^{-1}$ in DPPC monolayer films to show the effect of film compression and cation interaction on the phosphate vibrations of the phospholipid headgroup. An additional illustration of the sensitivity of the IR external reflectance method to the ionic state of the hydrophilic headgroups of monolayers was shown for the case of 1,2-dimyristoyl-*sn*-glycero 3-phosphatidic acid (DMPA) (87), a molecule with a net negative charge at neutral pH. It was found that the phosphate vibrations in this molecule could easily track the state of protonation of the phosphatidic acid headgroup because of the wide pH dependence of the nature of the vibrations of the PO_4 group.

One study has appeared (95) in which the interaction of subphase Ca^{2+} with a binary mixture of neutral and acidic lipids has been studied by external reflection IR spectroscopy. In this case the monolayer was composed of an equimolar mixture of DPPC with 1,2-dipalmitoyl-*sn*-glycero-3-phosphoserine (DPPS); DPPS has a net negative charge at neutral pH and strongly binds Ca^{2+}. The presence of ionic Ca^{2+} in the subphase induced an acyl chain ordering in the binary mixture that was not observed in the case of pure DPPC alone. It was found that Ca^{2+} did not induce a phase separation in the DPPC:DPPS monolayer, as it does in bulk phase mixtures, but rather the components of the mixed monolayer appeared to retain their miscibility.

The structure-function relationships in isolated and model systems that mimic pulmonary surfactant have been studied with the external reflectance IR technique. In the first study of its kind, natural pulmonary surfactant isolated from bovine lungs was studied in situ at the A-W interface (96). It was found that under surface pressure conditions that correspond to large lung volumes in vivo, the surfactant exists mostly in an ordered conformation. Upon film compression, a weakly cooperative phase transition was noted for the surface film. This work demonstrates that ability of the IR reflectance method to obtain structural information on the surfactant in physical states that directly relate to those in vivo.

Others studies have used model monolayer films composed of binary mixtures of PG and PC lipids in order to test the "squeezing-out" hypothesis of pulmonary surfactant function (97). In this case the PC component was composed of DPPC containing completely perdeuterated acyl chains (i.e., DPPC-d_{62}). In this fashion, the C—H stretching vibrations (due to the PG component) and the C—D vibrations (due to the PC component) could be simultaneously monitored. The relative intensities of these two vibrational bands as a function of surface pressure could then be used to determine the fractional concentration of each component in the mixed monolayer. It was found that when PG lipids containing unsaturated acyl chains were selectively added to the mixed monolayer, these lipids were excluded, or squeezed out, from the monolayer at high surface pressures. This phenomenon did not occur for PG lipids containing saturated acyl chains, confirming the earlier ATR-IR results on the same system (45,46).

A trend in the use of external reflectance IR spectroscopy has been to study peptide monolayer films at the air-water interface (83,84,99,100). Because of the unique refractive index dispersion properties of water in the region 1700–1600 cm^{-1}, there exists a problem with acquiring the spectra of peptide and protein monolayer films in the structurally important amide I spectral region. Therefore, an IR monolayer sampling accessory has been designed that allows the use of small quantities of D_2O as the substrate (102). IR spectra of model synthetic peptides at the A-W interface were obtained using this device. The results presented in the paper suggested that although the spectra of β-sheet peptides are readily observable, the situation for α-helical and mixed conformations is complicated by uncertainties due to H-D exchange, among other factors.

Studies of peptides at the air-water interface have used both PM-IRRAS (83,84) and an updated static IRRAS accessory (78) to acquire the IR spectra of peptide monolayers. The quality of these spectra has increased greatly over the past few years, as shown in Fig. 20 for PM-IRRAS spectra of synthetic peptides at the air-water interface. The level of detail apparent in these spectra is now sufficient for the calculation of structural and orientation information for these peptide films (84,100).

IV. SUMMARY

Reflection infrared spectroscopy has proved to be a versatile and powerful technique for the spectroscopic characterization of biophysical monomolecular films at solid or liquid interfaces. Advances in instrumentation and sampling methodology have enabled researchers to develop these techniques into methods that have a submonolayer sensitivity that was unheard of a decade ago. A detailed examination of the underlying physical and optical principles of internal and external reflection spectroscopy enables

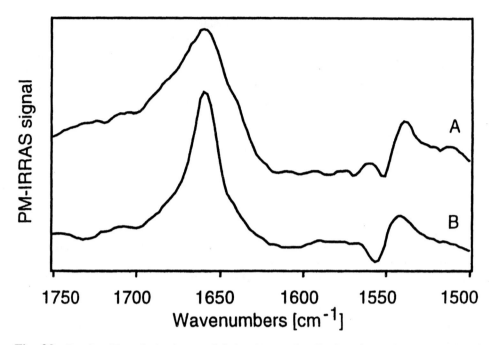

Fig. 20 In situ IR polarization-modulation external reflection-absorption spectrum of synthetic hydrophobic (A) and amphipathic (B) peptides at the air-water interface. (From Ref. 84, © 1989 Biophysical Society, used with permission.)

one to understand the orientational and geometric origins of the spectra with a high degree of precision and enables a more detailed interpretation of the experimental spectroscopic data. This has allowed even greater insight into not only the study of monolayer structural and conformational analysis (for which IR spectroscopy has most widely been used) but also the study of monolayer spatial and orientational analysis, which has led to increased insights into molecular organization in two dimensions.

ACKNOWLEDGMENT

This work was supported by the U.S. Public Health Service through National Institutes of Health grant GM40117.

REFERENCES

1. CM Knobler. Recent developments in the study of monolayers at the air-water interface. Adv Chem Phys 77:397, 1990.
2. CM Knobler, RC Desai. Phase transitions in monolayers. Annu Rev Phys Chem 43:207, 1992.
3. HM McConnell. Structures and transitions in lipid monolayers at the air-water interface. Annu Rev Phys Chem 42:171, 1991.
4. P Dutta, JB Peng, B Lin, JB Ketterson, M Prakash, P Georgopoulos, S Ehrlich. X-ray diffraction studies of organic monolayers on the surface of water. Phys Rev Lett 58:2228, 1987.

5. M Seul, P Eisenberger, HM McConnell. X-ray diffraction by phospholipid monolayers on single-crystal silicon substrates. Proc Natl Acad Sci USA 80:5795, 1983.

6. K Kjaer, J Als-Nielsen, CA Helm, LA Laxhuber, H Möhwald. Ordering in lipid monolayers studied by synchrotron x-ray diffraction and fluorescence microscopy. Phys Rev Lett 58:2224, 1987.

7. CA Helm, H Möhwald, K Kjär, J Als-Nielsen. Phospholipid monolayer density distribution perpendicular to the water surface. A synchrotron x-ray reflectivity study. Europhys Lett 4:697, 1987.

8. CA Helm, H Möhwald, K Kjaer, J Als-Nielsen. Phospholipid monolayers between fluid and solid states. Biophys J 52:381, 1987.

9. K Kjaer, J Als-Nielsen, CA Helm, P Tippman-Krayer, H Möhwald. Synchrotron x-ray diffraction and reflection studies of arachidic acid monolayers at the air-water interface. J Phys Chem 93:3200, 1989.

10. D Jacquemain, S Grayer-Wold, F Leveiller, M Deutsch, K Kjaer, J Als-Nielsen, M Lahav, L Leiserowitz. Two-dimensional crystallography of amphiphilic molecules at the air-water interface. Angew Chem Int Ed Engl 31:130, 1992.

11. H Kuhn, D Möbius, N Bucher. Spectroscopy of monolayer assemblies. In: A Weissberger, BW Rossiter, eds. Physical Methods in Chemistry. Vol I. Part 3B. New York: Wiley-Interscience, 1972, p 577.

12. GG Roberts. An applied science perspective of Langmuir-Blodgett films. Adv Phys 34:475, 1985.

13. A Fischer, M Lösche, M Möhwald, E Sackmann. On the nature of the lipid monolayer phase transition. J Phys Lett 45:785, 1984.

14. G Picard, G Munger, RM LeBlanc, R LeSage, D Sharma, A Siemarczuk, JR Bolton. Fluorescence lifetime of chlorophyll *a* in pure and mixed Langmuir-Blodgett films. Chem Phys Lett 129:41, 1986.

15. JD Swalen, JF Rabolt. Characterization of orientation and lateral order in thin films by Fourier transform infrared spectroscopy. In: JR Ferraro, LJ Basile, eds. Fourier Transform Infrared Spectroscopy. Vol 4. Orlando, FL: Academic Press, 1985, p 283.

16. JF Rabolt, JD Swalen. Structure and orientation in thin films: Raman studies with integrated optical techniques. In: RJH Clark, RE Hester, eds. Spectroscopy of Surfaces. Advances in Spectroscopy, Vol 16. London: Wiley, 1988, p 1.

17. NJ Harrick. Surface chemistry from spectral analysis of totally internally reflected radiation. J Phys Chem 64:1110, 1960.

18. J Fahrenfort. Attenuated total reflection: A new principle for the production of useful infrared reflection spectra of organic compounds. Spectrochim Acta 17:698, 1961.

19. DR Scheuing. Fourier Transform Infrared Spectroscopy in Colloid and Interface Science. Vol 447. Washington, DC: American Chemical Society, 1991.

20. NJ Harrick. Internal Reflection Spectroscopy. New York: Wiley, 1967.

21. BW Gregory, S Thomas, S Stephens, RA Dluhy, LA Bottomley. Preparation of atomically smooth germanium substrates for infrared spectroscopic and scanning probe microscopic characterization of organic monolayers. Langmuir 13:6146, 1997.

22. WN Hansen. Surface chemistry by reflection spectroscopy. Prog Nucl Chem 11:3, 1972.

23. WN Hansen. Electric fields produced by the propagation of plane coherent electromagnetic radiation in a stratified medium. J Opt Soc Am 58:380, 1968.

24. WN Hansen. Internal reflection spectroscopy in electrochemistry. In: RH Muller, ed. Advances in Electrochemistry and Electrochemical Engineering. Vol 9. New York: Wiley-Interscience, 1973, p 1.

25. PH Axelsen, WD Braddock, HL Brockman, CM Jones, RA Dluhy, BS Kaufman, FJ Puga. Use of internal reflectance infrared spectroscopy for the in situ study of supported lipid monolayers. Appl Spectrosc 49:526, 1995.

26. B Jasse, JL Koenig. Orientational measurements in polymers using vibrational spectroscopy. Macromol Sci Rev Macromol Chem C17:61, 1979.

27. PH Axelsen, MJ Citra. Orientational order determined by internal reflection infrared spectroscopy. Prog Biophys Mol Biol 66:227, 1996.

28. UP Fringeli, HH Günthard. Infrared membrane spectroscopy. In: E Grell, ed. Membrane Spectroscopy, Molecular Biology, Biochemistry and Biophysics. Vol. 31. Berlin: Springer-Verlag, 1981, p 270.

29. E Goormaghtigh, J-M Ruysschaert. Polarized attenuated total reflection infrared spectroscopy as a tool to investigate the conformation and orientation of membrane components. In: R Brasseur, ed. Molecular Description of Biological Membranes by Computer-Aided Conformational Analysis. Boca Raton, FL: CRC Press, 1990, p 285.

30. UP Fringeli. In situ infrared attenuated total reflection membrane spectroscopy. In: FM Mirabella Jr, ed. Internal Reflection Spectroscopy. Theory and Applications. New York: Marcel Dekker, 1993, p 255.

31. LK Tamm, SA Tatulian. Infrared spectroscopy of proteins and peptides in lipid bilayers. Q Rev Biophys 1997.

32. S Krimm, J Bandekar. Vibrational spectroscopy and conformation of peptides, polypeptides, and proteins. Adv Protein Chem. 38:181, 1986.

33. DG Cameron, RA Dluhy. FT-IR studies of molecular conformation in biological membranes. In: RM Gendredu, ed. Spectroscopy in the Biomedical Sciences. Boca Raton, FL: CRC Press, 1986, p 53.

34. MS Braiman, KJ Rothschild. Fourier transform infrared techniques for probing membrane structure. Annu Rev Biophys Biophys Chem 17:541, 1988.

35. M Jackson, HH Mantsch. Biomembrane structure from FT-IR spectroscopy. Spectrochim Acta Rev 15:53, 1993.

36. JLR Arrondo, A Muga, J Castresana, F Goñi. Quantitative studies of the structure of proteins in solution by Fourier transform spectroscopy. Prog Biophys Mol Biol 59:23, 1993.

37. E Goormaghtigh, V Cabiaux, J-M Ruysschaert. Determination of soluble and membrane protein structure by Fourier transform infrared spectroscopy. I. Assignments and model compounds. In: HJ Hilderson, GB Ralston, eds. Physiochemical Methods in the Study of Biomembranes, Subcellular Biochemistry. Vol 23. New York: Plenum, 1994, p 329.

38. E Goormaghtigh, V Cabiaux, J-M Ruysschaert. Determination of soluble and membrane protein structure by Fourier transform infrared spectroscopy. II. Experimental aspects, side chain structure, and H/D exchange. In: HJ Hilderson, GB Ralston, eds. Physiochemical Methods in the Study of Biomembranes. Subcellular Biochemistry, Vol 23. New York: Plenum, 1994, p 363.

39. E Goormaghtigh, V Cabiaux, J-M Ruysschaert. Determination of soluble and membrane protein structure by Fourier transform infrared spectroscopy. III. Secondary structures. In: JH Hilderson, GB Ralston, eds. Physiochemical Methods in the Study of Biomembranes. Subcellular Biochemistry, Vol 23. New York: Plenum, 1994, p 404.

40. M Jackson, HH Mantsch. The use and misuse of FTIR spectroscopy in the determination of protein structure. Crit Rev Biochem Mol Biol 30:95, 1995.

41. UP Fringeli. The structure of lipids and proteins studied by attenuated total reflection (ATR) infrared spectroscopy. II. Oriented layers of a homologous series: Phosphatidylethanolamine to phosphatidylcholine. Z Naturforsch. C Biosci 32C:20, 1977.

42. L Ter-Minassian-Saraga, E Okamura, Umemura. Fourier transform infrared–attenuated total reflection spectroscopy of hydration of dimyristoylphosphatidylcholine multi-bilayers. Biochim Biophys Acta 946:417, 1988.

43. E Okamaura, J Umemura, T Takenaka. Orientation studies of hydrated dipalmitoyl-phosphatidylcholine multibilayers by polarized FTIR-ATR spectroscopy. Biochim Biophhys Acta 1025:94, 1990.

44. W Hübner, HH Mantsch. Orientation of specifically ^{13}CO labelled phosphatidylcholine multilayers from polarized attenuated total reflection FT-IR spectroscopy. Biophys J 59:1261, 1991.

45. FR Rana, AJ Mautone, RA Dluhy. A combined infrared and ^{31}P NMR spectroscopic method for determining the fractional composition in Langmuir-Blodgett films of binary phospholipid mixtures. Appl Spectrosc 47:1015, 1993.

46. FR Rana, AJ Mautone: RA Dluhy. Surface chemistry of binary mixtures of phospholipids in monolayers. Infrared studies of surface composition at varying surface pressures in a pulmonary surfactant model system. Biochemistry 32:3169, 1993.

47. BW Gregory, RA Dluhy, LA Bottomley. Structural characterization and nanometer-scale domain formation in a model phospholipid bilayer as determined by infrared spectroscopy and scanning tunneling microscopy. J Phys Chem 98:1010, 1994.

48. WF DeGrado, GF Musso, M Lieber, ET Kasier, FJ Kezdy. Biophys J 37:329, 1982.

49. TC Terwilliger, L Weissman, D Eisenberg. The structure of melittin in the form I crystals and its implication for melittin's lytic and surface activities. Biophys J 37:353, 1982.

50. H Vogel, F Jänig, V Hoffman, J Stüpel. The orientation of melittin in lipid membranes. A polarized infrared spectroscopic. Biochim Biophys Acta 733:201, 1983.

51. H Vogel, F Jähnig. The structure of melittin in membranes. Biophys J 50:573, 1986.

52. H Vogel. Comparison of the conformation and orientation of alamethicin and melittin in lipid membranes. Biochemistry 26:4562, 1987.

53. J Brauner, R Mendelsohn, F Prendergast. Attenuated total reflectance Fourier transform infrared studies of the interaction of melittin, two fragments of melittin, and *d*-hemolysin with phosphatidylcholines. Biochemistry 26:8151, 1987.

54. S Frey, LK Tamm. Orientation of melittin in phospholipid bilayers. A polarized attenuated total reflection infrared study. Biophys J 60:922, 1991.

55. AJ Weaver, MDBJW Kemple, R Mendelsohn, SFG Prenderga. Fluorescence, Cd, attenuated total reflectance (ATR) FT-IR, and ^{13}C NMR characterization of the structure and dynamics of synthetic melittin and melittin analogs in lipid environments. Biochemistry 31:1301, 1992.

56. MS Briggs, DG Cornell, RA Dluhy, LM Gierasch. Spectroscopic studies of signal peptides in phospholipid monolayers: Conformations induced by lipids suggest possible initial steps in protein export. Science 233:206, 1986.

57. DG Cornell, RA Dluhy, MS Briggs, LM Giersach. Conformation and orientation of a signal peptide interacting with phospholipid monolayers. Biochemistry 28:2789, 1989.

58. LK Tamm, SA Tatulian. Orientation of functional and nonfunctional PTS permease signal sequences in lipid bilayers. A polarized attenuated total reflection infrared study. Biochemistry 32:7720, 1993.

59. B Pastrana-Rios, AJ Mautone, R Mendelsohn. Fourier transform infrared studies of secondary structure and orientation of pulmonary surfactant SP-C and its effect on the dynamic surface properties of phospholipids. Biochemistry 30:10058, 1991.

60. C Vandenbussche, A Clerx, T Curstedt, J Johansson, H Jornvall. Structure and orientation of the surfactant-associated protein C in a lipid bilayer. Eur J Biochem 203:201, 1992.

61. JE Baatz, KL Smyth, JA Whitsett, C Baxter, DR Absolom. Structure and functions of a dimeric form of surfactant protein SP-C: A Fourier transform infrared and surfactometry study. Chem Phys Lipids 63:91, 1992.

62. G Vandenbussche, A Clerx, M Clerx, T Curstedt, J Johansson, H Jornvall, J Ruysschaert. Secondary structure and orientation of the surfactant protein SP-B in a lipid environment: A Fourier transform infrared spectroscopy study. Biochemistry 31:9169, 1992.

63. DL Allara, RG Nuzzo. Spontaneously organized molecular assemblies. 2. Quantitative IR spectroscopic determination of equilibrium structures of solution adsorbed *n*-alkanoic acids on an oxidized aluminium surface. Langmuir 1:52, 1985.

64. DL Allara, A Baca, CA Pryde. Distortions of band shapes in external reflection infrared spectra of thin polymer films on metal substrates. Macromolecules 11:1215, 1978.

65. RA Dluhy, DG Cornell. In-situ measurement of the infrared spectra of insoluble monolayers at the air-water interface. J Phys Chem 89:3195, 1985.

66. MD Porter, TB Bright, DL Allara. Quantitative aspects of infrared external reflection spectroscopy: Polymer/glassy carbon interface. Anal Chem 58:2461, 1986.

67. JA Mielczarski, RH Yoon. Fourier transform infrared external reflection study of molecular orientation in spontaneously adsorbed layers on low-absorption substrates. J Phys Chem 93:2034, 1989.

68. M Born, E Wolf. Principles of Optics. New York: Pergamon, 1975.

69. OS Heavens. Optical Properties of Thin Solid Films. New York: Dover, 1965.

70. RA Dluhy. Quantitative external reflection infrared spectroscopic analysis of insoluble monolayer spread at the air-water interface. J Phys Chem 90:1373, 1986.

71. HA Pearce, N Sheppard. Sur Sci 59:205, 1976.

72. WN Hansen. Reflection spectroscopy of adsorbed layers. Symp Faraday Soc 4:27, 1970.

73. JA Mielczarski. External reflection infrared spectroscopy at metallic, semiconductor, and nonmetallic substrates. 1. Monolayer films. J Phys Chem 97:2649, 1993.

74. H Brunner, U Mayer, H Hoffman. External reflection infrared spectroscopy of anisotropic adsorbate layers on dielectric substrates. Appl Spectrosc 51:209, 1997.

75. LJ Fina, YS Tung. Molecular orientation of monolayers on lipid substrates: Optical model and FT-IR methods. Appl Spectrosc 45:986, 1991.

76. JT Buontempo, SA Rice. Infrared external-reflection spectroscopy of adsorbates on dielectric substrates: Determining adsorbate orientation in Langmuir monolayers. J Chem Phys 98:5825, 1993.

77. A Gericke, AV Michailov, Hühnerfuss. Polarized external infrared reflection-absorption spectrometry at the air/water interface: Comparison of experimental and theoretical results for different angles of incidence. Vib Spectrosc 4:335, 1993.

78. CR Flach, A Gericke, R Mendelsohn. Quantitative determination of molecular chain tilt angles in monolayer films at the air-water interface: Infrared reflection/absorption spectroscopy of behenic acid methyl ester. J Phys Chem B 101:58, 1997.

79. RA Dluhy, NA Wright, PR Griffiths. In situ measurement of the FT-IR spectra of phospholipid monolayers at the air-water interface. Appl Spectrosc 42:138, 1988.

80. D Blaudez, T Buffeteau, JC Cornut, B Desbat, N Escafre, M Pezolet, JM Turlet. Polarization-modulated FT-IR spectroscopy of a spread monolayer at the air/water interface. Appl Spectrosc 47:869, 1993.

81. D Blaudez, T Buffeteau, JC Cornut, B Desbat, N Escafre, M Pezolet, JM Turlet. Polarization modulation FTIR spectroscopy at the air-water interface. Thin Solid Films 242:146, 1994.

82. LA Nafie, DW Vidrine. Double modulation Fourier transform spectroscopy, in Fourier Transform Infrared Spectroscopy, Vol. 3, J.R. Ferraro, and L.J. Basile, Eds., Academic Press, New York, pp., 1982, 83.

83. I Cornut, B Desbat, JM Turlet, J Dufourcq. In-situ study by polarization modulated Fourier transform infrared spectroscopy of the structure and orientation of lipids and amphipathic peptides at the air-water interface. Biophys J 70:305, 1996.

84. M Boncheva, H Vogel. Formation of stable polypeptide monolayers at interfaces: Controlling molecular conformation and orientation. Biophys J 73:1056, 1997.

85. ML Mitchell, RA Dluhy. In-situ FT-IR investigation of phospholipid monolayer phase transitions at the air/water interface. J Am Chem Soc 110:712, 1988.

86. RD Hunt, ML Mitchell, RA Dluhy. The interfacial structure of aqueous phospholipid monolayer films. An infrared reflectance study. J Mol Struct 214:93, 1989.

87. RA Dluhy. Langmuir revisited: Using infrared reflectance spectroscopy to determine monolayer structure at the air-water interface. In: DG Cameron, ed. 7th International Conference on Fourier Transform Spectroscopy. Vol 1145 Bellingham, WA: SPIE, 1989, p 22.

88. RA Dluhy, DG Cornell. Infrared reflectance spectroscopy as a probe of monolayer structure at gas-liquid and gas-solid interfaces. In: DR Scheuing, ed. Fourier Transform Infrared Spectroscopy in Colloid and Interface Science. ACS Symposium Series, Vol 447. Washington DC: American Chemical Society, 1991, p 192.

89. JT Buontempo, SA Rice. Infrared, external-reflection spectroscopic studies of phase transitions in Langmuir monolayers of heneicosanol. J Chem Phys 98:5835, 1993.

90. JT Buontempo, SA Rice, S Karaborni, JI Siepmann. Differences in the structures of relaxed and unrelaxed Langmuir monolayers of heneicosanol: Dependence of collective molecular tilt on chain conformation. Langmuir 9:1604, 1993.

91. JT Buontempo, SA Rice. Infrared external reflection spectroscopic studies of phase transitions in Langmuir monolayers of stearyl alcohol. J Chem Phys 99:7030, 1993.

92. A Gericke, H Hünerfuss. In situ investigation of saturated long-chain fatty acids at the air/water interface by external infrared reflection-absorption spectrometry. J Phys Chem 97:12899, 1993.

93. A Gericke, J Simon-Kutscher, H Hhnerfuss. Influence of the spreading solvent on the properties of monolayers at the air/water interface. Langmuir 9:2119, 1993.

94. A Gericke, J Simon-Kutscher, Hühnerfuss. Comparison of different spreading techniques for monolayers at the air/water interface by external infrared reflection-absorption spectroscopy. Langmuir 9:3115, 1993.

95. CR Flach, JW Brauner, R Mendelsohn. Calcium-ion interactions with insoluble phospholipid monolayer films at the A/W interface—External reflection-absorption IR studies. Biophys J 65:1994, 1993.

96. RA Dluhy, KE Reilly, RD Hunt, ML Mitchell, AJ Mautone, R Mendelsohn. Infrared spectroscopic investigations of pulmonary surfactant. Surface film transitions at the air water interface and bulk phase thermotropism. Biophys J 56:1173, 1989.

97. B Pastrana-Rios, CR Flach, JW Brauner, AJ Mautone, R Mendelsohn. A direct test of the "squeeze-out" hypothesis of lung surfactant function. External reflection FT-IR at the air/water interface. Biochemistry 33:5121, 1994.

98. B Pastrana-Rios, S Taneva, KMW Keough, AJ Mautone, R Mendelsohn. External reflection absorption infrared spectroscopy study of lung surfactant proteins SP-B and SP-C in phospholipid monolayers at the air/water interface. Biophys J 69:2531, 1995.

99. CR Flach, JW Brauner, JW Taylor, RC Baldwin, R Mendelsohn. External reflection FT-IR of peptide monolayer films in-situ at the air/water interface—Experimental design, spectra-structure correlations, and effects of hydrogen-deuterium exchange, Biophys J 67:402, 1994.

100. A Gericke, CR Flach, R Mendelsohn. Structure and orientation of lung surfactant SP-C and dipalmitoylphosphatidylcholine in aqueous monolayers. Biophys J 73:492, 1997.

101. RA Dluhy, ML Mitchell, T Pettenski, J Beers. Design and interfacing of an automated Langmuir-type film balance to an FT-IR spectrometer. Appl Spectrosc 42:1289, 1988.

102. CR Flach, JW Brauner, R Mendelsohn. Coupled external reflectance FT-IR miniaturized surface-film apparatus for biophysical studies. Appl Spectrosc 47:982, 1993.

22

Fluorescence Microscopy for Studying Biological Model Systems: Phospholipid Monolayers and Chiral Discrimination Effects

K. J. Stine

University of Missouri–St. Louis, St. Louis, Missouri

I. INTRODUCTION

The application of flourescence microscopy to the study of monolayers of membrane lipids and their interaction with proteins and other molecules introduced in the subphase beneath the film has been fruitful and relevant to improving the understanding of membrane processes and potentially to the assembly of biomolecular electronic devices using the Langmuir-Blodgett (LB) technique (1–4). In this chapter, practical advice to the experimentalist wishing to establish an effort in fluorescence microscopy studies of Langmuir films will be presented. In addition, both general and more recent results for phospholipid monolayers will be surveyed to provide an entry into the large literature in this area. Aspects of chiral discrimination in phospholipid monolayers will be discussed, selected aspects of the interaction of subphase molecules or enzymes with phospholipid monolayers described, and a few selected biotechnological applications mentioned. It is hoped that the survey will indicate to the potential user the scope of problems and systems addressable using the technique.

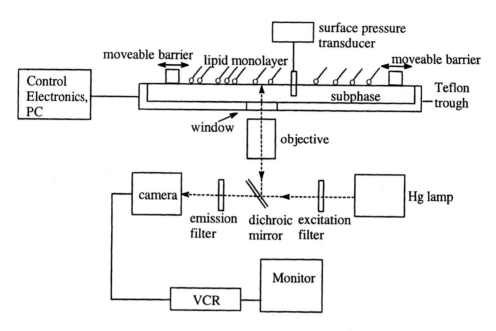

Fig. 1 Schematic of an apparatus for fluorescence microscopic investigations of lipid monolayers, shown for a generic inverted microscope configuration.

II. FLUORESCENCE MICROSCOPY IN THE STUDY OF LIPID MONOLAYERS

A. Experimental Apparatus for Fluorescence Microscopy

The most straightforward manner in which to establish an apparatus for fluorescence microscopy investigations of Langmuir films is to build the system around a commercial fluorescence microscope. A schematic diagram of the general features of an apparatus for fluorescence microscopic study of monolayers is shown in Fig. 1. The experimenter needs to decide between an upright "metallurgical" microscope and an inverted microscope. The choice has consequences for the design of the trough. The author used an upright microscope to simplify the trough design and simplify the necessary machining. If an inverted microscope is chosen, provision must be made for a window at the bottom of the trough for viewing the monolayer. The advantage of the inverted microscope choice is that the area above the trough remains free for barriers, the surface pressure transducer, or dippers for LB transfer. Additional spectroscopic apparatus such as fiber optics for collecting the fluorescence spectrum of the monolayer, micromanipulators for trapping and probing individual domains electrically or mechanically, or optics for simultaneous observation by Brewster angle microscopy (BAM) can be positioned above the trough. A comprehensive design for a simultaneous fluorescence–polarized fluorescence–BAM system has been described by Zasadzinski and coworkers (5) and used to study a lung surfactant system. A slight disadvantage of the inverted configuration is that it restricts the experimenter to viewing fewer locations on the film surface.

    The surface textures of Langmuir films are often heterogeneous and it can be informative to be able to translate the trough in the x and y directions and observe different areas, at least within a several square centimeter area. For example, while the *overall* surface fraction of two phases at fixed conditions of temperature and surface density in a single-component Langmuir film in equilibrium in a two-phase coexistence region is governed by the thermodynamic lever rule (6), one field of view may reveal uniformity and another field of view may show domains of both phases. If a trough is placed on an upright microscope, the microscope stage can be translated to examine different regions of the surface film. The microscope should be acquired with a mercury lamp and lamp power supply and a selection of filter cube assemblies to allow use of the expected fluorescent labels One or more long-working-distance objectives of different magnifications ($10\times$, $20\times$, and $50\times$ would be typically available) should be purchased to allow variation in the field of view for instances in which the monolayer shows large structures as well as cases in which the domains are smaller in size.

The trough to be placed on the microscope stage can be either designed and built inhouse or purchased commercially. A commercial trough can be intended for either an upright or an inverted microscope. The typical commercial unit comes equipped with a machined Teflon trough with several capabilities, including (a) two barriers for symmetric compression of the monolayer film at a controlled rate, (b) tubing beneath or around the trough for circulation of water for temperature control, (c) a surface pressure transducer, (d) a temperature sensor, and (e) software for the control of surface pressure or surface area and the measurement of surface pressure isotherms. The degree of sophistication of the software may be of concern depending on the application to special experiments, such as monitoring monolayer area while maintaining a constant surface pressure or measurements of surface pressure versus time. Troughs for fitting on the stage of the microscope can be purchased from a number of current vendors including Nima (CTC Technologies in the United States) and Kibron. Of course, it is possible to construct home-built troughs at a lower cost. The level of sophistication achieved will depend on the amount of time invested in machining, electronics, and computer software interfacing. It is important that the system be placed on a vibration-free and level surface. Although it is not necessary to purchase expensive platforms with active electronic feedback isolation of vibrations, a heavy marble table (a balance table is a good choice) or a moderate size optical table will be appropriate. Vibrations will affect both the image quality and the resolution of the surface pressure readings, and a tilt of the surface will lead to problems with water spillage out of the trough. The area around the trough should be covered to protect the monolayer from air currents and accumulation of airborne impurities. A Plexiglas enclosure or use of a laminar flow hood is appropriate. In a study of a type of spread interfacial monolayer not yet well investigated, a trough design appropriate for studying phospholipids at an oil-water interface by fluorescence microscopy was reported (7).

Another important decision is the choice of camera for detection of the monolayer image. The author's system, assembled in early 1990, employs an SIT (silicon-intensified-target) tube camera, but the use of CCD (charge-coupled device) cameras has become more popular and is preferable because the CCD is more rugged. SIT cameras are tube-type devices equipped with an additional image intensifier that amplifies the number of incident photons, whereas CCD cameras are solid-state devices based on the generation of electron-hole pairs in individual semiconductor pixels with the accumulated charges subsequently read out of the pixel array and processed into an image. The CCD will be more robust, and SIT cameras can easily be damaged by accidental exposure to even modest light levels. ISIT (intensified SIT) cameras have another stage of amplification and achieve sensitivity to ultralow light levels, with some loss of lateral resolution; it is our recommendation that an SIT camera is a better choice if one must use a tube-type camera. The intensities of the pixels in the CCD depend on the number of bits in the gray scale; a larger number of bits will give the image a greater dynamic range for covering bright to dim features. For imaging Langmuir films by fluorescence microscopy, a standard 8 bits or a 0–255 gray scale is adequate; it should not be necessary to have the greater number of bits employed when CCDs are used in spectroscopy, although CCDs with 12 bits and 16 bits are common. The light level for fluorescence from a monolayer doped with < 1% of a fluorescent tag is in the low to very low region and a CCD camera coupled to an image intensifier is needed; it is also possible to use thermoelectrically or cryogenically cooled CCD cameras. With CCDs, lateral image resolution depends on the pixel array; a square CCD chip with 512×512, 1024×1024, or more pixels is needed. The experimenter will most likely wish to view images of the monolayer on a monitor and record them on video cassette recorder (VCR) tape for later retrieval. Continuous marking of the image by a commercial time and date generator placed between the camera and VCR is recommended.

B. Image Acquisition and Data Analysis

It is generally the practice to record fluorescence images of the Langmuir film throughout the course of the experiment or at selected times. Later retrieval of selected images for detailed analysis or labeling and preparation for presentation can be aided by an image analysis system. The most economical choice is a personal computer (PC)–based frame grabber board so that tapes can be viewed and images captured and stored in files in a common picture format (jpg, gif, tif, etc.). Images can be enhanced and their contrast improved using various software filters applied to the image gray scale. Image analysis software packages are commercially available and there is a free package from the National Institutes of Health (NIH). Units accomplishing this in real time as the monolayer is viewed are also available . These units allow real-time contrast enhancement, brightness enhancement, and integration, a process in which the displayed image is a composite of a certain number of frames. Integration does not work well with Langmuir films in most conditions, as the domains tend to drift slowly on the water surface, especially if the barrier is moving. If it is necessary to observe the same domains over a long period of time, as can be desirable in studies of enzyme or protein interaction with monolayers, then a Teflon collar with a slit can be placed over the objective and through the interface to trap the same domains and prevent them from slowing drifting away. For phospholipid + cholesterol mixed films, quantitative analysis of the domain shapes for the cholesterol-rich phase in terms of harmonic analysis of the domain contours was performed and compared with theoretical models of liquid domain shape dynamics; the necessary image analysis algorithms have been reviewed by Seul et al. (8).

C. Polarized Fluorescence Microscopy

The technique of polarized fluorescence microscopy was introduced in studies of DPPC doped with nitrobenzoxadiazol-dipalmitoylphosphatidylethanolamine (NBD-DPPE) (9). In this experiment, the monolayer is illuminated at an oblique angle with a laser beam polarized with respect to the surface normal, such as the 488-nm line of an argon ion laser. If the fluorophore of the probe is at a fixed orientation with respect to the chain axis of the probe, then the overlap between the electric field of the laser beam and the transition moment of the fluorophore will depend on the chain tilt. If the laser polarization and the angle of incidence are fixed, regions of different tilt will fluoresce with different intensities. A particularly striking effect can be seen if one splits the laser beam and illuminates the same field of view first from the left side of the trough and then from the right side of trough at the same angle. On switching between these two illuminations, regions appearing dark when illuminated from the right will appear bright when illuminated from the left, and vice versa. Images obtained in this manner were presented in studies of the orientational order within DPPC and DPPE monolayers. This technique relies upon the incorporation of some of the probe into the ordered phase and its alignment with the neighboring molecules. In the case of DPPC monolayers, the method revealed molecular tilt paralleling the borders of the spiral domains. It has been the author's experience that a probe can orient itself and show fluorescence anisotropy in one lipid and then not show the effect in a monolayer of a different lipid under conditions in which other methods clearly indicate that the film is expected to be in a tilted or anisotropically ordered condensed phase. It is also possible to observe fluorescence anisotropy by simply placing polarizers oriented perpendicular to one another on a slider that fits into most microscopes (10). For unambiguous studies of molecular orientation within domains, BAM is a preferable technique.

D. Sample Preparation and Handling

The study of a single- or mixed-component Langmuir film by fluorescence microscopy, without any proteins or biomolecules in the subphase, is generally straightforward. A source of ultrahigh-purity water (10^{18} MΩcm resistivity and <10 ppb organic content) such as a Millipore system is essential for monolayer studies. The Π-A isotherms and domain textures are very sensitive to surface-active impurities. Solvents used to prepare spreading solutions, or to clean the trough, should be high-performance liquid chromatography (HPLC) grade or better. The trough will need to be cleaned thoroughly, first with a solution of high-purity sulfuric acid; then with high-purity water, high-purity organic solvents; and finally with more high-purity water. To determine whether the trough is clean enough, it is possible to measure the isotherm for pure subphase, compress the barriers close together, and look for a change in Π of < 0.1 mN m^{-1}.

The lipids should be dissolved in a spreading solvent at millimolar concentration. Suitable spreading solvents are those that spread rapidly on the water surface, are not soluble in the subphase, and evaporate well after being spread, but do not pose storage problems for the lipid solution by being too volatile. Chloroform is the best spreading solvent; hexane and hexane-ethanol (9:1) are also good choices. A study of the subtle effects of spreading solvent choice on the structure of fatty acid monolayers employing infrared reflection-absorption spectroscopy (IRRAS) has been reported (11). In order to add the fluorescent probe, prepare a stock solution of the probe in the same solvent and then add a small volume of this to the main lipid solution to achieve the desired probe concentration. Solutions containing fluorescent probe should be stored in dark brown bottles, wrapped so as to prevent slow photobleaching from ambient light. A probe concentration of 0.5 mol % may be a good initial value; if the contrast is very good in the images, then this can be reduced or, if needed, increased. It would be prudent to make some observations at different probe concentrations, going as low as allowed by the camera employed, to see if there are any concentration-dependent changes in the domain structures. Depositing the appropriate volume of the spreading solution on the trough surface to achieve the desired initial area per molecule should be done using a microsyringe, and the spreading solution should not come in contact with any plastic elements as these will leach out surfactants. Solution concentrations should be such as to avoid spreading more than $\sim 100\,\mu L$ of solution. It is useful to observe the surface pressure versus time during and after the spreading of the solution. The surface pressure will increase dramatically on spreading and then decay back to equilibrium as the solvent evaporates and the molecules equilibrate on the surface; this is likely to take 15–30 min.

The key feature of fluorescence microscopy studies of Langmuir films is the addition to the monolayer of a small amount (typically 0.1–2.0 mol %) of a fluorescently labeled lipid that will partition itself between surface phases of different surface density so that optical contrast between these two phases is achieved. Proper behavior of the probe is crucial for these studies; if the probe is equally soluble in the two phases or segregates from the monolayer, no useful images or misleading images will be obtained. For example, NBD-DPPE distributes equally between the expanded and condensed phases of DPPC, providing no image contrast under unpolarized illumination, whereas NBD-PC partitions into the expanded phase and provides contrast between the two phases. However, introduction of NBD-DPPE into DPPC monolayers is useful for observations with polarized excitation of the fluorescence because the orientation of the probe in the condensed phase enhances the fluorescence when the electric vector of the incident light is aligned near the transition moment of the NBD-DPPE and striking fluorescence anisotropy effects can be seen (9).

A number of suitable probes are commercially available. The most popular derivatives for monolayer studies have been the NBD-labeled (NBD = 7-nitro-2,1,3-benzoxadiazol-4-yl) phospholipids and amines. In a study comparing the behavior of NBD-labeled cholesterol and NBD-labeled phosphatidylcholine with regard to their partitioning between coexisting phases in phospholipid and phospholipid + cholesterol mixed monolayers; both were found to behave similarly and act as impurities (12). Figure 2 shows the structure of some common fluorescent probe lipids. The emission filter, dichroic mirror, and excitation filter used in the microscopic must be appropriate for the chosen probe. These are typically available as a cube containing all three elements. The fluorescent labeling of specific proteins requires covalent attachment using protocols from the literature.

E.   Relative Merits of Fluorescence and Brewster Angle Microscopy

Brewster angle microscopy is discussed in another section of this chapter. It has been the author's experience that higher quality images of better lateral resolution (~ 1–$2\,\mu m$) can be obtained by fluorescence microscopy than can be obtained by earlier BAM models (~ 6–$8\,\mu m$) of the standard design; however, improved and more expensive BAM systems with specialized optics have become available (Nanofilm Technologie GmbH). The establishment of a system for fluorescence microscopy requires less in the way of machining and optical design compared with that required to construct a BAM system; however, a home-built BAM can be had at a fraction

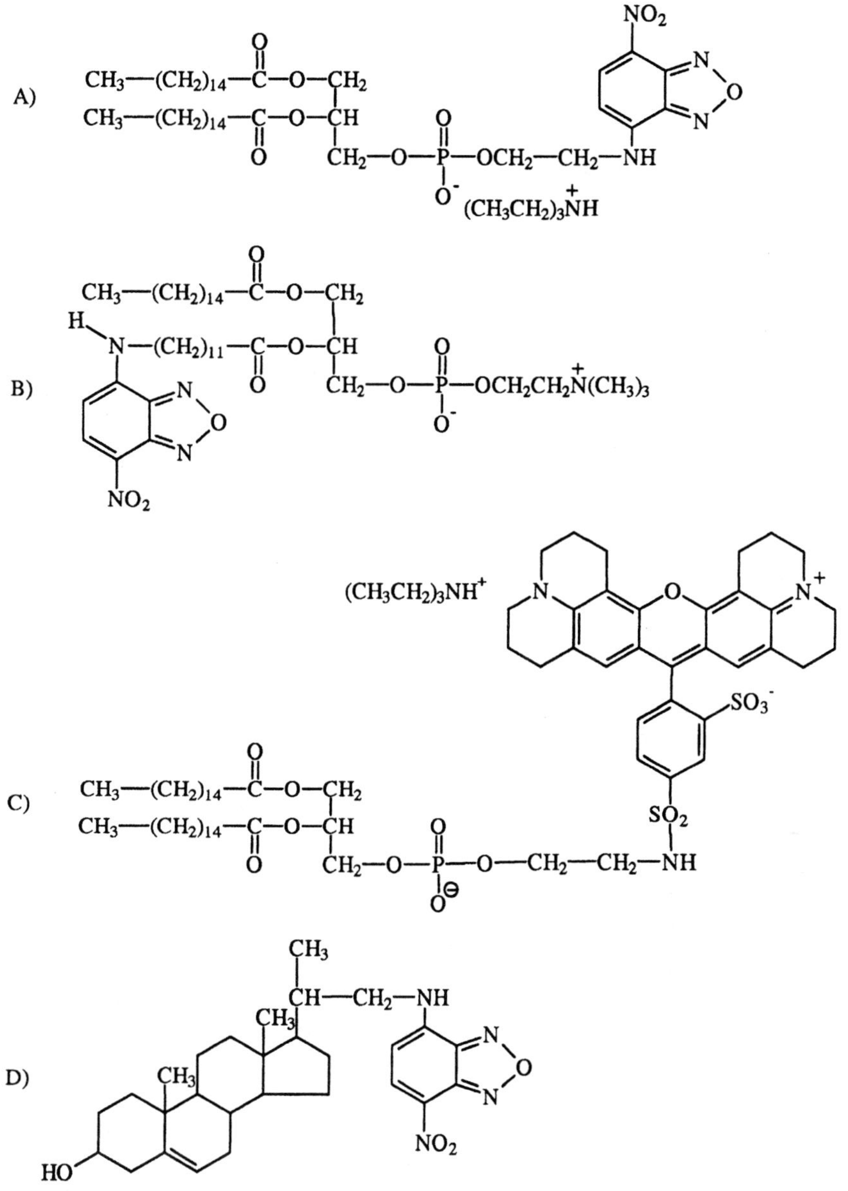

Fig. 2 The molecular structures and absorbance and emission wavelengths in a selected solvent for a number of commonly employed fluorescent lipid probe molecules: (A) *N*-(7-nitrobenz-2-oxa-1,3-diazol-4-yl)-1,2-dihexadecanoyl-*sn*-glycero-3-phosphoethanolamine, triethylammonium salt (NBD-PE), $\lambda_{abs} = 463$ nm, $\lambda_{em} = 536$ nm (methanol); (B) 2-(12-(7-nitrobenz-2-oxa-1,3-diazol-4-yl)amino)dodecanoyl-hexadecanoyl-*sn*-glycero-3-phosphocholine (NBD-C$_{12}$-HPC), $\lambda_{abs} = 465$ nm, $\lambda_{em} = 534$ nm (ethanol); (C) *N*-(Texas Red sulfonyl)-1,2-dihexadecanoyl-*sn*-glycero-3-phosphoethanolamine, triethylammonium salt (Texas Red DHPE), $\lambda_{abs} = 582$ nm, $\lambda_{em} = 601$ nm (methanol); (D) 22-(*N*-(7-nitrobenz-2oxa-1,3-diazol-4-yl)amino)-23,24-bisnor-5-cholen-3β-ol, $\lambda_{abs} = 469$ nm, $\lambda_{em} = 537$ nm (chloroform). All of these are available from Molecular Probes (Eugene, OR), along with numerous others.

of the cost of the commercial BAM systems. High lateral resolution is of particular significance for discerning the shapes and the shape changes of monolayers of lipids exhibiting "small" domain sizes below about $20\,\mu$m, as is often the case with phospholipids. Domain borders appear more distinctly with fluorescence microscopy than with BAM.

BAM offers the clear advantage of dispensing with the need for fluorescent probes; however, fluorescence microscopy allows dual-labeling experiments such as those in which a protein in the subphase is tagged with a fluorescent probe emitting at a wavelength different from that of the fluorescently tagged lipid dispersed in the monolayer. In such experiments, the use of two sets of filters (commonly available as cubes inserted into the microscope) allows alternate imaging of the domains of protein near the interface and of the domain structure of the lipid monolayer. Dual labeling has also been employed to study a lung surfactant system (13). The method of fluorescence recovery after photobleaching (FRAP) applied to monolayers can provide measurements of the diffusion coefficient of a probe molecule in different monolayer states (14, 15).

BAM experiments in which the angle of the analyzer in front of the CCD camera is varied can unambiguously give information on lipid chain tilt in different sections of domains. Such information can also be obtained from polarized fluorescence observations provided that the orientation of the chromophore in the probe lipid is directly coupled to the orientation of the surrounding molecules and the probe lipid is slightly soluble in the condensed lipid phase. BAM has obvious advantages for the study of adsorbed layers, mono- or multilayers of unlabeled proteins, monolayers containing nanoparticles, and other systems in which the behavior of probe lipids would be either problematic or uninformative.

III. FLUORESCENCE MICROSCOPY OF PHOSPHOLIPID MONOLAYERS: DOMAIN STRUCTURES AND CHIRALITY RELATED EFFECTS

A. Structural and Thermodynamic Aspects

One aspect of the experimental study of monolayers concerns unraveling the effect of stereochemistry on structure and thermodynamics (16). Generally, these studies of chiral discrimination have concerned comparing the behavior of films of one enantiomer of a compound with both of the racemate and films of the opposite enantiomer. Most studies have focused on the case of one chiral center, generally in the headgroup, although some efforts on molecules with two chiral centers have been reported; one can readily imagine the number of possible subtle effects to be investigated if molecules with multiple chiral centers are studied as a set of diastereomers. For one chiral center two cases can be defined: homochiral discrimination and heterochiral discrimination. In the case of homochiral discrimination, there is a preferential interaction between like enantiomers, whereas heterochiral discrimination occurs when there is a preferential interaction between opposite enantiomers. In the case of strong chiral discrimination effects of monolayers, differences in Π-A isotherms between enantiomers and racemic films, differences in domain structures, and differences in molecular-level properties such as lattice packing and infrared (IR) spectra are expected by analogy with three-dimensional chiral discrimination in molecular solids (17,18).

There has been no evidence for strong chiral discrimination in the common membrane phospholipids studied to date in terms of unambiguously measured differences between the Π-A isotherms of racemic and enantiomeric films; however, evidence for chiral discrimination has been reported for the Π-A isotherms of a triple-chain phosphatidylcholine [3-*O*-hexadecyl-2-(2-hexadecylstearoyl)-glycero-*sn*-1-phosphocholine] at 5°C with the racemic film being slightly less compact than the enantiomeric film. The evidence for chiral discrimination in this system comes primarily from synchrotron X-ray diffraction data (19). The enantiomer was found to form an oblique lattice, and the racemic film formed a centered rectangular lattice at low surface pressure; the difference disappeared at higher surface pressure. For DPPC, X-ray data showed an oblique structure for both the enantiomer and the racemate, established by ordering of the glycerol backbones (20). In dipalmitoylphosphatidyl ethanolamine (DPPE), only the enantiomer exhibited a chiral lattice structure (21). In a related study, monolayers of 1-hexadecylglycerol showed an oblique lattice for the enantiomeric films and a centered rectangular lattice for the racemic films (22).

In contrast to phospholipids, studies of *N*-acylamino acids (23–25) show large differences between the Π-A isotherms of enantiomeric and racemic films. Amphiphiles in which the chirality is coupled to a strong interaction such as hydrogen bonding between the headgroups seem to show significant differences in Π-A isotherms and other properties. The more buried nature of the chiral center in phospholipids leads to smaller differences in the packing with neighboring molecules than occurs in molecules with a chiral center(s) bearing functional groups directly associating with those of nearest neighbor molecules.

B. The Influence of Chirality on Lipid Domain Shapes

In the pioneering work of McConnell and coworkers, fluorescence microscopy was employed to examine the domain morphologies of Langmuir films of common phospholipids. Although differences between the Π-A isotherms of enantiomeric and racemic DPPC or DPPE were not observed, visually dramatic differences in the shapes of the domains of the condensed phase were observed. Domains of the condensed phase of R-DPPC appeared as three armed spirals with arms curving exclusively counterclockwise, whereas the domains for S-DPPC curved exclusively clockwise (26). Domains of rac-DPPC exhibited branching but no net curvature. In Fig. 3, polarized fluorescence micrographs of R-DPPC with 2 mol% NBD-DPPE probe are shown; in this figure, the spiral nature of the domains is evident and comparison of the two images demonstrates the presence of long-range molecular orientational order within the domains (9). Domains of S-DPPC would be similar in appearance, but the curvature would be of the opposite sense. These results of polarized fluorescence microscopy observations clearly indicated that the lipid molecules have a uniform tilt orientation parallel to the sides of the domains. These studies were conducted using the probe NBD-DPPE (nitrobenzoxadiazole-labeled DPPE; see Fig. 2), which partitioned into and became oriented in the condensed phase. Later position-dependent electron diffraction studies of LB films of two phosphatidyl-ethanolamines clearly showed that the molecular orientation follows the turning of the domain (27).

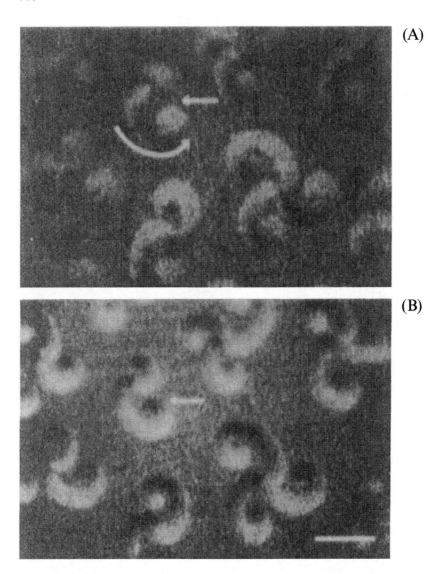

Fig. 3 Fluorescence micrographs of an R-DPPC monolayer doped with 2 mol % NBD-DPPE. The micrographs were taken with oblique illumination from above the monolayer with a p-polarized laser beam (488 nm) from the left (A) and from the right (B). Note that the fluorescence intensities for the two directions of illumination are complementary. The straight arrow points to a boundary between domains. The curved arrow shows the curvature of the domain. The bar represents 10 μm. (From Ref. 27.)

The equilibrium shapes of condensed lipid domains have been treated in great detail theoretically (28). An important feature of the phospholipid domains is that the observed shapes appeared to represent equilibrium configurations. The shapes of lipid domains are understood in terms of the competing effects of electrostatic repulsions between the oriented headgroup dipoles, tending to favor elongated or noncircular domains, and the line tension around the peripheries of the domains tending to favor compact, circular forms. For the films of enantiomeric DPPC, it is expected that the structural environments on the two long sides of the domain will not be equivalent and the line tension will be anisotropic, i.e., $\lambda_L \neq \lambda_R$, where λ_L is the line tension of the "left" side of the domain and λ_R is the line tension of the "right" side of the domain. Such an anisotropy in the line tension promotes a curving of the domain in a direction so as to reduce the total length of the side of higher line tension, resulting in a spiral domain. Given the mirror symmetry of the packing of the two opposite enantiomers, it is thus expected that the curvature of the domains will be unique to the enantiomeric form of the lipid. Sensitivity of the detailed nature of the domain shape distribution of DPPC monolayers to compression rate has been noted by Klopfer and Vanderlick (29) and indicates that "equilibrium" of the domain shapes may not be truly achieved on the time scales employed in typical monolayer studies (minutes to hours).

Studies of other biologically relevant phospholipid systems using fluorescence microscopy have flourished and revealed a further variety of domain shapes and shape transitions, also discussed in terms of theories invoking electrostatic forces and line tension. Prominent among these are studies of DMPC + cholesterol mixed monolayers (30), which show a two-dimensional critical point between a denser cholesterol-enriched fluid phase and phase of lower density less rich in cholesterol. In this system, labyrinth-like domain patterns denoted as the "stripe" phase have been observed and accounted for theoretically. Such patterns have also been observed in other single-component monolayers (31,32). Alterations in the headgroup charge effect the observed domain shapes significantly; for example, studies of long-chain phosphatidic acids showed elongation of the condensed phase domains on cooling as the subphase was made basic (33), due to enhanced electrostatic repulsion between the headgroups and a resultant tilting of the alkyl chains.

A significant amount of work in the area of chiral discrimination of Langmuir monolayers as studied by fluorescence microscopy does not concern membrane lipids but clearly warrants mention. Monolayers of *N*-acyl amino acids including *N*-stearoylvaline, *N*-stearoylserine methyl ester, and *N*-myristoylalanine were examined (23–25). In these studies, macroscopic domain chirality was shown to be a stereospecific expression of the molecular chiral recognition and ordering due to amide-amide hydrogen bonding. Related studies have also appeared for a chiral imidazole amphiphile (34). Although these molecules are not membrane lipids, the observations are relevant to the assembly of supramolecular aggregates such as tubules, fibers, and vesicles from synthetic surfactants for various applications and for fundamental study of self-assembly. Chiral segregation has also been observed in atomic force microscopy (AFM) studies of LB films of a racemic tetracylic alcohol on mica (35); in these studies, oblique lattice structures of opposite orientation were clearly seen in AFM images.

C. Chirality Effects in Mixed Monolayers of Phospholipids and Cholesterol

The addition of cholesterol to a phospholipid monolayer has been observed to amplify the chiral nature of the domains for DPPC. As observed by Weis and McConnell (36), the addition of small amounts (1–4%) of cholesterol to DPPC results in thinner, more elongated domains, increasing the length of the interface by a factor of 10–100 or more. For the pure enantiomers, the consequence is more obviously curved spiral structures. The large effect of small amounts of cholesterol is believed to arise from its action as a

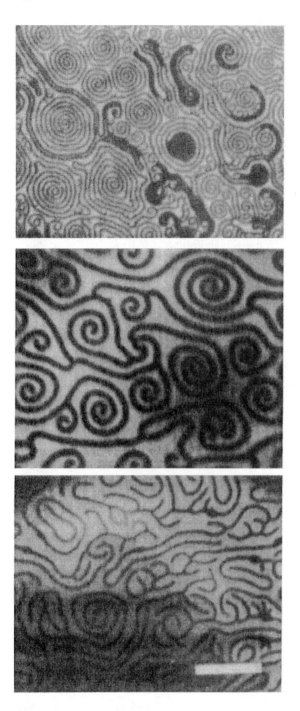

Fig. 4   Fluorescence micrographs showing the influence of the absolute configuration of DPPC on the domain curvature and the effect of cholesterol on the domains. The monolayers are composed of R-DPPC (top), S-DPPC (middle), and equimolar R-DPPC and S-DPPC (bottom). The monolayers contain 2 mol % cholesterol and 2 mol % NBD-PC. The bar represents 30 μm. (From Ref. 46.)

"line-active" molecule, localizing at the interface between the expanded and condensed phases. The reduction of the line tension by the line-active nature of cholesterol results in a shape change toward elongated domains. A reduction of the line tension for a domain of oriented chiral molecules, without loss of the anisotropy of the line tension, will result in more dramatically visible chiral shapes. In Fig. 4, micrographs from unpolarized fluorescence microscopy for R-DPPC, S-DPPC, and rac-DPPC with 2 mol; % cholesterol and 2 mol % NBD-PC probe are shown. In these images, one can clearly see how the reduction in line tension has thinned and elongated the condensed phase domains into stripes that coil counterclockwise for S-DPPC, clockwise for R-DPPC, and have no net curvature for rac-DPPC.

Given the variety of natural membrane lipids and steroids, further investigations of mixed monolayers have been pursued. Fluorescence microscopy was used to determine the relative strengths of the phospholipid-cholesterol interaction by comparing mixed monolayers of cholesterol with DPPC and *N*-palmitoylsphingomyelin (37). The characteristics of the condensed-phase domains differed between these two systems. The domains of the mixed films with the sphingomyelin were smaller and exhibited stable shapes up to higher surface pressures than those formed with DPPC, along with other consistent quantitative differences, such as a narrower size distribution. A study of mixed monolayers of DPPC and a number of sterols showed that 3-keto sterols do not form condensed domains with DPPC (38). A study of mixed monolayers of 20–33 mol % cholesterol with dialkylphosphatidylcholines with chain lengths from C_{10} to C_{16} showed domain formation for all chain lengths (39).

IV. PROTEIN-MONOLAYER INTERACTIONS

Studies of protein–lipid monolayer interactions employing fluorescence microscopy have provided great insight into membrane-protein recognition processes and have contributed practically to the development of methods for the organization of proteins at functionalized interfaces. Studies in the area can be partly divided into three categories: (a) the direct observation of the action of lipase enzymes on lipid monolayers, (b) the study of specific recognition of proteins by monolayers of lipids bearing functional groups recognized or bound by the protein, and (c) the interaction of proteins with lipid monolayers driven by nonspecific interactions.

A. Experimental Methods for the Direct Observation of Enzyme Action on Monolayers

In order to study the interaction of subphase proteins, enzymes, or other molecules with a monolayer, additional facilities and different trough designs are desirable. One method is to compress the monolayer to a certain Π and then inject the protein beneath the film from a syringe. These studies have been pursued extensively and are discussed in this volume by R. Verger, who is responsible for the introduction of the "zero-order" trough, especially appropriate for such studies. This type of trough has been employed to study enzyme action on monolayers and, more recently, the sequestering of cholesterol and other steroids from mixed monolayers with phospholipids by cyclodextrins introduced into the subphase (40). The stereoselectivity of enzyme action has been observed using the zero-order trough technique in a study

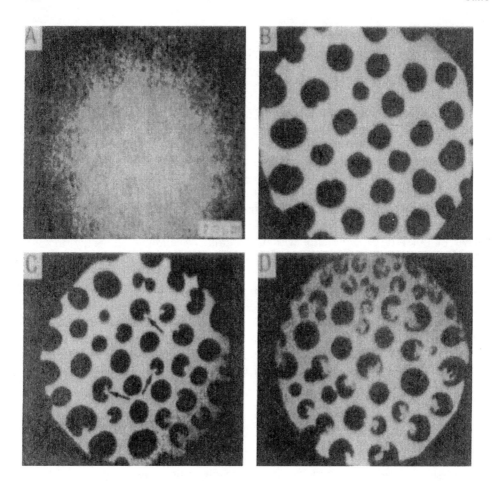

Fig. 5 Fluorescence micrographs of an L-α-DPPC monolayer phase transition region after slow compression and subsequent lipid hydrolysis by injection of phospholipase A$_2$ into the aqueous subphase. Image (A) taken at time zero (immediately after enzyme injection) through the fluorescein filter to observe the homogeneous subphase signal from fluorescein-labeled phospholipase A$_2$. Image (B) also at time zero, but imaged through the sulforhodamine filter in the L-α-DPPC phase transition region. Image (C) at 20 min after enzyme injection imaged through the sulforhodamine filter. Arrows indicate attack of phospholipase A$_2$ at lipid domain edges. Image (D) shown at 40 min after enzyme injection. The scale bar in (A) is 20 μm. The temperature was 30°C and the surface pressure 22 dyne cm^{-1}. (From Ref. 61.)

of the action of 23 different animal and microbial lipases against monolayers of dicaprin isomers (41).

B. Fluorescence Observation of the Stereospecific Action of Phospholipases on Phospholipid Monolayers

The use of fluorescence microscopy to follow the action of phospholipase A$_2$ against phospholipid monolayers revealed unique information about the process (42). Figure 5 shows fluorescence images of a monolayer of L-DPPC beneath which phospholipase

A_2 has been injected in which the action of the enzyme can be visually observed. This is an excellent example of a dual-labeling experiment; the monolayer contains a sulforhodamine-labeled long-chain amine preferentially soluble in the liquid-expanded phase of the film and the enzyme has been labeled with fluorescein isothiocyanate. As can be seen in Fig. 5, the enzyme action begins at the domain border and works inward. The enzyme action is complete after 4–5 h. At the time, imaging of the fluorescence from the tagged enzymes shows that the enzyme forms large aggregates at the water-air interface. A further study by Maloney and Grainger (43) of the DPPC + lyso-DPPC + palmitic acid ternary "products" monolayer indicated that lateral phase separation could create domains with net negative charge to which the enzyme could become electrostatically bound. In another study, the action of phospholipase A_2 against 1-caproyl, 2-palmitoylphosphatidylcholine was studied; with the caproic acid product readily desorbed into the subphase, very little enzyme induced domain structure was observed (44).

The action of cholesterol oxidase against mixed films of DMPC and cholesterol revealed that the enzyme penetrated the monolayer in the regions of lower surface density and that the cholestenone product resulted in a more fluid monolayer with dissipation of the boundary between the condensed and expanded phases (45). Subsequent studies included examination of the rate of cholesterol oxidation in ternary mixed monolayers of cholesterol with sphingomyelins and phosphatidylcholines by cholesterol oxidase introduced into the subphase (46). Comparison of mixed monolayers of cholesterol with DPPC and C_{16} alkylether-linked analogues surprisingly showed stronger association of cholesterol with the alkylether-linked phosphatidylcholines (47). This indicates the sensitivity of monolayer component associations with steroids to modest changes in molecular structure.

C. Fluorescence Microscopy Studies of Specific Protein–Functionalized Lipid Monolayer Interactions

The specific binding of the protein streptavidin by biotinylated lipids in monolayers is a model system for specific recognition of proteins by interfaces and their two-dimensional crystallization. A series of biotinylated lipids were prepared and streptavidin labeled with sulforhodamine was introduced into the subphase (48,49). Large shifts were observed in Π-A isotherms due to specific binding of streptavidin to the monolayer and were easily distinguished from much smaller changes arising from nonspecific adsorption of streptavidin with its sites occupied when biotin was dissolved in the subphase. Fluorescence microscopy showed the formation of regular two-dimensional protein crystals at low surface pressure where the biotin lipid–streptavidin complex retained lateral mobility. The size, morphology, and crystallization kinetics were dependent on surface pressure and the nature of the biotinylated lipid. Polarized fluorescence showed the two-dimensional protein crystals to be optically anisotropic (50). Electron microscopic analysis of the bound biotin lipid–streptavidin complex transferred onto grids confirmed the crystalline structure and showed the presence of two empty binding sites on the other side of the protein. At the water-air interface, the two binding sites could be filled with a second spacer lipid and a second layer of streptavidin assembled. This was confirmed using fluorescence microscopy examinations in a dual-labeling experiment; the first protein layer was labeled with sulforhodamine and the second with fluorescein (51).

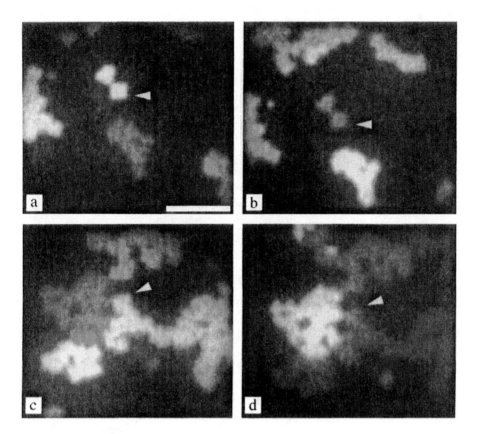

Fig. 6 Fluorescence micrographs of streptavidin aggregates grown on a dioleoyliminodiacetic acid (DOIDA)-Cu monolayer on a buffered subphase [20 mM 3-(*N*-morpholino)-propanesulfonic acid, 250 mM NaCl, pH 7.8] imaged through polarizing filters. Images on the left and right were taken under perpendicular orientations of the polarizers. White arrowheads indicate equivalent points on the monolayer in each pair of images. Scale bar = 100 μm. (From Ref. 72; micrograph kindly provided by Prof. F. Arnold.)

A new approach for targeting of proteins to the water-air interface has been developed by Arnold and coworkers (10,52–54). Synthetic lipids with iminodiacetic acid (IDA) headgroups such as 1,1'-[[9-[2,3-bis[(*Z*)-octadec-9-enyloxy]propyl]-3,6,9-trioxanonyl]imino]diacetic acid (DOIDA), which has an IDA group attached to a glycerol substituted with two unsaturated chains by a triethylene glycol spacer, were prepared (10). The IDA headgroup binds a Cu^{2+} ion, forming a neutral complex with two coordination sites free and available to bind to surface histidines on proteins introduced into the subphase beneath the monolayer. The lipid is spread "premetalated," Cu^{2+} having been introduced into the spreading solution. The Cu^{2+} in the subphase inhibits protein crystallization at the interface. The films were found to serve as templates for the crystallization of streptavidin and observed by fluorescence microscopy. In Fig. 6, streptavidin aggregates grown under a DOIDA-Cu monolayer are shown under polarized fluorescence observation. These aggregates took about

8 h to grow and thus required great care in maintaining a clean and well-controlled trough environment. The aggregates appear as square "tiles," often connected to each other, and the inversion of contrast seen on changing the polarization of the illuminating light by 90° demonstrates their crystalline nature. In this example, binding occurs to a histidine located on the streptavidin surface near the binding pocket for biotin. This approach has the advantage of greater generality; other proteins bound to monolayers of IDA lipids in this manner include myoglobin, which has 11 histidines, and cytochrome b_5 engineered to contain a hexahistidine tag at the C-terminus.

D. Fluorescence Microscopy Studies of Nonspecific Protein–Phospholipid Monolayer Interactions

In this context, we use the term "nonspecific" to mean that a particular binding or "molecular recognition" process is not involved in the association of the protein with the monolayer. Cytochrome c was found to be attracted electrostatically to the surface of a dimyristoylphosphatidic acid monolayer (55), slowly adsorbing onto and penetrating the monolayer and then being squeezed out of the monolayer above a surface pressure of 22 dyne cm^{-1}. Using labeled protein and NBD-DPPE as the probe in the monolayer, it was observed that the protein preferred the liquid-expanded phase and remained beneath the film after being squeezed out. The sugar-binding protein concanavalin A was found to aggregate around the domain boundaries in the coexistence region of DMPE monolayers (56). The specific binding of Con A to a DPPC monolayer containing a small amount of the glycolipid recognized by Con A was detected by changes in Π and by fluorescence microscopy. It was postulated that the Con A–glycolipid conjugate was line active because of the observed formation of sections of high curvature along the domain boundaries.

V. FUTURE PROSPECTS

Many studies of the interactions between enzymes and monolayers, between drugs or peptides and monolayers, and of mixed monolayers of lipids with other biomolecules such as steroids, ion channels, and gangliosides are still reported relying on information derived from surface pressure data alone. Given the continuing interest in the understanding of membrane functions for fundamental reasons and in the applications of tailored membrane-like structures in biotechnology, supplementing more of these studies with direct observation of the domain morphologies of the films would lead to improved characterization. Fluorescence microscopy is relatively easy to apply and a significant literature discussing the implications of different domain morphologies for the molecular ordering exists to aid in qualitative interpretation. An example along these lines is the study of anesthetic (procaine)–phospholipid monolayer interactions conducted by Cadenhead and coworkers (57). Arguments about membrane fluidity, miscibility of monolayer components, and phase structure can be made more definitive by the use of fluorescence microscopy combined with traditional characterization methods. Studies of chiral discrimination and two-dimensional stereochemical effects in synthetic lipids represent an area in which many details remain unexplored. The future utilization of the technique will be determined by its value as applied to biotechnological and membrane chemistry problems.

REFERENCES

1. KJ Stine. Investigations of monolayers by fluorescence microscopy. Microsc Res Tech 27:439, 1994.
2. H Möwald. Surfactant layers at water surfaces. Rep Prog Phys 56:653, 1993.
3. F Rondelez. Fluorescence methods at the liquid-vapor interface. In: D Langevin, ed. Light Scattering by Liquid Surfaces and Complementary Techniques. New York: Marcel Dekker, 1991, pp. 405–424.
4. CM Knobler. Seeing phenomena in flatland: Studies of monolayers by fluorescence microscopy. Science 249:870, 1990.
5. MM Lipp, KC Lee, A Waring JA Zasadzinski. Fluorescence, polarized fluorescence, and Brewster angle microscopy of palmitic acid and lung surfactant protein B monolayers. Biophys J 72:2783, 1997.
6. BG Moore, CM Knobler, S Akamatsu, F Rondelez. Phase diagram of Langmuir monolayer of pentadecanoic acid: quantitative comparison of surface pressure and fluorescence microscopy. J Phys Chem 94:4588, 1990.
7. M Thoma, H Möhwald. Phospholipid monolayers at hydrocarbon/water interfaces. J Colloid Interface Sci 162:340, 1994.
8. M Seul, MJ Sammon, LR Monar. Imaging of fluctuating domain shapes: Methods of image analysis and their implementation in a personal computing environment. Rev Sci Instrum 62:784, 1991.
9. VT Moy, DJ Keller, HE Gaub, HM McConnell. Long-range molecular orientational order in monolayer solid domains of phospholipid. J Phys Chem 90:3198, 1986.
10. K Ng, DW Pack, DY Sasaki, FH Arnold. Engineering protein-lipid interactions: targeting of histidine-tagged proteins to metal-chelating lipid monolayers. Langmuir 11:4048, 1995.
11. A Gericke. J Simon-Kutscher, H Hühnerfuss. Comparison of different spreading techniques of monolayers at the air/water interface by external infrared reflection-absorption spectroscopy. Langmuir 9:3115, 1993.
12. JP Slotte, P Mattjus. Visualization of lateral phases in cholesterol and phosphatidylcholine monolayers at the air/water interface—A comparative study with two different reporter molecules. Biochim Biophys Acta 1254:22, 1995.
13. K Nag, S Taneva, J Perez-Gil, A Cruz, KMW Keough. Combinations of fluorescently labeled pulmonary surfactant proteins SP-B and SP-C in phospholipid films. Biophys J 72:2638, 1997.
14. R Peters, K Beck. Translational diffusion in phospholipid monolayers measured by fluorescence microphotolysis. Proc Natl Acad Sci USA 80:7183, 1983.
15. S Kim, H Yu. Lateral diffusion of amphiphiles and macromolecules at the air/water interface. J Phys Chem 96:4034, 1992.
16. PL Rose, NG Harvey, EM Arnett. Chirality and molecular recognition in monolayers at the air-water interface. In: D Bethell, ed. Advanced in Physical Organic Chemistry. Vol 28. New York: Academic Press, 1993, pp 45–170.
17. EL Eliel, SH Wilen. Stereochemistry of Organic Compounds. New York: Wiley-Interscience, 1994.
18. J Jacques, A Collet, SH Wilen. Enantiomers, Racemates, and Resolutions. New York: Wiley, 1981.
19. F Bringezu, G Brezinski, P Nuhn, H Mohwald. Chiral discrimination in a monolayer of a triple-chain phosphatidylcholine. Biophys J 70:1789, 1996.
20. G Brezinski, A Dietrich, B Struth, C Bohm, WG Bowman, K Kjaer, H Möhwald. Influence of ether linkages on the structure of double-chain phospholipid monolayers. Chem Phys Lipids 76:145, 1995.
21. C Bohm, H Möhwald, L Leserowitz, J Als-Nielsen, K Kjaer. Influence of chirality of the structure of phospholipid monolayers. Biophys J 64:553, 1993.

22. E Scales, G Brezinski, H Möhwald, VM Kaganer, WG Bouwman, K Kjaer. Chirality effects on 2D phase transitions. Thin Solid Films 284/285:56, 1996.

23. KJ Stine, J-Y Uang, SD Dingman. Comparison of enantiomeric and racemic monolayers of *N*-stearoylserine methyl ester by fluorescence microscopy. Langmuir 9:2112, 1993.

24. P Nassoy, M Goldman, O Bouloussa, F Rondelez. Spontaneous chiral segregation in bidimensional films. Phys Rev Lett 75:457, 1995.

25. H Hühnerfuss, V Neumann, KJ Stine. Role of hydrogen-bond and metal complex formation for chiral discrimination in amino acid monolayers studied by infrared reflection-absorption spectroscopy. Langmuir 10:2561, 1996.

26. RM Weis, HM McConnell. Two-dimensional chiral crystals of phospholipid. Nature 310:47, 1984.

27. SW Hui, H Yao. Electron diffraction studies of molecular and orientation in phospholipid monolayer domains. Biophys J 64:150, 1993.

28. HM McConnell. Structures and transitions in lipid monolayers at the air-water interface. Annu Rev Phys Chem 42:171, 1991.

29. KJ Klopfer, TK Vanderlick. Isotherms of dipalmitoylphosphatidylcholine monolayers: Features revealed and features obscured. J Colloid Interface Sci 188:220, 1996.

30. NY Morgan, M Seul. Structure of disordered droplet domain patterns in a monomolecular film. J Phys Chem 99:2088, 1995.

31. M Yoneyama, A Fujii, S Maeda, T Murayama. Light-induced bubble-stripe transitions of gaseous domains in porphyrin Langmuir monolayer. J Phys Chem 96:8982, 1992.

32. KJ Stine, DT Strattman. Fluorescence microscopy study of Langmuir monolayers of stearylamine. Langmuir 8:2509, 1992.

33. WM Heckl, DA Cadenhead, H Möhwald. Cholesterol concentration dependence of quasi-crystalline domains in mixed monolayers of the cholesterol-dimyristoylphosphatidic acid system Langmuir 4:1352, 1988.

34. JH van Esch, RJM Nolte: H Rongsdorf, G Wildburg. Monolayers of chiral imidazole amphiphiles: Domain formation and metal complexation. Langmuir 10–1955, 1994.

35. CJ Eckhardt, NM Peachey, DR Swanson, JM Takacs, MA Khan, X Gong, J-H Kim, J Wang, RA Uphaus. Separation of chiral structures in monolayer crystals of racemic amphiphiles. Nature 362:614, 1993.

36. RM Weis, HM McConnell. Cholesterol stabilizes the crystal-liquid interface in phospholipid. J Phys Chem 89:4453, 1985.

37. JP Slotte. Lateral domain formation in mixed monolayers containing cholesterol and dipalmitoylphosphatidylcholine or *N*-palmitoylsphingomyelin. Biochim Biophys Acta 1235:419, 1995.

38. JP Slotte. Effect of sterol structure on molecular interactions and lateral domain formation in monolayers containing dipalmitoylphosphatidylcholine. Biochim Biophys Acta 1237:127, 1995.

39. JP Slotte. Lateral domain heterogeneity in cholesterol/phosphatidylcholine monolayers as a function of cholesterol concentration and phosphatidylcholine acyl chain length. Biochim Biophys Acta 1238:118, 1995.

40. H Ohvo, JP Slotte. Cyclodextrin-mediated removal of steroids form monolayers: Effects of sterol structure and phospholipids on desorption rate. Biochemistry 35:8018, 1996.

41. E Rogalska, S Nury, I Dochet, R Verger. Lipase steroeselectivity and regioselectivity toward three isomers of dicaprin: A kinetic study by the monomolecular film technique. Chirality 7:505, 1995.

42. DW Grainger, A Reichert, H Ringsdorf, C Salesse. Hydrolytic action of phospholipase-A_2 in monolayers in the phase transition region: Direct observation of enzyme domain formation using fluorescence microscopy. Biochim Biophys Acta 1023:365, 1990.

43. KM Maloney, DW Grainger. Phase-separated anionic domains in ternary mixed lipid monolayers at the air-water interface. Chem Phys Lipids 65:31, 1993.

44. KM Maloney, M Grandbois, DW Grainger, C Salesse, KA Lewis, MF Roberts. Phospholipase-A$_2$ domain formation in hydrolyzed asymmetric phospholipid monolayers at the air/water interface. Biochim Biophys Acta 1235:395, 1995.

45. JP Slotte. Direct observation of the action of cholesterol oxidase in monolayers. Biochim Biophys Acta 1259:180, 1995.

46. P Mattjus, JP Slotte. Does cholesterol discriminate between sphingomyelin and phosphatidylcholine in mixed monolayers containing both phospholipids? Chem Phys Lipids 81:69, 1996.

47. P Mattjus, R Bittman, JP Slotte. Molecular interaction and lateral domain formation in monolayers containing cholesterol and phosphtidylcholines with acyl- and alkyl-linked C16 chains. Langmuir 12:1284, 1996.

48. SA Darst, M Ahlers. PH Meller, EW Kubalek, R Blankenburg, HO Ribi, H Ringsdorf, RD Kornberg. Two-dimensional crystal of streptavidin on biotinylated lipid layers and their interactions with biotinylated macromolecules. Biophys J 59:387, 1991.

49. R Blankenburg, P Meller, H Ringsdorf, C Salesse. Interaction between biotin lipids and streptavidin in monolayers: Formation of oriented two-dimensional protein domains induced by surface recognition. Biochemistry 28:8214, 1989.

50. M Ahlers, R Blankenburg, DW Grainger, P Meller, H Ringsdorf, C Salesse. Specific recognition and formation of two-dimensional streptavidin domains in monolayers: Applications to molecular devices. Thin Solid Films 180:93, 1989.

51. M Ahlers, M Hoffman, H Ringsdorf, AM Rourke, E Rump. Specific interaction of desthiobiotin lipids and water-soluble compounds with streptavidin. Markromol Chem Marcromol Symp 46:307, 1991.

52. DW Pack, G Chen, KM Maloney, C-T Chen, FH Arnold. A metal-chelating lipid for 2D protein crystallization via coordination of surface histidines. J Am Chem Soc 119:2479, 1997.

53. DW Pack, FH Arnold. Langmuir monolayer characterization of metal chelating lipids for protein targeting to membranes. Chem Phys Lips 86:135, 1997.

54. DW Pack, K Ng, KM Maloney, FH Arnold. Ligand-induced reorganization and assembly in synthetic lipid membranes. Supramol Sci 4:3, 1997.

55. J Peschke, H Möhwald. Cytochrome *c* interaction with phospholipid monolayers and vesicles. Colloids Surf 27:305, 1987.

56. H Haas, H Möhwald. Specific and unspecific binding of concanavalin A at monolayer surfaces. Thin Solid Films 180:101, 1989.

57. B Asgharian, DA Cadenhead, M Tomoaia-Cotisel. An epifluorescent microscopy study of the effects of procaine on model membrane systems. Langmuir 9:228, 1993.

23

Scanning Force Microscopy in Biology

D. J. Keller
University of New Mexico, Albuquerque, New Mexico

I. INTRODUCTION

The scanning force microscope is one example of a larger class of instruments called scanning probe microscopes that create images in a fundamentally new way. Unlike conventional light or electron microscopes, which use lenses and focused radiation to form images at a distance, scanning probe microscopes create images by scanning the sample at close range with a sharp sensor tip. The various kinds of probe microscopy—scanning force microscopy (SFM), scanning tunneling microscopy (STM), near-field scanning optical microscopy, (NSOM), scanning ion conductance microscopy, (SICM), etc.—use different kinds of probe tips and exploit different kinds of tip-sample interactions. The SFM uses a sharp point mounted on a flexible cantilever and measures the force of interaction between tip and sample. The STM uses a sharp conductive tip and measures a small electrical current between tip and sample. The NSOM uses a tip made from a sharpened optical fiber and measures the absorption, emission, and scattering of light by the sample.

For biological applications, the most important properties of SFM are its very high resolution (angstroms to nanometers) and its ability to form images in physiological buffer solutions. The SFM thus combines the main advantage of light microscopy (ability to observe active samples in solution) with that of electron microscopy (high resolution). The SFM is also a force-measuring device, and information on forces in biological systems is itself directly useful. The emerging major areas of application for SFM reflect this unique combination of capabilities. For example, several reports have shown that the SFM can visualize the operation of individual active proteins and protein complexes as they carry out their functions. Examples include the digestion of DNA by nucleases (1), the transcription of

DNA by RNA polymerase (2), and the conformational changes accompanying the opening and closing of membrane channels (3). A second major area of application is the imaging of membranes and membrane proteins. Lipid bilayers are flat, thin, and adsorb easily to common substrates. The membrane acts as a stabilizing matrix for membrane proteins, which enhances high-resolution imaging, especially if they form two-dimensional crystals. Images of membrane proteins in two-dimensional crystals show the highest resolution to date for any biological sample (4).

A third major area of application uses the SFM's highly sensitive force sensor to measure forces on the molecular scale. This ability has been exploited to investigate the force needed to rupture biomolecular complexes [biotin-avidin complexes, antibody-antigen complexes, DNA base-pairing interactions, etc. (5–7)] as a function of environmental conditions. Two other variations of SFM are friction imaging and compliance imaging, in which image contrast corresponds to surface material properties (tip-sample friction or tip-sample compliance, respectively) rather than structure. Such properties are usually more sensitive to chemical composition than simple topography and are widely used as a means of identifying and investigating specific sample features. Finally, an area with great untapped potential is the use of the SFM to follow the time dynamics and time fluctuations of single biomolecules as they carry out their function.

The purpose of this chapter is to review the most basic principles and methods of SFM imaging. Section II is a brief outline of probe microscope instrumentation and the main modes of SFM imaging. Section III describes the origin and properties of the forces of interaction between the probe tip and the sample. These forces are responsible for both the high resolution of the SFM and some of its limitations. Section IV is an overview of the most common experimental techniques for imaging biological molecules. Section V is an introduction to probe microscope imaging theory. The chapter concludes with a short perspective on the strengths and weaknesses of SFM at its present stage of development.

II. SFM INSTRUMENTATION

A. Major Instrumental Components

All scanning force microscopes contain four major components: (a) a tip-cantilever-deflection system, (b) a feedback system, (c) a piezoelectric positioner, and (d) a computer that stores the data and controls the instrument (8). The first component, the tip-cantilever-deflection system, is the heart of the instrument (Fig. 1a). The probe tip is the primary imaging device (its role is analogous to the objective lens in a light microscope), and its properties determine the resolution and other characteristics of the images. It is mounted on a flexible cantilever, which deflects slightly when forces act on the tip. The deflection is measured by, in this case, a laser beam reflected from the back of the cantilever into a two-segment photodiode. This setup, called an optical lever, is the most common method for measuring cantilever deflections. It is simple and robust but can be very sensitive: deflections of less than 1 Å are easily detected. Other methods for measuring cantilever deflections include interferometry, capacitance, and tunneling current (8). Some of these have greater sensitivity than the optical lever, but the limiting factor in most SFM measurements is thermal noise rather than detector sensitivity, so the simplicity and reliability of the optical lever are advantages.

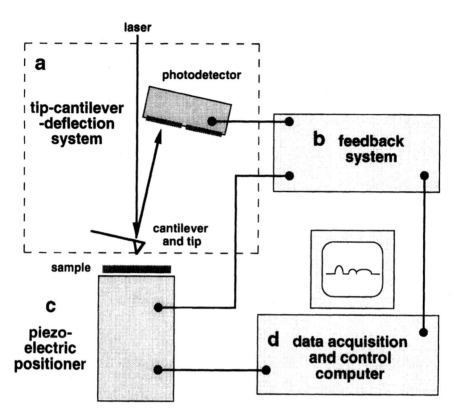

Fig. 1 SFM instrument diagram with four major components: (a) the tip-cantilever-deflection system; (b) the feedback system; (c) the piezoelectric positioner; and (d) the data acquisition and instrument control computer. All probe microscopes have components b, c, and d. They differ only in the type of probe-sensor utilized (component a in this diagram).

The force acting on the tip, F, is proportional to the cantilever deflection, Δx, by $F = -k\Delta x$, where k is the force constant of the cantilever (usually given in units of N/m or nN/nm). The ability of the SFM to measure small forces is thus limited by its ability to measure small deflections. For a typical force constant of 0.1 nN/nm and a minimum measurable deflection of 0.5 Å, the tip-sample force is about 5 pN. For comparison, the force required to stretch a typical chemical bond by 0.1 Å is about 500 pN, so the forces measured in SFM experiments are truly atomic. In practical experiments considerably larger forces (≥ 100 pN) are used because of noise and attractive tip-sample forces.

With good experimental design, the largest source of noise in SFM measurements is fluctuation of the cantilever caused by thermal energy. The mean square amplitude of thermal fluctuations can be estimated from the equipartition theorem and depends only on the cantilever's force constant, k, and temperature, T:

$$\langle \Delta x^2 \rangle = \frac{k_{\mathrm{B}}T}{k} \tag{1}$$

At room temperature and for a force constant of 0.1 nN/nm, the root-mean-square fluctuation $\sqrt{\langle \Delta x^2 \rangle}$, is about 2 Å. This is larger than the minimum deflection that can be measured by the optical lever, so the sensitivity of the SFM is limited by intrinsic thermal noise rather than by instrumental design.

As might be expected, the shape and properties of the tip play an important role in the quality of SFM images. Commercial SFM cantilevers are usually made of silicon nitride or silicon, by semiconductor microfabrication methods. They come in two shapes: a single-arm "diving board" or a double-arm V shape. Most varieties have a pyramidal tip at the free end, but some are provided without tips for specialty experiments or for users who make their own tips. The commercial tips are hydrophilic (they have a surface film of SiO_2) and vary in sharpness from about 1000 Å to about 100 Å end diameter. There is often considerable variation in tip size and shape within a single batch. The sharper tips generally yield higher resolution but are also more fragile and can damage delicate biological samples. It is possible to modify the as-received commercial tips by adding an electron beam–deposited tipette with a scanning electron microscope (9). Such "e-beam" tips are hydrophobic and have relatively uniform shapes with end diameters of about 200 Å.

The signal from the cantilever–optical lever is sent to the feedback circuit (Fig. 1b), which controls the piezoelectric positioner on which the sample rests. The piezoelectric positioner or "piezo" is usually a hollow tube of piezoelectric ceramic with several electrodes coated onto its surface. By applying voltages to various combinations of these electrodes, the piezo can move in the x, y, or z direction with sub-angstrom precision. During imaging, triangle-wave voltages from the computer drive movements in the x and y directions, and a voltage from the feedback circuit controls movements in the z direction. The triangle-wave voltages cause the sample to scan in a raster pattern back and forth underneath the tip, so that each point in the field of view is profiled. The feedback circuit monitors the signal from the cantilever–optical lever and continually adjusts the height of the sample to cancel any changes in tip-sample interaction. The voltage needed to maintain constant interaction is proportional to the height of the sample at each point. The computer stores this height information as a function of xy position, thus creating the most basic SFM image. Most SFMs can display such topographic maps in several different ways: as a gray-scale image, a false color image, a three-dimensional surface, etc.

B. Modes of Operation

SFM instruments can operate in several imaging modes: *contact mode*, in which the tip is in continuous contact with the sample surface; *tapping mode*, in which the tip makes intermittent contact with the sample; and *noncontact mode*, in which the tip oscillates a short distance above the sample and makes no contact at all.

1. Contact Mode Imaging

Contact mode images can be collected on samples in high vacuum, in ambient air, in nonaqueous liquids, in aqueous physiological buffers, in concentrated electrolyte solutions, or in any other fluid medium. When an optical lever is used to measure cantilever deflection, the instrument is set up so that the laser reflects from the end of the cantilever and hits the photodetector. The normalized difference, $(V_{top} - V_{bottom})/(V_{top} + V_{bottom})$, in output of the top and bottom photodiode segments is the primary signal sent to the feedback circuit. In *topographic contact mode*, the feedback circuit adjusts the height of the sample to keep this signal constant, that is, to keep the cantilever at a constant deflection with a constant force of contact beneath the tip.

In most SFMs the feedback circuit can be adjusted for fast response (high gain) or slow response (low gain). In topographic mode the feedback is set for fast response, and the sample is scanned slowly, so the tip follows the sample surface almost perfectly. In *deflection mode*, the feedback circuit is set slow and the tip is scanned rapidly, so that the feedback cannot cancel all cantilever deflections. The computer stores these uncanceled deflections, and the SFM image is a map of cantilever deflection versus *xy* position. Deflection mode is used mainly for very high resolution images of hard samples with low relief (such as well-ordered, atomically smooth substrates), which are not easily damaged by high forces. The deflection map has high sensitivity, which helps make shallow subangstrom features more visible. Slight variations of deflection mode, in which the feedback gains are set low even on samples with high relief, are also sometimes useful for highlighting features with low relief in the presence of features with high relief.

In a contact mode experiment the tip slides over the sample, which generates frictional forces. These lateral forces cause the cantilever to twist or buckle, which deflects the optical lever in different directions than would a vertical force. This allows lateral forces, which contain useful information, to be distinguished from vertical forces and measured separately. By using a four-element photodetector that is split both in the vertical direction and in the horizontal direction, a friction signal, $(V_{\text{right}} - V_{\text{left}})/(V_{\text{right}} + V_{\text{left}})$, can be collected at the same time as the topographic signal. Differences in friction usually indicate differences in chemical composition or structure, so friction imaging gives SFM some chemical sensitivity. In many applications the topographic image and friction image are collected simultaneously and displayed side by side. This allows parts of the sample that differ in frictional properties to be correlated with topographic features.

On soft and resilient samples it is sometimes possible to measure differences in mechanical compliance from one feature to another (10,11). When the tip is pressed into a soft sample the surface yields slightly, and the deflection of the cantilever is less than it would be on a hard surface. If Δh is the change in the height of the sample and Δx is the corresponding deflection of the cantilever, then $\Delta x = \Delta h/(1 + k_{\text{c}}/k_{\text{s}})$, where k_{c} is the force constant of the cantilever and k_{s} is the effective force constant of the compliant surface. The measured deflection, Δx, is thus smaller on a soft part of the sample (small k_{s}) than on a hard part (large k_{s}), and the elastic properties of the sample can be mapped. Like lateral force images, compliance maps are often compared with topographic maps to identify the chemical or mechanical properties of specific topographic features.

2. Tapping Mode and Non-Contact Mode Imaging

Contact mode is usually best for creating high-resolution images, but it has one great disadvantage: the frictional forces generated by the continuous pressure between tip and sample can be very destructive. When imaging in ambient air, the sample surface is almost always coated with a thin layer of adsorbed water that forms a meniscus between the tip and the sample and generates a large, attractive capillary force. This means that it is not possible to make contact mode images at arbitrarily low force: the minimum practical force is the value of the capillary attraction itself. For a typical capillary force of 10 nN and a contact area of $1 nm^2$, the pressure beneath the tip is $10^{10} Pa$, or about 100,000 atmospheres. This high downward pressure generates large frictional forces that can damage or sweep aside many biological samples and usually damage the tip as well.

Tapping mode and noncontact mode were developed to overcome this difficulty. Both use an oscillating tip to minimize contact and hence preserve delicate samples and sharp tips. The cantilever is usually driven into oscillation by a small dedicated piezo element somewhere in the cantilever mount or sometimes by oscillating the sample with the main positioning piezo. In tapping mode the amplitude of oscillation is relatively large (from a few nanometers to a few dozen nanometers), and the sample height is adjusted so that the tip lightly taps the surface at the bottom of each oscillation. The fact that the tip makes direct physical contact with the sample means that resolution is nearly as high as in contact mode. In both cases the forces responsible for imaging, direct van der Waals repulsions between atoms of the tip and atoms of the sample, are very short ranged and hence capable of high resolution. But because the tip is only in brief contact with the sample, lateral forces are effectively eliminated and the sample need only be robust enough to survive the less harmful vertical force of the taps.

The tip-sample interaction at each tap affects the oscillation of the cantilever, usually by reducing its amplitude and changing its phase. The primary feedback signal in tapping mode is generated by rectifying and filtering the AC component of the raw optical lever signal into a DC level that is proportional to the amplitude of cantilever oscillation. In most cases the tip sticks to the sample on each tap, which is why a large amplitude of oscillation is needed: if the amplitude is too small the tip will be captured, ending the experiment.

Noncontact mode is similar to tapping mode, but the amplitude of oscillation is usually much smaller, on the order of an angstrom or less, and the tip never makes direct contact with the sample. Instead, the oscillating tip is brought close to the surface (typically within a few nanometers), where long-range dispersion forces or electrostatic forces become significant. The presence of this external force field alters the effective equation of motion of the cantilever: the gradient of the force field adds to the effective force constant: $k_{effective} = k_{cantilever} - dF/dz$, where F is the long-range force field due to the sample, and z is the height of the tip above the sample. This shifts the cantilever resonance frequency and alters both the amplitude and phase of oscillation. As in tapping mode, these changes are converted to a DC voltage that is used as the primary feedback signal. Because the tip never comes in contact with the sample and the quantity actually measured is the force gradient rather than the force itself, the interaction between tip and sample can be extremely small, and any well-defined surface, no matter how soft, can in principle be imaged. However, because the forces responsible for imaging are long ranged, resolution in noncontact mode is lower than in contact or tapping mode. Also, as described in the next section, the presence of attractive forces makes the tip unstable when it comes too close to the surface. This can make noncontact imaging unpredictable or difficult to distinguish from tapping mode.

III. TIP-SAMPLE INTERACTIONS

A. Measuring Tip-Sample Interactions: Force Curves

The most important forces for SFM imaging may be grouped into five major categories, roughly in order of decreasing range: capillary forces, dispersion forces, electrostatic forces, structure forces, and contact/elastic forces. The first three types are relatively long ranged (nanometers to micrometers) and are usually attractive. The latter two are short ranged (angstroms to nanometers) and mainly repulsive. In contact mode, where the inertia of the cantilever is negligible, the sum of the repulsive and attractive forces must exactly balance the force due to cantilever bending at every point on the surface

$$F_a + F_r - k\Delta x = 0 \tag{2}$$

where F_a and F_r are the attractive and repulsive forces, respectively, k is the force constant of the cantilever, and Δx is the cantilever deflection from its equilibrium height, h_{tip0}: $\Delta x = h_{tip} - h_{tip0}$. In tapping or noncontact mode, inertial forces due to accelerations of the cantilever are also present.

The primary method by which tip-sample forces are measured in the SFM is called a *force curve*. When taking a force curve the xy scan and the feedback circuit are disabled, and the sample is raised until it touches the tip and the cantilever bends and is then retracted again. The force curve is a plot of cantilever deflection versus tip-sample separation, as shown in Fig. 2. At the beginning of the curve, in the region labeled A in Fig. 2, the cantilever is unbent and the sample is far from the surface. As the sample rises, long-range attractive forces (most often capillary or electrostatic) begin to act on the tip, and it deflects downward slightly, although this deflection is often too small to see. At point B the attractive force overcomes the ability of the cantilever to prevent motion, and the tip suddenly jumps to make contact with the surface. The existence of this "snap-down" means that it is not possible to control the height of the tip when it is too close to the surface: the tip always becomes unstable at small tip-sample separations. The point at which the tip becomes unstable depends on the stiffness of the cantilever and the size and nature of the attractive interactions. When the surface is covered with a liquid film, snap-down occurs when tip and liquid layer meet and a meniscus forms. In the absence of a liquid layer, snap-down occurs when the gradient of the attractive force (electrostatic or dispersion) exceeds the force constant, k, of the cantilever. Typical snap-down distances are from a few nanometers to an angstrom or two. With a very stiff cantilever ($k > 10$ N/m or so) and a clean surface with no capillary layer, snap-down may be completely absent.

Snap-down is one of the main sources of difficulty for noncontact SFM. If a soft cantilever is used to increase sensitivity to weak tip-sample forces, the tip must be kept well above the snap-down height at all times. This means that the interactions responsible for image formation must be relatively long ranged and will not yield high resolution. On the other hand, if a stiff cantilever is used to avoid the snap-down problem, sensitivity is lost and the tip may not detect weak noncontact forces. In this case the feedback circuit will lower the tip until it begins to tap the sample, and a nominally noncontact experiment becomes a tapping mode experiment by default. It is not always easy to tell whether the SFM is operating in noncontact or in tapping mode.

At point C the cantilever is unbent, and the only forces acting in the tip are attractions and repulsions from the sample itself. This is the point where the tip would have first touched the sample if not for attractive forces. In region D the cantilever bends upward as the sample continues to rise. On this part of the curve, the change in the deflection of the cantilever is identical to the upward movement of the sample

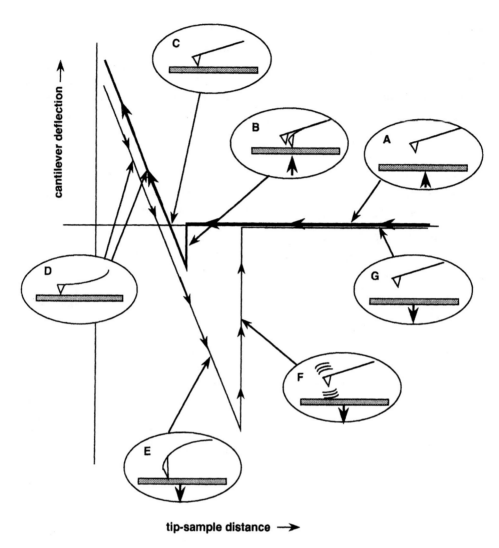

Fig. 2 Standard force curve. In a standard force curve the x axis is the difference in the heights of the tip and the sample, $s = h_{tip} - h_{sample.}$, and the y axis is either cantilever deflection, $\Delta x = h_{tip} - h_{tip0}$, or the negative of cantilever force, $-F = k\Delta x$.

(assuming a hard sample surface), so that the slope of the curve, $\Delta x / \Delta s$, must be unity. This fact is routinely used to calibrate the optical lever detector (or whatever other detection scheme is used). If the detector has already been calibrated and the slope of the curve is less than unity (or if the curve is not linear in this region), the sample is soft and is deforming under the pressure of the tip. This effect can be used to investigate sample elastic properties and is the basis for compliance imaging (10,11) (Sec. II.B.1).

After rising to its maximum height, the sample changes direction and begins to retract. In region D the force curve follows the same path as on the upward part of the cycle, but in region E the tip continues to deflect downward, creating a large *adhesion triangle* that is not present in the advancing part of the curve. The adhesion triangle is caused by the same attractive forces that caused snap-down, which now pin the tip to the surface. As the sample retracts the cantilever bends further and further downward until the bending force exceeds the maximum adhesion force and the tip snaps up at point *F.* The size of the adhesion triangle is a good measure of attractive forces at the surface and hence of the pressure beneath the tip when imaging in contact mode. If the height of the triangle is $\Delta x_{\text{adhesion}}$, the force under the tip for an unbent cantilever is $k \, \Delta x_{\text{adhesion}}$. In tapping mode the tip has inertia as well as potential forces, so the adhesion force is a *minimum* estimate of the momentary force exerted on the sample during each tap. Because adhesion force is directly responsible for damage to tip and sample in either contact or tapping mode, the adhesion triangle is one of the main reasons for taking force curves.

Not all force curves look like the one in Fig. 2. If there is a strong repulsive force between tip and sample, most often due to electrostatic repulsion, the curve will not have a snap-down region and may not have an adhesion triangle. The shape of the curve in the region near the surface can then be used to quantify and investigate the nature of the tip-sample forces. In other cases the force curve can have multiple adhesion triangles. This always indicates the presence of something soft and sticky on the surface and tip, so that there is a long neck of "glue" between the tip and sample. Each of the several adhesion triangles is caused by breaking some part of this glue neck, and the tip is not fully separated from the surface until all the triangles have been passed.

Sticky surfaces are always bad for imaging experiments but are often the main subject of interest in force measurement experiments with the SFM. If the stickiness is caused by an interesting interaction, say, because an antibody is adsorbed to the surface and its antigen is adsorbed to the tip, multiple adhesion triangles may contain unique information. Such curves have been determined to measure the force required to rupture single ligand-receptor interactions (5–7).

B. Origin and Properties of Tip Sample Forces

1. Capillary Forces

The capillary force arises when a meniscus forms between the tip and a liquid film on the sample surface. It is a sum of two contributions, both of which ultimately originate in the surface tension at the interface between two fluids (12). The first contribution, F_{st}, is due to the surface tension at the line of contact between the tip and the liquid:

$$F_{\text{st}} = 2\pi R \gamma \cos \phi \tag{3}$$

where R is the radius of the ring of contact, γ is the surface tension, and ϕ is the angle between the surface at the contact point and the z axis (Fig. 3). The second contribution is called the Laplace force and is due to the curvature of the liquid-air interface in the meniscus region. The curved interface creates a pressure difference, $P_{\text{in}} - P_{\text{out}}$, between the capillary layer and the bulk medium that pulls the tip downward. An estimate of the Laplace force is given by (12)

$$F_{\text{L}} = 4\pi R \gamma \cos \theta \tag{4}$$

where θ is the tip-interface contact angle.

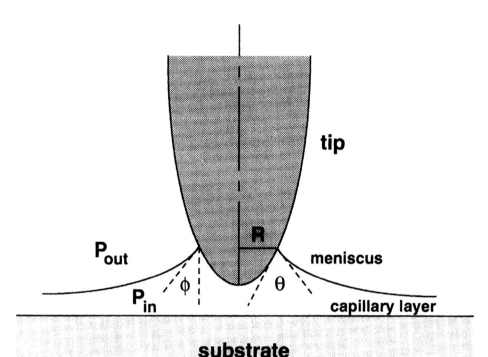

Fig. 3 Meniscus between the tip and liquid layer at the sample surface: R is the radius of the ring of contact, ϕ is the angle between the interface and the vertical axis at contact, and θ is the angle between the interface and the tip surface. The concave curvature of the interface means that the pressure inside the layer, P_{in}, is lower than the ambient pressure, P_{out}, giving rise to a downward Laplace force on the tip.

The sum of both contributions, F_{st} and F_L, can be quite large, even for a reasonably sharp tip: for $\phi = \theta = 0$, $R = 20$ nm, $\gamma = 72$ dyne/cm (air-water interface), the total capillary force, $F_c = F_{st} + F_L$, is 27 nN. For a contact area of, say, 1 nm^2, this corresponds to a pressure of about 3×10^{10} Pa (3×10^5 atm), large enough to flatten or destroy almost any biomolecule (or tip). Forces of this magnitude are easily generated in actual imaging experiments if steps are not taken to eliminate capillary layers.

Even a very thin liquid layer, such as a molecular monolayer of water, is enough to generate a large capillary force. When imaging in ambient air, water layers are almost always present, and adhesion forces are difficult to avoid. When low force is needed (as in virtually all experiments with biological molecules) it is common practice either to dehumidify the sample by imaging in a dry box or to immerse the sample in a liquid that is miscible with water. Even these measures may not be sufficient to eliminate capillary forces: unless the sample has been carefully cleaned and has not been exposed to ambient conditions for long, organic material will be present that will generate smaller but still significant tip-sample adhesion. The best results are obtained when the substrate has been carefully cleaned (or freshly cleaved, in the case of a molecularly layered substrate like mica or graphite), and imaging is done in a liquid that is a good solvent for both water and organics, such as propanol. For samples compatible with these conditions, it is often possible to reduce capillary forces effectively to zero.

Biological samples are usually destroyed by such "good" solvents, so the best choices are imaging in clean buffer solutions in a carefully cleaned fluid cell or, for imaging in air, placing the entire SFM into a closed dry box with a steady, slow flow of dry nitrogen. When using mica substrates, which are exceedingly hydrophilic and hard to dry out, we have found it necessary to put a drying tube filled with solid P_2O_5 into the gas inlet line to get completely dry conditions and to allow up to an hour for surface water to evaporate.

Finally, a capillary or surface adhesion force sometimes exists between the probe tip and the sample molecules themselves. In mild cases this can simply be tolerated. In severe cases there is little that can be done besides changing the chemical composition of the tip surface.

2. Dispersion Forces

The dispersion force is a nearly universal interaction between any two polarizable bodies caused by correlated, fluctuating induced dipoles. For example, if the polarizable bodies are molecules the dispersion force is the familiar van der Waals attraction. If the two bodies are larger—in the present case we are interested in the interaction between the probe tip and the sample substrate—the interactions are less familiar but still have the same fundamental origin: the presence of a random dipole in one body, generated by thermal energy or a quantum fluctuation, creates an electric field that induces a corresponding dipole in the other body. Because the two dipoles are correlated, their interaction does not average to zero, even though the average dipole in each body is zero. Virtually all materials are polarizable to some extent, so the dispersion interaction is always present and almost always attractive. The only exception, as will be shown in the following, is when two bodies of very different properties interact through a medium with intermediate properties, e.g., an air bubble interacting through water with a metal. In cases like this it is possible to find conditions were the dispersion interaction is zero or even repulsive, but in all normal cases the dispersion interaction is attractive. The dispersion force is usually not large (compared with capillary or electrostatic forces) but can cause snap-down instabilities and tip-sample adhesion even when capillary and electrostatic forces are negligible. The dispersion force therefore often sets the lower limit of imaging forces in contact and tapping mode.

The dispersion interaction between two arbitrarily shaped bodies is very difficult to calculate exactly (12). But the calculations are greatly simplified if each body is thought of as a collection of small polarizable volumes and if it is assumed that these volumes interact in a pairwise-additive fashion. The dispersion potential energy between two point polarizable volumes separated by a distance r is $v_{\text{disp}}(r) = -C/r^6$, which can be recognized as the $1/r^6$ part of the Lennard-Jones potential (12). For simple geometries it is possible to integrate this potential over all pairs and find the total interaction between the bodies. Comparison with more elaborate calculations shows that this approximation is very good for separations smaller than about a micrometer. At larger separations the time required for an electromagnetic disturbance to travel between bodies allows dipoles to become decorrelated (the so-called retardation effect), and the true dispersion interaction is smaller than would be pre-

Table 1 Dispersion Interaction Between a Tip and a Planar Surface

	Energy of Interaction	Force	Force at Contact (nN) ($d = 3\,\text{Å}, A = 10^{-20}\,\text{J}$)
parabole	$w(d) = \dfrac{-AR}{6d}$	$F(d) = -\dfrac{AR}{6d^2}$	$0.19\,\text{nN}$ $R = 10\,\text{nm}$
cone	$w(d) \cong -\dfrac{A\tan^2\gamma}{6}\left[\ln\dfrac{d}{h}\right]$ for $\dfrac{d}{h} \ll 1$	$F(d) \cong -\dfrac{A\tan^2\gamma}{6}\dfrac{1}{d}$	$0.055\,\text{nN}$ $\gamma = 45°$
pyramid	$w(d) \cong -\dfrac{2A\tan^2\gamma}{3\pi}\left[\ln\dfrac{d}{h}\right]$ for $\dfrac{d}{h} \ll 1$	$F(d) \cong -\dfrac{2A\tan^2\gamma}{3\pi}\dfrac{1}{d}$	$0.070\,\text{nN}$ $g = 45°$
cusp	$w(d) \cong -\dfrac{A}{24R^2}\left(\dfrac{1}{2}h^2 - 4dh\right)$ for $\dfrac{d}{h} \ll 1$	$F(d) \cong -\dfrac{A}{8R^2}\left[h - \left(\dfrac{13}{3}\right) + 4\ln\dfrac{h}{d}\right]d$	$4.1 \times 10^{-7}\,\text{nN}$ $h = 3\,\mu m, R = 3\,\mu m$
hollow pyramid	$w(d) = -\dfrac{2A\tan^2\gamma}{3\pi}\ln\left(\dfrac{d}{d+t}\right)$ for $\dfrac{d}{h} \ll 1$	$F(d) = -\dfrac{2A\tan^2\gamma}{3\pi}\left(\dfrac{t}{d(d+t)}\right)$	$0.070\,\text{nN}$ $\gamma = 45°, t = 0.5\,\mu m$
cylinder	$w(d) = -\dfrac{Ax^2}{12d^2}$ for $\dfrac{d}{h} \ll 1$	$F(d) = -\dfrac{Ax^2}{6d^2}$	$620\,\text{nN}$ $x = 0.1\,\mu m$

dicted by pairwise additivity. Table 1 shows the results of calculations made assuming pairwise additivity for a variety of SFM tips with standard shapes interacting with a flat, planar substrate. The Hamaker constant, A, which appears in these formulas is related to the constant C in $v_{\text{disp}}(r)$:

$$A = \pi^2 \rho_1 \rho_2 C \tag{5}$$

where ρ_1 and ρ_2 are the densities of the two interacting bodies. For a given pair of materials it is possible to estimate the value of A from dielectric constants and refractive indexes (12):

$$A \cong \frac{3}{4} k_B T \left(\frac{\varepsilon_{tip} - \varepsilon_{med}}{\varepsilon_{tip} + \varepsilon_{med}} \right) \left(\frac{\varepsilon_{sub} - \varepsilon_{med}}{\varepsilon_{sub} + \varepsilon_{med}} \right)$$

$$+ \frac{3h\nu_e}{8\sqrt{2}} \left[\frac{\left(n_{tip}^2 - n_{med}^2 \right) \left(n_{sub}^2 - n_{med}^2 \right)}{\left(n_{tip}^2 + n_{med}^2 \right)^{1/2} \left(n_{sub}^2 + n_{med}^2 \right)^{1/2} \left[\left(n_{tip}^2 + n_{med}^2 \right)^{1/2} \left(n_{sub}^2 + n_{med}^2 \right)^{1/2} \right]} \right]$$

$$\tag{6}$$

where ε_{tip}, ε_{sub}, and ε_{med} are the dielectric constants of the tip, substrate, and the medium in which they are immersed, respectively, and n_{tip}, n_{sub}, and n_{med} are the corresponding refractive indexes. The quantity ν_e is an electronic transition frequency that is chosen to make the best overall match with experiment, usually around 3×10^{15} Hz. Typical values of A range from 10^{-20} to 10^{-19} J (12). From Eq. (6) it can be seen that whenever the dielectric constants and refractive indexes of the two interacting bodies are both smaller than or both larger than the dielectric constant and refractive index of the medium, A will be positive, and the dispersion force will be attractive. When one of the two bodies has ε and n less than the medium and the other has ε and n greater than the medium, A will be negative and the dispersion force will be repulsive. If either body has values of ε and n that match the medium, the dispersion force will be zero. It is therefore possible in principle to cancel or even reverse the dispersion force. In practical situations, where the nature of both the substrate and the medium is dictated by other considerations, this is usually difficult to do.

All the force laws in Table 1 have much weaker dependence on tip-sample distance than the $1/r^6$ for the interaction of point particles. The tips with sharp ends (the cone, pyramid, and cusp) have force laws that vary more slowly with d than the blunt shapes (paraboloid, cylinder) and have smaller calculated contact forces. This reflects the fact that the surface interacts most strongly with the bottom part of the tip, so shapes that have only a small mass near the surface interact less, even though they may widen out rapidly at their tops. In particular, the cusp-shaped tip has a force law that is essentially linear for small values of d, which is too weak to cause snap-down. This is somewhat misleading, however, because no real tip can be cusp shaped all the way to the very end. The magnitude of the calculated adhesion force varies from negligibly small to many nanonewtons, in agreement with general experience. Because pairwise additivity was assumed in arriving at the formulas in Table 1, more elaborate tip shapes can be treated by combining several of these basic shapes and then adding the forces acting on each piece.

3. Electrostatic Forces

Electrostatic forces depend on the presence of free, unbalanced charge on tip, sample, or both. Electrostatic forces can be very strong and very long ranged. At their strongest, they dominate all other forces, but in most real situations they are small or modest even when considerable free charge is present, due to screening by countercharge. When imaging in air, static charges can build up on the tip as it slides over the surface or can be present on the sample when it is initially placed in the SFM. If the charge on the sample is distributed uniformly, the resulting electric field is nearly uniform over even macroscopic distances, and the force acting on the tip is almost independent of tip height. A strong, very slowly varying force is the hallmark of electrostatic interactions in air. In many SFMs the sample or tip is grounded to reduce buildup of static charge.

Electrostatic forces are most common and most interesting when imaging in aqueous solutions, in which tip, substrate, and sample molecules may have ionized groups on their surfaces. In ionic solution these charges are surrounded by Debye-Hückel atmospheres, which strongly screen all electrostatic interactions. The main result is that although electrostatic interactions are long ranged in air and other neutral media, they are quite short ranged in ionic solution. The Debye-Hückel screening length is

$$l_D = \sqrt{\varepsilon \varepsilon_0 k_B T / (2000 N_A e^2 I)} \tag{7}$$

where ε is the dielectric constant of the solvent (water, in this case), ε_0 is the permittivity of free space (8.85×10^{-12} coul^2/(Jm)), N_A is Avagadro's number, e is the charge of an electron (in coulombs), I is the ionic strength, $I \equiv \frac{1}{2}\sum_{\text{ionic species}} z_i^2 c_i$, and z_i and c_i are the valence and concentration (in molar units) of the ith ion. For typical buffer concentrations of about 0.1 M, the screening length is about 1 nm, so electrostatic interactions are significant only when the tip is quite close to the sample.

4. Structure Forces

Structure forces are caused by changes in the organization of solvent molecules near sample surfaces. In the neighborhood of a liquid-solid interface, the statistical packing of liquid molecules often changes dramatically. This is especially true for polar molecules such as water near a charged surface like mica: the water molecules are strongly attracted to the surface and tend to adopt an ordered, solid-like arrangement. This ordering extends for a few molecular diameters into the bulk liquid. The same thing may also happen near the tip surface, especially if it is charged, as standard silicon nitride tips are. (They are thought to have a thin film of SiO_2 on their surfaces.) When tip and sample approach each other, the two ordered layers collide and produce forces. The variation in force with tip-sample distance can be complicated and may even oscillate between attractive and repulsive forces at different tip-sample separations (12). Structure forces are quite short ranged and quite weak and have been observed only rarely in SFM experiments.

5. Contact/Elastic Forces

Contact forces, as the name implies, are the van der Waals repulsions that occur when the tip and sample are in direct contact. They are very short ranged (subangstroms) and, if the tip is atomically sharp and attractive forces are small, give the SFM atomic resolution. Usually, however, attractive forces are large enough to damage a single-atom tip very quickly or to cause elastic deformations of the sample that reduce resolution. The main contact interactions are then spread over at least a few atoms at the tip apex, and resolution is in the range from a few angstroms to a few nanometers.

IV. SAMPLE PREPARATION AND IMAGING

Now that the basics of instrumentation have been covered, we turn attention to practical methods for imaging biological molecules by SFM. The emphasis in this section is on imaging individual biomolecules, but methods for imaging membranes, whole cells, two-dimensional crystals, and other large samples are similar. Only molecules that are well adsorbed to a solid substrate can be imaged by SFM. Molecules that are not bound or are so loosely adsorbed that they move under the influence of the scanning tip will be invisible. The main need is therefore to find ways to attach the sample molecules to a substrate.

The most common methods are similar to those used in electron microscopy. For most types of samples, the overall preparation and imaging process can be divided into four steps: (a) preparing a solution containing the molecule or molecules of interest, (b) depositing the molecule on a suitable substrate, (c) postdeposition treatment (drying, rinsing, fixation, etc.), and (d) SFM imaging. The first step, preparing a sample solution, is no different than preparing any other biochemical solution, except that the concentration of the biomolecules, the pH of solution, the type of buffer, the ionic strength, and the presence or absence of polyvalent metal ions may all affect the adsorption of the target biomolecule to the substrate and hence may have a large effect on the outcome of the experiment. The second step, depositing the target molecules, is usually just a matter of exposing the substrate to the sample solution for a short time. If the substrate and the solution conditions (concentration, pH, etc.) have been well chosen, the target molecules will spontaneously adsorb and will remain adsorbed throughout postdeposition treatments and imaging. The third step, postdeposition treatment, is not needed in some cases (for example, when imaging native molecules under buffer) and usually involves only simple rinses or short drying steps. The fourth step is the imaging itself.

A. Some Substrates and Their Properties

For high-resolution SFM imaging the substrate should be atomically smooth and clean, should strongly bind the molecules of interest, and should be easy to prepare and use. Several common materials satisfy most of these requirements, including mica, polished glass, annealed gold, and graphite.

1. Mica

Mica is by far the most commonly used substrate for biological samples. Crystals of mica have a layered structure that cleaves easily without special equipment (see Appendix A). The cleaved surface is atomically flat, chemically clean, and carries a negative charge in solution. Mica can be purchased from electron microscope supply companies and used directly. All grades and varieties seem to be useful, but high-grade red or ruby mica seems to give the most consistent results.

The 1,0,0 planes in the mica crystal contain negatively charged silanol groups and positively charged potassium counterions. When these planes are exposed to water in the cleaved crystal the potassium ions dissolve away, leaving behind a lattice of negatively charged pockets. Adsorption of biomolecules to this highly charged surface is often controlled by electrostatic interactions: positively charged molecules usually adsorb strongly, but negatively charged molecules, especially nucleic acids, do not. A standard way to overcome this problem is to add a small, doubly charged cation such as Mg^{2+} to the sample solution. If the cation is small enough to fit into mica's charged pockets, the surface gains a net positive charge. The literature discusses methods by which mica can be quantitatively loaded with magnesium (13), but millimolar concentration of the cation in the sample buffer is sufficient for good adsorption in many applications (14). A wide variety of metal cations have been investigated for their ability to aid binding of nucleic acids (15,16).

Adsorption to mica is also strongly influenced by the total ionic strength and the pH of the imaging solution. As ionic strength increases, electrostatic interactions are increasingly screened out. This usually decreases binding by positive molecules and increases binding by negative molecules. As pH is lowered, acidic groups are protonated and binding to normal, negatively charged mica decreases for positive molecules and increases for negative molecules. Adsorption of biomolecules to mica may be inhibited by certain common buffer components [glycerol, for example (13,15)] and it is sometimes necessary to replace each buffer component systematically until the "surface blocking" material (or materials) is found.

The fact that binding to mica depends on solution conditions can sometimes be used to advantage. One example is the imaging of transcription of DNA by RNA polymerase. During transcription the RNA polymerase complex must move along the DNA template, but for good SFM imaging both the DNA and RNAP must be well adsorbed to the substrate and hence immobilized. These seemingly incompatible requirements were satisfied by first depositing transcription complexes on the substrate in an imaging buffer that strongly favors adsorption of both DNA and RNAP and taking an initial SFM image. Then the buffer was changed to a reaction buffer that favors adsorption of RNAP but not DNA, allowing the reaction to proceed. Then the reaction was stopped by reintroducing the imaging buffer, and a new SFM image was taken to record the changes. By alternating buffers in this way, a time-lapse "movie" of the transcription process was obtained (2).

2. Glass Coverslips

Glass coverslips of the type used for light microscopy are widely used as SFM substrates. They are not atomically smooth; they typically have an RMS surface roughness of a few angstroms, enough to hide small molecules, but are flat enough for most applications with proteins or other large molecules. Like mica, many biomolecules spontaneously adsorb to glass. A clean glass surface contains silanol groups that carry a negative charge, but the charge density is much smaller than on mica, especially at low pH, and the adsorption process seems to be less dominated by electrostatic interactions. New coverslips are often coated with a thin film of oil and must be thoroughly cleaned before use. Glass surfaces can also be chemically functionalized with silanes such as aminopropyltriethoxysilane (APTES) (15,17,18), which can give greater control over surface properties and greatly enhance surface adsorption. But such surface layers can also add to surface roughness and cause fouling and capillary interactions with the tip.

Glass substrates are commonly used for SFM imaging of membrane bilayers. Methods for adsorbing uniform bilayers on glass have been worked out (e.g., 19), and the bilayer makes an atomically smooth background against which to view membrane proteins. The deposition methods are essentially the same as for proteins or nucleic acids, but the substrate is exposed to a suspension of protein-containing lipid vesicles instead of a simple solution of biomolecules. Adsorption is much faster if the lipids are chosen to be in the fluid state at the deposition temperature. Very soft, fluid lipid mixtures such as egg PC are often invisible to the SFM because they are so fluid that the tip penetrates to the substrate and plows through them. Even so, such membranes can act as matrices for imaging membrane proteins. The deposition and distribution of membranes doped with a small amount of fluorescent lipid can be monitored independently by epifluorescence light microscopy.

3. Gold

Gold is widely used as a substrate for scanning tunneling microscopy (STM) because it is highly conductive, atomically flat, chemically inert, and quite easy to prepare. Gold surfaces in contact with electrolyte solutions have been extensively characterized by electrochemical methods. They are also easy to functionalize chemically using the well-known thiol chemistry (20,21). In SFM experiments a conductive substrate is not usually necessary and gold is less often used, but it is still common in situations where conductivity is required for reasons other than the imaging itself (e.g., electrochemical experiments) or when other substrates are not adequate for some reason.

Atomically flat gold substrates can be prepared either by flame annealing or by evaporation onto a smooth substrate in vacuum. The flame annealing process is quite simple and fast and requires only minimal equipment (see Appendix B). It produces millimeter-scale crystalline 1,1,1 surfaces that are atomically flat except for occasional steps. Its primary disadvantage is that the surfaces come in the form of flat facets on small spheres, which can be hard to mount and handle in some experiments (Appendix B). In the vacuum evaporation method, gold atoms are coated onto a heated surface (usually mica, glass, or silicon), where they form wide, crystalline, 1,1,1 plateaus. The atomically flat parts of the evaporated surfaces are much smaller than the flame-annealed surfaces (typically about 0.1 μm), but the substrate is much easier to handle. The gold surface is chemically clean immediately after preparation by either method but quickly becomes contaminated when exposed to air and must be used immediately.

One advantage of a conductive substrate, even for SFM imaging, is the ability to control deposition and surface properties by electrochemistry. By setting up the SFM experiment in an electrochemical cell and applying an external potential, a conductive substrate can be charged and its chemical composition and surface properties altered in a controlled, reproducible way (22). The deposition of charged biomolecules can be monitored externally by current measurements. Electrodeposition also produces very strong bonds between the sample molecules and the substrate, which is usually an advantage in SFM imaging.

4. Graphite

Another useful conductive substrate is crystalline graphite, often called highly oriented pyrolytic graphite or HOPG. HOPG is very conductive, very hydrophobic, very inert, and, like mica, is a layered material that can be cleaved easily. The cleaved surfaces are infinite sheets of fused benzene rings, with a honeycomb pattern of carbon atoms. Like flame-annealed gold and mica, HOPG surfaces are atomically flat over macroscopic distances, except for step edges.

In the early days of probe microscopy, HOPG was the most popular substrate for STM of biomolecules. Then it was discovered that some reported images on HOPG were artifacts: even in the absence of biological samples, HOPG can display features that are easily mistaken for biomolecules by the unwary (23). HOPG has since been out of favor. Despite this history HOPG has a unique combination of properties and is the substrate of choice for some types of experiments, if used with caution and appropriate control experiments.

B. Imaging in Air

Once the sample has been deposited on a suitable substrate, it is ready for imaging. The two main options for imaging biomolecules are imaging in air and imaging in buffer. Imaging in air requires that the sample be dried onto the substrate surface, which means that the molecules have lost most of their water, and buffer salts, dissolved carbonates, and other solution impurities are deposited on and around them. One positive consequence of this is that the molecules are often much harder and more resistant to damage and movement by the tip. In solution, contact mode imaging of most biomolecules is impossible: they are destroyed or swept aside even at the smallest practical imaging forces.* But the opposite is true in air: most biomolecules can be reliably imaged in contact mode and are not visibly damaged after many imaging passes. Likewise, in solution it is sometimes difficult to find conditions under which the sample adsorbs strongly enough to avoid being knocked loose by the tip. When imaging in air, adsorption must only be strong enough to keep the sample in place during the drying process. This is apparently a much weaker requirement, and deposition problems are less important in air than in solution.

On the other hand, the height of dried samples is consistently less than expected, suggesting that they collapse upon removal of water. Damage can also be done by surface tension at the edges of the last droplets of water to evaporate from the sample. These problems are also common in electron microscopy, and experience shows that in most cases the main features of the molecules are preserved. Distortions caused by drying and surface tension can sometimes be reduced by fixing the sample before deposition, by adding around 1% electron microscopy (EM) grade glutaraldehyde (available from EM supply companies) to the sample solution and waiting a few minutes for the cross-linking reaction to take place. Glutaraldehyde cannot be used with TRIS buffers however and sometimes seems to interfere with adsorption to mica. Another method commonly used in EM preparations that also works for SFM is to fix at the rinse step by adding glutaraldehyde to the rinse water.

Resolution in air is usually limited by surface salts, sample deformation, and sample damage. Although dried samples are robust, resolution is generally no better than in solution: the small surface details are covered by salts, destroyed by surface tension or structural changes caused by drying, or flattened by high pressure under the tip. Strong capillary forces also make it difficult to use sharp tips. When capillary forces are large, the best images are often obtained with tips that are somewhat blunter than the sharpest ones available. In most SFM experiments these factors limit resolution to a few nanometers.

Capillary forces are the dominant attractive interactions between tip and sample in air. Even when the sample has been carefully dried and kept under zero humidity, organic films and other debris create capillary interactions that set the lower limit of imaging force. Clean buffers and highly purified biomolecules can reduce capillary forces but never eliminate them completely.

Most substrates used with biological samples are hydrophilic and will attract a thin film of water in ambient air. Even samples that have been carefully dried and

---

* There are exceptions to this rule, however, most notably the very high resolution images obtained on two-dimensional crystals of membrane proteins in solution.

appear dry to the eye will be wet on the microscopic scale. An easy way to avoid large capillary forces (and partial hydration of the sample) is to place the SFM in a dry box with a slow flow of dry gas. In many cases it is possible to watch the sample dry by taking force curves at intervals. When the surface is really dry the adhesion triangle should diminish dramatically.

If the sample molecules are pure (especially if there are no contaminant proteins in the sample) and adsorb strongly to mica, deposition is simple: the substrate is exposed to a solution of biomolecules for a short time (typically a few minutes), rinsed in a low-salt solution (or distilled water) to remove excess buffer salts, and dried in a gentle stream of clean, dry air or nitrogen. A good deposition procedure yields a reproducible, uniform distribution of sample molecules at the desired coverage. To avoid misidentifications and artifacts, it is crucial that the deposition procedure be reproducible, that the features identified as the sample molecules are plentiful in the images, and that their surface coverage changes appropriately when deposition conditions are changed. It is often necessary to vary solution parameters (sample concentration, ionic strength, pH, divalent cations) and deposition time systematically until reliable conditions are found. Air-dried samples usually cannot be kept for more than a few hours after deposition because of contamination buildup.

The details of the actual imaging experiment vary with the type of SFM instrument and will not be described here except for a few general notes: (a) When imaging in contact mode, always choose the softest cantilever available: even relatively tough, dried biomolecules are still easily damaged in contact mode. Likewise, the imaging force should always be carefully minimized before scanning. In tapping mode, such tip-induced damage is a smaller problem but it is still useful to minimize both tapping force and the amplitude of tapping oscillation. (b) At high magnification the sample may drift rapidly in the field of view (in the xy directions and in the z direction), which makes optimum imaging difficult. Drift is usually caused by temperature differences in the various parts of the SFM, which are in turn caused by warmth from fingers, the motors and laser in the SFM itself, and air currents in the room. The easiest way to reduce the problem is to enclose the SFM after mounting the sample and then wait for thermal equilibrium to be reached (about 20 min to an hour) before beginning to image.

C. Imaging in Buffers

Imaging in buffer has several inherent advantages over imaging in air: a biomolecule immersed in physiological buffer is more likely to be in its native conformation than is a dried molecule; changes in the system brought about by chemical reactions, the binding of another molecule, or by changes in the solution environment can be observed directly, on the same field of molecules; because there are no dried salts, the bare protein surface is accessible, and resolution is limited only by the capabilities of the SFM instrument itself. On the other hand, molecules in buffer solution are much more easily deformed or knocked loose from the surface by the probe tip than in air, and contact mode imaging is usually not possible. Even in tapping mode, molecules are often damaged, especially after multiple imaging passes.

SFM imaging in buffer requires a fluid cell that allows the sample, cantilever, and tip to be immersed in liquid. In one common design, the upper surface of the fluid cell is a rectangle of glass with an optical window for the laser beam of the optical lever. The tip is mounted on the underside of the window, and a soft rubber O-ring creates a seal between the sample substrate and the glass and defines the enclosed volume of the cell. Small fluid ducts in the glass allow the enclosed space to be filled externally. To prevent capillary forces it is important that the fluid cell be kept clean. The buffer environment prevents capillary forces due to water films, but organic contamination from dirty fluid cells can still contribute to tip-sample adhesion.

Deposition of sample molecules for imaging in buffer is essentially the same as for imaging in air without the rinse and drying steps: the substrate is exposed to the sample solution for a period of time and then placed in the fluid cell. As already discussed, the strength of deposition is affected strongly by the nature of the sample, the type of substrate, and the solution conditions.

Tapping mode imaging in liquid is quite different from imaging in air. In air, the oscillating cantilever experiences relatively little frictional drag and the force constant is usually high.* This means that the cantilever will have a large, sharp resonance at frequency $v_0 = (1/2\pi)\sqrt{k/m}$, where k is the force constant and m is the effective mass of the cantilever. For cantilevers designed to be used in air, the resonance frequency is typically in the hundreds of kilohertz. At frequencies near resonance the amplitude of oscillation is sensitive to small perturbations, so the interaction between tip and sample can be small and still be strong enough for imaging.

In liquids an oscillating cantilever experiences very strong hydrodynamic drag, and all its dynamical properties are altered. The cantilevers most commonly used for liquid imaging have force constants of about 0.4 N/m and, in air, have a sharp resonance at about 40 kHz. In water, the same cantilevers have a very broad resonance at about 10 kHz. The broadening of the resonance is the expected result of extra frictional drag, but the large decrease in resonance frequency can be explained only if the effective mass of the cantilever is much larger in water. Hydrodynamic theory shows this to be a reasonable explanation: a sphere moving in viscous liquid is surrounded by a large, circulating fluid flow field that reacts back on the sphere, causing it to behave as if its mass were larger (24). The presence of such a flow field surrounding SFM cantilevers in liquids is verified by the fact that whenever a cantilever approaches a substrate surface, its dynamical properties begin to change when it is still many micrometers from the sample. This long-range interaction between tip and sample is really the effect of the flow field interacting with the sample and usually causes still more viscous drag and further broadening of the resonance as tip and sample are brought together.

The main consequence of these hydrodynamic effects is that the sensitivity made possible by the sharp resonance in air is absent when imaging in water. From experience, the best results are obtained with cantilevers that have a thin profile and a moderately stiff force constant. Cantilevers with thick arms are more strongly affected by hydrodynamics. Cantilevers with small force constants can be hard to drive into oscillation and are relatively insensitive to tip-sample interactions.

---

* Stiff cantilevers are used for tapping mode in air because the capillary force would otherwise pin the tip to the surface. Stiff cantilevers usually do more damage to biological samples and are a disadvantage.

V. IMAGE INTERPRETATION FOR PROBE MICROSCOPY

As with all forms of microscopy, the images produced by probe microscopes requires interpretation. The purpose of any theory of imaging is to aid interpretation by clarifying the relationship between the features of the image and the properties of the sample. In this section three closely related theories of probe microscope imaging are outlined: the envelope theory (25), the dilation theory (26), and the Legendre transform theory (27). All three are geometric theories; that is, they answer the question: "What surface results when one hard object (the tip) is traced over another (the sample) and the position of the first is recorded at each point?" Because the tip, sample, and image are treated as hard, geometric objects, the theories explain only the most prominent and characteristic features of probe microscope images and ignore subtler effects such as deformation of the sample by the pressure of the tip or quantum effects on the atomic scale. Mathematically, they are classified as local, nonlinear theories. This is in sharp constrast to the theories that are most useful for light and electron microscopes, which are based on nonlocal, linear operations such as convolution and Fourier transformation. The difference in mathematical properties reflects fundamental differences in physics between probe microscopes and lens-based microscopes. As will be shown in the following, one consequence of these differences is that there is no entirely satisfying definition of resolution for probe microscopes.

A. Envelope Theory

The envelope theory begins by dividing the sample surface into a collection of sharp spikes, as in Fig. 4a (25). Any sample can be decomposed in this way, by matching the heights of the spikes to the height of the surface at each point. The spikes are infinitely sharp, so they act as an array of ideal probe tips, which create multiple (inverted, reversed) images of the tip (Fig. 4b), one at each location on the sample (Fig. 4c). The image is then the envelope or top surface formed by all such tip profile surfaces (Fig. 4d). In areas where the sample is flat, the image surface coincides with the true sample surface, but in areas where the sample is steep, the image differs significantly from the sample. As expected, sharp, narrow tips create faithful images, and blunt tips create distorted images. The most common distortions are broadening of sharp peaks and ridges and filling in of narrow holes and grooves (Fig. 4d).

Some of these tip distortions can be systematically removed by another envelope process (25). Instead of placing inverted and reversed tips on the sample surface, normally oriented tip surfaces are placed at each point on the image surface. The envelope formed by all such tip surfaces is then the reconstructed surface. If the image has no experimental noise and if the tip shape is known exactly, the reconstructed surface is identical to the true sample surfaces at all points where the tip touched the original surface. In sharp interior corners or narrow holes and cracks, where the tip cannot actually touch, the reconstruction process replaces the image with a bridging segment of the tip surface. The bridging surface is the highest possible surface consistent with the information contained in the image. In almost all cases the true surface lies somewhere below.

Finally, by exchanging the roles of the tip and sample, it is possible to reconstruct the shape of the probe tip itself from an image of a known sample. In this case, the sample acts as the effective tip, and the envelope formed by placing sample surfaces at all points on the image surface is the unknown tip surface. By making use of

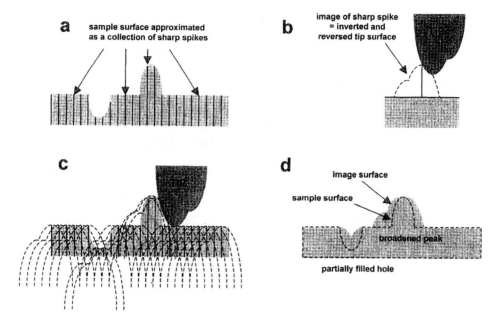

Fig. 4 (a) Any sample surface can be thought of as a collection of infinitely sharp spikes. (b) Each spike acts effectively as an infinitely sharp imaging tip and creates an inverted, reversed image of the probe tip. An asymmetric tip shape has been chosen for this example to show that the tip image is reversed. (c) When imaging a sample surface each spike creates an inverted, reversed probe tip image. (d) The final image is the envelope surface formed from the probe tip images. Note the characteristic distortions caused by the relatively blunt, asymmetric tip used in this example.

the general properties of probe microscope images, Villarrubia (26,28) showed that tip shape can also be estimated from images of an appropriately chosen random surface. This avoids the difficult problem of finding a sample that is well characterized on the nanometer scale and is one of the few practical methods of determining tip shapes without directly viewing each tip in a transmission electron microscope.

B. Dilation Theory

The dilation theory is very similar to the envelope theory but treats the tip, sample, and image as sets of geometric points filling up three-dimensional volumes rather than as surfaces (26). The dilation of set 2 by set 1 is the *union* of many copies of set 1, each copy being placed at a point inside set 2 (Fig. 5a). Informally this means that set 1 acts as a paintbrush and set 2 as a guide. The shape that results from dilation is created by sweeping the center point (or any other fixed point) of the paintbrush over every point inside the guide. In probe microscope imaging, set 1 is the inverted and reversed probe tip, and set 2 in the sample itself. The image is then created by dilating the sample with the inverted and reversed tip. It can be shown formally that dilation produces the same surface as the envelope theory.

The reconstruction process makes use of a related operation called erosion. Erosion of set 2 by set 1 means first taking the complement of set 2 (that is, forming the set of all points *not* in set 2), then dilating this with set 1, and then taking the

a DILATION

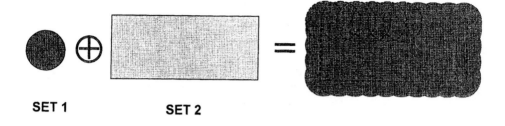

SET 1   SET 2

b EROSION

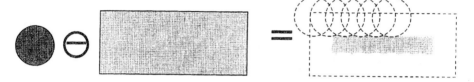

Fig. 5 (a) Dilation of a rectangle by a circle. The dilated set includes all points touched by the circle as it sweeps through every point inside the rectangle. (b) Erosion of a rectangle by a circle. The eroded set includes only points *not* touched by the circle as it sweeps through all points *outside* the rectangle.

complement of the result. Informally, this amounts to using set 1 like an eraser around the edges of set 2, as in Fig. 5b. The reconstructed sample surface is created by eroding the image (set 2) with the tip (set 1).

Finally, in analogy with the envelope theory, the shape of the probe tip can be reconstructed from an image of a known sample by eroding the image with the sample surface itself. The dilation theory is essentially identical to the envelope theory, but the algebra of dilations and erosions has some advantages in formal calculations.

C. Legendre Transform Theory

Unlike the envelope and dilation theories, which represent the tip, sample, and image as abstract sets of points, the Legendre transform theory represents the tip, sample, and image surfaces as explicit two-dimensional functions: $t(x, y)$ for the tip, $s(x, y)$ for the sample, and $i(x, y)$ for the image. This has the advantage that the usual tools of analysis can be brought to bear, and results can be stated in the form of explicit equations. But it also makes the theory somewhat more complex and less intuitive. The fundamental result of the theory is the Legendre transform rule, which says that the Legendre transform of the sample surface is the sum of the Legendre transforms of the image and tip surfaces (27):

$$L[s(x, y)] = L[i(x, y)] + L[t(x, y)] \tag{8}$$

where $L[s(x, y)]$, $L[i(x, y)]$ and $L[t(x, y)]$ are the two-dimensional Legendre transforms of the sample, image, and tip, respectively. The two-dimensional Legendre transform of any function $f(x, y)$ is defined by

$$L[f(x, y)] \equiv b(m_x, m_y) = f(x, y) - m_x x - m_y y \tag{9}$$

where m_x and m_y are partial derivatives of s along the x and y directions, respectively, and the variables x and y are expressed in terms of the slopes, m_x and m_y (so that the transform is a function only of m_x and m_y) by inverting the derivatives:

$$m_x = \frac{\partial f(x, y)}{\partial x} \quad \text{and} \quad m_y = \frac{\partial f(x, y)}{\partial y} \tag{10}$$

Geometrically, the Legendre transform represents the function $f(x, y)$ as the envelope of a family of planes tangent to the original curved surface. The *inverse* Legendre transform is then

$$L^{-1}[b(m_x, m_y)] \equiv f(x, y) = b(m_x, m_y) + m_x x + m_y y \tag{11}$$

with m_x and m_y written in terms of x and y using $\partial b / \partial m_x = -x$ and $\partial b / \partial m_y = -y$. Using inverse Legendre transforms, Eq. (8) can be used to calculate any one of the three surface functions, $s(x, y)$, $i(x, y)$, and $t(x, y)$, from the other two. The Legendre transform is local and nonlinear, so Eqs. (8) through (11) compactly express the local, nonlinear character of probe microscope imaging.

Using Eq. (8) it is possible to derive the *curvature rule*, which says, in general, that the inverse curvature tensor (strictly speaking, the inverse of the second derivative matrix) of the sample surface is the sum of the inverse curvature tensors of the image and tip surfaces. For the special case in which both sample and tip have parabolic surfaces, this rather abstract result simplifies: if the sample surface has radius R_s, and the tip surface is has radius R_t, the curvature rule says that the image will also be parabolic, with radius $R_i = R_s + R_t$ (see Fig. 6). This simple special case applies for most tips and most isolated surface features, which are usually approximately parabolic on the parts of their surfaces that come into contact.

The curvature rule then provides a simple way to estimate the true size of isolated features (such as protein molecules) that have been exaggerated by tip distortions. The image of a protein of radius R_s profiled by a parabolic tip of radius R_t is approximately a parabola with radius of curvature $R_s + R_t$:

$$z(x, y) = h - \frac{x^2 + y^2}{2(R_s + R_t)} \tag{12}$$

where z is the height of the image at (x, y), h is the height of the feature, and x and y are position coordinates in the plane of the sample. The width of the image at the level of the substrate ($z = 0$) is therefore given by

$$h = \frac{(w/2)^2}{2(R_s + R_t)} \tag{13}$$

IMAGE SURFACE, R$_i$

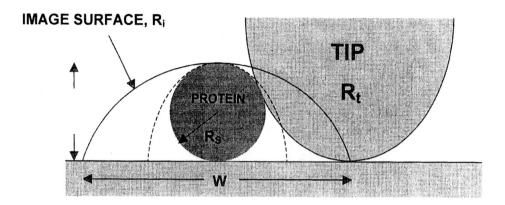

$$R_i = R_s + R_t$$

Fig. 6 Estimating the size of a parabolic feature imaged by a parabolic tip. Spherical features are approximately parabolic at points near their tops where the probe tip touches. For parabolic surfaces the radius of curvature (more precisely, the second derivative) is a constant. The curvature rule then says that the image is also parabolic and that its radius is the sum of radii of tip and sample. The radius of the feature can then be estimated from the measured width at the level of the substrate, *w*.

Solving for the true sample radius, we then have

$$R_s = \frac{(w/2)^2}{2h} - R_t \tag{14}$$

which gives the true protein radius, R_s, as a function of the measured width, *w*, and the tip radius, R_t, For example, suppose the SFM image shows a protein molecule adsorbed to a mica substrate with height 5 nm and width 22 nm, and the tip is known or estimated to have an end radius, R_t, of 10 nm. Using Eq. (14) the estimated lateral radius of the protein is then 5.6 nm, suggesting that the protein is slightly anisotropic in shape.

D. Resolution in Probe Microscopy

One way to define the resolution of any microscope is to use an analogue of the Rayleigh criterion from optical microscopy. Consider a sample containing two very sharp spikes, as in Fig. 7a. The probe microscope image of the two spikes has two maxima separated by a minimum or "dimple" halfway between. When the spikes are far apart, the minimum is deep, but as they move closer the minimum becomes shallower, until it disappears entirely when the spikes coincide. In an ideal, noise-free image, the dimple allows the two underlying spikes to be identified even when they are very close. By in a real image, which always contains some noise, the dimple will disappear when it becomes shallower than the noise level. The probe microscope analogue of the Rayleigh resolution length, L_R, is then the separation, *s*, of the spikes when the root-mean-square (rms) vertical noise is equal to the dimple depth. With

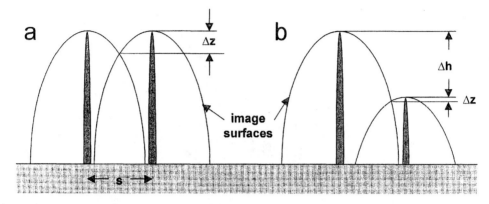

Fig. 7   (a) Probe microscope analogue of the Rayleigh resolution length. Two sharp spikes of equal height are resolved if the depth of the dimple between them, Δz, is larger than the root-mean-square image noise. The smallest distance, s for which this remains true is the Rayleigh resolution length. (b) The Rayleigh resolution length depends on the relative height of the two spikes and hence is not a property of the imaging system alone.

this definition, L_R is related to the tip size and the noise level by (29).

$$L_R = \sqrt{8R\Delta z_{rms}} \tag{15}$$

where a parabolic tip shape has been assumed, R is the radius of curvature of the tip, and $\Delta z_{rms} = \sqrt{\langle \Delta z^2 \rangle}$ is the root-mean-square image noise in the vertical direction. For a tip radius of 10 nm and rms noise of 0.5 nm, the resolution length according to Eq. (15) is 6.3 nm, in rough agreement with experience for SFM imaging under these conditions.

The Rayleigh analogue is a useful definition but has the awkward property that it applies only to features of equal height. As shown in Fig. 7b, the dimple depth depends on the height difference, Δh, between the spikes. When Δh becomes large, the dimple becomes shallow and can even disappear. Thus, features that are resolved when they are of equal height may not be resolved when they are of unequal height, and estimates of resolution obtained by measuring the separation of features may vary from place to place in a single image. Another well-known manifestation of this effect is that relatively blunt tips can sometimes yield images with very high apparent resolution on flat samples with features of similar height and yield poor apparent resolution on other, rougher samples.

VI.   SUMMARY AND CONCLUSIONS

SFM imaging of biomolecules is nearing the stage where most experiments are routine. The main secrets of good imaging are (a) a clean, well characterized sample, (b) careful attention to deposition conditions, and (c) carefully minimized imaging force. Samples that are pure and adsorb well to substrate, and for which the tip-sample interactions are weak, are easily imaged at high resolution.

The information contained in probe microscope images is often complementary to the information in light or electron micrographs. Probe microscopes see only the surface of the sample, and contrast is usually topographic, while in light microscopy contrast is related to the sample's optical properties and in electron microscopy to its electron density. These differences mean that features and properties easily visible in probe micrographs may be invisible in light or electron micrographs and vice versa. In particular, features with small height differences from their surroundings are visible in SFM images, even if they have no optical or density differences. A common example is a small DNA-binding protein such as Cro, which is usually difficult to image by electron microscopy but is prominent in SFM images (30). In general, more is learned by imaging with both SFM and electron microscopy than from either technique alone.

Scanning force microscopy is a young technique, with many unexplored directions for further development. The SFM's ability to visualize molecular movements and measure forces as functions of time has only begun to be exploited. Improvements in control of tip-sample forces would allow increases in resolution, perhaps to the atomic level. Special probe tips with chemically modified surfaces that interact specifically with sample features would give SFM added chemical sensitivity. Measurement of electrostatic interactions between tip and sample would allow mapping of molecular charge distributions in solution. As these and other improvements are realized, the basic theory and methods outlined in this chapter are likely to remain as the foundation of an expanding field.

VII. APPENDICES

A. Cleaving Mica or Graphite

1. Attach the substrate (good-quality ruby mica or highly oriented pyrolytic graphite) to a metal shim or other sample mount.
2. Cut a piece of adhesive tape about 5 cm long, and double back one end to form a handle.
3. Gently press the sticky end of the tape onto the substrate so that it adheres uniformly. Sharp tools or strong pressure can cause subsurface damage to the crystals and should be avoided.
4. Pull gently upward and slightly backward until the tape peels away, bringing the top layer of substrate with it.
5. Inspect both cleaved surfaces. The layer of substrate stuck to the adhesive tape should be continuous and complete, so that none of the original top surface remains on the cleaved substrate. Both sides should be smooth although there are always irregularities visible on the macroscopic scale that are too large to be a problem on the microscopic scale. With mica, loose flakes are also sometimes a problem.
6. If the first cleave did not produce an acceptable surface, cleave a second or third time.
7. Immediately deposit the sample.

B. Flame-Annealed Gold Balls

1. Cut approximately 1–2 inches of gold wire (0.08 inch diameter, >99.99% purity), and hold with clamping tweezers.

2. Prepare a reducing flame with a hydrogen-oxygen microtorch. The reducing environment of the flame prevents the formation of surface oxides.

3. Thrust the end of the wire into the reducing part of the flame until enough wire has melted to form a droplet of about 1 mm, and quickly remove from flame. Leave a short piece of wire to act as a handle.

4. Allow the droplet to cool, and visually inspect it for atomically flat 1,1,1 surfaces, which should appear as eye-shaped, flat facets. Under ideal conditions, when the droplet is a single crystal throughout, the facets will appear at regular angular intervals on the surface of the spherical droplet. If any facets have dark, rough slag patches, avoid them. Sometimes slag can be floated to the base of the droplet by holding the wire droplet-downward during annealing. If it seems impossible to form clean balls, the gold may not be sufficiently pure.

5. If no good facets can be found, it is sometimes possible to remelt the existing ball. Otherwise it will be necessary to repeat steps 1–4 with a fresh section of wire.

6. Immediately place the ball in a holder, and deposit sample. Balls that are exposed to air for more than a few minutes have altered surface properties that indicate the formation of contaminant layers.

A narrow-gauge syringe needle glued to a metal shim makes an excellent holder for flame-annealed balls. The wire can be bent and wedged into the bore of the needle, and the slanting, hollow, needle tip then cradles the ball. The facet of interest can be rotated to a level orientation with a pair of tweezers. When setting up the sample for imagining the SFM tip must be carefully aligned with the facet. A stereomicroscope and a micrometer-fitted sample mount make this task much easier.

REFERENCES

1. C Bustamante, DA Erie, D Keller. Biomolecular imaging by scanning force microscopy. Curr Opin Struct Biol 4:750–760, 1994.
2. S Kasas, NH Thompson, BL Smith, HG Hansma, X Zhu, M Guthold, C Bustamante, ET Kool, M Kashlev, PK Hansma. *Escherichia coli* RNA polymerase activity observed using atomic force microscopy. Biochemistry 36:461–468, 1997.
3. DJ Muller, CA Schoenenberger, F Schabert, A Engel. Structural changes in native membrane proteins monitored at sub-nanometer resolution with the atomic force microscope: A review. J Struct Biol 119:149–157, 1997.
4. DJ Muller, GN Buldt, A Engle. Force-induced conformational change of bacteriorhodopsin. J Mol Biol 249:239–243, 1995.
5. M Rief, M Gautel, F Oesterhelt, JM Fernandez, HE Gaub. Reversible unfolding of individual titin immunoglobulin domains by AFM. Science 276:1109–1112, 1997.
6. GU Lee, LA Chrisey, RJ Colton. Direct measurement of the forces between complementary strands of DNA. Science 266:771–773, 1994.
7. EL Florin, VT Moy, HE Gaub. Adhesion forces between individual ligand-receptor pairs. Science 264:415–417, 1994.
8. D Sarid. Scanning Force Microscopy with Applications to Electric, Magnetic, and Atomic Forces. New York: Oxford University Press, 1991.
9. D Keller, D Deputy, A Alduino, K Sharp. Vertical-walled tips for scanning force microscopy of steep or soft samples. Ultramicroscopy 42–44:1481–1484, 1992.

10. M Radmacher, M Fritz, CM Kacher, JP Cleveland, PK Hansma. Measuring the viscoelastic properties of human platelets with the atomic force microscope. Biophys J 70:556–576, 1996.

11. M Radmacher, M Fritz, JP Cleveland, DA Walters, PK Hansma. Imaging adhesion forces and elasticity of lysozyme adsorbed on mica with the atomic force microscope. Langmuir 10:3809–3814, 1994.

12. J Israelachvili. Intermolecular and Surface Forces with Applications to Colloidal and Biological Systems. San Diego: Academic Press, 1991, Chap 11.

13. RW Keller, DJ Keller, D Bear, J Vasenka, CJ Bustamante. Electrodeposition procedure for *Escherichia coli* RNA polymerase onto gold and deposition of *Escherichia coli* RNA polymerase onto mica for observation with scanning force microscopy. Ultramicroscopy 42–44:1173–1180, 1992.

14. C Bustamante, J Vasenka, CL Tang, WA Rees, M Guthold, RW Keller. Circular DNA molecules imaged in air by scanning force microscopy. Biochemistry 31:22–26, 1992.

15. M Bezanilla, S Manne, DE Laney, YL Lyubchenko, HG Hansma. Adsorption of DNA to mica, silylated mica and minerals: Characterization by atomic force microscopy. Langmuir 11:655–659, 1995.

16. HG Hansma, DE Laney. DNA binding to mica correlates with cationic radius: Assays by atomic force microscopy. Biophys J 70:1933–1939, 1996.

17. M Fritz, M Radmacher, JP Cleveland, MW Allersma, RJ Stewart, R Gieselmann, P Janmey, CF Schmidt, Hansma. Imaging globular and filamentous proteins in physiological buffer solutions with tapping mode atomic force microscopy. Langmuir 11:3529–3535, 1995.

18. YL Lyubchenko, BL Jacobs, SM Lindsay, A Stasiak. Atomic force microscopy of nucleoprotein complexes. Scanning Microsc 9:705–727, 1995.

19. S Singh, P Turina, CJ Bustamante, DJ Keller, R Capaldi. Topographical structure of membrane-bound *Escherichia coli* F_1F_0 ATP synthase in aqueous buffer. FEBS Lett 397:30–34, 1996.

20. A Kamura, NL Abbot, E Kim, HA Biebuyck, GM Whitesides. Patterned self-assembled monolayers and mesoscale phenomena. Accounts Chem Res 28:219–226, 1995.

21. GE Poirier. Characterization of organosulfur molecular monolayers on Au(111) using scanning tunneling microscopy. Chem Rev 97:1117–1127, 1997.

22. NJ Tao, JA DeRose, SM Lindsay. Self-assembly of molecular superstructures studies by in situ scanning tunneling microscopy: DNA bases on Au(111). J Phys Chem 97:910–919, 1993.

23. CR Clemmer, TP Beebe. Graphite: A mimic for DNA and other biomolecules in scanning tunneling microscope studies. Science 251:640–642, 1991.

24. LD Landau, EM Lifshitz. Course of Theoretical Physics. Vol 6. Fluid Mechanics. Oxford: Pergamon, 1986, pp 88–98.

25. D Keller, F Franke. Envelope reconstruction of scanning probe microscope images. Surf Sci 294:409–419, 1993.

26. JS Villarrubia. Morphological Estimation of tip geometry for scanned probe microscopy. Surf Sci 321:287–300, 1994.

27. D Keller. Reconstruction of STM and AFM images distorted by finite size tips. Surf Sci 253:353–363, 1991.

28. JS Villarrubia. Scanned probe microscope tip characterization without calibrated tip characterizers. J Vac Sci Technol B 14:1518–1521, 1996.

29. C Bustamante, D Keller. Scanning force microscopy in biology. Phys Today 48:32–38, 1995.

30. DA Erie, GL Yang, HC Schultz, C Bustamante. DNA bending by Cro protein in specific and non-specific complexes: Implications for protein site recognition and specificity. Science 266:1562–1566, 1994.

24

Cryo-Transmission Electron Microscopy

D. Danino and Y. Talmon
Technion-Israel Institute of Technology, Haifa, Israel

I. INTRODUCTION

Understanding the structure of a material system is often the key to uncovering, controlling, and improving its properties. Full characterization requires direct information, namely images. In most cases this information is on the supramolecular level, describing how molecules arrange to form clusters of various sizes and shapes. Cryo-transmission electron microscopy (cryo-TEM) refers to direct imaging of a fluid specimen thermally fixed into a solid or quasi-solid state. This is potentially one of the most useful techniques in the study of microstructured fluid systems, as it provides direct high-resolution images of the assemblies making up the system; thus the interpretation of data is not model dependent. Today, cryo-TEM is applied in the study of a wide variety of complex fluid systems in academia and industry. In fact, some companies use the technique not only for research but also for quality control and production-related troubleshooting.

Cryo-TEM can elucidate the nature of the basic building blocks that make up the systems, covering a wide range of length scales (from few nanometers to several micrometers), including fine details. In addition, coexistence of many different assemblies present in the examined systems is quite easily observed in the micrographs. In contrast, with indirect methods interpretation of experimental data is very complicated in systems containing more than one type of aggregate. Direct images are always useful, but in some cases they provide the only way to prove a suggested or a theoretically predicted model, as was demonstrated in the case of branched micelles (1).

However, one should realize that cryo-TEM is not a strictly quantitative technique. Although it is the technique of choice to determine the structural building blocks of complex fluid systems, the quantitative data should come from other techniques, such as small-angle X-ray scattering (SAXS), small-angle neutron scattering (SANS), or nuclear magnetic resonance (NMR). Microstructural information obtained from cryo-TEM can be a good basis for a physical model required for interpretation of indirect methods data.

In the following sections the basics of cryo-TEM are described, followed by a broad review of the applications of the technique in the study of biological and biological-model systems. More comprehensive discussions on the technique and its use to other systems can be found elsewhere (2,3).

II. PRINCIPLES OF CRYO-TEM

Complex fluid systems contain a high water content and other volatile components. To examine these samples in the TEM it is necessary to lower the vapor pressure of these systems to make them compatible with the high vacuum in the microscope column, typically better than 10^{-4} Pa. Also, any supramolecular motion must be arrested, to prevent blurring of the recorded image. TEM specimens must be thin, not thicker than about 250 nm. Thicker specimens give rise to inelastic electron scattering that deteriorates image quality.

A. Thermal Fixation and Vitrification

Reduced vapor pressure and arrested motion can be achieved by chemical or "physical" (thermal) fixation. Chemical fixation involves addition of an alien chemical substance to the sample. However, microstructured fluids are very sensitive to changes in composition. Obviously, addition of compounds such as a stain or fixative, followed in some cases by a chemical reaction between the fixative and the specimen, and often drying of sample, alter the original microstructure of the studied system, thus making chemical fixation unsuitable for the study of microstructured fluids.

The method of choice is thermal fixation, i.e., the conversion of liquid specimens into a solid or quasi-solid state via ultrafast cooling. This is done by rapidly plunging the specimen into a suitable cryogen. Thermal diffusivities are larger than mass diffusivities and therefore thermal fixation is much more rapid than chemical fixation. Also, thermal fixation eliminates the addition of an alien compound to the system, as is the case in chemical fixation.

Cryo-TEM is based on the ability to cool the specimen fast enough to vitrify it, thereby preserving its original microstructure. The cooling rate needed for vitrification of water is on the order of 100,000 K/s, as estimated theoretically (4) and measured experimentally (5). When cooling is not fast enough, crystalline (hexagonal or cubic) ice forms, which may lead to optical artifacts, to mechanical damage to the microstructure, and to redistribution of solutes and their spatial organization. Solutes are expelled from the growing ice lattice and are deposited either in the ice grains or often at grain boundaries. Additional information can be found in reviews on direct-imaging cryo-TEM (2,3) and references therein.

To achieve the high cooling rate required for vitrification we maximize the surface area-to-volume ratio of the specimen. The geometry of choice is that of a thin film. Thin films are also required because of the limited penetration power of even high-energy electrons. As already stated, regular applications require specimen thickness less than about 250 nm. For high-resolution images, thinner samples are preferable. Most direct-imaging samples display areas of a wide thickness range. Thin vitrified films may also be obtained by sectioning thermally fixed bulk specimens using a cryo-ultramicrotome [e.g., (6,7)]. Today, microscopes operating at 200, 300, and 400 kV are found in many universities and research institutes. However, although these instruments are capable of imaging thicker specimens, image interpretation becomes increasingly difficult with specimen thickness because of superposition of information resulting from the high depth of focus of the TEM.

A proper cryogen is required to vitrify the specimen successfully. The cryogen has to be at a low temperature, far below its boiling point, to avoid formation of a gas film around the specimen during quenching. Such a film reduces the heat transfer coefficient considerably. The cryogen should also have a high enough thermal conductivity. Liquid nitrogen is a poor cryogen because of the narrow temperature between its freezing and boiling points. Practically, liquid ethane cooled to its freezing point by liquid nitrogen is the best cryogen (its normal boiling point is about 100 K higher). Another advantage of ethane is that crust that forms during sample preparation sublimes quite readily in the microscope, mostly during specimen transfer. (Liquid propane is another possible coolant but because of the lower vapor pressure of the solid it is rather difficult to get rid of its crust.)

B. Interaction Between Electron Beam and Cryo-Specimens

To achieve good imaging through cryo-TEM, electron beam–specimen interactions should be recognized and understood. The most important interactions are discussed briefly here. Additional references for more detailed discussions of the subject are mentioned in the text.

Transmission electron microscopy is an optical technique, in many ways similar to light microscopy. Image formation in the TEM is the result of specific interactions between the electron beam and the sample it traverses. To obtain a meaningful image these interactions need to be different in areas of different structure or composition, reflected in image contrast.

Useful image information is the result of elastic scattering of incoming electrons by the specimen. Electrons deflected by large enough angles are blocked by the microscope objective lens aperture, thus reducing locally the signal for that particular area. Contrary to elastic scattering, inelastically scattered electrons do not contribute to the image information, but they produce a background "fog." Some modern microscopes have electron energy filters that operate in the range of 80 to 120 kV and are used to filter out inelastically scattered electrons. Such filters improve contrast and allow the study of thick specimens.

1. Contrast

Image contrast is defined, in general, as the difference in signal intensity between two regions of the picture, divided by a reference intensity. Several different contrast mechanisms, crystalline contrast, mass-thickness contrast, and phase contrast, contribute to the image contrast. In the study of vitrified hydrated samples the most important contrast mechanism is phase contrast, which is based on converting wave phase differences into amplitude differences. Phase contrast is achieved in the TEM by underfocusing the objective lens (8). Thus, regions of different inner electron potential relative to vitreous ice, such as micelles and vesicles, are made clearly visible. However, changing the amount of underfocus changes the so-called microscope transfer function, an expression of changes in the image induced by the microscope optics. This may lead to filtering out certain spatial frequencies (details of certain sizes) while overemphasizing others, leading to loss of resolution and possible distortion of the true microstructure. To rule out focus-related optical artifacts one needs to record a series of images at increasing underfocus to assess its effect on the recorded images.

Mass-thickness contrast, a result of density and thickness differences from one area to another, is usually limited in biological specimens because they are made mainly of light elements: hydrogen, carbon, nitrogen, and oxygen. Some heavier elements such as sulfur, chlorine, and phosphorus found in some phospholipids or proteins may produce better contrast. Crystalline contrast is of minor importance because direct cryo-TEM specimens are typically vitreous.

2. Radiolysis

Electron beam radiation damage, or radiolysis, is an inevitable consequence of electron beam–specimen interaction required for image formation. It is a result of free radical chain reactions started by ionization of specimen molecules by the high-energy electron beam. Radiolysis is especially severe in the presence of water and much more so in vitrified specimens that contain significant amounts of organic compounds. A comprehensive discussion of the subject is given in Talmon (9). In general, the goal is to reduce radiation damage as much as possible. Modern microscopes are equipped to perform "minimal dose exposure," through which the electron exposures for recording a micrograph are reduced to less than 1000 electrons per nm^2, four or five orders of magnitude lower than typical exposures used to study "beam-resistant" materials such as metal and ceramics. The advancement of digital microscopy, i.e., recording the image by a charge-coupled device (CCD) camera, has made low-dose imaging easier and more effective. In some cases selective etching can enhance contrast (10–12). Also, one can apply beam etching after the image has been recorded at low electron exposure to determine whether certain structural elements are part of the bulk of the specimen or are just surface contamination.

III. SPECIMEN PREPARATION AND IMAGING

In general, cryo-TEM is a very delicate and time-consuming technique, based on many successive steps, each crucial for achieving reliable results. In the following section we present the main devices, sample preparation procedures, and important parameters that constitute experimental microscopy work.

A. Controlled-Environment Vitrification Systems

The basic apparatus used for specimen preparation is the controlled-environment vitrification system (CEVS), which we have developed and continuously upgraded over the years. Samples are prepared in the CEVS at well-controlled temperature in the range of 0 to above 80°C and at relative humidity. This ensures microstructural preservation of the specimen during preparation. The original system was first described in detail by Bellare et al. (13). Briefly, the CEVS is a polycarbonate closed chamber, heated by a high-power light bulb controlled by an on-off temperature controller connected to a sensor in the chamber and by a rheostat controlling the voltage to the lamp. Cooling is achieved by filling an insulated reservoir mounted on the outside of the chamber back wall with a refrigerant (e.g., ice-water or liquid nitrogen). To saturate the air inside the chamber, two large sponges placed in beakers in the back of the chamber are wetted with water or another volatile liquid. A small fan circulates the air over the sponges, to minimize relative humidity and temperature gradients. The specimen is held by a tweezer mounted on a spring-loaded plunger and is manipulated from outside the chamber through two rubber septa on the side walls of the chamber. Upon triggering the plunger mechanism, a trapdoor placed at the bottom of the chamber opens, and the specimen is propelled into the cryogen reservoir placed underneath the CEVS.

Specimen preparation in the CEVS is rather simple: a drop, typically 3 to 5 μL, is applied onto a perforated carbon film supported on a TEM copper grid held by a tweezer. The drop is then blotted by a piece of filter paper wrapped on a thin metal strip to produce a thin specimen film. After blotting, the specimen is plunged into the cryogen and vitrified. Finally, the vitrified sample is transferred under liquid nitrogen to the "working station" of a cooling holder, where it is loaded into the special holder and transferred with it into the microscope. For some applications it is advantageous to use a "bare grid," without a perforated film. In those cases one should use grids of 700 mesh or finer.

Several variations of the CEVS have been built during the past 10 years. The apparatus described by Chestnut et al. (14) was designed to introduce an on-the-grid temperature jump by shining an intense light beam on the grid prior to plunging. Temperature jumps of 30–60°C are achieved with exposure times of 150–450 ms. Much improved temporal resolution has been achieved in a more recent design based on heating the specimen by a flash just before it hits the cryogen. The light is produced by a flash tube that can produce temperature jumps of 70°C in a few milliseconds (5). The most recent design is the computer-controlled flow-through CEVS (FT-CEVS) (15) shown in Fig. 1. Exact control of temperature and humidity in the FT-CEVS is achieved by a continuous flow of two streams of humid or dry gas, air or nitrogen, through a small cylindrical chamber. The gas is continuously discharged through a small hole in the center of the chamber bottom. This hole is also used for plunging the specimen. A personal computer controls the temperature and humidity of the two gas streams, monitors the specimen temperature, changes temperature and humidity in the chamber according to the set program, and eventually triggers the plunging mechanism. The small volume of the chamber (0.3 L compared with 1.7 L of the CEVS), its small thermal inertia, and the continuous flow of gas through it make the system ideal for performing controlled rapid changes in either the temperature or the relative humidity for on-the-grid processing (OTGP) and time-resolved cryo-TEM.

In OTGP experiments a thin liquid film (e.g., of a low-viscosity precursor to a gel phase) is prepared, and a new phase is then formed on the grid (e.g., a gel) by inducing a physical or a chemical change such as heating, cooling, controlled drying, pH change, or mixing of reactants (5,16–19). A recent example of OTGP is thermally induced phase separation and thermal gelation performed in the FT-CEVS (20). Such an on-the-grid process may be stopped before it reaches completion, by plunging the grid into the cryogen, to capture intermediate structures. A series of such experiments

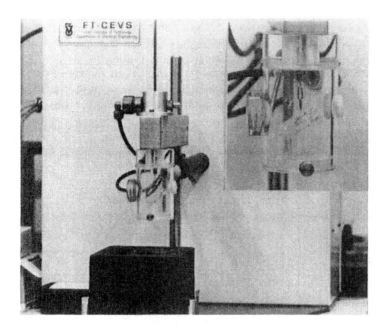

Fig. 1   The flow-through controlled environment vitrification system (FT-CEVS). Inset shows the specimen preparation chamber itself.

with increasing OTGP durations results in "time-resolved cryo-TEM" (TRC-TEM) (16,17).

A different route of direct-imaging cryo-TEM is freeze-fracture cryo-TEM (FF-TEM). This method is based on fracturing a frozen specimen and preparing a metal replica of the fracture surface by vapor deposition of a heavy metal that enhances contrast and carbon, which affords mechanical stability to the replica. Afterward, the sample is melted away and the replica washed, dried, and examined in the microscope at room temperature. Thin enough (a few tens of nanometers) replicas can easily be made by controlling the evaporation and deposition processes. A comparison of these two techniques can be found in Refs. 21 and 22.

B.   Blotting

The most crucial step in sample preparation is thinning of the specimen via blotting, which determines the film thickness and quality. The exact amount of blotting and its mode, i.e., how much shear is applied on the liquid and whether it is done from the back and/or from the front of the grid, depend on the nature of the fluid. The operator has to adjust the blotting based on his or her experience with that particular system or a similar one.

An automatic computerized blotting device has been developed by Frederik and coworkers (P. Frederik, private communication). The solution to be studied is equilibrated in a controlled chamber at the required temperature. Following that, the tweezer holding the grid is dipped into a reservoir of the solution and two arms equipped with filter papers squeeze the grid, thereby producing a thin film. This is followed by plunging of the grid into the cryogen. The amount of blotting in this device is determined by the time the filter papers are held against the specimen and the number of times it is applied.

Complex flow patterns develop in the liquid when the sample is blotted to its final thickness. Usually, a biconcave film forms over holes of the perforated carbon film, with thin areas usually in the center and thicker areas anchored at the edges of the holes. This effect causes larger objects to be selectively pushed to thicker areas, while smaller objects are preferentially left in the thinner domains, leading to size segregation. The thinnest areas may be completely devoid of suspended objects. Large objects such as vesicles and liposomes are squeezed during sample thinning and may be totally expelled from the specimen.

As the liquid drop applied onto the grid is thinned to its final thickness, it is subjected to very high shear rates. The velocity of the blotting paper surface is in the range of a few cm/s, and the final thickness of the film is tens to hundreds of nanometers, so shear rates on the order of 10,000 reciprocal seconds and more may be expected. Such high shear rates may cause gross microstructural changes in complex fluids, for example, increase of size of threadlike micelles (23), alignment of structures (24), or even phase transformation (25–28). The formation of a thin film also promotes very high mass transfer rates, which may lead to composition changes due to loss of volatiles if 100% relative humidity is not maintained around the specimen. The problem of size segregation is linked to local concentration of the dispersed aggregates close to hole edge. Several studies demonstrated the formation of new phases on the grid by locally higher concentrations produced during blotting (18,29).

C. Imaging

The vitrified samples must be kept under liquid nitrogen until they are examined in the microscope to prevent warming leading to water crystallization and rearrangement of material. In typical cryo work, transfer of specimen into the microscope is done in a cryo-holder cooled to at least $-170°C$ with liquid nitrogen. Special helium-cooled specimen holders operating at about $-269°C$ provide reduced electron beam radiation damage (30,31).

Imaging of cryo-TEM specimens is difficult because of the low inherent contrast of the specimens and their sensitivity to the electron beam. Correct imaging strategy takes these factors into account and optimizes the conditions to make most of the possible information available to the user. For most applications a microscope operating at an accelerating voltage of 100 kV, an optimum between reasonable penetration power, which goes up with the voltage, and contrast, which goes down with it, is sufficient. The vacuum system of the microscope needs to be kept in excellent condition at all times. An "anticontaminator," consisting of large cooled surfaces installed in the microscope column as close as possible to the specimen, is also needed to trap molecules of residual volatiles and prevent them from condensing on the specimen. Images are typically recorded at magnification of 20,000 to 140,000, the minimum needed to obtain the desired structural information, using a sensitive photographic film or a CCD camera. CCD cameras offer ease of use, better low-dose operation, and immediate postmicroscopy image processing, which is very important in the study of cryo-specimens that typically show very little contrast. We regularly use low dose exposure to minimize radiation damage. Phase contrast may be enhanced through defocus of the microscope objective lens.

Surface contamination, which may arise, for example, from water vapor that condenses on the specimen, may take different forms, sometimes resembling the structure one is trying to image. Contaminants may be introduced into the microscope column through leaks, from the photographic film in the microscope camera chamber, and at the introduction of the specimen holder into the microscope through the air lock. Good microscopy practice, keeping the instrument at top performance at all time, is essential.

IV. APPLICATIONS FOR BIOLOGICAL SYSTEMS

A. Vesicles

Vesicles and liposomes have been extensively studied by cryo-TEM because of their similarities to biological membranes. Phospholipids, which are a major component of membranes and one of the three major components in bile, disperse in water to form vesicles or liposomes. Phospholipid vesicles are important models for cell membranes and contribute to our understanding of the role of specific lipids and the biochemistry and biophysics of transport and receptor proteins (32). In addition, phospholipid vesicles have enormous potential as drug carriers, in packaging hemoglobin as an emergency blood substitute, as a formulation for skin care products, for cosmetics, and as carriers for viral or other antigens.

Phospholipids undergo a temperature-dependent, reversible, order-disorder transition that is associated with the "melting" of the hydrocarbon chain region of the lipid bilayer. This transition depends on several parameters such as the fatty acid chain length and the extent of saturation, the degree of hydration, and the nature of the polar head. A characteristic appearance of liposomes is observed, according to whether they are maintained before vitrification at a temperature above or below their melting temperature. This effect was demonstrated in a time-resolved cryo-TEM study of unilamellar dimyristoyl phosphatidylcholine (DMPC) vesicles (33). Perfect smooth vesicles and a faceted-rippled surface were seen when vitrified above and below the transition temperature, respectively. Similar results were obtained in DPPC (dipalmitoyl PC) samples (34). Ripple structures were found in DHPC (diheptanoyl PC) and DPPC mixtures below the main transition temperature (35). Disclinations were detected in the ripple structures. Above the main transition temperature discoid structures were seen in this system and in DHPC-sphingomyelin mixtures as well (36).

In the ternary system DMPC, water, and graniol (a branched long-chain biological alcohol derived from oil-soluble vitamins, a cosurfactant), a phase consisting of bilayer sheets in the form of tubular vesicles was suggested (37). Tubules were found in bulk DMPC samples vitrified at temperatures close to room temperature (28.5°C) (38). Coexistence of tubules with closed unilamellar vesicles and bilamellar structures was observed in several synthetic sodium di(polyprenyl) phosphates, which can in principle be regarded as primitive membrane-forming phospholipids (39) and in surfactants such as SHBS (sodium 8-phenyl-*n*-hexadecyl-*p*-sulfonate) as well (35,40). In dihexadecyl phosphate dispersions open and folded bilayer fragments rather than smooth vesicles were the dominating structure after sonication at high temperature and vitrification from room temperature (41). As in other systems mentioned, the discontinuous curvature was attributed to the fact that vesicles were vitrified below the chain melting temperature. Very long tubular vesicles similar to those observed in PC samples were also seen, coexisting with threadlike micelles, in dilute aqueous solutions of phosphoglucolipids from membranes of the bacterium *Acholeplasma laidlawii* (42).

Mui et al. (43), examined the morphological consequences of transbilayer transport of dioleoylphosphatidylglycerol (DOPG) in large unilamellar vesicles of dioleoyl PC DOPC-DOPG between the inner and outer layers by exposing the system to pH gradients. A progression of morphologies leading eventually from invaginated vesicles to the formation of long tubules was observed, as DOPG was transported to the outer monolayer. Tubules were also seen in response to outward DOPG transport in large unilamellar vesicles of DOPC-DOPG-cholesterol. When DOPG was transported from the outer monolayer to the inner monolayer, reverse behavior was noticed. These morphological changes were interpreted by an elastic bending theory of the bilayer.

Drugs can be loaded into preformed liposomes. Doxorubicin, an anticancer drug, loaded into the interior of vesicles by applying ammonium sulfate gradients, was studied by Lasic et al. (44). Whereas initial lecithin-cholesterol-PEG (polyethylene glycol) vesicles were small, round, and unilamellar, the vesicles containing doxorubicin were ellipsoidal and contained dark strips, usually aligned with the major semiaxis of the ellipse. This study provided direct evidence of the incorporation of doxorubicin into the liposomes and showed that the high encapsulation efficiency of doxorubicin into liposomes, obtained by the ammonium sulfate gradient loading method, is due to the formation of a gel-like precipitate in the liposome interior.

B.   Membrane Solubilization and Reconstruction

Many studies have focused on understanding the interactions and mechanisms of membrane solubilization and reconstitution. Experimental laboratory work generally focuses on solubilization of vesicles formed by one type of amphiphilic molecules by the addition of another type that tends to form spheroidal micelles (reconstitution is the inverse process). The transition from vesicles to micelles (and vice versa) is not sharp; it involves formation of several different intermediates at different ratios of the two types of amphiphiles.

Solubilization of phospholipid vesicles by surfactants is considered to occur in three main stages. At low surfactant/PC ratio the surfactant incorporates into the bilayer membrane of the vesicles, until the PC bilayer is saturated by the surfactant. High surfactant/PC ratios induce the formation of small mixed micelles. At intermediate ratios various intermediate structures coexist. The concentrations at which the microstructural changes occur and the nature of intermediate structures formed depend on the geometry and charge of the surfactant and lipid molecules and environmental parameters such as temperature and concentration (45).

Cryo-TEM is a powerful technique for the study of the structural steps of such mixed systems, as it can quite easily follow the transitions, detect coexisting intermediates, and provide direct information on the intermediates shape and size. Indeed, a very significant contribution to the structural study of surfactant-PC systems has been made by cryo-TEM, as described in the following.

Vesicle-to-micelle transition in egg-PC–octylglucaside (OG) mixtures was studied by Vinson et al. (46). Cryo-TEM micrographs showed that upon OG addition, originally dialyzed vesicles decrease in size considerably at first. With further increase in surfactant concentration, small vesicles were observed coexisting with open vesicles, pieces of lamellar fragments, and long cylindrical micelles. The open vesicles were all larger than the intact ones. The transition from vesicles to cylindrical micelles occurred by a separation of the lamellar sheets into strings of material. At higher OG/egg-PC ratios the dominant structures were cylindrical micelles, and eventually, upon addition of excess OG, spheroidal micelles formed.

OG was also used to solubilize cholesterol-rich membranes of diglycerol hexadecyl ether ($C_{16}G_2$) containing small amounts of dicetyl phosphate (47). An initial increase in the vesicle size followed by a decrease in their size and formation of spheroidal micelles were observed with increasing OG concentration. However, threadlike micelles that are usually seen in solubilization of lipid vesicles were not observed in this study. The absence of threadlike micelles was also suggested in solubilizing vesicles of a dimeric surfactant with either a conventional or a dimeric micelle-forming surfactant (48).

Structural symmetry of the vesicle-to-micelle and micelle-to-vesicle transitions was shown to exist in the sodium cholate–egg-PC system (49,50). Addition of low cholate concentrations to egg-PC vesicles resulted in an increase of vesicle size. In the coexistence region, at a higher surfactant/egg-PC ratio, open vesicles, large bilayer sheets, and long flexible sometimes branched threadlike micelles, evolving from the edges of the bilayer sheets, were observed, as seen in Fig. 2. Eventually, at high surfactant concentrations only small spheroidal mixed micelles were noted in the electron micrographs. In the micelle-to-vesicle transition short threadlike mixed micelles at the higher cholate concentrations, long threadlike micelles at intermediate concentrations, and then a mixture of threadlike mixed micelles, lamellar fragments, and close vesicles with large pores at the lower cholate concentrations were observed.

The effect of Triton X-100 on sonicated lecithin small unilamellar vesicles (SUVs) was studied by Edwards et al. (51). Time-resolved cryo-TEM experiments

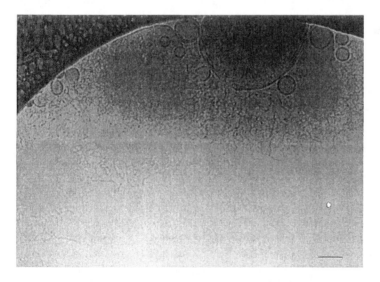

Fig. 2 Electron micrograph of a vitrified aqueous mixture containing 9 mM lecithin and 6.5 mM cholate, showing large membrane patches with many cylindrical mixed micelles emerging simultaneously from the edges. (From Ref. 49.)

showed that the formation of the large vesicles occurred in about 2 min, and breakdown to mixed micelles at high surfactant concentrations was completed within a few seconds. The same group also studied the structural transitions induced by addition of the nonionic surfactant octaethylene glycol *n*-dodecyl monoether ($C_{12}E_8$) to lecithin SUVs (52). Similar morphologies were observed in solubilizing lecithin vesicles with the cationic surfactant cetyltrimethylammonium chloride (CTAC) (53).

Silvander et al. (54) solubilized unilamellar lecithin vesicles by three different alkyl sulfates ($C_{10}SO_4$, $C_{12}SO_4$ and $C_{14}SO_4$). The three surfactants induced vesicle growth at subsolubilizing concentrations and a transformation into small globular lipid–surfactant mixed micelles at high surfactant concentration in the presence of NaCl. In the coexistence region of the shortest surfactant (C_{10}), open and closed vesicles appeared with long, entangled, branched threadlike micelles. With both C_{12} and C_{14} surfactants, vesicles and open sheets of holey bilayers, but no threadlike micelles, were observed before the bilayers solubilized into small globular micelles. Upon decreasing the salt concentration, intact and open vesicles coexisted with bilayer fragments.

The effect of ionic strength on the self-assembly in mixtures of lecithin and sodium cholate was studied by Meyuhas et al. (55). Vesicles about 200 nm in diameter were seen by cryo-TEM in samples with high NaCl concentrations as compared with vesicles about 50 nm in diameter observed in the absence of salt. Threadlike micelles were also seen in the presence of salt. With the zwitterionic bile salt derivative CHAPS (3-[(3-cholamidopropyl)dimethylammonium]-1-propansulfonate) only a few large vesicles and membrane fragments were observed by cryo-TEM. Threadlike micelles existed in CHAPS-lecithin solutions, some of them branched, whereas very few threads, if any, were observed in the presence of salt.

These studies demonstrate the unique contribution of direct-imaging cryo-TEM when several structures coexist and where data obtained by quantitative indirect methods such as SAXS or SANS are difficult to interpret.

C. Phase Transitions and Membrane Fusion

Common examples of cellular events that involve fusion of membranes are exocytosis, endocytosis, and infection by enveloped viruses. Understanding the mechanisms of membrane fusion requires identification of intermediate structures formed during the process. Freeze-fracture TEM has revealed three intermediate fusion structures in artificial membrane systems (56,57): particles that reflect interlamellar attachments (ILAs), particles that reflect joining (inverted micellar intermediates, IMIs), and particles that reflect fusion, the formation of an aqueous channel between two compartments that were previously separated by their limiting membranes. ILAs have relatively long lifetime compared with IMIs (5). ILAs represent the fission stage at which an aqueous connection is established between the contents of two vesicles. IMIs represent the fusion stage at which two membranes become continuous.

Many lipids in all biomembranes reveal a tendency to form inverted nonbilayer lipid structures such as the inverted hexagonal and cubic mesophases. This transition from a lamellar structure to the inverted phases involves an extreme change in the topology of the lipid-water interfaces. It has been suggested that the intermediate structures formed during lamellar-to-inverted hexagonal and lamellar-to-inverted cubic transitions are related to the structures that mediate membrane fusion (e.g.,

57,58). Determination of the mechanisms of these phase transitions requires high resolution, both spatial and temporal (i.e., visualization of transient structures with a short lifetime), that can be achieved by direct imaging cryo-TEM.

Siegel et al. (16) applied an on-the-grid pH jump using time-resolved cryo-TEM and directly visualized the evolution of inverted phases in several phospholipid liposomal aggregates. They identified ILAs, with the predicted structure and dimensions. They suggested that the same ILAs appear to assemble the inverted cubic phase from the lamellar phase by formation of an ILA lattice and to mediate membrane fusion of aggregated liposomes, in agreement with the earlier freeze-fracture studies. To capture and identify the initial intermediates formed, on a time scale of milliseconds, Siegel et al. (5) performed temperature-jump experiments, using the TRC-TEM, on large unilamellar vesicles of DOPC, *N*-monomethylated DOPE (DOPE-Me), and phosphatidylethanolamine. The results obtained in this study disagree with the mechanism of IMI formation but support the theoretical proposed transition through the formation of stalk intermediates (59). Stalks are catenoidal connections between the interfaces of opposed bilayers; they are the lowest energy intermembrane structures that have been proposed. TRC-TEM experiments were performed on large unilamellar vesicle suspensions of dipalmitoleoyl phosphatidylethanolamine (DiPoPE) as well (60). Beautiful micrographs showing different intermediates that formed during successive pH jumps provide further support for the formation of stalk intermediates in the lamellar-to-hexagonal phase transition. An example of these structures is given in Fig. 3.

Frederik et al. (61) studied the formation of intermediate structures during fusion between artificial membranes. The two investigated systems were cardiolipin-PC mixtures in which fusion was triggered by addition of Ca^{2+} ions and cholesterol-DOPE-DOPC solutions in which fusion was triggered by heating. Both mixtures exhibit a complex phase behavior in which lamellar systems with lipidic particles and a honeycomb or a cubic phase coexist. In another series of experiments (62) they identified ILAs. The ILAs shape and size correlate well with results of freeze-fracture experiments that were conducted in parallel and with the results of Siegel et al. (16). Structures characteristic of the transition from lamellar to cubic and from lamellar to inverted hexagonal mesophases were observed. In some micrographs, coexistence of bilayers and inverted hexagonal and cubic phases were seen.

During cellular infection enveloped animal viruses fuse with either the plasma membrane or the endosomal membrane. Membrane fusion is accomplished through the specific action of viral glycoproteins, or spike proteins (63). Influenza virus, one of the most extensively studied enveloped animal viruses, enters the cell by receptor-mediated endocytosis (64). In mixtures of unilamellar vesicles and influenza virus, lowering the pH induces a conformational change in the hemagglutinin spikes (62), and diffusion of the viral spike proteins into the liposomal membrane is observed (65). In phosphatidylserine-containing vesicles to which annexin V binds in the presence of calcium ions, electron micrographs showed a change from spherical vesicles into faceted ones (34).

D. Bile and Model-Bile Systems

Bile salts are the major component of bile. Because of their inflexible ring structures they are unusual surfactants having a hydrophilic and a hydrophobic face. The phase

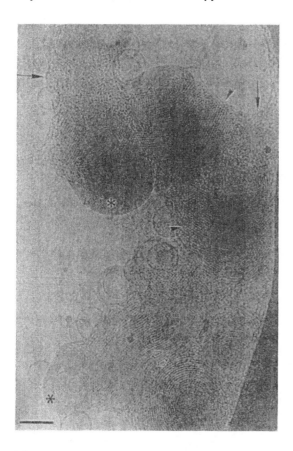

Fig. 3 TEM image of DiPoPE sample at pH 5, 31°C. Solid arrows, clusters of intermembrane connections; hollow arrows, lattices of dark striations; arrowheads, regions of hexagonal symmetry. The striations and hexagonal symmetry are always associated, and termed quasi-hexagonal domains. Asterisk, multilamellar vesicles. The interbilayer water spacings are too small to be clearly resolved. (From Ref. 60.)

behavior and aggregation structures in mixtures of bile salts and surfactants have been studied using various techniques. Dilute aqueous solutions of sodium cholate (NaC) and sodium desoxycholate (NaDOC) were studied in the presence of various alkyl trimethylammonium bromide (C_nTAB) surfactants, with and without added salt (66,67). Monodisperse globular micelles were observed in the system of NaC-NaCl and $C_{14}TAB$, and under the same conditions threadlike micelles were seen in the presence of $C_{16}TAB$. Spheroidal micelles formed with $C_{18}TAB$ without salt, and globular micelles coexisting with disklike structures were observed at increasing salt concentration. Increasing NaC concentration led to the formation of small globular micelles (67). In NaDOC-CTAB mixtures flexible threadlike micelles, sometime branched, were observed in the dilute micellar phase and in the two-phase region, as a function of bile salt mole fraction (66). Threadlike mixed micelles were also seen at low equimolar composition, whereas at a higher equimolar composition flocks or clumps of more compact material appear. In the dilute two-phase region a three-dimensional network of the interwoven threadlike aggregates appeared.

Cholesterol monohydrate crystals constitute the microscopic building blocks of cholesterol gallstones. Understanding the process of cholesterol crystallization in bile and characterization of the dynamic microstructural evolution of nucleating bile are essential for understanding gallstone formation. Mixtures of bile salts, phospholipids, and cholesterol provide model systems for following cholesterol crystallization in bile. Cryo-TEM studies, usually combined with light microscopy (LM) observation, can cover the whole range of microstructures from a few nanometers up to several micrometers, revealing the structures present from the early stages of this process to the appearance of platelike cholesterol monohydrate crystals. Experimental studies have also compared synthetic bile model systems with the more complex native human bile.

Kaplun et al. (68) studied the microstructures formed as a function of time during cholesterol crystallization in human bile samples. In a 1-day-old gallbladder bile samples from a gallstone patient, spheroidal micelles and small spherical vesicles were observed, similar to structures previously found in a model system composed of egg-PC–sodium cholate (49). Electron-dense elongated structures (probably microcrystals) and very thin long structures were also seen. In hepatic bile spheroidal micelles, elongated structures and vesicles of various shapes and sizes were found 1 day after collection. Part of the bilayer was missing in some vesicles, indicating a process of vesicle formation. The same types of structures were seen 3 days after collection, together with cylindrical tubes. Six days after collection concentric tubes and closed and opened vesicles were seen, and after 9 days large vesicles and lamellar fragments were found.

The sequence of appearance of aggregates prior to the formation of microcrystals is very similar to that found by indirect methods in bile models studied earlier (69,70). Thus, the cryo-TEM images provide strong support for the previously proposed mechanism (e.g., 71,72) that describes biliary cholesterol crystal formation as being a consequence of cholesterol reorganization within aggregates of cholesterol-rich lecithin-cholesterol vesicles. The polymorphism found is reproduced in simplified model systems of cholesterol, egg yolk lecithin, and sodium taurocholate (73).

Another simple bile model system, composed of lecithin-cholesterol vesicles mixed with sodium cholate at concentrations similar to those found in supersaturated human gallbladder bile, was studied by Fudim-Levin et al. (74). In specimens vitrified 2 min after dilution small spheroidal micelles were seen, supporting the hypothesis that a micellar phase forms immediately after mixing the vesicles with cholesterol. In samples vitrified 6 min after preparation, some lipid vesicles were observed. These cryo-TEM results combined with LM, turbidity, and ^{1}H-NMR measurements also support the sequence of events preceding the appearance of cholesterol crystals proposed previously (71). Elongated threadlike micelles that were found in egg-PC–sodium cholate samples (49) were not observed in any of these model systems. The results suggest that their formation is probably inhibited by the presence of cholesterol.

E. DNA Complexes, Viruses, and Biological Compounds

Studies of the shape of supercoiled DNA in solution are important for the precise determination of the mechanical properties of DNA and for understanding biological mechanisms that involve bending, twisting, and looping of DNA. Direct imaging cryo-TEM micrographs proved that naturally supercoiled (plasmid pUC18) DNA molecules in solution adopt an interwound superhelix form (75,76). Also, the presence of magnesium ions in the solution, which support protein-DNA interactions in typical in vitro reactions, strongly affect the supercoiled DNA shape; the diameter of the interwound superhelix decreased from about 12 to 4 nm upon addition of magnesium salt to solution. Natively supercoiled pUC9 molecules taking the form of a tight superhelix induced by the presence of $MgCl_2$ were also seen.

The phospholipid membrane and glycoprotein spikes seen in vitrified samples of LaCrosse virus demonstrate the preservation of original microstructure during vitrification (77), as opposed to staining and drying, in which the virions were collapsed and destroyed. In a micrograph taken at about the Scherzer defocus the membrane double layer was resolved. The structure of the envelope of Semliki Forest virus (SFV), one of the simplest icosahedral enveloped RNA viruses, was reconstructed to 3.5 nm resolution (78). The geometry of the surface lattice, the shape of the trimeric spikes, and their arrangement on the lipid bilayer were visualized by direct imaging cryo-TEM. Three-dimensional reconstruction of the virus was performed. The three-dimensional structure based on cryo-electron micrographs was also reconstructed for a complex between an intact virus, rotavirus, and Fab fragments of a neutralizing monoclonal antibody (79).

Fujiyoshi (30) used a helium-cooled cryo-stage to image fluid microstructures at up to 0.3 nm, very high resolution in comparison with about 2 nm in standard cryo-TEM work. Electron diffraction patterns from a frozen crystal of transfer RNA (tRNA) were obtained and vitrified complexes of rec A proteins and DNA molecules were observed. In intact influenza A virus a layer of moderate electron density was observed inside a single phospholipid layer of a high contrast. The high resolution made it possible to identify each layer of the lipid bilayer and also membrane defects, as seen in the micrograph presented in Fig. 4.

The papovavirus family consists of two types, both of which include human pathogens. These viruses contain a single circular molecule of double-stranded DNA (dsDNA) associated with histones or histonelike proteins inside an unenveloped, icosahedral capsid (see Ref. 80 and references therein). The structures of bovine papillomavirus type 1 (BPV-1) and human papillomavirus type 1 (HPV-1) found by cryo-TEM were similar, both displaying nearly spherical outer profiles with a highly uniform diameter. Capsomers are clearly discernible, especially at the particle peripheries. Three-dimensional reconstructions of the two viruses computed from digitized electron micrographs were sufficient to define the capsomer symmetry and provide basic information about the nature of intercapsomer contacts. Similar studies were performed for two members of the polyoma genus, polyoma virus and simian virus 40 (81,82), and for pyruvate dehydrogenase complexes (PDCs), which are among the largest enzymes known (83).

Biosurfactants constitute an interesting group of biomaterials. Several classes of these materials are known. In aqueous solutions, biosurfactants reduce surface tension and can interact with organic and inorganic solutes and affect the aqueous dispersion of these solutes. The morphology of one type of biosurfactant, a rhamnolipid biosurfactant produced by *Pseudomonas aeruginosa* ATCC 9027, was studied by Champion et al. (84). Cryo-TEM micrographs showed the morphology as a function of pH, changing from lamellar to vesicular and finally

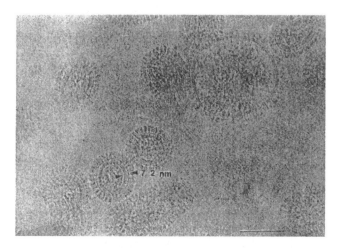

Fig. 4   Intact influenza A virus. A layer of moderate electron density, as seen between arrowheads, is observed inside a single phospholipid bilayer of high contrast. (From Ref. 30.)

to micellar structure as pH is increased. Cadmium seems to stabilize rhamnolipid vesicle structures, and octadecane favors the micellar structure. Micrographs of another biosurfactant, surfactin (produced by *Bacillus subtilis*), revealed spherical, ellipsoidal, and threadlike micelles with an inhomogeneous size distribution at pH equal or larger than 7. Addition of NaCl and $CaCl_2$, to surfactin solutions induced transformation from cylindrical micelles to small ellipsoidal and spheroidal micelles (31).

   Peptide- and protein-based tubular structures are quite common in nature (85). Tobacco mosaic virus, bacterial pili, and microtubulines are a few examples of natural tubular assemblies. Peptide nanotubes are formed by the self-assembly of flat, ring-shaped peptide subunits made up of alternating D- and L-amino acid residues. Its self-aggregation is directed by the formation of an extensive network of intersubunit hydrogen bonds (85). Cryo-TEM was used in parallel with several indirect techniques to characterize and assign the structural features of the crystals. Highly ordered nanotube crystals were obtained from several peptide subunits. Typical striations running along the long axis of the crystal were seen in cryo-TEM images. Cryo–electron diffraction patterns that often extended to 0.12 mm resolution were indicative of the highly ordered specimens.

   Low density lipoprotein (LDL) is a macromolecular assembly that serves as a primary transport particle for cholesterol in human plasma. It has been associated positively with the development of atherosclerosis. The glycoprotein apolipoprotein B-100 (apoB) is one of the largest monomeric proteins known. The structure of the human LDL particles and the disposition of apoB at its surface were investigated by Spin and Atkinson (86) using cryo-TEM. Electron micrographs show that LDL is a quasi-spherical particle with a region of low density (lipid) surrounded by a ring of high density believed to represent apoB. This ring is composed of several large regions of high-density material that may represent protein superdomains. Analysis of LDL images reveals that areas of somewhat lower density connect these regions, in some cases crossing the projectional interiors of the LDL particles.

Carrier-free delivery systems are based on the principles of dispersing pure drug in water, either as solid particles or as liquid droplets. Model systems of a hydrophobic drug such as emulsions of cholesteryl acetate particles stabilized by lecithin and bile salts were characterized by cryo-TEM in combination with indirect scattering methods (87). Drops, platelets, and a few vesicles were seen in the different systems studied. Particles with a composition similar to that of LDL were found to be large spherical globules covered by small vesicles. Dark spheres representing the emulsion particles coexisting with unilamellar phospholipid vesicles were also found in commercial Intralipid emulsions made of lecithin and soybean oil in an aqueous solution of glycerol (88). Such fat emulsions are used clinically as a source of energy in parenteral nutrition.

Tropomyosin is a 40-nm long α-helical, coiled-coiled protein that forms, together with troponin, a calcium-sensitive regulatory switch in many muscles. Highly hydrated tropomyosin Bailey crystal is made up of supercoiled filaments similar in structure to those that wind around the actin filaments in muscle (89). Electron micrographs of vitrified tropomyosin clearly show the separation of the two coiled-coiled filaments, as well as the woven appearance of the crossover regions. The crossover regions are especially accentuated due to the high density of protein in this region. Distortions in the crystal lattice were also seen in the micrographs. Tropomyosin-troponin Bailey cocrystals were somewhat larger, and some of the troponin binding sites were not occupied. Density maps and structure factors of the crystals that were generated using image processing were in excellent agreement with a corresponding projection of the three-dimensional structure determined from X-ray crystallography.

Hepatitis B surface antigen (HBsAg) particles serve as the immunogen of hepatitis B subunit vaccine (HBV). Cryo-TEM was combined with image processing to resolve the structure of recombinant HBsAg particles prepared from mammalian Chinese hamster ovary (CHO) cells that resemble those derived from hepatitis B carriers in their lipid and protein composition (40:60). Roughly spherical vesicles with nonuniform membranes are seen in the micrographs. Spheroidal micelle-like particles, possibly proteins excluded from the particles, are also observed. Previous biochemical analysis had suggested that the particles are either leaky vesicles or lipoprotein-like structures with a neutral lipid core surrounded by a phospholipid, cholesterol, and lipid shell. Image processing on selected particles showed the existence of holes in the particle membrane, suggesting that the particles are leaky vesicles (90). Areas of higher optical density corresponded to protein-rich domains, and areas of lower optical density were richer in phospholipid.

The development of gene therapy and genetic engineering depends on reliable and efficient systems for the introduction of DNA into target cells. Association structures formed by DNA plasmids and liposomes composed of synthetic cationic surfactants and phospholipids have been successfully employed as gene carriers in transfection assays. The cationic surfactants give the vesicle a positively charged surface, which promotes strong interactions between vesicles and plasmids. Those complexes are supposed to carry a positive net charge, which facilitates their approach to the cell surface. Gustafsson et al. (91) found that addition of DNA to DOPE vesicles in mixtures of CTAC, DODAB (dioctadecyldimethylammonium bromide), or DOTAP [1,2-bis(oleoyloxy)-3-(trimethylammonio)propane] leads to spontaneous formation of discrete complexes that appear as clusters of aggregated dense particles. Low ratios of CTAC to DOPE stabilized the vesicles, and large amounts of CTAC solubilized the lipid into micelles. Free plasmids or protruding DNA strings appeared at high DNA/lipid ratios. Binding and entrapment of DNA into aggregated multilamellar structures were found at low ratios. It also appears that the choice of surfactant does not affect the morphology of the DNA-lipid complexes.

Cationic lipid-DNA complexes were further studied by Templeton et al. (92). Using cryo-TEM, they studied the morphology and colloidal properties of DOTAP:cholesterol-DNA liposome complexes. In the absence of DNA, extruded DOTAP:cholesterol liposomes were usually invaginated, consisting of lipids with excess surface area. Some vesicles were completely invaginated with two concentric lamellae and a small orifice with an approximate diameter of 50 nm (a "vase structure"). Others consisted of bilamellar and unilamellar tubular shapes, unilamellar erythrocyte shapes, and unilamellar bean shapes. Multilamellar liposomes were not observed in electron micrographs. When extruded DOTAP:cholesterol liposomes were complexed with DNA at optimal concentrations, the DNA was localized in the liposome interior between two lipid bilayers. This structure could account for the high efficiency of gene delivery in vivo and for the broad tissue distribution of the DNA:liposome complexes. Extruded DOTAP:DOPE liposomes also formed a few vase structures; however, their orifices were larger, and many spheres were formed. In addition, little or no DNA assembled in the liposomes, and the DNA was frequently found on the outside of these liposomes. DODAB-containing liposomes did not form any invaginated structures. The internalization of DNA within invaginated structures is a unique feature of extruded DOTAP liposomes and had not been observed in other cryo-TEM studies of complexed DNA:liposomes (91,93). Internalization of DNA within liposomes has been observed only in large multilamellar vesicles (93). A model for the assembly of DNA:liposome complexes was proposed in which the DNA adsorbs onto the invaginated and tubular liposomes via electrostatic interactions. A micrograph of a vitrified DNA:liposome sample is shown in Fig. 5. Note the large liposome encapsulating a thick DNA complex.

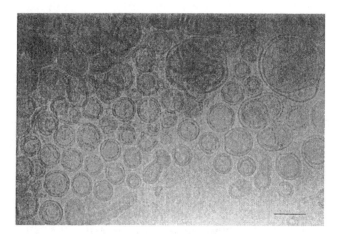

Fig. 5   Electron micrograph of a 5 mM DOTAP:DOPE liposome mixture (molar ratio of 1:1), mixed with 81 μg of DNA in a final volume of 200 μL. The DNA-to-liposome charge ratio is 0.5.

REFERENCES

1. D Danino, Y Talmon, H Levy, G Beinert, R Zana. Branched threadlike micelles in an aqueous solution of a trimeric surfactant. Science 269:1420, 1995.
2. Y Talmon. Transmission electron microscopy of complex fluids: The state of the art. Ber Bunsenges Phys Chem 100:364, 1996.
3. Y Talmon. Cryogenic temperature transmission electron microscopy in the study of surfactant systems. In: BP Binks, ed. Modern Characterization Methods of Surfactant Systems. New York: Marcel Dekker, 1999, p. 147.
4. DR Uhlmann. A kinetic treatment of glass formation. J Noncryst Solids 7:337, 1972.
5. DP Siegel, WJ Green, Y Talmon. The mechanism of lamellar-to-inverted hexagonal phase transitions: A study using temperature-jump cryo-electron microscopy. Biophys J 66:402, 1994.
6. I Sabanay, T Arad, S Weiner, B Geiger. Study of vitrified, unstained frozen tissue sections by cryoimmunoelectron microscopy. J Cell Sci 100:227, 1991.
7. S Weiner, T Arad, I Sabanay, W Traub. Rotated plywood structure of primary lamellar bone in the rat: Orientations of the collagen fibril arrays. Bone 20:509, 1997.
8. DL Misell. Image Analysis, Enhancement of Interpretation. Amsterdam: North-Holland, 1978.
9. Y Talmon. Electron beam radiation damage to organic and biological cryospecimens. In: RA Steinbrecht, K Zierold, eds. Cryotechniques in Biological Electron Microscopy. Berlin: Springer Verlag, 1987, p 64.
10. Y Talmon. The study of latex IPNs by cryo-TEM using radiation-damage effects. In: D Klempner, KC Frisch, eds. Advances in Interpenetrating Polymer Networks. Lancaster, PA: Technomic, 1990, p 141.
11. A Hass-Bar Ilan, I Noda, LA Schechtman, Y Talmon. Cryo-TEM and DSC characterization of latexes stabilized with surface active block oligomers. Polymer 33:2043, 1992.
12. K Mortensen, Y Talmon. Cryo-TEM and SANS microstructural study of pluronic polymer solutions. Macromolecules 28:8829, 1995.
13. JR Bellare, HT Davis, LE Scriven, Y Talmon. Controlled environment vitrification system (CEVS). J Electron Microsc Tech 10:87, 1988.
14. MH Chestnut, DP Siegel, JL Burns, Y Talmon. A temperature-jump devise for time-resolved cryo–transmission electron microscopy. Microsc Res Tech 20:95, 1992.
15. Y Fink, Y Talmon. The flow-throu controlled environment vitrification system. Proceedings 13th International Congress on Electron Microscopy, 1994, pp 37–38.
16. DP Siegel, JL Burns, MH Chestnut, Y Talmon. Intermediates in membrane fusion and bilayer/nonbilayer phase transitions imaged by time-resolved cryo–transmission electron microscopy. Biophys J 56:161, 1989.
17. A Sein, JF van Breemen, JBFN Engberts. Emergence of a lyotropic lamellar phase: Surfactant-aqueous contact experiments examined with a cryo–transmission electron microscope. Langmuir 11:3565, 1995.
18. D Danino, Y Talmon, R Zana. Aggregation and microstructure in aqueous solutions of the nonionic surfactant $C_{12}E_8$. J Colloid Interface Sci 186:170, 1997.
19. J-F Ménétret, W Hofmann, RR Schröder, G Rapp, RS Goody. Time-resolved cryo-electron microscopic study of the dissociation of actomyosin induced by photolysis of photolabile nucleotides. J Mol Biol 219:139, 1991.
20. M Goldraich, Y Cohen, Y Talmon. Gelation in EHEC/H_2O and EHEC/SDS/H_2O systems studied by light microscopy, cryo-TEM and SAXS, in preparation.
21. L Bachmann, Y Talmon. Cryomicroscopy of hydrated specimens: Direct imaging versus replication. Ultramicroscopy 14:211, 1984.

22. S Chiruvolu, E Naranjo, JA Zasadzinski. Microstructure of complex fluids by electron microscopy. In: CA Herb, RK Prud'homme, eds. Structure and Flow in Surfactant Solutions. ACS Symposium Series 578. Washington, DC: American Chemical Society, 1994, p 86.

23. DC Roux, JF Berret, G Porte, E Peuvrel-Disdier, P Linder. Shear-induced orientations and textures of nematic wormlike micelles. Macromolecules 27:1681, 1995.

24. TM Clausen, PK Vinson, JR Minter, HT Davis, Y Talmon, WG Miller. Viscoelastic micellar solutions: microscopy and rheology. J Phys Chem 96:474, 1992.

25. I Wunderlich, H Hoffmann, H Rehage. Flow birefringes and rheological measurements on shear induced micellar structures. Rheol Acta 26:532, 1987.

26. FS Bates, KA Koppi, M Tirell, K Almdal, K Mortensen. Influence of shear on the hexagonal-to-disorder transition in a diblock copolymer melt. Macromolecules 27:5934, 1994.

27. B Lu, X Li, HT Davis, LE Scriven, Y Talmon, JL Zakin. Effect of chemical structure on viscoelasticity and extensional viscosity of drag reducing cationic surfactant solutions. Langmuir, 14:8, 1998.

28. D Danino, Y Talmon, R Zana. Cryo-TEM of thread-like micelles: on-the-grid microstructural transformations induced during specimen preparation. Colloid and Surfaces, submitted.

29. JL Burns, Y Cohen, Y Talmon. The structure of cubic mesomorphic phases determined by low-temperature electron microscopy and small-angle x-ray scattering. J Phys Chem 94:5308, 1990.

30. Y Fujiyoshi. High resolution cryo-electron microscopy for biological macromolecules. J Electron Microsc 38:S97, 1989.

31. A Knoblich, M Matsumoto, R Ishiguro, K Murata, Y Fujiyoshi, Y Ishigami, M Osman. Electron cryo-microscopic studies on micellar shape and size of surfactin, an anionic lipopeptide. Colloids Surf B5:43, 1995.

32. A Walter. Vesicle-micelle transitions of surfactant phospholipid systems. In: BP Gaber, KRK Easwaran, eds. Biomembrane Structure and Function—The State of the Art. Schenectady, NY: Adenine Press, 1992, p 21.

33. R Groll, A Böttcher, J Jäger, JF Holzwarth. Temperature dependent intermediate structures during the main phase transition of dimyristoyl phosphatidylcholine vesicles: A combined iodine laser–temperature jump and time resolved cryo-electron microscopy study. Biophys Chem 58:53, 1996.

34. PM Frederik, MCA Stuart, H Andree, CPM Reutelingsperger, EJ Boekema, KNJ Burger, AJ Verkleij. Lipidic polymorphism, lipid protein interaction and membrane fusion as studies by cryo-electron microscopy. Electron Microscopy 3 EUREM, Granada, Spain, 1992, pp 113–114.

35. PK Vinson, JR Bellare, HT Davis, WG Miller, LE Scriven. Direct imaging of surfactant micelles, vesicles, discs, and ripple phase structures by cryo–transmission electron microscopy. J Colloid Interface Sci 142:74, 1991.

36. PK Vinson. Cryo–electron microscopy of microstructures in complex fluids. Doctoral dissertation, University of Minnesota, 1990.

37. S Chiruvolu, HE Warriner, E Naranjo, SHJ Idziak, JO Rädler, RJ Plano, JA Zasadzinski, CR Safinya. A phase of liposomes with entangled tubular vesicles. Science 266:1222, 1994.

38. GT Barnes, GA Lawrie, BJ Battersby, SM Sarge, HK Cammenaga, PB Schneider. Dimyristoyl phosphatidylcholine: Equilibrium spreading behaviour. Thin Solid Films 242:201, 1994.

39. V Birault, G Pozzi, N Plobeck, S Eifler, M Schmutz, T Palanché, J Raya, A Brisson, Y Nakatani, G Ourisson. Di(polyprenyl) phosphates as models for primitive membrane constituents: Synthesis and phase properties. Chem Eur J 2:789, 1996.

40. DD Miller, JR Bellare, DF Evans, Y Talmon, BW Ninham. Meaning and structure of amphiphilic phases: Inferences from video-enhanced microscopy and cryotransmission electron microscopy. J Phys Chem 91:674, 1987.
41. L Hammarström, I Velikian, G Karlsson, K Edwards. Cryo-TEM evidence: Sonication of dihexadecyl phosphate does not produce closed bilayers with smooth curvature. Langmuir 11:408, 1995.
42. D Danino, A Kaplun, G Lindblom, L Rilfors, G Orädd, JB Hauksson, Y Talmon. Cryo-TEM and NMR studies of a micelle-forming phosphoglucolipid from membranes of *Acholeplasma laidlawii* A and B. Chem Phys Lipids 85:75, 1997.
43. BLS Mui, H-G Döberiner, TD Madden, PR Cullis. Influence of transbilayer area asymmetry on the morphology of large unilamellar vesicles. Biophys J 69:930, 1995.
44. DD Lasic, PM Frederik, MCA Stuart, Y Barenholz, TJ McIntosh. Gelation of liposome interior. A novel method for drug encapsulation. FEBS 312:255, 1992.
45. D Lichtenberg. Micelles and liposomes. In: M Shinitzki, ed. Biomembranes, Physical Aspects. Weinheim: VCH, 1993, p 63.
46. PK Vinson, Y Talmon, A Walter. Vesicle-micelle transition of phosphatidylcholine and octyl glucoside elucidated by cryo–transmission electron microscopy. Biophys J 56:669, 1989.
47. M Seras, K Edwards, M Almgren, G Carlson, M Ollivon, S Lesieur. Solubilization of nonionic monoalkyl amphiphile-cholesterol vesicles by octyl glucoside: Cryo-transmission electron microscopy of the intermediate structures. Langmuir 12:330, 1996.
48. D Danino, Y Talmon, R Zana. Vesicle-to-micelle transformation in systems containing dimeric surfactants. J Colloid Interface Sci 185:84, 1997.
49. A Walter, PK Vinson, A Kaplun, Y Talmon. Intermediate structures in the cholate–phosphatidylcholine vesicle–micelle transition. Biophy J 60:1315, 1991.
50. A Kaplun, A Walter, Y Talmon. Symmetry of the vesicle-to-micelle and micelle-to-vesicle transitions in mixtures of sodium cholate and egg phosphatidylcholine, to be submitted to Biophys J.
51. K Edwards, M Almgren, JR Bellare, W Brown. Effects of Triton X-100 on sonicated lecithin vesicles. Langmuir 5:473, 1989.
52. K Edwards, M Almgren. Solubilization of lecithin vesicles by $C_{12}E_8$. Structural transitions and temperature effects. J Colloid Interface Sci. 147:1, 1991.
53. K Edwards, J Gustafsson, M Almgren, G Karlsson. Solubilization of lecithin vesicles by a cationic surfactant: Intermediate structures in the vesicle-micelle transition observed by cryo-transmission electron microscopy. J Colloid Interface Sci 161:299, 1993.
54. M Silvander, G Karlsson, K Edwards. Vesicle solubilization by alkyl sulfate surfactants: A cryo-TEM study of the vesicle to micelle transition. J Colloid Interface Sci 179:104, 1996.
55. D Meyuhas, A Bor, I Pinchuk, A Kaplun, Y Talmon, MM Kozlov, D Lichtenberg. Effect of ionic strength on the self-assembly in mixtures of phosphatidylcholine and sodium cholate. J Colloid Interface Sci 188:351, 1997.
56. AJ Verkleij, B Humbel, D Studer, M Müller. 'Lipidic particles' systems as visualized by thin-section electron microscopy. Biochim Biophys Acta 812:591, 1985.
57. AJ Verkleij. Lipidic intramembranous particles. Biochim Biophys Acta 779:43, 1984.
58. H Ellens, DP Siegel, D Alford, P Yeagle, L Boni, L Lis, PJ Quinn, J Bentz. Membrane fusion and inverted phases. Biochemistry 28:3692, 1989.
59. DP Siegel. Energetics of intermediates in membrane fusion: Comparison of stalk and inverted micellar intermediate mechanisms. Biophys J 65:2124, 1993.
60. DP Siegel, RM Epand. The mechanism of lamellar-to-inverted hexagonal phase transitions in phosphatidylethanolamine: Implications for membrane fusion mechanisms. Biophys J 77:3089, 1997.

61. PM Frederik, MCA Stuart, AJ Verkleij. Intermediary structures during membrane fusion as observed by cryo-electron microscopy. Biochim Biophys Acta 979:275, 1989.

62. PM Frederik, KNJ Burger, MCA Stuart, AJ Verkleij. Lipidic polymorphism as observed by cryo-electron microscopy. Biochim Biophys Acta 1062:133, 1991.

63. J White, M Kielian, A Helenius. Membrane fusion proteins of enveloped animal viruses. Q Rev Biophys 16:151, 1983.

64. KS Matlin, H Reggio, A Helenius, K Simons. Infectious entry pathway of influenza virus in a canine kidney cell line. J Cell Biol 91:601, 1991.

65. KNJ Burger, G Knoll, PM Frederik, AJ Verkleij. Influenza virus mediated membrane fusion: The identification of fusion intermediates using modern cryotechniques. In: JAF Op den Kamp, ed. Dynamics and Biogenesis of Membranes. NATO ASI Series. Berlin: Springer-Verlag, 1990, p 185.

66. MS Vethamuthu, M Almgren, W Brown, E Mukhtar. Aggregate structure, gelling, and coacervation within the L_1 phase of the quasi-ternary system alkyltrimethylammonium bromide–sodium desoxycholate–water. J Colloid Interface Sci 174:461, 1995.

67. MS Vethamuthu, M Almgren, G Karlsson, P Bahadur. Effect of sodium chloride and varied alkyl chain length on aqueous cationic surfactant-bile salt systems. Cryo-TEM and fluorescence quenching studies. Langmuir 12:2173, 1996.

68. A Kaplun, Y Talmon, FM Konikoff, M Rubin, A Eitan, M Tadmor, D Lichtenberg. Direct visualization of lipid aggregates in native human bile by light- and cryo-transmission electron microscopy. FEBS Lett 340:78, 1994.

69. D Lichtenberg, S Ragimova, A Bor, S Almog, C Vinkler, M Kalina, Y Peled, Z Halpern. Stability of mixed micellar bile models supersaturated with cholesterol. Biophys J 54:1013, 1988.

70. FM Konikoff, DS Chung, JM Donovan, DM Small, MC Carey. Filamentous, helical, and tubular microstructures during cholesterol crystallization from bile. Evidence that cholesterol does not nucleate classic monohydrate plates. J Clin Invest 90:1155, 1992.

71. D Lichtenberg, S Ragimova, A Bor, S Almog, C Vinkler, Y Peled, Z Halpern. Stability of mixed micellar systems made by solubilizing phosphatidylcholine-cholesterol vesicles by bile salts. Hepatology 12:149S, 1990.

72. Y Peled, Z Halpern, B Eitan, G Goldman, FM Konikoff, T Gilat. Biliary micellar cholesterol nucleates via the vesicular pathway. Biochim Biophys Acta 1003:246, 1989.

73. A Kaplun, FM Konikoff, A Eitan, M Rubin, A Vilan, D Lichtenberg, T Gilat, Y Talmon. Imaging supramolcular aggregates in bile models and human bile. Micros Res Tech 39:85, 1997.

74. E Fudim-Levin, A Bor, A Kaplun, Y Talmon, D Lichtenberg. Cholesterol precipitation from cholesterol-supersaturated bile models. Biochim Biophys Acta 1259:23, 1995.

75. M Adrian, B ten Heggeler-Bordier, W Wahli, AZ Stasiak, A Stasiak, J Dubochet. Direct visualization of supercoiled DNA molecules in solution. EMBO J 9:4551, 1990.

76. J Dubochet, M Adrian, I Dustin, P Furrer, A Stasiak. Cryoelectron microscopy of DNA molecules in solution. Methods Enzymol 211:507, 1992.

77. Y Talmon, BVV Prasad, JPM Clerx, GJ Wang, W Chiu, MJ Helwett. Electron microscopy of vitrified-hydrated LaCrosse virus. J Virol 61:1121, 1987.

78. RH Vogel, SW Provencher, C-H von Bonsdorff, M Adrian, J Dubochet. Envelope structure of Semliki Forest virus reconstructed from cryo-electron micrographs. Nature 320:533, 1986.

79. BVV Prasad, JW Burns, E Marietta, MK Estes, W Chiu. Localization of VP4 neutralization sites in rotavirus by three-dimensional cryo-electron microscopy. Nature 343:476, 1990.

80. TS Baker, WW Newcomb, NH Olson, LM Cowsert, C Olson, JC Brown. Structures of bovine and human papillomaviruses. Analysis by cryoelectron microscopy and three-dimensional image reconstruction. Biophys J 60:1445, 1991.

81. TS Baker, J Drak, M Bina. Reconstruction of the three-dimensional structure of simian virus 40 and visualization of the chromatin core. Proc Natl Acad Sci USA 85:422, 1988.

82. TS Baker, J Drak, M Bina. The capsid of small papove viruses contains 72 pentameric capsomers: Direct evidence from cryo-electron microscopy of simian virus 40. Biophys J 55:243, 1989.

83. T Wagenknecht, R Grassucci, GA Radke, TE Roche. Cryoelectron microscopy of mammalian pyruvate dehydrognase complex. J Biol Chem 266:24650, 1991.

84. JT Champion, JC Gilkey, H Lamparski, J Retterer, RM Miller. Electron microscopy of rhamnolipid (biosurfactant) morphology: Effects of pH, cadmium, and octadecane. J Colloid Interface Sci 170:569, 1995.

85. JD Hartgerink, JR Granja, RA Milligan, MR Ghadiri. Self-assembling peptide nanotubes. J Am Chem Soc 118:43, 1996.

86. JM Spin, D Atkinson. Cryoelectron microscopy of low density lipoprotein in vitreous ice. Biophys J 68:2115, 1995.

87. B Sjöström, A Kaplun, Y Talmon, B Cabane. Structures of nanoparticles prepared from oil-in-water emulsions. Pharm Res 12:39, 1995.

88. M Rotenberg, M Rubin, A Bor, D Meyuhas, Y Talmon, D Lichtenberg. Physico-chemical characterization of Intralipid™ emulsions. Biochim Biophys Acta 1086:265, 1991.

89. D Cabral-Lilly, GN Phillips Jr, GE Sosinsky, L Melanson, S Chacko, C Cohen. Structural studies of tropomyosin by cryoelectron microscopy and x-ray diffraction. Biophys J 59:805, 1991.

90. D Diminsky, Y Barenholz, M Goldraich, Y Talmon. Hepatitis B surface antigen particles: Composition, structure and immunogenicity, in preparation.

91. J Gustafsson, G Arvidson, G Karlsson, M Almgren. Complexes between cationic liposomes and DNA visualized by cryo-TEM. Biochim Biophys Acta 1235:305, 1995.

92. NS Templeton, DD Lasic, PM Frederik, HH Strey, DD Roberts, GN Pavlakis. Improved DNA: liposome complexes for increased systemic delivery and gene expression. Nature Biotechnol 15:647, 1997.

93. PM Frederik, MCA Stuart, PHH Bomans, WM Busing, KNJ Burger, AJ Verkleij. Perspective and limitations of cryo-electron microscopy. From model systems to biological specimens. J Microsc 161:253, 1991.

Index